UNMANNED AIRCRAFT SYSTEMS

UNMANNED AIRCRAFT SYSTEMS

UNMANNED AIRCRAFT SYSTEMS EDITORS

Ella Atkins

Department of Aerospace Engineering, University of Michigan, Ann Arbor, MI, USA

Aníbal Ollero

Universidad de Sevilla and Scientific Advisory Department of the Center for Advanced Aerospace Technologies, Seville, Spain

Antonios Tsourdos

School of Aerospace, Transport & Manufacturing and Centre for Autonomous and Cyber-Physical Systems, Cranfield University, Cranfield, UK

ENCYCLOPEDIA OF AEROSPACE ENGINEERING EDITORS-IN-CHIEF

Richard Blockley

Aerospace Consultant, Cranfield University, Cranfield, UK
Former Head of Technical Programmes, BAE Systems, Farnborough, UK

Wei Shyy

Hong Kong University of Science and Technology, Hong Kong, P. R. China

WILEY

Library of Congress Cataloging-in-Publication Data is available for this title.

ISBN: 978-1-118-86645-0

A catalogue record for this book is available from the British Library.

Front cover: Drone image from John Lawson, Belhaven/Gettyimages

Typeset in 10.25/12.5pt TimesLTStd-Roman by Thomson Digital, Noida, India

Printed and bound in Singapore by Markono Print Media Pte Ltd

1   2016

CONTENTS

CONTRIBUTORS

Brandon R. Abel
*International Center for Air Transportation,
Massachusetts Institute of Technology, Cambridge,
MA, USA*

Domenico Accardo
University of Naples "Federico II", Napoli, Italy

José Joaquin Acevedo
*Grupo de Robótica, Visión y Control, Universidad de
Sevilla, Seville, Spain*

Florian-Michael Adolf
*German Aerospace Center (DLR), Department of
Unmanned Aircraft, Institute of Flight Systems,
Braunschweig, Germany*

Jessica Alvarenga
*Ritchie School of Engineering and Computer Science,
DU Unmanned Research Institute, University of Denver,
Denver, CO, USA*

Brian M. Argrow
*Department of Aerospace Engineering Sciences,
Research and Engineering Center for
Unmanned Vehicles, University of Colorado Boulder,
Boulder, CO, USA*

Begoña C. Arrue
*Grupo de Robótica, Visión y Control, Universidad
de Sevilla, Seville, Spain*

Ella M. Atkins
*Department of Aerospace Engineering, University of
Michigan, Ann Arbor, MI, USA*

Randal W. Beard
*Department of Electrical and Computer Engineering,
Brigham Young University, Provo, UT, USA*

Yunfeng Cao
*College of Astronautics, Nanjing University of Aeronautics
and Astronautics, Nanjing, China*

Jesús Capitán
*Grupo de Robótica, Visión y Control, Universidad
de Sevilla, Seville, Spain*

Philip B. Charlesworth
Airbus Group Innovations, Newport, UK

Wen-Hua Chen
*Department of Aeronautical and Automotive Engineering,
Loughborough University, Loughborough, UK*

Yang Quan Chen
*School of Engineering, University of California,
Merced, CA, USA*

Matthew Coombes
*Department of Aeronautical and Automotive
Engineering, Loughborough University,
Loughborough, UK*

Mary L. Cummings
*Humans and Autonomy Laboratory, Duke University,
Durham, NC, USA*

Dan DeLaurentis
*School of Aeronautics and Astronautics, Purdue University,
West Lafayette, IN, USA*

Pedro F.A. Di Donato
Department of Aerospace Engineering, University of Michigan, Ann Arbor, MI, USA and National Civil Aviation Agency–Brazil (ANAC), São, José dos Campos, Brazil

Haibin Duan
School of Automation Science and Electrical Engineering, Beihang University, Beijing, P.R. China

John T. Economou
Centre for Defence Engineering, Defence Academy of the United Kingdom, Cranfield University, Swindon, UK

Gary J. Ellingson
Mechanical Engineering Department, Brigham Young University, Provo, UT, USA

Paul G. Fahlstrom
United States Army Materiel Command, Huntsville, AL, USA

Farhan A. Faruqi
Information Processing and Human Sciences Group, Combat and Mission Systems, WCSD, Defence Science and Technology Organisation, Edinburgh, South Australia

Giancarmine Fasano
University of Naples "Federico II", Napoli, Italy

Karen Feigh
Cognitive Engineering Center, Georgia Tech, Atlanta, GA, USA

C.E. "Noah" Flood
CAVU Global LLC, Purcellville, VA, USA

Michael S. Francis
United Technologies Research Center, East Hartford, CT, USA

Seng Keat Gan
Australian Centre for Field Robotics, The University of Sydney, Sydney, Australia

Alessandro Gardi
RMIT University, Melbourne, Australia

Thomas J. Gleason
Gleason Research Associates, Inc., Columbia, MD, USA

R. John Hansman
International Center for Air Transportation, Massachusetts Institute of Technology, Cambridge, MA, USA

Inseok Hwang
School of Aeronautics and Astronautics, Purdue University, West Lafayette, IN, USA

Mario Innocenti
Munitions Directorate, Eglin Air Force Base, Air Force Research Laboratory, FL, USA

Pantelis Isaiah
Faculty of Aerospace Engineering, The Technion—Israel Institute of Technology, Haifa, Israel

Stéphane Kemkemian
Thales Airborne Systems, Elancourt, France

Seungkeun Kim
Department of Aerospace Engineering, Chungnam National University, Daejeon, Republic of Korea

Trevor Kistan
RMIT University, Melbourne, Australia and THALES Australia, Melbourne, Australia

Daniel P. Koch
Mechanical Engineering Department, Brigham Young University, Provo, UT, USA

Cheolhyeon Kwon
School of Aeronautics and Astronautics, Purdue University, West Lafayette, IN, USA

Jack W. Langelaan
Department of Aerospace Engineering, The Pennsylvania State University, University Park, PA, USA

Nicolas Léchevin
Department of Mechanical and Industrial Engineering, Concordia University, Montreal, Quéebec, Canada

Christopher W. Lum
William E. Boeing Department of Aeronautics & Astronautics, University of Washington, Seattle, WA, USA

Douglas M. Marshall
*TrueNorth Consulting LLC, Grand Forks, ND, USA
and De Paul University College of Law, Chicago, IL, USA*

David W. Matolak
Department of Electrical Engineering, University of South Carolina, Columbia, SC, USA

Iván Maza
Grupo de Robótica, Visión y Control, Universidad de Sevilla, Seville, Spain

Timothy W. McLain
Mechanical Engineering Department, Brigham Young University, Provo, UT, USA

Luis Merino
Grupo de Robótica, Visión y Control, Universidad Pablo de Olavide, Seville, Spain

Antonio Moccia
University of Naples "Federico II", Napoli, Italy

Linas Mockus
School of Aeronautics and Astronautics, Purdue University, West Lafayette, IN, USA

Eric Mueller
NASA, Moffett Field, CA, USA

Myriam Nouvel
Thales Airborne Systems, Elancourt, France

Paul W. Nyholm
Mechanical Engineering Department, Brigham Young University, Provo, UT, USA

Hyondong Oh
Department of Aeronautical and Automotive Engineering, Loughborough University, Loughborough, UK

Aníbal Ollero
Universidad de Sevilla and Scientific Advisory Department of the Center for Advanced Aerospace Technologies, Seville, Spain

Martina Orefice
Air Transport Sustainability Department, CIRA Italian Aerospace Research Center, Capua, Italy

Charles H. Patchett
School of Engineering, University of Liverpool, Liverpool, UK

Lorenzo Pollini
Department of Information Engineering, University of Pisa, Pisa, Italy

Amy Pritchett
Cognitive Engineering Center, Georgia Tech, Atlanta, GA, USA

Camille A. Rabbath
Department of Mechanical and Industrial Engineering, Concordia University, Montreal, Québec, Canada

Matthew R. Rabe
International Center for Air Transportation, Massachusetts Institute of Technology, Cambridge, MA, USA

Subramanian Ramasamy
RMIT University, Melbourne, Australia

Francisco J. Ramos
UAS Ground Segment Department, Airbus Defence & Space, Getafe, Spain

James M. Rankin
Avionics Engineering Center, School of Electrical Engineering and Computer Science, Russ College of Engineering and Technology, Ohio University, Athens, OH, USA

Keith A. Rigby
BAE Systems, Warton Aerodrome, Preston, UK

Matthew J. Rutherford
Ritchie School of Engineering and Computer Science, DU Unmanned Research Institute, University of Denver, Denver, CO, USA

Roberto Sabatini
RMIT University, Melbourne, Australia

Daniel P. Salvano
Aviation Consultant, Safety, Certification and CNS Systems, Haymarket, VA, USA

A. Savvaris
Centre for Cyberphysical Systems, Institute for Aerospace Sciences, Cranfield University, Cranfield, UK

Corey J. Schumacher
711 HPW/RH, Wright-Patterson AFB, Ohio, OH, USA

Pau Segui-Gasco
Centre for Autonomous and Cyber-Physical Systems, SATM, Cranfield University, Cranfield, UK

Madhavan Shanmugavel
School of Engineering, Monash University Malaysia, Selangor, Malaysia

Tal Shima
Faculty of Aerospace Engineering, The Technion—Israel Institute of Technology, Haifa, Israel

Hyo-Sang Shin
Centre for Autonomous and Cyber-Physical Systems, SATM, Cranfield University, Cranfield, UK

Brandon J. Stark
School of Engineering, University of California, Merced, CA, USA

Chun-Yi Su
Department of Mechanical and Industrial Engineering, Concordia University, Montreal, Quéebec, Canada

Salah Sukkarieh
Australian Centre for Field Robotics, The University of Sydney, Sydney, Australia

Shigeru Sunada
Department of Aerospace Engineering, Osaka Prefecture University, Osaka, Japan

Hiroshi Tokutake
Department of Aerospace Engineering, Osaka Prefecture University, Osaka, Japan

Christoph Torens
German Aerospace Center (DLR), Department of Unmanned Aircraft, Institute of Flight Systems, Braunschweig, Germany

Giulia Torrano
Air Transport Sustainability Department, CIRA Italian Aerospace Research Center, Capua, Italy

Antonios Tsourdos
School of Aerospace, Transport & Manufacturing and Centre for Autonomous and Cyber-Physical Systems, Cranfield University, Cranfield, UK

Dai A. Tsukada
William E. Boeing Department of Aeronautics & Astronautics, University of Washington, Seattle, WA, USA

Joseph J. Vacek
Department of Aviation, University of North Dakota, Grand Forks, ND, USA

Kimon P. Valavanis
Ritchie School of Engineering and Computer Science, DU Unmanned Research Institute, University of Denver, Denver, CO, USA

Antidio Viguria
Center for Advanced Aerospace Technologies (CATEC), Seville, Spain

Vittorio Di Vito
Air Transport Sustainability Department, CIRA Italian Aerospace Research Center, Capua, Italy

Nikolaos I. Vitzilaios
Ritchie School of Engineering and Computer Science, DU Unmanned Research Institute, University of Denver, Denver, CO, USA

David O. Wheeler
Department of Electrical and Computer Engineering, Brigham Young University, Provo, UT, USA

an White
tre for Autonomous Systems and Cyber-Physical
ems, Cranfield University, Cranfield, UK

e Xu
*ralian Centre for Field Robotics, The University of
ney, Sydney, Australia*

g A. Yakimenko
*duate School of Engineering and Applied Science,
al Postgraduate School, Monterey, CA, USA*

Andy Yu
*School of Aeronautics and Astronautics, Purdue University,
West Lafayette, IN, USA*

Greg L. Zacharias
Charles River Analytics, Cambridge, MA, USA

FOREWORD

The *Encyclopedia of Aerospace Engineering*, first published in 2010, represents a singular attempt to capture the aerospace community's ever-expanding collective body of knowledge into an easy-to-use, cohesive, universal reference framework.

The past few years have marked rapid growth in aerospace systems and technology – as new and innovative designs and applications come to the fore, new ways of thinking about old challenges emerge, and as existing technology and systems have continued to evolve in new and exciting directions. This growth has been especially dynamic in the field of unmanned aircraft systems (UAS).

No longer solely the tools of the military, UAS have experienced a cost and capability revolution, performing important missions across many fields – agricultural sensing, infrastructure inspection, scientific research, and logistics – with significant implications for the research and development enterprise. The new complementary technologies involving intelligent systems are continually changing how we think about the capabilities and applications of UAS technology and how it will continue to transform our lives.

The absence of the human payload and its associated systems has inspired and delivered remarkable innovations in our industry. Yet unmanned systems still face extraordinary challenges to deliver comparable situational awareness to the operator, and we are all aware of the potential threats to safety and security associated with the widespread availability and increasing affordability of small-scale, remotely piloted aircraft. To address these challenges and to leverage these associated innovations fully, we need ongoing access to information. We therefore welcome this addition to the *Encyclopedia of Aerospace Engineering* as both timely and comprehensive, covering a remarkable range of UAS issues from platform technology, autonomy, security, and fail-safe systems through to integration with manned aviation and the regulatory and legal regimes – all critical pieces of knowledge if we are to continue developing the UAS enterprise to its fullest potential.

The year 2016 marks both the 150th anniversary of the Royal Aeronautical Society (RAeS) and the 85th anniversary of the American Institute of Aeronautics and Astronautics (AIAA). With a combined membership of more than 50,000 aerospace professionals, our two organizations celebrate these milestones and our members' never-ending quest for knowledge and solutions not just to the problems and challenges of today but also of the next impossible thing. It is our members who evolve UAS technology to even greater capabilities and uses than that exist today.

Aerospace make the world safer, more connected, more accessible, and more prosperous. We hope that the addition of this volume to the *Encyclopedia* continues this trend and is as professionally valuable and influential to its readers – and the industry – as were the preceding volumes.

As we write, there is perhaps no issue more timely in aviation than unmanned aircraft systems. That is why it is our pleasure to jointly commend to you this new contribution to the aerospace engineering body of knowledge.

Mr. James Maser
President, American Institute of Aeronautics and Astronautics
and
Vice President, Operations Program Management,
Pratt & Whitney, East Hartford, CT, USA
and
Dr. Chris Atkin
President, Royal Aeronautical Society
and
Professor of Aeronautical Engineering,
City University London, UK

ley *Encyclopedia of Aerospace Engineering* offers the
ce and robotics communities a series of accessible
s covering all disciplines of the Aerospace field. While
yclopedia is regularly updated to ensure currency, the
also decided to pursue new key volumes in important
erging Aerospace areas. This volume covers the tech-
operations, and policy challenges associated with both
nd large unmanned aircraft systems (UAS).

ll UAS operating at low altitudes are rapidly proliferating
s ranging from hobby to surveillance and package
. Configurations range from traditional fixed-wing air-
the popular multirotor helicopter or multicopter offering
dented maneuverability. Plastic and composite materi-
-cost manufacturing processes, and capable embedded
and processors support both the fully piloted and fully
nous flight. Motors powered by lithium–polymer batte-
mass-produced at low cost, yet further improvements in
l energy storage and power requirements are essential to
 small UAS range and endurance. This UAS volume
s essential background in UAS configurations and
em design with respect to aerodynamic, structural, pro-
, and power system considerations as well as avionics,
nication, sensing, control, and planning functions.

ause traditional manned aircraft have always relied
le onboard pilot or crew to assimilate information and
afety-critical decisions, UAS necessarily introduce a
r of new challenges in control, communication, and
ation management. What sensing and control strate-
 effective for the spectrum of UAS configurations and
is, and what level of decisional autonomy is required
 desired? How do remote operators maintain situa-
awareness, how can the ground–air link be ensured
and reliable, and what protocols are appropriate in lost
uations? How will UAS sense and avoid each other
nned aircraft? What functions should be implemented
d and which in the ground station? Small UAS may be
ially organized in multiagent teams to simplify mis-
ordination and handling in the National Airspace
 (NAS), the National Air Traffic Services (NATS),
ler air traffic control systems. The chapters in this
 covers the spectrum of sensors, guidance, navigation,
ntrol algorithms, and mission-level decision-making
ms offering UAS the ability to autonomously execute
 plans and effectively coordinate actions with other

UAS. Remaining challenges in ensuring secure, safe, reliable, and robust UAS operation are also discussed.

The number of UAS operations per day is expected to
quickly exceed the number of manned operations. Furthermore, these operations will routinely occupy the low-altitude
airspace not commonly used for manned aircraft today. Small
maneuverable UAS can be launched and recovered from
almost any site and flown in cluttered areas. These factors
introduce a variety of new concerns related to airspace access
and policy, privacy, and social/legal issues. What restrictions
should be placed on UAS operations based on overflown rural
to urban property as well as airspace class? How can policy and
law balance the desire to capitalize on new UAS capabilities
while respecting privacy concerns and ensure acceptable
levels of risk exposure to overflown people and property?
This UAS volume offers chapters on UAS airspace access
requirements and associated policy issues. These chapters
outline capabilities and needs for standards and processes
enabling UAS safety certification and security. As camera-
equipped UAS operate "just over our backyards," new privacy
and airspace ownership and control issues have emerged that
are still under discussion. Chapters in this volume also outline
privacy, social, and legal issues in the context of legal precedent and emerging community concerns.

As is evident from the diverse technology, operations, and
policy content in this volume, UAS are truly "multidisciplinary systems" that offer exciting new mission capabilities but
that also challenge traditional aviation assumptions regarding
operational norms and personnel roles. UAS are motivating
us to rethink information handling while truly offering
everyone low-cost access to the sky.

Ella Atkins
Department of Aerospace Engineering,
University of Michigan, Ann Arbor, MI, USA

Aníbal Ollero
Universidad de Sevilla and Scientific Advisory
Department of the Center for Advanced Aerospace
Technologies, Seville, Spain

Antonios Tsourdos
School of Aerospace, Transport & Manufacturing and
Centre for Autonomous and Cyber-Physical Systems,
Cranfield University, Cranfield, UK

PART 1

Introductory

Chapter 1
UAS Uses, Capabilities, Grand Challenges

Michael S. Francis

United Technologies Research Center, East Hartford, CT, USA

INTRODUCTION

Unmanned aircraft have existed for almost as long as the human quest to achieve manned flight. Early unmanned heavier-than-air gliders, built by such notables as George Cayley and Otto Lilienthal, were used to pioneer the technologies required of early manned aircraft. With the powered variants that followed, the twentieth century is littered with a rich history of unmanned aircraft that were created to support an ever-increasing number of missions and applications, many driven by military needs and opportunities (Holder, 2001; Keane and Carr, 2013; Newcome, 2004). The last several decades have witnessed an even more explosive increase in unmanned aircraft of all shapes and sizes, including an increasingly large number intended for civil and commercial applications. Despite this long history, the unmanned aircraft revolution is arguably still in its infancy. To better understand this assertion, it is helpful to review the technological origins of these systems.

The technological roots of the heavier-than-air, manned aircraft are firmly implanted in the industrial age. But these machines evolved considerably over their first century of existence due, in part, to the infusion of technologies that would eventually underpin the information age. Electronics, solid state devices, microprocessors, data storage, and sensors of many types would find their way into aviation systems at virtually all levels of system architecture and operation. Later, the advent of digital communications technology and introduction of the satellite-based global positioning system (GPS) added vital elements that would further enable practical, low-cost, remote operation of these platforms.

The dramatic increases in information technology over the last several decades, coupled with concomitant decreases in the size and cost of enabling electronics, would help usher in the era of affordable unmanned air systems, or UAS, that we know today. The levels of innovation and discovery that have spurred recent growth in UAS capabilities can be expected to continue. With no abatement to Moore's law in sight and new fundamental advances such as quantum computing forecast for the not-too-distant future, the information revolution is showing few signs of slowing down. From a technology perspective, UAS can be viewed as a bellwether "marriage" of the industrial age and the information revolution.

Despite the push from this high-power technology "engine" and the enthusiasm of its many proponents, UAS capability has not yet seen widespread acceptance and adoption, tempered by a number of factors that can be associated with societal "inertia." These include cultural and regulatory inhibitions, legal precedents, and infrastructure constraints. And while these factors have impacted other industries and markets, their effects appear to be especially prominent for this disruptive entrant to the aviation arena.

Unmanned Aircraft Systems. Edited by Ella Atkins, Aníbal Ollero, Antonios Tsourdos, Richard Blockley and Wei Shyy.
© 2016 John Wiley & Sons, Ltd. ISBN: 978-1-118-86645-0.

2   USES – MISSIONS AND APPLICATIONS

2.1   Early evolution

It is not surprising that development progress in unmanned aircraft was slow during the first half of the twentieth century when the technology of electronics was in its infancy and modern digital systems were essentially nonexistent. Their early operational adaptation significantly lagged that of their manned counterparts. The first entrants were experimental and supported high-risk research and development activities aimed primarily at establishing the feasibility of manned flight. The first unmanned operational designs were intended for military applications, serving as aerial targets or actual weapons.

Perhaps the first invention to have major impact on the viability of unmanned aircraft was the multi-axis gyroscope, introduced by Elmer Sperry almost a decade after the Wright brothers first flew. Sperry's invention has been cited by some as the single most significant enabler for unmanned aviation as we know it today (Newcome, 2004). Despite this important advance, early unmanned aircraft lacked the level of navigational precision necessary to reliably accomplish military objectives. Perhaps the greatest limitation for these early systems was the lack of "intelligence" required to accomplish complex missions in challenging and uncertain combat environments. Not only did these vehicles lack an onboard pilot, but also the technologies necessary for providing access to adequate *remote* intelligence (operators) did not yet exist. As a result, the operational footprint remained limited, with a focus largely on launch-and-leave concepts that were best suited to weapons and other expendable system applications. The Kettering "Bug" and Sperry–Curtiss aerial torpedo (circa 1916–1917) and the World War II vintage German V1 "buzz bomb" are examples of this trend from different eras. Over much of the twentieth century, unmanned platforms continued to serve in these roles, as end-of-life aircraft were converted to aerial targets, and high-tech cruise missile designs further expanded the arsenal of expendable, one-way platforms. It was much later that the term *unmanned air vehicle*, or UAV for short, would be employed to differentiate the reusable platforms from the expendable variants.

The first widespread use of reusable unmanned aircraft in an operational environment came during the course of the Vietnam War. The Ryan *Firebee*, originally developed for aerial target applications, was adapted to serve as an information-gathering platform in what we would call today an intelligence, surveillance, and reconnaissance – or ISR role. Launched from a large mother ship, often a C-130 transport,

the Ryan AQM-34 "Lightning Bug" was configured to perform a preprogrammed mission over a scripted route followed by a parachute recovery into friendly territory. It was a true unmanned air system, by today's definition (Keane and Carr, 2013). In contrast to the high-speed, turbojet powered Firebees developed in the United States, the Israelis introduced the first low-speed, real-time surveillance UAV during the 1973 Yom Kippur war. In both cases, these systems were introduced to achieve specific tactical objectives and retained during the conflict solely for those purposes.

Radio frequency communications technology necessary to achieve remote unmanned aircraft operation was explored and tested as early as the eve of World War I, but deemed impractical. The idea was reintroduced on the eve of World War II with some success. But the concept of remotely operated "drones," as they had then come to be called, never found a niche for a role in the broader conflict. It was a subsequent key technology development, the introduction of inexpensive solid-state radios in the 1950s that "kick started" the era of modern radio-controlled aircraft in the United States. Now familiar to present day hobbyists, this was also the first introduction of remotely operated aircraft technology to a larger non-military marketplace.

It was not until the introduction of modern, compact, high performance computer technology that the contemporary UAV became attractive enough to its user community to earn a permanent place in the defense inventory. In the 1980s, the Defense Advanced Research Projects Agency, or DARPA, began developing a new class of low-cost, long endurance unmanned aircraft that could be employed in a variety of ISR missions. The agency's preoccupation with information technology and its role in the broader information revolution at that time helped motivate adoption of the UAV as an ideal "test case." As these systems matured and gained recognition, they were embraced by a variety of government customers including the intelligence agencies and the military departments.

2.2   Dull, dirty, and dangerous

The first Gulf War provided the first large-scale operational opportunity to test unmanned air systems in a realistic military environment. A number of them, such as the Predator medium-altitude long-endurance (MALE) UAV, gained notoriety for their ability to provide persistent, real-time streaming video imagery to remote operational command posts, including the Pentagon, during the actual course

tions. At the time, this capability was viewed as a game
...ger in modern warfare.

...AVs in a variety of sizes with similar capabilities
...ged to support the allied war-fighting enterprise at
...lly all levels of command. From the portable, hand-
...hed, locally controlled Raven peeking over-the-hill for
...ill Marine Corps squad to the medium-altitude, wide
...urveillance Predator – operated remotely by a conti-
...l US-based ground crew, these systems gained broad
...tance by their user communities. These systems have
...proliferated in large numbers as a result of the conflicts
...q and Afghanistan. The introduction of the even more
...le Global Hawk increased operational altitudes to
...d 60 000 ft and endurance timelines to in excess of
...A variety of intermediate-sized unmanned aircraft such
...Shadow and Scan Eagle Tactical UAVs, among others,
...also introduced to further expand battlefield ISR to
...echelons of command to a level never before seen in
...at.

...e Reaper, a weaponized variant of the Predator UAV,
...ded a unique capability never before seen in armed
...ct. Combining persistent surveillance and precision
...ing with near instantaneous lethal response, these plat-
...served as ultimate "standoff snipers," demonstrating an
...cedented level of precision engagement coupled with
...al collateral damage. Despite these unique capabilities,
...at UAS have struggled to gain acceptance across the
...er military community.

Emergence of civil and commercial applications

...bly, the first highly visible civil (government-sponsored,
...ilitary) applications of modern UAS were in the pursuit
...entific understanding. In the mid-to-late 1990s, the US
...nal Aeronautics and Space Administration (NASA)
...d a number of then fledgling UAS developers to
...nstrate very high altitude, long-endurance civil UAS
...its Environmental Research and Sensor Technology
...ST) program. ERAST was focused on developing
...ilities for probing the upper atmosphere and providing
...pportunity for *in-situ* measurements and remote-sensing
...tion that space-based sensors could not achieve. Air-
...uch as Aurora Flight Sciences' Pegasus and Aeroviron-
...s pioneering solar-powered HELIOS, among others,
...developed and flown as part of that effort. In recent
...UAS have also been employed for studying a variety
...ospheric phenomena ranging from hurricanes to super
...understorms and incipient tornadoes (Figures 1 and 2)
...n *et al.*, 2011).

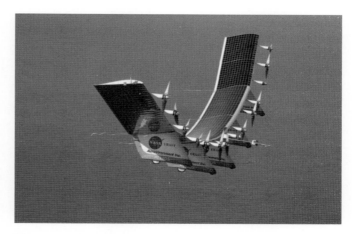

Figure 1. HELIOS unmanned air vehicle.

While UAS have become a staple in modern warfare, their
application to nonmilitary missions has risen dramatically in
just the past few years. And despite the rich operational
history of several now well-known UAS models over the past
two decades of conflict, it is a new generation of platforms
and technology that have captured the public's attention and
interest.

Attempts to develop the so-called micro air vehicle
(McMichael and Francis, 1997) reach back to the mid-
1990s. But the recent, rapid ascendance and public embrace
of these systems has been facilitated by the emergence of a
number of new key technology elements that do not have
their roots in defense technology. These include:

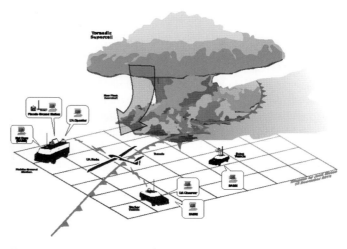

Figure 2. Operational architecture for tornadic supercell region
penetration and assessment during Vortex II campaign. (Reproduced
with permission from Wiley, 2011. © Wiley.) (Courtesy J. Elston and
B. Argrow, University of Colorado.)

- small, inexpensive, inherently stable, operator-friendly "quad-rotor" air vehicles;
- low cost, compact imaging video sensors; and
- low cost, portable control stations equipped with digital wireless radio connectivity and intuitive digital operator interface.

These small and compact systems have underpinned the explosion in interest in UAS applications. The inherent stability and straightforward control of the signature multi-rotor platforms enable novice operators to easily control or manage their aircraft trajectories within line of sight. Public interest in social media, coupled with the fascination for flight and the low-cost entry to own and operate these systems have put them in high demand for commercial and recreational users. From realtors trying to carve out a new approach to selling property to infrastructure managers wanting to inspect otherwise inaccessible areas to a host of video enthusiasts simply trying to capture a "new perspective," small UAS have created both interest and controversy across the public domain. But while the small, easy-to-operate low-end platforms have played a significant role in increasing public interest in UAS, they do not possess the range, endurance, speed, or payload capacity of their larger, fixed-wing counterparts. A diverse array of these larger vehicles stands ready to further expand the spectrum of mission and applications.

In a recent publication addressing the civil and commercial marketplace, the Association of Unmanned Vehicle Systems International (AUVSI) highlighted a diverse range of applications; encompassing wildfire mapping; agricultural monitoring; disaster management; telecommunications; thermal infrared power line surveys; law enforcement; weather monitoring; aerial imaging/mapping; television news coverage; sporting events; moviemaking; environmental monitoring; oil and gas exploration; and freight transport (Jenkins and Vasigh, 2013). That report also predicts precision agriculture and public safety to be the two most impactful areas of commercial/civil use in the United States over the coming decade. Many of these projected applications exploit the remote sensing legacy of contemporary UAS, so successfully demonstrated by earlier military systems. This array of ISR-like applications has also been energized by the proliferation of very low cost, miniature commercial imaging sensors that have flooded the cell phone and tablet computer markets. But UAS can also be expected to be employed in an array of other uses, including the transport of cargo as an example.

Like the sensor–shooter combination demonstrated by the Reaper UAS, the on-platform integration of sensors with other payload elements affords an opportunity for further expansion of missions and applications. As an example, the combination of multispectral imaging with real-time nutrie pesticide dispensing can potentially take "precision agricu ture" to another level. Similarly, real-time infrared imagi with concurrent fire suppressant application could great improve the ability to mitigate incipient wildfires. Althoug the remote sensing capability adds great value by itself, t ability to integrate it with a timely response/action mech nism greatly increases the utility of the resultant system and host of its applications.

3   EMERGING CAPABILITIES AND A LOOK AHEAD

Today's UAS, even the small inexpensive variety, posse attributes that would be the envy of their early radio-controll predecessors. New capabilities that improve operational ve satility seem to emerge on a regular basis. Auto takeoff, au land, and waypoint navigation, coupled with highly stable a controllable air vehicle designs, are now commonplace even the emerging commercial marketplace.

And while information age by-products have added si nificant new enabling capabilities, they have also spurr new developments on the industrial age side of the equatio For example, the search for more effective and efficie propulsion methodologies, new structural constructs, a new materials may be more extensively exploited in t unmanned systems community than the traditional mann aircraft segment.

3.1   Expanding the design space and operational envelope

The elimination of human presence on board the air vehic affords an opportunity to introduce new attributes and missi capabilities to the aircraft, and, to some extent, redefine flig as we know it. To better understand this opportunity, it is use to examine the benefits and constraints of the tradition onboard human presence (pilot, crew, and passengers).

In the early days of manned flight, the pilot performed the functions necessary to control and manage the aircra The achievement of meaningful mission objectives und complex circumstances without the aid of an onboard hum was virtually unfathomable. The pilot's eyes and ears we primary sensors, hands and feet served as primary effecto and the human brain was the integrated flight and missi computer responsible for everything from basic maneuv execution to comprehensive mission management. In add tion to direct sensing, the pilot was responsible for da interpretation and information synthesis related to all aspe f aircraft operation. Today, many of these requirements are located to automated systems, allowing the pilot and crew • focus on top-level supervisory tasks.

In contrast, the accommodation of human presence on bard the aircraft has proven an ever more daunting and source consuming task, as aircraft operate in domains far ore demanding and complex compared to that experienced y the early aviation pioneers. The addition of pressurized stems with oxygen has become indispensable for high-titude long-range operations. Other constraints imposed by e human anatomy have limited the way flight is prosecuted .g., coordinated bank to turn) and has significant impact in e design and configuration of the aircraft. The need for ecific orientation with respect to the gravity vector has mited the way vertical takeoff or landing (VTOL) flight is hieved and has significant impact on the design of those rcraft.

Onboard human presence impacts flight vehicle design in any ways. It constrains platform acceleration in all axes, d limits vehicle endurance. The human factor has impacted e approach to reliability and safety from both design and erational perspectives. The need to maintain human func-onality and performance within the volume and weight onstrained confines of the cockpit has necessitated extensive d often costly training and proficiency regimens for the rcrew community. Onboard human presence has also voked the addition of unique infrastructure, and diversely nging from specialized training simulators to search, res-e, and other support capabilities that come into play in the ent of aircraft mishaps. The on-aircraft interface between uman and machine has become quite complex, encompass-g everything from integrated control effectors, and com-ex displays to power-consuming environmental systems at enable crew comfort and survival during flight at all hievable altitudes and airspeeds.

While today's unmanned aircraft take full advantage of eir capability for remote operation, few designs fully ploit the absence of human presence. This potential to large the air vehicle design and operational envelopes is gnificant. A number of attributes that could be more fully vantaged in that regard include the following:

Extreme Endurance: This ability of a platform to stay aloft for periods that well exceed normal crew limitations has been demonstrated to a large extent in current opera-tional systems. The attribute is a key performance driver for ISR mission systems such as those depicted here. 24-h endurance capabilities are commonplace for larger plat-forms and are rapidly becoming possible for their smaller, tactical-size counterparts. Designers are currently focused on week-long operation, with some experimental systems

(a) Northrop Grumman *Global Hawk*

(b) General Atomics *Predator B*

Figure 3. Long-endurance aircraft examples.

attempting even longer durations. Future missions such as aerial cell-phone relay and Internet distribution platforms will benefit greatly from these capabilities (Figure 3).

- *Small Size/Scale:* The ability to build and operate aircraft incapable of physically accommodating an adult human presence has already been realized. Small UAS (sUAS), such as the AAI Shadow UAS, have proliferated in military missions for over a decade and performed admir-ably in a variety of tactical roles. They are enablers for a variety of civil and commercial applications ranging from highway/bridge infrastructure inspection to precision agriculture. Even smaller variants, so-called "micro air vehicles" have begun to appear in real operational roles, although the very smallest, such as the Aerovironment Hummingbird is still in the experimental stage. With linear dimensions not exceeding 15 cm in any axis, they may prove extremely useful in highly confined spaces, such as building or even pipe interiors. With further miniaturization, they may even prove useful in internal bio-medical applications – exploring the interior of the human anatomy (Figure 4).

- *Extreme Maneuvering Capability:* Unmanned platforms are, in principal, capable of sustaining accelerations and

(a) Textron-AAI *Shadow UAV*

(b) Aerovironment *Hummingbird* Micro Air Vehicle

Figure 4. Small UAVs. (Courtesy AeroVironment, Inc.)

• *Arbitrary Orientation:* Unlike manned aircraft, the phy ical orientation of an unmanned air vehicle in any pha of flight need never be dictated by human physiolog limitations. In theory, it can be completely arbitrar limited only by physics-of-flight considerations and mi sion needs. An example illustrating the potential utility this attribute is found in the tail-sitter concept, who roots go back to the early 1950s in a then-impractic manned design (Chana and Coleman, 1996; Taylor ar Michael, 1977). A modern UAS variant (the Sikorsk VTOL X-Plane) is an aircraft designed to pioneer th prospect of runway independent launch and recove while achieving fixed wing-competitive cruise speed This class of aircraft has the potential to revolutioni high-priority transport – enabling, for example, tl retrieval of cargo from a ship at sea and its rapid transpo to even remote, unimproved areas without the need for runway or any other form of terrestrial transportation the process (Figure 6).

• *Unique Configurations:* Unmanned aircraft designe have already provided numerous examples of unconve tional configurations ill suited to manned flight. Co temporary vehicle control technology has enabl innovative designs that capture the best of fixed ar rotary wing designs in a relatively simple mechanic package. The lack of need for a conventional cockp coupled with other attributes mentioned above can rest in novel shapes and configurations more germane niche missions or unique flight environments. Exampl of these include the 1998-vintage Cypher UAV and as yet untested dual-free wing concept capab of morphing from the biplane configuration (showi operating at near zero-speed hover conditions, to tandem-winged, tailless orientation that could fly extremely high speeds because of its low aerodynam drag (Figure 7).

• *Unconventional Launch and Recovery:* Novel appr aches for launching and recovering UAS have increas significantly, especially for smaller air vehicles. Assist rail launch capability and net or tether recovery techniqu

forces limited only by structural considerations, operating well beyond the tolerance of any human pilot. Turning accelerations up to approximately 30 g's – the limit where modern turbojet engines begin to experience "out-of-round" geometric distortion, might be possible without the introduction of other new technology. Such capabilities could revolutionize modern air combat, even with the introduction of more agile air-to-air weapons. Structural morphing capability, such as that depicted in this DARPA concept could be exploited to increase maneuver accelerations in future unmanned combat aircraft (Figure 5).

Figure 5. Extreme maneuvering may be enabled by morphing structure. (Reprinted with permission from Defense Advanced Resear Projects Agency. Approved for Public Release, distribution unlimited.)

Figure 6. Arbitrary orientation – example: the Sikorsky VTOL X-Plane Concept. (Reproduced with permission from Chris Van-Buiten, 2016. © Sikorsky Innovations.)

(a) Cypher UAV

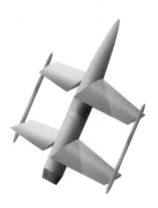

(b) Dual Free Wing UAV Concept

Figure 7. Unique configurations.

are employed on operational aircraft such as the Insitu Scan Eagle. Larger systems may find similar opportunities as in the depicted shipboard concept where a high-performance, unmanned UCAV is tube-launched like a cruise missile and later recovered shipside, following a near vertical, high angle-of-attack approach. An articulating, conforming porous arresting structure is configured to match the aircraft's approach orientation. The concept eliminates the need for conventional, often heavy landing gear, improving the vehicle's range-payload performance. More broadly, the concept reduces the need for conventional aircraft carrier operations, while simultaneously providing airpower projection capabilities to other surface ship types. Future novel launch and recovery techniques may well enable UAS operations in otherwise impractical civil and commercial environments as well (Figure 8).

Attritability: The notion of an "attritable" (limited life, yet reusable) vehicle design is unrealistic for manned aircraft, but the capability could become a practical option for a number of unmanned vehicle applications. This airplane equivalent of a reusable but otherwise throw-away styrofoam cup could fill a gap between the long-life, fatigue-limited manned aircraft designs and the single-use missile/projectile configurations in common use today.

The very first DARPA UCAV platform concept depicted here invoked that capability. During peacetime, manned combat aircraft can spend in excess of 90% of their flight hours for aircrew training and proficiency. For UAS, there is no "seat-of-the-pants" benefit to flying these aircraft during training missions unless operational synergies with close proximity manned aircraft are also sought. For all applications that require highly intermittent flight operations or those that involve lengthy downtime periods, limited life aircraft designs could offer significant life-cycle cost advantages. Missions such as *in situ* sensing of toxic or radiation clouds, where the vehicle may have to be disposed following a mission, as well as other missions into dangerous environmental conditions, such as extreme weather or other threats may be ideal for attritable aircraft. However, cost savings derived from attritable designs are likely limited, since these aircraft must possess the necessary levels of safety and reliability to conduct missions in shared common airspace (Figure 9).

Figure 8. Novel shipboard launch and recovery of high-performance UAV.

Figure 9. "Attritable" aircraft concept icon – DARPA's Unmanned Tactical Aircraft.

3.2 Autonomy

Automation has already contributed to a number of operational improvements for UAS by providing more time flexibility for the remote human crew to assess and act on information as the mission unfolds. The continued advance of modern computing power is opening the door to ever higher levels of autonomous operation, where the human element is fully relieved of many minor decisions and becomes essentially supervisory in nature.

Fully autonomous flight remains an elusive objective for UAS proponents and will likely remain so for the foreseeable future. The leap from today's automation to tomorrow's autonomy is not a small step. An automated system is constrained to operate within prespecified bounds, with anticipated and preprogrammed alternatives available in the event of non-nominal circumstances. Most automation today is centered on basic, prescriptive flight functions such as, for example, vehicle control (e.g., fly-by-wire control) or navigational route execution (so-called way-point navigation). These advances have greatly improved the ability for the remote crew to interact intermittently in controlling the aircraft. However, much remains to be done in this domain. For example, many systems today limit UAS operation to one vehicle by one operator. Studies have been conducted to illustrate the possibility of managing multiple aircraft with a single human operator, if the level of supervisory interaction is high enough (Ruff, Narayanan and Draper, 2002). For the latter to occur, the level of autonomy at the vehicle and system levels must increase dramatically.

Increasing the level of autonomy of an unmanned air system requires more than adding functionality in the form of new tasks or increasing task levels. A truly autonomous system would be capable of identifying and assessing a broad range of mission-level conditions and then adapting, as needed, to accomplish necessary tasks as the mission unfolds. It would be capable of brokering solutions that account for multiple objectives and circumstances that may have impact on the aircraft in its mission over several time scales (epochs) simultaneously. For example, an aircraft may be faced with a short-term requirement to avoid an unexpected obstacle, while coping with a potential threat just over the horizon, and while also facing a change to its overall mission endgame objective. A capable *autonomous mission manager* must cope with all of these circumstances simultaneously, while projecting an acceptable solution and executing successful outcomes throughout the mission timeline. The system would be capable of dealing with a broad range of variables, ranging from traditional well-defined physical parameters to less objective conditions, such as evolving environmental changes or even less predictable threats, such as those imposed by a human adversary.

The metric that best separates an autonomous system from a highly automated one is the ability to cope with the unknown – the condition or situation which was not considered in the system's "in-the-box" design. The ability to manage these kinds of contingencies will define the level of autonomy in future UAS. These future autonomous systems must be capable of learning from their experience, for it is that trait and the ability to adapt as a result that enables this behavior in the first place. Coping with the statistical probabilities associated with the operational environment, and adapting to conditions in a manner that will improve performance, mission success, safety, or other desired objectives are key behaviors that the autonomous system must master.

Robotics and artificial intelligence remain hot topics with seemingly limitless applications – from biomedical devices to driverless cars and unmanned air systems to domestic robots that can do the family laundry. Many of the challenges associated with advancing this disciplinary arena for the broader robotics community are well documented (Hager *et al.*, 2015).

GRAND CHALLENGES AHEAD

Despite the tremendous potential of unmanned air systems across a range of economically beneficial and compelling applications, the obstacles to their successful introduction and implementation are significant. UAS today face a number of constraints that technology alone cannot overcome. Many are rooted in competing legacy systems and methods, as well as in institutional, regulatory, and cultural precedents that minimally assure a lengthy transition to an acceptable, productive future state. As a result, economic limitations for these systems are no longer centered on the cost of hardware and software. The fundamental inhibitions to ownership and operation can be found in the lack of acceptable regulatory infrastructure to guide their operations, combined with institutional conservatism in dealing with companion liability, insurability, legal issues, along with the concomitant consequences of negative public perceptions.

The "Grand Challenges" are those that require a coordinated, integrated approach to collectively address all these issues, technical and otherwise, in a manner that will enable UAS of all types to reach their full potential.

.1   Access to the airspace

Today, limited access to the airspace is the dominant barrier to the realization of the full economic potential that can be derived from UAS capabilities. Most of today's operational requirements that regulate UAS operations in the common airspace are rooted in the regulatory precedents set by and for manned systems over decades with an evolved operational paradigm centered on pilot capabilities and behavior. In the manned aircraft, the pilot is omnipresent – assumed able to assess in-flight circumstances from a cockpit perspective and react to them virtually instantaneously. This is not the case with the modern UAS.

An array of real-world constraints and limitations is responsible for this dichotomy. These include wireless connectivity issues, including communications latency; environmental factors; and human frailties that can become exaggerated in the quest to provide the continuous human presence. The latter set ranges from situation awareness limitations imposed by the finite number and types of sensor and information sources to fundamental limits to human attention spans. More subtle factors associated with human cognition may also play a role. These constraints can be less significant in some operational circumstances, for example, short-duration flights within visual line of sight between the aircraft operator in reasonably good weather conditions. The problem can become acute in long-range beyond-line-of-sight operations, especially in adverse weather and/or in airspace crowded by aircraft or other physical obstacles. The need for UAS to project a continuous "crew presence" – able to respond with no delay, replete with a fault-free wireless connection between platform and remote crew simultaneously – represents its most demanding requirement and its greatest vulnerability.

In keeping with the slow evolution of manned flight prevalent over the last half-century, these rules have been slow to change despite the emergence of new or improved technologies designed to enhance reliability and safety. To the impatient drone entrepreneur, progress in integrating these systems into the common airspace appears glacial across much of the breadth of the international landscape.

Larger UAS must compete to share already crowded airspace with the manned platforms that have set the precedent and expectations for flight safety. Small UAS, in contrast, are pioneering access to a new region of airspace largely unfamiliar to both pilots and regulators. This low-altitude, obstacle-rich environment, ranging from below approximately 150 m down to the "blades of grass" adjacent to the earth's surface, presents a variety of challenges to remote operations. These include people – often transiting near vehicle flight routes, personal property adjacent to and along those routes, and other hazards, including nearby trees, buildings, and other obstacles.

The most difficult situations will likely involve operations in urban canyons, where traditional navigation sources like

GPS are intermittent or unavailable. The most demanding of these environments have rarely, if ever been encountered by larger manned aircraft. They present a new set of challenges for the regulatory communities and the public, as well. Ironically, it is this most complex set of environments that the smallest, least capable platforms and systems (size, weight, and power) have chosen to invade.

4.2   The quest for *trust*

The arguably greatest challenge and impediment to UAS acceptance and mission proliferation lies in gaining *trust* in the behavior of these technologically advanced systems. This need extends to the manned aircraft-dominated user community, an outdated and often incompatible regulatory system needed to support and promote their operation, and most importantly, a skeptical public.

Although UAS technologies have made significant strides over the past several decades, their vulnerabilities are well known to most. A century of manned aviation evolution has set high expectations for safety and reliability yet to be matched by the unmanned community. Growing prospects for cyber-physical security threats in recent years have added to public skepticism. Along with growing concerns over the illicit use of UAS and their prospects for violating individual privacy, resistance to their broad introduction has been significant (Tam, 2011; Watts, Ambrosia, and Hinkley, 2012). Many of these concerns are directed at the system users, and especially at their intent and integrity. They are likely to be resolved through a combination of properly defined regulatory constraints, coupled with adequate education of potential users and the public as well.

A more immediate concern that has long-term implications over the continued evolution of UAS revolves around the issue of trust in *intelligent software*. This turns out to be a problem for manned and unmanned systems alike. And it has its roots in a long evolved methodology for developing trust in physical systems.

Traditional rigorous hard science-based evaluation methods created to assure developer, user, and even public confidence in engineering products such as airplanes are not likely to prove adequate for the certification of future intelligent unmanned air systems. And software is the culprit. As software-based approaches and processes have proliferated within the aviation ecosystem, their collective verification, validation, and certification (VV&C) has proven to be perhaps the most significant factor to date to impact aircraft affordability. Current VV&C techniques based on, for example, FAA-referenced DO-178 B/C and comparable standards continue to consume an ever-increasing proportion of aircraft

development budgets. Prospects for their application future advanced unmanned air systems could prove eve less successful.

The incompatibility of today's software and system VV&C regulations with future, more fully autonomou systems represents a major obstacle to the advance towar more capable UAS. As is the case with hardware, softwar verification and validation techniques rest heavily on a tes ing philosophy that is comprehensive and a companio methodology that is thorough. In hardware, scientific law and principles underpinned by years of research have bee used to derive the transfer functions that relate input stimu to quantifiable output expectations, with predictable err limits. This is not the case for software. The substitute for th elusive transfer function has been exhaustive testing of eve logic path that exists within the man-made code. As softwar has become more capable and complex, this testing proce has become more imposing and costly, in many case pressing on the limits of system affordability. The bas construct, which served so well in gaining engineerin confidence in the early days of software definition an development, has become a significant burden in the nea explosive advance of the information revolution.

The software test philosophy has affected all curre generation aeronautical systems, due to the sheer complexi associated with the large number of system interactions th the software must reflect. Recently, suggestions to redefi the verification and validation (V&V) processes based o methodologies that rely on model-based design and form methods have provided some near-term hope for reductio in testing. But these tools today have limited to no utility an supporting the development of *intelligent software*.

The intelligent software that will enable true autonomo functionality will be capable of adaptation to emergin mission and environmental conditions, potentially exhibitin attributes such as emergent behavior and other nondetermi istic features. These are systems capable of learning in t course of operations and applying that knowledge to futu situations. Current bottoms-up methods for software evalua tion based on or that assume inherent determinism a incompatible with these intelligent systems. The fundament issue that must be resolved is not only related to the curre approach to VV&C, but also to the very attributes an characteristics that must define the *intelligent software* itse

It is interesting to contrast the methodology employed develop trust in human-authored software (today's VV& procedures) with the seemingly much simpler and qui different process used to certify true *human software*, th is, the pilot. That latter interaction between pilot and exam iner is usually a relatively brief encounter, involves most high-level logic associated with complex mission-base cenarios, invokes the desire for flexible, acceptable out-omes, and takes place in the domain language of flight, as pposed to some foreign language (i.e., software code) nfamiliar to the principals. The dialog between student nd certifier is less about precision than it is about decisions nd judgment. And it explores the learning acquired by the udent as the mission unfolds, along with the behavior it vokes. Future *intelligent software* may need to possess some f the same traits to permit a very different approach to V&C from what we know it to be today.

.3   Integration

ltimately, the development of a methodology that addresses ie certification of and trust in an integrated *man–machine* ystem, where both elements are considered together in chieving acceptable operational results, is essential. The aditional methodology of dealing with the machine and uman operator separately made sense in the industrial age here *all* system intelligence was provided by the human, nd the exclusively hardware-based machine was solely the roduct of hard science-based engineering. That is certainly o longer the case for even today's modern systems, and the istinction will continue to blur as more and more of the itelligence resides in and is endemic to the machine.

The nature of the interaction with human supervisory perators will begin to evolve based on our understanding f *human–machine intelligence* integration. A system that ptimizes this interaction in a manner so that the integrated ystem performance well exceeds that of the independent um of its parts" will likely prove to be a significant iallenge for some time to come.

SUMMARY

espite a century plus of slow evolution, the unmanned air ystems revolution is technologically still in its infancy. ontinuing advances in computing power will enable ever iore capable systems – exploiting enhanced logic and :nsing to achieve more versatile platforms that enable ew and diverse missions with an economic leverage as ood as any to emanate from the revolution in robotic ystems. The integrated regulatory, legal, social, and cultural ndscape poses the greatest array of impediments to this advance, but an ever-increasing and compelling array of capabilities and applications appears to have the edge in shaping the future of this upstart niche in aviation and aerospace.

REFERENCES

Chana, W. and Coleman, J. (1996) World's First VTOL Airplane Convair/Navy XFY-1 Pogo, SAE Technical Paper 962288. doi: 10.4271/962288.

Elston, J.S., Roadman, J., Stachura, M., Argrow, B., Houston, A., and Frew, E. (2011) The tempest unmanned aircraft system for *in situ* observations of tornadic supercells: design and VORTEX2 flight results. *J. Field Robot.*, 28(4), 461–483.

Hager, G.D., Rus, D., Kumar, V., and Christensen, H. (2015) *Toward a Science of Autonomy for Physical Systems*, Computing Community Consortium, Ver. 1.

Holder, B. (2001) *Unmanned Air Vehicles – An Illustrated Study of UAVs*, Schiffer Publishing Ltd, Atglen, Pennsylvania.

Jenkins, D. and Vasigh, B. (2013) *The Economic Impact of Unmanned Aircraft Systems Integration in the United States*, Association of Unmanned Vehicle Systems International (AUVSI), Alexandria, Virginia.

Keane, J.F. and Carr, S.S. (2013) *A Brief History of Early Unmanned Aircraft*. Johns Hopkins Applied Physics Laboratory Technical Digest, 32(3), 558–571.

McMichael, J. and Francis, M. (1997) The micro air vehicle – toward a new dimension in flight, *Unmanned Systems*, vol. 15, Association of Unmanned Vehicle Systems International (AUVSI), Alexandria, Virginia, 10–17.

Newcome, L.R. (2004) *Unmanned Aviation – A Brief History of Unmanned Air Vehicles*, American Institute of Aeronautics and Astronautics (AIAA), Reston, Virginia.

Ruff, H.A., Narayanan, S., and Draper, M.H. (2002) Human interaction with levels of automation and decision-aid fidelity in the supervisory control of multiple simulated unmanned air vehicles. *Presence – Teleoperators and Virtual Environments*, vol. 11, MIT Press, Cambridge, UK, 335–351.

Tam, A. (2011) Public perception of unmanned aerial vehicles, *Aviation Technology Graduate Student Publications*. Paper 3, Purdue University e-Publications, 2011, Available at http://docs.lib.purdue.edu/atgrads/3.

Taylor, J.W. and Michael, J.H. (1977) *Jane's Pocket Book of Research and Experimental Aircraft*, Collier Books, New York.

Watts, A.C., Ambrosia, V.G., and Hinkley, E.A. (2012) Unmanned aircraft systems in remote sensing and scientific research: classification and considerations of use. *Remote Sensing*, 4(6), 1671–1692.

PART 2
Missions

Chapter 2

Remote Sensing Methodology for Unmanned Aerial Systems

Brandon J. Stark and Yang Quan Chen

School of Engineering, University of California, Merced, CA, USA

1 INTRODUCTION

Unmanned aerial systems (UASs) have rapidly developed into a promising tool for remote sensing applications across a wide range of disciplines, from archeology to wildlife conservation. They can be designed and customized to fulfill a spectrum of characteristics and capabilities, such as low-altitude flying, long endurance, high maneuverability, and automated flight controls. But the UAS is simply the platform from which the target data are acquired. Unfortunately, with the multitude of UASs and combinations of sensing equipment, it can be a daunting challenge to determine the correct or cost-effective solution. The development of a thorough project methodology is an effective tool for addressing this challenge.

Section 2 of this chapter provides a guide to developing an effective methodology for UAS-based applications. Section 3 identifies several core attributes across three major types of remote sensing applications to guide the development of a methodology and influence equipment choices. Finally, in Section 4, imaging equipment attributes are discussed to provide guidance in their selection. While there are a multitude of different types of UASs and sensors, the chapter will utilize small UASs (<55 lb) and optical-based remote sensing as an example, although the overarching message is applicable for any UAS and sensing technique.

2 UAS REMOTE SENSING METHODOLOGY

It is far too easy for an application or project to be proposed with a UAS without a clear concept of the necessary methodology to address the problem. While public interest has fostered technological innovation, literature has been sparse of general methodology approaches for the unique challenges of UASs. Instead, UAS research is saturated with specific application with specialized workflows and methodologies unique for the immediate application. It has become necessary to promote methodology for the development of new applications and mature UASs.

An important challenge for the UAS project developers is to translate layman statements such as "Let's use a drone to improve land management practices" into "Let's use a remote sensing platform carrying radiometrically calibrated optical imagers in the visible and near-infrared (NIR) spectra for the bare ground classification of a 10 square mile area with a desired optical resolution to discern the endemic

Unmanned Aircraft Systems. Edited by Ella Atkins, Aníbal Ollero, Antonios Tsourdos, Richard Blockley and Wei Shyy.
© 2016 John Wiley & Sons, Ltd. ISBN: 978-1-118-86645-0.

population of Meadowfoam (*Limnanthes alba*)." The first statement is a wishful goal; the second introduces the methodology necessary to ensure a successful application and that the initial development and equipment purchases will lead to an effective solution.

An effective methodology defines the end goal, the activity, the implementation of the activity, the measurement of progress, and the success of the project. It provides a guideline for solving the targeted problem with specific tasks, components, and metrics. An incomplete or poorly defined project methodology can lead to development delays, spiraling costs, purchases of incorrect equipment, or complete project failure. In practice, many project developers find it useful to formulate a project methodology in terms of a series of questions such as the following (as adapted from Bhatta (2013)):

- What is the purpose of the project?
- What is the stated goal of the project?
- Is the goal quantitative or qualitative?
- Does this project utilize the scientific method or the technological method?
- What objects or events are the desired outcomes related to?
- Are there specific relationships found within the object or event of interest that can be utilized or must be taken into consideration?
- What data are necessary to address the problem?
- How should the data be collected?
- What procedures should be used to analyze the data?
- Are there available models/procedures sufficient to analyze the data?
- Does it require developing new models/procedures?
- What efforts must be undertaken to ensure the validity and reliability of the project?
- What ethical issues need to be addressed?

Addressing the questions above and/or other clarifying questions about the proposed project is designed to help form connections between goal and implementation and identify specific methods that will enable the successful completion of the application or the project.

The first step in any project is to understand the goal with the intended purpose of narrowing down the language to actionable items. Simple classifications such as separating the goal between quantitative goals and qualitative goals are often useful in this regard. This step often requires a thorough understanding of the desired goal that may not always align with the wording of the stated goal. For example, a project with a purpose of "improving crop yield" utilizes language that implies a qualitative goal, but in practice would require quantitative goals such as "improve yield by 5%," which implies accurate measurements to be achievable.

The method or body of techniques of the project is another example of a way to provide guidance to the development of an effective methodology. For UAS remote sensing applications, the scientific method and the technological or engineering method are the most common. Whereas the scientific method strives to advance knowledge, the technological method addresses specific problems or issues. If the scientific method is about *knowing*, then the technological method is about *applying* (Bhatta, 2013). The two methods may overlap at times and utilize similar approaches and equipment, but the differences play a role in the development of a UAS remote sensing methodology.

The scientific method can be described as a set of techniques based on empirical and measurable evidence with principles of reasoning and inquiry to arrive at new knowledge. It is a cycle of observations, refining hypotheses, and testing, until a thoroughly vetted understanding can be presented as knowledge. Environmental research UAS applications typically fall under this category and assume that the technical capability of the UAS-based remote sensing is sufficient. In contrast, the technological method is an application of research, directed at a specific target goal or a desired state. In this approach, the enabling technical capability is the target end goal. Validation and testing become methods to measure progress rather than part of the implementation. In some projects, both methods may be employed, such as answering a scientific inquiry while developing the underlying technical capability. Clarifying the goals and the methods of the project can help put realistic targets and progress metrics within the context of the project end goals, and prevent cost-control problems from inadequate detail planning.

Examining the relationships of the desired objects and events of the goal is another aspect of forming a methodology. Keeping track of strong correlations and dependencies can be valuable. In some cases, the target goal, for example, "measuring chlorophyll content," might show a positive correlation with a reflectance ratio calculation known as normalized difference vegetation index (NDVI; Jones and Vaughan, 2010). Thus, utilizing NDVI might be an effective method. However, a thorough methodology may identify that NDVI also shows a strong correlation with a leaf area index (Jones and Vaughan, 2010), which may complicate the desired goal measurements if the influences of the two correlations cannot be separated.

Understanding the goals and ways that the desired data can be collected provides some guidance for equipment, software, and workflow requirements, but selecting the right pieces can still be a significant challenge. There are a wide variety of platforms, sensor packages, software solutions at an equally wide range of costs, and capabilities already commercially offered, but even still many researchers and developers end up implementing their own custom solutions (Stark, Smith, and Chen, 2013). This application-centric approach, choosing equipment based on the specific requirements of the application, is common, given the narrow and specialized applications proposed. However, this drives up costs and delays projects when incorrect equipment is purchased or developed.

Once the project's data goal is selected, the data must be collected. Data collection strategies vary significantly based on equipment, although there are plenty of examples of the use of modified equipment (Chabot and Bird, 2012; Jang and Kim, 2008; Jensen, Baumann, and Chen, 2008). However, one of the major challenges for remote sensing applications of UASs is the lack of standardized processing procedures. As many developers and researchers have discovered, specialized workflows are often necessary to process their data. Unfortunately, this poses problems in addressing whether or not the results of the project were valid and reliable. It is not an uncommon problem however, especially for remote sensing operations where different data generating processes can create data that may not be comparable with other sources (Trishchenko, Cihlar, and Li, 2002). Sections 3 and 4 provide guidance on selecting what type of data should be collected and how to collect it.

Ethical and legal issues are significant topics that require addressing with an effective methodology. The current legal environment, especially in the United States, is particularly challenging to traverse. However, it is important that UAS applications are developed with the legal restrictions and limitations in mind and understand how they may affect the data collection process and feasibility of the proposed goal. A challenge may arise from addressing privacy concerns. A common technique is to employ a "Privacy by Design" approach (Cavoukian, 2009), incorporating privacy considerations into the technology and methodology that addresses it at all stages: data collection, data management, data dissemination.

CORE CONCEPTS IN UAS REMOTE SENSING APPLICATIONS

In the following section, several core concepts are identified to provide guidance in the selection of the necessary data requirements and its influence on UAS and sensor selection. While there are many unique solutions in UAS remote sensing applications, there are some common equipment and workflow implementations that are useful to refer to when analyzing the data goal for a proposed project.

UAS remote sensing applications can be grouped by data goals into three major categories: detection or counting applications, identification or localization applications, and analysis applications. Detection or counting applications are focused on detecting or counting targets. Unlike the other types of applications, the data in these applications are in the form of contrasts, such as person versus not-person. Identification or localization applications are focused at understanding the contextual information associated with a target. Rather than looking for a herd of cattle, the size and location of the herd is vital to the application. Analysis applications require further investigation of the data and contextual information to create calibrated and meaningful or actionable information, although these applications can be very complex to establish. In general, the increasing complexity of the application is proportional to the costs, both in time and money.

3.1 Detection/Counting Applications

The detection or counting of targets is a common and valuable wide-area monitoring UAS application. Conceptually, the goal of such applications is simple: to find the existence of the desired target. The significant challenge is to determine the optimal way to separate the target from the rest of the scene, either of which could be static or moving. The target is the primary goal, thus the accuracy of the separation or classification is paramount to success rather than the accuracy of the image or other measurements. The separation or classification of the target can be accomplished in any variety of ways by focusing on finding specific characteristics such as color, texture, or shapes that are unique for the target. Additional contextual information, such as location, time of day, or *a priori* knowledge, may also be valuable for improving the accuracy, and could require the use of data fusion techniques or statistical modeling to reduce errors. However, in contrast to the accuracy requirement of the detection, the collection of the characteristic or contextual information is reliant on precision or the repeatability or reproducibility of the collection of the information. This is an important distinction to make because it may affect equipment choice. For example, if the goal is to find hogs on property (Hirsch, 2013), a thermal camera is an effective tool, but the temperature measurement of the hog is not of value, only the contrast of hot and cold. A lower cost precise thermal imager may be utilized rather than a more expensive accurate thermal imager.

Identifying the characteristics or contextual information necessary for detection influences equipment selection. Many characteristics such as texture and shape often require a high spatial resolution to discern small features. Contextual information, such as location, size, and depth, can be inferred from motion determined from images with a high temporal resolution such as individual frames in a video. Automated low-level control found in many cameras and video systems such as color balance, autofocus, aperture, and shutter speed control can be effective at maintaining the visibility of the image for characteristics to be discerned.

The time sensitiveness of the application also plays a role in the equipment selection, more so in detection and counting applications than the others (Doherty and Rudol, 2007). Often an immediate reaction is desired at the detection of a target, such as returning home or changing search patterns. This level of visual feedback into the system often requires real or near real-time communication and systems with a high frame rate are best suited (Peschel and Murphy, 2013). The desire to have an independently operated imaging system often requires the same level of visual feedback as well. For these reasons, video systems are more common for detection and counting applications where immediate visual feedback is prioritized over image quality and resolution.

The processing of the data can be automated or manually done with a human operator. Automated machine vision algorithms have been utilized and demonstrated widely, although human operator monitoring are commonplace. Search and rescue operations, especially, are staffed with human operators due to scene complexity and ease of implementation (Woods et al., 2004).

However, there are specific challenges to detect and counting applications. For automated machine vision systems, the data processing increases significantly with image resolution, but too low of a resolution limits the ability to discern details such as texture. Human operators who monitor real-time video also have a number of challenges, as documented by studies on human factors for search and rescue operations with teleoperated robotics (Murphy, 2004). Operator fatigue and sensory overloads are common issues that lead to decreased detection and counting accuracy (Freed, Harris, and Shafto, 2004). Long operations may be limited by UAS platform capabilities, proper selection of the desired platform is another key for success (Stark, Smith, and Chen, 2013). In addition, the data bandwidth of the video system is often much greater than the rate of detection, leading to a significant amount of wasteful redundant data. From that challenge, it is important to recognize the value of optimal path planning and optimal sensing strategies (Chao and Chen, 2012).

3.2 Identification/Localization Applications

In many situations, the characteristic or contextual information of a target is a part of the data goal. This transforms the application into an identification or localization application, where instead of asking "is it there?" the question is "what is it?" Characteristic or contextual information commonly includes location and surroundings, but may also include size, time, color, or texture. These attributes often require a higher spatial resolving capability of imagery, though not necessarily always a faster temporal resolution. The addition of this information enables the classification or identification of a number of items such as plants, animals, vehicles, or sustained damage. However, the challenge of classification introduces the need for repeatability and consistency from image to image.

A wide variety of sensor equipment can be utilized for identification or localization applications. Video systems can be utilized effectively as described in firefighting efforts (Ambrosia et al., 2003; Hinkley and Zajkowski, 2011). Digital cameras can often provide a higher resolution and many are affordable solutions where real time is not necessary. Other specialized equipment such as thermal imagers, multispectral imagers, or hyperspectral imagers are also effective equipment though are often a costly investment. Remote sensing applications may also utilize nonimaging sensors for air quality measurements and the inclusion of localization data enables the creation of detailed spatial maps.

Whereas some detection applications can be accomplished without specialized equipment, identification and localization applications often require contextual information to be stored during image collection and additional processing to fully utilize it. UAS payloads may employ camera systems with embedded Global Positioning Systems (GPSs) to record image locations. Photogrammetry software such as Pix4D (Pix4D SA, www.pix4d.com) and Ensomosiac UAV (MosaicMill Inc., www.mosaicmill.com) are commonly a part of the workflow.

The tracking of moving targets is another common UAS application that combines the challenge of object identification and localization (Ren and Beard, 2004). Challenges such as multiobject tracking may require the use of real-time data downlinks or significant onboard computing power. As with detection and identification applications, the use of auxiliary processing and data fusion algorithms may be useful for improving results at the expense of cost and complexity.

3.3 Analysis Applications

Analysis applications are typically complex and require significant development and a strong methodology. While identification applications ask "what is it?" analysis applications are designed around the question "what does it mean?" In essence, they are designed for the purpose of transforming remote sensing data into meaningful or actionable intelligence. The counting application will return with the information that there are 12 trees in the grove. The identification application will return with the location and size of each tree. The analysis application will generate the data to make estimations on the health of the trees and how much fruit will be produced.

In analysis applications, often the data produced is not the image, but rather a 2D map of the optical sensor measurements. As such, sensor calibration, radiative transfer models, ground control points, and bias corrections are standard elements of the analysis application workflow in an effort to relate sensor measurements to physical features. Commercially available point and shoot digital cameras may not always be well suited for these applications as they typically lack the ability to record sensor measurements. Multispectral cameras and hyperspectral imagers are commonly implemented and have demonstrated effectiveness in agricultural applications such as crop monitoring (Berni *et al.*, 2009) and environmental applications such as invasive weed monitoring (Rasmussen *et al.*, 2013).

The value of calibrated imaging equipment can be interpreted in the spectral reflectance of grass, dry grass, and brown sandy loam (Figure 1) (Baldridge *et al.*, 2009). Live vegetation, including grass, has a distinctive pattern of spectral reflectance or the amount of light that is reflected. Vegetation typically appears green to the human eye because it reflects more light in the green spectrum (0.53–0.58 μm) than red or blue. Most vegetation is also highly reflective in the near-infrared spectrum that is in the range of 0.7–1.0 μm, beyond what the human eye can see. An imaging system that can measure the reflectance of an object at multiple wavelengths would be able to very clearly determine the difference between grass and dry grass, which has a different spectral signature as depicted in Figure 1. However, if a sensor was uncalibrated and suffered from an unknown bias, the different materials may be separated, but not identified. The following section examines this issue in more detail.

4 UAS IMAGING EQUIPMENT

The development of an effective UAS remote sensing methodology requires knowledge of various equipment available and their capabilities. Rather than focusing on specific technological metrics, the following discussion focuses on the common qualities of selected imaging equipment types. Without specifying existing imaging resolutions or shutter speeds, it is still valuable to examine the different defining

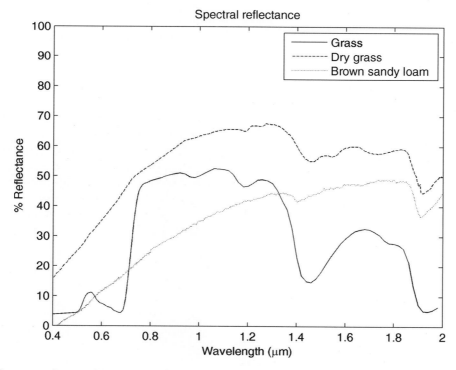

Figure 1. Spectral reflectance of grass, dry grass, and brown sandy loam (Baldridge *et al.*, 2009).

aspects and how they dictate the remote sensing workflows and best practices. The following section examines common UAS payloads such as video systems, digital cameras, and calibrated digital imagers with a discussion of the implementation strategies and methodology development. Additional equipment, such as thermal imagers, have been found to be significantly useful (Stark, Smith, and Chen, 2014), but are outside the scope of this section.

4.1 Video Systems

Video systems can be a simple payload to integrate into a UAS. It can be as simple as affixing a small HD video recorder to the UAS but also as advanced as a remotely operated gimbaled video system with real-time communication and control. The wide range in capabilities does enable project developers the ability to decide on the best system, balancing performance and cost with functionality.

Image quality and resolution vary significantly with quality and price, although, in general, they are not at the same level as digital cameras. However, the key aspect of video systems is the high frame rate rather than optical quality. For human viewers of live or recorded video, the implied motion visible from the rapid progression of frames provides significant contextual information such as movement direction, relative size and orientation of visible objects, and object depth that are difficult to discern from still imagery at lower frame rates.

With machine vision algorithms and automated processing, the high frame rate enables superior object tracking and coverage area with faster moving vehicles. The use of a controllable gimbal system provides improved situational awareness for human operators (Peschel and Murphy, 2013), a valuable capability for search and rescue operations, although at the cost of added complexity. While video systems typically have a lower image resolution than digital cameras, the use of a narrow field of view lens or a controllable zoom lens can enable a similar high spatial resolution at the tradeoff of a smaller viewing area.

Implementation of video systems into a project workflow is straightforward. Typically, they do not require preflight calibration or image correction as the information goal is to obtain visual references of objects or of characteristic information. Setting up ground control points can be utilized for postprocessing georeferencing. Depending on the desired autonomy, video processing can be done onboard or on the ground, though typically the computer power is greater on the ground.

4.2 Digital Cameras

Digital cameras are effective for many UAS operations that require high spatial resolution but do not require immediate visual feedback or a high frame rate. Many cameras, even those that are commercially available, have advanced automated features such as automatic focus, color balance, white balance, and image stabilization that ensure excellent pictures are generated. Overall, digital cameras provide excellent resolution for quantitative measurements of many characteristics such as small features and object texture, making them ideal for identification or localization applications. The additional contextual information, such as known ground control points or recording the position the picture was taken in, can enable accurate spatial measurements as well. With a sufficient coverage, a mosaic can be generated from the set of pictures over the targeted area (Figure 2). Combined with the contextual information, this enables high-resolution georectified orthophotos that can be used for applications such as mapping fire damage (Hinkley and Zajkowski, 2011) and rangeland management (Laliberte et al., 2010). In the example orthophoto, the discoloration of soil is apparent in the area surrounding the water tower located on the right side of the orthophoto, which was caused by sediment leakage.

The pictures generated can also be used with a photogrammetry technique of generating 3D surface models from aerial images (Figure 3). Utilizing sufficiently overlapping pictures, image points from a structure-from-motion (SfM) algorithm are matched together to generate pixel depths and stitched together to form a digital surface model. These digital surface models have been presented as both accurate

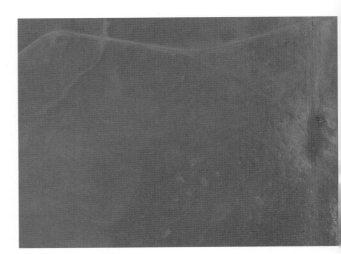

Figure 2. Example orthophoto.

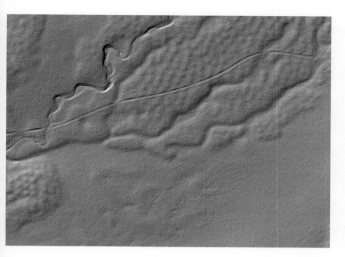

Figure 3. Example digital surface model (hillshaded for clarity).

and precise (Rock, Ries, and Udelhoven, 2011) enough to be utilized for applications such as modeling river topology (Javernick, Brasington, and Caruso, 2014) and mapping ice flows (Whitehead, Moorman, and Hugenholtz, 2013). In Figure 3, the digital surface model depicts the abundance of sediment mounds that characterize the formation of the seasonal vernal pools in the Merced Vernal Pool and Grassland Reserve.

While digital cameras have a number of advantages, they are less suited for applications where immediate responses or quantitative spectral measurements are needed. The automated features that enable high-quality pictures obscure accurate reflectance radiation measurements by dynamically adjusting color, light, and introducing artifacts through lossy compression.

4.3   Calibrated Digital Imagers

Quantifiable spectral measurements are a powerful analytical tool and the basis for most satellite remote sensing applications. While satellites suffer from low spatial resolution, low temporal resolution, and atmospheric interference, UASs can be utilized to counter these issues.

Calibrated systems are designed to provide accurate radiometric measurements, typically of the radiation emanating from the surface (Jones and Vaughan, 2010). Rather than looking at images in terms of colors, images are comprised of the intensity of energy received at particular wavelengths. Whereas a red object may appear slightly pink or orange depending on the time of day, camera orientation, or camera settings, a calibrated system is designed to isolate only the

reflectance of an object and provide a consistent measurement across multiple settings and viewings.

4.3.1   Digital cameras as calibrated imagers

Digital cameras can be utilized as radiometrically calibrated imagers, although additional procedures are required for calibration. In Figure 4, an example workflow for using digital cameras as calibrated digital imagers is depicted. Field data collection is often a necessity for most workflows for radiometric calibration. Camera identification is also a process done prior to the flight operation, although this may not be necessarily prior to each flight. Lens calibration calibrates for the optical qualities. Flat-field calibration provides for adjustments from nonuniform image collection (vignetting, nonlinear response, and dead pixels). Spectral sensitivity enables radiometric data to be collected for spectral signature matching, which often requires ground control points and spectral control points. The data processing workflow includes the integration of metadata for spatial processing and raw band separation to adjust for band-to-band registration.

Digital cameras can also be modified to measure reflectance at the near-infrared spectrum. The CMOS- and CCD-based imaging sensors used for commercial cameras are also sensitive to the NIR spectrum, although normally NIR blocking filters are installed for regular pictures. Removal of this filter restores the NIR sensitivity, although it can be mixed with the red light spectrum. The installation of a NIR pass filter such as Hoya R72 (Hoya Filters, hoyafilters.com) blocks out the red spectrum to enable NIR measurements. Other solutions utilize a red notch filter, blocking only visible red while allowing visible blue and green and NIR (LDP LLC, www.maxmax.com). On some cameras, the blue channel is also marginally sensitive for NIR. In those cases, it is possible to install a blue notch filter. This has the intended effect of blocking the visible blue wavelengths on the blue channel while still allowing the NIR wavelengths to be measured on the blue channel.

4.3.2   Multispectral and hyperspectral imagers

Imaging equipment that specialize in measuring the reflected radiation at specific wavelengths are either considered multispectral or hyperspectral imagers. Multispectral imagers are typically only a handful of selected wavelengths, while hyperspectral generate upward of 60 channels of selected wavelengths, typically at much narrower bands than multispectral.

Advances in technology have led to the feasibility of the use of multispectral imagers such as those developed by Tetracam (Tetracam Inc., www.tetracam.com) and

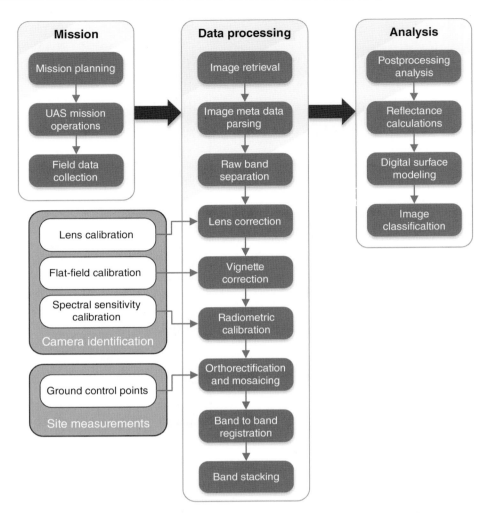

Figure 4. UAS analysis workflow for converted digital cameras.

MicaSense (MicaSense Inc., www.micasense.com). For applications that rely on spectral signatures of targets, often these systems are a necessity. A variety of agricultural applications such as crop water stress (Zarco-Tejada *et al.*, 2013) and identifying citrus greening disease (Garcia-Ruiz *et al.*, 2013) have demonstrated the effectiveness of these systems for both multispectral and hyperspectral imaging.

Many of the implementation strategies of calibrated digital cameras can be similarly applied to these calibrated imagers. As with other optical systems, corrections such as background noise, radial distortion, and vignetting are required for accurate radiometric measurements (Del Pozo *et al.*, 2014). Multispectral sensors, based on CMOS or CCD sensors, utilize a wide range of spectral sensitivity of sensors and optical bandpass filters such as those commercially sold

by Androver (Androver Inc., www.androver.com) or Edmund Optics (Edmund Optics Inc., www.edmundoptics. com). The advantage of these specialized sensors is the quality of the spectral measurements. While calibrated camera systems have broadband spectral responses, the specialized imagers are capable of measuring specific spectrum as described in the following section.

4.3.3   Spectral sensitivity

An understanding of spectral sensitivity is an important quality for proper measurement of reflected radiation. For optical imaging systems, a simplified model of the measured light radiation for each channel or band can be described as the integration of the camera's sensitivity, scene illumination, and the scene's reflectance over the spectral range as shown in the following equation:

$$I_{k,x} = \int_{\lambda_{\min}}^{\lambda_{\max}} C_k(\lambda)L(\lambda)R_x(\lambda)\, d\lambda$$

where k is the channel, x is the spatial position, I is the measured intensity, $C_k(\lambda)$ is the imager sensitivity for band k, $L(\lambda)$ is the spectral power distribution of the illuminate, and $R_x(\lambda)$ is the spatial reflectance of point x. $C_k(\lambda)$ of the imager sensitivity can be measured or estimated through a variety of means (Jiang *et al.*, 2013). The illumination can be measured or estimated with existing solar models. The goal for most analysis application involves solving $R_x(\lambda)$ given $I_{k,x}$, which is a challenge due to the low intrinsic dimensionality. However, the solution for $R_x(\lambda)$ can be approximated when the camera sensitivity is sufficiently narrow, as with multi-spectral or hyperspectral imaging sensors.

When the channels or bands are not sufficiently narrow, a common solution utilizes colored panels or objects with a known spectral response. To calibrate scene illumination, *in situ* measurements either concurrently with the imagery or immediately prior or after are used (Clemens, 2012). The calibration of the imager with known reflectance values ensures an accurate ratio between bands rather than accurate radiation measurements.

Although the intended effect of calibrated imagers is to provide satellite-like measurements of particular wavelengths, in practice the differences in spectral sensitivity of the imagers pose a challenge for a unified data set. The following plots of the spectral sensitivity of a Canon 600D digital camera (Figure 5), Tetracam Mini-MCA6 (Figure 6),

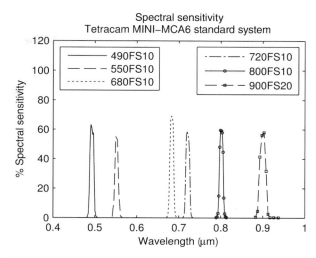

Figure 6. Spectral sensitivity of standard filters of a Tetracam Mini-MCA6 Standard System. (Reproduced with permission from Tetracam, 2016. © Tetracam Inc.)

and the Landsat 8 Satellite (Figure 7) depict the significant variation. For common calculations such as NDVI, the differences in spectral sensitivities of the imaging systems can have significant differences in the final calculations even with satellite systems (Trishchenko, Cihlar, and Li, 2002). As these differences play a large role in the accuracy of the data, care should be taken in the proper selection of the sensor sensitivity to the desired data goal.

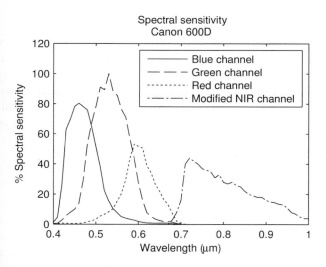

Figure 5. Spectral sensitivity for a Canon 600D camera. Modified NIR channel on a second camera (Jiang *et al.*, 2013).

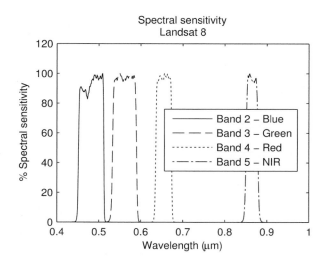

Figure 7. Spectral sensitivity of Landsat 8 (NASA, http://landsat.gsfc.nasa.gov/?page_id=7195).

5   CONCLUSION

The use of UASs as a remote sensing tool has a number of significant advantages to complement existing technology and methodology. However, as new capabilities are developed, there is a need for describing how to utilize and capitalize them efficiently. As more and more applications are developed and described, UAS methodology will mature and effective projects will be the norm. For many applications, such as those based around detection or identification applications, existing technology is capable. While it is tempting to use UASs as a direct replacement for satellites for analysis applications, there are additional challenges that need to be addressed, especially toward accurate spectral measurements. However, the future is bright for UAS remote sensing applications, and sooner than later the use of UASs will become regular and mature.

REFERENCES

Ambrosia, V.G., Wegener, S.S., Sullivan, D.V., Buechel, S.W., Dunagan, S.E., Brass, J.A., Stoneburner, J., and Schoenung, S.M. (2003) Demonstrating UAV-acquired real-time thermal data over fires. *Photogramm. Eng. Remote Sensing*, **69**(4), 391–402.

Baldridge, A., Hook, S., Grove, C., and Rivera, G. (2009) The ASTER spectral library version 2.0. *Remote Sens. Environ.*, **113**(4), 711–715.

Berni, J., Zarco-Tejada, P.J., Suárez, L., and Fereres, E. (2009) Thermal and narrowband multispectral remote sensing for vegetation monitoring from an unmanned aerial vehicle. *IEEE Trans. Geosci. Remote Sens.*, **47**(3), 722–738.

Bhatta, B. (2013) *Research Methods in Remote Sensing*, Springer.

Cavoukian, A. (2009) *Privacy by Design: The 7 Foundational Principles*, Information and Privacy Commissioner of Ontario, Ontario, Canada.

Chabot, D. and Bird, D.M. (2012) Evaluation of an off-the-shelf unmanned aircraft system for surveying flocks of geese. *Waterbirds*, **35**(1), 170–174.

Chao, H. and Chen, Y. (2012) *Remote Sensing and Actuation Using Unmanned Vehicles*, John Wiley & Sons, Inc., Hoboken

Clemens, S.R. (2012) Procedures for correcting digital camera imagery acquired by the AggieAir Remote Sensing Platform. Masters thesis. Utah State University.

Del Pozo, S., Rodríguez-Gonzálvez, P., Hernández-López, D. and Felipe-García, B. (2014) Vicarious radiometric calibration of a multispectral camera on board an unmanned aerial system. *Remote Sens.*, **6**(3), 1918–1937.

Doherty, P. and Rudol, P. (2007) A UAV search and rescue scenario with human body detection and geolocalization. *AI 2007: Advances in Artificial Intelligence*, Springer, Berlin, pp. 1–13.

Freed, M., Harris, R., and Shafto, M. (2004) Human-interaction challenges in UAV-based autonomous surveillance. *Proceedings of the 2004 Spring Symposium on Interactions between Humans and Autonomous Systems Over Extended Operations.*

Garcia-Ruiz, F., Sankaran, S., Maja, J.M., Lee, W.S., Rasmussen, J., and Ehsani, R. (2013) Comparison of two aerial imaging platforms for identification of Huanglongbing-infected citrus trees. *Comput. Electron. Agric.*, **91**, 106–115.

Hinkley, E.A. and Zajkowski, T. (2011) USDA forest service – NASA: unmanned aerial systems demonstrations – pushing the leading edge in fire mapping. *Geocarto Int.*, **26**(2), 103–111.

Hirsch, J. 2013 *Hunting wild pigs with drones.* April 18, 2013. Available at http://modernfarmer.com/2013/04/hunting-pigs-with-drones/ (accessed September 23, 2014).

Jang, S. and Kim, J. (2008) Survey of electro-optical infrared sensor for UAV. *Aerospace Technol. Trends*, **6**(1), 124–134.

Javernick, L., Brasington, J., and Caruso, B. (2014) Modeling the topography of shallow braided rivers using structure-from-motion photogrammetry. *Geomorphology*, **213**, 166–182.

Jensen, A.M., Baumann, M., and Chen, Y.Q. (2008) Low-cost multispectral aerial imaging using autonomous runway-free small flying wing vehicles. *2008 IEEE International Geoscience and Remote Sensing Symposium (IGRASS 2008)*, IEEE, pp. V-506–V-509.

Jiang, J., Liu, D., Gu, J., and Susstrunk, S. (2013) What is the space of spectral sensitivity functions for digital color cameras? *2013 IEEE Workshop on Applications of Computer Vision (WACV 2013)*, IEEE, pp. 168–179.

Jones, H.G. and Vaughan, R.A. (2010) *Remote Sensing of Vegetation: Principles, Techniques, and Applications*, Oxford University Press.

Laliberte, A.S., Herrick, J.E., Rango, A., and Winters, C. (2010) Acquisition, orthorectification, and object-based classification of unmanned aerial vehicle (UAV) imagery for rangeland monitoring. *Photogramm. Eng. Remote Sensing*, **76**(6), 661–672.

Murphy, R.R. (2004) Human–robot interaction in rescue robotics. *IEEE Trans. Syst. Man. Cybern. C Appl. Rev.* **34**(2), 138–153.

Peschel, J.M. and Murphy, R.R. (2013) On the human–machine interaction of unmanned aerial system mission specialists. *IEEE Trans. Hum. Mach. Syst.*, **43**(1), 53–62.

Rasmussen, J., Nielsen, J., Garcia-Ruiz, F., Christensen, S. and Streibig, J.C. (2013) Potential uses of small unmanned aircraft systems (UAS) in weed research. *Weed Res.*, **53**(4), 242–248.

Ren, W. and Beard, R. (2004) Trajectory tracking for unmanned air vehicles with velocity and heading rate constraints. *IEEE Trans. Control Syst. Technol.*, **12**(5), 706–716.

Rock, G., Ries, J. and Udelhoven, T. (2011) Sensitivity analysis of UAV-photogrammetry for creating digital elevation models (DEM). *Proceedings of Conference on Unmanned Aerial Vehicle in Geomatics.*

Stark, B., Smith, B., and Chen, Y.Q. (2013) A guide for selecting small unmanned aerial systems for research-centric applications. *Proceedings of Research, Education and Development of Unmanned Aerial Systems (RED-UAS2013).*

Stark, B., Smith, B., and Chen, Y.Q. (2014) Survey of thermal infrared remote sensing for unmanned aerial systems. *Proceedings of the International Conference on Unmanned Aerial Systems (ICUAS)*, pp. 1294–1299.

Trishchenko, A.P., Cihlar, J., and Li, Z. (2002) Effects of spectral response function on surface reflectance and NDVI measured with moderate resolution satellite sensors. *Remote Sens. Environ.*, **81**(1), 1–18.

Whitehead, K., Moorman, B. and Hugenholtz, C. (2013) Brief communication: low-cost, on-demand aerial photogrammetry for glaciological measurement. *Cryosphere*, **7**(6), 1879–1884.

Woods, D.D., Tittle, J., Feil, M., and Roesler, A. (2004) Envisioning human–robot coordination in future operations. *IEEE Trans. Syst. Man Cybern. C Appl. Rev.*, **34**(2), 210–218.

Zarco-Tejada, P.J., Gonzalez-Dugo, V., Williams, L., Suarez, L., Berni, J.A., Goldhamer, D. and Fereres, E. (2013) A PRI-based water stress index combining structural and chlorophyll effects: assessment using diurnal narrow-band airborne imagery and the CWSI thermal index. *Remote Sens. Environ.*, **138**, 38–50.

Chapter 3

Autonomous Parachute-Based Precision Delivery Systems

Oleg A. Yakimenko

Graduate School of Engineering and Applied Science, Naval Postgraduate School, Monterey, CA, USA

1 INTRODUCTION

Aerial cargo delivery using conventional uncontrolled round canopies has been around for quite a while. It has been and still is a major method for delivering a variety of payloads to the areas that are otherwise hard to get to. For various reasons, an aircraft that carries cargo in its cargo bay may need to fly high forcing a high-altitude deployment of parachute systems. Canopy opening usually occurs right after a cargo parachute system release. The inevitable consequence of this high-altitude deployment/high-altitude opening approach is that uncontrolled parachute system remains at the mercy of the winds all way down resulting in large misses of the intended point of impact (IPI). Introduction of controlled gliding parachutes, ram-air parafoils, to

Unmanned Aircraft Systems. Edited by Ella Atkins, Aníbal Ollero, Antonios Tsourdos, Richard Blockley and Wei Shyy.
© 2016 John Wiley & Sons, Ltd. ISBN: 978-1-118-86645-0.

skydivers in the late 1960s (Jalbert, 1966) immediately suggested their use for cargo delivery as well. While earlier systems could only rely on the ground-based radio navigational aids (beacons), traditionally used in aviation, availability of the satellite-based Global Positioning System (GPS) eliminated the last barrier toward full autonomy of aerial payload delivery systems and resulted in the development of the family of different-weight systems capable of delivering and accurately placing payload with very little prior knowledge of the winds. The recently published monograph (Yakimenko, 2015) describes the current state of precise aerial delivery from the standpoint of modeling and design of control algorithms and this chapter represents a quintessence of this work providing the overall look at the problem and making some key points.

2 CONCEPT OF OPERATIONS AND KEY REQUIREMENTS

Figure 1a shows a typical precision aerial delivery system (PADS) rigged with a cargo bag. It consists of parachute system sitting atop the airborne guidance unit (AGU), which in turn sits atop a cargo bag. Figure 1a also features a paper honeycomb beneath a cargo bag to attenuate impact and plywood skidboard. In rigged PADS, a paper honeycomb also separates AGU from a cargo. Shown in Figure 1a is the A-22-type container delivery system (CDS) consisting of a sling assembly, cargo cover, and four suspension webs connecting the container to the parachute. Other CDS may utilize pallets or an adjustable nylon cloth and webbing

(a)

(b)

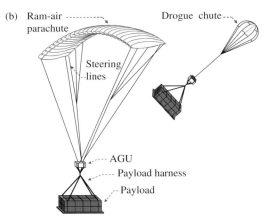

Figure 1. Rigged (a) and deployed (b) PADS. (Reproduced with permission from AIAA, 2016.)

container. These CDS may have different dimensions to accommodate bundles of food, water, medicine, ammunition, and so on. CDS can also accommodate carrying rigid inflatable and foldable boats, motorcycles, snowmobiles, wheeled vehicles, and other bulky equipment.

Upon release of PADS at the so-called computer air release point (CARP), computed based on the best knowledge of the winds, and a short freefall phase, a parachute canopy fully opens (Figure 1b) (Lingard, 1995). The release from an airplane may involve arming AGU manually/remotely or using a static line. As shown in Figure 1b, extraction of the heavier PADS out of an air carrier is done using a drogue chute. It also helps stabilizing the system before a main canopy deploys. A fully deployed PADS consists of the canopy, suspension lines, payload harness, and AGU. Different PADS designs may also include the slider, risers, and other auxiliary equipment such as reefing, drogue, and brake release cutters (Lingard, 1995). AGU includes sensors, computer, and electrically driven actuators (motors) to pull the steering lines down.

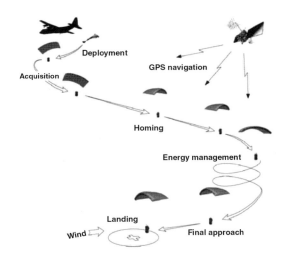

Figure 2. General representation of a sequence of events from PADS deployment till soft landing. (Reproduced with permission from AIAA, 2016).

Figure 2 shows a schematic diagram of a typical airdrop mission (Wegereef and Jentink, 2003). Once the GPS signal is received and PADS position relative to IPI determined (navigation task), AGU computes the best course of actions (guidance task) and produces and executes the servos control commands (control tasks). Directional control of a typical parafoil-based PADS (aka steering) is realized by pulling down the control lines attached to the outer portion (spanwise) of the trailing edge (TE) differentially (asymmetrically). First, PADS steers toward a desired landing zone (aka drop zone (DZ)). When DZ is reached, energy management (EM) phase is commenced. Lastly, the final approach maneuver (also referred to as a terminal guidance phase) is conducted to land at IPI, preferably against the winds, using a flare maneuver (simultaneous symmetric pull down of the steering lines) to reduce the sink rate ensuring soft landing. If DZ is reached with enough altitude excess, the touchdown precision depends on the terminal guidance algorithm. That is where the constantly varying surface winds take a toll. As such AGU tries to constantly estimate the current altitude winds and propagate this prediction to the lower altitudes.

Some PADS utilize a two-stage architecture. In this case, a high-speed parafoil, the first stage, with a good wind penetration capability brings PADS to some point above IPI (or slightly upwind IPI if the surface winds are known) where a standard uncontrollable round canopy, the second stage, is deployed.

These days a majority of PADS use a ram-air parafoil (Figure 3a). However, the earlier research also involved other concepts (Figure 3b–f) (Eilertson, 1969). Even though these

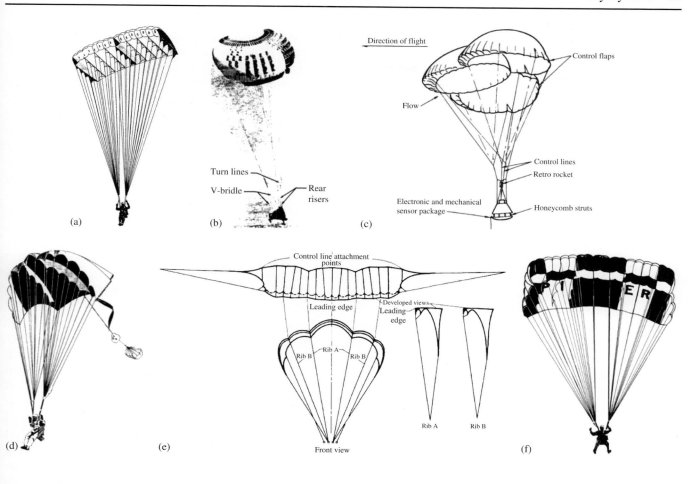

Figure 3. Steerable parachute concepts. (a) Parafoil. (b) Parasail. (c) Cloverleaf. (d) Parawing. (e) Sailwing. (f) Volplane. (Reproduced with permission from AIAA, 2016.)

other solutions exhibit lesser glide ratio (GR), some of them are still considered as a viable option in a variety of applications.

Based on the success of the earlier research using ram-air parafoils, a joint US Army/US Air Force program, Joint Precision Airdrop System (JPADS), was established in 1997 to explore the way to cardinally improve accuracy of aerial cargo delivery. The Air Force was responsible for developing the mission planning computer (JPADS-MP) that would forecast winds over DZ based on all wind data available in the vicinity of DZ in the recent past and then produce a CARP that would minimize the landing error for conventional nongliding payload delivery systems. The Army was responsible for developing different-weight PADS and common-architecture AGU to support a variety of parachute sizes and designs and ensure a flare maneuver prior to impact. As of 2011, the five JPADS categories were defined as microlight (ML) ($\sim$5–70 kg), ultralight (UL) ($\sim$100–300 kg), extra-

light (XL) ($\sim$300 kg to 1.1 tons), light (L) ($\sim$2.3–4.5 tons), and medium (M) ($\sim$4.5–19 tons) (JPADS, 2014).

One of the major requirements to JPADS is to have a substantial standoff range from which PADS can still reach IPI. This range is obviously proportional to the product of lift-to-drag ratio (L/D) (equal to GR in no-wind conditions) and deployment altitude. Winds aloft can either extend this range or shrink it. The slower the rate of descent, the greater the effect. Typically, PADS has at least 2.5:1 GR, and therefore, if deployed from 11 km altitude above the surface level, may reach IPI from a standoff distance of up to about 25 km. (The X:1 format for GR shows explicitly how much horizontal displacement X can be obtained per unit of vertical displacement.)

Another major requirement is to have a touchdown accuracy (IPI miss distance) less than 100 m circular error probable (CEP). These days the smaller PADS are capable of achieving 10 m CEP, while the larger PADS are only able to deliver payloads with about 300 m CEP.

Speaking of touchdown accuracy, it should be noted that it has two dimensions. The first dimension is a system accuracy e_{sys}, which is computed as the touchdown error for all PADS released from an air carrier. The second dimension is a terminal accuracy, e_{term}, which is computed for only those systems that reached the DZ area and had enough altitude to execute the terminal guidance phase (Brown and Benney, 2005). While the latter one is defined entirely by perfection of guidance, navigation, and control (GNC) system, its ability to mitigate the effect of the unknown surface winds and perform a flare maneuver if appropriate, the first one may be affected by inability to reach the DZ area or reaching it with no altitude excess because of one of the following factors: launch acceptability region (LAR) being computed with major errors or CARP being missed by air carrier, parachute opening failure, damaged or twisted control lines, and inoperable or malfunctioning AGU. The difference between system accuracy and terminal accuracy can also be cast as PADS reliability expressed as a probability of PADS reaching the DZ area from which a guided approach and landing can successfully be completed, P_{DZ}. If $P_{DZ} = 1$, then $e_{sys} = e_{term}$. In most of the cases however, $P_{DZ} < 1$ and as a result $e_{sys} > e_{term}$.

Distinguishing between two accuracies (or assessing P_{DZ}) is a tricky thing. The desire to limit the number of successful airdrops to improve terminal guidance leads to a degraded value of PADS reliability. Statistical analysis performed on the results of massive PADS airdrops allows revealing P_{DZ} in a more systematic way. As an example, Figure 4 shows a typical distribution of PADS touchdown error measured in the units of median radial error (MRE). Compared to the circular normal distribution (CND), distributions for three same-weight-category PADS shown in Figure 4 feature a "heavy tail," so that the chance of PADS landing beyond a three-MRE circle around IPI is about two orders of magnitude higher compared to what would be in the case of CND. Situation shown in Figure 4 is typical for any guided system

in general – While improving the MRE (or CEP) drastically compared to an unguided analogous one, the error distribution does not follow CND anymore.

For three different PADS shown in Figure 4, MRE is different (not shown here) and the system accuracy would be judged based upon its values. System reliability could be judged upon the number of "outliers" (the number of data points greater than three MREs). For three PADS shown in Figure 4, P_{DZ} could be estimated as 0.81, 0.76, and 0.88, respectively. Throwing outliers away and reassessing MRE for the remaining data points yield terminal accuracy for each PADS.

Finally, one more major requirement to JPADS is that airdrop equipment should be recovered and reused unless cost per drop unit is negligible. The desired relative cost for PADS was set to $6 per pound of payload delivered ($13.22 per kilogram). This requirement has a major effect on the range of applications PADS can be used in because recovering them is not always practical or possible at all. Hence, decreasing the cost of PADS but ensuring the same or even better accuracy is the main stream in the development of next-generation PADS.

3   PADS FAMILY AND STEADY-STATE PERFORMANCE

The Orion PADS developed in the early 1990s by SSE Inc. of Pennsauken, NJ was the first commercially available system with high-GR parafoils supplied by Pioneer Aerospace of Melbourne, FL (Allen, 1995). It was capable of utilizing different ram-air parafoils from 30 to 680 m^2 and delivering 90 kg to 16 tons payload. Upon exiting, a drogue chute was deployed to stabilize attitude and velocity. As the drogue deploys, AGU separates from the payload. After a preset time, or at a preset altitude, the main canopy opens, and AGU takes control. Orion's AGU had all functionality the later PADS had. The GS-750 parafoil-based Orion (with the 14 m wingspan and 70 m^2 surface area) demonstrated the desired 100 m CEP accuracy and was the first PADS fielded to the US Department of Defense. The larger 45 m wingspan version of Orion PADS capable of steering a 16 metric tons payload was adapted by NASA as the recovery system for the International Space Station X-38 experimental Crew Return Vehicle (CRV) and was demonstrated at Precision Airdrop Technology Conference and Demonstration (PAT-CAD) event held in Yuma, AZ, in 2001.

Later on, within the JPADS program, several different-weight self-guided PADS were developed. They were demonstrated during five PATCAD in the United States and two Precision Airdrop Capability Demonstration events conducted in France. Overall almost 700 airdrops, primarily

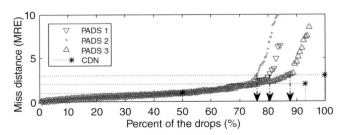

Figure 4. Typical miss distance distribution for the same-weight JPADS category. (Reproduced with permission from AIAA, 2016.)

Table 1. PADS representatives by JPADS weight categories.

Microlight	Ultralight	Extralight	Light	Medium
		2 K Screamer	10 K Screamer	
200 CADS	500 Panther	2 K Panther	10 K CADS	
160 Snowbird	500 Pegasus	2 K Sherpa Ranger	10 K Sherpa Provider	
150 Mosquito	500 MicroFly	2 K FireFly	10 K DragonFly	30 K MegaFly
				42 K Gigafly
Onyx ML	Onyx UL		5 K Para-Flite	
5 Mosquito	300 SPADeS	1 K SPADeS	1 K SPADeS	
5 Snowflake	500 AGAS	2 K AGAS	5 K AGAS	

using a Lockheed C-130 Hercules air carrier, were executed. Among other payload delivery systems, the self-guided PADS included Affordable Guided Airdrop System (AGAS) developed by Natick, Vertigo Incorporated, and Capewell Components LLC; Buckeye PADS developed by Southwest Research Incorporated; CADS PADS developed by Cobham Public Ltd., UK; MegaFly, DragonFly, FireFly, and MicroFly PADS developed by Airborne Systems; Onyx PADS developed by Atair Aerospace Inc.; Panther PADS developed by Pioneer Aerospace Corporation/Aerazur; Para-Flite PADS developed by Para-Flite Inc.; ParaLander PADS developed by European Aeronautic Defence and Space Company, Pegasus PADS developed by FXC Corporation, Inc., Screamer PADS developed by Strong Enterprises; Sherpa, SnowGoose, and Snowbird PADS developed by Mist Mobility Integrated Systems Technology Inc. (MMIST), Canada; SNCA PADS developed by NAVOCAP, France; Snowflake PADS developed by the University of Alabama in Huntsville/Naval Postgraduate School/Arcturus UAV; SPADeS PADS developed by Dutch Space, The Netherlands; and Mosquito PADS developed by STARA Technologies Inc. Table 1 shows a distribution of these self-guided PADS among different JPADS weight categories with a number in front of the name specifying maximum weight in pounds.

Four of 27 aforementioned systems are shown in Figures 5 and 6 as representative examples of the ML, UL, EL, and M JPADS weight categories (PATCAD, 2005; Sego, 2001; JPADS, 2014).

One unique design, AGAS, mentioned in Table 1, was conceived to bridge the gap between expensive high-GR PADS and relatively inexpensive uncontrolled (ballistic) round parachutes. Slight modification of rigging system for a standard round-canopy-based payload delivery system and addition of AGU allowed achieving a limited steering authority casting this system as PADS. More specifics on this system will be given in the last section of this chapter.

Also demonstrated at PATCAD events were two powered PADS that might be cast as an unmanned aerial vehicles

(a)

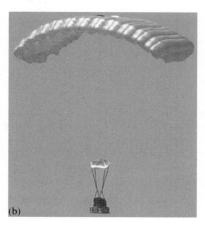

(b)

Figure 5. Mosquito (a) and Pegasus (b) PADS. (Reproduced with permission from AIAA, 2016.)

(UAVs). The first one was the SnowGoose PADS by MMIST, which is now fielded and used by the US Special Operations Command (Figure 7a). The second one was a Buckeye powered paraglider (PPG). These days PPG is commonly used in university research because it allows gaining easily some altitude as a powered UAV (Figure 7b) and then shutting its engine down to emulate the behavior of PADS.

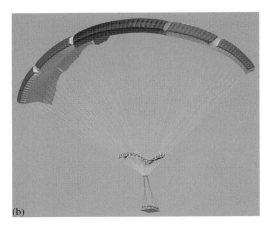

Figure 6. Sherpa (a) and GigaFly (b) PADS. (Reproduced with permission from AIAA, 2016.)

Figure 7. SnowGoose UAV (a) and an example of PPG (b). (Reproduced with permission from AIAA, 2016.)

Table 2. Reported and computed properties of representative different-weight PADS.

PADS	m (ton)	S (m^2)	c (m)	b (m)	AR	m/S (kg m^{-2})	GR, X:1	V_h (m s^{-1})	V_v (m s^{-1})	TR ($^\circ$ s^{-1})
SPADeS	0.16	34	4	8.6	2.2	5	3.3	11.7	3.6	47
DragonFly	4.5	325	11	30	2.7	14	3.9	18.0	4.6	10
MegaFly	11.8	836	16	52	3.3	14	3.6	21.0	5.9	5

Table 2 presents some parameters of three different-weight PADS to show their spread. It shows a nominal payload mass m, canopy area S, chord c, span b, aspect ratio (AR), wing loading (WL), GR, maximum turn rate (TR), horizontal V_h and vertical V_v components of airspeed in a steady gliding flight. The TR corresponds to the maximum safe differential (asymmetric) TE deflection affecting about a quarter of the aft part of a chord.

To supplement this table, Figure 8 shows TR versus canopy area as a log–log graph superimposed on prediction for different-size canopies obtained for WL of 4.9 kg m^{-2} in

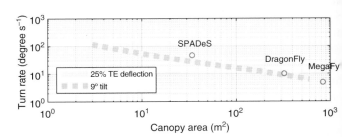

Figure 8. Control authority versus canopy area. (Reproduced with permission from AIAA, 2016.)

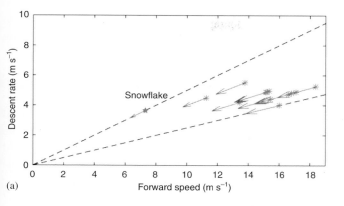

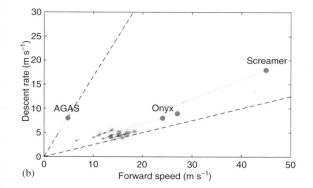

Figure 9. Rate of descent versus forward speed for one-stage PADS (a), and for all PADS (b). (Reproduced with permission from AIAA, 2016.)

(Goodrick, 1984). Figure 8 also shows TR that would be achieved using another method for a directional control, alternative to that of TE deflection. This alternative would involve tilting the resultant aerodynamic force vector into the turn by slightly deflecting the inboard tip and thus producing a net lateral pressure differential across the span. Data in Figure 8 correspond to 9° tilt for canopies of all sizes.

Figure 9a shows a spread of gliding performance (in no-wind conditions) for the most parafoil-based one-stage systems listed in Table 1 graphically. For comparison, Figure 9b adds some unique systems, specifically the round-canopy-based AGAS and the two two-stage PADS, Onyx and Screamer, featuring a higher WL, which results in a higher airspeed, but still about the same GR as all one-stage systems.

To better understand the relationship between the GR and airspeed components, we should write down two equations for the forces acting on PADS horizontally and vertically:

$$L \sin(\gamma) - D \cos(\gamma) = 0, \quad mg - L \cos(\gamma) - D \sin(\gamma) = 0 \tag{1}$$

Here γ is the glide angle (negative to the flight path angle), $L = L_c + L_l + L_s$ and $D = D_c + D_l + D_s$ are the total system

lift and drag with contributions from canopy, suspension lines, and payload (also referred to as store), $m = m_s + m_c + m_e$ is the total mass of the system composed of the mass of store, canopy with suspension lines, and air entrapped inside ram-air canopy, and g is the acceleration due to gravity. The first equation further yields

$$\tan(\gamma) = (L/D)^{-1} = (C_L/C_D)^{-1} = GR^{-1} \tag{2}$$

where $C_L = L/(QS)$ is the lift coefficient and $C_D = D/(QS)$ is the drag coefficient ($Q = 0.5\rho V^2$ is the dynamic pressure determined by the air density ρ and the airspeed V). The higher the GR, the smaller the glide angle and therefore the greater the gliding range for a given height loss.

Substituting Equation 2 in Equation 1 allows defining the airspeed of PADS in a steady gliding flight as

$$V = \left(\frac{2g}{\rho}\frac{m}{S}\frac{1}{\sqrt{C_L^2 + C_D^2}}\right)^{0.5} = \left(\frac{2g}{\rho}\frac{m}{S}\frac{1}{C_L\sqrt{1 + GR^{-2}}}\right)^{0.5}$$

$$\approx \left(\frac{2gm}{\rho S C_L}\right)^{0.5}(1 - 0.25\,GR^{-2}) \tag{3}$$

This equation shows that the airspeed is dependent on air density, WL, and aerodynamic characteristics of parachute (which implicitly depends on the trim angle of attack). An airspeed down along the glide path defined by Equation 2 will increase with increasing altitude, WL, and GR.

The horizontal and vertical components of the airspeed vector are

$$V_h = V \cos(\gamma), \quad V_v = V \sin(\gamma) \tag{4}$$

Substitution of Equations 2 and 3 in Equation 4 yields

$$V_h \approx \left(\frac{2gm}{\rho S C_L}\right)^{0.5}(1 - 0.75 GR^{-2}),$$

$$V_v \approx \left(\frac{2gm}{\rho S C_L}\right)^{0.5}(1 - 0.75 GR^{-2})GR^{-1} \tag{5}$$

which means that under the same conditions, the increase of GR results in a slight increase of horizontal component of airspeed and decrease of its vertical component. Equation 5 computed for the design C_L of 0.5 and GR of 2–4 for varying WL at sea level are shown graphically in Figure 10. Vertical arrows show the direction of parameter variation while increasing GR (decreasing glide angle). As seen, data of Figure 10 correlate with that of Figure 9.

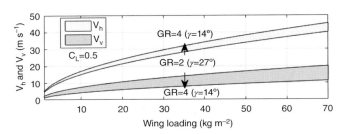

Figure 10. Ram-air parachute airspeed vector components versus WL.

4   MODELING

4.1   Governing equations

For the purpose of mission planning and trajectory optimization, the PADS model can be represented by its kinematic equations:

$$\begin{bmatrix} \dot{x} \\ \dot{y} \\ \dot{z} \end{bmatrix} = \begin{bmatrix} V_h \cos(\chi_a) \\ V_h \sin(\chi_a) \\ V_v \end{bmatrix} + \begin{bmatrix} w_x \\ w_y \\ w_z \end{bmatrix} \qquad (6)$$

In these equations, x, y, and z represent the PADS coordinates in the local tangent plane $\{n\}$, w_x, w_y, and w_z are the components of the wind vector $\mathbf{W}$, and χ_a is the heading angle, that is, the angle from North to the projection of airspeed vector $\mathbf{V}_a$ onto a horizontal plane. In practice, for the simple models, the sideslip angle β is usually neglected, so χ_a is considered to be the same as the yaw angle ψ, which defines an orientation of the longitudinal axis of the body coordinate frame $\{b\}$ with respect to North, i.e. $\chi_a \approx \psi$.

Obviously, assuming PADS model described by Equation 6, the horizontal and vertical motion can be decoupled. In this case, the first two equations

$$\begin{bmatrix} \dot{x} \\ \dot{y} \end{bmatrix} = \begin{bmatrix} V_h \cos(\psi) \\ V_h \sin(\psi) \end{bmatrix} + \begin{bmatrix} w_x \\ w_y \end{bmatrix} \qquad (7)$$

represent a three degree-of-freedom (DoF) model with x and y being the states and yaw angle, serving as a control input. To account for yaw angle dynamics, we can write

$$\dot{\psi} = K_\psi \delta_a \qquad (8)$$

where δ_a represents an asymmetric TE deflection, and K_ψ is the gain. More sophisticated model may assume second-order dynamics:

$$\begin{bmatrix} \dot{\psi} \\ \ddot{\psi} \end{bmatrix} = \begin{bmatrix} 0 & 1 \\ 0 & T_\psi^{-1} \end{bmatrix} \begin{bmatrix} \psi \\ \dot{\psi} \end{bmatrix} + T_\psi^{-1} \begin{bmatrix} 0 \\ K_\psi \end{bmatrix} \delta_a \qquad (9)$$

where T_ψ is the time constant.

The last equation in Equation 6 can be augmented with an additional term increasing the descent rate while turning compared to that of a straight gliding flight:

$$\dot{z} = V_v + w_z + k_{v\dot{\psi}}|\dot{\psi}| \qquad (10)$$

Here, $k_{v\dot{\psi}}$ is the weighting coefficient that can be determined from flight data. This equation ties maneuvering in the horizontal plane with the total altitude loss and should definitely be accounted for during the EM and terminal guidance phases.

For the purpose of trajectory optimization, Equation 7 should be rewritten to exclude time:

$$\begin{bmatrix} x'_h \\ y'_h \end{bmatrix} = \frac{-V_h}{V_v + w_z} \begin{bmatrix} \cos(\chi_a) \\ \sin(\chi_a) \end{bmatrix} - \frac{1}{V_v + w_z} \begin{bmatrix} w_x \\ w_y \end{bmatrix} \qquad (11)$$

Here x'_h and y'_h are derivatives with respect to altitude h ($h = -z$). Equations 8 and 9 then take the following form:

$$\psi'_h = \frac{-1}{V_v + w_z} u_c \qquad (12)$$

$$\begin{bmatrix} \psi'_h \\ \dot{\psi}'_h \end{bmatrix} = \frac{-1}{V_v + w_z} \begin{bmatrix} 0 & 1 \\ 0 & T_\psi^{-1} \end{bmatrix} \begin{bmatrix} \psi \\ \dot{\psi} \end{bmatrix} - \frac{T_\psi^{-1}}{V_v + w_z} \begin{bmatrix} 0 \\ K_\psi \end{bmatrix} \delta_a \qquad (13)$$

While this model captures all features important from the standpoint of mission planning, it might be necessary to involve one more state, the roll angle ϕ, and account for effect of the angle of attack (AoA) α. In this case, the 4DoF model can be written as follows (Jann, 2004):

$$\begin{bmatrix} \dot{u} \\ \dot{\psi} \\ \dot{w} \end{bmatrix} =$$
$$\begin{bmatrix} m^{-1}(L(\alpha)\sin(\alpha) - D(\alpha)\cos(\alpha)) - w\dot{\psi}\sin(\phi) \\ u^{-1}(g\sin(\phi) + w\dot{\phi})\cos^{-1}(\phi) \\ m^{-1}(-L(\alpha)\cos(\alpha) - D(\alpha)\sin(\alpha)) + g\cos(\phi) + u\dot{\psi}\sin(\phi) \end{bmatrix} \qquad (14)$$

In this equation, u and w are the components of the ground-speed vector $\mathbf{V}$ ($\mathbf{V} = \mathbf{V}_a + \mathbf{W}$) expressed in $\{b\}$. The kinematic equations then take the following form:

$$\begin{bmatrix} \dot{x} \\ \dot{y} \\ \dot{z} \end{bmatrix} = {}^n_b\mathbf{R} \begin{bmatrix} u \\ 0 \\ w \end{bmatrix} \qquad (15)$$

where $_b^n\mathbf{R}$ is the rotation matrix (from $\{b\}$ to $\{n\}$). Roll angle dynamics due to asymmetric TE deflection can be either modeled as a first-order system

$$T_\phi\dot{\phi} + \phi = K_\phi\delta_a \qquad (16)$$

or neglected

$$\phi = K_\phi\delta_a \qquad (17)$$

The 6DoF model accounts for three translational and three rotational degrees of freedom, and includes the angular velocity vector $\boldsymbol{\omega} = [p, q, r]^T$ (defined in $\{b\}$) and the v component of the groundspeed vector $\mathbf{V}$. Translational and rotational kinematics are described as

$$\left[\ \right] = {}_b^n\mathbf{R}\mathbf{V}, \qquad \begin{bmatrix} \dot{\phi} \\ \dot{\theta} \\ \dot{\psi} \end{bmatrix} = \begin{bmatrix} 1 & \sin(\phi)\tan(\theta) & \cos(\phi)\tan(\theta) \\ 0 & \cos(\phi) & -\sin(\phi) \\ 0 & \sin(\phi)\sec(\theta) & \cos(\phi)\sec(\theta) \end{bmatrix}\boldsymbol{\omega}$$
$$(18)$$

The dynamic equations are cumbersome, but may be presented in a compact form as

$$\dot{\mathbf{V}}^* = \mathbf{A}^{-1}\left(\mathbf{B}(\boldsymbol{\omega}, \phi, \theta, \psi, \mathbf{W})\mathbf{V}^* + \begin{bmatrix} \mathbf{F}_a + \mathbf{F}_g \\ \mathbf{M}_a \end{bmatrix}\right) \qquad (19)$$

where $\mathbf{V}^* = [u, v, w, p, q, r]^T$, $\mathbf{F}_a$ and $\mathbf{M}_a$ are the aerodynamic force and moment vectors, $\mathbf{F}_g$ is the gravitational force vector (all in $\{b\}$). Apart from the matrix $\mathbf{B}$ depending on the states and the wind vector, both matrices $\mathbf{A}$ and $\mathbf{B}$ depend on geometric and mass properties of PADS (the distance between the origin of $\{b\}$ and canopy-fixed coordinate frame $\{p\}$, rigging angle μ defining a rotation of $\{p\}$ with respect to $\{b\}$, and PADS total mass), apparent mass tensor $\mathbf{I}_{am}$ and apparent inertia tensor $\mathbf{I}_{ai}$ (to be considered next).

While the 6DoF model includes three inertial position components of some point in the body frame $\{b\}$ and three Euler orientation angles of $\{b\}$ with respect to $\{n\}$, the higher fidelity models also include additional Euler angles defining orientation of payload (coordinate frame $\{s\}$) relative to $\{b\}$. The number of DoF depends on payload rigging geometry. The 7DoF might be needed in the case the canopy harness is attached to payload using four risers, as shown in Figure 5a. The extra DoF in this case describes payload yaw angle relative to $\{b\}$. A two-riser scheme (Figures 5b, 6b, and 7) would call for adding the payload yaw and pitch angles (8DoF). To fully describe a single-riser (swivel) scheme (Figure 6a) allowing all three Euler angles (including a bank angle of payload relative to $\{b\}$) to have their own dynamics, a 9DoF model would be required (Gorman and Slegers, 2011).

4.2 Apparent mass and inertia

While Equation 19 explicitly includes aerodynamic and gravity forces and moments driving translational and rotational dynamics, there is one more force that should be taken into account. When PADS glides in the air, it sets the air around it into a motion. In turn, this motion introduces pressure forces on PADS that are called the apparent mass pressures. The magnitude (effect) of these pressures is inversely proportional to the mass ratio representing the ratio of a mass of PADS to an air mass displaced or associated with it, $M_r = m\rho^{-1}S^{-1.5}$. For a large PADS, M_r can fall as low as about 0.5 and therefore must be accounted for.

The apparent mass force

$$\tilde{\mathbf{F}}_{am} = -\left(\mathbf{I}_{am}\begin{bmatrix} \dot{\tilde{v}}_x \\ \dot{\tilde{v}}_y \\ \dot{\tilde{v}}_z \end{bmatrix} + \mathbf{S}(\tilde{\boldsymbol{\omega}})\mathbf{I}_{am}\begin{bmatrix} \tilde{v}_x \\ \tilde{v}_y \\ \tilde{v}_z \end{bmatrix}\right) \qquad (20)$$

and apparent inertia moment

$$\tilde{\mathbf{M}}_{ai} = -\left(\mathbf{I}_{ai}\begin{bmatrix} \dot{\tilde{p}} \\ \dot{\tilde{q}} \\ \dot{\tilde{r}} \end{bmatrix} + \mathbf{S}(\tilde{\boldsymbol{\omega}})\mathbf{I}_{ai}\begin{bmatrix} \tilde{p} \\ \tilde{q} \\ \tilde{r} \end{bmatrix} + \mathbf{S}(\tilde{\mathbf{V}}_a)\mathbf{I}_{am}\begin{bmatrix} \tilde{v}_x \\ \tilde{v}_y \\ \tilde{v}_z \end{bmatrix}\right) \qquad (21)$$

act at the centroid of PADS apparent mass. For ellipsoidal canopy shape, this centroid roughly coincides with a volumetric canopy centroid. Equations 20 and 21 expressed in the rotating parafoil coordinate frame $\{p\}$ involve the canopy airspeed angular velocity vectors $\tilde{\mathbf{V}}_a = [\tilde{v}_x, \tilde{v}_y, \tilde{v}_z]^T$ and $\tilde{\boldsymbol{\omega}} = [\tilde{p}, \tilde{q}, \tilde{r}]^T$. In Equations 20 and 21, $\mathbf{S}$ is a skew symmetric matrix and apparent mass and inertia matrices, $\mathbf{I}_{am}$ and $\mathbf{I}_{ai}$, assume the diagonal form:

$$\mathbf{I}_{am} = \text{diag}([A, B, C]), \quad \mathbf{I}_{ai} = \text{diag}([I_A, I_B, I_C]) \qquad (22)$$

For a typical arched wing with elliptical, noncambered cross section in potential flow with t and a being the thickness and arc, respectively, the basic expressions for six apparent mass terms in Equation 22 can be written as

$$A = k_A\rho\frac{\pi}{4}t^2b, \quad B = k_B\rho\frac{\pi}{4}(t^2 + 2a^2)c, \quad C = k_C\rho\frac{\pi}{4}c^2b \qquad (23)$$

$$I_A = k_A^*\rho\frac{\pi}{48}c^2b^3, \quad I_B = k_B^*\rho\frac{4}{48\pi}c^4b, \quad I_C = k_C^*\rho\frac{\pi}{48}t^2b^3 \qquad (24)$$

For a typical PADS with $AR = bc^{-1} = 3$, arc-to-span ratio $a^* = ab^{-1} = 0.15$, and relative thickness $t^* = tc^{-1} = tb^{-1}AR = 0.15$, the coefficients in Equations 23 and 24 were estimated in Lissaman and Brown (1993) as

$$k_A = 0.899, \quad k_B = 0.34, \quad k_C = 0.766, \quad k_A^* = 0.630,$$

$$k_B^* = 0.961, \quad k_C^* = 1 \tag{25}$$

Utilizing the computational fluid dynamics (CFD) code for the same wing and accounting for the shape of the wing tips, these coefficients were corrected in Barrows (2002) as

$$k_A = 0.945, \quad k_B = 0.614, \quad k_C = 0.806, \quad k_A^* = 0.589,$$

$$k_B^* = 0.784, \quad k_C^* = 0.953 \tag{26}$$

Both sets of coefficients are relatively close to each other (the difference in k_B is most likely caused by the tip geometry) and are used in Equations 23 and 24 to model apparent mass forces and moments in Equations 20 and 21.

4.3 PADS aerodynamics

When fully inflated, the ram-air parachute resembles a low-AR wing. The airfoil-shaped ribs are sewn chordwise between the low-permeability upper and lower surfaces at a number of spanwise intervals to form a series of cells. While most early parafoil wings used the modified Clark YM-18 airfoil with a maximum thickness of $0.18c$, nowadays other low-speed airfoils, like NASA LS1-0417, are used as well. The wing shape is maintained by ram-air pressure entering through the opening over the entire length of parafoil leading edge (Jalbert, 1966).

Suspension lines are attached to alternate ribs and connect the parachute to payload. To reduce the number of suspension lines (and consequently reduce the drag) but still maintain the chordwise profile of the lower surface suspension lines are usually cascaded. All lines in the spanwise direction have the same length $R = [0.6; 1]b$ chosen to ensure stability). This results in arc-anhedral shape.

The earlier theoretical studies on ram-air parachute aerodynamics were undertaken in the early 1970s. A comprehensive bibliography on this subject up to the late 1990s is given in Goodrick (1975) and Lingard (1995). Some of the latest developments, including application of extended lifting-line theory, CFD, and coupled CFD-FSI (fluid–structure interaction), are highlighted in Yakimenko (2015).

Comprehensive wind tunnel testing of ram-air parafoils was conducted by Nicolaides (1971) and Ware and Hassell (1969) at the University of Notre Dame $2\,\text{ft} \times 2\,\text{ft}$ and

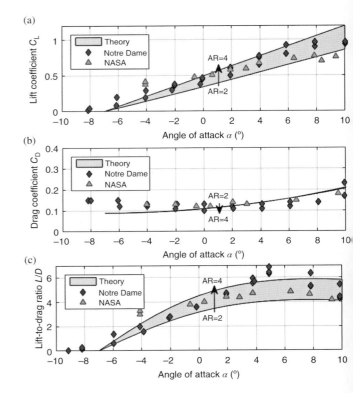

Figure 11. Theoretical lift (a) and drag (b) coefficients along with their ratio (c) for ram-air wings with $AR = 2$–4.

Langley $30\,\text{ft} \times 60\,\text{ft}$ low-speed wind tunnels. Parafoil model sizes ranged from $84\,\text{cm}^2$ to $14\,\text{m}^2$ and AR ranged from 0.5 to 3. Figure 11a and b represents some of the data obtained in these experiments with an $AR = 3$ parafoil overlaid over theoretical lift and drag coefficients, C_L^c and C_D^c, obtained for parafoils with $AR = 2$–4 with the zero-lift AoA α_0 of $-7°$ and stall AoA of about $10°$ (Lingard, 1995). The vertical arrow in these figures represents the direction of parameter change while increasing AR (the lift curve slope increased while the drag curve did not vary much). As seen from these figures, theoretical data match experimental data fairly well. Figure 11c represents a comparison between the theoretical and experimental values of GR.

Arc anhedral, described by the anhedral angle $\varepsilon = 0.25/(R/b) \neq 0$ (spanwise downward angle from horizontal to a parafoil tip), causes the reduction of the total wing lift proportional to $\cos^2(\varepsilon)$, while wing's drag is not affected explicitly. However, other elements of PADS introduce an additional drag so that the overall drag coefficient of PADS is considerably larger than that of Figure 11b resulting in about 40–50% decrease of GR. Among the largest contributors of that additional drag are the suspension lines. Figure 12 shows

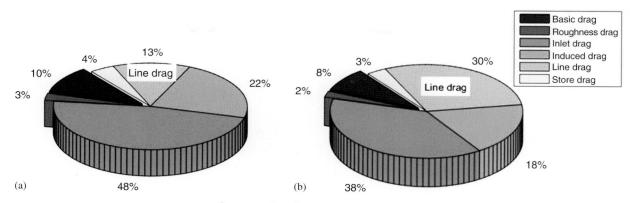

Figure 12. Typical drag contributions for a $30\,\mathrm{m}^2$ (a) and $300\,\mathrm{m}^2$ (b) ram-air parachutes. (Reproduced with permission from AIAA, 2016.)

examples of total drag contributions for the small and large ram-air parafoils (Lingard, 1995).

Figure 12 shows that open airfoil nose, inlet, contributes the most. Specifically, its contribution is on the order of $0.5\overline{h}$, where $\overline{h} = h_{\mathrm{in}}/c$ is the relative inlet height. Contribution of basic airfoil drag is about 0.015 and contribution of surface irregularities and fabric roughness is on the order of 0.004 (Ware and Hassell, 1969).

Assuming a drag coefficient for a line related to its lengthwise area (perpendicular to the flow) being 1, the drag coefficient for n suspension lines almost perpendicular to the airspeed vector and having an average length R and a diameter d is proportional to the ratio of the lines area and canopy area, $C_{\mathrm{D}}^{l} = nRd/S$. Increasing the size of PADS and keeping the same n/S ratio makes C_{D}^{l} being proportional to R, which in turn is proportional to b (to keep the R/b ratio the same). That is why Figure 12 features an increase of relative contribution of suspension lines drag with the size of PADS. In practice, increase of the canopy size leads to increase of the volume of trapped air inside it, and results in PADS center of gravity shifting toward canopy. Hence, larger PADS require an increase of R/b ratio compared to that of smaller PADS.

The lift, drag, and pitch moment coefficients of payload (store) depend of its geometry. For a rectangular parallelepiped and small AoA, they can roughly be represented as

$$C_D^s \approx C_D^*(k + (1-k)\alpha^2)\overline{S}^s, \quad C_L^s \approx \alpha\overline{S}^s, \quad C_m^s \approx 0.2\alpha\overline{S}^s \tag{27}$$

where $k = S_{\mathrm{fr}}S_{\mathrm{bot}}^{-1}$ is the ratio of the front and bottom areas of payload, $\overline{S}^s = S^{\mathrm{bot}}S^{-1}$ is the relative bottom area, and AoA is in radians. For cubic solid payload, a drag coefficient C_D^* is 1.05. For parallelepiped geometries other than cube, it may vary from 0.9 to 1.8 depending on payload's

cross-sectional (front) AR. Dependencies of $C_{\mathrm{D}}(\beta)$, $C_Y(\beta)$, and $C_n(\beta)$ are described by similar equations with S^{bottom} replaced with S^{side} and the values of coefficients for $C_Y(\beta)$ and $C_n(\beta)$ decreased.

Compared to the rigid wings, the theoretical studies (including CFD) can only produce some rough estimates of aerodynamic parameters. The use of suspension lines and the flexible nature of parafoils make it difficult to handle them in the wind tunnel and almost impossible to introduce the sideslip angles. That is why employing combined CFD–FSI simulations to account for a canopy deformation might be the most reliable resource of aerodynamic data. Canopy deformation occurs both spanwise and chordwise, affects anhedral arc radius and projected planform in flight, and also causes spanwise negative twist (because of uneven aerodynamic loading). Deflecting TE causes even more changes of planform. Figure 13 shows what happens to a low-AR double-cell rectangular planform parafoil in a brake regime with both TE down (see more details in Yakimenko (2015)). For these canopies, such deformations may cause up to about 15% reduction in span and 5% reduction in root chord, which results in about 20% reduction in a projected planform area and 10% reduction in AR. Simultaneous (up to 20%) change of the lift and drag coefficients, however, leaves GR unaffected.

Ultimately, PADS system identification (SID) can be conducted using the flight tests results. However, compared to other aerial vehicles, SID of PADS is a real challenge. Flight test instrumentation includes the GPS receiver, which records its 3D position and velocity vector components in $\{n\}$ at 1–5 Hz, and inertial measurement unit (IMU), which adds measurements of accelerations, Euler angles and angular rates at up to 100 Hz. It may also include air data sensors to record surrounding air parameters along with components of an airspeed vector. AGU records actuator positions. More

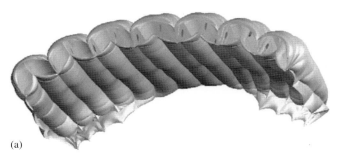

(a)

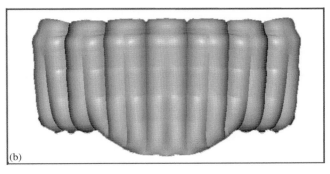

(b)

Figure 13. Parawing shape in a brake regime (a) as compared to the nominal constructed shape (b). (Reproduced with permission from AIAA, 2016.)

or less recent wind data are usually provided by dropsondes. Otherwise, airspeed and components of the wind vector must be estimated as well.

However, in most PADS configurations, all these data describe the motion of AGU only. If AGU resides atop payload, these data describe the motion of payload. Developing a high-fidelity model additionally requires information about relative motion of payload with respect to parachute canopy. To address this issue, onboard instrumentation package may include uplooking camera installed atop payload. After a proper calibration, image processing algorithms allow estimating both relative position and relative orientation of parachute with respect to payload (camera). In addition to this, these days a miniature IMU can be sewn into canopy at multiple locations to measure a variety of parameters of various canopy parts at the same time. If the airdrop is recorded using multiple ground cameras, spread apart around DZ, post-flight image processing may also allow estimating positions of both canopy and payload.

As a result, while flight test data are routinely used to validate the low-fidelity and linear models, not many attempts were made to date to perform SID on the high-fidelity PADS models. The most comprehensive ones were based on a single-criterion error method used to validate a 6DoF model of the ALEX PADS featuring 19 varied parameters (mostly coefficients of aerodynamic and control

derivatives), multicriteria identification technique to investigate an 8DoF model of the Pegasus PADS that included 33 varied parameters and eight adequacy criteria, and the Extended Kalman filter approach applied in attempt to estimate 55 parameters of the 6DoF model of the prototype of X-38 Crew Return Vehicle (CRV) (Yakimenko, 2015).

4.4 Effect of the control inputs

So far, no consideration was given to the effect of symmetric, δ_s, and asymmetric, δ_a, control inputs on horizontal and vertical components of airspeed vector (excluding Equation 10). Turn dynamics assumed a symmetric linear dependence of a steady-state yaw rate $\dot{\psi}_{ss}$ versus δ_a. That is where SID using flight test data becomes useful.

Assuming normalized inputs, so that $\delta_s \in [0;1]$ and $\delta_a \in [-1;1]$, the regression analysis performed on several PADS reveals that the dependence of V_h, V_v, and GR on the control inputs can accurately be described by the same-form quadratic relationship:

$$\xi = \xi_0(1 - p_\delta \bar{\delta}_{s;a}^2) \qquad (28)$$

In this equation ξ_0 is the value at $\bar{\delta} = 0$ and parameter p_δ shows the decrease in the value at full actuator deflection ($\bar{\delta} = 1$). Depending on the size of PADS and trimming conditions (parafoil rigging), these parameters vary as follows: For GR $\xi_0 \in [3;4]$ and $p_\delta \in [0.2;0.4]$, for V_h $\xi_0 \in [11;23]$ m s^{-1} and $p_\delta \in [0.24;0.43]$, and for V_v $\xi_0 \in [3.6;6.8]$ m s^{-1} and $p_\delta \in [-0.8;0.24]$. The spread of the decrease factor p_δ for these parameters is shown graphically in Figures 14 and 15.

As seen, a full symmetric TE deflection causes about 20% decrease of V_v and about a doubled decrease of V_h. As a result, a full symmetric TE deflection usually leads to about 20–40% decrease of GR. This decrease is gradual, so at $\bar{\delta}_s \approx 0.3$ it is only 1/10 of it, meaning that for the relatively small $\bar{\delta}_s$ the effect may be considered negligible. An asymmetric TE deflection happens to have a much stronger effect. While having almost no effect on V_h, because of banking it has a strong effect on V_v causing its increase. This increase can be quite substantial (up to 80%).

The effect of asymmetric TE deflection on $\dot{\psi}_{ss}$ can essentially be modeled with the quadratic regression as well. Slightly modified to allow for the asymmetry and nonzero TR at $\bar{\delta}_a = 0$, it takes the following form (Figure 16):

$$\dot{\psi}_{ss} = K_\psi \text{sign}(\bar{\delta}_a)\bar{\delta}_a^2 + \dot{\psi}_0 \qquad (29)$$

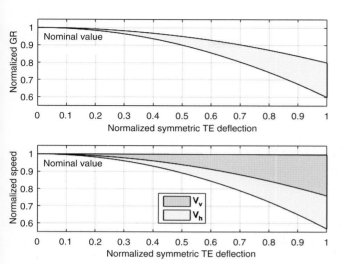

Figure 14. Variation of GR, V_v, and V_h versus $\bar{\delta}_s$.

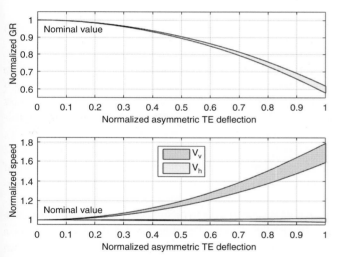

Figure 15. Variation of GR, V_v, and V_h versus $\bar{\delta}_a$.

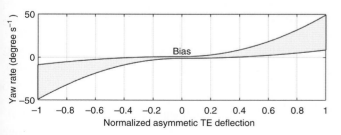

Figure 16. Variation of $\dot{\psi}_{ss}$ versus $\bar{\delta}_a$.

Introduction of nonlinear control efficiency compensates for some canopy deformation effects while deflecting TE as described earlier. The values of K_ψ vary from 10 to $50°\,s^{-1}$ (the smaller the PADS, the more the agility), which matches data of Figure 8. Due to canopy asymmetry and errors in a nominal control lines setting, Equation 29 allows for a bias. Usually, it is on the order of a couple of degrees per second. As seen from Figure 16, small deflections lead to a sluggish response. Moreover, if their steering lines have a sag, then PADS may exhibit inverse reaction in response to small δ_a. That is why the nominal control lines setting usually assumes some tension, that is, $\delta_{s0} > 0$. That also brings PADS to the range of a higher sensitivity to control inputs, so that the linear control models can be used. (The negative effect however is that because of this tension in the control lines, pulling them down takes a little longer time than releasing.)

4.5 Linearized models and stability

Linearized models are used to study PADS stability and develop controllers. These models have the standard form:

$$\delta\underline{x} = \mathbf{A}\delta\mathbf{x} + \mathbf{B}\delta\mathbf{u} \qquad (30)$$

where $\mathbf{A}$ and $\mathbf{B}$ are the state and input matrices, and $\delta\mathbf{x}$ and $\delta\mathbf{u}$ are variations of state and control vectors with respect to a nominal flight. Usually, longitudinal and lateral–directional states are decoupled. Depending on application, the state and the control vector could be as simple as

$$\mathbf{x}^{\text{lon}} = [u, w, q, \theta]^T, \quad \mathbf{x}^{\text{lat}} = [v, p, r, \phi, \psi]^T \qquad (31)$$

(to study stability of a 6DoF model), all way up to

$$\mathbf{x}^{\text{lon}} = [u, w, q, \theta, q_s, \theta_s]^T \quad \text{and}$$

$$\mathbf{x}^{\text{lat}} = [v, p, r, \phi, \psi, p_s, r_s, \phi_s, \psi_s]^T \qquad (32)$$

(to capture a payload motion in the 9DoF model). (In Equation 32, additional parameters with subindex s describe relative dynamics of a store.) The corresponding control vectors are

$$\mathbf{u}^{\text{lon}} = \delta_s \quad \text{and} \quad \mathbf{u}^{\text{lat}} = \delta_a \qquad (33)$$

For a powered PADS, the control vector for the longitudinal channel would also include an engine throttle setting:

$$\mathbf{u}^{\text{lon}} = [\delta_T; \delta_s]^T \qquad (34)$$

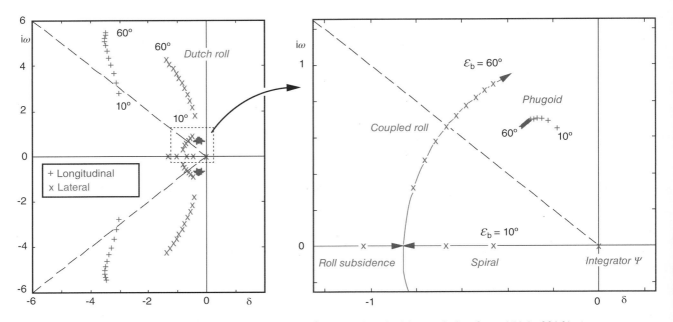

Figure 17. PADS model root locus while varying $\varepsilon_b = 10-60°$. (Reproduced with permission from AIAA, 2016.)

Varying PADS parameters and using a standard root locus technique applied to the state matrix **A** of a linearized model allows exploring system dynamics graphically. For example, Figure 17 shows root loci while varying the angle $\varepsilon_b = 2\varepsilon$ (double anhedral angle), which changes the R/b ratio from 2.9 to 0.5 (Jann, 2004). As seen, increasing canopy's curvature (shortening suspension line length) results in increase of the frequency of oscillations of AoA (short-period mode) and sideslip angle (Dutch-roll mode). It also leads to a decrease of damping of these two modes as well as a coupled roll-spiral mode (roll angle). Hence, increasing canopy's curvature (decreasing line length R) has a destabilizing effect. A very small canopy curvature (anhedral angle) results in the coupled roll-spiral mode split into two aperiodic modes, roll subsidence mode and spiral mode, traditional for an aircraft.

5 PADS GNC

All self-guided PADS rely on AGU GNC software to gather information about current PADS position relative to IPI, estimate PADS parameters and winds, and develop and continuously adjust guidance strategy generating the corresponding control inputs to follow this strategy. Hence, as in the case of any autonomous vehicle, AGU software consists of the navigation, guidance, and control blocks. For most of modern PADS, AGU's sensor suite includes GPS receiver and barometric altimeter only; some AGUs may also include

IMU. Larger PADS also employ ultrasound or laser altimeters to sense the height above the ground to execute a flare maneuver. Future AGU may include optical sensors enabling navigation in the GNP-denied environment.

Compared to an aircraft, PADS are unpowered, much slower, and underactuated. Landing PADS would be similar to executing an engine-off maneuver using a rudder only. Also, compared to an aircraft PADS are very vulnerable to the wind. Hence, along with the traditional task of understanding current PADS position with respect to IPI, determining the current wind as long as the current values of V_h, V_v, and TR (which in general are not known upfront because ideally PADS should be able to use different parafoils/rigging schemes and different loads) is the primary goal of the navigation block.

Guidance block accepts this information to provide with the best possible solution allowing getting from the current PADS position to IPI. The control block ensures stable flight behavior, satisfactory tracking of the trajectory generated by the guidance block, and timely implementation of the flare maneuver for the soft landing. The outer loop of the control block generates its commands in terms of a desired yaw rate $\dot\psi_c$, and the inner loop then converts these commands into the desired inputs to two motors.

In PADS utilizing a one-point swivel attachment, AGU is suspended above the confluence point (like in Figure 6a). Some other PADS utilize a two-point attachment scheme and may have AGU suspended above the confluence point

(Figure 6b) or sitting atop a payload (Figure 5b). Using a four-point attachment, like in Figure 5a, allows having AGU atop a payload as well. Having AGU atop a payload seems to be a simpler solution, while AGU suspended between the canopy and payload provides a very flexible attachment point for a variety of payload sizes and shapes and, what is probably even more important, simplifies interface between AGU and canopy as well as feedback control by measuring the states of canopy rather than payload.

5.1 Maneuver-based guidance

The first detailed GPS-based guidance strategy was published for the NASA Spacewedge PADS (Figure 18a) (Sim *et al.*, 1994). The transitions between phases were controlled by onboard barometric altimeter measuring altitude above IPI. The surface wind direction and speed as well as IPI barometric altimeter settings were provided upfront. Upon approaching the DZ area with a certain altitude excess, which completes the homing phase, PADS enters a standard holding pattern aligned with the surface wind. The length of this pattern was chosen to produce about 150 m altitude loss per one pattern loop. When altitude falls below 90 m, PADS continues with the landing pattern gliding along the downwind leg passing IPI and then proceeding with the base and final approach legs. The initiation of a base turn starts at 45–60 m based on the surface wind. The flare maneuver is initiated when onboard height sensor (ultrasonic or laser) senses a height of 8 m above ground.

The major pitfall of this approach was in determining a relative PADS position based on a GPS signal only. Back then (before the accuracy-degrading selective availability "jitter" was removed), it could only provide an accuracy of ±100 m (against today's ±10 m). Hence, further refinement of this scheme involved development of an integrated GPS/IMU navigation system. Control algorithms included online estimate of the current winds.

For the subsequent PADS, guidance strategy was refined to substitute a rectangular EM pattern with the circular one (Figure 18b). In this latter approach, the turn radius while in the EM pattern was initially fixed. Later, however, it was made adjustable (resulting in a different altitude loss per one loop) to be able to enter the final approach phase at the correct altitude and direction (as shown in Figure 18b) (Jann, 2004). Alternative versions of this guidance scheme included shifting the holding pattern downwind to allow more time for possible final corrections (Figure 18c).

These days, many PADS in different weight categories still use the circular pattern of Figure 18b. Specifically, this guidance strategy was employed by single-stage ML and UL

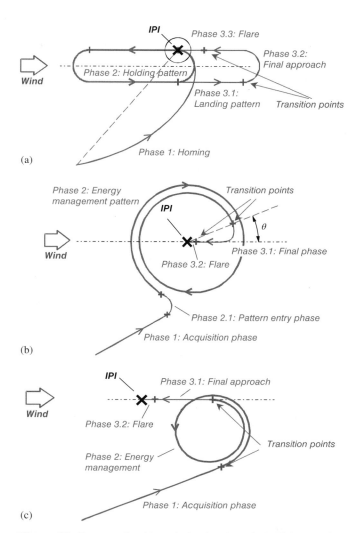

Figure 18. Rectangular (a) and circular (b and c) EM strategies. (Reproduced with permission from AIAA, 2016.)

Mosquito and Onyx PADS. It is also used by the first stage of the two-stage EL and L Screamer PADS with a following deployment of the second stage on the final approach phase elevated to accommodate the altitude loss while deploying and dereefing the second-stage round canopy.

Instead of a circular EM pattern, many other PADS utilize a figure-8 (lemniscate) pattern (Figure 19) oriented transverse to the desired landing direction. Compared to a circular pattern downwind of IPI (Figure 18c), figure-8 pattern is more stable and ensures a simpler exit because most of the time PADS flies either perpendicular to the wind direction or upwind (while the circular EM pattern has a substantial downwind portion). Most figure-8-based algorithms utilize a maneuver-based EM, that is, a sequence of a predefined control inputs while in the pattern, others establish this

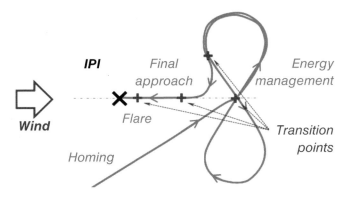

Figure 19. Figure-8 EM strategy.

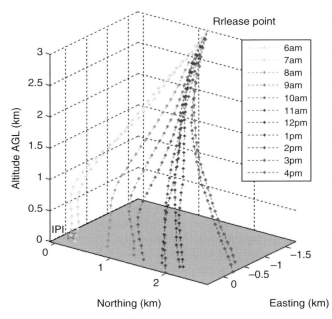

Figure 20. Uncontrolled trajectories with the different wind profiles. (Reproduced with permission from AIAA, 2016.)

maneuver in $\{n\}$ using a set of four way points (WPs) (Jann, 2004). The location of these WPs is varied based on the current 3D position relative to IPI and wind estimates.

5.2 Accounting for the variable winds

Obviously, variable winds aloft and surface winds play a crucial role in both system accuracy and terminal accuracy. Some preliminary knowledge of the winds aloft is necessary to generate a proper LAR and assign a reliable CARP within it; some knowledge of the surface winds is essential for the terminal guidance phase.

As an illustration, Figure 20 shows the winds at the same location collected hourly, starting from 6 a.m. in the morning. By looking at these data, there is no doubt that atmospheric data change throughout a day drastically, so even an hour difference could result in a completely different wind profile. Actual temperature, pressure, and density (versus altitude) change throughout the day as well, and for sure differ from parameters of the standard atmosphere model, especially near the surface.

As mentioned in Section 2, JPADS-MP was developed to have a better wind forecast around the DZ area. It uses all available data, including NOAA's national operational weather forecasts, wind sounding balloons, wind data derived by the airdrop aircraft and on-scene *in situ*, wind and weather data observations from hand-launched dropsondes, and produces a high-fidelity, high-resolution 3D grid of winds, pressure, and density valid at the intended drop time. It also utilizes the local topographic data since it drives the atmospheric flow in the lower layers of the atmosphere, especially in complex, rugged terrain (Wright, Benney, and McHugh, 2005). Knowing these winds, $\mathbf{W}(h)$, allows offsetting CARP computed for a release altitude h^* for a specific

PADS (characterized by an estimate $\hat{V}_v(h)$) in no-wind conditions by a horizontal vector $\mathbf{d}_W^*$:

$$\mathbf{d}_W^* = \int_{h_{IPI}}^{h^*} \frac{\hat{\mathbf{W}}(h)}{\hat{V}_v(h)} dh \tag{35}$$

During the actual descent, AGU attempts to estimate the current winds and adjust guidance strategy based on these most recent estimates.

Decisions made during the terminal guidance phase are made upon some knowledge of the surface-layer winds. These decisions explicitly affect the PADS terminal accuracy and that is why knowing surface-layer winds is very important. Modeling the near-surface winds in GNC algorithms may be based on different assumptions:

$$W^{cons}(h) = const\,(h), \quad W^{lin}(h) = (\hat{W}_H - \hat{W}_{surf})H^{-1}h + \hat{W}_{surf}$$

$$W^{log}(h) = \frac{\hat{W}_H - \hat{W}_{surf}}{\ln(H+1)} \ln(h+1) + \hat{W}_{surf} \tag{36}$$

In these models, $\hat{W}_H$ is the wind estimate available at altitude H and $\hat{W}_{surf}$ is the wind estimate at IPI (if available).

Figure 21 shows two examples of the wind profiles from the ground up to 750 m above IPI. They are split into

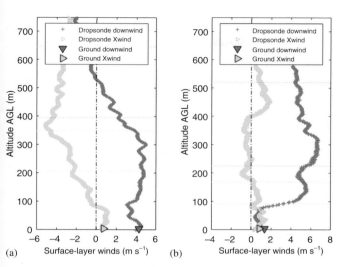

Figure 21. Examples of the surface-layer winds. (Reproduced with permission from AIAA, 2016.)

downwind and crosswind components relative to the desired direction of landing. Shown with a triangle marker are measurements provided by the surface winds measurement device and collected several minutes after the entire $W(h)$ profile was obtained (using a dropsonde). Horizontal lines depict decision-making altitudes. These two profiles show all kinds of problems associated with the wind modeling. The wind profile of Figure 21a demonstrates that at $h \approx 550$ m wind changes its direction, so whatever assumptions were made above this altitude became obsolete. It also shows the

crosswind component dying toward the ground. Starting from $h \approx 300$ m all way down, the downwind component can be modeled as a constant. If surface wind data were available to PADS when it was at $h \approx 300$ m, it would definitely add confidence to the validity of this model. On the contrary, the wind profile of Figure 21b exhibits no major changes down to $h \approx 100$ m, when the winds suddenly die. If the constant model were assumed, it would result in a huge overshot. Again, knowing the surface winds and assuming a linear model could probably help in this case as well.

Figure 22 features a miniature portable IPI station based on the Kestrel pocket weather tracker paired wirelessly with a Blackberry cell phone (via Bluetooth interface) to allow transmitting the IPI coordinates and real-time ground atmospheric data to a descending PADS (in this case, Snowflake PADS AGU also used a cell phone paired to autopilot). If GSM network is not available, a common RF-based station could be used (Bourakov, Yakimenko, and Slegers, 2009). Figure 22a features a situation when the landing direction is determined by a portable IPI station orientation. Alternatively, the weather station may be mounted on a vane to broadcast a current surface wind direction as well. That would allow PADS to land exactly into the wind. Introduction of this weather station alone allowed Snowflake PADS to mitigate situations as shown in Figure 21b when the winds estimated right before the final base turn to IPI and happened to be way too different compared to the current surface winds. Mounting this miniature portable IPI station onto a moving platform enabled a novel capability of landing onto a

Figure 22. Miniature surface weather station (a), landing onto a moving platform (b), and miniature MAXMS dropsonde (c). (Reproduced with permission from AIAA, 2016.)

nonstationary IPI. Experimenting with the Snowflake PADS also demonstrated a capability to deploy a miniature dropsonde shown in Figure 22b allowing measuring the entire $W(h)$ profile rather than just a surface wind and incorporating this profile into terminal guidance decisions.

5.3   Optimal precision placement guidance

Most of guidance algorithms accommodate the best-known wind profile by planning a trajectory in the wind-fixed coordinate system with the IPI position shifted according to Equation 35. In this case, while executing a terminal guidance maneuver, these algorithms have to somehow accommodate the changes in the surface-layer wind. The only remedy in this case is to start a final approach leg a little bit high and execute a side-wave maneuver if actual winds happen to be weaker than predicted.

Ideally, mitigation of a constantly changing situation (discrepancy between the actual and desired positions caused by unmodeled winds and variations in PADS dynamics) could be accomplished by continuously solving the two-point boundary value problem with a fixed time. PADS kinematics in the horizontal plane with a yaw rate being a bounded control input can be described with as little as three differential equations (Equations 7 and 8). The initial condition of PADS at some point "A" right after exiting the EM pattern would be described by $\mathbf{x}(0) = [x_0, y_0, \psi_0]^T$, and the final condition, point "B", by some point on the final approach leg close to the flare initiation point $\mathbf{x}(t_f) = [x_f, y_f, \psi_f]^T$.

One attempt to address this problem using a classical calculus of variations resulted in a creation of bank of solutions computed off-line that was stored on a memory card and then used in AGU. This bank of optimal trajectories, lookup table, allowed choosing a specific terminal flight path (a sequence of TR commands) while entering a terminal area based on the current conditions defined by altitude, along-track and cross-track positions with respect to IPI and heading. Each of lookup table trajectories either hits the target or, if that is not possible from the given initial state, minimizes a function of position and heading error at impact (Carter *et al.*, 2007).

Direct methods of calculus of variations that search for the optimal solution within a set of parametrized candidate solutions represent much more robust choice allowing to conduct a real-time trajectory optimization/correction. To date, two different approaches were implemented and published. One approach was developed by Draper Laboratory and tested for the FireFly, DragonFly, and MegaFly PADS. Another approach was developed and tested for the Snowflake PADS. Both approaches are schematically shown in Figure 23 and proved to be very effective.

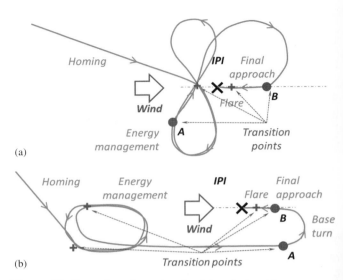

Figure 23. Examples of quasi-optimal terminal guidance.

The Draper Lab approach assumes homing toward IPI and executing the figure-8 EM upwind of IPI. When PADS descents low enough (in Figure 23a, it corresponds to some point A), the terminal phase J-hook maneuver is commenced. During this maneuver, the band-limited guidance (BLG) method is used to compute the commanded heading rate as a function of current relative position, heading, and heading rate with the goal of reaching a prescribed final position, heading, and heading rate at point B.

BLG relies on parametrization of a candidate heading rate profile:

$$\psi'_h(h) = \sum_{k=0}^{M} \psi'_k \sin(\xi_k)/\xi_k \tag{37}$$

with $\xi_k = \pi(h - k\Delta h)/\Delta h$. Varied parameters ψ'_k represent TR at consecutive multiples of Δh and ensure that the control system will be able to track the heading rate command profile accurately (i.e., where specific PADS dynamics is incorporated). This form allows explicitly restricting the heading rate profiles by the frequency that is significantly less than the bandwidth of closed-loop system (Carter *et al.*, 2009). The trajectory is obtained by integrating Equation 37 and substituting results back to Equation 7. The optimized performance index consists of a weighted sum of the squared miss distance and heading error at point B (Figure 23a). Utilizing a finite-horizon control, a new optimal solution is generated from the current point to the same final point B every several seconds.

Another direct-method-based approach is based on the inverse dynamics in the virtual domain (IDVD) method (Figure 23b). In this case, after homing to a rectangular EM pattern upwind of IPI and executing at least two full EM pattern loops allowing to accurately estimate the current winds, PADS follows a standard aircraft landing pattern (Slegers and Yakimenko, 2011). The base turn initiation point (point A in Figure 23b) is based on the wind estimates continued throughout the downwind leg. Using the IDVD method, AGU constantly updates the TR command based on the optimal solution sought among parametrizations of the PADS coordinates that use some scaled abstract argument $\bar{\tau} = \tau/\tau_{\mathrm{f}} \in [0; 1]$:

$$x(\bar{\tau}) = P_1(\bar{\tau}) = a_0^1 + a_1^1 \bar{\tau} + a_2^1 \bar{\tau}^2 + a_3^1 \bar{\tau}^3 + b_1^1 \sin(\pi\bar{\tau}) + b_2^1 \sin(2\pi\bar{\tau})$$
$$y(\bar{\tau}) = P_2(\bar{\tau}) = a_0^2 + a_1^2 \bar{\tau} + a_2^2 \bar{\tau}^2 + a_3^2 \bar{\tau}^3 + b_1^2 \sin(\pi\bar{\tau}) + b_2^2 \sin(2\pi\bar{\tau})$$
$$(38)$$

The coefficients a_i^η and b_i^η ($\eta = 1, 2$) in Equation 38 are defined by the boundary conditions set for up to the second-order derivative at $\tau = 0$ and $\bar{\tau} = 1$. The yaw angle ψ is then found from Equation 7, which using Equation 38 becomes

$$\psi(\bar{\tau}) = \tan^{-1}\left(\frac{\lambda(\bar{\tau})y'_{\bar{\tau}}(\bar{\tau}) - \hat{w}_y}{\lambda(\bar{\tau})x'_{\bar{\tau}}(\bar{\tau}) - \hat{w}_x}\right) \qquad (39)$$

Equation 39 uses the speed factor $\lambda(\bar{\tau})$ allowing mapping the virtual domain of parameter $\bar{\tau}$ to the physical time domain to ensure a constant forward speed. In this scheme, parameter τ_{f} then is the only varied parameter; however, for more flexibility, the list of varied parameters could easily be extended to include some or all final states as long as they belong to the glide path leading to IPI.

6 OTHER DEVELOPMENTS

6.1 Glide slope angle control

Improving touchdown performance for large PADS featuring higher V_{h} and slower TR compared to those of smaller PADS is still a challenge. One solution to this problem might be to control V_{v} to be able to increase it during the final approach, right before touchdown. In this case, PADS would always arrive to the DZ area a little high and then follow the common strategy employed by human jumpers. This approach is also referred to as a glide slope (angle) control.

One technique includes augmenting a parafoil-based PADS with a drogue using a two-point drogue bridle attachment at AGU and payload swivel as shown in Figure 24a (Moore, 2012). This concept assumes a one-time step decrease of the overall GR at an accurately determined deployment point on the final approach. Another scheme includes continuous control of GR via changing PADS rigging angle and therefore AoA using a variable AoA

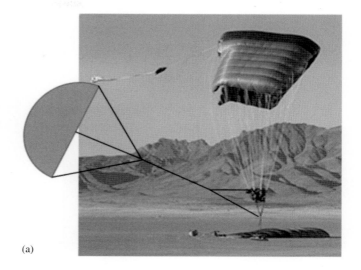

(a)

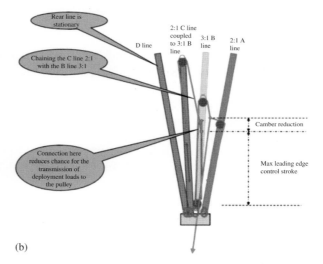

(b)

Figure 24. Drogue add-on (a) and VACS concepts (b). (Reproduced with permission from AIAA, 2016.)

control system (VACS) (Figure 24b). The entire pulley riser system would be removed from the deployment load path during canopy opening by a slackened control line. Figure 24b shows an example of varying the length of three out of four line groups (Moore, 2012). A differential AoA change in this case could induce differential lift (Figure 8). Hence, while paraglider operators (where this scheme originally came from) typically employ this so-called speed system in combination with the TE control, VACS needs no TE deflection control at all.

Another technical solution to control the glide slope angle is to use the upper surface aerodynamic spoilers. To enable it, a spanwise slit should be introduced across a number of cells in the center section of the upper surface of the canopy

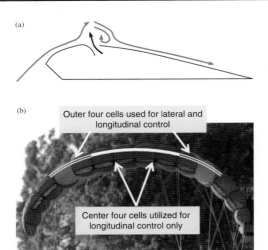

Figure 25. Airflow bubble acting like a spoiler (a) and slit spoilers locations (b). (Reproduced with permission from AIAA, 2016.)

(Figure 25a) (Higgins, 1979). The cells that contain a slit have a control line attached to the leading edge of the slit. These lines pass through the bottom skin of parafoil and run down to AGU. The same type of aerodynamic controllers spread farther apart from midsections along the upper surface of parafoil that can produce a sufficient TR control (when actuated differentially) and therefore eliminate the need for the TE deflection control as well (Figure 25b) (Ward and Costello, 2013).

6.2 Reduced cost PADS

To this end, Figure 26 shows an achieved relative cost for the different-weight PADS presented in Section 3 against the desired $6 per pound value. As seen, currently the objective value can only be met by L (and M) systems. However, UL and XL systems, that have much broader usage, are far away from meeting it. The ratio of the AGU and parafoil costs is about the same, 6 to 4, for all JPADS weight categories

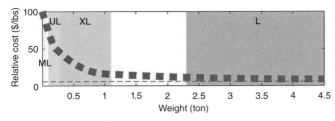

Figure 26. Relative cost of the current different-weight PADS-compared to the required objective value.

(because larger systems require larger motors and batteries). Hence, there are developments to bring down the cost of both PADS components. While the alternative approaches to steer a parafoil without using TE, discussed in the previous section, contribute to reducing the size (cost) of AGU, this section explores alternative methods of aerial payload delivery using modified round and cross canopy systems, even at the cost of slightly degraded precision performance compared to that of PADS. That is because traditional flat canopies, like G-14, G-12, and G-11, are 3-4 times cheaper compared to the same-weight-category ram-air parafoils and cruciform parachutes are about 20 times cheaper.

One of the first successful attempts of this kind was the development of AGAS in the late 1990s (Dellicker, Benney, and Brown, 2001). The idea was to use a standard G-12 type canopy and utilize four variable-length risers to disturb the shape of parachute canopy and therefore provide a limited horizontal control authority. The canopy disturbance was achieved by lengthening one or two adjacent risers from the nominal 6 to 8 m (30% lengthening). Both CFD analysis and flight tests proved that lengthening a single riser leads to about 0.5:1 GR, while lengthening two adjacent risers results in up to 0.8:1 GR. These modest (compared to those of PADS) values allowed developing a robust GNC that ensured very good touchdown accuracy given that AGAS is deployed within LAR computed based on the relatively accurate winds profile (that is, when the development of the JPADS-MP had begun). Released at CARP somewhere within LAR, AGAS was capable of steering toward some nominal trajectory, defined in $\{n\}$ and originated at IPI, compensating for wind variations and unmodeled dynamics. Figure 27a shows an example of such a nominal trajectory (the thick central curve) and a set of Monte Carlo simulations originating within LAR and employing the high-fidelity AGAS model with the developed control algorithm. In practice, during the very first tests, three AGAS deployed from 3 km above IPI level exceeded the threshold requirement of 100 m CEP landing 56, 76, and 78 m away from IPI, while two standard (uncontrolled) G-12-based systems released at the same time as AGAS, were blown for more than 1 km away from IPI. Figure 27b features two AGAS with the two adjacent and single riser lengthened, which gives an idea about canopy shape disturbance.

Further development of this idea may involve even cheaper cruciform-type canopies, featuring good drag to canopy area ratio and good static and dynamic stability characteristics. These parachutes are composed of two identical cloth rectangles, crossed and joined to each other at the square intersection to form a flat surface having four equal arms. Suspension lines are attached to the outer edges of the four arms (Figure 28a).

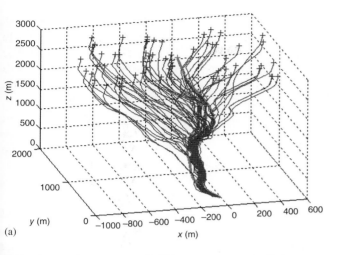

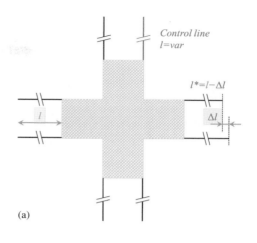

Figure 27. Monte Carlo simulation for GNC algorithm verification (a), and two AGAS steering toward the same nominal trajectory (b). (Reproduced with permission from AIAA, 2016.)

Figure 28. Layout of a controlled cruciform parachute (a) and subscale tests (b). (Reproduced with permission from AIAA, 2016.)

Fixed shortening or lengthening of one of the suspension lines by Δl (Figure 28a) puts parachute to spin (Figure 28b). Deflecting the adjacent line allows controlling this spin (Figure 29). Deflecting both lines evenly creates a horizontal force that can be used in a manner similar to that of AGAS. Preliminary tests on a small prototype exhibited a 0.4:1 GR. Compared to AGAS, this scheme utilizes a single (as opposed to four) control and allows a directional control (while AGAS may only steer in one of eight directions).

Partially connecting two- and three-canopy assemblies at the skirt, as shown in Figure 30, and pulling one side of the skirt down allow steering a cluster of canopies. Figure 30a and b shows two snapshots of a flight test of the prototype two-canopy assembly composed of one-quarter-scale G-12 canopies featuring a steady-state glide and turn, respectively (Lee and Buckley, 2004). In these initial tests, the encouraging values of up to 0.9:1 GR and a 2.5° s⁻¹ TR were achieved.

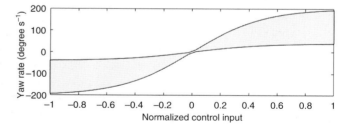

Figure 29. Yaw rate control authority for a cruciform parachute.

One more seemingly inexpensive approach includes varying the descent rate of circular parachutes. It can be done continuously, by reefing and disreefing of the skirt resulting in changes to the effective drag area (which might be more appropriate for small canopies due to the power demands of the actuation system), or in a one-time action, by delayed canopy deployment. Figure 31 presents a general idea of

Figure 30. Steady glide (a) and turn (b) of a steerable cluster of two parachutes. (Reproduced with permission from AIAA, 2016.)

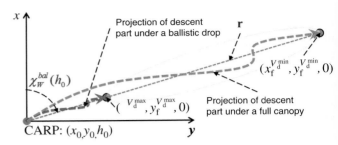

Figure 31. Reachability area using a canopy reefing control.

varying reefing level, PADS has some range it can land within (shown as a shaded area). A projection of all possible touchdown points initiated from the current 3D point with a constant descent rate V_d is represented by the line

$$x_f(V_d) = x_0 + \frac{h_0}{V_d} W^{bal}(h_0) \cos\left(\chi_W^{bal}(h_0)\right),$$

$$y_f(V_d) = y_0 + \frac{h_0}{V_d} W^{bal}(h_0) \sin\left(\chi_W^{bal}(h_0)\right) \qquad (40)$$

where $W^{bal}(h_0)$ and $\chi_W^{bal}(h_0)$ are the magnitude and direction of the so-called ballistic winds computed for an altitude h_0. The technical implementation of a reversible reefing system for circular canopies can vary. To this end, several reefing techniques were investigated (Fields and Basore, 2015). These techniques included the four reefing strategies stemmed from a single control line approach, in which a single control line can manipulate the parachute size/shape directly while carrying only a small portion of the suspension line load. This way the descent rate is inversely proportional to the control line length (linearly proportional to the level of reefing).

7   CONCLUSION

The touchdown accuracy of uncontrolled round parachutes deployed from high altitudes depends entirely on some knowledge of underlying air column characteristics. They include the vertical profiles of air density defining the height above DZ and winds. Since neither of these parameters is known precisely, a touchdown error for these systems is measured in hundreds of meters. Introduction of a gliding parachutes and capability to use GPS as a major sensor for controlling them resulted in reducing a touchdown error of aerial payload delivery by an order of magnitude. A variety of GNC algorithms developed in the past two decades made these PADS fully autonomous with a very little interface

using a variable reefing (Fields, 2013). With a canopy fully opened (reefing control line fully extended), PADS has a slowest descent rate and therefore glides with the winds for a longer time. As the reefing control line is reeled in, the effective canopy drag area is reduced and the descent rate increases. This results in a shorter gliding distance. By required only to set up an upcoming mission. The efforts on improving PADS performance, including extending standoff distances, further reducing a miss distance even in the GPS-denied environment and reducing overall costs of PADS operations, still continue. The latest developments incorporate using terrain maps to be able to deliver payloads to a complex rugged-terrain DZ, and networking between multiple PADS, which enables using massive airdrop with collision avoidance. Other research is conducted in the area of precise aerial placement of a remote sensor grid or delivering payloads in the urban environment. Started as a program to improve precision of aerial payload delivery, JPAD program developed a new type of unmanned aerial vehicle equipped to address the challenges of traditional and new applications.

REFERENCES

Allen, R.F. (1995) Orion advanced precision airborne delivery system. *Proceeding of the 13th AIAA Aerodynamic Decelerator Systems Technology Conference and Seminar*, AIAA-1995-1539-CP, Clearwater Beach, FL.

Barrows, T.M. (2002) Apparent mass of parafoils with spanwise camber. *J. Aircr.*, **39**(3), 445–451. (Also available as Barrows, T.M. (2001) Apparent mass of parafoils with spanwise camber. *Proceeding of the 16th AIAA Aerodynamic Decelerator Systems Technology Conference and Seminar*, Boston, MA, May 21–24.)

Bourakov, E.A., Yakimenko, O.A., and Slegers, N.J. (2009) Exploiting a GSM network for precise payload delivery. *Proceedings of the 20th AIAA Aerodynamic Decelerator Systems Technology Conference*, AIAA, Reston, VA.

Brown, G. and Benney, R. (2005) Precision aerial delivery systems in a tactical environment. *Proceedings of the 18th AIAA Aerodynamic Decelerator Systems Technology Conference and Seminar*, AIAA, Reston, VA.

Carter, D., George, S., Hattis, P., McConley, M., Rasmussen, S., Singh, L., and Tavan, S. (2007) Autonomous large parafoil guidance, navigation, and control system design status. *Proceedings of the 19th AIAA Aerodynamic Decelerator Systems Technology Conference and Seminar*, AIAA-2007-2514, Williamsburg, VA.

Carter, D., Singh, L., Wholey, L., Rasmussen, S., Barrows, T., George, S., McConley, M., Gibson, C., Tavan, S., and Bagdonovich, B. (2009) Band-limited guidance and control of large parafoils. *Proceedings of the 20th AIAA Aerodynamic Decelerator Systems Technology Conference and Seminar*, AIAA-2009-2981, Seattle, WA.

Dellicker, S., Benney, R., and Brown, G. (2001) Guidance and control for flat-circular parachutes. *J. Aircr.*, **38**(5), 809–817.

Eilertson, W.H. (1969) *Gliding Parachutes for Land Recovery of Space Vehicles: Case 730*, NASA-CR-108990, Bellcomm Inc., Washington. DC.

Fields, T. (2013) A descent rate control approach to developing an autonomous descent vehicle, PhD thesis, University of Nevada, Reno, NV.

Fields, T. and Basore, N. (2015) Reversible control line reefing system for circular parachutes. *Proceeding of the 23rd AIAA Aerodynamic Decelerator Systems Technology Conference and Seminar*, Daytona Beach, FL.

Goodrick, T. (1975) Theoretical study of the longitudinal stability of high performance gliding airdrop systems. *Proceedings of the 5th AIAA Aerodynamic Decelerator Systems Conference*, AIAA-1975-1394, Albuquerque, NM.

Goodrick, T. (1984) Scale effects on performance of ram air wings. *Proceedings of the 8th AIAA Aerodynamic Decelerator and Balloon Technology Conference*, AIAA 84–0783, Hyannis, MA.

Gorman, C.M. and Slegers, N.J. (2011) Comparison and analysis of multi-body parafoil models with varying degrees of freedom. *Proceeding of the 21st AIAA Aerodynamic Decelerator Systems Technology Conference and Seminar*, Dublin, Ireland.

Higgins, M.W. (1979) Control system for ram air gliding parachute. U.S. Patent 4,175,722.

Jalbert, D. (1966) Multi-cell wing type aerial device. U.S. Patent 3,285,546.

Jann, T. (2004) *Modellierung, Identifizierung und Autonomes Fliegen eines Gleitfallschirm-Last-Systems*, PhD dissertation, DLR-FB-2004-33, Institute of Flight Mechanics, DLR Braunschweig, Germany.

JPADS (2014) JPADS: making precision airdrop a reality. *Defense Industry Daily.* www.defenseindustrydaily.com/jpads-making-precision-airdrop-a-reality-0678/

Lee, C.K. and Buckley, J. (2004) Method for steerable clustered round parachutes. *J. Aircr.*, **41**(5), 1191–1195.

Lingard, J.S. (1995) Precision aerial delivery/ram-air parachute system design. *Proceeding of the 13th AIAA ADS Conference and Seminar*, Clearwater Beach, FL.

Lissaman, P.B.S. and Brown, G.J. (1993) Apparent mass effects on parafoil dynamics. *Proceedings of the 12th RAeS/AIAA Aerodynamic Decelerator Systems Technology Conference and Seminar*, AIAA-1993-1236, London, England.

Moore, J. (2012) White Paper: Variable Angle of Attack Control System (VACS) for Gliding Aerial Delivery, Response to Request for Information prepared for InSitech by JM Technologies Inc.

Nicolaides, J.D. (1971) *Parafoil wind tunnel tests*. AFFDL-TR-70-146, U.S. Air Force Flight Dynamics Lab, Wright-Patterson Air Force Base, Ohio.

PATCAD (2005) Precision Airdrop Technology Conference and Demonstration, Final Report, U.S. Army RDECOM, Natick, MA.

Sego, K., Jr. (2001) Development of a high glide, autonomous aerial delivery system – Pegasus 500 (APADS). *Proceedings of the 16th AIAA Aerodynamic Decelerator Systems Technology Conference*, AIAA-2001-2073, Boston, MA.

Sim, A.G., Murray, J.E., Neufeld, D.C., and Reed, R.D. (1994) Development and flight testing of a deployable precision landing system. *J. Aircr.*, **31**(5), 1101–1108.

Slegers, N. and Yakimenko, O. (2011) Terminal guidance of autonomous parafoils in high wind-to-airspeed ratios. *Proc. Inst. Mech. Eng. G J. Aerosp. Eng.*, **225**(3), 336–346.

Ward, M. and Costello, M. (2013) Autonomous control of parafoils using upper surface spoilers. *Proceedings of the 22nd AIAA Aerodynamic Decelerator Systems Technology Conference*, AIAA, Reston, VA.

Ware, G.M. and Hassell, J.L., Jr. (1969) *Wind Tunnel Investigation of Ram Air Inflated All-Flexible Wings of Aspect Ratio 1.0 to 3.0*, NASA TM SX-1923.

Wegereef, J. and Jentink, H. (2003) SPADES: a parafoil delivery system for payloads until 200 kg. *Proceedings of the 17th AIAA Aerodynamic Decelerator Systems Technology Conference and Seminar*, Monterey, CA.

Wright, R., Benney, R., and McHugh, J. (2005) Precision airdrop system. *Proceedings of the 18th AIAA Aerodynamic Decelerator Systems Technology Conference and Seminar*, AIAA, Reston, VA.

Yakimenko, O. (Ed.) (2015) *Precision Aerial Delivery Systems: Modeling, Dynamics, and Control*, Progress in Astronautics and Aeronautics, AIAA, Arlington, VA. ISBN: 978-1-62410-195-3.

Chapter 4
Networked Multiple UAS

Philip B. Charlesworth
Airbus Group Innovations, Newport, UK

1 INTRODUCTION

Unmanned aerial systems (UAS) are widely percieved as a key emerging technology. Many applications have emerged for single UAS (1):

- Land management, including forestry and agriculture, vegetation, and livestock monitoring
- Commercial, including crop dusting, surveying, and broadcasting
- Earth science, including cloud and aerosol measurements, meteorology, contaminant measurement, glacier and ice-sheet monitoring, extreme weather monitoring, and wildlife census

- Homeland security including coastal patrol, forest fire mapping, and emergency communications

Networks of multiple UAS may be required for many reasons. The most common reason is to form UAS sensor networks with the aim of collaborating on a common task. Another reason for networking UAS is to use one or more platforms as an intermediate data relay to overcome terrain or other obstructions. There is a growing desire to use UAS as elements in ad hoc communications networks.

Such networks are conceptually similar to terrestrial networks; however, there are some significant differences. A key benefit of multi UAS networks is the mobility of the nodes, allowing the network to physically reconfigure itself in response to changing demand. Changes in location and attitude of each platform can obstruct the radio path between platforms, resulting in changes to the network topology. The main performance parameters for UAS networks are similar to terrestrial networks: available bandwidth, range, availability, latency and other quality of service (QoS) metrics.

Flight paths can be planned to optimise against these QoS metrics, avoid threats or forbidden regions, or make best use of available onboard resources such as fuel or stored electrical power. This dynamic behavior makes UAS networks different to their terrestrial equivalents. The mobility of UAS may be constrained by other limitations such as the physical performance of the platforms or the airspace regulatory environment in which they operate.

The design of multi-UAS networks requires an understanding of basic radio principles. This chapter starts with a short introduction to radio link modelling. It then considers some common topologies for line-of-sight (LOS) and beyond line-of-sight (BLOS) communications with aircraft. Selection of the appropriate antenna, and correctly locating it on

Unmanned Aircraft Systems. Edited by Ella Atkins, Aníbal Ollero, Antonios Tsourdos, Richard Blockley and Wei Shyy.
© 2016 John Wiley & Sons, Ltd. ISBN: 978-1-118-86645-0.

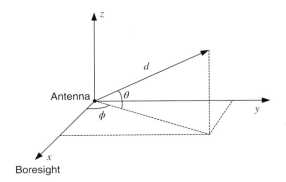

Figure 1. Coordinate system for antennae.

the airframe, is a key decision so this area is explored in greater depth. Finally, the chapter describes the state of the art in multi-UAS networks.

2 PRINCIPLES OF RADIO LINKS

A transmitter that radiates a power P_t equally in all directions at a frequency f Hz will generate a finite amount of power at a distance d m. The received power P_r can be calculated by using the well-known Friis Equation 2. In this equation, c is the speed of light.

$$P_r = P_t \left(\frac{c}{4\pi f d} \right)^2 \tag{1}$$

Sometimes Equation 1 is presented in terms of the wavelength $\lambda = \frac{c}{f}$ rather than the frequency f. Equation 1 can be rearranged to show the amount of loss that occurs between the transmitter and receiver antenna as the signal passes through the free space. This loss, known as the free space path loss L_{fs}, is simply the ratio of the received and transmitted powers.

$$L_{fs} = \frac{P_t}{P_r} = \left(\frac{4\pi f d}{c} \right)^2 \tag{2}$$

Practical antennae do not transmit and receive equally in all directions but tend to favor some directions, either by design or as a consequence of their location. This preference for transmission and reception in certain directions is referred to as the gain of the antenna and is defined as the ratio of the power in a particular direction over the power that would occur if the antenna was isotropic.

Figure 1 shows the coordinate system that is commonly used for specifying antenna gain. An antenna reference direction known as the boresight is aligned with the x-axis.

The boresight is often the direction of maximum gain and is generally aligned with the x-axis. The direction of an object from the boresight can be defined in terms of the two angles, ϕ and θ. The antenna radiation pattern is commonly annotated $G(\phi, \theta)$ to indicate that gain changes with direction. It is common for antenna manufacturers to specify radiation pattern in two orthogonal planes, usually the $x-y$ plane and the $x-z$ plane.

Equation 1 can be modified to include the use of directional transmitter and receiver antennae whose gains are denoted $G_t(\phi_t, \theta_t)$ and $G_r(\phi_r, \theta_r)$, respectively:

$$P_r = P_t G_t(\phi_t, \theta_t) G_r(\phi_r, \theta_r) \left(\frac{c}{4\pi f d} \right)^2 \tag{3}$$

For links that are modulated with data, the required link quality can be expressed in one of several forms. Most commonly it is expressed as a minimum receiver power P_r that will satisfy a specific signal-to-noise ratio (SNR) at the receiver input. Sometimes, it is specified as a required SNR that is normalized to provide a measure of independence from the choice of modulation scheme. When normalized to 1 Hz of bandwidth and 1 bit s^{-1}, it is expressed as the ratio of the energy in one bit E_b to the noise power in 1 Hz N_0. Equation 3 can be modified to calculate the ratio E_b/N_0 as follows:

$$\frac{E_b}{N_0} = \frac{P_t G_t(\phi_t, \theta_t) G_r(\phi_r, \theta_r)}{T_{sys} R_b K} \left(\frac{c}{4\pi f d} \right)^2 \tag{4}$$

In Equation 4, T_{sys} is the equivalent nose temperature of the receiver in Kelvin, R_b is the data rate in bits s^{-1}, and K is Boltzmann's constant 1.38×10^{-23} (J K^{-1}).

The boresight gain of an antenna with an effective area A_{eff} and efficiency η can be calculated from Equation 5. In Equation 5, the effective area A_{eff} is related to the physical area of a reflector or other aperture antenna. Further details on the design and theory of antennae can be found in Ref. 3.

$$G = \frac{4\pi A_{eff}}{\lambda^2} \eta \tag{5}$$

The main frequency bands for UAV communications networks are VHF, UHF and SHF. The limits of these frequency bands are defined by Article 2 of the ITU Radio Regulations (4) and summarized in Table 1.

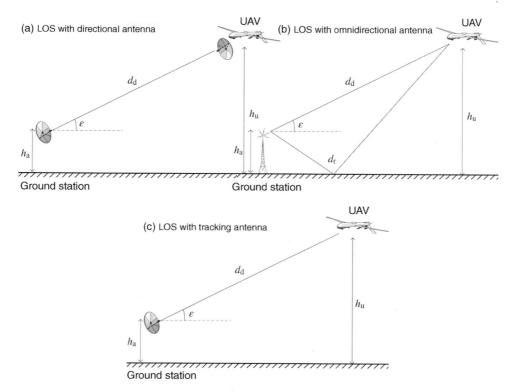

Figure 2. Line-of-sight radio links.

Table 1. Radio frequency bands.

Frequency band	Frequency limits	ITU band number
VHF	30–300 MHz	8
UHF	300–3000 MHz	9
SHF	3–30 GHz	10

3 AIR-TO-GROUND COMMUNICATIONS

Air-to-ground communications can be categorized as line-of-sight (LOS) or beyond line of sight (BLOS). LOS communications require exist when the UAV is above the radio horizon of the ground terminal. This does not always imply visibility of the UAV as the refraction and diffraction of radio waves can facilitate communications beyond the optical horizon, particularly in UHF and SHF bands.

The most effective communications system has directional antennae at both ends of the link as shown in Figure 2a. If the product $G_t (\phi_t, \theta_t) G_r (\phi_r, \theta_r) > 1$, it enables a corresponding reduction in power P_t or increase in data rate R_b, both of which are desirable. As the gain of either antenna increased more of the RF power is focused into a narrower beam. This generates a need for a tracking system to accurately point the antennae. On a small UAV, a tracking system can require significant amounts of internal space and electrical power.

The simplest case of a LOS path is a connection between a ground station with omnidirectional antennae at the ground station and on the UAV, as shown in Figure 2b. The performance of a system with omnidirectional antennae is relatively independent of the attitude or location of the aircraft, so no tracking or antenna pointing is required. This comes at the cost of reduced antenna gain so power must be increased or data rate reduced to satisfy the link budget. Furthermore, it opens the possibility of multiple RF paths and destructive interference of signals, inter-symbol interference, and other significant problems.

Between these two extremes is a compromise in which the UAV has an omnidirectional antenna and the ground station has a tracking antenna. In this system, the product $G_t (\phi_t, \theta_t) G_r (\phi_r, \theta_r) > 1$ so the power and data rate are better than the system with omnidirectional antennae. The omnidirectional antenna on the UAV allows changes in aircraft attitude without affecting link availability. Using a directional ground antenna introduces a tracking requirement; however, there are fewer constraints on space and power on ground than on the aircraft.

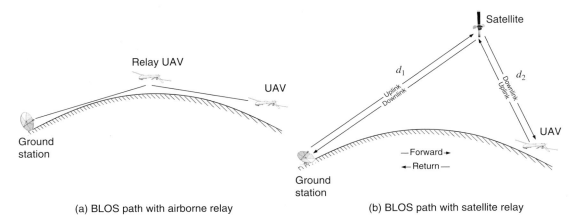

(a) BLOS path with airborne relay   (b) BLOS path with satellite relay

Figure 3. Beyond line-of-sight radio links.

At low elevation angles ε the signal path approaches the geometric horizon. Paths beyond the horizon are generally referred to as beyond line-of-sight (BLOS).

Radio paths can be supported beyond the geometric horizon by atmospheric refraction or diffraction over ridges, but such paths have high losses and are unreliable. It is normal to use airborne or satellite relay to provide reliable communications over BLOS paths, as shown in Figure 3.

In its simplest form, airborne or satellite relay amplifies the received signal to a power suitable for rebroadcast. A frequency shift and tight filtering are required to prevent the system from oscillating. This type of rebroadcast system is sometimes called a "bent pipe." There are many satellite communications textbooks that describe how to calculate link budgets for bent pipe systems (5).

There is relatively little space for satellite antennae in a UAV airframe. This places some limitations on the data rate that can be achived over satellite links. Low frequency systems such as Inmarsat (L band, 1.2 GHz) can support data rates up to a few hundred kbit s^{-1}. Systems in X band (7/8 GHz) and Ku band (12/14 GHz) can achieve some useful transmit and receive gain, leading to data rates of a few Mbit s^{-1}.

4   AIR-TO-AIR COMMUNICATIONS

The geometry of air-to-air radio paths suggests that they are range limited by the earth bulge, and that the maximum range occurs when the radio path just clears the horizon of both UAVs. Altitude changes along the LOS path between two aircraft result in changes of radio refractive index. This has the effect of refracting the radio signal toward the earth, causing the actual radio path appears to pass over an earth with a radius that can be $k = 4/3$ of the actual earth radius.

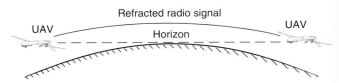

Figure 4. Air-to-air radio links.

This k factor can change from 2/3 to 4/3 depending on weather conditions, but is generally >1. This can help to extend the range of air-to-air radio signals.

5   ANTENNA TYPES AND LOCATIONS

Several antenna types are required to satisfy the different communications roles. This section considers some common antenna types for use on UAVs, their uses, and the optimal locations on the airframe.

5.1   Omnidirectional antennae

Wire, blade, and patch antennae tend to exhibit omnidirectional radiation patterns. Examples of these antennae can be seen in Figure 5.

The simplest and cheapest antenna type is a wire antennae. Wire antennae tend to have omnidirectional radiation patterns, resulting in low gain but good all-round coverage. Wire antennae are mainly used at VHF/UHF bands and have a relatively narrow bandwidth. These are normally $\lambda/4$ long monopoles that use the airframe as a ground plane.

Wire antennae have poor rigidity and tend to bend at high speeds, distorting their polar pattern. The blade antenna is a more aerodynamic form of omnidirectional antenna. The radiation pattern can be distorted by the sweep of the antenna.

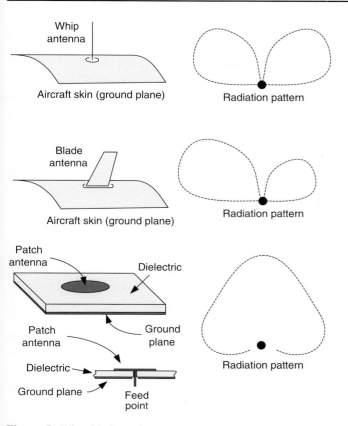

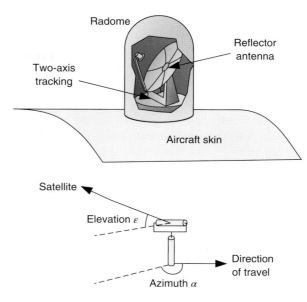

Figure 6. Dish antenna in a radome.

Figure 5. Wire, blade, and patch antennae.

Blade antennae are useful at VHF/UHF bands and can be made with a wider bandwidth than wire antennae.

Wire and blade antennae have a signal null that is aligned with the radiating element, generally at right angles to the ground plane. Patch antennae do not have these nulls and give good all-round cover. Patch antennae can be made to be conformal to the aircraft skin, resulting in very low drag, and can also be load bearing components of the airframe. They are useful but relatively complex antennae and are useful at UHF/SHF bands where their wide bandwidth can be exploited.

5.2 Directional antennae

Directional antennae are mainly used in the high UHF and SHF frequency bands where their narrow beamwidths and high gain can be fully exploited. The most common use for directional antennae on UAVs is for satellite links, usually at Ku band or above. Reflectors are often made from carbon fibre and other composites to reduce weight.

The gain of an antenna can be calculated using Equation 5. A parabolic reflector with diameter D, the effective area A_{eff}

is approximately the same as the physical area of the dish, hence a form of the equation can be derived in terms of the reflector diameter.

$$G = \frac{(\pi D)^2}{\lambda^2}\eta \qquad (6)$$

The beamwidth θ_{h} of a parabolic reflector can be approximated to (5):

$$\theta_{\text{h}} \approx 70\frac{\lambda}{D} \qquad (7)$$

Equation 6 shows that the boresight gain of a dish is proportional to $\left(\frac{D}{\lambda}\right)^2$, and Equation 7 shows that beamwidth is proportional to $\left(\frac{D}{\lambda}\right)^{-1}$. The dimensionless term D/λ defines the diameter of the dish as a number of wavelengths. At Ku band, for the types of satellite antenna used in UAVs, $D/\lambda \approx 50$.

A tracking system is required to ensure that the antenna is pointed in the desired direction while the aircraft manoeuvres. An azimuth and elevation (az–el) tracking system such as the one shown in Figure 6 allows the antenna to track the satellite across a hemisphere. When the satellite passes through zenith, the az–el tracker can require a high rate of turn, a problem that can be resolved by using X–Y trackers.

Directional antennae are usually installed in a radome to protect the antenna and control aerodynamic drag.

5.3 Phased arrays

Individual antennae can be placed in a grid, or repeating pattern, to form array antennae. Varying the phase and power to individual antennae allows the radiation pattern to be altered. Phased arrays are commonly used to steer lobes in the radiation pattern, creating gain in a wanted direction, or to create nulls that can be used to reject interference or jamming.

Beamforming with planar phased arrays is relatively straightforward, particularly if the radiation pattern is fixed over long-time intervals. Beam steering, that is the changing of the radiation pattern over time, requires regular recalculation of the power and phase for all elements. The processing requirement is affected by many factors, particularly the number of elements in the array, the complexity of the radiation pattern, and the frequency at which recalculation is required.

Planar arrays are not always suited to the aerodynamics of a UAV. Conformal arrays can be shaped to follow the skin of the UAV, allowing low drag antennae. The aerodynamic advantages of conformal arrays come at the cost of additional processing complexity. The far-field contributions of each element need to allow for its phase and polarization. Furthermore, some elements may be unusable because of body obstruction.

Phased arrays offer a potentially useful alternative to reflector antennae, especially in high-performance UAVs, but can be expensive end complex to develop and implement.

5.4 Antenna locations

The location of an antenna on the airframe is a compromise between many factors:

- Required coverage
- Body masking
- Effect on aerodynamics
- Effect on centre of gravity (CoG)
- Cable routes
- Structural requirements

A common problem with locating antennae is that the signal path can be obstructed by the airframe in certain directions. Diffraction and reflections from the wing, tail, and fuselage can create complex interference patterns, so it is difficult to accurately predict the radiation pattern of an antenna. Simple obstruction models such as that shown in Figure 7, which shows a blade antenna located on the underside of the starboard wing of a UAV, can provide a satisfactory estimate of the coverage of an omnidirectional

antenna. It can be seen that there is unobstructed coverage beneath the aircraft; however, the wing, tail, and forward fuselage obstruct a large portion of the sky.

Antennae for GPS and satellite communications need to be located so that they have a clear field of view of the sky. Most locations on the upper surfaces of a winged aircraft are suitable for the GPS antenna.

Directional satellite antennae can be heavier because of the tracking mechanism, dish, and the need to mount the electronics very close to the antenna. It is common to mount small satellite antennae on the spine of the UAV, and larger dishes forward of the wing. These locations minimize obstruction by the tail. Mounting satellite antennae on rotary aircraft if more problematic. Ideally, the antenna should be clear of the disc of the rotors; however, this is seldom achieved because it tends to move the CoG away from the rotor shaft. There is also the practical problem that the interruption of the blades amplitude modulates the RF signal.

Air-to-air and air-to-ground communications normally use omnidirectional blade antennae. Antennae located on the spine of the aircraft can give good coverage of the horizon, making this location ideal for air-to-air networks. Ventral, or underwing, locations are preferable for air-to-ground networking or short-range communications.

Mission antennae are located to give a field of view that is appropriate to their purpose. Surveillance and radar antennae are usually located on the forward or ventral areas, depending on the mission. For some applications performance can be improved by locating the antennae in wingtip pods, increasing their separation. For rotary UAVs, the antenna is sometimes deployed in flight below the UAV and retracted before landing. This is a convenient way of deploying a radar from a UAV for hemispheric coverage.

The wide field of view from the tail has established it as a prime location for mission antennae on both winged and rotary aircraft. It is an excellent location for airborne communications antennae, and also threat warning systems on military UAVs.

Figure 8 shows some commonly used locations for antennae on winged and rotary UAVs.

6 UAS NETWORKS

The use of UAS as communications nodes was first proposed by Pinkney in 1996 (6). The focus has remained on the use of single UAS for many years. There is an emerging trend toward the use of multiple UAS in networks. The primary use is to provide intermediate relays for UAS that are required to operate beyond line-of-sight. Multi-UAS sensor networks are currently the preserve of academic research, but may start to

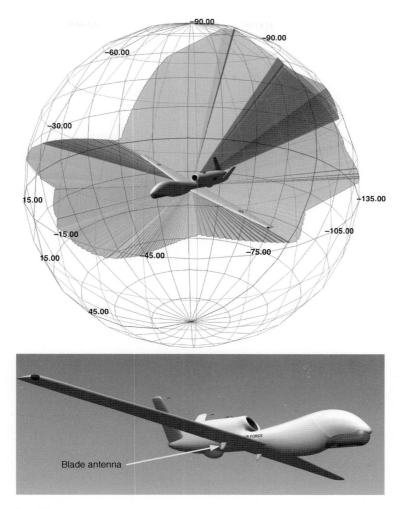

Figure 7. Signal masking by the airframe.

become a reality with the development of lightweight, radio-aware, routers.

Pinkney's proposal for a UAS communications node was based around a simple transparent transponder payload. The X band transponder was intended to extend the range of high-capacity trunk radios while providing simple data broadcast (7). Common concepts and technologies for using UAV in military communications systems have yet to emerge. Current systems are based on either a UAV that circles a common point, such as Northrop Grumman's Battlefield Area Communications Node (BACN), or provision of a communications payload as a secondary system on a surveillance UAV.

Concepts for using multiple communications UAVs in networks are still evolving. There are good cost justifications as the cost of a single high-altitude long-endurance (HALE) platform is significantly greater than the cost of several medium-altitude long-endurance (MALE) UAVs.

Furthermore, a multiple UAS system offers operational advantages as it is inherently more resilient to single failures. Notwithstanding these advantages, little work has been done to develop networks of large UAS.

There are many demonstrations of networks of small UAS. The University of Colorado' "Ad hoc UAV-Ground Network (AUGNet)" experiment with small UAVs equipped with IEEE 802.11b systems (8) demonstrated the principle that UAVs could improve connectivity in ad hoc networks. University of Klagenfurt has used small UAVs to provide range extension for a sensor (9). Bekmezci and others have introduced the idea of flying ad hoc networks (FANET) as an airborne equivalent to mobile ad hoc networks (10). These exemplar projects reflect a wide interest in the development of networks of small UAS.

All these systems used omnidirectional antennae for payload communications to remove the issues of accurate tracking. The protocol stacks are generally

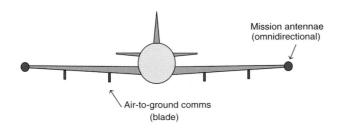

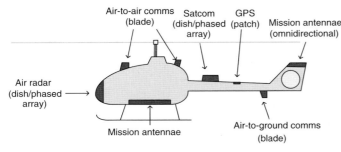

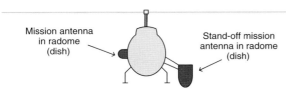

Figure 8. Antenna locations.

location planning task. There is extensive research available on planning the paths of individual UAS but relatively little on coordinating multiple UAS. Algorithms for coordinating UAS need to allow for the possible failure to fully distribute a common plan to all agents, thus some degree of autonomous planning is required onboard each UAS. One method for the autonomous generation of UAS locations has been suggested by Charlesworth, but there remains considerable scope for original work in this field.

7 CONCLUSIONS

This chapter has explained some of the basic issues that need to be considered when designing multi-UAS networks. This includes the modeling of single radio paths and the choice and optimal location of antennae. It has described the state-of-the-art in multi-UAS networking.

NOTATION

P_r	Received power (W)
P_t	Transmitter power (W)
c	Speed of light $\approx 3 \times 10^8$ (m s^{-1})
f	Transmitter frequency (Hz)
d	Slant range between the transmitter and receiver (m)
$G_t(\phi, \theta)$	Antenna gain profile of the transmitter (dimensionless)
$G_r(\phi, \theta)$	Antenna gain profile of the receiver (dimensionless)
K	Boltzmann's constant 1.38×10^{-23} (J K^{-1})
R_b	Information rate on the link (bit s^{-1})
T_{sys}	Effective receiver noise temperature (Kelvin)
$G(\phi, \theta)$	Antenna gain profile (dimensionless)
A_{eff}	Effective aperture of the antenna (m^2)
λ	Wavelength (m)
θ_h	Half-power beamwidth of an antenna (degrees)
D	diameter of a circular dish antenna (m)

ABBREVIATIONS

AODV	Ad hoc On-Demand Distance Vector
BACN	Battlefield Area Communications Node
BLOS	Beyond line-of-sight
CoG	Centre of Gravity
FANET	Flying ad hoc networks
HALE	High-altitude long endurance

unmodified 802.11b/g using AODV and similar routing protocols.

Multi-UAS networks have an implicit requirement to optimise the locations of all the platforms against some aspect of the current task. In a trivial case of range extension, it could be that the optimal solution is to locate all the platforms in a straight line, space so that the required SNR is just exceeded. In real situations, there are many other factors to be considered, for example, terrain obstructions may render some links unusable. As a minimum, there must be sufficient usable paths to form a minimum spanning tree between all active UAS.

Hazardous situations such as military operations or forest fires may introduce no-go areas, further complicating the

ITU	International Telecommunications Union
LOS	Line-of-sight
MALE	Medium-altitude long endurance
QoS	Quality of Service
SHF	Super high frequency
SNR	Signal-to-noise ratio
UAS	Unmanned Aerial System
UHF	Ultra high frequency
VHF	Very high frequency

REFERENCES

Balanis, C.A. (2005) *Antenna Theory: Analysis and Design*, John Wiley & Sons, Inc.

Bekmezci, I., Sahingoz, O.K., and Temel, (2013) Flying ad-hoc net-works (fanets): a survey. *Ad Hoc Netw.*, **11**(3): 1254–1270.

Brown, T.X., Argrow, B., Dixon, C., Doshi, S., Thekkekunnel, R.G., and Henkel, D. (2004) Ad hoc UAV ground network (AUGNet). *In AIAA 3rd Unmanned Unlimited Technical Conference.*

Charlesworth, P.B. (2014) Using non-cooperative games to coordinate communications UAVs. *Globecom 2014 Workshop—, Wireless Networking and Control for Unmanned Autonomous Vehicles (GC14 WS—, Wi-UAV)*, Austin.

Cox, T.H., Nagy, C.J., Skoog, M.A., and Somers, I.A. (2004) Civil UAV capability assessment. Technical report, NASA.

Friis, H.T. (1946) A note on a simple transmission formula. *Proceedings IRE*, Vol. **34**, pp. 254–256.

International Telecommunications Union. *Radio Regulations* (2012) Volume **1**, Article 2.

Maral, G. and Bousquet, M. (2011) *Satellite Communications Systems: Systems, Techniques and Technology.* John Wiley & Sons, Inc.

Pinkney, F.J., Hampel, D., and DiPierro, S. (1996) Unmanned aerial vehicle (UAV) communications relay. Milcom.

Pinkney, F., Hampel, D., DiPierro, S., Abbe, B., and Sheha, M. (1997) *UAV Communications Payload Development.* Milcom: IEEE.

Yanmaz, E., Kuschnig, R., and Bettstetter, C. (2013) Achieving air-ground communications in 802.11 networks with three-dimensional aerial mobility. *In INFOCOM, 2013 Proceedings IEEE*, IEEE, pp. 120–124.

Chapter 5
Weapons Integration

Keith A. Rigby

BAE Systems, Warton Aerodrome, Preston, UK

1 INTRODUCTION

One of the key differences between civilian and military aircraft is that many military aircraft have the ability to carry and release weapons. From the earliest days of aviation when the pilot would drop simple bombs by hand, engineers have striven to develop the capability to accurately deliver weapons against targets reliably and safely. The integration of weapons onto aircraft requires multidisciplinary capabilities.

Unmanned Aircraft Systems. Edited by Ella Atkins, Aníbal Ollero, Antonios Tsourdos, Richard Blockley and Wei Shyy.
© 2016 John Wiley & Sons, Ltd. ISBN: 978-1-118-86645-0.

The release of a weapon whether it is a forward-fired missile or a downward-ejected store, such as a fuel tank from either an externally mounted pylon or from an internal bay, creates issues such as the ability to achieve safe separation and the ability of the aircraft structure to withstand the imparted loads. The complexity of weapons integration is increased when the requirements for priming and aiming are considered.

This chapter will cover the various aspects of weapons integration, primarily from the viewpoint of systems integration. This will be done by following a typical timeline covering preparation, loading, selection, targeting, priming, aiming, deployment and post-release actions for a number of different weapon types of increasing complexity. Safety considerations will also be addressed.

2 ISSUES FOR SYSTEM DESIGN AND INTEGRATION

As with any system design, a structured, top–down approach is essential. However, for weapons integration, the higher level requirements will also include aeromechanical aspects such as the desired launch envelope, carriage life, the number of weapons that can be carried, influences of other weapon and store types to be carried on the same sortie, and so on.

This chapter predominantly covers the system integration aspects of weapons integration. Therefore, in relation to the top-level requirements for the capability to be delivered, there is a need to segment individual requirements to aircraft subsystems. The segmentation process may use software-based requirements management tools, as these assist in validation and verification of the system implementation against the requirements. The actual segmentation will depend on the system architecture and is therefore aircraft-specific. However, the requirements could be segmented into, for example, mission-critical and safety-critical processes; navigation, targeting requirements; aerodynamic requirements; and so on.

At the lower system design levels, there is a need to understand the operation of the weapon and the data exchanges with the aircraft. This is documented in an interface control document (ICD), which is agreed between the aircraft and weapon design organizations. The ICD defines all information relevant to the integration such as mechanical attachments, electrical signal sets, data structures, timeline (a detailed temporal sequence of data and state transitions required for the aircraft to operate the weapon), environmental data, aerodynamic data, and so on. Negotiation of the ICD can be a significant activity, particularly when a new weapon is being developed in parallel with integration to the platform.

Following implementation of the requirements in the aircraft subsystems, integration testing will be undertaken in the systems integration laboratory. Employing either weapon simulators or inert weapons with operational electronics, integration testing tests that all the subsystems are working together to control the weapon. Such testing identifies problems that need to be corrected during an iteration of the subsystem design and implementation.

Following successful systems integration on the ground, the complete system will be flight-tested, leading to the eventual live firing of weapons against representative targets.

3   TYPES OF WEAPON

Many different types of weapon can be found in the inventory of air forces around the world. These will range from ballistic bombs, through bombs with added control mechanisms that improve accuracy, to powered weapons having a greater range (air-to-air and air-to-surface missiles).

Ballistic bombs are unguided. These weapons are inherently inaccurate, since their delivery requires the launch aircraft to perform complex aiming calculations to determine the release point that will maximize the probability of hitting the target.

Smart bombs, for example, those with laser guidance kits or global positioning system (GPS) assisted inertial navigation systems, have a degree of maneuverability that can be exploited to increase terminal accuracy. For these weapons, the launch aircraft must align the weapon's navigation system prior to launch.

The standoff range of ballistic and guided bombs is dictated by the kinetic and potential energy of the weapon at release, which are, in turn, dependent on a number of factors, not least being the launch altitude and aircraft speed. In order to increase the standoff capability, missiles contain a source of propulsion such as a turbojet or rocket motor. While increasing the effective range of the missile, a propulsion system can also reduce the time for the missile to engage its target. This is a key factor for an air-to-air missile. Powered air-to-ground weapons also employ either a rocket motor (e.g., for a high-speed short-range weapon) or turbojets (e.g., for a long-range cruise missile).

In addressing the weapon integration problem space, this chapter uses these examples to explore the differences in the integration of each weapon type.

4   BALLISTIC BOMBS

4.1   Physical preparation

All bombs are made up of a number of constituent parts, these being the explosive warhead, the fuze (see STANAG 4187 for bomb fuze design requirements), a tail arrangement, and possibly a height sensor that is used to detonate the bomb prior to impact. The main body of the bomb comprises the warhead to which various tail units can be fitted depending on mission requirements (i.e., a slick bomb will have its aerodynamic drag minimized; a retarded bomb will have a high aerodynamic drag).

When released, a slick bomb will follow a ballistic trajectory until the fuze either detects an impact or is notified by the height sensor that the air burst height has been achieved. The bomb fuze is typically programmable by switches located on the fuze for setting arming and initiation delays. Slick bombs would typically be released from a medium altitude (around 3000 m) to maximize the standoff range.

Retarded bombs are used for low-level releases where the retarding mechanism reduces the velocity of the bomb, so that the aircraft can escape the debris zone when the bomb detonates.

4.2   Aircraft attachment

NATO aircraft and bombs generally use common "hook-and-eye" mechanical attachments defined by STANAG 3726. The hook mechanism is part of a release unit or bomb rack operated either by the initiation of a pyrotechnic cartridge (the product of which is a high-pressure hot gas) or by the

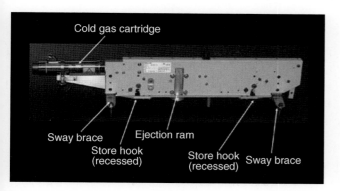

Figure 1. Example of a bomb rack.

release of compressed gas (usually purified air in a so-called cold gas system). In both systems, in addition to opening the hooks, the gas is used to operate pneumatic rams that push the weapon away from the aircraft, so aiding aerodynamic separation. Older systems or systems for very light stores may be operated by energizing a solenoid to open the hooks. On these systems, the store falls away from the aircraft under the influence of gravity. Figure 1 shows a typical bomb rack.

The weapon is mechanically attached to the aircraft by closing the hooks to grab the eyes fitted to the weapon (bale lugs) and then lanyards are attached that will remove locking pins from the fuzing mechanism when the bomb separates from the aircraft. Modern fuzes contain a thermal battery (a chemical device initiated by a pyrotechnic squib) that is activated by a fuzing power supply switched through to the fuze by the aircraft on release.

4.3   Targeting

Broadly, two types of targeting are employed; preplanned and target of opportunity (TOO). With a preplanned target, the target type and location is known prior to the aircraft sortie enabling the optimum method of attack to be devised. During the mission planning stage, the approach route to the target, the attack, and the egress route can be meticulously planned and rehearsed such that the probability of destroying the target can be maximized.

Targets of opportunity are targets that are identified while the aircraft is in flight and therefore do not have a mission plan to follow. The location of the target must therefore be fixed using either onboard sensors or from a third party relaying target coordinates to the aircraft. The aircrew must enter the target coordinates into the attack system, so that the aircraft can be steered to the weapon release point. Targeting using only the aircraft's own sensors to identify the target's location will inherently result in a Target Location Error

(TLE – an error expressed in three dimensions: latitude, longitude and height). The navigation and attack systems need to be highly accurate if TLE is to be minimized.

The accurate delivery of unguided ballistic bombs relies on the accuracy of the aircraft's navigation system to determine the aircraft's exact position, altitude and three-dimensional velocities in order to minimize targeting errors. Modern aircraft have multi-sensor systems where the navigation solution is derived using techniques that may include a combination of inertial sensors, radar, GPS, visual fixing systems, Kalman filtering (a mathematical approach to linear filtering and prediction), etc. The aircraft's weapon aiming system will compute the impact point of a bomb if released from its current location in the sky. This calculation will be repeated continually at a rate largely dependent on the processing power available in the weapon aiming computer and any other tasks that the system needs to undertake. The bomb is released when the weapon aiming system determines that the Continuously Calculated Impact Point (CCIP) overlays the target's position. The system will have processing delays, data transmission delays, and system latencies that must also be accounted for in the aiming solution. The weapon aiming system therefore needs to advance the release point to account for these delays. However, as the weapon aiming calculations are cyclic and take a finite time to complete, then the point where the CCIP overlays the target exactly may in reality, occur part way through a processing cycle. It is therefore necessary for the designer of the weapon aiming algorithms to undertake statistical analysis to determine the error in the true release point and the calculated release point solution so that this too can be factored into the overall calculation, thereby minimizing system-induced errors.

4.4   Release

As a ballistic bomb is un-guided, when it is released from the aircraft it will fall in accordance with the physical laws of motion. While theoretically this provides a level of predictability, in reality there are many factors that influence the bomb's trajectory. These include the aircraft speed and attitude at the point of release, the downward ejection force imparted by the aircraft bomb rack, wind velocity and direction (which itself will vary from the release point throughout the trajectory), and the air density profile from release to the target. The effects of aerodynamic drag will also influence the weapon aiming solution. Much effort is expended during the design of the weapon aiming algorithms to ensure that an approximation of the bomb's trajectory can satisfy all the variables such that miss distance is minimized. The actual impact point of a weapon will have a Gaussian distribution about the mean impact point with ideally, the mean impact point coinciding with the target's location. In practice it rarely does.

For low-level releases of retarded bombs there is the added complication that the aircraft must be clear of the blast debris when the bomb detonates. Since bomb fragments could have a velocity in the region of 1000 ms^{-1} the maximum energy boundary of such fragments will need to be included in the weapon aiming computations so that the aircraft can remain safe for the given set of release conditions.

At a predetermined time after release the bomb's fuze will arm in readiness for the conditions required for detonation to be achieved (e.g., height above the ground or impact).

5   SMART BOMBS

5.1   Physical preparation

A smart bomb consists of a guidance kit that is attached to a ballistic bomb. The earliest bombs that could be considered to be in this category are laser-guided bombs that were developed in the USA during the 1960s. Other early bombs such as the GBU-15 employed a video sensor, movable control surfaces, and a radio link with the launch aircraft that enabled the bomb to be steered onto its target. Following the 1991 Gulf War, the USA identified the need for higher-precision weapons that would have improved kill efficiencies. These truly smart bombs such as the Joint Direct Attack Munition family of weapons use a relatively low-cost GPS-aided inertial unit coupled to a simple flight control system to guide the bomb to a target's GPS coordinates.

The inclusion of electronics means that the weapon has to include a thermal battery power source. The battery is initiated just prior to launch, either by an electrical fuzing supply switched through to the bomb when the bomb rack hooks are opened, or by the use of a smart weapon interface such as that defined by MIL-STD-1760 (see Section 10).

When released, a smart bomb will initially follow a ballistic trajectory. The guidance system enables the weapon to be steered toward its target within certain constraints. The control surfaces can also provide a level of aerodynamic lift that will provide some extension to the weapon's range. Indeed, some weapons can be fitted with wing kits to extend the range still further. However, as most smart bombs are unpowered, their range is limited by their potential and kinetic energy at launch.

Fuzes used by a smart bomb are identical to those used in ballistic bombs. However, as fuze technology develops, new fuzes are being developed specifically for smart bombs.

5.2   Aircraft attachment

Smart bombs utilize the standard mechanical and fuzing attachments used by ballistic bombs. However, as a smart bomb is likely to contain electronics such as navigation and flight control systems, the primary system interface will be as defined in standards such as MIL-STD-1760.

5.3   Targeting

With the guidance capability of smart bombs it is possible to trade energy at release with route to the target. Clearly, if the weapon has sufficient energy at release and sufficient maneuverability, then it is feasible to hit the target from a variety of predefined direction and azimuth and elevation angles. This complicates the mission planning activity but does provide a greater level of flexibility for the attack.

Where the accurate delivery of unguided ballistic bombs requires the aircraft to be in a specific location in the sky with a defined attitude and speed, a smart bomb does not have such stringent constraints. In effect the aircraft need only be in a defined three-dimensional volume in the sky. This volume will vary with altitude, speed, distance from the target, and weapon performance. This overcomes the need for the aircraft to calculate the CCIP. However, there is still a need for the aircrew to know when the smart bomb can be released such that it will hit the target. The volume in the sky where, if released, the bomb will reach the target is known as the launch basket or Launch Acceptability Region (LAR – see Figure 2).

While Figure 2 shows the LAR in relation to a point target, it can also be shown as a projected "footprint" on the ground (in effect, an inversion of Figure 2). This would then show all the possible impact points if the weapon was "launched now".

The LAR is dependent on the actual release conditions and will change as these parameters vary. The LAR is therefore dynamic, requiring continuous calculation that places greater demands on the weapon aiming system's processing power, particularly if a number of smart bombs are to be released in a single attack against dispersed targets.

The primary method of defining the LAR is to employ a six degree-of-freedom (6-DOF) model that can predict the weapon's trajectory from release to impact for a given set of conditions. The 6-DOF model uses the dynamics of the weapon movements (body rates, angles, etc.), the environmental conditions (temperature, air density, wind speed, etc.), and the weapon's flight control system dynamics to predict the weapon's trajectory. The 6-DOF model is considered to provide a true representation of the dynamics of the

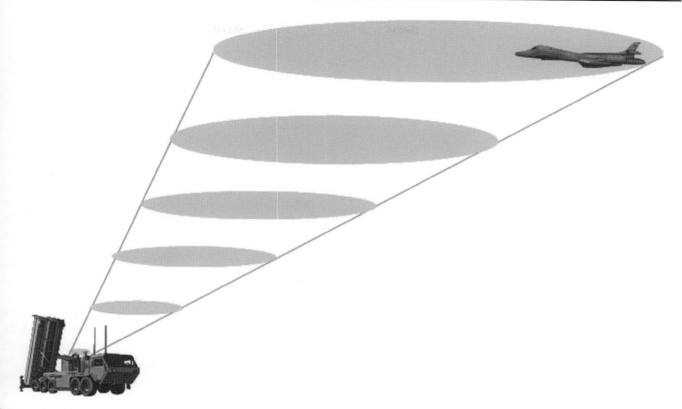

Figure 2. A launch acceptability region.

weapon. However, the continuous processing of a 6-DOF model in an airborne computer would be prohibitively complex. It is for this reason that the models are simplified to give a good approximation of the true weapon performance. Two methods are commonly used, these being the dynamic LAR and the parametric LAR.

A dynamic LAR employs the equations of motion based on the physical characteristics of the weapon in a similar way to a 6-DOF model. However, the equations are simplified usually by considering a reduced number of degrees of freedom, typically a 3-DOF model. The parametric model is matched to the output of the 6-DOF model but uses an approximation such as "least squares fit".

LAR displays are often simplified as shown in Figure 3. In this example, markers for the maximum and minimum In-Range and In-Zone LARs are displayed to the crew. In the figure, the plane of the LAR is indicated to show how the markers displayed to the crew are constructed. Here, LAR cross-track limits may also be displayed to the crew (as shown) in order to improve overall situational awareness.

When combining LARs for several weapons, the resulting LAR will be of a significantly reduced volume. Here, the update rate of the combined LAR may become critical as minor changes in parameters for all individual LAR

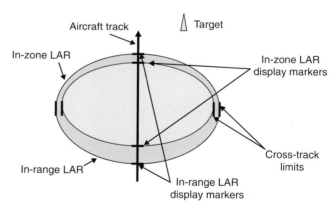

Figure 3. Typical aircraft LAR display.

computations can cause dramatic changes in the combined LAR. Displaying multiple LARs to the aircrew therefore places significant demands on the display computer and graphics generator.

As weapons are developed in isolation then it is not uncommon for an aircraft's weapon aiming computer to hold several different LAR algorithms for different weapons. This means that should a particular algorithm need to be modified, it can be very expensive.

To address this situation the US Air Force tasked the Society of Automotive Engineers to develop a common approach to LAR algorithms. The intention was to have a single algorithm for all unpowered smart weapons that could be changed simply by inserting different coefficients into the weapon aiming computer. Such an approach is documented in AIR5682, AIR5712, and AIR5788.

5.4   Release and guidance

Smart bombs will guide themselves to their designated target using a navigation solution produced by its own systems and following a glide path dictated during mission planning. For optimum performance, a smart bomb would be released when inside the LAR. However, should there be a reason why the launch aircraft cannot achieve the LAR required to satisfy the desired impact conditions, then it may be possible to trade reduced maneuverability with release from a greater stand-off distance from the target. It is for this reason that two LAR's may be calculated (see Figure 3). The first is the so-called "In Zone" LAR where the weapon, if released, will attack the target with the desired impact conditions. The second is the "In Range" LAR where the weapon, if released, will have suffi- cient energy to reach the target but where the trajectory contains the minimum of maneuvers, trading increased stand-off range for reduced end-game maneuverability.

Smart bombs can be used to attack targets of opportunity but their accuracy is dictated by the accuracy of precision code GPS. This places a greater reliance on the sensors used to locate the target to minimize TLE.

6   COMPLEX AIR-TO-GROUND WEAPONS

One of the main shortcomings of smart bombs is that their maximum range is limited to around 15 to 20 nautical miles (depending on launch altitude). In certain scenarios smart weapons with a greater stand-off range are required. This is achieved in two ways:

(i)   With a range extension kit (normally a deployable wing kit that provides additional lift, enabling a greater glide range).

(ii)   With a source of propulsion such as a small turbo-jet or rocket motor.

A smart bomb fitted with a wing kit is operated in an almost identical way as the weapon without a wing kit. However, due to the greater stand-off range, the LAR algorithms need to be modified to account for the greater footprint. The weapon may also need to have an increased capacity energy source to power weapon subsystems for the full duration of the extended flight time.

A smart weapon with a propulsion system such as a turbo- jet typically has a large stand-off range and is classed as a cruise missile.

Once launched, a typical cruise missile is completely auton- omous guiding itself to the target using a range of techniques such as GPS navigation and scene matching algorithms. Scene matching uses a sensor to match the visible scene during the end game with a digital image loaded into the weapon's guidance system. The error between the sensed "real" scene and the stored image is used to make final course corrections, thereby improving the terminal accuracy of the weapon.

Weapons with a rocket motor have a limited range but are able to prosecute target engagements quickly. Such weapons would typically include a seeker based on either laser, optical, or radar technology, depending on the target types being engaged and the operation of the overall system (e.g. whether it includes a human in the guidance loop or has an autonomous "fire-and-forget" capability).

7   AIR-TO-AIR MISSILES

7.1   Aircraft attachment

Air-to-air missiles are designed for the specific purpose of attacking airborne threats and typically consist of a body tube housing a sensor, guidance electronics, fuze, warhead, and rocket motor. They can either be launched by ejection from the launcher or fired along a rail. For a rail-launched missile, the mechanical attachments are defined by STANAG 3842AA. For an eject-launched missile, there are currently no standardized mechanical interfaces. Some missiles can either be eject-launched or rail-launched.

Air-to-air missiles have traditionally employed bespoke electrical interfaces, although the adoption of MIL-STD- 1760 is bringing a level of standardization to such weapons. The analog interface of the AIM-9 missile family has also been adopted by many short-range missiles in order to improve interoperability and increase export potential through commonality.

7.2   Targeting

There are two primary methods of targeting air-to-air missiles that are defined by the type of target sensor employed, these being infra-red (IR) and radar.

7.2.1 Ir Sensor

Early air-to-air missiles employed a single element IR sensor to detect the heat emitted from an enemy fighter's engines. This relatively simple seeker technology suffered from the problem that it would detect any source of heat including the sun or heat reflected from the launch aircraft's fuselage, thereby causing the missile to lock on to a false target. To overcome this problem modern IR sensors use imaging IR focal plane arrays. This enables the missile control algorithms to discriminate between a real aircraft target and false or spoof targets (e.g. IR countermeasure flares).

7.2.2 Radar sensor

Radar sensors have a significantly greater range than IR sensors and therefore afford the ability for a fighter aircraft to increase its stand-off distance from its target. The increased use of jamming as a defensive mechanism has led to radar sensors employing counter-countermeasures such as providing a "home on jam" capability where the sensor acts in a passive mode to identify the relative bearing of the jamming signal. This enables the missile's guidance electronics to maneuver, home in on, and attack the source of the jamming signal.

7.3 Release and guidance

There are a number of engagement modes in which air-to-air missiles can be used. These are "Lock-on-Before-Launch" (LOBL) and "Lock-on-After Launch" (LOAL). For a SRAAM with an IR seeker, a LOBL engagement could be achieved by simply maneuvering the aircraft so that it is pointing directly at the target (so-called boresight). This has a major disadvantage against targets that are themselves maneuvering, as the pilot has to use all his flying skills to keep the target on boresight for a sufficient period to enable missile lock. However, with the advent of modern computers it has been possible to set up scan patterns where the IR sensor is moved in a defined pattern to search for a potential target. When a target is detected, missiles such as the AIM-9 Sidewinder generate an audio tone that is passed through the aircraft's communication system to the pilot's headset. The audio tone is then modulated to inform the pilot of missile lock status.

With a highly agile and maneuvering aircraft target, the boresight and scan pattern methods of aiming SRAAM's is no longer viable. Missiles such as the Sidewinder employ an extended acquisition mode, whereby the IR sensor can be slaved to the launch aircraft's radar, enabling high off-boresight target acquisition. However, the use of radar can warn enemy aircraft that they are being tracked. Infra-red Search and Track (IRST) systems passively detect and track targets, with data produced by such systems used to point the missile's sensor at the target prior to launch.

Highly agile missiles such as the IRIS-T and the AIM-132 ASRAAM can be targeted at off-boresight angles that are greater than the sensor's gimbal limits. To exploit this capability, a helmet-mounted sight is used to detect where the pilot is looking (at the target aircraft) and feeds this data into the missile's guidance computer. Post launch, the missile will then maneuver until the target is within its sensor's field of view. This gives modern SRAAMs a LOAL capability.

Missiles with an IR seeker detect the heat signature of jet engines and determine if there is an off-boresight error angle caused by the missile not being pointed directly at the target. Any error is used as an input to a closed loop control system enabling the guidance electronics to alter the deflection of control surfaces to maneuver the missile and to correct the off-boresight error.

Radar guided missiles can operate in both LOBL and LOAL modes. In a LOBL mode, as with IR-guided sensors, the aircraft's radar is used to detect a target and prime the missile's radar so that it can acquire the target prior to launch. After launch, the missile may receive target position updates through a post-launch data link. Radar guided missiles may also contain an inertial navigation system to enable a LOAL mode to be employed. Here, the missile is given target coordinates prior to launch. As the target may be travelling at high speed and could be maneuvering, a post-launch data link is used to update the missile with the target's current position. On nearing the target, the missile will then activate its radar seeker to home onto the target. This method of operation enables a relatively stealthy attack to be prosecuted with the missile's radar only being used for end-game guidance. This provides a very high probability of kill (P_k).

In a modern military aircraft, the weapon aiming system will also be calculating the P_k. To do this effectively, the weapon aiming computer must employ a high-fidelity missile fly-out model based on the real performance of the missile. Fire cues will be given to the pilot in order to maximize the P_k. The fly-out model must be run repeatedly during the engagement and the fire cue updated in near real-time. As the fly-out model could, in its purest form, be a 6-DOF model the computer processing power required to provide a real-time update of the fire cue can be very demanding. In addition, in the modern air-to-air engagement, the window for a successful prosecution of an attack can depend on split-second timing. The overall P_k of the system therefore depends not only on the performance of the missile but also that of the aircraft sensor and computing systems.

The use of a helmet-mounted sight can reduce pilot workload during an engagement but can also increase delays in the system, which need to be countered by high processing speed.

It follows, therefore, that the integration of the components of an air-to-air engagement system is a complex but essential task if the aircraft is to have a leading edge capability.

7.4 End-game

It is likely that the target aircraft will be aware that it is being attacked and so will be taking evasive action by maneuvering and deploying countermeasures such as IR flares or radar reflecting chaff. Modern missiles use complex algorithms to counter such countermeasures and are themselves agile.

When the missile guidance system detects that it is closing in on its target, a proximity sensor will be activated. Many types of proximity sensor have been used ranging from simple radio frequency systems to more complex laser range detectors. On detecting the target the warhead will detonate.

8 RELEASING WEAPONS FROM WEAPON BAYS

With the need for low observability, several modern aircraft have the provision to carry weapons in internal bays. This increases the aeromechanical problems associated with weapons release. Downward ejected weapons need to be ejected with sufficient force to penetrate the air flow in and around the bay. Forward-fired missiles need to be lowered into the air flow before firing, and for a LOBL capability, a means of transferring target data to the weapon over the interface with the aircraft needs to be considered.

9 STORES MANAGEMENT SYSTEMS

An essential element of the aircraft system is the Stores Management System (SMS). The SMS manages the weapon load-out and controls the safe release and jettison from the aircraft.

From a safety and certification viewpoint it is essential that an aircraft only releases a store when intended. This appears to be an obvious requirement but it is the primary driver in the design of the armament system and it is this requirement that adds complexity to the design of the SMS.

The first electrical release systems were based on relays that, when energized by the Bomb Aimer pressing the release button, would switch current to the bomb rack, causing it to open. To some extent, this provided a safe system as the relay contact provided an air gap that would prevent the bomb rack being operated. However, a short-circuit failure in the Bomb Aimer's button or a similar failure in the relay would mean that the bomb could be inadvertently released. This drove the design of systems with multiple breaks in the bomb rack firing chain such that a single failure, on its own, could not cause an inadvertent release. However, a significant drawback was that there were more components that could fail in a safe manner (i.e. open circuit) and therefore the system was less reliable. Another significant drawback was that the accuracy of the release point was determined by the skill of the Bomb Aimer and his reaction time. For ballistic bombs, this increased the release point error and therefore the terminal accuracy.

The introduction of a weapon aiming computer meant that the Bomb Aimer was now committing to release a weapon and it was the computer that was actually generating the signal to close the fire relays. This basic concept has evolved into the systems in today's aircraft.

A modern SMS is required to control the firing and release of many different types of weapons with varying release options and modes. This has driven the system design to include software, and for improved reliability and life, relays have been replaced by semiconductor switches. However, the same basic safety principle of "no single failure shall cause an unintended release" and the availability principle that "no single failure shall prevent a release when intended" still hold true. These principles are captured, for example, by the United Kingdom Defence Standard 00-970. Figure 4 shows how these design principles could be implemented.

This figure shows a simple twin-channel system operating from dual power supplies. The release circuits are initiated from a multiple-pole release button. When both release circuits are operating correctly, each circuit switches its own upper switch in the channel's fire supply (the power that is used to operate the ERU) with the lower switch being controlled by the other channel. However, if Built-In Test (BIT) circuitry detects a fault in a channel, full authority to switch the fire supply is handed to the good channel. This implementation is protected against a single failure either causing an unintended release or preventing an intended release.

A modern SMS also has to determine its own weapon inventory. Many smart weapons have the ability to tell the aircraft what type of weapon they are. This simplifies the logistics of preparing an aircraft for a mission and enables the aircraft to know exactly what is loaded on which station. This can be particularly important, as this information may be required by the aircraft's flying control system to alter

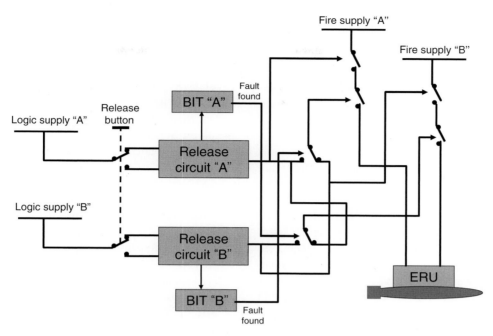

Figure 4. Simplified SMS architecture.

performance parameters. Also, as stores are released, it may be important for continued controlled flight that the distribution of heavy stores is controlled such that their release does not impose an unstable condition on the aircraft (e.g., many heavy stores loaded on one wing with very few loaded on the other).

The modern SMS may also be required to control the safe (unarmed) jettison of stores from the aircraft under emergency conditions (e.g., an engine flame-out on takeoff) when the mass of the aircraft may need to be quickly reduced.

All these demands make a modern SMS a complex system. Multi-channel systems that are designed to maintain integrity and availability are common. Such systems employ high-integrity software and are generally designed to be immune to electromagnetic interference (to ensure the system remains safe at all times).

10   WEAPON INTERFACE STANDARDS

With the advent of smart weapons, the electrical interfaces employed were often bespoke and usually optimized for the specific weapon. This led to aircraft being required to provide many different interfaces. Aircraft pylons quickly became congested with wiring, dictating the need for aircraft to undergo a role change if a specific weapon type was to be operated.

This situation led the US Department of Defense to develop a common interfacing standard for future smart stores (Military Standard 1760 (MIL-STD-1760)). MIL-STD-1760 defines an aircraft electrical interconnection system using a standard connector and providing a flexible signal set that contains various power supplies, media for transferring analog signals such as audio and video, a dual-redundant data bus for controlling the store, and a safety discrete signal. MIL-STD-1760 is the primary interface standard used by current smart weapons.

As current military requirements demand physically smaller weapons, the MIL-STD-1760 connector has been deemed to be too large and require too high a break force during ejected stores releases. This led the US Air Force to approach the SAE to develop a new standard for miniature munitions (munitions in the 25 kg–100 kg class). The Miniature Mission Store Interface is defined by AS5725 and maintains the principles of MIL-STD-1760 in that it provides a standard signal set containing power, data, and GPS interfaces.

The prevalence of the Sidewinder missile has driven more recent SRAAM designs to adopt the same analog interface as a necessity to capture weapon sales to existing Sidewinder customers and to reduce aircraft integration costs by enabling their operation on an unmodified system interface. The Sidewinder analogue interface has therefore become a de facto standard for SRAAMs, although the need for increased capability means that it is now being supplemented by a digital interface based on MIL-STD-1760.

11   FUTURE SYSTEMS

So what does the future hold for weapon systems?

Weaponization of Unmanned Air Systems (UAS) is becoming reality. While the "no single failure" principles are still very much applicable, the removal of the pilot from the aircraft brings new challenges for the safe integration of weapons.

UAS are also driving the need for even smaller weapons (in the sub 3 kg class), and for these, new interfaces are required. The SAE has developed AS5726 (Interface for Micro Munitions).

Also, future weapons may not depend on kinetic kill. The development of weapons employing high-powered lasers and microwaves is being investigated and one system, the US Airborne Laser Programme, is in advanced development.

While these new weapon types bring new challenges for the weapons integrator, the basic principles for safe, available systems that can deliver a military effect with high precision still hold true.

ACKNOWLEDGMENTS

The author would like to thank a number of people for their contributions in either providing information or images for this chapter or taking the time to review and critique draft manuscripts. Of note are Rod Robinson and Dennis Griffin.

RELATED ARTICLE

Chapter 18

FURTHER READING

STANAG 4187 – Fuzing Systems: Safety Design Requirements.

STANAG 3726 – Bail (Portal) Lugs for the Suspension of Aircraft Stores.

MIL-STD-1760 – Aircraft/Store Electrical Interconnection System.

STANAG 3842AA – Rail Launched Missile/Launcher Mechanical Interface.

Def-Stan-00-970 – Design and Airworthiness Requirements for Service Aircraft.

AS5725 – Interface Standard, Miniature Mission Store Interface.

AS5726 – Interface for Micro Munitions.

AIR5682 – Common Launch Acceptability Region Approach Interface Control Document.

AIR5712 – Common Launch Acceptability Region Approach Rationale Document.

AIR5788 – Common Launch Acceptability Region Truth Data Generator Interface Control Document for the CLAR Approach.

MIL-HDBK-1760A – Aircraft/Store Electrical Interconnection System.

2001-01-02951; Rigby KA: The Role of Standardisation on the Journey to "Plug & Play" Weapons. Paper 2001-01-02951, Society of Automotive Engineers (www.sae.org).

2001-01-2952; Provenza, J & Benedick, F: Development of a Standard Electrical Interface for Miniature Munitions. Society of Automotive Engineers (www.sae.org).

2001-01-2953; Clark, D, Faust, A & Jones, A: Common Launch Acceptability Region Task Group. Society of Automotive Engineers (www.sae.org).

2001-01-2954; Gregory, DA: The Generic Aircraft-Store Interface Framework. Society of Automotive Engineers (www.sae.org).

2004-01-3113; Millett, SB: Mission Data Exchange for the Netted Future. Society of Automotive Engineers (www.sae.org).

PART 3
Airframe Configurations

Chapter 6
Classes and Missions of UAVs

Thomas J. Gleason[1] and Paul G. Fahlstrom[2]
[1]*Gleason Research Associates, Inc., Columbia, MD, USA*
[2]*United States Army Materiel Command, Huntsville, AL, USA*

ACRONYMS

UAV	unmanned aerial vehicle
AV	air vehicle
VTOL	vertical takeoff and landing
NIR	near-infrared

1 OVERVIEW

We here describe a representative sample of unmanned aerial vehicles (UAVs), including some of the earlier designs that had a large impact on current systems. The range of UAV sizes and types now runs from air vehicles (AVs) small enough to land on the palm of your hand to large lighter-than-air vehicles.

Much of the early development of UAVs was driven by government and military requirements, and the bureaucracies that manage such programs have made repeated efforts to establish a standard terminology for describing various types of UAV in terms of the capabilities of the air vehicles. While the "standard" terminology constantly evolves and occasionally changes abruptly, some of it has come into general use in the UAV community and is briefly described.

Finally, we attempt to summarize the applications for which UAVs have been or are being considered, which provide a context for the system requirements that drive the system designs.

The material in this chapter is based on material published in *Introduction to UAV Systems* (Fahlstrom and Gleason, 2012).

2 EXAMPLES OF UAVS

We attempt here to provide a broad survey of the many types of UAVs that have been or are being designed, tested, and fielded throughout the world. The intent of this survey is to introduce those who are new to the UAV world to the wide variety of systems that have appeared over the few decades since the revival of interest in UAVs began in the 1980s.

There are a variety of guides to UAVs available and a great deal of information is posted on the Internet. We use *The Concise Global Industry Guide* (Kemp, 2011) as a source for quantitative characteristics of current UAVs and a variety of open source postings and our own personal files for information on systems no longer in production.

As a general organizing principle, we will start with the smallest UAVs and proceed to some that are the size of a corporate jet. The initial efforts on UAVs in the 1970s and 1980s concentrated on AVs that had typical dimensions of 2 or 3 m (6.6–9.8 ft), partly driven by the need to carry sensors

Unmanned Aircraft Systems. Edited by Ella Atkins, Aníbal Ollero, Antonios Tsourdos, Richard Blockley and Wei Shyy.
© 2016 John Wiley & Sons, Ltd. ISBN: 978-1-118-86645-0.

and electronics that at that time had not reached the advanced state of miniaturization that has since become possible. In more recent years, there has been a growing interest in extending the size range of UAVs down to insect-sized devices at one extreme and up to medium air transport sizes at the other end.

Initially, some of the motivation for smaller UAVs was to make them man portable so that a soldier or a border guard could carry, launch, and control a model-airplane-sized UAV that allows him or her to take a look over the next hill or behind the buildings that are in front of him or her. Further miniaturization, to the size of a small bird or even an insect, is intended to allow a UAV to fly inside a building or perch unnoticed on a window sill or roof gutter and provide a look inside the building or into a narrow street.

More recently, miniaturization down to dimensions of a few centimeters and weights of a few ounces has been motivated by a desire to make UAVs that can be mass produced at consumer-level prices and easily can be transported and launched from any small open space. This allows them to be marketed to hobbyists and various commercial users who use them for a rapidly growing list of personal, business, and government applications.

The realm of small UAVs is one in which there is no competition from manned vehicles. It is unique to vehicles that take advantage of the micro-miniaturization of sensors and electronics to allow humans to view the world from a flying vehicle that can go places that are not accessible to anything on a human scale.

The motivation for larger UAVs is to provide long endurance at high altitudes with the ability to fly long distances from a base and then loiter over an area for many hours using a larger array of sensors to search for something or keep watch over some area. Increasingly, in the military arena, the larger UAVs also provide a capability to carry a large weapons payload a long distance and then deliver it to the destination area.

There now is increasing talk about performing missions such as heavy air transportation, bombers, and even passenger transportation with unmanned systems. Whatever may be the outcome of those discussions, it is likely that there eventually will be unmanned systems of all sizes.

In the following sections, we use intuitive size classes that are not in any sense standardized but are convenient for this discussion.

2.1   Very small UAVs

For the purposes of this discussion, "very small UAVs" range from "micro" sized, which are about the size of a large insect, up to an AV with dimensions of the order of a 30–50 cm (12–20 in.). There are two major types of very small UAVs. One type uses flapping wings to fly like an insect or a bird and the other uses a more or less conventional aircraft configuration, usually rotary wing for the micro size range. The choice of flapping wings or rotary wings often is influenced by the desire to be able to land and perch on small surfaces to allow surveillance to continue without having to expend the energy to hover. Another advantage of flapping wings is covertness, as the UAV may look a lot like a bird or insect and be able to fly around very close to the subjects of its surveillance or perch in plain view without giving away the fact that it is actually a sensor platform.

Examples of very small UAVs include the Israeli IAI Malat Mosquito, which is an oval flying wing with a single tractor propeller; the US Aurora Flight Sciences Skate, which is a rectangular flying wing with twin tractor engine/propeller combinations that can be tilted to provide "thrust vectoring" for control; and the Australian Cyber Technology CyberQuad Mini, which has four ducted fans in a square arrangement.

The Mosquito wing/fuselage is 35 cm (14.8 in.) long and 35 cm (14.8 in.) in total span. It uses an electric motor with batteries and has an endurance of 40 min, and claims a radius of action of about 1.2 km (0.75 mi). It is hand or bungee launched and can deploy a parachute for recovery.

The Skate fuselage/wing has a wingspan of about 60 cm (24 in.) and length of 33 cm (13 in.). It folds in half along its centerline for transport and storage. It has twin electric motors on the leading edge that can be tilted up or down and allow vertical takeoff and landing (VTOL) and transition to efficient horizontal flight. There are no control surfaces, with pitch control being accomplished with thrust vectoring by tilting the motor/propeller assemblies and yaw control by varying the speed of the two propellers. It can carry a payload of 227 g (8 oz) with a total takeoff weight of about 1.1 kg (2 lb).

The CyberQuad Mini has four ducted fans, each with a diameter of somewhat less than 20 cm (7.8 in.), mounted so that the total outside dimensions that include the fan shrouds are about 42 cm × 42 cm (16.5 in.). The total height including the payload and batteries, which are located in a fuselage at the center of the square, is 20 cm (7.8 in.). This AV resembles a flying "toy" called the "Parrot AR Drone" currently marketed for about $300 US. The CyberQuad Mini includes a low-light level solid-state camera or thermal camera and a control system that allows fully autonomous waypoint navigation. The toy has two onboard cameras, one facing forward and one facing down, and is controlled much like a video game from a portable digital device such as a tablet computer or a smart phone.

The quad-rotor configuration was an important milestone in the explosion of mass-produced UAVs intended for the consumer and commercial markets. By the time that this is

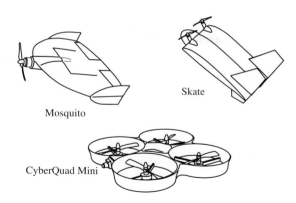

Figure 1. Very small UAVs.

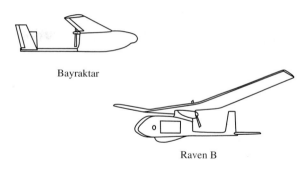

Figure 2. Small UAVs.

being written it has evolved to include configurations with up to eight rotors, with or without rotor guards or ducting.

Drawings of these UAVs are shown in Figure 1.

2.2 Small UAVs

What we will describe as "small UAVs" have at least one dimension of greater than 50 cm (16 in) and go up to dimensions of a meter or two. Many of these UAVs have the configuration of a fixed-wing model airplane and are hand launched by their operator by throwing them into the air much as we launch a toy glider.

Examples of small UAVs include the US AeroVironment Raven and the Turkish Byraktar Mini, both conventional fixed-wing vehicles. There are also a number of rotary-wing UAVs in this size grouping, but they are basically scaled-down versions of the medium rotary-wing systems discussed in the following section and we do not offer an example in this group.

The RQ-11 Raven is an example of a UAV that is in the "model airplane" size range. It has a 1.4 m (4.6 ft) wingspan and is about 1 m (3.3 ft) long. It weighs only a little less than 2 kg (4.4 lb) and is launched by being thrown into the air by its operator in much the same way that a toy glider is put into flight. It uses electrical propulsion and can fly for nearly an hour and a half. The Raven and its control station can be carried around by its operator on his/her back and can carry visible, near-infrared (NIR), and thermal imaging systems for reconnaissance as well as a "laser illuminator" to point out target to personnel on the ground. (Note that this is not a laser for guiding laser-guided weapons, but more like a laser pointer, operating in the NIR to point things out to people using image-intensifier night-vision devices.)

The latest model, the RQ-11B Raven, was added to the US Army's Small UAV program in a competition that started in 2005. Built by AeroVironment, the Raven B includes a number of improvements from the earlier Raven A, including improved sensors, a lighter Ground Control System, and the addition of the onboard laser illuminator. Endurance was improved as was interoperability with battlefield communication networks.

The Bayraktar Mini UAV was developed by Baykar Makina, a Turkish company. It is a conventionally configured, electrically powered AV somewhat larger than the Raven, with a length of 1.2 m (3.86 ft), wingspan of 2 m (6.6 ft), and weight of 5 kg (10.5 lb) at takeoff. It is advertised to have a spread spectrum, encrypted data link, which is a highly desirable, but unusual, feature in an off-the-shelf UAV. The data link has a range of 20 km (12.4 mi), which would limit the operations to that range, although it may depend on the local geography and where the ground antenna is located.

The Bayraktar Mini has a gimbaled day or night camera. It offers waypoint navigation with GPS or other radio navigation systems.

Despite its slightly greater size and weight, it is launched much like the Raven. It can be recovered by a skidding landing on its belly or with an internal parachute. It is fielded with small army units and has been heavily used by the Turkish Army since it became operational in about 2006. Drawings of these examples are shown in Figure 2.

2.3 Medium UAVs

We describe a UAV as "medium" if it is too large to be carried around by one person and still smaller than a light aircraft. (As with all of these informal size descriptions, we do not claim rigorousness in this definition. Some attempts at standardized and universal classifications of UAVs are described later.)

The UAVs that sparked the present resurgence of interest, such as the Pioneer and Skyeye, are in the medium class. They have typical wingspans of the order of 5–10 m (16–32 ft) and carry payloads of from 100 to more than

200 kg (220–440 lb). There are a large number of UAVs that fall into this size group. The Israeli–US Hunter and the UK Watchkeeper are more recent examples of medium-sized, fixed-wing UAVs.

There are also a large number of rotary-wing UAVs in this size class. A series of conventional helicopter with rotor diameters of the order of 2 m (6.4 ft) have been developed in the United Kingdom by Advanced UAV Technologies. There are also a number of ducted-fan systems configured much like the CyberQuad Mini but having dimensions measured in meters instead of centimeters.

Finally, we mention the US Boeing Eagle Eye, which is a medium-sized VTOL system that is notable for using tilt-wing technology.

The RQ-2 Pioneer is an example of an AV that is smaller than a light manned aircraft but larger than what we normally think of as a model airplane. It was for many years the workhorse of the stable of US tactical UAVs. Originally designed by the Israelis and built by AAI in the United States, it was purchased by the US Navy in 1985. The Pioneer provided real-time reconnaissance and intelligence for ground commanders. High-quality day and night imagery for artillery and naval gun-fire adjustment and damage assessment were its prime operational missions. The 205 kg (452 lb), 5.2 m (17 ft) wingspan AV had a conventional aircraft configuration. It cruised at 200 km h^{-1} and carried a 220 kg (485 lb) payload. Maximum altitude was 15 000 ft (4.6 km). Endurance was 5.5 h. The ground control station could be housed in a shelter on a High Mobility Multipurpose Wheeled Vehicle (HMMWV) or truck. The fiberglass air vehicle had a 26 hp engine and was shipboard capable. It had piston and rotary engine options.

The Pioneer could be launched from a pneumatic or rocket launcher or by conventional wheeled takeoff from a prepared runway. Recovery was accomplished by conventional wheeled landing with arrestment or into a net. Shipboard recovery used a net system.

The BAE Systems Skyeye R4E UAV system was fielded in the 1980s and is roughly contemporary with the Pioneer, with which it has some common features, but the air vehicle is significantly larger in size, which allows expanded overall capability. It uses launchers similar to the Pioneer but does not have a net-recovery capability. It uses a ground control station similar in principle to that of the Pioneer.

The Skyeye air vehicle is constructed of lightweight composite materials and is easy to assemble and disassemble for ground transport because of its modular construction. It has a 7.32 m (24 ft) wingspan and is 4.1 m (13.4 ft) long. It is powered by a 52 hp rotary engine (Teledyne Continental Motors) providing high reliability and low vibration. Maximum launch weight is 570 kg (1257 lb) and it can fly for 8–10 h and at altitudes up to 4600 m (15,000 ft). Maximum payload weight is about 80 kg (176 lb).

Perhaps the most unique feature of the Skyeye when it was fielded was the various ways in which it could be recovered. The Skyeye has no landing gear to provide large radar echoes or obstruct the view of the payload. The avoidance of a nose wheel is particularly significant as a nose gear often obstructs the view of a payload camera looking directly forward, precluding landing based on the view through the eyes of the camera. However, it can land on a semiprepared surface by means of a retractable skid located behind the payload. This requires one to control the landing by observing the air vehicle externally during its final approach. This is particularly dangerous during night operations.

The landing rollout, or perhaps more accurately the "skid-out," is about 100 m (322 ft). The Skyeye also carries either a parafoil or a parachute as alternative recovery systems. The parafoil essentially is a soft wing that is deployed in the recovery area to allow the air vehicle to land much slower. The parafoil recovery can be effective for landing on moving platforms such as ships or barges. The parachute can be used as an alternative means of landing or as an emergency device. However, using the parachute leaves one at the mercy of the vagaries of the wind, and it primarily is intended for emergency recoveries. All of these recovery approaches are now offered in various fixed-wing UAVs, but having all of them as options in one system still would be unusual.

The RQ-5A Hunter was the first UAV to replace the terminated Aquila system as the standard "Short Range" UAV for the US Army. The Hunter does not require a recovery net or launcher, which significantly simplifies the overall minimum deployable configuration and eliminates the launcher required by the Skyeye. Under the appropriate conditions, it can takeoff and land on a road or runway. It utilizes an arresting cable system when landing, with a parachute recovery for emergencies. It is not capable of net recovery because it has a tractor ("puller") propeller that would be damaged or broken or would damage any net that was used to catch it. It also has a rocket-assisted takeoff option to allow launch to occur when no suitable road or runway is available.

The Hunter is constructed of lightweight composite materials, which afford ease of repair. It has a 10.2 m (32.8 ft) wingspan and is 6.9 m (22.2 ft) long. It is powered by two four-stroke, two-cylinder (v-type), air-cooled Moto Guzzi engines, which utilize fuel injection and individual computer control. The engines are mounted in-line, tractor and pusher, giving the air vehicle twin engine reliability without the problem of unsymmetrical control when operating with a single engine. The air vehicle weighs approximately 885 kg (1951 lb) at takeoff (maximum), has an endurance of about 12 h, and a cruise speed of 120 knots.

The Hermes 450/Watchkeeper is an all-weather, intelligence, surveillance, target acquisition and reconnaissance UAV. Its dimensions are similar to the Hunter. The Watchkeeper is manufactured in the United Kingdom by a joint venture of the French company Thales and the Israeli company Elbit Systems. It has a weight of 450 kg (992 lb) including a payload capacity of 150 kg (331 lb).

The Watchkeeper became operational with British forces late in 2014 or 2015.

A series of rotary-wing UAVs called the AT10, AT20, AT100, AT200, AT300, and AT1000 have been developed by the UK firm Advanced UAV Technology. They are all conventionally configured helicopters with a single main rotor and a tail boom with a tail rotor for yaw stability and control. The rotor diameters vary from 1.7 m (5.5 ft) in the AT10 to about 2.3 m (7.4 ft) for the AT1000. Speed and ceiling increase as one moves up the series, as does the payload capacity and payload options. All are intended to be launched by vertical takeoff and all claim the ability for autonomous landings on moving vehicles.

The Northrop Grumman MQ-8B Fire Scout is an example of a conventionally configured VTOL UAV. It looks much like a typical light helicopter. It has a length of 9.2 m (30 ft) (with the blades folded so that they do not add to the total length), height of 2.9 m (9.5 ft), and a rotor diameter of 8.4 m (27.5 ft). It is powered by a 420-shaft hp (shp) turbine engine. The Fire Scout is roughly the same size as an OH-58 Kiowa light observation helicopter, which has a two-man crew and two passenger seats. The Kiowa has a maximum payload of about 630 kg (1389 lb), compared to the 270 kg (595 lb) maximum payload of the Fire Scout, but if one takes out the weight of the crew and other things associated with the crew, the net payload capability of the Fire Scout is similar to that of the manned helicopter.

The Fire Scout was tested by the US Army and Navy for a variety of missions that are similar to those performed by manned helicopters of a similar size. It was fielded in 2013 and 2014.

The tilt-rotor Bell Eagle Eye was developed during the 1990s. It uses "tilt-wing" technology, which means that the propellers are located on the leading edge of the wing and can be pointed up for takeoff and landing and then rotated forward for level flight. This allows a tilt-wing aircraft to utilize wing-generated lift for cruising, which is more efficient than rotor-generated lift, but still to operate like a helicopter for VTOL capability.

The Eagle Eye has a length of 5.2 m (16.7 ft) and weighs about 1300 kg (2626 lb). It can fly at up to about 345 km h^{-1} (190 knots) and at altitudes up to 6000 m (19 308 ft).

Some of these UAVs are shown in Figure 3.

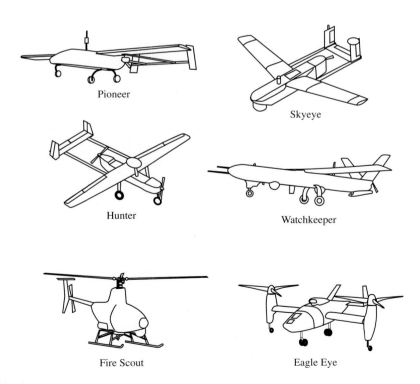

Pioneer

Skyeye

Hunter

Watchkeeper

Fire Scout

Eagle Eye

Figure 3. Medium UAVs.

2.4 Large UAVs

Our informal size groupings are not finely divided and we will discuss all UAVs that are larger than a typical light manned aircraft in the group called "large."

This includes, in particular, a group of UAVs that can fly long distances from their bases and loiter for extended periods to perform surveillance functions. They also are large enough to carry weapons in significant quantities. The lower range of such systems includes the US General Atomics Predator A, which has a significant range and endurance, but can carry only two missiles of the weight presently being used. The limitation to two missiles is serious as it means that after firing the two missiles that are on board, the UAV either has lost the ability to deliver weapons or must be flown back to its base to be rearmed. For this reason, a second generation of UAVs designed for missions similar to that of the Predator, including a Predator B model, called the "Reaper" has now appeared that is larger and able to carry many more weapons on a single sortie.

The Cassidian Harfang is an example of a system much like the Predator A and the Talarion, also by Cassidian, is an example of the emerging successors to the Predator A.

At the high end of this size group, an example is an even larger UAV designed for very long range and endurance and capable of flying anywhere in the world on its own, the US Northrop Grumman Global Hawk.

There are a number of specialized military and intelligence systems for which information available to the public is very limited. An example of this is the US Lockheed Martin Sentinel. Little or no authoritative information is available on these systems and we leave it to the reader to explore what is available on the Internet.

The MQ-1 Predator A is larger than a light single-engine private aircraft and provides medium altitude, real-time surveillance using high-resolution video, infrared imaging, and synthetic aperture radar. It has a wingspan of 17 m (55 ft) and a length of 8 m (26 ft). It adds significantly higher ceiling (7620 m or 24 521 ft) and longer endurance (40 h) to the capabilities of the smaller UAVs. GPS and inertial systems provide navigation, and control can be via satellite. Speed is 220 km h^{-1} (119 knots) and the air vehicle can remain on station for 24 h, 925 km (575 mi) from the operating base. It can carry an internal payload of 200 kg (441 lb) plus an external payload (hung under the wings) of 136 kg (300 lb).

The Harfang UAV is produced by Cassidian, which is subsidiary of the French company EADS. It is about the same size as the Predator and is designed for similar missions. The configuration is different, using a twin-boom tail structure.

There are a variety of possible payloads. Its stated performance is similar to that of the Predator, but it has a shorter endurance of 24 h. It takes off and lands conventionally on wheels on a runway. Control can be via satellite.

Talarion is under development by Cassidian as a second-generation successor to the Predator/Harfang class of UAVs. It uses two turbojet engines and can carry up to 800 kg (1764 lb) of internal payload and 1000 kg (2205 lb) of external payload with a ceiling of over 15 000 m (49 215 ft) and speeds around 550 km h^{-1} (297 knots).

The RQ-4 Global Hawk is manufactured by Northrop Grumman Aerospace Systems. It flies at high altitude and utilizes radar, electro-optical, and infrared sensors for surveillance applications. It uses a turbofan engine and appears to have a shape that reduces its radar signature, but is not a "stealth" aircraft. It is 14.5 m (47 ft) long with a 40 m (129 ft) wingspan and has a maximum weight at takeoff of 1460 kg (3219 lb). It can loiter at 575 km h^{-1} (310 knots) and has an endurance of 32 h. It has a full set of potential payloads and it appears that it is routinely controlled via satellite links.

The RQ-170 Sentinel is reported to be a stealthy AV manufactured by Lockheed Martin. No data are officially available, but based on pictures recently in the press it appears to be a flying wing configuration much like the B-2 bomber and to be in the medium-to-large size class, with a wingspan of around 12–13 m (38–42 ft).

Some of these large UAVs are illustrated in Figure 4.

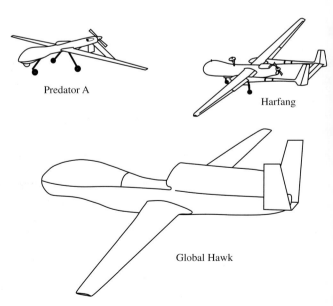

Predator A

Harfang

Global Hawk

Figure 4. Large UAVs.

3 EXPENDABLE UAVS

Some expendable UAVs are not designed to return after accomplishing their mission. In the military world, this often means that they contain an internal warhead and are intended to be crashed into a target destroying it and themselves. This type of expendable is not really a UAV, but rather a missile or munition of some sort. There is a considerable area of overlap between guided missiles and UAVs, as illustrated by the fact that the first "UAVs" of the aviation era were mostly guided weapons.

Another group of expendables is designed so that it can (and should) be recovered if possible, but is inexpensive and small enough that there can be many spares available, so that it can have a very high loss rate. We consider this type of "expendable" to belong to the general family of UAVs.

The electric motor-powered Raven, described in Section 2.2.2, is an example of a recoverable, but expendable UAV. It is hand launched and uses a hand-carried ground control station. The Raven is used to conduct reconnaissance missions out to about 5 km and is recoverable, but if it does not return or crashes during landing, the loss is considered acceptable.

4 CLASSES OF UAV SYSTEMS

It is convenient to have a generally agreed upon scheme for classifying UAVs rather like the classification of military aircraft in general into such classes as transport, observation, fighter, attack, cargo, and so on.

4.1 Classification by range and endurance

Shortly after being appointed the central manager of US military UAV programs, the Joint UAV Program Office defined classes of UAVs as a step toward providing some measure of standardization to UAV terminology. They were as follows:

- *Very Low Cost Close Range:* Required by the Marine Corps and perhaps the Army to have a range of about 5 km (3 mi) and cost about $10 000 (in the 1980s) per air vehicle. This UAV system fits into what could be called the "model airplane" type of system and its feasibility with regard to both performance and cost had not been proven but since has been demonstrated by systems such as the Raven and Dragon Eye.
- *Close Range:* Required by all of the services but its concept of operation varied greatly depending on the

service. The Air Force usage would be in the role of airfield damage assessment and would operate over its own airfields. The Army and Marine Corps would use it to look over the next hill, and desired a system that was easy to move and operate on the battlefield. The Navy wanted it to operate from small ships such as frigates. It was to have a range of 50 km (31 mi), with 30 km (19 mi) forward of the forward line of friendly troops. The required endurance was from 1 to 6 h depending on the mission. All services agreed that the priority mission was reconnaissance and surveillance, day and night.

- *Short Range:* The short-range UAV was also required by all of the services and, like the close-range UAV, had the day/night, reconnaissance and surveillance mission as a top priority. It had a required range of 150 km (93 mi) beyond the FLOT, but 300 km (186 mi) was desired. The endurance time was to be 8–12 h. The Navy required the system to be capable of launch and recovery from larger ships of the Amphibious Assault Ship and Battleship class.
- *Mid Range:* The mid-range UAV was required by all the services except the Army. It required the capability of being ground or air launched and was not required to loiter. The latter requirement suggested that the air vehicle was a high-speed deep penetrator and, in fact, the velocity requirement was high subsonic. The radius of action was 650 km (404 mi) and it was to be used for day/night reconnaissance and surveillance. A secondary mission for the mid range was the gathering of meteorological data.
- *Endurance:* The endurance UAV was required by all services and, as the name suggested, was to have a loiter capability of at least 36 h. The air vehicle had to be able to operate from land or sea and have a radius of action of approximately 300 km (186 mi). The mission was day/night reconnaissance first, and communications relay second. Speed was not specified, but it had to be able to maintain station in the high winds that will be experienced at high altitudes. The altitude requirement was not specified, but it was thought probably to be 30 000 ft (9.14 km) or higher.

This classification system has been superseded. However, some of the terminology and concepts, particularly the use of a mix of range and mission to define a class of UAV, persists today and it is useful for anyone working in the field to have a general knowledge of the terminology that has become part of the jargon of the UAV community.

The following sections outline some of the more recent terminology used to classify UAVs. Any government-dictated classification scheme is likely to change over time to meet the changing needs of program managers, and the reader is advised to search the literature on the Internet if the current standard of government classification is needed.

4.2   The tier system

A set of definitions that has become pervasive in the UAV community stems from an attempt to define a hierarchy of UAV requirements in each of the US services. The levels in these hierarchies were called "tiers" and terms such as "tier II" often are used to classify a particular UAV or to describe a whole class of UAVs.

The tiers are different in each US service, which can lead to some confusion, but they are listed below with brief descriptions:

US air force tiers

Tier not specified: small/micro UAV.

Tier I: low altitude, long endurance.

Tier II: medium altitude, long endurance. An example is the MQ-1 Predator.

Tier II+: high altitude, long-endurance conventional UAV. Altitude: 60 000–65 000 ft (19 800 m), less than 300 knots (560 km h^{-1}) airspeed, 3000 nautical mile (6000 km) radius, 24 h time-on-station capability. Tier II+ is complementary to the Tier III aircraft. An example is the RQ-4 Global Hawk.

Tier III−: HALE low-observable UAV. Same as the Tier II+ aircraft with the addition of "stealth" features. An example is the RQ-3 DarkStar.

Marine corps tiers

Tier not specified: micro UAV. An example is the Wasp.

Tier I: mini UAV. An example is the Dragon Eye.

Tier II: an example is the RQ-2 Pioneer.

Tier III: medium range – an example is the Shadow.

Army tiers

Tier I: small UAV. An example is the RQ-11A/B Raven.

Tier II: short-range tactical UAV. Role filled by the RQ-7A/B Shadow 200.

Tier III: medium-range tactical UAV.

4.3   Commercial and consumer UAVs

Until relatively recently UAVs actually in use were applied almost solely to military and government applications. The types of UAVs most talked about for commercial applications were similar to light or ultra-light manned aircraft and

expected to be of similar cost and to require runways for takeoff and landing.

There were, of course, radio-controlled model airplanes flown by hobbyists, but they required at least semiskilled operators, had very limited payloads capability and endurance, and generally not considered candidates for commercial applications.

Starting around 2010, there was a convergence of many technologies that made it feasible to design and produce small and very small UAVs that could be sold in quantity at prices that were easily affordable by hobbyists or potential commercial users and that required little skill to fly.

Multirotor, electric motor UAVs became available on the consumer market. They were controlled using inexpensive remote control consoles or even by an "App" on a smart phone or tablet computer, carried high-definition video cameras, were transportable in a private automobile, and could be launched and recovered from any small open space. The age of the "drones" had arrived.

By the time that this is written, it is estimated that there are as many as a million of these "drones" in private hands and sales are expected to continue at an accelerating rate.

As discussed in the next section, the availability of these inexpensive, highly maneuverable, easy-to-fly platforms, capable of carrying high-definition video cameras, and other sorts of sensors has resulted in an explosion of actual and proposed commercial applications.

5   MISSIONS

Defining the missions for UAVs has always been a difficult task because (i) there are so many possibilities and (ii) there have never previously been enough systems in the field to develop all of the possibilities. This is not to say that the subject has not been thought about, because there have been repeated efforts to come up with comprehensive lists of missions as part of classification schemes. All such lists tend to become unique to the part of the UAV community that generates them and they all tend to become out-of-date as new mission concepts continually arise.

The arrival of commercial/consumer UAVs has led to a positive explosion of proposed missions. Because there are now inexpensive UAVs available with which to attempt those missions, and in the absence of effective regulation, many of those proposals have been tried out and demonstrated by users for whom the UAV is just another tool. They require only a little more skill on the part of the user than other common tools, such as hand-held video cameras, but offer new possibilities limited only by the imagination of the user.

The list of commercial applications already underway includes, but is far from limited to,

- taking overhead pictures of real estate for sale, in which the real-estate agent may own and operate his or her own UAV;
- inspections of inaccessible roofs (or bridges, towers, chimneys, power lines, steeples, trees, etc.) using a high-definition video camera that can be positioned a few feet from the area to be inspected and can adjust its point of view up, down, or to the side at will;
- surveying/inspecting fields of crops;
- spraying crops with pesticides;
- observing scenes of accidents, flooding, landslides, or other incidents;
- providing video feeds for "breaking news" reports;
- performing precision three-dimensional surveys using sterioscopic cameras and precision locating of the platform at low altitiude over the area to be surveyed;
- delivery of drugs or other supplies to remote areas;
- search for people missing in rough terrain;
- monitoring forest areas for wildfires;
- sensing of pollutants near their source;
- providing flexible and dynamic camera positioning for cinema and television video production;
  competitive UAV racing, which revives the thrill of the manned air races of a century ago and adds the excitement and challenge of complex and difficult obstacle courses and of many bloodless crashes.

In addition, there are major commercial development programs underway to implement at least the following missions:

Fast, same-day delivery of everything from pizzas to general merchandise.
- Search and rescue using a UAV that is capable of landing and carrying a person found (not a pilot) back to its base.
- Self-piloted personal aircraft.
- Long-endurance, high-altitude platforms for communications relays, serving as local area satellites for telephone or digital broadband services.

These lists can only suggest the breadth and scope of ctual and proposed applications. Any such list will be ncomplete by necessity and out-of-date by the time that it an be printed.

.1   Military versus civilian missions

wo major divisions of missions for UAVs are civilian and nilitary, but there is significant overlap between these two in the area of reconnaissance and surveillance, which a civilian might call search and surveillance or observation, which is the largest single application of UAVs in both the civilian and military worlds.

The development of UAVs has been led by the military and there are other areas along recognized as potential military missions that also have civilian equivalents. These include atmospheric sampling for radiation and/or chemical agents, providing relays for line-of-sight communications system, and meteorological measurements.

An area of interest to both the military and civilian worlds is to provide a high-altitude platform capable of lingering indefinitely over some point on the earth that can perform many of the functions of a satellite at lower cost and with the capability of landing for maintenance or upgrade and of being redeployed to serve a different part of the world whenever needed.

5.1.1   Military missions

A fundamental difference between military and civilian missions is that the military use of UAVs includes the direct delivery of lethal force. This began in 2001 as a single, rather specialized mission of delivering small, precision-guided munitions against terrorists. It has grown into a situation where virtually all combat missions for manned aircraft are at least under consideration for being performed by UAVs.

The initial antiterrorist mission created a new class of military aircraft that could loiter for long periods of time over suspected locations of possible targets in order to take advantage of fleeting opportunities to engage individual targets. This was an essential requirement of the mission that could only be met by UAVs. This mission has a number of significant distinctions from nonlethal missions in the areas of AV design and raises new issues related to the level of human control over the actions of the AV.

Of course, all missiles are "unmanned aerial vehicles," but we consider systems that are designed to deliver an internal warhead to a target and destroy themselves while destroying that target as weapons and distinguish them from vehicles that are intended to be recoverable and reused for many flights. Although there are areas in common between flying weapons and reusable unmanned aircraft, there are also many areas in which the design tradeoffs for weapons differ from those for the aircraft.

As of this writing, the primary form of armed UAV is an unmanned platform, such as the Predator, carrying precision-guided munitions and the associated target acquisition and fire-control systems such as imaging sensors and laser designators. This is evolving to include delivery of small guided bombs and other forms of dispensed munitions. These systems can be considered unmanned ground attack aircraft. The future seems to hold unmanned fighters and bombers, either as supplements to manned aircraft or as substitutes.

There is an ambiguous class of military missions in which the UAV does not carry or launch any weapons, but provides the guidance that allows the weapons to hit a target. This is accomplished using laser designators on the AV that "point out" the target to a laser-guided weapon launched from a manned aircraft or delivered by artillery. As we have seen, this mission was a primary driver for the resurgence of interest in UAVs in the US Army in the late 1970s. It remains a major mission for many of the smaller tactical UAVs in use by the military.

5.1.2   Shared missions

Among the core missions of UAVs for both military and civilian use are reconnaissance (search) and surveillance, which often are combined, but are different is important ways, as seen in the following definitions.

- *Reconnaissance:* The activity to obtain by visual or other detection methods information about what is present or happening at some point or in some area.
- *Surveillance:* The systematic observation of aerospace, surface or subsurface areas, places, persons or things by visual, aural, electronic, photographic, or other means.

Thus, surveillance implies long endurance and, for the military or law enforcement, somewhat stealthy operations that will allow the UAV to remain overhead for long periods of time. Because of the interrelationship between surveillance and reconnaissance, the same assets are usually used to accomplish both missions.

5.1.3   Land- and ship-based missions

There are both land- and ship-based missions in both the military and civilian worlds. A land-based operational base may be fixed or may need to be transportable. If it is transportable, the level of mobility may vary from being able to be carried in a backpack to something that can be packed up and shipped in large trucks or on a train and then reassembled at a new site over a period of days or even of weeks. Each of these levels affects the trade-offs between various approaches to AV size, launch and recovery methods, and almost every other part of the system design.

Ship-based operations almost always add upper limits to AV size. If the ship is an aircraft carrier, the size restrictions are not too limiting, but may include a requirement to be able to remove or fold the long, thin wings that are typical of long-endurance aircraft.

6   CONCLUSION

In conclusion, UAVs are divided into two major classes: military and civilian. These classes overlap conceptually, because there are core missions that are common to both the military and civilian world. However, almost all specific UAVs are intended for one or the other market because of the different environments in which they are required to operate.

The civilian UAV market is further divided between commercial and government applications that required relatively large and expensive UAVs and a rapidly growing market for small and inexpensive UAVs that have been applied to a rapidly growing list of applications and are generally referred to by the public as "drones."

Within these subsets, UAVs may be classified by size, by mission, or by a combination of size and mission.

The missions of UAVs, which were dominated by military applications until recently, are now rapidly expanding in both the military and civilian worlds. In the military, UAVs are moving into almost every area of military aviation. In the civilian world, the arrival of the commercial/consumer "drone" is creating altogether new applications and missions for aerial vehicles.

RELATED ARTICLE

Chapter 38

REFERENCES

Fahlstrom, P.G. and Gleason, T.J. (2012) *Introduction to UAV Systems*, John Wiley & Sons, Ltd., Chichester.

Kemp, I. (ed.) (2011) Unmanned vehicles, *The Concise Global Industry Guide*, Issue 19, The Shephard Press, Slough, Berkshire, UK.

Chapter 7
Launch of UAVs

Thomas J. Gleason[1] and Paul G. Fahlstrom[2]
[1]*Gleason Research Associates, Inc., Columbia, MD, USA*
[2]*United States Army Materiel Command, Huntsville, AL, USA*

ACRONYMS

UAV	unmanned aerial vehicle
AV	air vehicle
KE	kinetic energy
RATO	rocket-assisted takeoff
JATO	jet-assisted takeoff
HP	hydraulic/pneumatic
ESCO	Engineered Arresting Systems Corporation
VTOL	vertical takeoff and landing

1 OVERVIEW

Here, we primarily address the launch of small-to-medium unmanned aerial vehicles (UAVs), particularly using concepts that avoid the need for a runway, road, or large, open area. If a runway or road is always available for a particular UAV, then

Unmanned Aircraft Systems. Edited by Ella Atkins, Aníbal Ollero, Antonios Tsourdos, Richard Blockley and Wei Shyy.

the simplest and least expensive launch mode is to takeoff using wheeled landing gear. There still might be reasons for using one of the other techniques discussed here, but they would be based on some system-specific requirements.

Launch without a takeoff run often is referred to as a "zero-length launch." In fact, it is generally necessary to accelerate any fixed-wing air vehicle (AV) to some minimum controllable airspeed before releasing it from the launcher, and that cannot be done in zero distance. However, the use of catapults or rocket boosters can achieve a launch distance that is of the order of from one to a few times the length of the AV.

This chapter is based on material published in *Introduction to UAV Systems* (Fahlstrom and Gleason, 2012).

2 BASIC CONSIDERATIONS

The basic parameters to be considered for launch are straightforward and relate to physics. The formulae, which are interrelated, are given in the following:

Linear motion equation:

$$v^2 = 2aSn \tag{1}$$

Kinetic energy (KE) equation:

$$KE = \frac{1}{2}mv^2 \tag{2}$$

Equivalence of work and KE:

$$KE = FS \tag{3}$$

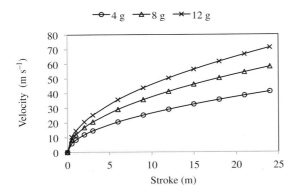

Figure 1. Velocity versus stroke.

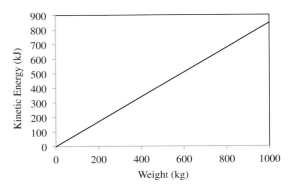

Figure 2. Kinetic energy versus velocity.

where v is the velocity, a is the acceleration (or deceleration), n is the efficiency factor, m is the total mass to be accelerated, F is the force, and S is the distance over which the force must be applied (launch distance, also called "stroke").

All real systems will have some variation in the acceleration during the stroke. The efficiency factor (n) is an empirical adjustment factor that takes this variation into account. If the acceleration were to be constant, of course, the value of n would be 1 and Equation 1 reduces to the familiar $v^2 = 2aS$.

These relationships are shown in Figure 1 as a plot of velocity versus stroke for three selected accelerations, expressed in units of g. For the sake of discussion, an $n = 0.9$ is included in Figure 1. One can see from Equation 1 that, for a given velocity, the loss of efficiency requires a longer stroke to launch the vehicle at the selected value of acceleration.

For ease of calculation, it is assumed that the UAV, we are interested in, has an all-up weight of 1000 lb (453.6 kg). For the current discussion, we will merely consider it "the weight" of the vehicle. Actually, the performance of a launcher must consider the "tare weight," not just the AV weight. The tare weight includes the weight of the AV and of all moving parts connected to the shuttle that carries the air vehicle up the launch rail. It is also assumed that the vehicle requires a launch velocity of 80 knots (41.12 m s^{-1}), true airspeed, there is no wind, and that the vehicle and its component parts can withstand an 8-g longitudinal acceleration or deceleration.

Referring to Figure 1, one can see that the launch stroke required for the assumed system with an acceleration of 8 g and an efficiency of 0.9 is about 12 m. Figure 2 is a plot of kinetic energy that must be provided to launch a vehicle of a given weight to an 80 knot flight speed. From this plot, we see that to launch the 1000 lb (453.6 kg) vehicle requires expending approximately 400 kJ of energy. Velocity is the key

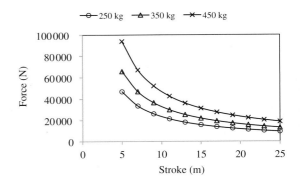

Figure 3. Force versus stroke for various vehicle weights.

factor in these calculations, since it is the velocity-squared factor that dominates the energy to be provided.

Once the required energy level is determined, and knowing the stroke necessary to limit acceleration to the selected value, it is easy to calculate the force that must be applied to the vehicle over the length of stroke to reach the launch velocity within the g limitation. Figure 3 is a plot of force versus stroke for the kinetic energy values for three masses. From this plot, we see that for a launch stroke of approximately 15 m and a 450 kg mass, the force required is about 30 000 N (about 6750 lb force). It is important to note that this theoretical force must be applied over the entire launch stroke; if not, then the actual stroke must be adjusted accordingly.

While we have accounted for some loss of efficiency in calculating the required stroke in Equation 1, we must now look at the force/stroke relationship for the particular power source to be used.

Remembering that the kinetic energy is the area under the force–stroke curve, Figure 4 shows the performance that results from the use of an elastic cord to drive the launcher. Typical of this type would be a bungee cord launcher. Force

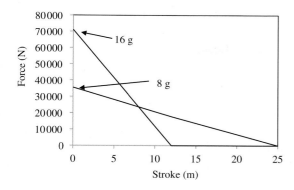

Figure 4. Force versus stroke for an elastic cord.

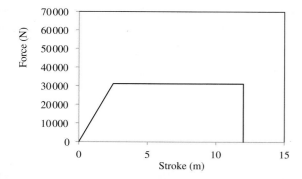

Figure 5. Force versus stroke for a pneumatic–hydraulic launcher.

and acceleration are high at the beginning of the stroke and decay as stroke proceeds. Obviously, the most desirable device would be the one that provides a constant force over the necessary stroke distance.

Practically speaking, it is possible to obtain constant (or near-constant) force over the stroke. However, to reach the desired force level quickly and efficiently, a rapid rate of change of applied force is necessary and frequently results in force overshooting the desired level. The overshoot, in turn, can lead to excessive "g" forces at the beginning of the stroke. To avoid an overshoot problem, the launcher design needs to allow time for a controllable buildup of forces that can be leveled out without significant overshoot. This requires a somewhat longer stroke in order to provide the required level of kinetic energy. Figure 5 shows a typical force–stroke plot for a pneumatic–hydraulic launcher with a tailored force that builds up to a desired level and then is constant for the remainder of the stroke.

As previously stated, the foregoing discussion is for basic theoretical considerations. These principles apply regardless of the means of launch. Of course, there are other practical considerations and they vary depending on the mechanical

equipment used. For example, to the casual observer, rocket launch appears to be "zero length," but in reality the rocket must impart the required force (as derived from the formulae presented) over the distance calculated, so although the mechanical part of the launch equipment may be "zero length," the UAV must ride the rocket thrust vector over the calculated distance.

3   UAV LAUNCH METHODS FOR FIXED-WING VEHICLES

There are many ways in which a UAV can be launched. Some are quite simple in concept, while others are very complex. A number of launch concepts are derived from full-scale aircraft experience, while others are peculiar to small unmanned vehicles.

Perhaps the simplest method is the "hand launch," derived from model airplane usage. This method is practical only for comparatively lightweight vehicles (under about 10 lb) having low wing loading and adequate power.

Also simple, but typically requiring a prepared surface, is normal wheeled takeoff.

Some UAVs, particularly target drones, are air-launched from fixed-wing aircraft. These UAVs typically have relatively high stall speeds and are powered by turbojet engines. Such vehicles frequently are also capable of being surface launched using rocket-assisted takeoff (RATO). The RATO launch method will be discussed in greater detail later, but generally requires that the launch force be applied over a significant distance in order to have the vehicle reach flying speed. For this application, the line of action of the propulsion force must be carefully aligned to ensure that no moments are applied to the vehicle, which might create control problems.

If one has available a smooth surface, even if too rough for a takeoff on the small wheels of a small UAV, truck launch is a low-cost practical approach. The larger wheels and suspension of even a small truck can allow driving it at takeoff speed despite gravel, washboard surfaces, or high grass that would make it impossible for a UAV smaller than a light aircraft to use the surface as a runway. The AV is held in a cradle that places it above the cab of the truck with its nose high to create the angle of attack for maximum lift. Once the airspeed is sufficient, the AV is released and lifts directly upward off its cradle into free flight. Driving a truck at over $27 \, \text{m s}^{-1}$ (60 mph) with a UAV and its supporting structure mounted above the roof might be exciting! Such an arrangement has been used and is illustrated in Figure 6.

One novel approach to UAV launch is a rotary system used with small target drones during World War II and by

Figure 6. Truck launch.

Flight Refueling Ltd. in the United Kingdom for their Falconet target drone. In this system, the UAV is cradled on a dolly that is tethered to a post centrally located within a circular track or runway. The engine is started with the UAV on the dolly. The dolly is released and circles the track picking up speed until launch velocity is reached. The UAV is then released from the dolly and flies off tangentially to the circle. While this system requires some interesting control inputs at the instant of release, it does work and is relatively easy to operate. The system does, however, require significant real estate and is not mobile.

Another launcher type that has been proposed in the past is the "flywheel catapult." This launcher uses the stored energy in a spinning flywheel to drive a cable system attached to a shuttle holding the UAV. The idea is that the flywheel can be brought up to speed slowly and when "launch" is called for, the flywheel engages a clutch attached to the power train (cables, etc.) and transfers its rotational energy to the UAV. Variations of this launcher type have used mechanical clutches and electromechanical clutches. While "flywheel" launchers have successfully been built for test and prototyping purposes, most have launched UAVs weighing no more than several hundred pounds and at comparatively low launch speeds. The problem with this concept is the operation of the clutch. Most clutch designs are not robust enough to withstand the rapid onset of energy transfer.

Large UAVs that use runways for conventional takeoffs and landings present some autopilot and control challenges, but otherwise require no special launch and recovery subsystems. The remainder of this discussion concentrates on smaller UAVs using less conventional approaches to launch and recover.

Many UAV launch systems have a requirement to be mobile, which means being mounted on a suitable truck or trailer. Generally, these systems can be categorized as either "rail" launchers or "zero-length" launchers. The material that follows addresses each type separately.

3.1   Rail launchers

A rail launcher is basically one in which the UAV is held captive to a guide rail or rails as it is accelerated to launch velocity. Although a rail launcher could use rocket power, some other propulsion force is usually utilized.

Many different designs of rail launchers have been used or proposed for use with UAVs. Bungee-powered launchers have been used for test operations, but this power source is limited to very lightweight vehicles. A typical example of bungee launcher is the one used to launch the Raven UAV in the United Kingdom (see Chapter 6 for descriptions of some of the UAS used as examples). For small AVs, the bungee launcher can be configured much like a large slingshot without a rail and may be hand held.

Most rail launchers used to launch UAVs in the 500–1000 lb weight class use some variation of pneumatic or hydraulic–pneumatic powered units.

3.2   Pneumatic launchers

Pneumatic launchers are those that rely solely on compressed gas or air to provide the force necessary to accelerate the UAV to flying velocity.

These launchers use compressed-air accumulators that are charged by a portable air compressor. When a valve is opened, the pressurized air in the accumulators is released into a cylinder that runs along the launch rail and pushes a piston down that cylinder. The piston is connected to an AV cradle that rides on the launch rail, sometimes by a system of cables and pulleys that can increase the force available at the expense of the stroke or increase the stroke at the expense of a smaller force. The cradle is initially locked in place by a latch. The unlatching process may use a cam to reduce the rate of onset of acceleration. At the end of the power stroke, the cradle is stopped using some type of shock absorbers and the AV flies off the cradle at sufficient airspeed to maintain flight.

Pneumatic launchers are satisfactory for relatively lightweight UAV launches, but operating at low ambient temperature can be troublesome. Using ambient air at low temperatures, it has been found that pollutants and moisture combined in the compressed air and adversely affect operation. The addition of conditioning equipment to solve the problem presents weight and volume problems.

Another novel pneumatic launcher concept is one using a "zipper" sealing free piston operating in a split cylinder. The cradle or other device, which imparts the driving force to the UAV, is connected to the free running piston. As the piston moves along the length of the cylinder, the sealing strap is displaced and reemplaced. The compressed air is held in a tank until "launch" is signaled. At that time, the compressed air is fed into the launch cylinder through a valve that modulates the onset of pressure to reduce initial shock loads and, in some cases, the valve regulates pressure throughout the stroke in an attempt to achieve constant acceleration. At the end of the power stroke, the piston either impacts a shock absorber or pressure that builds up ahead of the piston brings the piston to a halt.

This type of launcher would have the same drawbacks as exhibited by other pneumatic launchers described above. In one case, an attempt to use a "zipper seal" launcher was made after it had sat in rain and drizzle for several days. Although the prescribed prepressure was set, the launch velocity achieved was only about two-thirds of that predicted. After several additional attempts, the prescribed velocity was achieved. An investigation determined that moisture was sealing the tape ahead of the piston creating a back pressure, thus retarding the forward acceleration of the piston. Another possible problem with this type of launcher could be the proper mating of cylinder sections in the event the launcher needed to be folded for transportation.

A third type of pneumatic launcher is one that has been used with the Israeli/AAI Pioneer UAV. In this design, the compressed air, stored as before in a large tank, is discharged into an air motor, which in turn drives a tape spool. This spool, when powered, winds a nylon tape secured to the UAV with a mechanism that releases the end of the tape as the UAV passes over the end of the launch rails. This launcher has no shuttle; rather the UAV is equipped with slippers on the ends of small fins protruding from the fuselage, which ride in slots situated longitudinally along the launch rails. The air storage tank on this launcher contains enough volume to power several launches without refilling or repressurizing. Large tank volume and the effect of increased effective drum diameter as the tape is wrapped on the drum during launch results in a near constant launch acceleration rate, and hence relatively high efficiency.

So far as is known, this launcher was limited to use with UAVs weighing less than 500 lb, with launch velocities of less than 75 knots, and with sustained acceleration rates of 4 g or less. In any event, the launch stroke of units provided to the US Marines has a length of about 70 ft. Based on previous experience with purely pneumatic launchers, the authors would expect that while this launcher appears to operate satisfactorily in a temperate environment, problems could be encountered at low temperatures unless precompression dryers and/or coolers are employed to condition and dry the air. The adaptability of this type of launcher for higher weight UAVs and higher launch velocities is unknown, but the power requirements for these conditions would involve a significant increase in air motor size and the volume of air required.

3.3 Hydraulic/pneumatic launchers

The hydraulic pneumatic (HP) launcher concept has been successfully employed in a number of UAV programs.

Air vehicles weighing up to at least 555 kg (1225 lb) have been launched at speeds of up to 44 m s^{-1} (85 knots) with this type of launcher. Both full-sized and lightweight variants have been built by All American Engineering Company (now Engineered Arresting Systems Corporation (ESCO), a subsidiary of Zodiac Aerospace).

The basic HP launcher concept utilizes compressed gaseous nitrogen as the power source for launch. The nitrogen is contained within gas/oil accumulators. The oil side of the accumulator is piped to a launch cylinder, the piston rod of which is connected to the moving crosshead of a cable reeving system. The cable (in most cases a dual-redundant system) is routed over the end of the launch rail and back to the launch shuttle. The launch shuttle is held in place by a hydraulically actuated release system. After the UAV is placed upon the launch shuttle, the system is pressurized by pumping hydraulic oil into the oil side of the accumulators, thus pretensioning the cable reeving system and applying force to the UAV shuttle. When the pressure monitoring system indicates that the proper launch pressure has been achieved, the release mechanism is actuated to start the launch sequence. The release mechanism has a programmed actuation cycle that is designed to lessen the rate of onset of acceleration. After release, the shuttle and UAV are accelerated up the launch rail at an essentially constant rate of acceleration.

At the end of the power stroke, the shuttle engages a nylon arresting tape, which is connected to a rotary hydraulic brake, the shuttle is stopped and the UAV flies off. On some launchers, an optional readout is provided for launcher end speed. However, variations in end velocity rarely are more than ± 1 knot from the predicted value. Unlike purely pneumatic systems, the nitrogen precharge is retained, and except for rare leakage, seldom needs replenishment. This allows the use of dry, conditioned air or dry nitrogen in the charge and avoids the problems of using ambient air. The launch energy is provided by the pumps that transfer hydraulic fluid between the accumulators. This type of launcher has very low visual, aural, and thermal signature.

Figure 7 is a photograph of an HP-2002 launcher currently produced by ESCO. The HP-2002 is a light HP launcher rated to launch a 68 kg (150 lb) UAV at 35 m s^{-1} (68 knots) or a 113 kg (250 lb) UAV at 31 m s^{-1} (60 knots). It has a total weight, including a trailer, of 1360 kg (3000 lb). Other ESCO HP launchers can be used with AVs up to about 555 kg (1225 lb).

Figure 7. HP 2002 launcher. (Reproduced with permission from Engineering Arresting Systems Corporation.)

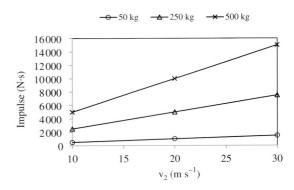

Figure 8. Energy require-ments for zero-length launcher.

3.4 Zero-length RATO launch of UAVs

A "zero-length" launcher does not use a rail. The AV rises directly from a holding fixture and is in free flight as soon as it starts moving.

One of the most common and most successful launch methods is RATO. Rocket assist dates back to the World War II era when it was used to shorten the takeoff roll required for large military aircraft; in those days, they were called JATO, for Jet Assisted Take-Off units, a term still occasionally used today. RATO launch has been routinely used for launching target drones for many years, and has been utilized for some of the USAF UAVs, such as Pave Tiger and Seek Spinner, for shipboard and ground launch of the US Navy Pioneer and for the US Marine Corps BQM-147 UAVs.

The following discussion presents several considerations pertinent to the design of RATO units for UAV applications. The information presented should only be used for preliminary approximations since many factors unique to the particular application and/or AV may significantly influence the RATO unit final design.

3.4.1 Energy (impulse) required

A RATO unit designer needs to know the mass of the AV to be accelerated and the desired AV velocity at RATO unit burnout. These two items determine the energy that must be imparted to the vehicle and will subsequently determine the size of the RATO unit. The required energy, or impulse, is calculated from the impulse momentum equation as follows:

$$I = m(v_2 - v_1) \qquad (4)$$

If the mass (m) is entered as kg and the velocity is expressed in $\mathrm{m\,s^{-1}}$, then the calculated impulse will be in

the units of N s. For a stationary launcher $v_1 = 0$. The above relationship can also be expressed graphically as shown in Figure 8. Note that this equation and figure assume that the mass of the RATO unit itself is small compared to the mass of the UAV, since the RATO unit must be accelerated along with the UAV. The RATO unit mass initially includes the mass of the motor grain, which burns during the acceleration. As a simple approximation to take this into account, one might add the mass of the RATO unit to that of the UAV and use the sum as the value of m in the equation.

For example, the Exdrone UAV had a mass of about 40 kg (neglecting the mass of the RATO unit) so for a velocity at RATO burn out of about $15\,\mathrm{m\,s^{-1}}$, it would lie slightly below the line for 50 kg at that value of v_2. This results in a required impulse of about 630 N s. The Pioneer is significantly heavier, with a mass of about 175 kg with a full set of sensors and for a velocity at burn out of about $40\,\mathrm{m\,s^{-1}}$ would require an impulse of about 7000 N s.

3.4.2 Propellant weight required

The amount of energy, or specific impulse, that a propellant can deliver depends primarily upon the type of propellant used and upon the efficiency of the rocket design.

Propellants range from high energy cast composites, such as polybutadiene binders with perchlorate oxidizers, to lower energy slow-burning ammonium nitrates, to extruded single- or double-base formulations. The propellant type will be selected by the designer depending upon the relative importance of such things as environmental conditions, age life requirements, smoke generation, burning rate, specific energy, processability, insensitivity to accidental ignition by artillery fragments and small arms, and cost. The "specific impulse" of a propellant is a measure of the amount of impulse that can be produced by burning a unit mass of the propellant. The units are impulse divided by weight, which comes out to lb(force) s $\mathrm{lb^{-1}}$ or $\mathrm{N\,s\,kg^{-1}}$. Specific impulse commonly is specified in English units. In general, propellants will deliver a specific impulse in the range of 180–240 lb s lb^{-1}.

$$W_p = \frac{I}{I_s P} \qquad (5)$$

Rocket design parameters that have an effect on motor efficiency include the operating pressure, the nozzle design, and to a lesser degree, the plenum volume upstream of the rocket nozzle. Simply dividing the required total impulse by the delivered specific impulse will provide an estimate of the total propellant weight required.

To estimate the overall weight of the RATO unit, one can use the approximation that the RATO unit will weigh roughly twice the propellant weight.

3.4.3 Thrust, burning time, and acceleration

A rocket's energy is delivered as the product of a force or thrust (F) over a finite time interval (from time t_1 to time t_2).

$$I = F(t_2 - t_1) \qquad (6)$$

Acceleration produced on an air vehicle with mass m (or weight w) can be expressed as

$$a = \frac{F}{m} = F\frac{g}{w} \qquad (7)$$

The maximum acceleration that a vehicle (and on-board subsystems) can withstand is very important and is usually dictated by the structural design of the airframe. Knowing the maximum acceleration and the vehicle weight, the thrust and burn time can be calculated using the above equations.

3.4.4 RATO Configuration

RATO units can be designed to interface with the AV in many different ways, depending on the design of the AV and location of the structural hard points. In some cases, more than one RATO unit is utilized. When a single RATO unit is used, it may be located behind the AV along its longitudinal axis, or it may be located below the vehicle fuselage. Where and how the RATO unit is mounted determines its size, its mounting attachment features and whether its nozzle is axial or canted. In any event, the RATO system is normally designed so that the resultant rocket thrust line passes through the AV center-of-gravity at the time of launch.

As mentioned earlier for "zero-length" RATO launch, the thrust direction must have an upward tilt to support the AV until it is moving fast enough to develop lift.

3.4.5 Ignition systems

RATO ignition systems can enter the rocket pressure vessel either through the head end or through the nozzle end. Either method is acceptable and can utilize initiators that can be shipped and stored separately, and installed in the field just prior to launch.

Several types of initiation have been used. These include a percussion primer actuated by an electrical solenoid for primary initiation and an electric squib built into a remotely actuated rotating safe/arm device. The Pioneer RATO unit used a dual-bridgewire, filter pin-initiator and the Exdrone RATO unit used a percussion-primer-actuated, shock-tube ignition system. Each ignition system was selected to comply with unique system and user requirements. As with munitions, the RATO ignition system will have to meet strict safety requirements to avoid unintentional ignitions.

3.4.6 Expended RATO separation

The flight performance of most AVs is very weight dependant. It is therefore, undesirable to carry along expended RATO unit launch hardware for the entire air vehicle flight. Consequently, expended hardware normally is separated from the AV by aerodynamic, mechanical, or ballistic means. Selection of the separation system will depend on how rapidly and in what direction the expended hardware must be jettisoned. Care must be exercised, since the falling RATO unit canister becomes an overhead safety hazard for personnel and equipment near the launch site.

3.4.7 Other launch equipment

Other launch equipment required for RATO launch includes a launch stand and usually an AV holdback/release system.

The launch stand positions the AV wings level and nose elevated at the desired launch angle. The angle of launch is unique to each specific AV. Normally, it is desirable to minimize the vehicle angle of attack during RATO unit burn. The launch stand may provide other features such as deck-tie-down provisions and RATO unit exhaust deflectors, and may also be collapsible or foldable for ease of transport.

The holdback/release mechanism provides a method of restraining the AV against wind gusts and the engine run-up thrust prior to launch; it also provides automatic release of the AV at the time of RATO unit ignition. Systems that have been used include a shear line release for the Pioneer and a ballistic cutter release for the Exdrone UAV.

4 VERTICAL TAKEOFF AND LANDING UAV LAUNCH

Vertical takeoff and landing (VTOL) UAVs, by virtue of their design, need little in the way of launch equipment, especially for ground-based operations. However, logic would dictate that for military operations mobility considerations would require that the VTOL UAV should be operated from a vehicle of some sort. This vehicle would contain devices to secure the UAV during ground transport, and would probably also contain checkout, start-up, and servicing equipment (such as service lifts).

The majority of the new class of commercial/consumer UAVs, commonly called "drones," are VTOL aircraft. They often are small enough to takeoff and land in any small, reasonably level patch of clear ground or pavement, requiring no supporting equipment of any sort. This characteristic is one of the key things that make these small UAVs easily useable by operators who are largely self-trained and view the UAV as a convenient tool to achieve some goal or as an interesting hobby.

5 AIR LAUNCH OF UAVS

As mentioned earlier, air launch of UAVs has been used ever since the first reconnaissance drones. This remains a proven way to launch fixed-wing UAVs.

At the time that this is being written, much of the known interest in air-launching of UAVs is in the military and many of the unmanned aircraft to be air-launched either are expendable aircraft that contain a warhead and are intended to blow themselves up or are semiexpendable reconnaissance and targeting systems that may be recovered in some way, but are considered successful if they perform one mission and then are lost. The former are actually weapons just like any guided missile, not properly considered UAVs. The latter are UAVs. At present, they often utilize specialized designs, such as wings that fold down along a cylindrical fuselage so that they can be ejected into the airstream in a tubular configuration with a drag parachute deployed to slow them down and stabilize them before the wings fold out and the begin controlled flight.

There is some talk of flying aircraft carriers, manned or unmanned, that could launch large numbers of small UAVs and refuel them in flight using the same sort of equipment that presently is used to refuel fighter aircraft (a trailing fuel line with a conical docking orifice into which the aircraft to be refueled inserts a refueling probe). It has been suggested that for relatively small fixed-wing UAVs the refueling system could capture the UAV and reel it in for an aerial recovery.

Launch of a rotary-wing UAVs from a fixed-wing aircraft presents a different set of problems. Unlike the launch of a fixed-wing UAV, where additional airspeed adds to stability, excess airspeed may lead to severe instability and tumbling of a rotary-wing UAV that is simply released into the airstream. This might be solved by using drag parachutes to slow the UAV down and stabilize it before it attempts to fly on its own.

Launch of a rotary-wing UAV from a larger, rotary-wing aircraft, manned or unmanned, would appear to be fairly simple. However, it would be necessary to ensure that the smaller UAV could make it out of the rotor downwash from the larger launch aircraft. This might be accomplished by lowering the UAV to be launched on a tether and maintaining enough forward airspeed to cause the tether to trail out of the downwash. Something similar might be done with a fixed-wing UAV that could be designed to go into a dive upon release until it achieved the necessary airspeed to level off and fly under its own power.

Based on recent experience, the whole area of air-launched UAVs seems likely to evolve fairly quickly, particularly if commercial applications using the inexpensive mass-produced small UAVs were to begin to appear.

6 CONCLUSIONS

A UAV developer has a number of launch options. He or she must evaluates the advantages and disadvantages of the various launch concepts to determine which is best for a particular AV and set of mission requirements and cost constraints. Above all, the designer should select the launch approach early in the design process, so that the incidence of problems associated with launch can be eliminated, or at least much reduced, by producing a total system design that integrates launch considerations into all aspects of the design.

RELATED ARTICLE

Chapter 6

ACKNOWLEDGMENT

We would like to thank Mr. Robert Veazy, who provided initial drafts of some of the material on launch concepts that are included in this chapter.

REFERENCE

Fahlstrom, P.G. and Gleason, T.J. (2012) *Introduction to UAV Systems*, John Wiley & Sons, Ltd, Chichester.

Chapter 8
Recovery of UAVs

Thomas J. Gleason[1] and Paul G. Fahlstrom[2]

[1]*Gleason Research Associates, Inc., Columbia, MD, USA*
[2]*United States Army Materiel Command, Huntsville, AL, USA*

ACRONYMS

AV	air vehicle
CARS	Common Automatic Recovery System
MARS	mid-air recovery system
SPARS	Ship Pioneer Arresting System
UAV	unmanned aerial vehicle
VTOL	vertical takeoff and landing

1 OVERVIEW

The simplest option for recovery is to land the unmanned aerial vehicle (UAV) just as one would land a manned aircraft

Unmanned Aircraft Systems. Edited by Ella Atkins, Aníbal Ollero, Antonios Tsourdos, Richard Blockley and Wei Shyy.
© 2016 John Wiley & Sons, Ltd. ISBN: 978-1-118-86645-0.

on a road, runway, smooth field, or carrier deck. For medium-to-large air vehicles (AVs), there are few other options, as the use of nets or parachutes becomes impractical. However, wheeled landing also has been used with many small-to-medium AVs because it often is the least expensive option.

When there is a requirement for "zero-length" recovery, there are a number of options available for small AVs. The most commonly used approaches are identified and discussed in this chapter. Undoubtedly, there are other schemes unique to specific UAVs and special mission requirements that will not be addressed.

This chapter is based on material published in *Introduction to UAV Systems* (Fahlstrom and Gleason, 2012).

2 CONVENTIONAL LANDINGS

The most obvious fixed-wing UAV recovery option parallels that of full-sized aircraft, that is, runway landing. For all but the smallest AVs, to utilize this option the UAV must be equipped with landing gear (wheels) and its control system must be able to perform the "flare" maneuver typical of fixed-wing aircraft. Experience has shown that directional control during rollout is extremely important, as is the requirement to have some sort of braking system.

One frequently used adaptation to the runway landing technique is to equip the UAV with a tail hook and position arresting gear on the runway. In this way, the need for directional control during rollout and for onboard brakes is minimized. This approach parallels carrier landing techniques.

There are two types of arresting-gear energy absorbers generally used. One is a friction brake that has a drum, around which is wound cable or tape that is connected in turn to the deck pendant (the cable or line which the UAVs tail hook engages is called a deck pendant even when used on a land runway). The second type is a rotary hydraulic brake, a simple water turbine with the rotor attached to a drum, around which is wound a nylon tape. As with the friction brake, the tape is attached in turn to the deck pendant. There is a significant difference between these two braking systems. With a friction brake, the retarding force is usually preset and fixed and the run-out (the distance it takes to arrest the UAV) varies depending on UAV weight and landing speed. Rotary hydraulic brakes, however, are considered "constant run-out" devices, and the UAV will always end up at approximately the same point on the runway even if the weight and landing speed vary. This statement only is true, of course, within limits. A rotary hydraulic brake system is configured for a "design point" of UAV weight and landing speed and variations of, say, 10–20% around that design point are readily accepted.

Skid landing has also been used successfully, notably by SkyEye (see Chapter 6 for descriptions of UAS used as examples), and does not require a paved surface. Any reasonably flat surface without large obstacles can be used. SkyEye uses a single skid equipped with a shock absorber and tracks to keep the UAV running straight. The engine is cut on touchdown and friction between the skid and landing surface brings the UAV to a halt. The use of the shock absorber eliminates the need for the flare maneuver; the UAV is merely set up with a low rate of descent and flies onto the landing surface.

A variation of the classic arresting gear recovery system is to attach a net to the purchase elements of the energy absorbers in lieu of the deck pendant. The net must be designed to envelop the UAV and distribute the retarding loads evenly on the airframe.

Very small AVs may be simply flown into the ground at a shallow angle and allowed to skid to a halt.

3 VERTICAL NET SYSTEMS

A logical outgrowth of runway arresting systems, both hook/pendant and net types, is the vertical net concept. In its basic form, the net is suspended between two poles with the net extremities attached to purchase lines, which are attached in turn to energy absorbers. The net usually also is suspended by a structure or lines to hold it above the ground and thus suspends the UAV within the net at the end of run-out.

Use of a net generally precludes the use of tractor propellers located at the front of the AV, as they are likely either to damage the net, be damaged by the net, or lead to damage to the engines from forces applied to the propeller and transmitted to the shaft and bearings of the engine. Depending on the configuration of the AV, even a pusher propeller may need to be shrouded to avoid these problems.

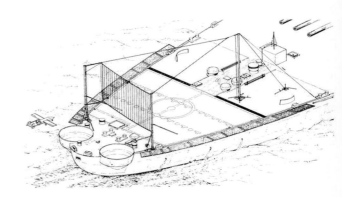

Figure 1. Pioneer installation on USS Iowa. (Reproduced with permission from Engineering Arresting Systems Corporation.)

The three-pole net recovery system used with Pioneer on the battleship Iowa and with the Lockheed Altair UAV, shown in Figure 1, predates most other vertical net systems. In the mid-1970s, the Teledyne Ryan STARS UAV system successfully utilized a three-pole vertical net system. This approach used a "purse-string" arrangement with a reeving system to create a pocket that captures the UAV as run-out proceeds. In some variations, a bungee is introduced to snap up the bottom edge of the net to insure that the UAV is securely ensnared. Another variation using four poles was used on early Israeli UAV systems.

Key to the success of net systems is the design of the net itself. Early net systems used simple cargo nets or, in at least one small UAV system, a tennis net. The net must properly distribute retarding forces on the UAV, and various means have been devised to accomplish this requirement. On the Pioneer/Iowa system, early tests using a multiple element net derived from full-scale aircraft work proved to be too heavy, and was easily displaced by the wind over the deck produced by the forward motion of the ship over the water. The final net configuration had a small (4.57 m × 7.62 m) conventional net at the "sweet point" or aiming point for recovery and triangular members leading to the four corners of the large net. This configuration provided low wind drag and presented a sufficiently large target for manual recovery.

A very important aspect to net design is how accurately the UAV can be flown into the net. Aquila and Altair used a very accurate automatic final approach guidance system, and variations of only a foot or so from the center of the net were normally achieved. This net was only approximately 4.57 m high × 7.62 m wide. The net on the Iowa, by comparison, was some 7.62 m high × 14.33 m wide. These larger dimensions were needed because Pioneer has a larger wingspan and final approach guidance was manual (under radio control with a human operator).

4   PARACHUTE RECOVERY

Parachute terminal recovery systems have a long history of use with target drones and other UAVs. The use of a parachute, of course, requires that the UAV have sufficient weight and volume capacity to accommodate the packed bulk of the parachute. Numerous parachute configurations have been used and they are usually designed to have a relatively low rate of descent in order to decrease the possibility of damage due to the impact with the ground or water. Some UAVs employ inflatable impact bags or crushable structures to attenuate loads on impact. Typical of this approach is the Teledyne Ryan Model 124, which utilized impact bags, and the BAE Systems Phoenix, which inverts to land on a crushable upper structure.

While there are many parachute designs and a wide variation in performance, the cross parachute design, shown in Figure 2, has been particularly successful, having a reasonably low packed bulk and excellent stability during descent. It also has the advantage of providing low forces during deployment.

One disadvantage of the standard parachute configuration is the lack of directional control after deployment. The decent of the UAV after parachute deployment is subject to the vagaries of the wind, and in high surface wind conditions a ground release mechanism is a necessity to avoid damage due to the UAV being dragged along the ground. Parachute deployment at very low altitudes is frequently programmed in order to reduce drift distance.

Parachute descent into the water requires that the internal systems of the UAV be protected against water immersion or that decontamination facilities be readily available for use after immersion, particularly when descent is made into salt water.

Gliding parachutes have been used in recent years to overcome some of the problems with standard parachutes. The parafoil, or ram air inflated parachute, has been shown to have considerable advantages. Parafoils can be directionally controlled using differential riser control and they trade off forward speed for a low rate of descent. Virtually pinpoint accuracy of landing has been demonstrated with manual control or with a homing beacon and an onboard sensor and control system. Parafoils exhibit essentially constant speed characteristics, so if thrust is provided by the UAV engine, the UAV suspended on a parafoil can be caused to climb, maintain level flight, or, with thrust decreased, descend. Directional control under power is generally excellent.

For shipboard recovery of parafoil-borne UAVs, some additional ship-based aids are desirable. A promising approach for the parafoil-borne UAV is to use a haul-down system. The UAV drops a line to the ship deck and a winch is used to haul down the UAV. Control of tension by the winch is considered necessary so that the parafoil and haul-down line are not overstressed due to ship motion. The sequence of operation for this approach is pictured in Figures 3 and 4. Some means of securing the landed UAV is also considered necessary, particularly when shipboard operations at sea states of up to four are considered. (Additional discussion of shipboard operational limitations is contained in Sections 5 and 7.)

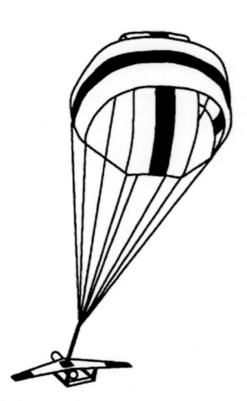

Figure 2. Cross parachute. (Reproduced with permission from Engineering Arresting Systems Corporation.)

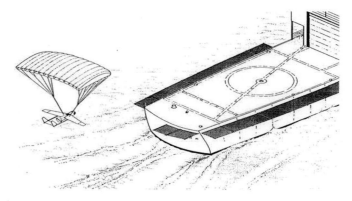

Figure 3. Parafoil recovery. (Reproduced with permission from Engineering Arresting Systems Corporation.)

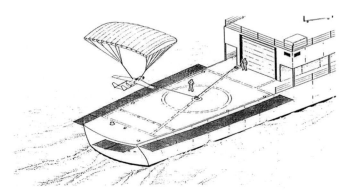

Figure 4. Parafoil recovery with winch. (Reproduced by permission of Engineering Arresting Systems Corporation.)

5   VTOL UAVS

A number of VTOL UAVs have been developed and fielded. These range from a pure helicopter to vectored-thrust devices and tilt-wing aircraft. Any of the VTOL UAVs have the distinct advantage of providing for a low relative velocity between the UAV and the ship deck or landing pad. This leads to a low relative energy transfer requirement.

For land operations, the landing pad is stationary, and if the UAV has good enough navigation capacity, it can be programmed to perform an independent automatic landing at the desired location at the center of the landing pad. At most, it might need an active altimeter to ensure a soft touchdown. For small VTOL UAVs, a manual or automated landing on any open area of grass or pavement is the normal mode of operation.

VTOL UAVs for shipboard operations require a more elaborate system to deal with the irregular motion of the landing pad in three dimensions. This includes a final approach guidance system, an airframe suitable for shipboard operation, and suitable capture equipment to insure that once landed on the deck, the UAV is secure and will not adversely affect other ship operations.

Figures 5 and 6 show two possible VTOL recovery techniques in a generic manner. Both involve a launch and recovery platform on the deck of the ship. The platform is on a track that allows it to be moved in and out of a hanger area without risk of sliding around on a heaving deck. There might be multiple platforms with sidings off the main track within the hanger to allow multiple AVs to be stored and retrieved from within the hanger.

Figure 5 shows a concept where the UAV drops a tether line and hook, which engages a snag line. A clever arrangement of blocks and lines (not shown in detail) allows the snag

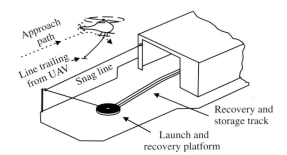

Figure 5. VTOL recovery by tether.

line to be reeled in by the recovery system to connect the tether line to a winch, similar to the winch recovery associated with parafoil recovery in Figure 4. The tether line then is winched in while tension is maintained on the line by the AV. When the AV is in contact with the platform, automatic securing devices lock it down and the AV engine is shut down.

Figure 6 shows a tetherless concept in which the AV is automatically landed on the platform using a closed-loop control system based on sensors located on the landing deck. The sensors provide precision information on the location of the AV relative to the recovery platform, which is used to implement a tightly closed loop with the autopilot of the AV and make a precision landing on the platform despite its motion in three dimensions on a heaving deck.

Figure 7 provides more detailed view of a recovery platform showing a simple concept for securing the UAV after landing. In this concept, hook-shaped clamps are located on tracks on either side of the center of the platform and when the air-vehicle landing skids contact the platform the clamps slide in over the skids and clamp down to secure them to the platform.

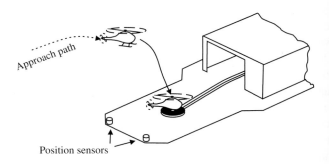

Figure 6. VTOL recovery by automatic landing.

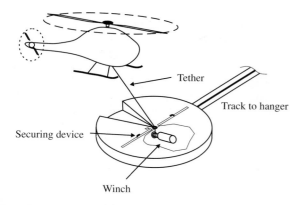

Figure 7. Launch and recovery platform.

6 MID-AIR RETRIEVAL

Use of a mid-air retrieval system (MARS) for UAV recovery provides the opportunity to perform the recovery operation away from the ship and then to fly the UAV as cargo down to the deck as is done with a normal helicopter operation. It would be possible to equip helicopters currently used for shipboard operation with a mission kit consisting of an energy-absorbing winch and a pole-operating auxiliary pod that would permit the helicopter to make the mid-air retrieval of a parachute-borne UAV. If the parachute utilized were a parafoil type, it would be possible to improve the performance of the recovery operation, since the helicopter pilot would not have to judge the vertical velocity of the parachute-borne UAV when affecting mid-air retrieval if the UAV continued to apply thrust after deployment of the parafoil so that it could continue in slow, powered flight. With power on the UAV would have a more gradual rate of descent than that of a conventional parachute, giving the helicopter pilot more time to make the recovery.

The sequence of operation of the classic MARS recovery is shown in Figures 8–10. For heavy (1100 lb) UAVs, the main parachute is jettisoned after engagement.

There is a wealth of experience in mid-air retrieval operations within the US Air Force and many thousands of successful recoveries of target drones, cruise missiles, and so on have been made. As an example, the reconnaissance drone program in Vietnam recorded a mission success rate of over 96%. This is one system that does not require a final approach guidance system. MARS requires, however, significant aircrew training, as well as a dedicated, especially configured aircraft. While MARS recoveries generally are only performed with good visibility and during daylight, some experimental night recoveries have been made by illuminating the parachute from below.

Figure 8. Mid-air retrieval. (Reproduced with permission from Engineering Arresting Systems Corporation.)

Figure 9. Mid-air recovery sequence – snagging. (Reproduced with permission from Engineering Arresting Systems Corporation.)

Figure 10. Mid-air recovery sequence – recovery. (Reproduced with permission from Engineering Arresting Systems Corporation.)

7 SHIPBOARD RECOVERY

Safety is the primary concern for UAV shipboard recovery. The type of UAV employed and the means of recovery must not endanger the ship or personnel aboard. This applies not only to the actual recovery action, but also to the recovery equipment installed on the ship, UAV handling, stowage, and all other aspects of the system. Other concerns are reliability and mission effectiveness. In most cases, space aboard ship is limited; therefore, a high degree of reliability is necessary so that large supply of spares is not required to keep the system operable. Mission effectiveness means that in addition to safely recovering the UAV, the system must be easily erected and operated by a minimum number of personnel, and must impose the minimum damage to the UAV being recovered. It must not require that the ship significantly deviate from its normal operating conditions in order to affect UAV recovery.

Unlike UAV operations over land, recovery at sea requires that the UAV must perform in spite of the motions of the ship in various sea states and in an environment that is very harsh. The relationship of sea state to ship motion is complex, and different classes of ship would have different reactions to the various sea state conditions. As an example, the pitch and roll rates of a battleship in a given sea state may be imperceptible compared to those of a frigate.

An example of the conditions under which UAV recovery operations might be conducted is contained in US Military Specification MIL-R-8511A, "Recovery Assist, Securing and Traversing System for LAMPS MK III Helicopter." This specification calls for maintenance of "all required functional characteristics" under stated conditions. The temperature range is −38 °C to +65 °C and exposure to relative humidity of 95% where the condensation takes the form of both water and frost. The ship motion conditions are

> When the ship is permanently trimmed down by the bow or stern as much as 5° from the normal horizontal plane, is permanently listed as much as 15° to either side of the vertical, is pitching 10° up or down from its normal horizontal plane, or is rolling up to 45° to either side of the vertical. Full system performance is not required when the ship roll exceeds 30°; however, exposure to ship rolling conditions up to 45° to either side of the vertical shall not cause loss of capability when rolling is reduced to 30° or less.

Complicating the effects of sea-state-induced ship motion on UAV recovery is the fact that most surface vessels create an air wake, or "burble" aft as a result of airflow past the various superstructures onboard. On some ships, this air burble can significantly affect UAV control as the UAV flies through the burble area on approach to the landing deck. Some data have been collected on this area of concern as an adjunct to insuring safe helicopter operation from the various ships. UAV designers should take these data into account when considering UAV shipboard recovery and plan to have adequate control when penetrating the burble area, or plan approaches to avoid the area.

Following the Gulf War, the US Navy retired their battleships, and lost the shipboard Pioneer UAV capability, which used the Ship Pioneer Arresting System (SPARS) vertical net recovery system. In order to maintain a shipboard UAV capability, the Navy had the SPARS equipment modified for installation on smaller ships designed to support amphibious operations. This system, designated SPARS III mounted the aft net support poles along the gunnels of the aft flight deck of the ship with the single forward pole mounted slightly off-center on the superstructure. The same basic geometry that was used for the battleship installation was maintained but the distance between the aft and forward poles was increased. This installation basically took up the entire flight deck, essentially preventing the deck from being available for helicopter operations. Following some initial problems with rigging, the system worked well. A very significant improvement in SPARS operations has been realized with the introduction of the Common Automatic Recovery System (CARS), which provides extremely accurate final approach guidance. With CARS available, smaller net sizes and overall smaller vertical net systems are feasible. This is important because one of the problems with the SPARS system relates to the area of the erected net and the drag the net imposes on the UAV during recovery.

This drag can affect the way the net envelopes the UAV and arrests it without dropping it out of the net.

Another improvement to the SPARS system has been proposed in which four poles are used to suspend the net system, thereby reducing the deck area used for the UAV operation. This modification, coupled with a simplified erection and lowering capability has the potential for making the ship deck more readily available for helicopter operations between UAV operations. To prevent the remote chance of a UAV undershooting the recovery net, a barrier net can be installed below the primary recovery net.

Vertical-net recovery systems can be adapted to virtually any fixed-wing UAV configuration that does not include tractor propellers. Coupled with CARS or some other automated control system, they provide an effective and reliable recovery system.

Finally, it must be recognized that ship captains are reluctant to have anything resembling a missile aimed at their ship. The UAV recovery method used must have demonstrated a high degree of reliability in its ability to recover the UAV without damage to the ship.

8 CONCLUSIONS

A UAV developer has a number of recovery options. He or she must evaluate the advantages and disadvantages of the various launch concepts to determine which is best for a particular AV and set of mission requirements and cost constraints.

The recovery options are not independent of the method by which the UAV is launched. If wheeled takeoff from a runway or road is used for launch then the same wheels that make that possible will make a wheeled landing possible. On the other hand, catapult launched AVs might need retractable

landing gear to avoid interference with the rail launch, and that would add cost and weight to the AV if a wheeled landing were desired. The launch and recovery trade-offs must be done at the same time as all the other basic trade-offs that determine the overall system configuration.

Above all, the designer should select the launch and recovery approaches early in the design process, so that the incidence of problems associated with launch and recovery can be eliminated, or at least much reduced, by producing a total system design that integrates launch and recovery considerations into all aspects of the design.

RELATED ARTICLE

Chapter 6

ACKNOWLEDGMENT

We would like to thank Mr. Robert Veazy, who provided initial drafts of some of the material on recovery concepts that is included in this chapter.

REFERENCE

Fahlstrom, P.G. and Gleason, T.J. (2012) *Introduction to UAV Systems*, John Wiley & Sons, Ltd, Chichester.

Chapter 9

Development of Centimeter-Sized Aerial Vehicles

Shigeru Sunada and Hiroshi Tokutake

Department of Aerospace Engineering, Osaka Prefecture University, Osaka, Japan

1 INTRODUCTION

In the 20th Century, a vehicle was just a transportation tool. However, we believe that in the future, a vehicle will become more than just a transportation tool; it will become a tool used in our everyday lives such as a mobile phone and a personal stereo. When we can acquire visual information from a high altitude with a vehicle, such ability will resemble that of a bird. To use a vehicle as a tool in everyday life, it must be small and lightweight. Slow flight is required. Recently, MAVs have been developed all over the world. However, to the best of our knowledge, no MAVs have been developed that can fly outdoors with low flight velocity. In this section, we consider centimeter-sized fixed-wing and rotary-wing UAVs with small weight and low flight velocity.

Unmanned Aircraft Systems. Edited by Ella Atkins, Aníbal Ollero, Antonios Tsourdos, Richard Blockley and Wei Shyy.
© 2016 John Wiley & Sons, Ltd. ISBN: 978-1-118-86645-0.

In Sections 2 and 3, the development of fixed-wing and rotary-wing UAVs is presented. In Section 4, the difficulties of the development of a flight controller for a UAV are explained, and some control methods are briefly introduced. In Section 5, the wing characteristics at a low Reynolds number and their effects on the flight stability of a fixed-wing UAV are stated as references for designing a centimeter-sized UAV.

2 DEVELOPMENT OF A FIXED-WING UAV

2.1 Overview of fixed-wing UAVs' configuration

UAV's design method has not been as sophisticated as manned vehicles. A wide variety of missions exist for UAVs, and such wide variation has caused many different kinds of designs. Pines and Bohorquez (2006) showed the trend of UAV and MAV designs and indicated that the relation between the wingspan and the gross take-off weight is linear on a log-log scale. Figure 1 shows the gross weight as a function of the wingspan of the fixed-wing UAV and was made by the present authors using Wilson's data (2007). In this figure, the x- and y-axes are logarithmic scale, and the electric powered UAVs and the non-electric ones are distinguished. This figure shows the followings: (a) When the wing span is larger than 10m, the non-electric UAVs are heavier than the electric UAVs. (b) When the wing span is less than 10m at the electric propulsion system is popular. The propulsion system of small UAVs with wingspans less than 1m are mainly electric because a non-electric propulsion system

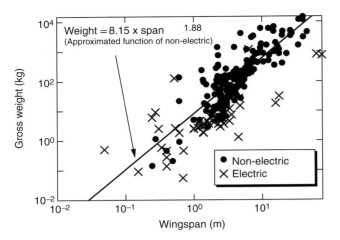

Figure 1. Wingspan-gross weight of UAVs. (Reproduced with permission from Wilson. 2007.)

with low weight and small size has not been developed. Therefore, an electric propulsion system must be used for a centimeter-sized UAV.

Figure 2 shows the relation between the wingspan and the wing loading of fixed-wing UAVs. We collected these data from the data sheets of unmanned vehicles. In this section, we use these figures to discuss the characteristics of fixed-wing UAVs.

The function of the masses m (kg) and the wingspans b (m) of the non-electric UAV were obtained from Figure 1 using the least mean square method by

$$m = 8.15 \times b^{1.88} \tag{1}$$

From this equation, wing loading mg/S and flight velocity V of the non-electric UAV are expressed by

$$\frac{mg}{S} = \frac{8.15\mathrm{AR}}{b^{0.12}}g, \qquad V = \sqrt{\frac{16.3\mathrm{AR}}{\rho C_L b^{0.12}}g} \tag{2}$$

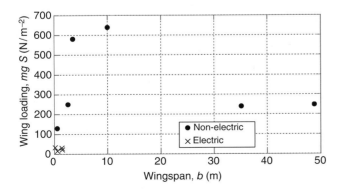

Figure 2. Wingspan-wing loading of UAVs.

Here, S, AR, g, ρ, and C_L are wing area, aspect ratio, gravitational acceleration, air density, and lift coefficient, respectively. In both equations, $\mathrm{AR}/b^{0.12}$ is included. The wing loading and forward velocity increase with the wingspan increase when $b < 10$ m, as shown in Figure 2 and by Pines and Bohorquez (2006). These relations are caused by the relation that the $AR/b^{0.12}$ increases with the increase of b. Equation 1 leads to the following relation:

$$\frac{T}{b^{1.88}} = \frac{8.15g}{L/D} \tag{3}$$

Here, T is the thrust, and L/D is the lift-to-drag ratio. When L/D is constant, this equation indicates that the required thrust divided by $b^{1.88}$ is a constant. Figure 1 shows when the wing span is larger than 10m, the non-electric UAVs are heavier than the electric UAVs. This means the slope between weight and span in Figure 1 is less than 8.15 in Equation 3, which is for the non-electric UAVs. And the thrust generated by the electric propulsion system is smaller than that by the non-electric system when they are mounted on an airplane with a similar wing span. This is perhaps caused by the fact that thrust per weight of an electric propulsion system is lower than that of a non-electric propulsion system. And the flight velocity of an electric powered UAV is less than that of a non-electric powered UAV and the airflame of the former is required to be lighter than that of the latter.

Now consider a situation where non-electric and electric UAVs encounter a vertical gust whose velocity is w_g. The vertical acceleration of UAV, $\ddot{Z}$, divided by the wingspan is as follows:

$$\frac{\ddot{Z}}{b} = \frac{\frac{1}{2}\rho V^2 S a \frac{w_g}{V}}{mb} = \frac{agw_g}{b}\sqrt{\frac{\rho}{2C_L(mg/S)}} \tag{4}$$

where a is the lift slope of UAVs. This equation indicates that the smaller the wing loading, the larger the acceleration caused by gusts when b, C_L, and a are common. As shown in Figure 2, the wing loading increases with the increase of wing span when $b < 10$m. And, as shown in Figure 1, the electric power system is more popular than the non-electric system when $b < 10$m. Then, small electric UAVs are strongly affected by a wind gust. Careful design is required for a small electric UAV to suppress the effect of a wind gust as much as possible.

2.2   Fixed-wing UAV developed in Japan

In Japan, the main aim of UAVs is environmental and disaster monitoring. A high-resolution camera with a sensor is mounted on the UAV. To mount the payload and to decrease the risk of crashes, fixed-wing UAVs have been developed with wingspans between 0.6 and 1.5 m, which are

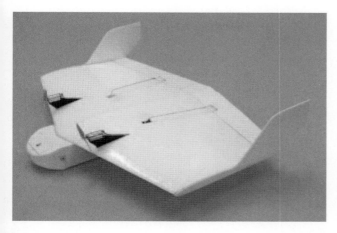

Figure 3. Fixed-wing UAV developed by Hitachi, Ltd and Kawada Industries Inc.

larger than MAVs. For example, Mitsubishi-Electric and the University of Tokyo developed a UAV with a 1.43-m span and a weight of 0.85 kg. Hitachi, Ltd and Kawada Industries, Inc., developed three UAVs of various sizes (Hitachi, 2003). The smallest has a 0.6-m span and weighs 400 g (Figure 3), the middle one has a 0.9-m span and weighs 1.3 kg, and the largest has a 1.51-m span and weighs 3 kg. The flight speeds of these UAVs are over 10 m s^{-1}. An ordinary wind does not greatly affect them because of their weights and flight speeds.

We developed a smaller and lighter fixed-wing UAV with a lower flight speed to investigate the possibility of an autonomous flight of a fixed-wing UAV with as little momentum as possible. The UAV, shown in Figure 4 (Fujinaga, Tokutake, and Sunada, 2007, 2008), has a 60-cm wingspan, an aspect ratio of 2.67, a cross section of airfoil CLARK-YS with a tip chord of 0.15 m, and weighs 270 g. It is designed to perform a slow flight with speed less than 10 m s^{-1} with a small airframe

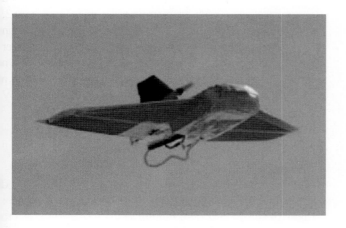

Figure 4. Fixed-wing UAV to investigate autonomous flight with as little momentum as possible.

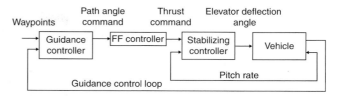

Figure 5. Longitudinal controller.

constructed of foamed styrol. To shorten its body length, it does not have horizontal tail wings. The longitudinal stabilities were achieved by S-cambered airfoil, and lateral-directional stabilities were achieved by vertical tail wing and swept-wing. Its center of gravity was set as far forward as we could manage. The sufficiency of the stabilities was attained by the flight control system. The vehicle dynamics are controlled using the elevon and thrust inputs. All the functions required for autonomous flight were installed.

This UAV's flight plan is given as waypoints, whose data are composed of positions and altitudes. The navigation and guidance controllers output the flight path and the bank angle commands to the inner loop. These commands are input to the stabilizing controller through the feed-forward filter. The closed-loop system of the longitudinal dynamics is shown in Figure 5. The longitudinal controller's design results are shown as an example.

The following are the linearized equations of the longitudinal dynamics:

$$\dot{x} = Ax + B_1 w + B_2 \delta_e, \quad x = \begin{bmatrix} u\, \alpha\, q\, \theta \end{bmatrix}^T$$
$$w = \begin{bmatrix} \alpha_g\, w_{th} \end{bmatrix}^T \tag{5}$$

Here, u is the forward velocity, α is the angle of attack, q is the pitch rate, θ is the pitch angle, and δ_e is the elevator input. The disturbance inputs are gusts in angle of attack α_g and disturbance derived from thrust input w_{th}. The input to the longitudinal stabilizing controller is the pitch rate.

This UAV model contains many uncertainties and disturbances because it flies at low velocities. This model is particularly affected by wind gusts. If a proportional-integral derivative (PID) controller is employed, the required performances and stabilities must be attained by trial and error. Therefore, μ synthesis was used after precise modeling of the structured uncertainties. A generalized plant was formulated from the following requirements and features:

1. stability derivatives contain uncertainties;
2. actuator dynamics are modeled as a second-order transfer function, and multiplicative uncertainties are found on the input side;

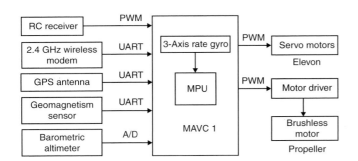

Figure 6. Block diagram of avionics system.

3. wind gusts are input;
4. sensor noises are input;
5. longitudinal disturbances exist that were derived from the thrust;
6. the deflection angles of the control surfaces are restricted;
7. the pitch rate, the roll rate, and the bank angle are restricted.

A controller is utilized to minimize the structured singular values of the formulated generalized plant by MATLAB and Robust Control Toolbox (The MathWorks, Inc., 2009).

The guidance system was constructed with a PID controller. The desired path angle was calculated from the positions and altitudes of the vehicle and the next waypoint. The desired heading angle was calculated from the positions of the vehicle and the next waypoint. The desired path angle and heading angle error are input to the PID controller to generate outputs to the stabilizing controller. The control gains were determined with numerical simulations. The guidance system outputs are the path angle commands to the longitudinal stabilizing controller and the bank angle commands to the lateral stabilizing controller. By stabilizing the airplane around the commands, it reaches the desired waypoints. The designed flight control system was implemented into the avionics system, which was constructed from the control board (MAVC 1) and several devices (Figure 6). Flight testing verified that the developed vehicle had the following abilities when the wind measured on the ground was less than 5 m s^{-1}: autonomous flight for passing the given waypoints and auto landing.

3 DEVELOPMENT OF A ROTARY-WING UAV

3.1 Centimeter-sized rotary-wing UAVs developed all over the world

A ducted fan called GTSpy (Figure 7) has been developed by the Georgia Institute of Technology (Johnson and Turbe,

Figure 7. GTSpy developed by Georgia Institute of Technology. (Reproduceed with permission from Johnson *et al.*, 2004.)

2006) that weighs about 3 kg and has a 30-cm rotor diameter. This ducted fan is based on Helispy developed by Micro Autonomous Systems. Micro Craft (Lipera *et al.*, 2001) has also developed a ducted fan iSTAR that weighs about 2 kg with about a 25-cm rotor diameter. Both ducted fans can make autonomous flights. A quadro-rotor type UAV called MD4-200 was developed by Microdrones (Microdrones GmbH, 2005) that weighs about 900 g with a 37-cm rotor diameter. The MD4-200, which can fit in a box whose side length is 1 m, can be utilized outdoors, and its flight duration is 20 min.

RMAX (Mettler, 2003) by Yamaha Motor Co. is a well-known unmanned helicopter that weighs about 80 kg with about a 3-m rotor diameter. This helicopter, which is used for spreading agricultural chemicals, is too large and heavy to be considered here. Since the centimeter-sized helicopters developed in Japan are radio controlled for amusement, they have no sensors except for a rate gyro. To the best of our knowledge, our cm-sized helicopter with autonomous flight is the only such model that has been developed in Japan.

Figure 8 shows the configuration of the rotary-wing UAV (Sunada and Tokutake, 2007; Ito *et al.*, 2006). The coaxial rotor's diameter is about 40 cm, and its total weight is about

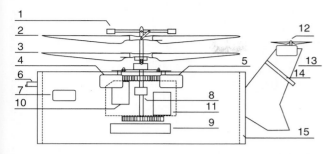

Figure 8. Configuration of rotary-wing UAV developed to eluci-date what is required not for injuring people. 1, stabilizer bar; 2, upper rotor; 3, lower rotor; 4, 5, servomotor; 6, GPS antenna; 7, battery; 8, altimeter; 9, flight controller; 10, 11, motor; 12, pitching propeller; 13 and 14, tails; 15, duct. (Reproduced with permission from Sunada and Tokutake, 2007. © AIAA.)

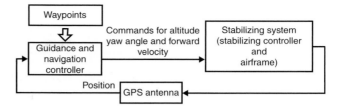

Figure 9. Guidance and navigation controller.

400 g, which is much lighter than the small-ducted fans developed by other groups (Guerrero *et al.*, 2003; Fleming *et al.*, 2003; Johnson and Turbe, 2006; Pines and Bohorquez, 2006; Wilson, 2007). Moreover, the present UAV's rotors are made of foamed styrol and break if they strike a person, minimizing the risk of injury. This rotary-wing UAV was developed to clarify the requirements for a rotary-wing UAV that does not injure people.

The coaxial rotor was originally designed for XRB, a toy model helicopter manufactured by Hirobo Co. Both its upper and lower rotors have two blades made of foamed styrol, which are driven by brushless motors. The cyclic pitch angle of the lower rotor is controlled by connecting this rotor to the two servomotors. The cyclic pitch angle of the upper rotor is controlled by a stabilizer bar when the vehicle has pitch and roll rates. The rotational speeds of the rotors are adjusted to control the yaw moment and the magnitude of thrust. A foamed styrol duct is attached to the vehicle below the center of gravity (CG). The vertical and horizontal tails stabilize the pitching and yawing motions for forward flight. The aft propeller on the vertical tail called the pitching propeller is about 13.5 cm in diameter and is driven by a DC motor. All the motors in this vehicle are driven by an 11.1-V, 720-mAh lithium-polymer battery mounted on it.

The following explains the significance of the duct and pitching propeller. A smaller vehicle is more strongly affected by the wind because the ratio of the parasite drag to the gravitational force generally increases with decreasing vehicle size. So, a smaller rotary-wing vehicle requires a highly stable attitude and the ability to generate sufficient thrust to control the position. To satisfy the first requirement, the present vehicle has a duct below its CG. Theoretically, it can be proven that the pitching and

translational motions are stabilized by an object below the center of gravity. Note that the duct should surround the rotors if higher flight efficiency is required (Pereira and Chopra, 2009). To satisfy the second requirement, the tip path plane of the coaxial rotor should be greatly inclined when the vehicle is either traveling forward or hovering under wind gusts. Since the vehicle's inclination due to the cyclic pitch angles is not sufficient to overcome the parasite drag, the present vehicle is equipped with a pitching pro-peller to generate a larger pitching moment regarding the CG to create larger forward thrust force.

The following sensors and actuators are connected to the flight control board (MAVC 1), whose details will be given in Section 4.3: Global Positioning System (GPS), altimeter, geomagnetism sensor, receiver, speed controllers, servomo-tors, modem. With this equipment, the flight control board weighs about 50 g. Depending on the preprogrammed flight plan, the flight duration is usually about 3–5 min. The flight controller can be divided into an inner loop and an outer loop. The outer loop consists of guidance and navigation controllers (Figure 9), and the inner loop consists of a stabilizing controller (Figure 10). The outer loop generates commands for altitude, yaw angle, and forward velocity as follows. The altitude command is preprogrammed. The yaw angle com-mand is a heading angle for a preprogrammed waypoint. The forward velocity command is proportional to the distance between the waypoint and a position measured using GPS. These commands are input into the inner loop, and then the stabilizing controller executes the desired state. The preprog-rammed altitude and waypoints can be changed from the ground base during flight by a wireless modem. The stabi-lizing controller comprises PID, PD, and P controllers. The gains of these controllers were adjusted by trial and error. In an emergency, the operator input can override the input of the stabilizing controller.

Many autonomous flight tests (Figure 11) were carried out to verify the flight control system. When the wind measured on the ground was less than 2 m s^{-1}, the rotary-wing UAV was controlled to avoid straying from the pro-grammed flight path.

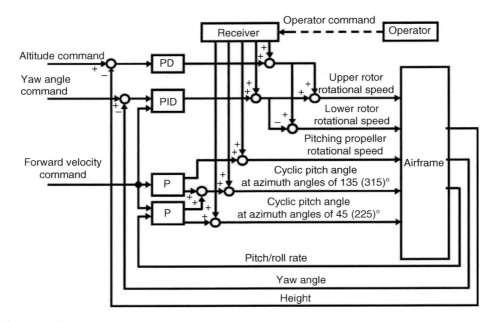

Figure 10. Stabilizing controller.

Figure 11. Photograph of developed rotary-wing MAV during test flight. (Reproduced with permission from Sunada and Tokutake, 2008. © The Japan Society for Aeronautical and Space Sciences.)

4   CONTROLLER DESIGN OF CENTIMETER-SIZED UAV

4.1   Control theory

In the development of UAVs, one often faces problems due to the constraint on the size and the weight of the equipments, which can be overcome by selecting an appropriate control theory and a high-performance and lightweight avionics system. In the following text, we briefly describe the constraints on the controller design of UAVs that reflect their size and weight.

- *Low sensor accuracy:* For UAV flight control, a small and lightweight rate gyro is definitely required. The typical micro electro mechanical systems (MEMS) rate gyro is less than 1 g (Analog Devices, Inc., 2008). However, its accuracy is lower than a fiber optic gyroscope that is used in large airplanes. In addition, the natural frequencies of micro aerial vehicle (MAV) flight dynamics are higher than those of a large airplane. The faster dynamics must be controlled with sensors of lower precision.

- *Low Micro Processing Unit (MPU) performance:* Due to the constraints on size, weight, and the equipment's power supply, a high-performance Micro Processing Unit (MPU) cannot be used. Low processing performance causes calculation delay and results in the deterioration of vehicle performance.

- *Large modeling error of flight dynamics:* The aerodynamics of centimeter-sized unmanned aerial vehicle (UAVs) at low Reynolds numbers has not been adequately investigated compared with those at high Reynolds numbers. The formulated vehicle dynamics, in general, have uncertainties, an issue that must be addressed.

The flight control of a centimeter-sized UAV is challenging because an inadequate dynamical model and low-performance equipment must be used. When flight tests

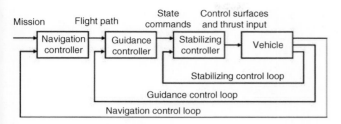

Figure 12. Flight control system.

show bad performance or instabilities, the controller should be able to be redesigned easily. In addition, the order of the designed controller should be low because the MPU performance is low. However, this presents many significant challenges, including development of an advanced sensor system, an actuator, and a controller design method.

The flight control system of a fixed-wing UAV (Figure 4) consists of a navigation controller, a guidance controller, and a stabilizing controller (Figure 12). The navigation and guidance controllers determine the commands given to the stabilizing controller to ensure that the vehicle follows the desired flight path.

Many controller design methods can be applied to stabilize a vehicle controller (Magni, Bennani, and Terlouw, 1997) that attains several design criteria. The typical methods used for flight controllers and their features are briefly mentioned in the following. Please refer to other sections for greater detail.

4.1.1 Elementary feedback control (McRuer, Ashkenas, and Graham, 1973)

The advantage of such a controller is that the time response and controller gain are directly related to each other and that the controller can be adjusted to attain certain requirements. Sometimes, the controller is designed without a dynamical model of a vehicle, for example, in a trial-and-error manner using a PID control method.

4.1.2 Eigenstructure assignments (Andry, Shapiro, and Chung, 1983)

In this method, the feedback gain is selected so that the eigenstructure of the closed-loop system becomes the desired one. Determining the desired eigenstructure is difficult because the relations between the time responses and the eigenstructure are often unclear. In addition, attaining robust stabilities is difficult with this method.

4.1.3 Linear quadratic regulator (LQR)/linear quadratic Gaussian (LQG)

Constant state feedback and observer gain are determined to minimize the quadratic form cost function. A closed-loop

system with LQR has both large gain and phase margins (Safonov and Athans, 1977). Using the design parameters of LQR, the time histories of vehicle response and control input can be shaped. However, designers encounter difficulties of controller adjustments when flight testing shows that the closed-loop system is unstable. In this case, the modeling process should be repeated to obtain a more precise vehicle model, increasing time and cost.

4.1.4 H_∞ controller (Zhou, Doyle, and Glover, 1995)

A full-order output-feedback controller is designed to decrease the H_∞ norm of a closed-loop system. The uncertainties are formulated, and a generalized plant is constructed. The robust stabilities are adjusted with the parameters of the uncertainties model.

4.1.5 μ Synthesis (Zhou, Doyle, and Glover, 1995)

Robust stabilities and performances are ensured for the structured uncertainties. Structured singular values are used to ensure the stabilities and performances for individual uncertainty blocks. Generally, the order of the obtained controller becomes very high.

4.2 Equipment

Since a computer on a vehicle cannot make continuous calculation, the controller needs to be discretized before implementation. However, conventional controller design methods usually depend on a continuous model, and the robust stabilities and performances of the closed-loop system cannot be guaranteed for the discretized controller. Therefore, the sampling rate must be as high as possible. There are three methods to obtain the discretized controller, that is, to discretize the continuous controller designed for the continuous dynamics using zero-order (or higher order) hold, to design the discretized controller with a digital control theory, and to design a discretized controller for the sampled data system. In digital control theory and control for the sampled data system, the performances and stabilities of the discretized controller are guaranteed. On the contrary, the discretized controller, which was originally designed for continuous dynamics, does not retain performance and stability. However, the control theory for continuous dynamics is more sophisticated than the others, and available examples are numerous. In addition, the discretizing process does not decrease the performance and stability very much if the sampling rate is sufficiently fast. In many cases, the controller was designed for continuous dynamics and was discretized and implemented.

Figure 13. MAVC 1.

Figure 14. MAVC 2.

The avionics system is a major weight component for centimeter-sized UAVs. Even if the airframe is small, the sensors and the flight computer must have the same functions as those of a larger vehicle. The flight control system simultaneously performs many processes such as calculation of guidance, navigation, stability control, sampling sensor output, transmitting flight data, and receiving commands from a ground base. However, since the UAV payload is often very small, all processes need to be performed by a single MPU. Hence, we developed a generalized control board specified for MAV control.

4.3   Flight control boards MAVCs 1 and 2

As examples, two flight controllers developed by the authors collaborating with a company (Y's Lab, Inc., 2008) are introduced here. Each unit comprises of a small, lightweight circuit board called Micro Aerial Vehicle Controller 1 (MAVC 1) and 2 (MAVC 2) with dimensions of 75 mm × 55 mm and 40 mm × 35 mm and weights of 29 g and 4.6 g, respectively (Figures 13 and 14). Table 1 shows the functions of the developed boards equipped with accelerometers and rate gyros (for MAVC 2, additionally geomagnetism sensor) into which many other devices can be connected. MAVC 1 is mounted on the above fixed-wing and rotary-wing UAVs.

5   WING CHARACTERISTICS AT A LOW REYNOLDS NUMBER AND FLIGHT STABILITY OF A FIXED-WING MAV

Some fixed-wing MAVs have wings with rectangular planforms and a low aspect ratio (AR ≈ 1) (Muller *et al.*, 2006) for the following reason. A gust relative to forward velocity, that is, a variation of the angle of attack, increases with a decrease of the vehicle's size. However, a lift acting on a wing with a rectangular planform with AR ≈ 1 increases with an increasing angle of attack between 0° and 45° at a low Reynolds number ($Re \approx 10^5$) (Muller *et al.*, 2006). Therefore, it does not stall over a wide range of angles of attack. On the other hand, a high aspect ratio wing often benefits at higher lift–drag ratios but has smaller power requirements. As a result, some studies have also investigated a high aspect ratio wing and an airfoil.

The C_D of a three-dimensional wing can be expressed by

$$C_D(\alpha) \approx C_{D_0}/C_{L=0} + \Delta_2(\alpha - \alpha_0)^2 \qquad (6)$$

where α_0 is a zero-lift angle of attack. The increase of $C_{D_0}/C_{L=0}$ with the decrease of the Reynolds number (Re) is mainly the increase of the skin friction drag. Laitone (1997)

Table 1. Functions of MAVCs 1 and 2.

	MAVC 1	MAVC 2
Size	75 mm × 55 mm	40 mm × 35 mm
Weight	29 g	4.6 g
PWM	8 ch	3 ch
Motor driver	–	1 ch
CCP	10 ch	3 ch
A/D	6 ch	1 ch
D/A	2 ch	–
I/O	16 ch	4 ch
UART	4 ch	2 ch
Rate gyro	3-Axis (onboard)	2-Axis (onboard)
Accelerometer	3-Axis (onboard)	3-Axis (onboard)
Geomagnetism sensor	1 ch	1 ch (onboard)
GPS	1 ch	Enabled (UART)
Rotary encoder input	1 ch	–
5 V output	3 ch	1 ch
Power supply	7.4–12 V	3.4–4.5 V

and Sunada et al., (2002, 2004) concluded that Δ_2 increases with decreasing Re when AR is about 5. Torres and Muller (2001) pointed out the increase when AR is about 1. Laitone states that this is the increase of the induced drag. (At a high Reynolds number, the term of Δ_2 corresponds to the induced drag because the increase of the profile drag with the increase of the angle of attack is much smaller than that of the induced drag at a high Reynolds number. Laitone probably describe the increase of Δ_2 as an increase of the induced drag in the same manner at a high Reynolds number.) On the other hand, Sunada et al. (2002, 2004) assumed that the difference of circulation distribution along the span from the elliptical one does not become much larger with the decrease of Re. They argued that the increase of Δ_2 is mainly caused by the increase of the profile drag, that is, the increase of the profile drag of an airfoil. If the assumption by Sunada et al. is relevant, the advantage of a high aspect ratio wing against a low aspect ratio wing becomes smaller (Sunada and Kawachi, 2004). It remains unclear which is dominant for the increase of Δ_2 between the increase of the induced drag and that of the profile drag. Measurements of the drag of wings with various aspect ratios including two-dimensional wings will resolve this issue.

The following is a similar problem with rotary wings: The power coefficient of a rotor at a high Reynolds number is given by (e.g., Bohorquez et al., 2003)

$$C_P = \kappa \frac{C_T^{3/2}}{\sqrt{2}} + \sigma \frac{C_{d0}}{8} \qquad (7)$$

where C_T, C_{d0}, κ, and σ are the thrust coefficient, the zero-lift drag coefficient, the induced power factor, and the rotor solidity, respectively. The first and second terms are induced power coefficient C_{Pi} and profile power coefficient C_{P0}, respectively. When the rotor performance at a low Reynolds number ($Re \approx 10^4$) is analyzed using Equation 7, the value of κ is far from 1 (Ramasamy, Johnson, and Leishman, 2008). However, using the blade element theory with the following drag coefficient of an airfoil, induced power factor $\kappa = \frac{C_{Pi}}{C_T^{3/2}/\sqrt{2}}$ is close to 1 and the obtained power is close to the measured one:

$$C_d(\alpha) \approx C_{d_0}/C_{L=0} + \delta_1(\alpha - \alpha_0) + \delta_2(\alpha - \alpha_0)^2 \qquad (8)$$

Results by Bohorquez et al. (2003) and Ramasamy, Johnson and Leishman (2008) suggest that the thrust and torque acting on a rotor can be estimated approximately by the same manner as at a high Reynolds number (by Blade element theory and the momentum theory) when a drag coefficient expressed by Equation 8 is used. The collective pitch is not

Table 2. Effects of increases of C_{D0} and Δ_2 on flight modes of a gliding swallowtail.

	Increase of C_{D0}	Increase of Δ_2
Lateral-directional motion		
Spiral mode	More stable	More unstable
Roll mode	Independent	Independent
Dutch roll mode	More stable	More unstable
Longitudinal motion		
Long period mode	More stable	More unstable
Short period mode	Independent	Independent

large, and the Reynolds number was about 2×10^4. Tsuzuki, Sato, and Abe (2007) also shows the thrust and torque can be estimated approximately by the same manner as at a high Reynolds number using Equation 4 when a collective pitch is roughly less than 10 deg. However, they also show the method underestimates them at a large collective pitch and the maximum figure of merit because of the leading edge vortex (Lentink et al., 2009). In their experiments, the Reynolds number was 4×10^3.

When an airplane becomes smaller due to its proportional shape, its flight stability is varied, caused by both different sizes and different aerodynamic coefficients as C_{D0} and Δ_2 due to those of Re. Note that flight stability is varied just by the variation of size even if the aerodynamic coefficients are not changed. We can understand this fact by referring to the equations of the motion of an airplane (e.g., Etkin, 1955). Okamoto, Sunada, and Tokutake (2009) focused the effects of the variations of C_{D0} and Δ_2 on the stability of each flight mode of a gliding swallowtail. Their results are summarized in Table 2, which shows that the effect of C_{D0} contradicts that of Δ_2. Therefore, the trade-off of these effects determines the effect of the variation of C_D due to Re on flight stability. The differences in the wing characteristics due to smaller Re might counter well-known qualitative data about meter-sized airplanes regarding flight stability. This possibility might also be observed in a future small aerial vehicle.

ACKNOWLEDGMENTS

The authors wish to acknowledge Professor Hao Liu, the section editor, for his fruitful comments on this chapter.

REFERENCES

Analog Devices, Inc. (2008) http://www.analog.com/.

Andry, A.N., Shapiro, E.Y. and Chung, J.C. (1983) Eigenstructure assignment for linear systems. *IEEE Trans.*, **AES-19**, 711–729.

Bohorquez, F., Samuel, P., Sirohi, J., Pines, D., Rudd, L. and Perel, R. (2003) Design, analysis and hover performance of a rotary wing micro air vehicle. *J. Am. Helicopter Soc.*, **48**, 80–90.

Etkin, B. (1955) *Stability and Control*, John Wiley & Sons, Inc., New York.

Fleming, J., Jones, T., Gelhausen, P. and Enns, D. (2003) Improving control system effectiveness for ducted fan VTOL UAVS operating in cross winds. AIAA paper. AIAA-2003-6514.

Fujinaga, J., Tokutake, H. and Sunada, S. (2007) Development of small unmanned aerial vehicle and flight controller design. AIAA Paper. AIAA-2007-6501.

Fujinaga, J., Tokutake, H. and Sunada, S. (2008) Guidance and control of a small unmanned aerial vehicle and autonomous flight experiments. *J. Jpn Soc. Aeronaut. Space Sci.*, **56**(649), 57–64 (in Japanese).

Guerrero, I., Londenberg, K., Gelhausen, P. and Myklebust, A. (2003) A powered lift aerodynamic analysis for the design of ducted fan UAVs. AIAA Paper. AIAA-2003-6567.

Hitachi, Ltd (2003) http://www.film.hitachi.jp/movie/movie615.html.

Ito, H., Tokutake, H., Sunada, S. and Sumino, H. (2006) Outside flight of a small and lightweight co-axial helicopter. *Proceedings of AHS International Meeting on Advanced Rotorcraft Technology and Life Saving Activities*, pp. T-243-2-1–T-243-2-4.

Johnson, E.N. and Turbe, M.A. (2006) Modeling, control, and flight testing of a small ducted-fan aircraft. *J. Guid. Control Dyn.*, **29**, 769–779.

Johnson, E.N., Schrage, D.P., Prasad, J.V.R. and Vachtsevanos, G.V. (2004) UAV flight test programs at Georgia Tech. AIAA Paper. AIAA 2004-6492.

Laitone, E.V. (1997) Wind tunnel tests of wings at Reynolds numbers below 700,000. *Exp. Fluids*, **23**, 405–409.

Lentink, D., Dickson, W.B., Van Leeuwen, J.L. and Dickinson, M.H. (2009) Leading-edge vortices elevate lift of autorotating plant seeds. *Science*, **324**, 1438–1441.

Lipera, L., Colbourne, J., Tischler, M.B., Mansur, M.H. and Patangui, P. (2001) The micro craft iSTAR micro air vehicle: control system design and testing. *Proceedings of Annual Forum of American Helicopter Society*, pp. 1998–2008.

Magni, J.F., Bennani, S. and Terlouw, J. (eds) (1997) *Robust Flight Control: A Design Challenge*, Springer, New York.

McRuer, D., Ashkenas, I. and Graham, D. (1973) *Aircraft Dynamics and Automatic Control*, Princeton University Press, Princeton.

Mettler, B. (2003) *Identification Modeling and Characteristics of Miniature Rotorcraft*, Kluwer Academic Publishers, Boston.

Microdrones GmbH 2005, http://www.microdrones.de./index.html.

Muller, T.J., Kellogg, J.C., Ifju, P.G. and Shkarayev, S.V. (2006) *Introduction to the Design of Fixed-Wing Micro Air Vehicles*, AIAA, Virginia.

Okamoto, M., Sunada, S. and Tokutake, H. (2009) Stability analysis of gliding flight of a swallowtail butterfly *Papilio xuthus*. *J. Theor. Biol.*, **257**, 191–202.

Pereira, J.L. and Chopra, I. (2009) Hover tests of micro aerial vehicle-scale shrouded rotors, part I: performance characteristics. *J. Am. Helicopter Soc.*, **54**, 012001.

Pines, D.J. and Bohorquez, F. (2006) Challenges facing future micro-air-vehicle development. *J. Aircr.*, **43**, 290–305.

Ramasamy, M., Johnson, B. and Leishman, J.G. (2008) Understanding the aerodynamic efficiency of a hovering micro-rotor. *J. Am. Helicopter Soc.*, October **53**, 412–428.

Safonov, M.G. and Athans, M. (1977) Gain and phase margin for multiloop LQG regulators. *IEEE Trans. Automatic Control*, **AC-22**(2), 173–179.

Sunada, S., Yasuda, T., Yasuda, K. and Kawachi, K. (2002) Comparison of wing characteristics at an ultralow Reynolds number. *J. Aircr.*, **39**(2), 331–338.

Sunada, S. and Kawachi, K. (2004) Effects of Reynolds number on characteristics of fixed and rotary wings. *J. Aircr.*, **41**(1), 189–192.

Sunada, S. and Tokutake, H. (2007) Outside flight of a small and lightweight helicopter. AIAA Paper. AIAA-2007-2789.

Sunada, S. and Tokutake, H. (2008) Developments of small electrically powered aircraft. *Aeronaut. Space Sci. Jpn*, **56**(651), 107–109 (in Japanese).

The MathWorks Inc., MATLAB R2009b and Robust Control Toolbox 3.4, 2009.

Torres, G.E. and Muller, T.J. (2001) Aerodynamic characteristics at low aspect ratio wings at low Reynolds numbers, in *Fixed and Flapping Wing Aerodynamics for Micro Air Vehicle Applications* (ed. T.J. Muller), AIAA, Virginia, pp. 115–141.

Tsuzuki, N., Sato, S. and Abe, T. (2007) Design guidelines of rotary wings in hover for insect-scale micro air vehicle applications. *J. Aircr.*, **44**(1), 252–263.

Wilson, J.R. (2007) UAV Worldwide Roundup 2007, Aerospace America, May 2007.

Y's Lab, Inc. (2008) http://www.yslab.co.jp/.

Zhou, K., Doyle, J.C. and Glover, K. (1995) *Robust Optimal Control*, Prentice Hall, Upper Saddle River.

PART 4
UAS Design and Subsystems

Chapter 10
Overview of UAS Control Stations

Francisco J. Ramos

UAS Ground Segment Department, Airbus Defence & Space, Getafe, Spain

1 INTRODUCTION

The use of unmanned aircraft systems (UASs) has grown exponentially in the last decades, and today they are used worldwide for a wide range of military and civil applications.

Traditionally, most of the available resources and information on UAS addressed the flying part of the system, providing less attention to other elements such as the control station. This aircraft-centered vision has progressively changed during the preceding years in favor of a more accurate system-wide approach. An example is the generalized adoption of the term "unmanned aircraft system" instead of "unmanned air vehicle" (UAV), in order to align the terminology with the concept of a complete system, composed not only of the aircraft itself but also of other elements such as the data link infrastructure or the control station (Trimble, 2005).

Unmanned Aircraft Systems. Edited by Ella Atkins, Aníbal Ollero, Antonios Tsourdos, Richard Blockley and Wei Shyy.
© 2016 John Wiley & Sons, Ltd. ISBN: 978-1-118-86645-0.

The UAS Control Station (UCS) is a key element of the unmanned aircraft system, which has evolved from the first remote control boxes and rudimentary cockpit implementations on ground to current systems based on modern hardware and software technologies, providing better human machine interfaces (HMI) and an extensive set of functions to support aircraft operation (Figure 1).

2 TERMINOLOGY AND DEFINITION

Although, as already mentioned, the term UAS seems to prevail in the past years, a common terminology is not yet established in the whole UAS community. For the overall system, UAS and UAV terms are usually mixed. Regarding control stations, the terms Ground Control Station and UAV Control Station are typically mentioned. On the other hand, the International Civil Aviation Organization (ICAO) has introduced the terms Remotely Piloted Aircraft System (RPAS), Remotely Piloted Aircraft (RPA), and Remote Pilot Station (RPS) in the regulatory domain, aiming at reinforcing the idea of a pilot in command (International Civil Aviation Organization, 2012).

The term UAS Control Station will be used throughout this chapter. The term unmanned air vehicle will be used to refer to the aircraft part of the UAS.

A UAS Control Station is then defined as the integrated set of elements that allow one or more operators to control the actions of one or more UAVs from a remote location.

3 CLASSIFICATION

UAS comprises a wide and heterogeneous set of systems with very different characteristics, from small quadcopters to big

Figure 1. NASA Global Hawk control station (inside view). The station provides several operator positions with keyboard, mouse, multiple displays, and intercom panels. Equipment such as computers and network devices is installed in the 19″ racks at the back. (*Source*: NASA Imagery Web Site http://www.nasa.gov/images/content/674938main_trailerinside.jpg)

High-Altitude Long Endurance (HALE) platforms, and these differences also affect the UCS. As in the case of terminology, to date no consolidated classification for UAS exists, and different organizations provide their own classifications (Maddalon *et al.*, 2013). Typical classifications are driven by the UAV characteristics (e.g., maximum takeoff weight, ceiling, or role). A basic UAS classification is provided in Table 1.

Table 1. UAS categories

UAS category	Description
Micro	Smallest available UAS systems. The UAV is small enough to be held in the palm of the hand, usually weighing less than a kilogram
Mini	UAS systems designed to be launched and operated by a single person. The UAV is small enough to be lifted by a human
Tactical	UAVs of several hundred kilograms, with an endurance of several hours and an operating radius of 200 km or less, based on line-of-sight (LOS) data links. They typically require more than one person to be deployed and operated
MALE	Medium-Altitude Long Endurance UAVs with endurance of about 24 h and long-range capability, generally used for operational reconnaissance
HALE	High-Altitude Long Endurance UAVs with endurance of a day or more and long-range capability, generally used for strategic reconnaissance

Based on this overall UAS classification, a specific UCS classification is proposed in Table 2 using system complexity and deployability needs as the main criteria.

4   MAIN DESIGN CHARACTERISTICS

4.1   Architecture

Although UCS architectures can be very different depending on their class, all modern implementations are essentially based on software applications running on top of computing platforms. A UCS provides two major interfaces: an interface with the controlled UAV(s), via data links, and an interface with UAS operator(s), based on electronic displays and control devices. Additional interfaces with external entities such as intelligence centers or air traffic control (ATC) can be included depending on the complexity of the system.

Regarding the data links and related equipment, some UAS architectures consider them as an independent part of the system, not belonging to the UCS, while others include ground data link equipment inside the UCS definition. Despite this conceptual decision, in many systems some data link parts are physically installed in the UCS.

An overview of different UCS architectures is provided in the following sections, which is based on their classification.

Table 2. UCS categories

UCS category	Description	Related UAS category
Portable	The UCS is based on a handheld device (rugged PC, tablet, or control pad) designed to be carried and operated by a single person, typically outdoors	Micro, mini
Deployable	The UCS is composed of a set of computers, displays, and controls that provides several operator positions (typically in the range of 2–5) to support missions carried out by two or more persons. It is usually installed on a container-based shelter to simplify deployment and to allow operation in a wide range of environmental conditions	Tactical, MALE
Distributed	The UCS is composed of several major elements that are positioned in different locations. Each of the elements provides one to several operator positions and can be deployed in a shelter or installed in existing facilities like a Command and Control Center. Elements can be interconnected by means of existing landlines or satellite links. A typical example is a UCS composed of a takeoff and landing element, deployed in the airfield, and a complete control element installed in a building far away from the operations theater	MALE, HALE

4.1.1   Portable architectures

Smaller UAS normally implements UCS solution as a single device containing computing resources and user displays and controls. Depending on the data link characteristics, the ground data link terminal(s) could be integrated in the main UCS body or deployed as a separate element. Portable architectures are then relatively simple, especially when compared with the complex systems designed for bigger UAS. As they are usually operated outdoors, many implementations rely on rugged portable devices such as tablets or laptops. An example is provided in Figure 2.

4.1.2   Deployable architectures

Deployable UCS architectures are designed to control medium and large UAVs, such as tactical and MALE platforms. These systems provide a wide set of functions and payload capabilities, and for most missions require more than one person for their operation. In some cases, a single UCS is also capable of controlling more than one aircraft simultaneously, which has additional impact on the total number of operators required. Deployable UCS is usually positioned on the field, particularly in the case of tactical UAS controlled by means of line-of-sight (LOS) data links.

Figure 2. Desert Hawk UAV Control Station, operated by a soldier of the British Army. The station consists of a briefcase containing a rugged laptop and several auxiliary devices. (*Source*: Picture by the UK Ministry of Defence, released under the Open Government License. Crown Copyright 2009.)

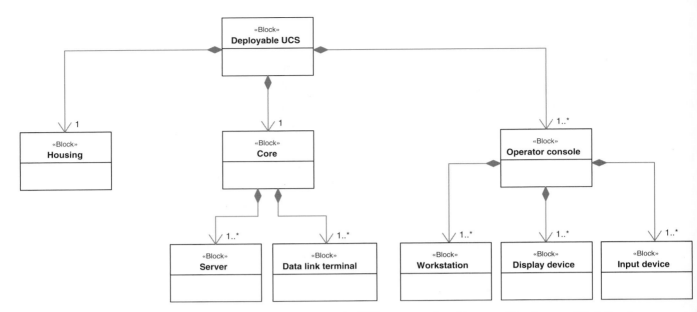

Figure 3. SysML™ block definition diagram showing deployable UCS conceptual architecture. The Systems Modeling Language, or SysML, is a general-purpose visual modeling language for systems engineering applications (Object Management Group, 2012).

A conceptual architecture of a typical deployable UCS is provided in Figure 3.

As depicted in the figure, the UCS is mainly composed of a core block, a set of one to *n* operator consoles, and a housing block.

The UCS core comprises computer servers, network infrastructure, and ground data link equipment. These shared resources are designed to support the functions of the console workstations and usually include data recording, database services, and data link management, among others. Depending on the different system needs and characteristics, part of the ground data link equipment, especially emitters, can be installed on the main UCS body (i.e., shelter) or in a separate element. The first option is more compact and can reduce UCS deployment time, while the second option provides more flexibility in the collocation of different elements (e.g., optimization of data link coverage area and installation of high-power radiating devices far from the personnel operating zones).

Operator consoles include one or several workstations connected to a set of displays and input devices. In missions with multiple operators involved, tasks are distributed by roles. An example of a common role division is the flight operator/payload operator pair, in which the first operator is in charge of the UAV flight and the second one controls the aircraft payload (e.g., an electro-optical/infrared sensors turret). Operator consoles are usually configurable to support different roles (Figure 4).

The UCS housing comprises the structural and ancillary parts required to provide a shelter container to the operators and equipment, including auxiliary services such as power distribution, environmental control, and lights, among others (Figure 5). In some cases, housing can be considered an optional part of the system, in case the UCS is installed in fixed existing facilities such as an office building. Some systems also integrate the UCS housing in a vehicle (e.g., an all-terrain truck) to simplify transport and deployment.

Systems designed in accordance with certification standards usually provide safety-driven architectures, in which flight critical functions are segregated from the rest of the system functions and implemented with some degrees of redundancy (see Section 4.5 for details).

4.1.3   Distributed architectures

Distributed architectures can be seen as an aggregation of multiple major elements, each of them similar to a single deployable UCS, which are installed in different locations and interconnected between themselves to allow a coordinated operation.

Examples of distributed UCS architectures are larger MALE and HALE systems, where the UCS is typically composed of two major elements: a "takeoff and landing" element deployed into the operations airfield and a "mission" element installed in a Command and Control center, far away from the operations theater. The first element is in charge of controlling takeoff and landing phases via LOS data links. Once the UAV is at cruise altitude, it is handed over to the second element, which controls the rest of the mission via satellite data links (Figure 6).

Figure 4. Example of a deployable UCS (Atlante GCS). *Interior view:* It shows flight and payload operator consoles. The consoles are equivalent and can be reconfigured for flight or payload roles during mission execution. They include several touchscreen displays, keyboard, mouse, a joystick for camera control (not used for flight), intercom, and radio panels. (*Source*: Reproduced with permission from Airbus Defence & Space. Airbus Defence & Space, Copyright 2014.)

Figure 5. Example of a deployable UCS (exterior view). The UCS is installed in a military shelter to facilitate transport. (*Source*: Reproduced with permission from Airbus Defence & Space. Airbus Defence & Space, Copyright 2014.)

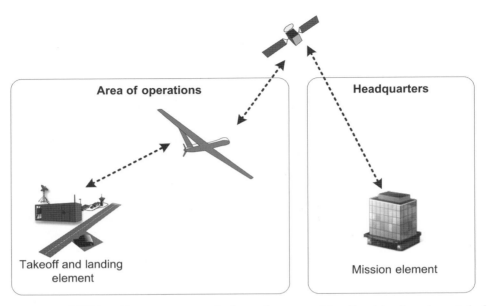

Figure 6. Example of distributed UCS architecture, composed of two elements: a takeoff and landing element deployed in the area of operations, and a mission element installed in the headquarters.

Distributed elements implement similar internal architectures, and most of the hardware and software components employed are the same. In the example mentioned above, the mission element is a "nonsheltered" version of the takeoff and landing element, with additional operator consoles dedicated to mission-related roles.

Communication links between elements are required to coordinate operations. These links could be part of the UCS system itself or rely on existing infrastructures such as satellite services or land networks.

4.2   Main Functions

UCS functions strongly depend on the overall UAS characteristics, but the following high-level functions can be generally found in most of the systems:

- *Aircraft Command and Control*: This function provides the displays and controls required to manage the UAV flight. Although some systems allow direct operator interaction with the aircraft control surfaces from the UCS, flight control in modern UAS is based on pre-planned flight routes and high-level outer loop commands such as heading, altitude, or speed changes. Due to the generalized use of automatic take-off and landing (ATOL) capabilities, in most of the cases takeoff and landing phases are also controlled by means of high-level commands (e.g., initiate, abort).

- *Aircraft Systems Monitoring*: Display of the onboard systems status in real time and related alerts in case of system malfunction. Most critical alerts are usually reinforced by means of acoustic indications.

- *Mission Planning and Replanning*: The UCS includes tools to prepare, check, and upload the mission plan to the aircraft, and in most of the systems it also allows to modify it during flight. Some systems also provide tools to allow mission planning outside the UCS (e.g., by means of software applications installed on a laptop). Mission planning tools strongly rely on support data such as maps, aeronautical databases, and terrain elevation models, not only for visualization but also for providing automatic checks to planned missions. Advanced tools also implement models to calculate aircraft and data link performances of the mission (e.g., fuel consumption, flight time, data link coverage, and signal strength).

- *Data Links Command and Control*: This function allows the operators to manage the UAS data links. It includes displays to check data links health (e.g., signal strength), and controls to activate, deactivate, and configure link parameters such as channels or transmission profiles.

- *Payload Command and Control*: UAS can include a wide range of payloads, which are constantly evolving to adapt to the different mission needs. UCS-related control functions are, in consequence, very different depending on the payloads to be managed. Most common payloads are EO/IR sensor turrets, where control functions are based on real-time video displays and a stick controller, or radars that provide Synthetic Aperture Radar (SAR) imagery displays or track management functions, depending on the supported radar modes. UCS payload functions tend to be modular to simplify integration of new payload types.

- *Exploitation and Dissemination of Payload Products*: As for payload command and control, exploitation functions strongly depend on the target payloads. For video- and imagery-based payloads, the UCS usually provides features such as snapshot capturing, image enhancement filters, or annotations. Dissemination functions are a common feature of military systems, allowing near real-time distribution of payload products to a remote intelligence center. In other applications such as photomapping, exploitation is usually performed out of the UCS after the flight based on the recorded data.
- *Voice Communications*: Mainly all deployable and distributed UCS solutions provide voice communications based on radio equipment. They are used to support operators' interaction with external entities such as ATC. Some systems also include other interfaces such as land or wireless phone lines.
- *Data Recording and Replay*: Recording and replay functions store data generated during flight (i.e., payload products, aircraft position, systems status, or voice communications), and allow their access in a structured way afterward. These functions are used to support mission debriefing and assessment, incident investigation, and offline data exploitation.

4.3 Human Factors

In the end, the main purpose of the UCS is to serve as the interface that allows the operators to interact with the whole system. Human factors are then a major aspect to be considered in UCS design. Until recent years, the focus in UAS development has been on the technology, and human factor aspects have not been explicitly handled or just addressed in an accessory way. Fortunately, latest designs have started to give human factor aspects the relevance they deserve. Human factors are also strongly considered in certification requirements (see Section 4.5 for details).

Physical ergonomics and cognitive ergonomics (Wilson and Corlett, 2005) are the most relevant human factor domains with respect to UCS design.

Physical ergonomics are related to anthropometric and biomechanical aspects of human–UCS interaction, such as accessibility to controls, visibility of displayed information, and workplace environmental conditions. Different human percentiles have to be considered to adapt the system to the whole range of potential users. Three-dimensional CAD tools and models are commonly used to support these activities (see Figure 7).

Cognitive ergonomics are concerned with mental processes such as workload, decision making, or stress. Operator task definition and allocation, and crew dimensioning, are tackled inside this scope. Automation is also a major topic related to cognitive ergonomics and workload. The final goal is to ensure that operators can perform their duties in all

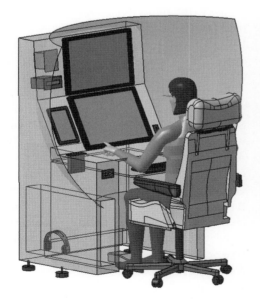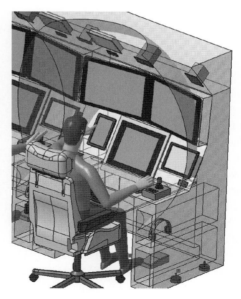

Figure 7. UCS ergonomics analysis supported by 3D CAD tool. (*Source*: Reproduced with permission from Airbus Defence & Space. Airbus Defence & Space, Copyright 2014.)

foreseeable normal and abnormal (e.g., degraded) scenarios, without comprising safety, and with an optimal support of automated functions.

Regarding HMI, current UCS implementations are based on electronic displays and a mixture of specific (e.g., sticks and custom hardware button panels) and general-purpose (e.g., keyboard, mouse, and touchscreens) input devices. But advances in HMI technologies are expanding the ways in which operators can interact with the system. Virtual/augmented reality, 3D displays, or speech and gesture recognition, among others, are progressively being incorporated in UCS designs, and its use is expected to grow in the following years.

Many civil and military standards on human factors and ergonomics are available as reference for the design. Some common examples are ISO 9241-210 (International Standards Office, 2010) and MIL-STD-1472G (United States Department of Defense, 2012), although additional standards exist.

4.4 Environmental Conditions

Many UCS systems are deployed on the field, exposed to adverse environmental conditions, including not only weather effects but also harsh electromagnetic environments (e.g., radar emitters in airfields or own UAS system data link signals). Vibrations and shocks are other aspects to be considered, particularly in the case of portable and deployable UCS systems.

Deployable military systems are normally subject to the most challenging requirements, as they must be able to operate worldwide in all foreseeable weather conditions. On the opposite side are systems designed to be operated indoors, where the expected operating conditions are relatively benign (i.e., office environments).

Design strategies to cope with environmental conditions are different depending on the UCS category. Portable UCS employs rugged enclosures and robust electronic components such as solid-state storage devices.

Deployable UCS systems, on the other side, are designed to ensure that conditions into the shelter are always maintained inside a comfortable range despite the external environment. This allows to alleviate the environmental requirements on the internal electronic equipment, which could be difficult to achieve for most complex devices. Isolation from the external weather conditions is provided by an environmental control system (ECS) that maintains temperature and in some cases humidity ranges inside the acceptable limits within the shelter enclosure. In some systems, the ECS is redundant to ensure operation even in

case of failure. Protection from electromagnetic emissions is also solved mainly at shelter level, using shielding materials such as honeycomb panels, embedded into the external and internal skin. Input and output signals based on wired cables are also filtered to avoid the propagation of external noise up to the connected equipment. Shocks and vibrations are normally mitigated by means of shock absorbers, installed in the joints of internal structures (e.g., equipment racks and operator consoles) with the shelter body. Standards like ASTM E1925 (ASTM International, 2010) provide guidance for developing deployable shelters.

4.5 Certification and Safety

One of the main challenges of UAS is their insertion into nonsegregated civil airspace. Until a clear and consolidated regulation is established, UAS operations will be significantly constrained to segregated airspaces and mainly military uses. Industry, certifying authorities, and UAS operators are making a big effort in this direction. Adaptation of existing standards and regulations and the definition of new ones are currently ongoing. There is a lot of literature on UAS certification and airspace integration, but few of these works address the UCS-related aspects (Gato and Ramos, 2013).

A common error in the development of UAS systems is to consider that the UCS is beyond the scope of certification standards, or only to take them into account at later phases of the project, when the impact of the required changes could be simply not affordable. As in the manned aircraft industry, the impact of a type certification process on the design must not be underestimated. Certification standards have to be considered from the very beginning of the development and applied to the overall system, including the UCS.

Civil airworthiness standards are still under development. To date, the most mature alternative is the military standard STANAG 4671 "Unmanned Aerial Vehicles Systems Airworthiness Requirements" (USAR) (NATO Standardization Agency, 2009), applicable for fixed wing UAS with an operating mass of 150 kg or more. This standard is based on the civil equivalent for manned aviation CS-23 (European Aviation Safety Agency, 2012), and it is also recognized by civil regulatory authorities like the European Aviation Safety Agency (EASA) as a reference airworthiness code for civil-type certification processes (European Aviation Safety Agency, 2009).

STANAG 4671 has been created to mirror as closely as possible the structure and content of CS-23. Requirements are grouped into nine interrelated subparts. There is a specific subpart dedicated to the UCS, although many transversal requirements are also applicable.

Major impact of STANAG 4671 on UCS design is related to safety assurance, where the strict development standards already in use in manned aviation (Society of Automotive Engineers, 1996a, 1996b) are required. These standards establish a set of iterative processes to assess and ensure system safety. In these processes, failure conditions are identified and allocated to system functions and components in an iterative way. Based on the criticality of these failure conditions, system components are categorized within one of the five possible Design Assurance Levels (DAL), ranging from DAL E (no safety effect) to DAL A (highest criticality, although STANAG 4671 only contemplates four levels for UAS, up to DAL B). Development processes imposed on the software (Radio Technical Commission for Aeronautics, 1992) and hardware (Radio Technical Commission for Aeronautics, 2000) components of the UAS take into account allocated DAL to set their objectives, which are progressively more demanding as level criticality increases. The described safety assurance processes have to be considered from the initial phases of the development as they will actually shape the overall UAS design.

The main architecture strategies to cope with safety assurance are segregation and redundancy. Segregation aims to ensure a proper isolation and allocation of most critical DALs to a subset of system components with well-defined boundaries and interfaces. This serves two major objectives: first, to guarantee that the functioning of the most critical functions cannot be adversely affected by malfunctions in the noncritical ones (lower DALs); and second, to optimize development effort and costs by setting the most appropriate DAL level to each component. Redundancy is used in most critical system components to cope with the strict failure probabilities required and to avoid single point of failure. Typical certifiable UCS architectures segregate aircraft control functions (with highest criticality in terms of safety) and mission functions (like payload control) and provide redundant implementations for the first ones.

STANAG 4671 also includes UCS requirements related to human factors, displays and controls, voice communications and recording, and power supply, among others.

4.6 Interoperability

One of the key aspects of UCS is the capability to interact with a wide range of external actors. UAS customers demand highly interoperable solutions, and manufacturers are working to continuously improve the interoperability of their systems (Defense Industry Daily, 2011).

Two major types of UCS interoperability can be considered: interoperability with external systems and interoperability with different UAV platforms. Interoperability with external systems is related to exchanges of information with other entities that can be consumers of the products generated by the UAS and/or providers of information used to support the UAS operation. Examples are military intelligence centers that provide tasking orders and receive elaborated results (e.g., imagery), cartography providers whose maps are used to support mission planning, or ATC that can receive the resulting flight plan. Interoperability with different UAV platforms is related to the concept of a common UCS, capable to control different types of UAVs and also adaptable to different types of payloads.

Interoperable solutions are based on the use of standardized interfaces and open architectures. Implementations of the first interoperability type described are generally more mature, as many of the related standards are well consolidated (e.g., map formats). On the other hand, out-of-the-box interoperability with different UAVs is still not widely available on the market. Most of the existing common UCS products are limited to UAVs of a single vendor's portfolio.

A common standard used in military systems is the STANAG 4586 "Standard Interfaces of UAV Control System (UCS) for NATO UAV interoperability" (NATO Standardization Agency, 2012), which covers both types of interoperability already described. This STANAG proposes a segregated architecture consisting of a Vehicle Specific Module (VSM), which isolates specific functions for a given UAV type, and a Core UAV Control System (CUCS), which provides generic functions potentially applicable to any UAV type. These two modules are communicated by means of a standardized Data Link Interface (DLI). The STANAG also specifies a standard interface for communicating with Intelligence Centers (or Command, Control, Communications, Computers, and Intelligence (C4I)), called Command and Control Interface (CCI), and proposes the implementation of a Command and Control Interface Specific Module (CCISM) to solve interconnection with legacy C4I systems noncompliant with CCI.

4.7 Security

When taking about overall UAS security, many works on the subject focus on communications security (or COMSEC) aspects, more related to the system data links. But the UCS, mainly being composed of interconnected computers and software, is more related to the Information Technology (IT) Security domain (sometimes also referred as computer security or cybersecurity).

The UCS makes an intensive use of information. It is the ground hub that receives, processes, stores, and disseminates mission-related information, which could be sensitive in many cases. In addition, complex UCS provides multiple and heterogeneous interfaces with external systems (sometimes even connected to the Internet), exposing potential vulnerabilities that could be exploited by malicious individuals. An unauthorized access to the UCS systems could lead to disclosure of sensitive information and even to exposure of UAV control functions. To avoid these risks, secured UCS systems implement several protection measures such as identification, authentication and access control mechanisms, use of firewalls, and encryption, among others.

Government agencies can also require formal certification processes to demonstrate that the UAS is compliant with a given set of security standards, like the Common Criteria for Information Technology Security Evaluation (Arrangement on the Recognition of Common Criteria Certificates in the Field of IT Security, 2012).

5   FUTURE TRENDS

Forecasts are obviously subject to errors, and this is especially true for UAS, being a new, nonestablished technology that is constantly evolving. Nevertheless, research efforts on the field of UAS and particularly on the UCS area depict some trends that seem to sketch the future capabilities of these systems. Some of them are described as follows.

- *Advanced Human Machine Interfaces*: Human factor is a key aspect of control stations and also a very active field of research. The latest HMI technologies such as immersive interfaces are being included to increase the operator situational awareness, to reduce mental workload, and to improve safety.
- *Automation*: Automation is probably the key research topic in UAS. Although most efforts are oriented toward providing automation on the UAV side, many autonomy functions are also applicable to UCS. Examples are automated mission planning, decision support tools, and human factor aspects of automation.
- *Multi-UAV Control*: Another active area of research is the control of several UAVs by a single operator. It comprises two different paradigms: conventional schemas based on individualized UAV tasks and the concept of UAV swarms, where a set of several UAVs is commanded as a whole. This capability, particularly in the case of swarms, is tightly related to automation.

6   CONCLUSIONS

Control stations are an essential part of unmanned air systems that have evolved from the first remote control boxes and rudimentary on-ground cockpits to current modern systems making intensive use of hardware and software technologies. They serve as the interface that allows the operators to interact with the complete UAS system and provide an extensive set of functions to support UAV operation. As in the case of UAV platforms, many UCS typologies exist, which range from the most simple portable devices used to manage mini-UAVs up to the complex, distributed systems designed for the large MALE and HALE UAVs. Human factors, certification, safety, interoperability, security, or environmental conditions are some of the key elements to be considered in the design of UAV Control Stations. UCS-related technologies are also an active field of research, and significant evolutions in HMI, automation, and multi-UAV operation are expected in the following years.

7   ACKNOWLEDGMENTS

The author wishes to acknowledge his colleagues from the different Airbus Defence & Space UAS related areas, especially from the UAS Ground Segment Department, for their support during the preparation of this chapter.

REFERENCES

Arrangement on the Recognition of Common Criteria Certificates in the field of IT Security (2012) Common Criteria for Information Technology Security Evaluation, version 3.1 revision 4, Cheltenham: CESG (UK CCRA member).

ASTM International (2010) *E1925-10 Specification for Engineering and Design Criteria for Rigid Wall Relocatable Structures*, ASTM, West Conshohocken.

Defense Industry Daily (2011) *It's better to share: breaking down UAV GCS barriers*. Available at http://www.defenseindustrydaily.com/uav-ground-control-solutions-06175 (accessed August 8, 2014).

European Aviation Safety Agency (2009) *E.Y013-01 Airworthiness Certification of Unmanned Aircraft Systems (UAS)*, EASA, Cologne.

European Aviation Safety Agency (2012) *CS-23, Certification Specifications for Normal, Utility, Aerobatic, and Commuter Category Aeroplanes, Amendment 3*, EASA, Cologne.

Gato, J.L. and Ramos, F.J. (2013) Certification of UAS – control stations perspective. *UAV World 2013*, Frankfurt, 5–7th November 2013.

International Civil Aviation Organization (2012) ICAO Amendment 43 to Annex 2 – Amendment to Definitions, Speed Variations, and Remotely Piloted Aircraft, Montreal: ICAO.

International Standards Office (2010) ISO 9241-210 Ergonomics of Human–System Interaction – Part 210: Human-Centred Design for Interactive Systems, ISO, Geneva.

Maddalon, J.M., Hayhurst, K.J., Koppen, D.M., Upchurch, J.M., Morris, A.T., and Verstynen, H.A. (2013) Perspectives on Unmanned Aircraft Classification for Civil Airworthiness Standards, Appendixes A–D. NASA Langley Research Center, Hampton.

NATO Standardization Agency (2009) STANAG 4671 (Edition 1) – Unmanned Aerial Vehicles Systems Airworthiness Requirements (USAR), NATO Standardization Agency, Brussels.

NATO Standardization Agency (2012) STANAG 4586 (Edition 3) – Standard Interfaces for UAV Control System (UCS) for NATO UAV Interoperability, NATO Standardization Agency, Brussels.

Object Management Group (2012) OMG Systems Modeling Language (OMG SysML™) Version 1.3, OMG, Needham.

Radio Technical Commission for Aeronautics (1992) DO-178B Software Considerations in Airborne Systems and Equipment Certification, RTCA, Washington.

Radio Technical Commission for Aeronautics (2000) DO-254 Design Assurance Guidance for Airborne Electronic Hardware, RTCA, Washington.

Society of Automotive Engineers (1996a) ARP 4761 Guidelines and Methods for Conducting the Safety Assessment Process on Civil Airborne Systems and Equipment, SAE, Warrendale.

Society of Automotive Engineers (1996b) ARP 4754 Certification Considerations for Highly-Integrated or Complex Aircraft Systems, SAE, Warrendale.

Trimble, S. (2005) *The Great Debate: UAV vs UAS*. Available at http://www.flightglobal.com/blogs/flight-international/2005/08/the-great-debate-uav-vs-uas (accessed August 8, 2014).

United States Department of Defense, (2012) MIL-STD-1472G Department of Defense Design Criteria Standard: Human Engineering, US DoD, Washington.

Wilson, J.R. and Corlett, N. (2005) *Evaluation of Human Work*, CRC Press, Boca Raton, p. **5**.

Chapter 11
Propulsion Systems

John T. Economou[1] and A. Savvaris[2]

[1]*Centre for Defence Engineering, Defence Academy of the United Kingdom, Cranfield University, Swindon, UK*
[2]*Centre for Cyberphysical Systems, Institute for Aerospace Sciences, Cranfield University, Cranfield, UK*

1 INTRODUCTION

The Wright brothers and their mechanical expert Charlie Taylor had decided for their powered flight to use an internal combustion engine that could deliver during thermal steady-state conditions a 12 hp power output at approximately 1000 RPM. The engine did not use spark plugs but instead used two contact points within the engine combustion area. The engine sparking energy required was supplied by the batteries while on land, whereas the magneto system supplied the sparking during the flight. The batteries compared to today's standards were much heavier and bulkier. Once the engine started, then the magneto system supplied the necessary energy and therefore the aircraft could takeoff without the batteries.

The magneto system operated at 10 V, 4A DC power and was driven from the running engine. A man was required on the aircraft to control the flight. In May 1834, the first successful example (Figure 1, milestone 1, MS1) of a usable working electric motor was demonstrated by Prussian Moritz Jacobi. However, at the time the battery technology was limited to ground applications due to its volume and mass

overheads. With the Wright brothers the battery volume and mass were an obvious technological barriers that offered a function that would perform on land however it was not practical as a means for propulsion in flight other than limiting its use to the initial engine sparking subsystem prior to takeoff.

Evidently from the 1850 (milestone 2, MS2), the internal combustion engine was developing as well as the electric motors. However, by today's standards, these were at a very early stage whereby reliability and longevity was still maturing. The big question behind these early propulsion developments is, how is it possible nowadays to have UAVs that were less than 20 g in 2012 (milestone 7, MS7), that carry useful payloads, and can have endurance utilizing electric motors of up to 25 min? What are the important technological developments within the last century that have resulted in these great achievements? Figure 1 illustrates the key developments that capture the key technological milestones that when integrated offer capabilities that otherwise would not be possible.

Meanwhile in the United States, the development of the Kettering Bug in 1918 contributed significantly in the development of the UAV navigation principles based on sensing and automatic actuation of the control surfaces with rudimentary sensors, including a gyroscope and an altimeter/barometer. These are impressive achievements for that time especially considering that the incorporation of feedback was critical for the successful use of autonomy in UAVs (milestone 4, MS4).

The Wright brothers' engine (milestone 3, MS3) was preset to operate at a given angular velocity hence there was no speed control. A milestone development of combustion engines was made by Frank Whittle in 1930 with the

Unmanned Aircraft Systems. Edited by Ella Atkins, Aníbal Ollero, Antonios Tsourdos, Richard Blockley and Wei Shyy.
© 2016 John Wiley & Sons, Ltd. ISBN: 978-1-118-86645-0.

development of the gas turbine (milestone 5, MS5). Electrical UAVs use very efficient motors thus converting most of the electrical source power into useful mechanical power. Historically, resistor networks could be used to divide the power delivered to the motor hence a form of speed control could be achieved.

This method would result in a bulky and inefficient speed control especially at low-to-medium speed requirements. Alternatively, a DC supply could be used by simply increasing with suitable contacts the supply voltage thus achieving higher motor velocities for given load torque. Another method therefore was necessary, which could reduce the packaging and increase operational efficiency. The important breakthrough was the development of the transistor that was first developed at Bell Laboratories in December 1947 (milestone 6, MS6). Figure 2 illustrates a simple velocity control schematic that captures the operational principle of an electric DC UAV thruster. The electrical power source can be either a battery, fuel cell (Pukrushpan *et al.*, 2004), supercapacitor (Miller, 2004) engine/generator based, or a combination of these systems (Moir and Seabridge, 2008). The choice of electrical motor is a DC-type brushed motor with electrical coil inductance L and coil armature resistance R. The motor is directly connected to the UAV propeller. Normally, the Electric Motor Speed Control is a pulsing system whereby it utilizes a power semiconductor switch similar to a power transistor that is pulsed at a suitable frequency for the given electric motor. The duty-cycle for

the same switch can be set either by the end user or by from a suitable UAV autopilot loop that sets the speed. Similar topologies to Figure 2 can be replicated for more complex motors such as brushless DC motors; however, the control algorithms and switching electronics for the semiconductor pulsing are more challenging since typically three phases are necessary to be managed rather than one.

1.1 Propulsion variants

UAV propulsion can be categorized into two main classes: (a) internal combustion engine (ICE) and (b) all-electric. Figure 3 shows these two main classes and their subcategories.

Larger UAVs such as, for example, the Predator (by General Atomics) use a reciprocating piston engine type of propulsion producing 115 hp. Similar propulsion technologies are used also for smaller UAVs such as the Pioneer (RQ-2B) producing 26 hp, and the Northrop Grumman (RQ-5A Hunter) that produces a power output of 57 hp. The graph in Figure 4 is a representation of several UAVs and shows how the vehicle gross mass and the propulsion power output correlate.

Figure 4 shows different propeller-based UAV propulsion technologies such as reciprocating piston engines (RPE) and Wankel rotary engines (WRE). Wankel engines operate at higher exhaust temperatures and are normally liquid cooled that inevitably adds complexity and weight (see

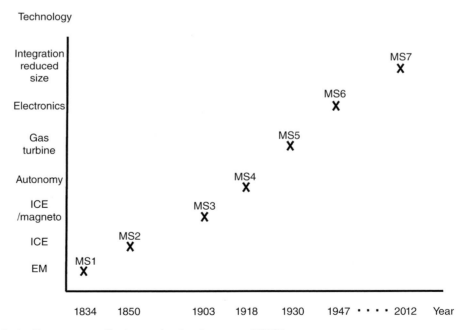

Figure 1. Technological milestones contributing to the development of UAVs.

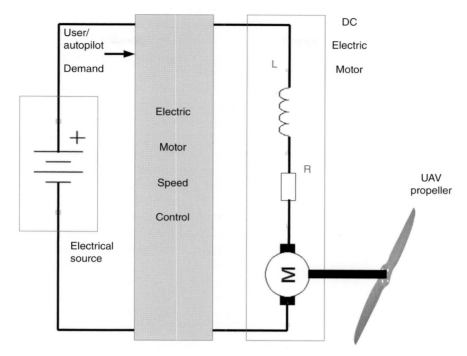

Figure 2. Rudimentary all-electric UAV propulsion topology using pulsing.

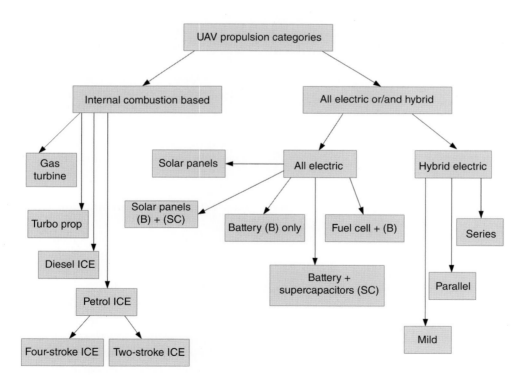

Figure 3. UAV classes based on the propulsion technology.

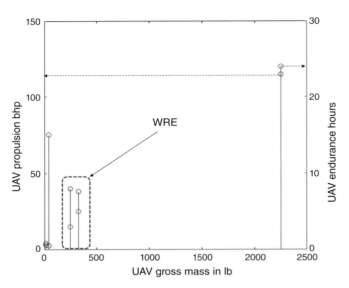

Figure 4. Examples of UAVs showing gross mass in relation to the propulsion bhp and UAV endurance.

US Department of Transportation, DOT/FAA/AR-09/11, 2009). An additional complexity with Wankel engines is the gradual burning of oil that for UAVs could affect and limit their range. Technically, the seals within WRE engines are important, since they assure that the no compression is lost thus power can be maintained throughout a mission.

In general, WRE are lighter when compared to four-stroke reciprocating piston engines.

Figure 5 shows the specific power P_s for the propulsion technology for five UAVs. The propulsion density is given from Equation 1:

$$P_s = \frac{\text{Propulsion output power}}{\text{Propulsion pack mass}} \quad (1)$$

$$E_s = \frac{\text{Propulsion energy}}{\text{Propulsion energy pack mass}} \quad (2)$$

Equation 2 allows the analysis of the UAV endurance via the specific energy E_s, whereby normally short-endurance UAVs can be designed to be more lightweight and are more popular in use.

Figure 4 relates the UAV bhp propulsion output rather than the propeller thrust. Hence, gas-turbine UAV engine examples are not included in Figure 4 because these normally generate directly thrust rather shaft power. Figure 4 shows three UAVs with RPE and two UAVs with WRE. Although many UAVs are omitted from Figure 4, it is generally the case that for extending endurance the UAV gross mass needs to increase significantly while also allowing a useful payload.

The five UAVs considered for this part are the MQ-1 Predator, ScanEagle A, Honeywell: MAV, RQ-7A Shadow 200, and the Sikorsky: Cypher. For relevant data, please see US Department of Transportation, DOT/FAA/AR-09/11 (2009).

Furthermore, to the internal combustion engine technologies, researchers have also investigated the hybrid electric options that incorporate the mixed benefits of both the engine and the advances in electrical thrusters and electronics/controls. In Merical *et al.* (2014), a small series hybrid electric propulsion system was designed and simulated using heavy fuels such as diesel, JP-5, and JP-8. The important improvement is that the internal combustion engine was decoupled from the propeller. The UAV propulsion design incorporated and integrated starter alternator that enabled restart of the engine after running in an electric-only mode. The series hybrid propulsion topology offers the advantage of utilizing the ICE at its optimum range for given flight conditions thus improving the UAV range. Consequently, as the number of propulsion subsystems increases the gross propulsion power increases, which can adversely reduce the useful UAV payload. Part of the UAV propulsion research is the reduction of mass and volume for all subsystems, thus maximizing the UAV payload while retaining the series hybrid propulsion benefits.

From Figure 5, the WRE can reach a specific power of $1.6 \, \text{bhp} \, \text{lb}^{-1}$ when compared to a specific power of approximately $1 \, \text{bhp} \, \text{lb}^{-1}$ for RPE. This is a significant advantage that for the same class of UAV can result in higher payload capability when WRE are used instead of RPE.

Rocket propulsion is generally a technology that can assist a UAV to takeoff (i.e., takeoff assist technology) and reach safe flight speeds without the need of long runways or complex infrastructure. The RQ-2B Pioneer UAV is an example whereby rocket propulsion is used as a takeoff assist. Assets that utilize rocket-only technology are normally single-use systems or systems that are capable of gliding.

1.2   Electrification propulsion variants

Miniaturization of electronics including sensor technologies has resulted in the design and development of many smaller size UAVs. Typical examples are the SoLong UAV that employs a brushless 3-phase motor as its propulsion incorporating a suitable gear reduction delivering $2.66 \, \text{W} \, \text{g}^{-1}$. With ICE-type UAVs as the fuel is combusted the UAV mass reduces as time progresses. However, with battery or battery/supercapacitor-type UAVs, the source mass does not reduce as time progresses but it remains constant. On the contrary, fuel cell- (FC) powered UAVs, assuming that these

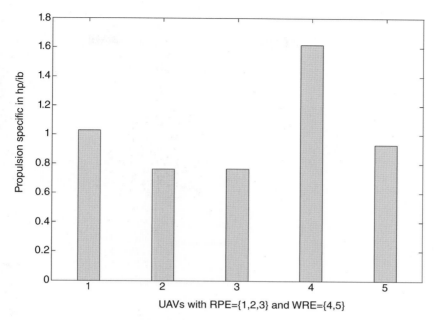

Figure 5. Propulsion-specific power in relation to RPE and WRE technologies.

utilize hydrogen, then the fuel is reacted with oxygen that in turn produces electricity. Normally, these types of UAV propulsion systems require a considerable mass budget for the hydrogen storage. Hence, when considering specific power calculations with electrical propulsion systems, it is very important to state the assumptions surrounding the claimed performance characteristics.

The SoLong UAV, for example, has a propulsion specific power of $800\,W/300\,g = 2.66\,W\,g^{-1}$ that is very impressive. The UAV vehicle gross mass to propulsion ratio is $12600/300 = 42$, whereas the MQ-1 Predator has vehicle gross mass to propulsion ratio of $2250/150 = 15$. The MQ-1 Predator delivers over 24 h flight time and the SoLong UAV delivers 48 h flight time with the ceiling being 25 000 ft for the Predator and 8000 ft for the SoLong UAV. Smaller UAVs such as the Lockheed Martin Desert Hawk can deliver normally up to 1 h flight time. Hence, depending on the role required the variants of propulsion technologies available today can offer a vast range of opportunities from pure electric solutions for smaller UAVs all the way to larger UAVs in which ICE type of technology is used.

UAV propulsion technologies defined in this chapter so far are "hard-based approaches" that have developed significantly over the last 20 years. The need to reduce the UAV size while increasing the useful payload for a range of missions is more challenging than ever before. It is therefore necessary to also consider "soft"-based approaches whereby software-based algorithms can add value into the overall UAV propulsion efficiency. Economou and Kladis

(2009) have shown soft-based methods for improving UAV endurance well beyond what is possible with simple use of existing technology. These methods are also known as "intelligent power management," which utilize suitable control methods that are fully programmable and in most cases would require the use of semiconductor switches for the activation/deactivation of certain power paths. Figure 6 shows a simplified example to illustrate how a UAV with multiple power sources, both conventional electric battery and three supercapacitor banks, powers the electric motor/propeller via a suitable pulsing speed control system (also refer to Figure 1).

Based on Figure 6, the UAV may require to takeoff or rapidly gain altitude, therefore requiring peak power delivery rather than steady-state power delivery. For these scenarios, the SC banks can provide the peak power for the necessary time, and when the UAV reaches the required altitude, then the main battery source can provide the steady-state power delivery. This potential use of power delivery to the electrical motor offers improved efficiency, thus allowing longer flight times.

In order to implement such a power delivery service, the use of semiconductor switches are necessary, for example, power electronics (PE), which will allow the power flow traffic around the UAV propulsion subsystems.

Figure 6 also shows within the context of UAVs the importance of the sensor signals, the control signals (low power), and the power line-related flows that relate to the electric motor requirements. The sensor signals allow the incorporation of

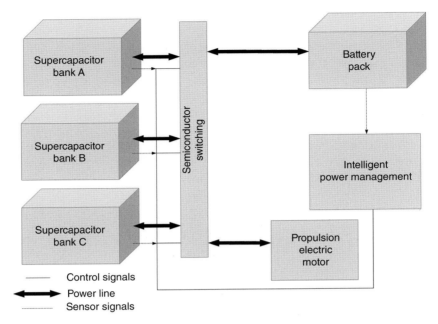

Figure 6. Multipower source UAV propulsion topology with "soft"-based approaches.

control (feedback) so that the "soft" part of the system (the intelligent power management) can decide, based on the user or the autonomous functions that the UAV may have, to switch on/off (charge and discharge) the appropriate set of capacitors while maximizing operational efficiency.

In Economou and Knowles (2011), a fuzzy hybrid modeling approach was utilized in order to model the parametric perturbations for an aerial vehicle electrical propulsion system. The method resulted in simulating a UAV under different operating conditions, such as altitude, thus relating to the motor parametric variations and the impact these changes had in the UAV state variables.

When small-size quadcopters are considered, then electrification of the propulsion systems offers better control of its thrusters. However, if one thruster fails, then the UAV (quadcopter) cannot be controlled effectively using simply individual actuator speed control. In Mueller and D'Andrea (2014), the authors discuss that for a quadcopter UAV to maintain normal flight it would normally require to be a multicopter utilizing, for example, six thrusters or more thrusters. Mueller and D'Andrea (2014) designed a control methodology whereby periodic solutions resulted in the quadrocopter maintaining a height around a position in space while having one thruster failure. In Segui-Gasco *et al.* (2013), the authors introduced a fault-tolerant propulsion method whereby three actuation mechanisms utilize gyroscopic torques, thrust vectoring, and differential thrusting.

The propulsion approach kept the UAV fault tolerant for up to two actuator failures.

1.3 Soft methods – intelligent power management and energy conservation

Traditionally, UAVs are remotely operated; however, based on current and future trends, different levels of autonomy have been defined, which define level one as "remotely guided" while level ten represents "fully autonomous flight." Based on the work shown in Economou and Kladis (2009), a methodology was developed whereby the propulsion system collaborates/interacts with the UAV guidance.

In Figure 7, a UAV consists of two electric motors. The UAV is autonomous with suitable guidance algorithms. The intelligent power management performs the functions described earlier. Figure 7 shows the conventional Guidance approach whereby the algorithms define the UAV propulsion requirements. This approach assumes that the UAV has suitable energy reserves to perform a given trajectory. However, this method is limited because it does not take into account the energy reserves. Figure 7 also demonstrates that with the Reflective Guidance method as developed by Economou and Kladis (2009) the algorithm incorporates from the start the energy reserves as well as the operator's trajectory requests.

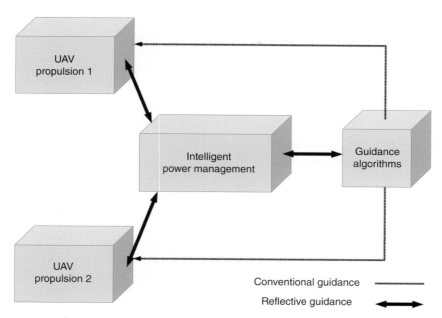

Figure 7. Intelligent power management propulsion and guidance collaboration (reflective guidance).

Figure 8 shows a three-dimensional space of a given example map (representing peak altitude around point 4 and low altitude around points 1 and 14). The operator can assign suitable waypoints and then request an energy optimum UAV path solution starting from a predefined waypoint and ending at a given end waypoint.

Thus, the overall mission path is influenced not only from the user requirements/operator but also from the UAV

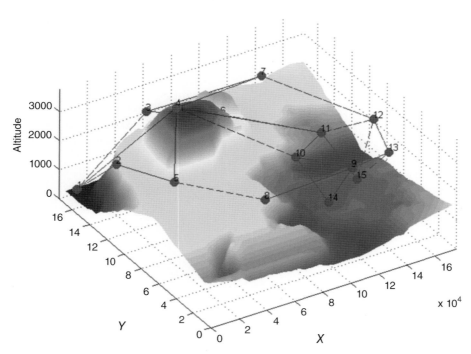

Figure 8. Example of UAV energy optimum path using intelligent power management propulsion and guidance collaboration (reflective guidance).

onboard availability of gross energy. This method results in a realistic flight plan that ensures UAV arrival to its destination with sufficient energy resources and utilization of the propulsion systems.

2　CONCLUSIONS

This chapter has explored the very first steps of aviation and the underpinning propulsion technologies, including the use of an internal combustion engine and also electrical propulsion solutions, and the technological challenges that engineers faced early on in the development of aerial vehicles. Nowadays, the rapid development of semiconductor technology that has resulted in the miniaturization of the control algorithms for the propulsion systems together with the development of light and compact sensors and improved efficiency of electrical power sources, electric motors, and engines including turbines has resulted in the design and development of much smaller UAVs being propelled by purely electrical thrusters whereas larger UAVs may be propelled by turbines or piston/Wankel internal combustion engines. The advancement of onboard processing power and effective communications have all contributed in impressive autonomous or/and command control UAVs that offer to the user unparalleled superiority. Furthermore the development of intelligent propulsion management algorithms has resulted in using the UAV propulsion systems' optimum or near-optimum conditions, thus resulting in improving range and endurance.

NOTATION AND NOMENCLATURE

AC	alternating current
bhp	brake horse power
DC	direct current
E_s	specific energy
EMSC	electric motor speed control
FC	fuel cell
hp	horsepower
ICE	internal combustion engine
MSi	milestone i-th
P_s	specific power
PE	power electronics
RPE	rotary piston engine
RPM	rotation per minute
SC	supercapacitor
SCB	supercapacitor banks
UAV	uninhabited aircraft/aerial vehicle
UAVP	uninhabited aircraft/aerial vehicle propulsion
WRE	Wankel rotary engine

REFERENCES

Economou, J.T. and Kladis, G. (2009) Intelligent Power Management for Autonomous UAV Systems. Systems Engineering for Autonomous Systems Defence Technology Centre, *Proceedings of the Fourth Annual Conference,* July 7–8, 2009, Edinburgh International Centre, Section C4.

Economou, J.T. and Knowles, K. (2011) Sugeno Inference Perturbation Analysis for Electric Aerial Vehicles, in *Electric Vehicles Modelling and Simulations* (ed. S. Soylu), Intech, Croatia, pp. 397–416.

Merical, K., Beechner, T., and Yelvington, P. (2014) Hybrid-electric, heavy-fuel propulsion system for small unmanned aircraft. *SAE Int. J. Aerosp.,* **7**(1), 126–134.

Miller, J.M. (2004) *Propulsion Systems for Hybrid Vehicles,* IEE Power & Energy Series 45, UK.

Moir, I. and Seabridge, A. (2008) *Aircraft Systems, Mechanical, Electrical, and Avionics Subsystems Integration,* 2nd ed., John Wiley & Sons, Inc., New York.

Mueller, M.W. and D'Andrea, R. (2014) Stability and control of a quadrocopter despite the complete loss of one, two, or three propellers. *IEEE International Conference on Robotics & Automation (ICRA),* Hong Kong Convention and Exhibition Center, May 31–June 7, Hong Kong, China.

Pukrushpan, J.T., Stefanopoulou, A.G., and Peng, H. (2004) *Control of Fuel Cell Power Systems: Principles, Modelling, Analysis and Feedback Design,* Springer.

Segui-Gasco, P., Al-Rihani, Y., Shin, H-S., and Savvaris, A. (2013) A novel actuation concept for a multi rotor UAV. *J. Intell. Robot Syst.,* doi 10.1007/s10846-013-9987-3.

US Department of Transportation, DOT/FAA/AR-09/11 (2009) Unmanned Aircraft System Propulsion Systems Technology Survey.

Chapter 12

Power Generation and Energy Management

Jack W. Langelaan

Department of Aerospace Engineering, The Pennsylvania State University, University Park, PA, USA

1 INTRODUCTION

Energy sources and management of available energy remain a key consideration in both aircraft design and mission planning for unmanned aircraft (indeed, this is key for manned aircraft as well). Small unmanned aerial vehicles (UAVs) are especially affected. Small size leads to lower Reynolds numbers, which results in lower aerodynamic efficiency. At the same time, payload and energy storage capacity tends to be lower, and there is a significant tradeoff between energy storage (i.e., range and endurance) and payload capacity. Further, the relatively low flight speeds of small UAVs means that winds aloft have a significant effect on range.

This chapter is not intended to be complete (indeed, one could fill several books on these topics). It should also be noted that technologies are rapidly improving (energy density of batteries and efficiency of solar cells are two notable examples), hence capabilities will soon outstrip the examples provided here. However, the methods for performance analysis and development of control and flight planning algorithms should remain applicable.

Internal combustion engines remain the primary source of power for UAVs larger than about 20 kg or those that have missions that require long range and/or endurance. Examples of fairly small aircraft include Scan Eagle, Blackjack, Aerosonde, and Pioneer; larger aircraft include Gnat and Watchkeeper, with turboprop and turbofan engines powering medium- and high-altitude aircraft such as Predator and Global Hawk, respectively.

Battery power is used almost exclusively in small (i.e., hand-launchable) aircraft such as the Raven and Puma. Many research UAVs (which tend to be purpose-built) are battery powered: the ease of use of battery power outweighs any endurance advantage that can be obtained from liquid-fueled engines, and the reliability of electric propulsion systems at small scale tends to be significantly better than fuel engines of comparable power.

Solar powered aircraft (both manned and unmanned) have been studied for several decades (Hall et al., 1985; Colozza, 1997). In 1974, the world's first entirely solar powered aircraft "Project Sunrise" flew over Bicycle Lake, California. A total of 28 flights were conducted before the aircraft was destroyed in turbulence, demonstrating the feasibility of solar powered flight.[1] Project Sunrise had a wingspan of 10 m, a maximum weight of 12 kg, and a wing loading of 1.2 kg m^{-2} (for comparison, typical high-performance sailplanes fly at a wing loading of approximately 50 kg m^{-2}, and the RQ-11 Raven UAV flies at a wing loading of approximately 5.5 kg

Unmanned Aircraft Systems. Edited by Ella Atkins, Aníbal Ollero, Antonios Tsourdos, Richard Blockley and Wei Shyy.

---

[1] http://www.projectsunrise.info

Table 1. Energy/fuel and power system for a selection of unmanned aircraft

Name	Wing span (m)	MTOW (kg)	Cruise altitude (m)	Engine type	Fuel	Mission
RQ-11 Raven	1.4	1.9	150 (AGL) 4200 (max.; MSL)	Electric	Battery	ISR
QinetiQ Zephyr	22	53	21 000	Electric	Solar	Research
Helios	75.3	929	29 523	Electric	Solar	Research
Scan Eagle	3.1	22	4570	Piston	Gasoline, JP-5, JP-8	ISR
RQ-21A Blackjack	4.8	61	6000	Piston	JP-5, JP-8	ISR
RQ-7 Shadow	4.3	170	4570	Rotary	Gasoline	ISR
Gnat 750	10.75	520	7600	Rotax	Gasoline	ISR
MQ-9 Reaper	20	4760	15 240	Turboprop	Heavy	ISR/strike
Global Observer	53	1800	16 700	Piston	Hydrogen	Research
RQ-4 Global Hawk	40	14 628	18 288	Turbofan	Heavy	ISR

m^{-2}). Since Project Sunrise, several solar powered aircraft have flown. Notable examples include Gossamer Penguin (1979), Solar Challenger (1980), SunSeeker (1990), Pathfinder (1995), Helios (2001), SoLong (2005), QinetiQ Zephyr (2008), and Solar Impulse (2015).

Fuel-cell-powered aircraft are still primarily a research tool, although commercial products are under development. In 2007, researchers at Georgia Institute of Technology demonstrated a fuel-cell-powered UAV (Bradley et al., 2007). In 2008, Boeing flew a manned fuel-cell-powered airplane (a modified Dimona motor glider, although this aircraft used a battery assist for takeoff and climb). In 2013, the US Navy Research Lab's Ion Tiger UAV flew for 48 h using a hydrogen fuel cell and liquid hydrogen (Stroman et al., 2014).

The remainder of this chapter begins with an overview of some currently available sources of onboard energy: liquid-fuel internal combustion, battery electric, solar-powered electric, and fuel cell electric are discussed. Next, flight planning tools to maximize energy efficiency in complex wind fields are described, with results of Monte Carlo simulations demonstrating the utility of the methods. Finally, this chapter discusses the utility of obtaining energy directly from the atmosphere via soaring flight.

2 ONBOARD ENERGY SOURCES AND DESIGN IMPLICATIONS

Not surprisingly, an aircraft's mission has a significant influence on the type of energy/power system that can feasibly be used. Table 1 shows a (noncomprehensive) selection of unmanned aircraft and their respective energy sources. Trends closely mirror those observed in manned aircraft: high-altitude, long-range missions tend to use turbofan engines; medium-altitude aircraft use turboprops or turbocharged piston engines; and low-altitude aircraft use piston or rotary engines. Very small aircraft (such as those intended for squad-level reconnaissance) are battery powered.

A key consideration is the energy density (energy per unit volume) and specific energy (energy per unit mass) of fuels/energy storage media typically used in aircraft. Energy properties of various fuels commonly used in aviation are provided in Table 2.

Table 2. Energy densities and specific energy of fuels typically used in aircraft (GFC, 2009). Data for "electricity" assumes a lithium-polymer cell

Fuel type	ρ (at 15°C) lb/gal (kg m^{-3})	Energy density BTU/gal (Wh L^{-1})	Specific energy BTU/lb (Wh kg^{-1})	Energy ratio to 87 octane mogas
Unlead mogas	6.09 (729.7)	115,000 (8903)	18,883 (12,202)	1.0
B100 bio-diesel	7.27 (871.1)	118,300 (9158)	16,272 (10,512)	1.0287
B20 bio-diesel	7.13 (854.4)	127,250 (9852)	17,847 (11,542)	1.1065
LH$_2$	0.567 (67.94)	34,644 (2682)	61,100 (39,517)	0.3013
Electricity (LiPo)	16.81 (2,014)	4676 (362)	278.6 (180)	33.7 kWh/gal
100LL avgas	6.02 (721.4)	120,000 (9290)	19933 (12,877)	1.0435
Jet A	6.76 (810.0)	135,000 (10,452)	19,970 (12,904)	1.1739
Diesel	7.09 (849.6)	129,500 (10,026)	18,265 (11,800)	1.1261

2.1   Combustion Engines

Internal combustion engines continue to be the primary source of power for aircraft larger than 20 kg. Heavy fuels such as diesel are an attractive energy source for many military UAVs because of commonality with other vehicles (e.g., trucks).

Derivations of range and endurance for propeller-driven aircraft can be found in any aircraft performance textbook. For completeness (and to show differences between aircraft whose mass varies during flight and constant mass aircraft such as battery-powered vehicles), the derivation is briefly repeated here. Assuming that specific fuel consumption (SFC) has units *weight/(power × time)* (which is equivalent to 1/*length*):

$$\dot{m}_f g = \mathrm{SFC}\frac{P_\mathrm{a}}{\eta_p} \tag{1}$$

Aerodynamic power is

$$P_\mathrm{a} = D v_\mathrm{a} = \frac{mg}{\frac{L}{D}} v_\mathrm{a} \tag{2}$$

Substituting and noting that $\dot{m}_f = -\dot{m}$,

$$-\frac{dm}{m} = \frac{\mathrm{SFC}}{\frac{L}{D}\eta_p} v_\mathrm{a} dt \tag{3}$$

thus

$$dt = -\frac{\eta_p \frac{L}{D}}{\mathrm{SFC} \cdot v_\mathrm{a}} \frac{dm}{m} \tag{4}$$

Recognizing that $\frac{L}{D} = \frac{C_L}{C_D}$ and $v_\mathrm{a} = \sqrt{\frac{2mg}{\rho S C_\mathrm{L}}}$,

$$dt = -\frac{\eta_p}{\mathrm{SFC}} \sqrt{\frac{\rho S}{2g} \frac{C_L^{\frac{3}{2}}}{C_D}} \frac{dm}{m^{\frac{3}{2}}} \tag{5}$$

Assuming constant altitude and constant lift coefficient, endurance is

$$t = \frac{\eta_p}{\mathrm{SFC}} \frac{C_L^{\frac{3}{2}}}{C_D} \sqrt{\frac{2\rho S}{m_0 g}} \left(1 - \frac{1}{\sqrt{1 + \frac{m_f}{m_0}}} \right) \tag{6}$$

where m_0 is the zero-fuel weight. Maximizing endurance thus implies flight at the condition that maximizes $\frac{C_L^{\frac{3}{2}}}{C_D}$ (i.e., the minimum power condition).

Range can be obtained from Equation 4. Assuming zero wind, $dR = v_\mathrm{a} dt$:

$$dR = -\frac{\eta_p \frac{L}{D}}{\mathrm{SFC}} \frac{dm}{m} \tag{7}$$

Assuming the aircraft flies at constant $\frac{L}{D}$, integrating gives the Breguet range equation:

$$R = \frac{\eta_p \left(\frac{L}{D}\right)}{\mathrm{SFC}} \ln\left(1 + \frac{m_f}{m_0} \right) \tag{8}$$

Note that this range is in zero-wind conditions. Later in this chapter the effect of wind on range will be discussed.

It is instructive to examine the relationship between specific fuel consumption and the efficiency of energy conversion. One must of course take appropriate care with units: in the Society of Automative Engineers (SAE) system, SFC is typically expressed as weight of fuel consumed per unit of time per unit of power; in SI it is expressed as mass of fuel consumed per unit of time per unit of power (i.e., lb/(hp hr) and g/kWh, respectively, with 1 lb/(hp hr) = 617 g/kWh).

The efficiency of energy conversion is

$$\eta_\mathrm{ec} = \frac{P_\mathrm{out}}{P_\mathrm{in}} = \frac{P_\mathrm{out}}{\dot{m}^\mathrm{f} e_\mathrm{f}} \tag{9}$$

where e_f is the specific energy of the fuel (see Table 2).

Now using the SI definition of SFC, the fuel flow rate $\dot{m}^\mathrm{f}$ is the product of specific fuel consumption and power output, $\dot{m}^\mathrm{fuel} = \mathrm{SFC} \cdot P_\mathrm{out}$.

$$\eta_\mathrm{ec} = \frac{P_\mathrm{out}}{\mathrm{SFC} \cdot P_\mathrm{out} e_\mathrm{f}} = \frac{1}{\mathrm{SFC} \cdot e_\mathrm{f}} \tag{10}$$

If the SAE definition of SFC is used (so that $\dot{m}^\mathrm{f} g = \mathrm{SFC} \cdot P_\mathrm{out}$), energy conversion efficiency is

$$\eta_\mathrm{ec} = \frac{g}{\mathrm{SFC} \cdot e_\mathrm{f}} \tag{11}$$

Table 3 shows typical fuel consumption for several internal combustion engines along with equivalent energy conversion efficiencies. Note that smaller engines tend to have significantly lower efficiency than larger engines (20% for 2.1 kW engines versus approximately 30% for 120 kW engines).

Electric motor efficiency is generally 90–95% and speed controller efficiency is similar. The efficiency of energy conversion for electrically powered aircraft is thus 81–90%, a factor of approximately 3 greater than the efficiency of

Table 3. Typical specific fuel consumption values for internal combustion engines (mogas assumed for efficiency calculation, unless noted)

Engine	Power [kW (hp)]	SFC (g/kWh)	(lb/(hp hr))	η_{ec}
Piston UAV engine	2.1 kW (2.9)	400	0.648	0.20
Rotary UAV engine	28 (38)	352	0.57	0.23
Rotary UAV engine	55 (75)	327	0.53	0.25
Ford EcoBoost	80 kW/L	247	0.40	0.33
Lycoming O-320 (100LL fuel)	118 (160)	259	0.42	0.30

internal combustion engines! The penalty for electrical power lies in the weight of batteries that must be carried, worsening the ratio of payload per unit weight and increasing the cost of climbing to cruise altitude. This is somewhat mitigated by the effect of increased wing loading on(L/D) at a given speed.

2.2   Battery Electric Power

From the standpoint of energy conversion efficiency, batteries are difficult (if not impossible) to match. However, specific energy remains a challenge: specific energy of readily available lithium-polymer batteries is a factor of 68 less than that of unleaded mogas (Table 2).

Battery power has several advantages over internal combustion engines: noise and emissions are significantly lower, and battery-powered aircraft are operationally simpler (although recharging a battery is significantly slower than refilling a fuel tank[2]). Further, battery energy density is improving steadily, current specific energy of lithium-sulfur batteries is nearly 400 Wh/kg (significantly greater than currently available lithium-polymer batteries); newer battery chemistries should give significant improvements as well. It can be further noted that the efficiency of internal combustion engines is fundamentally limited by thermodynamics, and continued improvement in battery technologies may eventually make internal combustion engines obsolete.

Energy consumption from the battery is

$$P_b = \frac{P_a}{\eta_{ec}\eta_p} \tag{12}$$

where η_{ec} is the efficiency of converting energy from the battery through the electronic speed controller and motor to the shaft. The total endurance is a function of power and energy available in the battery:

$$t_b = \frac{E_b}{P_b} = \frac{m_b e_b}{P_b} \tag{13}$$

thus

$$t_b = \frac{\eta_{ec}\eta_p}{P_a} e_b m_b \tag{14}$$

Again assuming steady horizontal flight, Equation (2) gives aerodynamic power, hence

$$t_b = \eta_{ec}\eta_p \frac{\frac{L}{D}}{v_a m g} e_b m_b \tag{15}$$

As before, $\frac{L}{D} = \frac{C_L}{C_D}$ and $v_a = \sqrt{\frac{2mg}{\rho S C_L}}$:

$$t_b = \eta_{ec}\eta_p \frac{C_L^{\frac{3}{2}}}{C_D} \sqrt{\frac{\rho S}{2 m^3 g^3}} e_b m_b \tag{16}$$

Rearranging and writing in terms of battery mass fraction (with $m = m_0 + m_b$):

$$t_b = \eta_{ec}\eta_p e_b \frac{C_L^{\frac{3}{2}}}{C_D} \sqrt{\frac{\rho S}{2 m_0 g^3}} \frac{\frac{m_b}{m_0}}{\left(1 + \frac{m_b}{m_0}\right)^{3/2}} \tag{17}$$

Specific range is the distance traveled per unit of energy consumed:

$$r_{sp} = \frac{\Delta s}{\Delta E} = \frac{ds}{dt}\frac{dt}{dE} = \frac{v_g}{P_b} \tag{18}$$

where $P_b = \frac{P_a}{\eta_{ec}\eta_p}$ is power consumed from the battery. Substituting aerodynamic power (Equation (2)):

$$r_{sp} = \frac{\eta_{ec}\eta_p\left(\frac{L}{D}\right)}{mg} \tag{19}$$

---

[2] Refueling a car at a gas station occurs at roughly 15 L min^{-1}: this is equivalent to 8 MW power input.

Clearly, specific range is maximized by flying at the speed that maximizes the product of propeller efficiency and $\frac{L}{D}$ and by minimizing weight.

In the case of a constant mass aircraft (such as battery-powered vehicles), maximum range is simply specific range multiplied by the total energy available:

$$R = r_{sp} e_b m_b \qquad (20)$$

Written in terms of battery mass fraction $\frac{m_b}{m_0}$, where m_0 is the "zero battery mass," (i.e. the mass of the aircraft with no battery),

$$R = \frac{e_b \eta_{ec} \eta_p \left(\frac{L}{D}\right)}{g} \frac{\frac{m_b}{m_0}}{1 + \frac{m_b}{m_0}} \qquad (21)$$

Not surprisingly, range increases linearly with increasing battery specific energy. Note that e_b/g has units of length, and $\frac{g}{e_b \eta_{ec}}$ is equivalent to specific fuel consumption for combustion aircraft.

Figure 1 shows the effect of fuel mass fraction m_f/m_0 and battery mass fraction m_b/m_0 on range and endurance. For reference, the Rutan Voyager has a fuel mass fraction of approximately 2.75 at takeoff, and a Boeing 747-400 has a fuel mass fraction of approximately 0.8 at takeoff. The Taurus G4 battery-powered aircraft had a battery mass fraction of 0.5.

There is a peak in endurance for battery-powered aircraft: when $\frac{m_b}{m_0} = 2$, the endurance of a battery-powered aircraft is maximized: increasing battery mass fraction increases range, but at a cost of decreased endurance. It is thus unlikely that very high battery mass fractions will be operationally useful. Note also that the peak is very "wide:" $\frac{m_b}{m_0} = 1$ gives 92% of the maximum possible endurance; $\frac{m_b}{m_0} = 1.5$ gives 98.5% of the maximum possible endurance.

2.3 Solar Power

Solar power is attractive because it is a "free" and easily replenishable source of energy. However, the usable amount of energy that can be absorbed from the Sun is currently fairly low (typical efficiency of solar cells used on UAVs is roughly 20%, but increasing steadily) and the energy obtained is dependent on the angle of solar incidence.

Details of solar power can be found in the literature, and Noth describes in detail the design and development of a small solar-powered aircraft (Sky-Sailor) (Noth, 2008). As a quick check on the feasibility of solar power for a given mission, it is helpful to develop an upper bound on allowable wing loading based on power required for level flight and the power that can be obtained from the Sun:

$$P_s = \frac{P_a}{\eta_{ec} \eta_p} + P_{systems} \qquad (22)$$

where the power available from the Sun is $P_s = I_s S_s \eta_s$, the product of Sun's intensity, the area of the solar cells, and the efficiency of converting solar energy to electricity. Writing S_s as a fraction of wing area ($S_s = c_s S$),

$$I_s c_s S \eta_s = \frac{mg}{\frac{L}{D} \eta_{ec} \eta_p} v_a + P_{systems} \qquad (23)$$

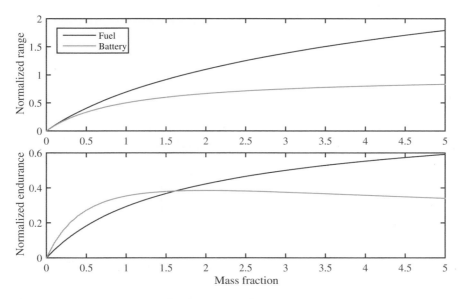

Figure 1. Comparison of normalized range and normalized endurance for combustion-powered and battery-powered aircraft.

Still assuming steady level flight,

$$I_s c_s S \eta_s = \frac{mg}{\eta_{ec}\eta_p} \frac{C_D}{C_L} \sqrt{\frac{2mg}{\rho S C_L}} + P_{\text{systems}} \qquad (24)$$

Rearranging slightly,

$$I_s c_s \eta_s = \frac{C_D}{C_L^{3/2}} \left(\frac{mg}{S}\right)^{3/2} \sqrt{\frac{2}{\rho}} \frac{1}{\eta_{ec}\eta_p} + \frac{P_{\text{systems}}}{S} \qquad (25)$$

hence the maximum possible wing loading is

$$\left(\frac{mg}{S}\right)^{3/2} = \left(I_s c_s \eta_s - \frac{P_{\text{systems}}}{S}\right) \eta_{ec}\eta_p \frac{C_L^{3/2}}{C_D} \sqrt{\frac{\rho}{2}} \qquad (26)$$

Note that this represents an upper bound: it does not directly include the effect of solar incidence angle on power. Clearly, the lower limit of wing area is that required to provide power to onboard systems such as the autopilot, servos, payload, and battery charging. From the standpoint of aircraft performance, the power required to drive systems is equivalent to a reduction in the energy that can be obtained from the solar cells, thus minimizing power requirements of onboard systems should be carefully considered during aircraft design.

Matching intuition, wing loading can increase as solar intensity increases (less wing area is needed to collect the required amount of power), and wing loading can increase as the net drivetrain efficiency increases.

Figure 2 shows limits of wing loading and aerodynamic efficiency as a function of altitude. Here $P_{\text{systems}} = 0$ (i.e., only power required to sustain level flight is considered). Further, solar intensity is assumed to be 1 kW m^{-2} (a generally accepted value for solar irradiance at sea level (Noth, 2008)); 90% of the wing's upper surface is covered in solar cells; and a net drivetrain efficiency of 20% (from solar cell through propeller) is assumed. Note that the plots assume constant net drivetrain efficiency: the effect of altitude on propeller efficiency is not directly included. For comparison, the figure also shows approximate data for a selection of sailplanes, hang gliders, and UAVs. Note that NASA's Helios solar UAV reached an altitude of 96 863 feet (29 524 m), and the SunSeeker Duo regularly flies well above 4 000 m altitude on solar power alone.

Figure 2 shows that the altitude that can be reached by a solar-powered aircraft is highly sensitive to wing loading. For wing loading greater than 10 kg m^{-2} (equivalent to a hang glider), it becomes challenging to achieve the required aerodynamic performance to enable high-altitude flight.

Since solar insolation varies significantly over the course of a day, and varies depending on latitude and time of year, aircraft designs typically aim for both lower than average insolation and include the capability to recharge batteries. The surface area of solar cells is thus significantly greater than that required purely to maintain flight, and thus wing loading is significantly lower than what is expected for "traditionally powered" aircraft with similar mission. The combination of low wing loading and a desire to fly above clouds generally leads to an aircraft designed for high-altitude missions.

2.4 Fuel Cells

The energy density of hydrogen makes fuel cells very attractive. Even when the mechanical systems are included, the specific energy of a gaseous hydrogen fuel cell drivetrain exceeds 1000 Wh kg^{-1} and specific energy of a liquid

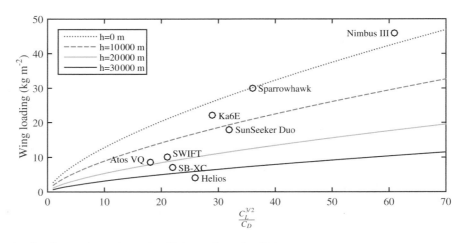

Figure 2. Allowable wing loading and aerodynamic efficiency for solar flight.

hydrogen fuel cell drivetrain exceeds $10\,000$ Wh kg^{-1}. Even when electrolyzer-based recharging systems are incorporated, net specfic energy greater than 320 Wh kg^{-1} is possible (Bradley et al., 2007). However, Bradley et al. note that a significant fraction of the total weight of small fuel-cell-powered aircraft is taken by the power system since weight does not scale particularly well as power is decreased (Bradley et al., 2007).

An interesting point to note is the relatively low energy density of liquid hydrogen (i.e., energy per unit volume). This suggests that hydrogen power may be more suited to low-speed, weight-sensitive vehicles such as Vertical Take-off and Landing aircraft rather than high-speed aircraft.

3 FLIGHT PLANNING FOR ENERGY MANAGEMENT

Both flight control and flight planning can aid in managing energy consumption during flight. Here "energy optimal" refers to flight at the condition that minimizes energy consumption per distance traveled, accounting for the effect of wind on that flight condition. This energy-optimal flight condition is the maximum specific range condition; for a propeller-powered aircraft in steady level flight in zero wind, this flight condition occurs when drag force is minimized (McCormick, 1995) (see also Equations (8) and (21)). In steady level flight, drag force is a function of weight and airspeed; thus as fuel is burned the minimum drag speed is reduced. However, winds can have a significant effect on the flight condition that maximizes specific range (Asselin, 1997; Mair and Birdsall, 1992).

3.1 Energy-Optimal Flight Speed

Referring to Figure 3, ground speed is affected by tailwind component and to a lesser extent the crosswind component (Langelaan et al., 2013):

$$v_g = \sqrt{v_a^2 \cos^2\gamma - w_c^2} + w_t \qquad (27)$$

The heading required to maintain the desired ground track is a function of airspeed and the crosswind component (the component of wind perpendicular to the ground track):

$$\psi_g = \psi + \sin^{-1}\frac{w_c}{v_a} \qquad (28)$$

Assume that flight is at constant altitude ($\gamma = 0$), hence $v_g = \sqrt{v_a^2 - w_c^2} + w_t$. It is thus clear that winds aloft will

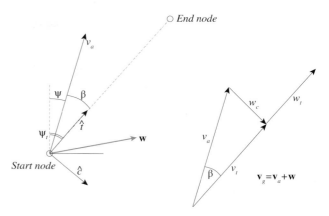

Figure 3. Track coordinate frames (left) and resolution of airspeed and wind vectors into the track coordinate frame (right). Positive angles are shown.

affect range; winds aloft will also affect the flight condition that maximizes range.

In the derivation here, assume that aircraft is propeller driven and powered by a liquid-fuel internal combustion engine. Specific range is the distance traveled for a given fuel consumption. Hence,

$$r = \frac{v_g}{m^{\cdot \text{fuel}}} \qquad (29)$$

where $m^{\cdot \text{fuel}}$ is the rate of fuel consumption.

In a propeller-powered aircraft fuel consumption depends on the power produced by the engine. In steady straight and level flight engine, power is

$$P = \frac{Dv_a}{\eta_p} \qquad (30)$$

where D is the drag force and η_p is propeller efficiency. Drag force is

$$D = \frac{1}{2}\rho v_a^2 S C_D \qquad (31)$$

where drag coefficient is typically expressed as a polynomial function of lift coefficient. In the absence of a detailed drag analysis, a second-order polynomial is generally assumed:

$$C_D = C_{D0} + \frac{C_L^2}{\pi R e_O} \qquad (32)$$

where C_L is the lift coefficient, R is the aspect ratio, and e_O is the Oswald's Efficiency, a parameter that combines span efficiency and the part of airfoil profile drag that depends on c_l^2.[3]

The rate of fuel consumption is

$$\dot{m}_{\text{fuel}} = \text{SFC} \cdot P = \text{SFC}\frac{Dv_a}{\eta_p} \tag{33}$$

Thus specific range is

$$r = \frac{\eta_p v_g}{\text{SFC}Dv_a} \tag{34}$$

To maximize range, one thus computes the flight condition that continuously maximizes specific range. Typically, two assumptions are made: (1) specific fuel consumption is approximately constant over the range of speeds typical of cruise flight; (2) propeller efficiency is approximately constant over the range of speeds typical of cruise flight (and during cruise propeller rpm is chosen to maximize propeller efficiency at that speed). If it is also assumed that wind speed is zero, then $v_g = v_a$ and (as before)

$$r_{w=0} = \frac{\eta_p}{\text{SFC} \cdot D} \tag{35}$$

In this case, maximum specific range occurs when the aircraft flies at its minimum drag speed. Note that this speed will vary continuously as fuel is burned.

Note that in addition to powering the propeller, the engine may also provide power to other aircraft systems, so that

$$P = \frac{Dv_a}{\eta_p} + P_{\text{systems}} \tag{36}$$

In this case, specific range becomes

$$r = \frac{v_g}{\text{SFC}\left(\frac{Dv_a}{\eta_p} + P_{\text{systems}}\right)} \tag{37}$$

This systems power decreases specific range and increases the speed to fly for maximum specific range.

The range maximizing airspeed can now be computed by solving the optimization problem:

$$v_a \, \text{maximize} \quad \frac{v_g}{\text{SFC}\left(\frac{Dv_a}{\eta_p} + P_{\text{systems}}\right)} \tag{38}$$

$$\text{subject to} \quad v_g = \sqrt{v_a^2 \cos^2 \gamma - w_c^2} + w_t \tag{39}$$

$$D = \frac{1}{2}\rho v_a^2 S\left(C_{D0} + \frac{C_L^2}{\pi A R e_O}\right) \tag{40}$$

$$C_L = \frac{2mg}{\rho v_a^2 S} \tag{41}$$

$$v_{a,\min} \leq v_a \leq v_{a,\max} \tag{42}$$

$$v_g > 0 \tag{43}$$

If propeller efficiency (as a function of advance ratio) is known, then one can include propeller rpm as a free variable in the optimization; similarly, if SFC (as a function of engine rpm or power) is known, then this can be included as a free variable.

Parameters for an example aircraft (a small tactical UAV) are provided in Table 4. These are representative of the RQ-21A Blackjack, but are not exact because only publicly available data were used. Resulting lift/drag ratio versus equivalent airspeed and shaft power required versus equivalent airspeed (at maximum takeoff weight) are presented in Figure 4.

Resulting speed to fly (normalized by the zero-wind maximum specific range speed, that is, the speed for best L/D) and range performance coefficient are plotted in Figure 5 as a function of tailwind component and crosswind component (both normalized by zero-wind maximum

---

[3] This formulation for drag is useful for basic performance analysis, but is less accurate when laminar flow airfoils are used on the aircraft. In that case, a detailed drag analysis will provide a significantly more accurate function of drag coefficient versus lift coefficient. Note also that Oswald's efficiency is distinct from (and should not be confused with) span efficiency.

Table 4. Nominal aircraft parameters and performance

Parameter	Value
Empty mass	42.6 kg (94 lb)
Max. fuel mass	18.6 kg (41 lb)
Wing area	1.44 m^2 (15.5 sq. ft)
Aspect ratio	16
C_{D0}	0.027
e_o	0.75
SFC (cruise)	0.421 g/Wh (0.69 lb/hp hr)
Propeller efficiency	0.8
$\frac{L}{D}\big|_{\max}$	18.7
$v_{w=0}^*$ (max. weight, zero wind)	26.0 m/s (50.5 kts)
$v_{w=0}^*$ (max. weight, zero wind and 350 W to payload)	28.1 m/s (54.6 kts)
v_a^{nom}	28.3 m/s (55 kts)

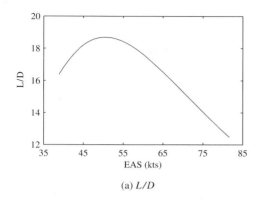

(a) *L/D*

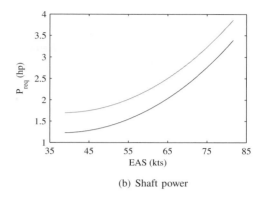

(b) Shaft power

Figure 4. Aircraft properties at maximum takeoff weight (135 lb). Light gray power curve includes 350 W to payload.

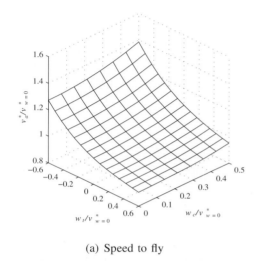

(a) Speed to fly

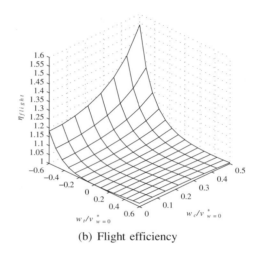

(b) Flight efficiency

Figure 5. Optimal speed to fly and flight efficiency for Integrator-class aircraft.

specific range speed). Flight efficiency (defined as a range performance coefficient) is defined as

$$\eta_{\text{flight}} = \frac{r_{\text{opt}}}{r^*_{w=0}} \tag{44}$$

where r_{opt} is specific range at the wind-optimal flight condition v^*_a and $r^*_{w=0}$ is specific range in wind when the aircraft is flying in its zero-wind maximum specific range condition.

Note that significant headwinds are required before a large effect on range performance coefficient (i.e., fuel consumption) is observed. Flying at the energy-optimal speed in a headwind of $0.34v^*_{w=0}$ gives roughly 3% improvement in fuel burn compared with flight at the zero-wind maximum range speed. This improvement is not large, but one must remember that it can be achieved with a software update in the flight control laws.

To assess the utility of energy-optimal flight, Monte Carlo simulations of flights over randomly generated flight paths in

randomly selected wind fields and altitudes were conducted. These simulations assess the improvement in fuel burn and time of flight obtained by flying at optimal speed versus nominal speed.

Wind forecasts generated for the 2011 Green Flight Challenge were used as representative wind fields for simulations. These were generated by the Numerical Weather Prediction Group of the Department of Meteorology at Penn State University using Weather Research Forecast, a state-of-the-art numerical weather prediction tool. At the Green Flight Challenge, these wind fields were used to compute energy-optimal trajectories and flight conditions; actual energy consumption measured during flight was well within 1% of predicted values.

Wind fields are shown in Figure 6. Note significant variations in both speed and direction throughout the region, and the variation in altitude and time.

In all cases, closed, triangular courses are used: it is assumed that the flight must start and end at the same point.

It is further assumed that the aircraft can compute the tailwind component and the crosswind component using GPS groundspeed, GPS track, airspeed, and heading measurements, where

$$w_c = v_a \sin(\psi_g - \psi) \tag{45}$$

$$w_t = v_g - v_a \cos(\psi_g - \psi) \tag{46}$$

A random triangular course is generated for each run by selecting waypoints and a flight altitude. The flight altitude depends on the windfield that has been randomly selected for that run (see Figure 6). To ensure reasonable courses, each segment must be at least 20 km in length and no more than 500 km in length. The total course length must be at least 250 km and no more than 1500 km.

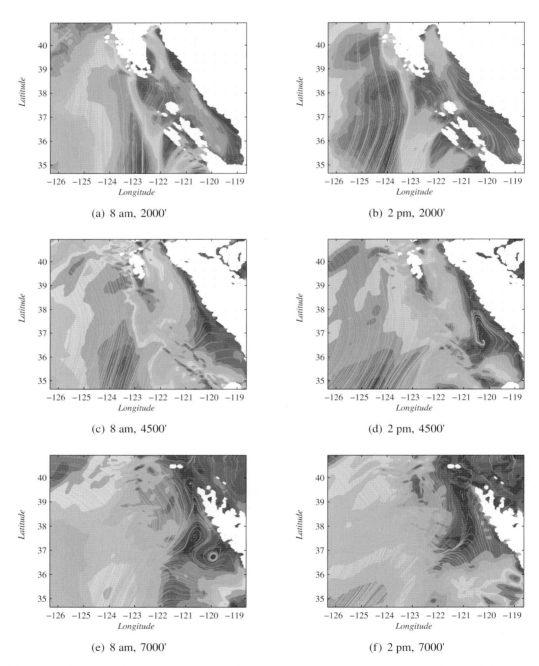

(a) 8 am, 2000'

(b) 2 pm, 2000'

(c) 8 am, 4500'

(d) 2 pm, 4500'

(e) 8 am, 7000'

(f) 2 pm, 7000'

Figure 6. Wind fields used for simulations. Streamlines show flow; contours show wind magnitude (dark gray = 0 kts; light gray = 34 kts); white means areas where terrain elevation is higher than the wind field altitude slice. Flow is generally from north to south.

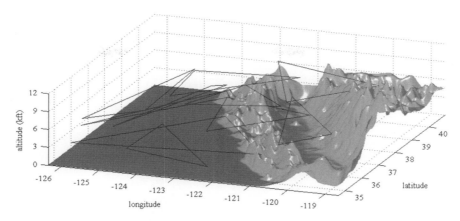

Figure 7. Sample of triangular courses used in simulations superimposed on terrain.

Terrain avoidance is not considered directly here. While turnpoints below terrain elevation were discarded as infeasible, a course leg that is below terrain over some portion was allowed. Winds for these parts of the flight were set to zero.

In all examined cases, four main parameters are compared:

- fuel burn fraction $\eta_{nom}^{opt} = \dfrac{m_{fuel}^{opt}}{m_{fuel}^{nom}}$;
- time of flight fraction $\tau_{nom}^{opt} = \dfrac{t^{opt}}{t^{nom}}$;
- average fuel burn per distance traveled;
- average ground speed.

3.2   Energy-Optimal Flight Versus Nominal Cruise Speed Flight

Before discussing Monte Carlo simulations, the effect of flight at the energy-optimal condition results of a single representative run with no wind measurement noise and nominal aircraft parameters is provided in Figure 8. For this run, the fuel burn fraction and time of flight fraction were

$$\eta_{nom}^{opt} = 0.923 \tag{47}$$

$$\tau_{nom}^{opt} = 0.773 \tag{48}$$

Thus flight at the minimum energy condition results in fuel savings of 7.7% over flight at the nominal flight speed and time savings of 22.7% over flight at the nominal flight speed. Most of the fuel and time savings occur during the second leg (from turnpoint 2 to turnpoint 3) when there is a significant headwind component: ground speed is higher when the aircraft is flying in its energy minimizing condition, thus the fuel consumption per distance traveled is lower, even though the consumption per hour is significantly higher.

Energy-optimal flight thus results in time savings as well as fuel savings. Monte Carlo simulations will show that in addition to savings in time and fuel, the variability in fuel consumption and time of flight is also reduced by flying at the energy-optimal flight condition.

Fuel burn fraction η_{nom}^{opt} and time of flight fraction τ_{nom}^{opt} for all runs are shown in Figure 9. Table 5 summarizes statistics of the results.

Interestingly, in addition to the energy-optimal flight condition showing lower average fuel consumption and higher average ground speed, the standard deviations of those quantities are also lower. Thus, there is less variability in both fuel consumption and ground speed when the aircraft flies at the energy-optimal condition.

3.3   Routing

In many cases, significant improvement in energy consumption can be realized by planning routes to exploit favorable winds or avoid adverse winds. Several methods for flight planning have been proposed: Rubio uses genetic algorithms (Rubio and Kragelund, 2003); Jardin uses neighborhood optimal control (Jardin and Bryson, 2001); Chakrabarty uses backward wavefront expansion to generate an energy map (Chakrabarty and Langelaan, 2011).

4   HARVESTING ATMOSPHERIC ENERGY

Long-endurance flight for medium- and high-altitude unmanned aircraft has become fairly common, with flights of tens of hours reported platforms such as Global Hawk and Predator. Aircraft with endurance of several days and even

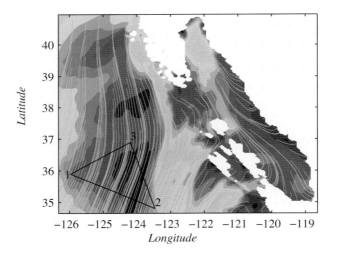

(a) Course flown

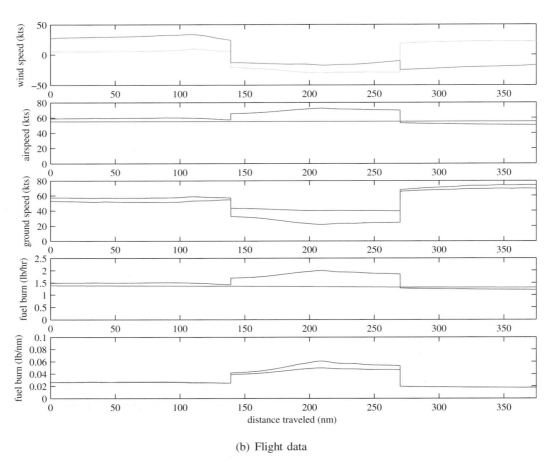

(b) Flight data

Figure 8. Speeds and fuel burn for single representative run. Course flown is shown in (a); flight data are shown in (b). Wind components (top plot) are tailwind (light gray) and crosswind (dark gray); flight data show nominal flight condition in dark gray and flight at energy-optimal condition in light gray.

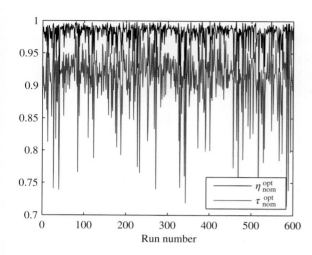

Figure 9. Fuel fraction and time fraction for each run. Note that both fuel and time are saved in all cases.

years are under development (AeroVironment's Global Observer (AeroVironment, Inc., 2010), DARPA's VUL-TURE program) and QinetiQ has demonstrated an endurance of 2 weeks with the Zephyr aircraft (QinetiQ, 2010). These vehicles are quite large (Predator's wingspan is roughly 15 m, Zephyr's wingspan is 22.5 m, GO's wingspan is 53 m, and the various VULTURE configurations have wingspans over 100 m), fly at medium to high altitude, and the endurance results from a combination of aerodynamic efficiency and the significant fuel supply which can be carried on board. With currently available sensors these vehicles are capable of very detailed surveillance from medium to high altitude as long as weather conditions are favorable. Low-level clouds, for example, will obscure ground targets from optical sensors. Thus, there is still a need for small, low-altitude (i.e., within the convective boundary layer) unmanned aircraft which can fly below clouds and can be deployed quickly by small groups of personnel. Flying large aircraft at low altitude makes them vulnerable to detection or destruction, while small vehicles are significantly more difficult to detect. Moreover, the low cost typically associated with small unmanned aircraft implies that riskier missions can be undertaken (e.g., flight in very windy or gusty conditions).

Table 5. Summary of results comparing fuel consumption and ground speed for constant airspeed versus optimal airspeed flight

Condition	Fuel consumption (lb/nm)		Ground speed (kts)	
	Mean	Std. dev.	Mean	Std. dev.
$v_a = 55$ kts	0.0263	0.00219	51.1	3.67
$v_a = v_a^*$	0.0257	0.00164	56.1	2.18

However, with these advantages come significant technical challenges. Any small robot (be it ground-, sea-, air-, or space-based) suffers from limited capacity for onboard energy storage and sensing payload. There is often an explicit trade between fuel (for increased mission duration) and payload (for increased data collection). Further, the low operating Reynolds numbers typical of small aircraft make it difficult to design a small vehicle with aerodynamic performance similar to its larger cousins. Small robotic aircraft, therefore, suffer both from reduced onboard energy capacity and from low aerodynamic performance. Together, these two factors greatly reduce the mission capabilities (and hence utility) of small robotic aircraft. Current endurance of small unmanned air systems is 1–2 h, greatly limiting the missions that can be performed. Battery technology is improving rapidly, but the capability to perform long-duration missions (such as persistent surveillance) will remain a limiting factor.

Underwater gliders such as Slocum (Webb et al., 2001), Spray (Sherman et al., 2001), and Seaglider (Eriksen et al., 2001) utilize variable buoyancy and trim to enable long transects with very little energy use. A similar technique has been proposed for exploration of Venus (Woolsey et al., 2005), but this is not practical for Earth-based unmanned aircraft (a very large volume of a light gas would be required, resulting in very high drag). However, appropriate flight techniques can permit extraction of energy from the atmosphere by heavier than air flight vehicles, greatly extending both range and mission duration of flight vehicles.

Significant energy is available from the atmosphere. Large birds and human sailplane pilots routinely exploit vertical air motion (lift) to remain aloft for several hours and fly hundreds of kilometers without flapping wings or the use of engines. Exploiting these long-duration vertical air motions has been an active area of research for manned glider flight for many years and is now becoming more active for small UAVs as well (Ostbo et al., 2004). Recent research in autonomous soaring has shown that enormous improvement in range and endurance is possible (Allen and Lin, 2007; Edwards, 2008).

Atmospheric energy can be harvested from three phenomena: (a) vertical air motion, (b) spatial wind gradients, and (c) temporal gradients (gusts). Energy harvesting from spatial gradients is known as *dynamic soaring*; energy exploitation from vertical air motion has become known as *static soaring* since the time scale of vertical air motion is long compared with vehicle dynamics.

Total energy is typically written in terms of airspeed (i.e., so that total energy is expressed in the frame of the airmass). Ignoring onboard stored energy (i.e., fuel or batteries),

$$E_{\text{tot}} = mgh + m\frac{v_a^2}{2} \qquad (49)$$

Dividing by weight gives specific energy:

$$e_{\text{tot}} = h + \frac{v_a^2}{2g} \qquad (50)$$

This has units of length and is sometimes known as energy altitude. The rate of change of specific total energy (specific power) is

$$\dot{e}_{\text{tot}} = \dot{h} + \frac{v_a \dot{v}_a}{g} \qquad (51)$$

Expressions for longitudinal aircraft dynamics are now required. Consider an aircraft located at $\mathbf{r}$ in an inertial frame I, where $\hat{x}^i$ and $\hat{z}^i$ define unit vectors (see Figure 10).

Using a common definition of stability axes, define $\hat{x}^s$ as a unit vector in the direction of airspeed (so that $\mathbf{v} = v_a \hat{x}^s$) and $\hat{z}^s$ opposite to lift. The velocity of the aircraft in the inertial frame is the sum of the velocity of the aircraft in the stability axes and the wind velocity:

$$\ddot{\mathbf{r}} = \mathbf{v} + \mathbf{w} \qquad (52)$$

$$\ddot{\mathbf{r}} = v_a \hat{x}^s + w_{i,x}\hat{x}^i + w_{i,z}\hat{z}^i \qquad (53)$$

Hence,

$$\ddot{\mathbf{r}} = \dot{v}_a \hat{x}^s + \boldsymbol{\omega}^s \times v_a \hat{x}^s + \dot{w}_{i,x}\hat{x}^i + \dot{w}_{i,z}\hat{z}^i \qquad (54)$$

Substituting $\boldsymbol{\omega}^s = \dot{\gamma}\hat{y}^s$, $\hat{x}^i = \hat{x}^s\cos\gamma + \hat{z}^s\sin\gamma$ and $\hat{z}^i = -\hat{x}^s\sin\gamma + \hat{z}^s\cos\gamma$ gives

$$\ddot{\mathbf{r}} = (\dot{v}_a + \dot{w}_{i,x}\cos\gamma - \dot{w}_{i,z}\sin\gamma)\hat{x}^s \\ +(-v_a\dot{\gamma} + \dot{w}_{i,x}\sin\gamma + \dot{w}_{i,z}\sin\gamma)\hat{z}^s \qquad (55)$$

Newton's law for aircraft longitudinal dynamics is

$$\mathbf{L} + \mathbf{D} + m\mathbf{g} + \mathbf{T} = m\ddot{\mathbf{r}} \qquad (56)$$

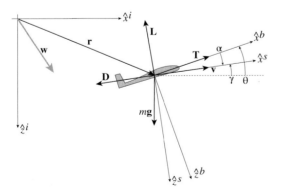

Figure 10. Reference frames for the longitudinal case. Positive rotations are indicated, so positive glideslope is upward and angle of attack is positive in the conventional sense.

where $\mathbf{L}$ is the lift vector, $\mathbf{D}$ is the drag vector, $\mathbf{g}$ is the acceleration due to gravity, $\mathbf{T}$ is the thrust vector, and m is the mass. Lift and drag are generally expressed in the stability frame, thrust is generally expressed in the body frame, and gravity is generally expressed in the inertial frame:

$$\mathbf{L} = -\frac{1}{2}\rho v_a^2 S C_{\text{L}}\hat{z}^s \qquad (57)$$

$$\mathbf{D} = -\frac{1}{2}\rho v_a^2 S C_{\text{D}}\hat{x}^s \qquad (58)$$

$$\mathbf{T} = \frac{1}{2}\rho v_a^2 S C_{\text{T}}\hat{x}^b \qquad (59)$$

$$\mathbf{g} = g\hat{z}^i \qquad (60)$$

and $\hat{x}^b = \hat{x}^s\cos\alpha - \hat{z}^s\sin\alpha$.

Thus

$$\dot{v}_a = \frac{qS}{m}(-C_D + C_T\cos\alpha) - \dot{w}_{i,x}\cos\gamma + \dot{w}_{i_z}\sin\gamma - g\sin\gamma \qquad (61)$$

$$v_a\dot{\gamma} = \frac{qS}{m}(C_L + C_T\sin\alpha) + \dot{w}_{i,x}\sin\gamma + \dot{w}_{i,z}\cos\gamma - g\cos\gamma \qquad (62)$$

where $q = \frac{1}{2}\rho v_a^2$.

For clarity, only longitudinal dynamics of unpowered flight will be considered further, thus $C_T = 0$. Substituting Equation (61) into Equation (51),

$$\dot{e}_{\text{tot}} = \dot{h} \\ + \left[-\frac{qS}{mg}C_D v_a - \dot{w}_{i,x}v_a\cos\gamma + \dot{w}_{i,z}v_a\sin\gamma - v_a g\sin\gamma \right] \qquad (63)$$

Recognizing that $\dot{h} = -\dot{z} = v_a\sin\gamma - w_{i,z}$,

$$\dot{E}_{\text{air}} = -w_{i,z} + \left[-\frac{qS}{mg}C_D v_a - \dot{w}_{i,x}v_a\cos\gamma + \dot{w}_{i,z}v_a\sin\gamma \right] \qquad (64)$$

Assuming a frozen gust field,

$$\frac{\mathrm{d}}{\mathrm{d}t}\mathbf{w} = \nabla\mathbf{w}\begin{bmatrix} \dot{x}_i \\ \dot{z}_i \end{bmatrix} = \begin{bmatrix} \frac{\delta w_x}{\delta x_i}\dot{x}_i + \frac{\delta w_x}{\delta z_i}\dot{z}_i \\ \frac{\delta w_z}{\delta x_i}\dot{x}_i + \frac{\delta w_z}{\delta z_i}\dot{z}_i \end{bmatrix} \qquad (65)$$

where $\nabla \mathbf{w}$ is the spatial gradient of the wind vector. Equivalently,

$$\frac{d}{dt}\mathbf{w} = \nabla \mathbf{w} \begin{bmatrix} v_a\cos\gamma + w_{i,x} \\ -v_a\sin\gamma + w_{i,z} \end{bmatrix} = \nabla \mathbf{w} \begin{bmatrix} v_a\cos\gamma \\ -v_a\sin\gamma \end{bmatrix} + \nabla \mathbf{w} \begin{bmatrix} w_{i,x} \\ w_{i,z} \end{bmatrix}$$

(66)

The rate of change of the total energy is therefore

$$\dot{e}_{tot} = -w_{i,z} - \frac{qS}{mg}C_D v_a - \mathbf{v}_a^T[\nabla \mathbf{w}]\mathbf{v}_a - \mathbf{v}_a^T[\nabla \mathbf{w}]\mathbf{w}$$ (67)

where $\mathbf{v}_a = [v_a\cos\gamma - v_a\sin\gamma]^T$.

The second term in the equation above is the energy lost due to drag. The first term is the total energy that can be gained (or lost) from vertical components of wind: note that $w_{i,z}$ is positive down, so upward components of wind will lead to energy gain.

We can also see the influence of wind gradients on airmass-relative total power, which is contained in the third and fourth terms of the right-hand side of Equation (67). The $\mathbf{v}_a^T[\nabla \mathbf{w}]\mathbf{v}_a$ term is likely to be larger in magnitude since airspeed is typically larger than wind speed. It will contribute to energy gain under two conditions: first, when $\nabla \mathbf{w}$ is negative and γ is negative; second, when $\nabla \mathbf{w}$ is positive and γ is positive. Note that for steady-state gliding flight, γ (the flight path angle with respect to the airmass) is always negative.

There are three main sources of vertical air motion: uneven solar heating of the ground, which produces convective instabilities known as thermals; long-period oscillations of the atmosphere, generally called wave; and orographic lift, where wind is deflected by the slopes of hills, ridges, and mountains. The time constant for these types of vertical air motion is long compared with the time constant of aircraft dynamics, with thermals lasting about 15 min, orographic lift lasting several hours, and wave lasting a few days. Typically, quasistatic approximations of aircraft dynamics are sufficient to permit energy exploitation from vertical air motion, and energy harvesting from these phenomena is known as static soaring.

4.1 Autonomous Static Soaring

The literature covering soaring flight is enormous, with many papers covering various aspects of the problem published in the *Journal of Technical Soaring*. It is thus not a surprise that the possibility of autonomous soaring should also be examined as a means of increasing range and endurance.

The first hardware demonstration of autonomous thermal soaring was performed by Michael Allen in 2007 (Allen and Lin, 2007). Since then, several researchers have developed controllers for thermal centering (Andersson et al., 2010), algorithms for path planning that explicitly trade exploration and exploitation of energy (Lawrance and Sukkarieh, 2009), and methods for thermal seeking (Bower et al., 2010).

Notwithstanding efforts put forth by competition soaring pilots, there is not yet a reliable device that can remotely sense the presence of updrafts: the only means of detecting a thermal is to fly into it. Human sailplane pilots use visual clues such as cumulus clouds and ground features that are likely to trigger thermals as a means of increasing the likelihood of finding thermals.

The autonomous thermal centering controller that was developed by Allen(Allen and Lin, 2007) and altered by Andersson (Andersson et al., 2010) has been shown to sufficiently center and maintain climb without the use of thermal mapping. This controller is based on Reichmann's thermal centering method (Reichmann, 1978). Reichmann's centering method consists of the following three rules (as illustrated in Figure 11):

1. As climb improves, flatten the circle (approximately 15–20° bank angle).
2. As climb deteriorates, steepen the circle (approximately 50° bank angle).
3. If climb remains constant, keep constant bank (approximately 25–30° bank angle).

Using the second derivative of energy ($\ddot{e}$) as a feedback term, this method was written as a control law:

$$\dot{\psi}_{cmd} = \dot{\psi}_{ss} - k\ddot{e}$$ (68)

where $\dot{\psi}_{ss}$ is the desired nominal steady-state turn rate and k is a gain (Andersson et al., 2009). This controller centers the aircraft's trajectory around the core of the thermal by adjusting the turn rate to follow a contour of constant $\dot{e}$. This control law has the built-in assumption that the thermal profile somewhat resembles a Gaussian, circular in shape with the highest updraft velocity at the core.

This method is proven stable under certain conditions (Andersson et al., 2010) but in practice can be difficult to implement successfully under broad weather conditions. It relies on the second derivative of the total energy ("energy acceleration"), which is not directly available from sensors. Obtaining energy acceleration from numerical derivatives of the total energy leads to a very noisy signal, and filtering this signal to an extent that it is useable for control leads to significant lag. Notwithstanding these issues, this approach has been successfully demonstrated in hardware.

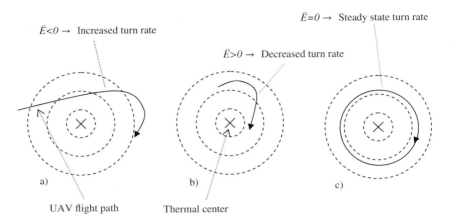

Figure 11. Reichmann's method of centering a thermal updraft using the total energy acceleration (Reichmann, 1978; Andersson et al., 2009).

Another approach is to estimate the location of the thermal core and its characteristic radius, and then compute the turn radius that maximizes energy gain. A flight controller can then ensure that the aircraft flies the correct radius, centered on the thermal core. This was successfully used in Dan Edwards' ALOFT autonomous glider during the Montague Cross Country Challenge (Edwards and Silverberg, 2010]).

Figure 12 shows a short segment from a 2 h long flight (using less than 5 min total motor run time) that used static soaring. The plot shows a map of vertical wind speeds (see Equation (67)) and aircraft flight path, color coded by the rate of change of total energy.

4.2   Dynamic Soaring

Dynamic soaring by albatrosses was first described by Lord Rayleigh in 1883 (Rayleigh, 1883; Rayleigh, 1889). Albatrosses and petrels are able to travel well over a thousand kilometers using dynamic soaring (Sachs, 2005), and human

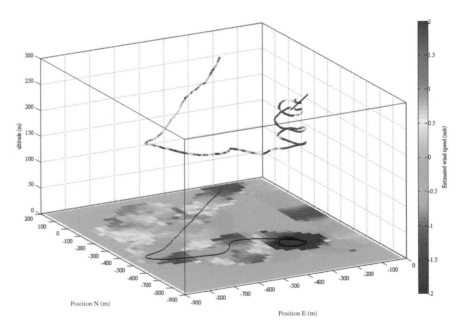

Figure 12. Segment of soaring flight showing mapped vertical winds and aircraft rate of change of the total energy. Figure courtesy Nathan T. Depenbusch and John J. Bird.

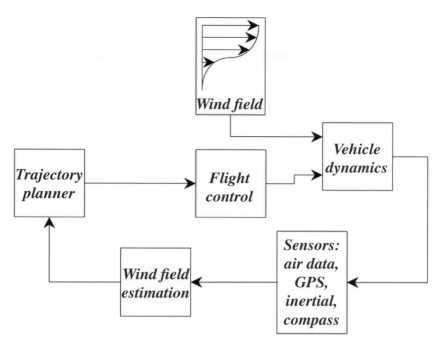

Figure 13. Top-level schematic of closed-loop dynamic soaring architecture (Bird et al., 2014).

radio-controlled glider pilots are able to reach speeds well over 300 mile/hour while dynamic soaring in the lee of ridges. As of September 2015, the speed record for dynamic soaring was 505 miles/hour.[4]

It should thus not be surprising that researchers have been examining dynamic soaring for autonomous aircraft as well (Boslough, 2002). Much of the research has focused on computing trajectories that maximize energy gain in a known wind field (Sachs and da Costa, 2003, 2006) or that ensure safety while provided at least an energy neutral trajectory (Flanzer et al., 2012).

However, wind fields are generally not known a priori, hence a dynamic soaring system should include a means for determining the wind field. Lawrance and Sukkarieh describe an approach based on Gaussian process regression (Lawrance and Sukkarieh, 2011); other approaches based on spline or polynomial approximations to wind field have also been proposed (Bird et al., 2014). A general architecture for a system to enable autonomous dynamic soaring is shown in Figure 13.

A wind field estimation block uses data from aircraft sensors (GPS/INS, air data and compass) to compute the wind velocity at the current aircraft location. This wind velocity information is then used to compute an estimate

of the wind field (i.e., the spatial variation in wind speed and direction). The wind field estimate is then used by the trajectory planner to compute a flight path that maximizes the change in the total energy over one cycle. If net energy change over one cycle is zero (or positive, meaning that energy is gained), then dynamic soaring is possible. Finally, a flight controller follows the computed trajectory.

Trajectory following control in the high wind and high-wind shear environments typical of dynamic soaring is nontrivial, although results of hardware in the loop simulations show promise. Further, significant improvements in terrain sensing must be implemented for dynamic soaring to be useable in any but the most restrictive of conditions.

5   CONCLUSION

Power generation and energy management remain a challenging aspect of both UAV design and UAV operations. Long endurance and long range still generally imply internal combustion engines, although steady improvements in electrical power and storage technologies (batteries and photovoltaic) suggest that these will become more prevalent in the future.

Operationally, flight control and flight planning can be used to manage energy usage and energy harvesting (both photovoltaic and atmospheric). Bank angle (to maximize

---

[4] http://www.rcspeeds.com/aircraftspeeds.aspx?rpt=LL
and https://www.youtube.com/watch?v=hFPJ6DUAY10

solar energy collection) can be controlled to maximize the total energy gain for solar-powered aircraft; range-maximizing airspeeds can be computed based on winds aloft. Further, knowledge of wind fields implies that route planning can be used to minimize energy usage.

REFERENCES

AeroVironment, Inc. (2010) *Global Observer.*

Allen, M.J. and Lin, V. (2007) *Guidance and Control of an Autonomous Soaring Vehicle with Flight Test Results. AIAA Aerospace Sciences Meeting and Exhibit, Reno, Nevada. AIAA Paper 2007-867, American Institute of Aeronautics and Astronautics.*

Andersson, K., Kaminer, I., and Jones, K. D. (2010) *Autonomous Soaring; Flight Test Results of a Thermal Centering Controller.* AIAA Guidance, Navigation and Control Conference, Toronto, Canada. AIAA Paper 2010-8034, American Institute of Aeronautics and Astronautics.

Andersson, K., Kaminer, I., Jones, K. D., Dobrokhodov, V., and Lee, D.-J. (2009) *Cooperative UAVs Using Thermal Lift to Extend Endurance.* AIAA Infotech@Aerospace Conference}, Seattle, Washington. AIAA Paper 2009-2043, American Institute of Aeronautics and Astronautics.

Asselin, M. (1997) *An Introduction to Aircraft Performance.* AIAA Education Series. American Institute of Aeronautics and Astronautics, Reston, VA.

Bird, J. J., Langelaan, J. W., Montella, C., Spletzer, J., and Grenestedt, J. (2014) *Closing the Loop in Dynamic Soaring. AIAA Guidance, Navigation and Control Conference,* National Harbor, Maryland. AIAA Paper 2014-0263.

Boslough, M. B. E. (2002) *Autonomous Dynamic Soaring Platform for Distributed Mobile Sensor Arrays.* Technical Report SAND2002-1896, Sandia National Laboratories, Sandia National Laboratories.

Bower, G. C., Flanzer, T. C., Naiman, A. D., and Saripalli, S. (2010) *Dynamic Environment Mapping for Autonomous Thermal Soaring.* AIAA Guidance, Navigation and Control Conference. AIAA Paper 2010-8031, American Institute of Aeronautics and Astronautics.

Bradley, T. H., Moffitt, B. A., Mavris, D. N., and Parekh, D. E. (2007) Development and experimental characterization of a fuel cell powered aircraft. *J. Power Sources,* **171**, 793–801.

Chakrabarty, A. and Langelaan, J. W. (2011) Energy-based long-range path planning for soaring-capable uavs. *J. Guidance Control Dyn.,* **34**(4), 1002–1015.

Colozza, A. J. (1997) *Effect of Date and Location on Maximum Achievable Altitude for a Solar Powered Aircraft.* Technical Report CR-202326, National Aeronautics and Space Administration.

Edwards, D. J. (2008) *Implementation Details and Flight Test Results of an Autonomous Soaring Controller. AIAA Guidance, Navigation and Control Conference,* Reston, Virginia. American Institute of Aeronautics and Astronautics.

Edwards, D. J. and Silverberg, L. (2010) Autonomous soaring: the montague cross-country challenge. *J. Aircraft,* **47**, 1763–1769.

Eriksen, C. C., Osse, T. J., Light, R. D., Wen, T., Lehman, T. W., Sabin, P. L., Ballard, J. W., and Chiodi, A. M. (2001) Seaglider: a long range autonomous underwater vehicle for oceanographic research. *IEEE J. Oceanic Eng.,* **26**(4), 424–436.

Flanzer, T. C., Bower, G. C., and Kroo, I. M. (2012) *Robust Trajectory Optimization for Dynamic Soaring.* AIAA Guidance, Navigation, and Control Conference,} Minneapolis, Minnesota, August 13–16, 2012.

Green Flight Challenge Team Agreement rev. 07.28.2009. Technical Report, CAFE Foundation, Santa Rosa, California.

Hall, D. W., Watson, D. A., Tuttle, R. P., and Hall, S. A. (1985) *Mission Analysis of Solar Powered Aircraft.* Technical Report CR-172583, National Aeronautics and Space Administration.

Jardin, M. R. and Bryson, A. E. (2001) Neighboring optimal aircraft guidance in winds. *J. Guidance Control Dyn.,* **24**(4), 710–715.

Langelaan, J. W., Chakrabarty, A., Deng, A., Miles, K., Plevnik, V., Tomazic, J., Tomazic, T., and Veble, G. (2013) Green flight challenge: Aircraft design and flight planning for extreme fuel efficiency. *J. Aircarft,* **50**(3), 832–846, doi: 10.2514/1.C032022.

Lawrance, N. R. J. and Sukkarieh, S. (2009) Wind Energy Based Path Planning for a Small Gliding Unmanned Aerial Vehicle. *AIAA Guidance, Navigation, and Control Conference* Chicago, Illinois, August 10–13, 2009.

Lawrance, N. R. J. and Sukkarieh, S. (2011) Autonomous Exploration of a Wind Field with a Gliding Aircraft. *J. Guidance Control Dyn.,* **34**(3), 719–733.

Mair, W. A. and Birdsall, D. L. (1992) *Aircraft Performance.* Cambridge Aerospace Series. Cambridge University Press, Cambridge, UK.

McCormick, B. W. (1995) *Aerodynamics, Aeronautics and Flight Mechanics,* 2nd edn, John Wiley & Sons, Inc., Hoboken, NJ.

Noth, A. (2008) *Design of solar powered airplanes for continuous flight.* PhD thesis. ETH Zurich.

Ostbo, M., Osen, P., Rokseth, G., Homleid, O., and Sevaldrud, T. (2004) *Exploiting Meteorology to Enhance the Efficiency and Safety of {UAV} Operations.* Technical Report 2004/00981, Norwegian Defence Research Establishment, Kjeller, Norway.

QinetiQ (2010) *Zephyr-unmanned aerial vehicle.* Available at http://www.qinetiq.com/media/news/releases/Pages/zephyr-14-days.aspx (accessed April 5, 2016).

Rayleigh, J. W. S. (1883) The soaring of birds. *Nature,* **27**, 534–535.

Rayleigh, J. W. S. (1889) The Sailing Flight of the Albatross. *Nature,* **40**, 34.

Reichmann, H. (1978) *Cross-Country Soaring.* Thomson Publications, Santa Monica, California.

Rubio, J. C. and Kragelund, S. (2003) *The trans-pacific crossing: long range adaptive path planning for uavs through variable wind fields. Proceedings of the 22nd Digital Avionics Systems Conference,* vol. 2, IEEE. Piscataway, New Jersey.

Sachs, G. (2005) Minimum shear wind strength required for dynamic soaring of albatrosses. *Ibis,* **147**, 1–10.

Sachs, G. and da Costa, O. (2003) Optimization of Dynamic Soaring at Ridges. *AIAA Atmospheric Flight Mechanics Conference*, Austin, Texas, August 11–14, 2003.

Sachs, G. and da Costa, O. (2006) *Dynamic Soaring in Altitude Region below Jet Streams.* AIAA Guidance, Navigation, and Control Conference, Keystone, Colorado, August 21–24, 2006.

Sherman, J., Davis, R. E., Owens, W. B., and Valdes, J. (2001) The autonomous underwater glider "spray". *IEEE J. Oceanic Eng.*, **26**(4), 437–446.

Stroman, R. O., Schuette, M. W., Swider-Lyons, K., Rodgers, J. A., and Edwards, D. J. (2014) Liquid hydrogen fuel system design and demonstration in a small long endurance air vehicle. *Int. J. Hydrogen Energy*, **39**, 11279–11290.

Webb, D. C., Simonetti, P. J., and Jones, C. P. (2001) SLOCUM: An underwater glider propelled by environmental energy. *IEEE J. Oceanic Eng.*, **26**(4), 447–452.

Woolsey, C., Hagerman, G., and Morrow, M. (2005) *A Self Sustaining Boundary Layer Adapted System for Terrain Exploration and Environmental Sampling.* Phase I final report, NASA Institute for Advanced Concepts.

List of Symbols

$\mathcal{R}$	aspect ratio
C_D	drag coefficient
C_L	lift coefficient
C_T	thrust coefficient
D	drag force
E	energy
E_b	energy in battery
E_{tot}	aircraft total energy
e_b	battery specific energy
e_f	specific energy of fuel
e_O	Oswald's efficiency
e_{tot}	aircraft specific total energy
g	acceleration due to gravity
h	height above ground
I_s	solar insolation
L	lift force
m	total mass
m_b	battery mass
m_f	fuel mass
m_0	zero-fuel weight (or zero-battery weight)
P_a	aerodynamic power
P_b	power drawn from battery
P_{in}	total power input
P_{out}	total power output
P_s	power drawn from solar cells
$P_{systems}$	power drawn by onboard systems
q	dynamic pressure ($\frac{1}{2}\rho v_a^2$)
r_{sp}	specific range
R	range
S	wing area
S_s	solar cell area
s	distance
SFC	specific fuel consumption
T	thrust force
t	time, endurance
v_a	air speed
v_g	ground speed
w_c	cross-wind component
w_t	tail-wind component

Greek symbols

γ	flight path angle
η_{ec}	energy conversion efficiency
η_{flight}	flight efficiency
η_p	propeller efficiency
η_s	solar cell efficiency
ρ	density of air
ψ	heading angle
$\psi_g \psi_g$	ground track

Chapter 13
Control System Mechanization

Yunfeng Cao,[1] Wen-Hua Chen,[2] and Matthew Coombes[2]

[1]*College of Astronautics, Nanjing University of Aeronautics and Astronautics, Nanjing, China*
[2]*Department of Aeronautical and Automotive Engineering, Loughborough University, Loughborough, UK*

1 CONTROL FUNDAMENTALS OF UAS

1.1 UAS and control systems

A UAS (unmanned aircraft system) control system aims to achieve control via aerodynamic control surfaces of the airplane and the thrust provided by the engines, which is normally controlled in its magnitude (Barnhart *et al.*, 2012). The functions of a UAS control system can be structured into both primary functions, such as pitch, roll, and yaw control, and secondary functions such as high lift, airbrake, and lift dump on the ground (Pratt, 2000), but only the primary functions are considered in this chapter. A typical composition of unmanned aerial vehicle (UAV) control systems is shown in Figure 1.

When developing a flight control system (FCS) for a UAV, it is common practice to assume that the UAV can be represented as a rigid body, defined by a set of body-axis coordinates. The rigid-body dynamics have six degrees of freedom given by three translations along and three rotations about the axes. All forces and moments acting on the UAV can be modeled within this framework. The forces and moments acting on the UAV vary substantially across its flight envelope. If we are able to control these forces and moments, we have the control of the UAV translational and rotational accelerations, and hence its velocities, attitude, and position.

1.2 Types of FCS

1.2.1 Stability augmentation systems

Stability augmentation systems (SAS) are a kind of automatic devices providing artificial stability for an airplane that has undesirable flying characteristics. A UAV has several inherent motion modes from its aerodynamic design. When the properties of these inherent motion modes do not comply with the requirements for performing various tasks and missions, the possible needs for improved damping and stability about all three axes can be achieved by an SAS.

The lower damping of the pitch short-period response may require the installation of a pitch damper system. Similarly, a yaw SAS may be required in a UAV to suppress its lightly damped yawing motion and the accompanying oscillatory roll motion due to Dutch roll motion. It is noted that a yaw damper system may be insufficient with some UAVs with large wing sweepback to suppress the effects of the cross-coupling between yaw and roll channels, and a roll damper system may also be necessary. Hence, three-axis SAS are installed in most high-performance UAVs.

It is important that the control authority exerted by a single-channel SAS is limited in order to avoid failure of SAS

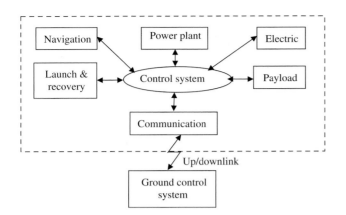

Figure 1. UAS system composition.

and actuator saturation. A diagram of a typical pitch SAS with pitch rate feedback is shown in Figure 2.

1.2.2   *Autopilot*

Different from manned aircraft, an autopilot is essential for UAV operation. In addition to attitude stability functions, an autopilot also provides the control of the UAV flight states (e.g., altitude, heading, and airspeed) such as following specified flight paths. An autopilot communicates with the ground control station (GCS) for receiving control commands by uplink, updates the flight state information by downlink, and sends out control inputs to the actuators on

UAVs. Figure 3 shows a typical attitude autopilot and how it interacts with an SAS.

1.3   UAS control architecture

Similar to FCS on manned aircraft, the cascaded control architecture is used on UAS, which is decomposed into two loops where reference signals for the inner loop are produced by controllers in the outer loop as shown in Figure 4.

The inner loop related to the rotation of UAV provides required attitude stability for UAV and achieves required attitude for performing a certain task. Typically, design of the inner loop aims at achieving high bandwidth and robust stability. The outer loop related to the translation is used for guidance, generating commands for the inner loop. It operates at a slower timescale than the inner loop. Furthermore, one of the main tasks of the inner loop is to provide adequate decoupling such that outer loop controllers may control each motion variable independently.

1.4   UAS control system design consideration

In developing UAS control systems, a number of key factors have to be taken into account (Pratt, 2000; Musial, 2008).

UAV dynamics. UAV dynamics have a direct impact on the stability. The resultant considerations affect the specifica-

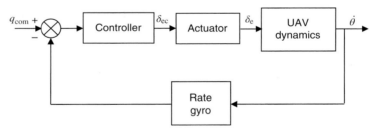

Figure 2. SAS using pitch rate feedback.

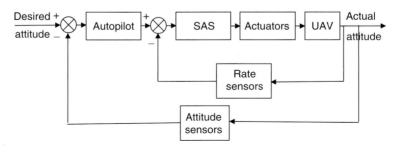

Figure 3. UAV autopilot functional block diagram.

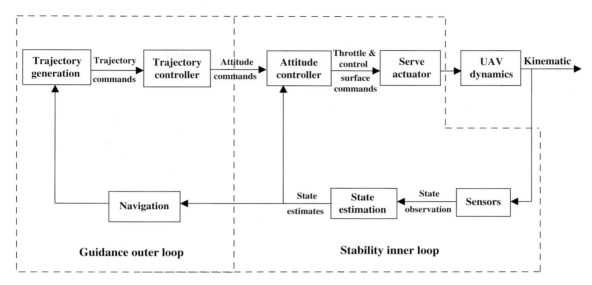

Figure 4. UAS control system architecture.

tions/choices of flight control computer processing power, the speed and authority of actuators, the computing speed of control laws, and the associated actuator drives. Highly agile unmanned compact aerial vehicles (UCAVs) may be designed to be unstable in order to achieve maneuverability, which implies high requirements on the flight control systems.

Safety. The control system design is greatly influenced by the results of hazard assessments. This may result in a certain redundancy in either system or sensor elements. In addition, the control laws may be segregated on flight control computer units necessitating mode logic and voting strategies to effect a change or a system reconfiguration. This has the undesirable effect of making the system more sophisticated, complex, and expensive.

Performance. The performance of the FCS in achieving the required stability or autopilot operations is specified by the overall index as a result of analyzing the overall mission requirements. The FCS performance is a function of the airframe dynamics, the equipment architecture, sensor configuration, and the actuation of the control surfaces. The selection of sensors plays a major role in achieving the performance. Signal accuracy, noise suppression, and time delay must be considered by the designer.

Physical constraints. The physical constraints on the size, weight, and power consumption of the equipment, coupled with the specification for the usage conditions, can influence the architecture. If the equipment needs to operate in a high-

vibration and high-temperature installation with the possibility of air contaminated with sand, salt, or seawater, the design must accommodate this while still ensuring maximum reliability.

Logistic support. The concepts of reliability, maintainability, testability, and manufacturability all have to be addressed as part of the FCS design. Reliability figures are usually included as part of the FCS performance specification in terms of the maximum number of failures per thousands of operating or flight hours. Therefore, the hardware design must use components that support the required reliability figures. Maintainability must be relatively simple and easy to accomplish by a trained operator. Testability is the concept of being able to diagnose the existence of faults by external or internal testing of the equipment.

Cost. The design process of FCS is a "V" iterative and a spiral up process where safety of flight is of paramount importance and the most cost-effective solutions must be applied. Clearly, a logical well-controlled design phase minimizes the number of late changes and the associated additional cost.

2   UAS CONTROL SYSTEM ELEMENTS

2.1   Sensors and its integration

Airborne sensors are vital for providing necessary information for control systems. Sensors and their data processing constitute one of the most complicated and expensive systems on UAV. Main purposes of airborne sensors (not including payload) in the UAS are to determine the attitude and position of the UAV in order to provide the required inputs to FCS or the GCS.

General requirements to UAV airborne sensors are basically similar to the requirements on other on-board systems for UAV although the cost and level of redundancy might be relaxed. They mainly include the following (Musial, 2008):

- Sensors must obey strict limitations regarding weight, size, power consumption, and cost.
- Sensors must be configured with standard interface in order to facilitate on-board information exchange and system integration.
- The bandwidth of sensors should match the performance of the UAS control system.

2.1.1   Inertial measurement unit

An on-board inertial measurement unit (IMU) classically consists of three angular rate gyros and three orthogonal accelerometers. Traditionally, high-grade IMUs were usually constructed with gyro-gimbaled platforms in constant orientation carrying the acceleration sensors. This primarily helps to minimize numerical integration errors because no rotational transforms are required. But this type of IMU is not quite suitable for all the UAVs due to size, weight, and cost issues. With processing technology advancing, these gimbaled IMUs are increasingly replaced by strap-down IMUs, with the sensors being fixed in the aircraft body axes and processing equipment taking care of rotational transforms. In spite of still slightly lower performance, this class of IMUs is widely used for UAVs, in particular for low-cost UAVs.

2.1.2   Global positioning system

The satellite-based Global Positioning System (GPS) provides the most obvious means of position measurement for UAV operating outdoors. Due to various error sources including "selective availability" (SA), the positioning accuracy of a conventional GPS may not be able to meet the application requirements of UAV. Differential GPS (DGPS) technology is now widely used to improve the accuracy of GPS and its integrity where correction signals from reference stations are used to correct the current GPS errors so that the accuracy could be significantly improved. Different from conventional DGPS, the so-called carrier-phase DGPS can even track carrier-phase differences between the reference and moving receivers. The absolute measuring accuracy of the carrier-phase DGPS is able to reach the level as low as centimeters. However, the existing carrier-phase DGPS

equipment is quite expensive and heavy, so the conventional DGPS is usually used for small UAVs.

2.1.3   Airspeed and altimeter

Air data systems provide accurate information on quantities such as pressure altitude, vertical speed, calibrated airspeed, true airspeed, Mach number, static air temperature, and air density ratio (McShea, 2010). These are derived from three basic measurements by sensors connected to probes including total pressure, static pressure, and air temperature. The static pressure of the free airstream is measured by an absolute pressure transducer connected to a suitable orifice located where the surface pressure is nearly the same as the pressure of the surrounding atmosphere. The total pressure is measured by means of an absolute pressure transducer connected to a pitot tube facing the moving airstream. It measures the impact pressure, which is the pressure exerted to bring the moving airstream to rest relative to the pitot tube, plus the static pressure. Airspeed can be derived from the impact pressure.

2.1.4   Other sensors

Magnetometer. The long-term online calibration of heading usually requires a magnetic field sensor, that is, a compass. While a 2-DOF compass is theoretically sufficient to determine the yaw angle with pitch and roll angles known, it is generally preferable to have a 3-DOF magnetic field sensor for better numerical handling and to decouple accelerometer and magnetometer errors as much as possible. A commonly used and fully sufficient class of magnetic field sensors is digital three-axis magnetometers.

Sonar. Both in the case of autonomous landing and for obstacle avoidance, sonar is a useful sensor to measure the altitude above ground. It is certainly the easiest and cheapest approach for obtaining this kind of information. Its main disadvantage consists in its limited range of only a few meters usually. It is possible if an additional "range" sensor for determining distances is included. This may be a laser range finder or stereo imaging equipment.

Laser range finder. Laser range finder emits a coded laser beam to determine the distance to a target. Distances are determined by reflecting the laser beam off a target object and measuring the round-trip time of light of the laser beam to and from the target object. The maximum measurable distances of any laser range finder are determined by several factors including the shape, size, reflectivity, and orientation of target as well as atmospheric conditions.

2.1.5 *Attitude heading reference systems and navigation*

It is clear that no single sensor can provide completely accurate vehicle status information. Therefore, multisensor integration is required in order to provide the on-board system with complementary, sometimes redundant, information for its location and navigation task. Many fusion technologies have been developed to fuse the complementary information from different sources into one representation format. The information to be combined may come from multiple sensors during a single period of time, or from a single sensor over an extended period of time. Integrated multisensor systems have the potential to provide high levels of accuracy and fault tolerance.

Attitude heading and reference systems. An attitude heading and reference system (AHRS) consists of a three-axis inertial measurement unit, a three-axis magnetic sensor, and an on-board processor that creates a virtual three-axis sensor capable of measuring heading (yaw), pitch, and roll angles of an object moving in 3D space.

Navigation systems. It is necessary for the operators to know where the UAV is at any moment. It may also be necessary for the UAV to "know" where it is if autonomous flight is required during the flight. This may be either as part or all of a preprogrammed mission or as an emergency return to base capability after system degradation. For fully autonomous operation, that is, without any communication between the GCS and the UAV, sufficient navigation equipment must be carried in the UAV. By combining the navigation information from a number of elements such as IMU and GPS, an integrated solution for realizable navigation of UAV could be achieved.

Multisensor integration and fusion also provide a system with additional benefits (Valavanis, 2007). These may include robust operational performance, extended spatial coverage, extended temporal coverage, an increased degree of confidence, improved detection performance, enhanced spatial resolution, improved reliability, and reduced ambiguity. Sensor fusion is essential for estimating nonmeasurable state variables and filtering, and also for performing system identification procedures. The main task of the data fusion module is to yield optimal estimates for key state variables using sensor measurements. State estimation techniques are also needed since almost all small UAVs are equipped with low-cost lightweight sensors prone to noisy measurements at different data rates and they are subject to considerable vibration and turbulent gusts. An excellent comparative overview of avionic systems and sensor fusion algorithms

of some of the early autonomous UAS projects was introduced in Saripalli *et al.* (2003).

2.2 Actuators

From the FCS point of view, there are two main types of actuators for UAV operation: aerodynamic control surface and propulsion (engine throttle). Propulsion is mainly used to control vehicle speed such as in airspeed holding autopilot. Aerodynamic control has a dominant influence in aircraft stability and maneuverability, which consists of two main parts: aerodynamic control surfaces and actuation systems as discussed below. Configuration of control surfaces and actuator is one of the key factors influencing FCS performance.

2.2.1 *Control surfaces*

UAV has a number of different control surfaces. The conventional control surfaces for a fixed-wing UAV may include some or all of the following:

- Ailerons: to control the roll angle.
- Elevator: to control the pitch angle (up and down).
- Rudder: to control the yaw angle (left and right).
- Throttle: to control the angle of throttle opening.

It must be noted that the type and use of a control surface has a significant impact on the requirements for the actuation system, in particular the actuator post-failure operation. The concepts such as reconfigurable vehicles and relaxed stability result in highly unstable UCAV to improve performance and agility.

2.2.2 *Actuation systems*

Traditionally, the aerodynamic control surfaces are actuated by mechanical or more widely hydraulic actuation systems. There is a quite significant dynamic loading on the control surface, particularly when UAV flies at a high speed. Recently, electromechanical actuation systems (EMAs) have received more attention. The advent of fly-by-wire technology has led to many actuators now having electrical signal (Pratt, 2000). Most flight control actuation systems on the existing UAVs are electromechanical actuation systems. According to the differences in output mechanism and motor work style, EMA has various categories, such as linear and rotary type, brushless and DC servo motor, and so on. A rotary brushless DC motor and PWM drive technology are mostly used currently on the UAV. If required, rotary motion can be converted to linear motion through a crank mechanism.

An actuator involved in FCS forms a closed servo loop, the so-called actuator loop. The basic performance criteria of the actuator loop considered in design and used are stall load, maximum rate capability, frequency response, dynamic stiffness, and failure transients. In addition, system reliability, size, weight, and installation details are also to be considered.

2.3   Flight control computer

When selecting flight control computers for a UAV, two basic sets of requirements are mainly considered (Musial, 2008): interface and computational power. Sensor and actuator components require a flight control computer with a great number of dedicated input and output features while sensor data processing often poses high demands on on-board computational power.

Usually, on-board systems may be designed around a single processing unit. The resulting architecture is more flat and less complex. This requires the use of a microcontroller-type processor providing suitable interfacing capabilities. The main advantage of this approach is that only a single software branch needs to be developed and maintained. Any time and data rate constraints can be addressed within a single program without suffering from the limitations imposed by an additional communications path. However, all requirements, both computational and interface-related, need to be met by the choice of a processor. This makes this choice much more difficult and considerably limits the variety of possible hardware platforms and operating systems.

Alternatively, a straightforward solution to this contradiction of requirements is to adopt multiple processing units: one or several more desktop-like processors for computationally expensive tasks, and one or several microcontroller-type processors for sensor and actuator interfacing. The main advantage of the multi-CPU approach is that each subtask is left to a processing unit that is best suited for it. The development of the higher level software can be conveniently performed solely on a computer with powerful operation systems such a desktop-like PC. As there are almost no requirements specifically related to the interfaces provided by the PC-type processor on board, the developer is free in choosing hardware and a suitable operating system. Meanwhile, the software running on the microcontroller is expected to be simple and to undergo only a very limited number of development cycles. However, it is worth noting that multiple processing platforms will not only increase the complexity of the control system structure but also increase the cost of development and maintenance of the control system.

3   FCS DEVELOPMENT PROCESS

As a typical safety-critical system, there is a long process in the development of FCS. It deals with not only stability and safety but also performance and robustness in the presence of a wide range of variations and uncertainties. A typical cycle of the development of FCS starts from aircraft dynamics analysis and handling quality study to design and simulation until implementation and eventually flight tests. Verification and validation of FCS is very important in guaranteeing that the desired performance is achieved within the full flight envelope.

3.1   Control system design

After developing flight dynamics models of a UAV, its stability and other dynamic performance will be investigated. Assessing against the performance specifications and flight profiles, both stability augmentation systems and autopilots will be investigated, and various control design methods could be used in designing FCS. PID controller design methods are among the most widely used ones. For small UAVs, a number of commercial autopilots are available such as ArduPilot and MicroPilot. It is not surprising that most commercial autopilots such as MicroPilot's MP2128 or Rotomotion's AFCSV2.5 UAV Controller are based on PID controllers. PID controller parameters may be easily adjusted allowing for online tuning when the model is not known exactly. On the other hand, when a more accurate model is available, classical design or optimization-based tuning strategies may be attempted in PID design. The main disadvantage of PID controllers relates to overlooking coupling among controlled variables, limiting bandwidth and agility that could be achieved. The control systems of UAVs can also be designed by other control techniques such as the linear quadratic regulator (LQR) and robust control.

Developing a full autopilot system from scratch is a relatively long and difficult process, as specialist knowledge is needed in electronic engineering, systems integration, and software engineering (Coombes *et al.*, 2012). By purchasing an open-source commercial off-the-shelf (COTS) autopilot system, much time and effort can be saved. ArduPilot is a hobbyist "do-it-yourself" autopilot meant for use on remote-controlled aircraft. It is capable of flying simple waypoints, or more importantly taking roll, pitch, rudder, and throttle commands from an external source. By using a communication protocol called MAVLink, any device or system can give ArduPilot these commands over a number of different connection types. As the low-level control is performed by this autopilot, it leaves the user the freedom to concentrate

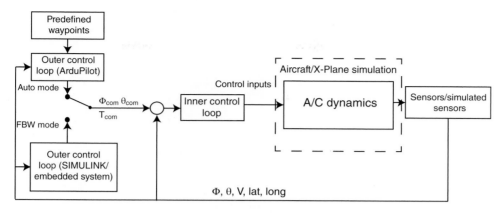

Figure 5. UAV control system based ArduPilot. (Reprinted with permission from Coombes *et al.* (2012). © 2012, IEEE.)

on the high-level control. Figure 5 shows an example of the UAV control system based ArduPilot.

3.2   Software-in-the-loop simulation

After initial simulation study and performance analysis under various operation conditions, the development of FCS moves to the implementation stage (Coombes *et al.*, 2012). It shall be realized that various assumptions or simplifications have been made in the FCS design. Furthermore, code development for real-time implementation of the designed FCS strategy is required as different computer languages from the design and development stage are usually used. Similar to the development of other embedded systems, two procedures known as software-in-the-loop (SIL) simulation and hardware-in-the-loop (HIL) simulation are common for testing the software and/or hardware of FCS under development. Putting algorithms through the full development cycle of SIL and HIL, and then flight testing is a good systematic debugging method, which significantly reduces risk. This process is shown in Figure 6. In the case of SIL simulation, only the

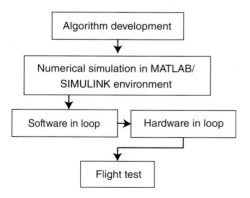

Figure 6. The development cycle for high-level control for UAS. (Reprinted with permission from Coombes *et al.* (2012). © 2012, IEEE.)

software normally running in the embedded system is tested. All hardware involved is replaced by a special emulation software layer that reads out or injects data at the lowest possible software interface level.

It should be pointed out that due to the coupling between embedded and emulation software and the difference in operation mode, pure SIL tests are most often performed on a PC-based platform instead of the embedded processor.

As an example, the SIL system for an open-source ArduPilot is shown in Figure 7. The open-source ArduPilot code has a desktop build, which means that, instead of the code being compiled on the Arduino on the ArduPilot, it is complied on a normal desktop computer running LINUX. This simulated ArduPilot communicates with a simulated aircraft in the X-Plane flight simulator using a plug-in that enables communication over a TCP/IP network connection using the MAVLink Protocol. The X-Plane aircraft model is controlled by the desktop build of ArduPilot, and, in turn, ArduPilot is controlled by an external system like SIMU-LINK or an embedded system running the users' high-level control algorithms. This SIL method enables development to be undertaken without any ArduPilot hardware.

3.3   Hardware-in-the-loop tests

SIL enables any bugs in the software to be ironed out (Coombes *et al.*, 2014). Then HIL brings to light bugs that come about due to the software's interaction with the hardware, and the wireless communication system. Finally, flight testing shows the functionality of the communications system at range and issues the system has in the real environment, such as wind. In the case of HIL tests, except for the embedded system (e.g., the autopilot for UAV), other components such as actuators and sensors should maximize the real physical parts, respectively. However, the HIL test usually cannot be followed to full extent. Certain parts of the hardware may have to

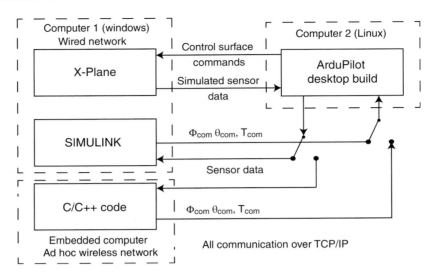

Figure 7. Flow diagram for ArduPilot SIL testing, used with SIMULINK and/or embedded system. (Reprinted with permission from Coombes *et al*. (2012). © 2012, IEEE.)

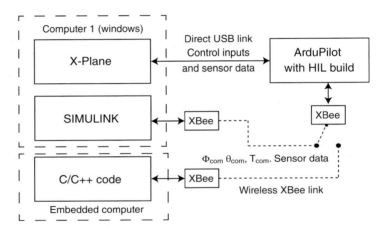

Figure 8. Flow diagram for ArduPilot HIL testing, used with SIMULINK and embedded system. (Reprinted with permission from Coombes *et al*. (2012). © 2012, IEEE.)

be left out such as the engine and some sensors; for example, GPS would require simulated satellites. Hence, HIL simulation environments will usually have to revert to SIL simulation for certain parts of the system under test.

As an example, shown in Figure 8, HIL is quite similar to the SIL but the actual ArduPilot hardware is used and sensor data are faked. The communication between X-Plane and ArduPilot is now done over a virtual serial connection, provided by the USB to serial chip. ArduPilot is now controlled over its telemetry port connected to SIMULINK, or a ground station using an XBee wireless serial transmitter, or to an embedded system directly.

When it comes time to conduct actual flight tests, there is no reconfiguration of software, hardware, or communications, apart from ArduPilot being mounted to the aircraft, and wired to the receiver and servos. As ArduPilot and MAVLink have

been abstracted to such a high level, throughout the whole development cycle, none of the ArduPilot code or high-level control ran externally needs to change at all to move to the next stage. All that changes is if the aircraft is real or simulated, where ArduPilot software is ran, and how the components of the system physically communicate with one another.

Comparing the above two methods of simulation test, the SIL has obvious advantages in many respects such as utility, unambiguity, flexibility, economy, and time, which are particularly relevant for small UAV development. Therefore, the HIL seems more desirable for high-performance or large UAVs, or for small UAVs that are not realizable during real field tests such as fault recurrence. Otherwise, SIL is more beneficial for small UAVs, and also should actually be implemented in every case of UAV development.

4 SOME PRACTICAL ISSUES

4.1 Fail-safe procedures for FCS

Fail-safe procedures are an important element in the operation of UAS. As a special flight control mode, they involve the whole flight process of a UAS and are only engaged when certain system components fail to work as desired. Usually fail-safe procedures have the following operation modes.

Pilot mode. The best ultimate fallback procedure with any kind of UAV is manual control, usually named as "direct chain control" or full manual remote control, where a ground operator controls aircraft actuators directly through the command and telemetry link. The operation by human remote control usually requires the smallest number of on-board components to work properly only. No sensors and autopilot might be involved at all. Therefore, the pilot mode is the fail-safe procedure applicable to the greatest subset of possible failure events. But this kind of control mode is closely related to the operator's personal experience with similar remote-controlled vehicles that provides a reliable basis for operating the UAV. This concerns both the very first flight control tests in an early phase of system development and regular verification of UAV's proper functions prior to autonomous operation.

Manual high-level control mode. As mentioned earlier, FCS usually consists of two loops: one of which is position control loop and the other is attitude control loop. Therefore, in cases of high-level sensor failure (e.g., GPS outage), the position control loop may be disabled and replaced by manual control inputs. A human pilot may command the desired orientation, or the desired velocity via remote control. Then, lower level or attitude control may work as usual. This kind of manual control is much easier than full manual remote-controlled piloting and requires that command and telemetry link work well.

Program control mode. In case of failure of the position sensor, for example, temporary GPS outage, upper level position control cannot be maintained at all. One best possible solution is to cut off position control loop and switch to program control mode. UAV can fly in accordance with the preset attitude and trajectory such as pull up and then return to base according to a preset route or fix the desired attitude of the lower control loop according to the current operating point. This kind of control also requires that command and telemetry link work well.

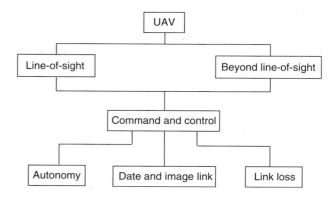

Figure 9. UAS C^3 function block.

Autonomous mode. In case of failure of the command and telemetry link, only best possible solution is autonomous mode. FCS automatically switches to autonomous mode after the true judgment of failure of command and telemetry link, and the UAV is guided by GPS according to a preset route and waits for both resting purposes and recovery.

4.2 Flight tests and communication with control station

Different from many other control applications, command, control, and communication (C^3) play a vital role in the control of UAS (see Figure 9). This is mainly because GCS is remotely located. The exchange of data and commands between the UAV and the GCS is essential in maintaining the safety of UAV. The uplink controls the activities of the UAV itself and the payload. This command and control link requires a sufficient degree of security to ensure that only authorized agents have access to the control mechanisms of the UAV (Stansbury *et al.*, 2009). The downlink transmits critical data from the onboard payload as well as the UAV health and status data to the operator and the analyst in the GCS. System health and status information must be delivered to the GCS without compromise, so a high degree of communication encryption technology is needed.

4.2.1 Wi-Fi communication

While the implementation of communication protocols and services is tedious, communication requirements in UAS involve real-time characteristics, the interconnection of heterogeneous component systems, and possibly the creation of a nondistributed abstract view of the system (Coombes *et al.*, 2014). Usually, off-the-shelf operating systems do not

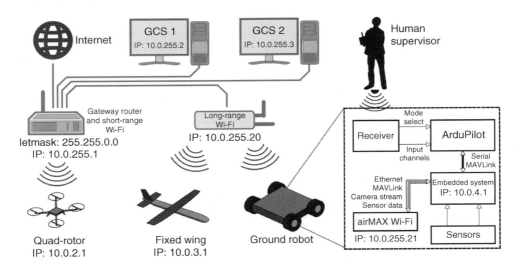

Figure 10. Example of network configuration. (Reprinted with permission from Coombes *et al.* (2014). © 2014, IEEE.)

provide any communications service that meets all these requirements. Therefore, the benefit from using an operating system tends to be limited for the lowest levels of communication. For a single communications port, for example, RS 232, the implementation of these layers is not so much costly to justify the use of an operating system just for this reason. In order to meet the more complex requirements mentioned, some middleware layer will usually have to be provided.

For short-range testing, a generic wireless router can be used as the gateway and wireless access point, with small-scale USB Wi-Fi adaptors used aboard the vehicles. This is sufficient to provide full 150 Mbit s^{-1} network capability, with the additional benefit of the minimal payload increase.

For long-range testing, more specialized equipment is required. The airMAX range of equipment from Ubiquiti Networks was found to be a promising practical solution, offering long-range, high-integrity wireless communications (Coombes *et al.*, 2014). To avoid adding extra complexity, airMAX devices connect directly to an Ethernet port on an embedded system and operate as a wireless bridge. As the Wi-Fi configuration is dealt with by the device, no additional drivers or software are required, allowing the devices to be both plug-and-play and interchangeable. For example, the airMAXPico station is small, has an omnidirectional antenna, and offers full bandwidth Wi-Fi up to 300 m range, making it suitable for use on board an aircraft. The airMAXNano station is larger, and is restricted by its 60° sector antenna, but offers greater range and bandwidth options, making it useful as the connection point on the ground. When used together, this setup provides a very reliable connection. An example network configuration is shown in Figure 10, where a network is based around a gateway router.

4.2.2   Communication protocol

As MAVLink is a protocol designed for serial communication, a method of encoding and decoding MAVLink data packets is required for network communication (Coombes *et al.*, 2014). This is achieved using MAVNode, a Node.js module that is a presentation of all vehicles and communication devices on the network. MAVNode utilizes the asynchronous, event-driven nature of Node.js to deal with messages from multiple systems via multiple communication channels efficiently.

In addition to handling MAVLink messages generated by systems such as an autopilot, MAVNode exposes a modified RESTful API to enable packets to be encoded/decoded over HTTP. All API interactions are handled via HTTP GET strings so as to be directly compatible with the MATLAB "urlread" command. A typical interaction between an application and MAVNode is shown in Figure 11. In this example, an application requests a data stream from MAVNode by issuing the HTTP request as shown. This requests data from system and component IDs since MAVNode is able to handle multiple systems. MAVNode responds with a plain text port number, which corresponds to the UDP port over which the requested data are sent. The application can now listen on this port for incoming data until it no longer requires it, informing MAVNode to terminate the stream.

5   SUMMARY

UAS control system mechanization is briefly discussed in this chapter. It provides a glance at control system composition, flight control system design and development, and

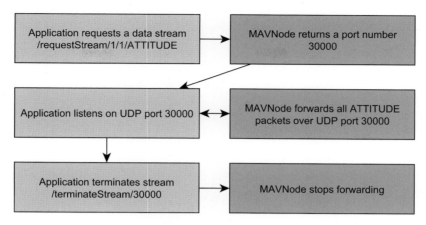

Figure 11. Application requesting data stream from MAVNode. (Reprinted with permission from Coombes *et al.* (2014). © 2014, IEEE.)

practical issues involved in UAV FCS development and operation. It is pointed out that flight control systems are a safety-critical function for UAV operation. It is quite complicated as it is composed of many different components such sensors and navigation, actuation systems, flight control computers, a number of control strategies (e.g., stability augmentation systems and autopilot), and aircraft dynamics, and these components interplay with each other. Attitude determination and navigation are very important in providing necessary information for FCS. For UAV, the communication/network between the UAV and its GCS is also heavily involved in the control system. FCS is a complex and interconnected combination of hardware and software implementation, and any malfunction or fault of each part may compromise the safety of the aircraft and the public. To reduce cost and development time, a widely used development procedure, including numerical simulation, software-in-the-loop simulation, and hardware-in-the-loop tests until real flight tests, is described. To address the safety concern, various fail-safe modes are introduced in UAV FCS development and operation. Depending on which component has a fault or malfunction, the FCS is able to switch to different fail-safe modes such as the manual control mode or the autonomous mode flight control.

Similar to many other technologies on UAS, control system mechanization is an evolving and important research area. New techniques and new capabilities affect the development of flight control systems; for example, automatic code generation may significantly reduce its development process.

REFERENCES

Barnhart, R.K., Hottman, S.B., Marshall, D.M., and Shappee, E. (2012) *Introduction to Unmanned Aircraft Systems*, CRC Press, New York.

Coombes, M., Eaton, W., McAree, O., and Chen, W.H. (2014) Development of a network enabled system for generic autonomous vehicles. *2014 UKACC International Conference on Control, July 9–11, 2014, Loughborough, UK*, pp. 621–627.

Coombes, M., McAree, O., Chen, W.H., and Render, P.M. (2012) Development of an autopilot system for rapid prototyping of high level control algorithms. *2012 UKACC International Conference on Control, September 3–5, 2012, Cardiff, UK*, pp. 292–297.

McShea, R.E. (2010) *Test and Evaluation of Aircraft Avionics and Weapon System*, AIAA Education Series, SciTech Publishers, Raleigh, NC.

Musial, M. (2008) *System Architecture of Small Autonomous UAVs*, VDM Verlag Dr. Müller Publications, Berlin.

Pratt, R.W. (2000) *Flight Control System – Practical Issues in Design and Implementation*, IEE Control Engineering Series 57, (eds. D.P. Atherton and G.W. Irwin, IEE, UK.

Saripalli, S., Roberts, J.M., Corke, P.I., Buskey, G., and Sukhatme, G.S. (2003) A tale of two helicopters. *2003 IEEE/RSJ International Conference on Intelligent Robots and Systems, October 27–31, 2003, Las Vegas, NV*, pp. 805–810.

Stansbury, S., Vyas, M.A., Timothy, A., and Richard, W. (2009) A survey of UAS technologies for command, control, and communication (C^3) . *J. Intell. Robot. Syst.*, **54**, 61–78.

Valavanis, K.P. (2007) *Advances in Unmanned Aerial Vehicles: State of the Art and the Road to Autonomy*, Springer.

PART 5
Autonomy

Chapter 14

Relative Navigation in GPS-Degraded Environments

David O. Wheeler,[1] **Paul W. Nyholm,**[2] **Daniel P. Koch,**[2] **Gary J. Ellingson,**[2]
Timothy W. McLain,[2] **and Randal W. Beard**[1]

[1]*Department of Electrical and Computer Engineering, Brigham Young University, Provo, UT, USA*
[2]*Mechanical Engineering Department, Brigham Young University, Provo, UT, USA*

1 INTRODUCTION

As processing, sensing, and battery technologies continue to develop, there are increased opportunities for unmanned air vehicles (UAVs) to contribute to society. Emerging applications include fire surveillance, search and rescue, infrastructure and agriculture monitoring, and the delivery of medical supplies to remote locations (Ambrosia *et al.*, 2011; Goodrich *et al.*, 2008; Herwitz *et al.*, 2002; Matternet, 2016). However, the majority of these applications require additional technology development and will likely be restricted to less populated environments. The integration of autonomous aircraft into mainstream life will depend in large part on the ability to safely and effectively operate in varied environments and with varied tasking. For example, current autonomous systems typically require external sensing or computation, such as a motion capture system, global

positioning system (GPS) localization, prior maps of the environment, or at least off-board sensor fusion and decision making. Other systems are accompanied with strong, limiting assumptions, such as a highly structured environment such as vertical walls, flat floors, stationary scenes, and so on. Small UAVs are also limited by size, weight, and power (SWaP) constraints. While autonomous flight is currently possible in specialized circumstances, the development of robust, real-time, onboard methods for autonomous control in cluttered, nonstructured environments without external or *a priori* information remains an open field of research.

A noted obstacle in reaching the navigation robustness necessary for the integration of UAVs in the national airspace is the heavy reliance on GPS. In 2010, the United States Joint Chief of Staff, Norton Schwartz, stated "It seems critical to me that the Joint Force should reduce its dependence on GPS-aided precision navigation and timing, allowing it to ultimately become less vulnerable, yet equally precise, and more resilient" (Schwartz, 2010). GPS not only provides global position estimates to constrain the drift introduced by noisy rate sensors like MEMS-based inertial measurement units (IMUs) but also provides a way to estimate ground speed and orientation, allowing the vehicle to estimate wind effects. A robust navigation solution cannot depend on accurate GPS measurements due to the varied sources of uncertainty presented in Table 1. Further, GPS measurements may be unavailable in the presence of GPS jammers, when shadowed by buildings or foliage, and simply cannot be applied indoors. Various sources report that GPS loss, even for a brief period, often results in catastrophic failure. As a result, GPS-denied navigation has become a strong emphasis of research over the last decade.

Unmanned Aircraft Systems. Edited by Ella Atkins, Aníbal Ollero,
Antonios Tsourdos, Richard Blockley and Wei Shyy.
© 2016 John Wiley & Sons, Ltd. ISBN: 978-1-118-86645-0.

Table 1. Sources of GPS uncertainty.

Type	Description
Multipath	Signal bounces before reaching receiver (false pseudo-range)
Number of satellites	Few visible satellites increases sensitivity to timing errors
Dilution of precision	Visible satellites are poorly spaced
Spoofing	Signal is locally recreated with false information
Atmospheric delays	Signal is delayed due to ionosphere and troposphere influences

Figure 1. The multirotor vehicle used for the implementation in this chapter. Details on the specific hardware are given in Table 2.

One important approach to GPS-denied navigation is known as simultaneous localization and mapping (SLAM). This approach involves estimating the vehicle's state relative to local landmarks by creating or adding to a map and localizing the vehicle within that map. This method allows for the direct use of relative measurements, such as visual odometry (VO), to estimate the vehicle's state. This chapter will serve as a tutorial outlining the basic components of a SLAM-based relative navigation solution. Although the framework, is applicable to other airframes and implementations, several of the relative navigation framework modules have been implemented on a multirotor aircraft and are presented as examples. Figure 1 and Table 2 describe the multirotor platform used in these implementations. Section 2 is an overview of the framework, while Sections 3 and 4 explain in detail the components of the relative front end and global back end. Conclusions are presented in Section 5.

Table 2. Hardware details.

Component	Description
Vehicle	Mikrokopter Hexacopter XL
Autopilot	Pixhawk
RGB-D camera	ASUS Xtion Pro Live
IMU	MicroStrain® 3DM-GX3®-15
Ultrasonic altimeter	MaxBotix MB1242
Processor	Intel Core i7-2710QE (2.1GHz × 4)

2 RELATIVE NAVIGATION FRAMEWORK

To drive an automobile safely, the driver should focus on the road ahead, rather than on their paper map, GPS unit, or communication device. The critical tasks of driving and responding to events on the road require fast and frequent attention. The tasks of navigating to the global destination, however, should be attended to opportunistically or perhaps even delegated to a passenger. Motivated by this observation, we propose using a relative navigation approach as shown in Figure 2, which decouples the relative, in-flight control from less critical global updates. Sections 2.1 and 2.2 outline the relative navigation framework, while Section 2.3 presents several scenarios that prove challenging for nonrelative frameworks.

2.1 Relative front-end overview

The core of the front end is a state estimation scheme (Section 3.2) that fuses sensor data, generally from a high rate IMU and infrequent exteroceptive sensors, such as an altimeter or a camera for visual odometry (Section 3.1). The state is estimated relative to a local, gravity-aligned coordinate frame (within about 1 m for a multirotor vehicle) known as a node frame. As the vehicle moves, new node frames are established and the transformation between node frames is stored as an edge (see Figure 3). A local path planner (Section 3.3) uses sensor data for obstacle avoidance and route planning and provides inputs to the aircraft's control (Section 3.4).

2.2 Global back-end overview

Decoupled from the flight-critical front end is the need for a globally accurate map. The back end, represented as a pose graph, is seeded with the node frames and edges created by the front end (Section 4.1). To eliminate accumulated drift error, place recognition algorithms efficiently compare the current image with all previous images (Section 4.2). When the vehicle returns to a previously visited location, known as a loop closure, an additional constraint is introduced to the set of edges, thereby over-constraining the map. Optionally, if GPS measurements are available, they can be added in the back end as constraints (Section 4.3). As desired, a robust optimization step minimizes constraint errors introduced by loop closures, GPS information, and odometry estimates (Section 4.4). With these refinements, the global map is then used by the high-level planner to fulfill global missions through a series of relative goals (Section 4.5). Subsequent

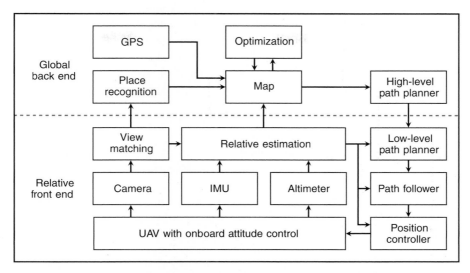

Figure 2. Relative navigation framework. The flight-critical elements of local estimation, path planning, and control are decoupled from global measurements. Odometry estimates, loop closures, and GPS measurements are optimized to form a globally consistent map making high-level missions possible.

loop closures are used to refine and improve the global map, making it sufficient for persistent, repeatable navigation. This global framework only interacts with the aircraft by influencing what low-level, relative goals are introduced, providing safer, more robust control in GPS-degraded environments.

2.3   Motivating scenarios

While working with respect to a single, inertial reference frame makes intuitive sense, the following scenarios highlight advantages to be gained from a relative framework. In each example, the UAV is assumed to be controlled with respect to its globally estimated position:

- A UAV loses GPS signal and receives IMU measurements only, for several minutes. Upon reacquiring GPS, the state estimate jumps drastically and the UAV is unable to recover.
- A multirotor vehicle moves from indoors to outdoors. Upon acquiring GPS signal for the first time its global state estimates, with respect to an arbitrary origin inside the building, will jump drastically.
- After a loop closure, a UAV's current estimated global state may jump significantly resulting in sudden, unintentional, and unpredictable vehicle motion.
- After flying for some time, the size of the optimization problem delays any updates to the UAV's estimated global state. The vehicle's control suffers as a result.

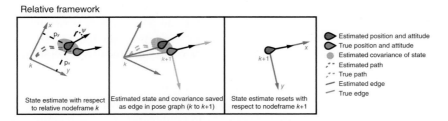

Figure 3. 2D illustration of estimated and true states relative to the current node frame. Roll and pitch (not shown) are inertially defined because the node frame z-axis is aligned with gravity. Position (p_x, p_y, p_z) and yaw (ψ) are defined relative to the current node frame, k. The state and covariance are estimated as usual with respect to this node frame. When a new node frame is declared, the estimated p_x, p_y, and p_z and ψ along with the associated covariance are saved as an edge in the global back-end pose graph. Node frame $k + 1$ is then defined as the current true state (with the roll and pitch removed), and the filter zeros p_x, p_y, and p_z and ψ and their corresponding covariance values. The result is that the state error and uncertainty are removed from the front end and delegated to the global back end.

- A vehicle receives an erroneous loop closure or GPS measurement. The estimated global state degrades without a method to later remove the effects of the outlying measurement.

3   RELATIVE FRONT END

The principal components of the relative front end, as introduced in Section 2.1, are explained in greater detail in the sections below.

3.1   Visual odometry

Visual odometry (VO) is the process of computing the motion of a camera by comparing the captured images (Nister, Naroditsky, and Bergen, 2004). VO algorithms fall into two general categories: appearance-based and feature-based. Appearance-based methods use information from all pixels in the images to compute motion, while feature-based methods use visually distinct features in the environment, such as corners, that are tracked from one image to the next. VO can be implemented using either monocular or stereo cameras, and more recently has been implemented using RGB-D cameras (e.g., Huang et al., 2011). In addition to providing a standard color image, RGB-D cameras, like the Microsoft Kinect sensor, provide a depth image that encodes how far the object imaged at each pixel is from the camera. With monocular VO the change in orientation of the camera can be computed, but the translation can be computed only up to an unknown scale factor. The depth information provided by stereo and RGB-D cameras enables the calculation of this scale factor.

Some VO algorithms compute the camera's motion between consecutive images, while others find the transformation to a chosen reference image, referred to as a key frame. While most VO algorithms could be used in the relative navigation framework, a key-frame approach fits best with the pose-graph SLAM paradigm. A new key frame is chosen once the vehicle has moved far enough that there is insufficient overlap between features in the current and key-frame images. The key-frame approach results in low drift in the VO estimates, especially for a hovering multirotor vehicle.

Good general tutorials on visual odometry are provided by Scaramuzza and Fraundorfer (2011) and Fraundorfer and Scaramuzza (2012). Additional approaches for VO using RGB-D cameras include those presented by Huang et al. (2011) and Zhang, Kaess, and Singh (2014), and a comparison between various methods is given by Fang and Scherer

(2014). Approaches utilizing laser scanners have also been widely explored in the literature (e.g., Grzonka, Grisetti, and Burgard, 2009; Bachrach et al., 2011), and techniques using point clouds from LiDAR or other sensors have been explored as well (e.g., Lui, Drummond, and Li, 2012). Increased UAV robustness is possible when relative measurements are incorporated from multiple sensors (Shen et al., 2014; Fabresse et al., 2014).

3.2   Estimation

To maintain stable flight, a filter must fuse available sensor data to provide robust estimates for attitude and velocity. Furthermore, many high-level goals require position to be estimated reliably. Several probabilistic methods have been developed to fuse a motion model with intermittent measurements for vehicle state estimation, including the complimentary filter, the particle filter, and the Kalman filter, where the latter is the most prevalent for UAV platforms. Due to the nonlinearity of UAV dynamics, the state uncertainty is often propagated by either using a second-order linearization by applying the analytic Jacobian of the dynamics (extended Kalman filter) or applying the nonlinear update to sigma points to reconstruct the state uncertainty (unscented Kalman filter (Wan and Van Der Merwe, 2000). While the following description outlines a Kalman filter approach, any estimation scheme could be used.

What most distinguishes the relative navigation approach from its conventional global counterparts is the decoupling of local state estimation from global states. Position and yaw states are defined relative to a local, gravity-aligned node frame. As illustrated in Figure 3, state estimates will drift from truth with the uncertainty estimated by the Kalman filter's covariance matrix. After moving a small distance, the current true state (with pitch and roll removed) is declared to be the origin of a new relative coordinate frame. The estimated position (p_x, p_y, p_z) and yaw (ψ), together with their associated covariance, are saved as an edge in the global back-end pose graph representing the estimated transformation between node frames with some uncertainty. These states are then replaced with states relative to the new node frame. Since the position and yaw states with respect to the new coordinate frame are now exactly zero, the associated covariance values are set to zero. This process can be thought of as augmenting and subsequently marginalizing the filter's states. The accumulated error and its accompanied uncertainty is effectively removed from the front-end filter and delegated to the back-end pose graph.

A multiplicative extended Kalman filter (MEKF) implementation of this relative estimation framework is presented by Leishman and McLain (2015). The filter estimates relative position, attitude (with relative yaw), body-fixed velocity, and gyroscope and accelerometer biases. The attitude of the vehicle is represented by a quaternion. While quaternions require four scalars to define a three degree of freedom rotation, they are computationally more efficient than Euler angles and avoid the singularity known as gimbal lock. When using quaternions however, the estimated attitude error, found in the typical update step of an extended Kalman filter (EKF), cannot simply be added to the current state estimate. A common approach is to multiply the attitude quaternion by the estimated attitude error, maintaining the quaternion norm, and earning the term multiplicative EKF (MEKF) (Lefferts, Markley, and Shuster, 1982). The underlying dynamics of the filter make use of the typical kinematic relationships (Beard and McLain, 2012) coupled with an enhanced rotorcraft drag model presented by Leishman *et al.* (2014). The velocity estimates are constrained by including, in conjunction with accelerometer measurements, drag terms that are proportional to the body-fixed forward and side velocities. In this implementation, gyroscope measurements are considered as inputs to the system (mechanization), while accelerometer measurements are used as updates. A feature-based VO for a forward-facing RGB-D camera, using FAST features (Rosten and Drummond, 2006) with BRIEF descriptors (Calonder *et al.*, 2010) was used (see Leishman, McLain, and Beard (2014) for details). Flight tests were performed using the hardware outlined in Table 2 and the MEKF described above. Figure 4 compares the MEKF's estimated forward and yaw states with truth as measured by a motion capture system. The discontinuities occur as the vehicle transitions from one node frame to the next.

3.3 Low-level path generation and following

The autonomous vehicle must be able to maneuver relative to the local environment without reliance on global state information. This is accomplished as local obstacles are determined from sensor information and goals with respect to the local node frame are received from the high-level path planner (Section 4.5). Low-level path planning typically takes the form of reactive obstacle avoidance, while the global path is calculated by the high-level path planner.

For example, one approach when using a LiDAR or RGB-D sensor is to simplify the path generation problem onto the 2D horizontal plane at the elevation of the vehicle. Obstacles are identified from the most recent depth information and saved with respect to the current node frame. A 2D path is defined relative to the current node frame that progresses toward the goal while avoiding obstacles. Obstacles, goal locations, and the current path are transformed into each new node frame. A path following approach, such as the one described in Beard and McLain (2012), can be used to provide a desired state to the controller.

Other approaches may be more appropriate given different vehicles and sensor information and may be extended beyond 2D. For example, if a fixed-wing aircraft is flying with a LiDAR then paths may be generated that account for the vehicle dynamics to avoid obstacles (e.g., Bry and Roy, 2011). This includes Dubins paths (Beard and McLain, 2012; Owen, Beard, and McLain, 2014) and vector fields for path following. A 3D multirotor approach could include biologically inspired steering algorithms, where obstacles are identified in the local frame and trajectories are chosen based on distance and bearing to obstacles and goals (Scherer *et al.*,

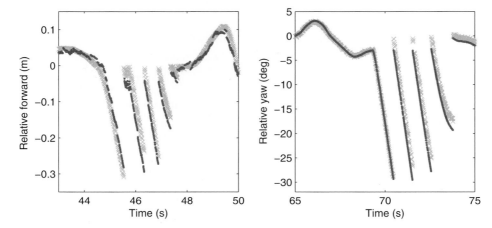

Figure 4. Performance plots for a MEKF implementation using the relative navigation framework. The estimated forward and yaw states (black) are compared to truth (gray). The discontinuities indicate a change in node frame.

2007). Other maneuver-based planning techniques generate a series of poses to smoothly navigate through the observed local environment while avoiding obstacles (Frazzoli, Dahleh, and Feron; 2005).

3.4   Control

Control approaches for both fixed-wing and multirotor vehicles have been widely explored in the literature and will not be reviewed in detail here. One of the key differences between a controller implementation for a relative framework to its global counterparts is that the desired states of the vehicle are expressed in the local node frame rather than in the global frame. However, because the actual state of the vehicle is also expressed in the local frame, the error between the actual and desired states will be the same as if both were expressed in the global frame. Therefore, many controllers designed to work in a global framework will also work well within the relative framework.

The position controller used in the example implementation for this chapter is adapted from the controller detailed by Ferrin *et al.* (2011), while attitude stabilization and control is achieved using the standard PID loops implemented on the autopilot. The position estimation and control are executed at approximately 100 Hz on the onboard computer, while the attitude estimation and control runs at a higher rate on the autopilot. The outputs of the position controller are roll, pitch, yaw rate, and thrust set points that become the inputs to the attitude controller. This architecture is illustrated in Figure 5.

4   GLOBAL BACK END

The principal components of the global back end, as introduced in Section 2.2, are explained in greater detail below.

4.1   Pose graph

To maintain a spatially consistent global map, the back end stores information about the vehicle's trajectory as a pose graph. A pose graph is a graphical representation of a vehicle's path that encodes global vehicle pose estimates as graph vertices and the relative transformation between two poses as graph edges. A pose graph representation is effective because efficient graph optimization algorithms have been developed and can be applied to refine pose estimates; additionally, the graph conveniently serves as an abstract map that can be used for 3D and human-readable map construction as well as high-level path planning.

Figure 6 depicts how a pose graph is constructed from vehicle odometry. In a relative framework, a pose graph vertex represents the global estimate of a node frame's translation and orientation. The graph edge connecting two consecutive poses encodes the relative transformation between two node frames and its respective covariance. The front-end estimator provides relative transformation edges to the back end. The transformations are compounded with all previous transformations to estimate the global position of the new node frame and a corresponding vertex is added to the graph.

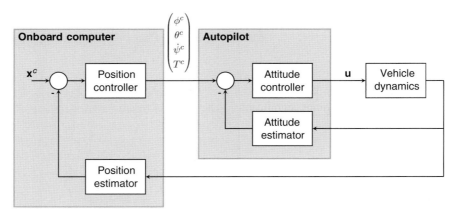

Figure 5. Control system architecture for the example implementation. The position estimation and control execute on the onboard computer, while attitude estimation and control run at a higher rate on the autopilot. In this figure $\mathbf{x}^c$ is the commanded position, $\mathbf{u}$ is the motor commands, and $(\phi^c, \theta^c, \dot{\psi}^c, T^c)^{\mathrm{T}}$ are the roll, pitch, yaw rate, and thrust commands sent from the onboard computer to the autopilot.

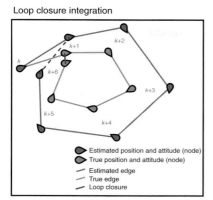

Loop closure integration

Estimated position and attitude (node)
True position and attitude (node)
— Estimated edge
— True edge
— Loop closure

Figure 6. A loop closure edge (dashed line) over-constrains the back-end pose graph. After optimization, the estimated node frame states (dark gray line) will more closely resemble truth (light gray line). The tear drops represent graph vertices (node frames), while lines represent graph edges. Each estimated and loop-closure edge also stores a covariance matrix representing the uncertainty in that transform. Over-constraining the graph in this manner enables the optimization routine to adjust the graph edges to minimize the total uncertainty in the pose graph (Section 4.4).

A pose graph is also capable of encoding measurements other than odometry. Loop closure measurements introduce edges between two existing, nonconsecutive nodes (Section 4.2). Similarly, GPS measurements can be added as edges between the vertex where the measurement was received and the origin of the global coordinate system (Section 4.3). Since the seminal work of Lu and Milios (1997), pose-graph SLAM has become the predominate method of SLAM in the literature. An excellent tutorial on pose-graph SLAM is given by Grisetti *et al.* (2010).

4.2 Place recognition

A vehicle's global position estimate will drift significantly over time in the absence of global updates. A common solution in the SLAM literature makes use of place recognition (PR). In place recognition, the current image is compared with previous images to determine if the vehicle has returned to the same location. While a naive solution would require significant computation and memory, not generally available with the size, weight, and power (SWaP) constraints of a UAV, the computer vision community has developed efficient vocabulary-based PR algorithms. As with VO algorithms, distinct image features are represented by mathematical descriptors. Using a large, representative training dataset of images, the most prominent, distinct feature descriptors are saved off-line and referred to as words in a vocabulary. During flight, any image can be succinctly represented by the set of nearest vocabulary

words found in the image. Images can be quickly compared using word occurrences, similar to many search engine algorithms (Sivic and Zisserman, 2006). Further work has improved place recognition performance in the presence of aliasing, where high correlation is found on noncorrelated images, like pictures of brick walls (Cummins and Newman, 2008).

In a relative navigation approach, each key frame image (Section 3.1) is passed through the place recognition software and archived. After a match is found, the images are recalled and visual odometry methods provide the estimated transformation from one node frame to another. The transform is communicated to the pose graph as an edge between nonconsecutive nodes, allowing for optimization and reduction of accumulated drift.

4.3 Intermittent GPS integration

With optimization, odometry and loop closures can be integrated to maintain a globally consistent, local map; the first node frame is considered to be the origin of a global coordinate system and the subsequent node frames are defined with respect to it. This type of map is useful for navigation but does not allow for high-level path planning techniques such as navigating to desired GPS coordinates. To position the globally consistent, local map on the earth's coordinate frame, GPS measurements are incorporated into the back end.

A virtual zero vertex is added to the graph to enable translation and rotation of the local map on the global plane (Rehder *et al.*, 2012). The virtual zero can be thought of as the origin of the Earth's global coordinate frame. A virtual constraint is added as an edge between the virtual zero vertex and another vertex on the graph. The virtual constraint has infinite covariance, meaning that there is no certainty about the relative transformation encoded by the edge. Adding a virtual zero and virtual constraint to the graph allows the map to translate freely within the global coordinate system.

Figure 7 depicts the process of adding a GPS measurement to the pose graph. First, when a measurement is received, an odometry edge and vertex is added to the graph representing the location where the measurement was received. This edge comes from the front end's current estimated state relative to the latest node frame. When the odometry has been established, an additional edge is created linking the virtual zero vertex and the newly created vertex. As with odometry and loop closures edges, the GPS edge encodes the relative transformation between the two vertices with its associated covariance. With GPS edges in place, the graph can be optimized to yield a globally consistent map that aligns with a known global coordinate system.

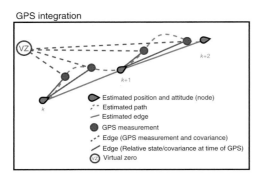

Figure 7. A virtual zero vertex is added to the pose graph and represents the origin of a global coordinate system. Edges between the virtual zero vertex and node frame vertices allow the map to be translated and oriented to its proper global location.

4.4   Map optimization

Because of erroneous odometry measurements, graph optimization is critical to maintaining a globally consistent map. A key advantage of using a pose graph to represent a map is that it can easily be formulated as a least-squares optimization problem where poses and edges become free variables and constraints, respectively. Least-squares optimization attempts to find the arrangement of poses that most likely results from a given set of odometry, loop closures, and GPS measurements. Many efficient graph optimization algorithms exist, some of which are popular open-source projects (Stachniss, Frese, and Grisetti, 2014).

Figure 8 shows a pose-graph map before and after optimization. The data was gathered using the relative front-end implementation described in Section 3.2 and the map was optimized with the popular g2o library (Kümmerle *et al.*,

2011). A multirotor aircraft was flown approximately 125 meters through a series of hallways, forming a loop. About 25 m of overlap exist in the path. It is clear that prior to optimization, errors accumulated and the global pose estimate drifted from truth. Optimization mitigates these errors and returns a globally consistent map that matches the flight path.

Intuitively, pose graph optimization can be thought of a mass spring system settling in a state of least energy. In this analogy, graph vertices are masses and edges are springs. A spring's stiffness corresponds to its edge's covariance, becoming increasingly stiff as the covariance approaches zero. In a thought experiment, one can visualize that when a graph constructed of only odometry edges is optimized, the output is simply the original graph because there are no additional loop closure or GPS springs to deform the system. When loop closures are added to the graph, additional stiff springs are added between nonconsecutive masses and the system settles into its new lowest energy state, resulting in a graph that more closely represents the true vehicle trajectory. Finally, in this example, the virtual zero and virtual constraint can be thought of as an infinitely long string tying the mass-spring system to the origin of the global coordinate system, as shown in Figure 9. This establishes a reference to the global coordinate system and allows the mass system to translate about the origin. GPS edges are then added as springs between the virtual zero and corresponding masses, pushing the map into its appropriate absolute position.

One drawback of least-squares optimization is that it is inherently sensitive to outliers. This means that false positive loop closures and erroneous GPS measurements have catastrophic effects on graph optimization. To compensate for these effects, dynamic covariance scaling (DCS) (Agarwal *et al.*, 2013), a variation of least squares that is robust to

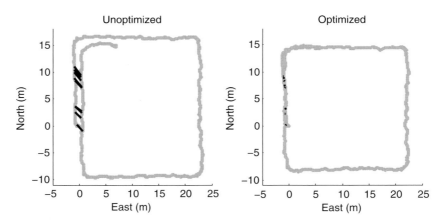

Figure 8. Map before (left) and after (right) optimization. Odometry and loop closure edges are shown in gray and black, respectively.

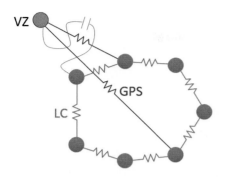

Figure 9. Illustration of mass spring optimization example. Springs represent edges in the pose graph, with the stiffness of the spring being analogous to the uncertainty associated with the edge (a stiffer spring corresponds to lower uncertainty). A virtual zero is established as the origin of the global coordinate frame and is tied to the mass spring system. GPS edges are added as springs between the virtual zero and the poses where a measurement is received. Map optimization is analogous to releasing the over-constrained mass spring system and allowing it to settle into its lowest energy state.

outliers, can be used. DCS can be thought of as giving no certainty to edges that are deemed as outliers. By using DCS, one is able to produce globally consistent maps even in the presence of erroneous loop closures and GPS measurements. Other methods exist for making graph optimization robust to outliers (Lee, Fraundorfer, and Pollefeys, 2013; Latif, Cadena, and Neira, 2013; Olson and Agarwal, 2013; Sunderhauf and Protzel, 2012). A comparison of several of these methods is given by Sunderhauf and Protzel (2013).

4.5 High-level path planning

The purpose of the high-level path planner is to transform global information, such as waypoints or obstacles, into the current node frame. As the relative estimator transitions from one node frame to the next, the high-level path planner passes up-to-date relative goals to the low-level path planner. These relative goals are the only way that the global back end influences the front-end control, effectively isolating the UAV from the effects of jumps in the global state estimate due to optimization or global measurements. The high-level path planner can be used for autonomous waypoint following, exploration and mapping, target tracking, or landing.

5 CONCLUSION

The integration and widespread use of unmanned air systems for many applications depends heavily on the ability of these aircraft to operate in a safe and reliable manner, often in the presence of degraded GPS signals. A relative SLAM-based approach has been presented as a viable solution for robust navigation in GPS-degraded environments. By decoupling the real-time local estimation and control from the global position estimation, the vehicle is able perform essential tasks such as stabilization and obstacle avoidance without being dependent on consistent and accurate global estimates. A pose-graph-based global back end allows techniques such as place recognition, loop closures, GPS integration, and map optimization to opportunistically improve global estimates. The separation between the relative front end and global back end also eliminates control issues that can arise when global estimates change as new information is incorporated. An example implementation has been demonstrated for a multi-rotor aircraft; however, the presented relative navigation framework is not implementation specific and could be adapted to other vehicles, sensors, and mission profiles.

REFERENCES

Agarwal P., Tipaldi G. D., Spinello L., Stachniss C., and Burgard W. (2013) Robust map optimization using dynamic covariance scaling, in *Proceedings of the IEEE International Conference on Robotics and Automation*, IEEE, pp. 62–69.

Ambrosia V., Wegener S., Zajkowski T., Sullivan D., Buechel S., Enomoto F., Lobitz B., Johan S., Brass J., and Hinkley E. (2011) The Ikhana unmanned airborne system (UAS) western states fire imaging missions: from concept to reality (2006–2010). *Geocarto Int.*, **26**, 85–101.

Bachrach A. G., Prentice S., He R., and Roy N. (2011) RANGE – robust autonomous navigation in GPS-denied environments. *J. Field Robot.*, **28**(5), 644–666.

Beard R. W. and McLain T. W. (2012) *Small Unmanned Aircraft: Theory and Practice*, Princeton University Press.

Bry A. and Roy N. (2011) Rapidly-exploring random belief trees for motion planning under uncertainty∗, in *2011 IEEE International Conference on Robotics and Automation (ICRA)*, IEEE, pp. 723–730.

Calonder M., Lepetit V., Strecha C., and Fua P. (2010) Brief: Binary robust independent elementary features, in *Computer Vision – ECCV 2010* (eds Daniilidis K., Maragos P., and Paragios N.), vol. 6314 of *Lecture Notes in Computer Science*, Springer, Berlin, pp. 778–792.

Cummins M. and Newman P. (2008) FAB-MAP: Probabilistic localization and mapping in the space of appearance. *Int. J. Robot. Res.*, **27**(6), pp. 647–665.

Fabresse F., Caballero F., Maza I., and Ollero A. (2014) Localization and mapping for aerial manipulation based on range-only measurements and visual markers, in *2014 IEEE International Conference on Robotics and Automation (ICRA)*, pp. 2100–2106.

Fang Z. and Scherer S. (2014) Experimental study of odometry estimation methods using RGB-D cameras, in *2014 IEEE/RSJ*

International Conference on Intelligent Robots and Systems, pp. 680–687.

Ferrin J., Leishman R. C., Beard R. W., and McLain T. W. (2011) Differential flatness based control of a rotorcraft for aggressive maneuvers, in *2011 IEEE/RSJ International Conference on Intelligent Robots and Systems,* IEEE, pp. 2688–2693.

Fraundorfer F. and Scaramuzza D. (2012) Visual odometry: part II: matching, robustness, optimization, and applications. *IEEE Robot. Automat. Mag.,* June, 78–90.

Frazzoli E., Dahleh M., and Feron E. (2005) Maneuver-based motion planning for nonlinear systems with symmetries, *IEEE Trans. Robot.,* 21, pp. 1077–1091.

Goodrich M. A., Morse B. S., Gerhardt D., Cooper J. L., Quigley M., Adams J. A., and Humphrey C. (2008) Supporting wilderness search and rescue using a camera-equipped mini UAV. *J. Field Robot.,* 25(1), 89–110.

Grisetti G., Kummerle R., Stachniss C., and Burgard W. (2010) A tutorial on graph-based SLAM. *IEEE Intell. Transp. Syst. Mag.,* 2(4), 31–43.

Grzonka S., Grisetti G., and Burgard W. (2009) Towards a navigation system for autonomous indoor flying, in *2009 IEEE International Conference on Robotics and Automation,* pp. 2878–2883.

Herwitz S. R., Johnson L. F., Arvesen J. C., Higgins R., Leung J., and Dunagan S. (2002) Precision agriculture as a commercial application for solar-powered unmanned aerial vehicles. *AIAA 1st Technical Conference and Workshop on Unmanned Aerospace Vehicles.*

Huang A. S., Bachrach A., Henry P., Krainin M., Fox D., and Roy N. (2011) Visual odometry and mapping for autonomous flight using an RGB-D camera, in *International Symposium on Robotics Research (ISSR).*

Kümmerle R., Grisetti G., Strasdat H., Konolige K., and Burgard W. (2011) g2o: a general framework for graph optimization, in *IEEE Int. Conf. Robot. Autom.,* pp. 3607–3613.

Latif Y., Cadena C., and Neira J. (2013) Robust loop closing over time for pose graph SLAM. *Robotics,* 32(14), 1611–1626.

Lee G. H., Fraundorfer F., and Pollefeys M. (2013) Robust pose-graph loop-closures with expectation-maximization, *2013 IEEE/RSJ Int. Conf. Intell. Robot. Syst.,* pp. 556–563.

Lefferts E., Markley F., and Shuster M. (1982) Kalman filtering for spacecraft attitude estimation. *IEEE Contr. Syst. Mag.,* 5(5), 417–429.

Leishman R. C., McLain T. W., and Beard R. W. (2014) Relative navigation approach for vision-based aerial GPS-denied navigation. *J. Intel. Robot. Syst.,* 74(1), 97–111.

Leishman R. C. and McLain T. W. (2015) multiplicative extended Kalman filter for relative rotorcraft navigation. *J. Aerosp. Inform. Syst.,* 12, pp. 728–744.

Leishman R. C., Macdonald J. C., Beard R. W., and Mclain T. W. (2014) Quadrotors and accelerometers: state estimation with an improved dynamic model. *IEEE Contr. Syst. Mag.,* 34(1), pp. 28–41.

Lu F. and Milios E. (1997) Globally consistent range scan alignment for environment mapping, *Auton. Robot.,* 4, 333–349.

Lui W. L. D., Tang T. J. J., Drummond T., and Li W. H. (2012) Robust egomotion estimation using ICP in inverse depth coordinates, in *2012 IEEE International Conference on Robotics and Automation,* IEEE, pp. 1671–1678.

Matternet (2016) Our company. available at https://www.mttr.net/company (accessed April 4, 2016).

Nister D., Naroditsky O., and Bergen J. (2004) Visual odometry, in *Proceedings of the International Computer Society Conference on Computer Vision and Pattern Recognition,* vol. 1, pp. 652–659.

Olson E. and Agarwal P. (2013) Inference on networks of mixtures for robust robot mapping. *Int. J. Robot.* 32(7), 826–840.

Owen M., Beard R. W., and McLain T. W. (2014) Implementing Dubins airplane paths on fixed-wing UAVs. *Handbook of Unmanned Aerial Vehicles,* Springer, The Netherlands, pp. 1677–1701.

Rehder J., Gupta K., Nuske S., and Singh S. (2012) Global pose estimation with limited GPS and long range visual odometry, in *Proceedings of the IEEE International Conference on Robotics and Automation,* IEEE, pp. 627–633.

Rosten E. and Drummond T. (2006) Machine learning for high-speed corner detection, in *Computer Vision – ECCV 2006* (eds Leonardis A., Bischof H., and Pinz A.), vol. 3951 of *Lecture Notes in Computer Science,* Springer, Berlin, pp. 430–443.

Scaramuzza D. and Fraundorfer F. (2011) Visual odometry: part I: the first 30 years and fundamentals. *IEEE Robot. Automat. Mag.,* December, 80–92.

Scherer S., Singh S., Chamberlain L., and Saripalli S. (2007) Flying fast and low among obstacles, in *2007 IEEE International Conference on Robotics and Automation,* pp. 2023–2029.

Schwartz N. (2010) The United States as an aerospace nation: challenges and opportunities, in *IFPA Fletcher Conf. on National Security Strategy and Policy,* Medford, MA.

Shen S., Mulgaonkar Y., Michael N., and Kumar V. (2014) Multi-sensor fusion for robust autonomous flight in indoor and outdoor environments with a rotorcraft mav, in *2014 IEEE International Conference on Robotics and Automation (ICRA),* pp. 4974–4981.

Sivic J. and Zisserman A. (2006) Video Google: Efficient visual search of videos, in *Toward Category-Level Object Recognition* (eds Ponce J., Hebert M., Schmid C., and Zisserman A.), vol. 4170 of *Lecture Notes in Computer Science,* Springer, Berlin, pp. 127–144.

Stachniss C., Frese U., and Grisetti G. (2014) *OpenSLAM.*

Sunderhauf N. and Protzel P. (2012) Switchable constraints for robust pose graph SLAM, *2012 IEEE/RSJ Int. Conf. Intell. Robot. Syst.,* pp. 1879–1884.

Sunderhauf N. and Protzel P. (2013) Switchable constraints vs. max-mixture models vs. RRR a comparison of three approaches to robust pose graph SLAM, *2013 IEEE Int. Conf. Robot. Autom.,* pp. 5198–5203.

Wan E., Van R., and Der Merwe (2000) The unscented Kalman filter for nonlinear estimation, in *The IEEE 2000 Adaptive Systems for Signal Processing, Communications, and Control Symposium (AS-SPCC),* pp. 153–158.

Zhang J., Kaess M., and Singh S. (2014) Real-time depth enhanced monocular odometry, in *2014 IEEE/RSJ International Conference on Intelligent Robots and Systems (IROS 2014),* pp. 4973–4980.

Chapter 15

Target Detection and Mission Planning Based on Pigeon-Inspired Optimization

Haibin Duan

School of Automation Science and Electrical Engineering, Beihang University, Beijing, P.R. China

1 INTRODUCTION

As with manned aircraft flights, mission planning is a critical element in successful mission performance. In the simplest case, a vehicle is supposed to fly over a disaster area such as wood fire or a large traffic accident and deliver high-quality sensor data such as images or videos. In this case, planning for this mission would require determination of flight paths to the monitored area in which the vehicle will loiter while monitoring the target point. More complicated missions may include several sub-missions with alternatives. This type of mission may put a premium on the ability to calculate times and fuel consumption so that all sub-missions can be accomplished on time and within the total endurance of the vehicle.

Compared to autonomous vehicles that perform solo missions, greater efficiency and operational capability can be realized from teams of autonomous vehicles operating in a coordinated fashion. It is easy to understand that a group of unmanned aerial vehicles (UAVs) is more capable than a single UAV, since the workload can be divided among the group. Using multiple UAVs cooperatively to accomplish a complicated task is a very challenging problem. In this, the mission assignment for multiple UAVs is key issue that attracts great attention and it has different objective functions according to concrete applications. Usually, multiple UAVs mission assignment can be formulated to be an optimization problem of the air-to-ground operations to maximize the effectiveness of UAV allocation. So planning function for multiple UAVs is also of great importance, such as mission assignment.

Inspired by the behavior of a swarm of pigeons, pigeon-inspired optimization (PIO) algorithm is proposed by Duan in 2014, which is a novel swarm intelligence algorithm. Pigeons have nice senses of orientation, which contributes to their superior ability to find the most efficient way to the destination. The main tools they use to find their way home are map, compass, and landmarks. Pigeons can sense the earth magnetic field and they take the sun position as a compass to form a map in their memories that guides them to the right direction. Meantime, pigeons also have the ability to recognize the landmarks they have met before so that they can obtain the best path to their destination. PIO algorithm completely reproduces these processes. The procedure of PIO can be divided into two parts: map and compass section and landmark section. In map and compass section, pigeons update their positions and velocities according to the global best position in each process of iteration. While in landmark section, pigeons update their positions using the best center

Unmanned Aircraft Systems. Edited by Ella Atkins, Aníbal Ollero, Antonios Tsourdos, Richard Blockley and Wei Shyy.
© 2016 John Wiley & Sons, Ltd. ISBN: 978-1-118-86645-0.

position in each iteration. Through these two parts of updates, pigeons will soon find the global best position of the history.

As image sensors have become more and more advanced, it has become imperative to design a satisfying target recognition system for UAVs in order to achieve autonomous reconnaissance and detection (Williamson *et al.*, 2009; Deng and Duan, 2014). The target recognition and detection method for UAVs has been investigated quite intensively in recent years. The objective of path planning is to search out an optimal or near-optimal flight path between an initial location and the desired destination under specific constraint conditions (Duan and Li, 2014). There are several considerations, including optimality, completeness, and computational complexity, for an ideal path planner. The pre-flight path planning in a large mission area is a typical large-scale optimization problem. Usually, multiple UAVs mission assignment can be formulated to be an optimization problem of the air-to-ground operations to maximize the effectiveness of UAV allocation. This chapter focuses on target detection and mission planning such as path planning and task assignment using PIO.

2   PIGEON-INSPIRED OPTIMIZATION

PIO algorithm is a new swarm intelligence algorithm inspired by the homing behaviors of pigeons, proposed by Duanin 2014 (Duan and Qiao, 2014). Population-based swarm intelligence algorithms have been widely accepted and successfully applied to solve many optimization problems. Unlike traditional single-point-based algorithms such as hill-climbing algorithms, a population-based swarm intelligence algorithm consists of a set of points (population) that solve the problem through information sharing to cooperate and/or compete among themselves. Exploration and exploitation is also the key issue for these meta-heuristic swarm intelligence algorithms. In recent years, there have been a lot of population-based swarm intelligence algorithms, such as ant colony optimization (ACO), particle swarm optimization (PSO) (Kennedy and Eberhart, 1995), artificial bee colony (ABC) algorithm, differential evolution (DE), and brain storm optimization (BSO). All the bio-inspired optimization algorithms are trying to simulate the natural ecosystem mechanisms, which have greatly improved the feasibility of the modern optimization techniques, and offered practical solutions for those complicated combinatorial optimization problems. Path planning is the problem of designing the path a vehicle is supposed to follow in such a way that a certain objective is maximized and a goal is reached (Ergezer and Leblebicioglu, 2013). Path planning is one of the most challenging issues of mission planning for UAVs, especially in complicated combating environments.

2.1   Natural behavior of pigeons

The word "pigeon" is derived from the Latin word "pipio," meaning "young cheeping bird." Pigeon is a type of very common and popular bird. The wild pigeon is found in coastal areas, and the feral pigeon is found almost exclusively in areas of human habitation. Pigeons were once widely used in military because of their homing behavior.

During the First and Second World Wars, pigeons especially contributed to the Australian, French, German, American, and UK forces. Pigeons have the special homing ability by which they are thought to use a combination of the Sun, the Earth's magnetic field, and landmarks to find their way around. Guilford argues that pigeons probably use different navigational tools during different parts of their journey (Guilford, Roberts, and Biro, 2004). Guilford and his colleagues developed a mathematical model that predicts when pigeons will swap from one technique to another. When pigeons start their journey, they may rely more on compass-like tools. While in the middle of their journey, they could switch to using landmarks when they need to reassess their route and make corrections.

Investigation of pigeons' ability to detect different magnetic fields demonstrates that the pigeons' impressive homing skills almost depend on tiny magnetic particles in their beaks. Exactly, there are iron crystals in pigeons' beaks, which can give birds a nose for north. Studies show that the species seem to have a system in which signals from magnetite particles are carried from the nose to the brain by the trigeminal nerve. Evidence that the Sun is also involved in pigeon navigation has been interpreted, either partly or entirely, in terms of the pigeon's ability to distinguish differences in altitude between the Sun at the home base and at the point of release. Recent researches on pigeon behaviors also show that the pigeon can follow some landmarks, such as main roads, railways, and rivers rather than head for their destination directly.

2.2   Mathematical model

2.2.1   Map and compass operator

Pigeons can sense the earth field by using magneto reception to shape the map in their brains. They regard the altitude of the sun as compass to adjust the direction. As they fly to their destination, they rely less on Sun and magnetic particles.

In the PIO model, virtual pigeons are used naturally. In this map and compass operator, the rules are defined with the position X_i and the velocity V_i of pigeon i, and the positions and velocities in a D-dimension search space are updated in each iteration. The new position and velocity of pigeon i at the tth iteration can be calculated with Equations 1 and 2.

$$V_i(t) = V_i(t-1) \cdot e^{-Rt} + \text{rand} \cdot (X_g - X_i(t-1)) \quad (1)$$

$$X_i(t) = X_i(t-1) + V_i(t) \quad (2)$$

where R is the map and compass factor, rand is a random number, and X_g is the current global best position, which can be obtained by comparing all the positions among all the pigeons. Figure 1 shows the map and compass operator model of PIO.

As shown in Figure 1, the best positions of all pigeons are guaranteed by using map and compass. By comparing all the flied positions, it is obvious that the right-centered pigeon's position is the best one. Each pigeon can adjust its flying direction by following this specific pigeon according to Equation 1, which is expressed by the thick arrows. The thin arrows are its former flying direction, which has relation to $V_i(t-1) \cdot e^{-Rt}$ in Equation 1. The vector sum of these two arrows is its next flying direction.

2.2.2 Landmark operator

When the pigeons fly close to their destination, they will rely on landmarks neighboring them. If they are familiar to the landmarks, they will fly straight to the destination. If they are far away from the destination and unfamiliar with the landmarks, they will follow the pigeons that are familiar with the landmarks.

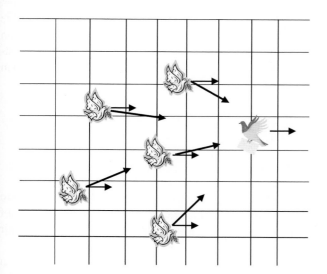

Figure 1. Map and compass operator model of PIO. (Reproduced with permission from H. Duan and P. Qiao. © Emerald Group Publishing Limited, 2014.)

In the landmark operator, half of the number of pigeons is decreased by N_P pigeonnum$_i\pi$ in every generation. However, the pigeons are still far from the destination, and they are unfamiliar to the landmarks. Let $X_c(t)$ be the center of some pigeon's position at the tth iteration, and suppose every pigeon can fly straight to the destination. The position updating rule for pigeon i at the tth iteration can be given by

$$N_P(t) = \frac{N_P(t-1)}{2} \text{pigeonnum}_i = \text{pigeonnum}_{i-1}/2 \quad (3)$$

$$X_c(t) = \frac{\sum X_i(t) \cdot \text{fitness}(X_i(t))}{N_P \sum \text{fitness}(X_i(t))} \text{xcenter}_i \quad (4)$$

$$= \text{xsum}_i/\text{pigeonnum}$$

$$X_i(t) = X_i(t-1) + \text{rand} \cdot (X_c(t) - X_i(t-1)) \quad (5)$$

where fitness() is the quality of the pigeon individual. For the minimum optimization problems, we can choose fitness$(X_i(t)) = 1/(f_{\min}(X_i(t)) + \varepsilon)$. For maximum optimization problems, we can choose fitness$(X_i(t)) = f_{\max}(X_i(t))$. For each individual pigeon, the optimal position of the Nc-thiteration can be denoted by X_p and $X_p = \min(X_{i1}, X_{i2}, \ldots, X_{iNc})$. Figure 2 shows the landmark operator model of PIO.

As shown in Figure 2, the center of all pigeons (the pigeon in the center of the circle) is their destination in each iteration. Half of all the pigeons (the pigeons out of the circle) that are far from their destination will follow the pigeons that are close to their destination, which also means that two pigeons may be at the same position. The pigeons that are close to their destination (the pigeons in the circle) will fly to their destination very quickly.

2.3 The procedure of basic PIO

The detailed procedure of basic PIO for UAV path planning can be described as follows and the pseudo-code of PIO are presented in Table 1.

Step 1: According to the environmental modeling, initialize the terrain information and the threaten information including the coordinates of threat centers, threat radiuses, and threat levels.

Step 2: Initialize parameters of pigeon-inspired optimization algorithm, such as solution space dimension D, the population size N_P, map and compass factor R, the number of iteration Nc$_1$ max and Nc$_2$ max for two operators, and Nc$_2$ max > Nc$_1$ max.

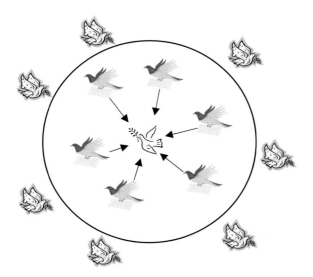

Figure 2. Landmark operator model. (Reproduced with permission from H. Duan and P. Qiao. © Emerald Group Publishing Limited, 2014.)

Step 3: Set each pigeon with a randomized velocity and path. Compare the fitness of each pigeons and find the current best path.

Step 4: Operate map and compass operator. First, we update the velocity and path of every pigeon by using Equations 1 and 2. Then we compare all the pigeons' fitness and find the new best path.

Step 5: If $Nc > Nc_{1max}$, stop the map and compass operator and operate next operator. Otherwise, go to step 4.

Step 6: Rank all pigeons according to their fitness values. Half of pigeons with bad fitness will follow those pigeons with high fitness according to Equation 3. We then find the center of all pigeons according to Equation 4, and this center is the desirable destination. All pigeons will fly to the destination by adjusting their flying direction according to Equation 5. Next, store the best solution parameters and the best cost value.

Step 7: If $Nc > Nc_{2max}$, stop the landmark operator, and output the results. If not, go to step 6.

The above programming steps of PIO algorithm can also be summarized as flowchart, as shown in Figure 3.

3   PIO FOR TARGET DETECTION

3.1   Problem formulation

In this chapter, the hybrid model of edge potential function (EPF) and Simulated Annealing Pigeon-Inspired Optimization

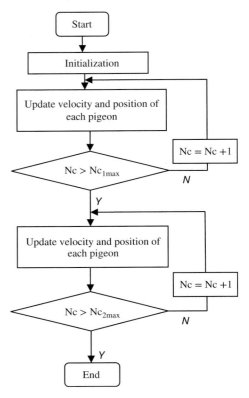

Figure 3. The implementation procedure of PIO. (Reproduced with permission from H. Duan and P. Qiao. © Emerald Group Publishing Limited, 2014.)

(SAPIO) algorithm is proposed to accomplish the target detection task for UAVs at low altitude (Li and Duan, 2014).

EPF is a newly developed shape-matching approach that was proposed by Dao, Natale, and Massa (2007). Similar to the electric potential generated by the electrostatic field, EPF is exploited to model the potential generated by edge structures of the image. The electric potential created by a set of point charges Q_i in a homogenous background, at a distance r from charges, can be shown as

$$v(\vec{r}) = \frac{1}{4\pi\varepsilon} \sum_i \frac{Q_i}{|\vec{r} - \vec{r}_i|} \tag{6}$$

where ε is the permittivity of the medium $\vec{r}$ and $\vec{r}_i$ are the observation point and charge locations, respectively.

In the EPF model, the ith edge of the image at coordinates (x_i, y_i) is equivalent to the charge point $Q_{eq}(x_i, y_i)$. The edge potential produced by a set of edge points $Q_{eq}(x_i, y_i)$ can be presented as follows:

$$EPF(x, y) = \frac{1}{4\pi\varepsilon_{eq}} \sum_i \frac{Q_{eq(x_i, y_i)}}{\sqrt{(x - x_i)^2 + (y - y_i)^2}} \tag{7}$$

Table 1. The pseudo-code of PIO.

Pigeon-Inspired Optimization algorithm

Input:
N_P: Number of individuals in pigeon swarm
D: Dimension of the search space
R: The map and compass factor
Search range: the borders of the search space
Nc_{1max}: The maximum number of generations that the map and compass operation is carried out
Nc_{2max}: The maximum number of generations that the landmark operation is carried out
Output:
X_g: The global optima of the fitness function f
1 **Initialization:**
Set initial values for Nc_{1max}, Nc_{2max}, N_P, D, R and the search range
Set initial path X_i and velocity V_i for each pigeon individual
Set $X_p = X_i$, $Nc = 1$
Calculate fitness values of different pigeon individuals
X_g:arg min$[f(X_p)]$
2 **Map and Compass operations**:
for Nc :1 to $Nc_{1\,max}$ **do**
for i :1 to N_P **do**
while X_i is beyond the search range do
calculate V_i and X_i according to Equations 1 and 2
end while
end for
Evaluate X_i, and update X_p and X_g
end for
3 **Landmark operations:**
for Nc: $Nc_{1\,max} + 1$ to $Nc_{2\,max}$ **do**
while X_p is beyond the search range **do**
rank all the available pigeon individuals according to their fitness values
$N_P = N_P/2$
keep half of the individuals with better fitness value, and abandon the other half
X_c:average value of the paths of the remaining pigeon individuals
calculate X_i according to Equation 5
end while
Evaluate X_i, and update X_p and X_g
end for
4 **Output:**
X_g is output as the global optima of the fitness function f

where ε_{eq} is the permittivity of the image to be matched. Three kinds of EPF models were proposed by Dao, Natale, and Massa (2007) and the Windowed EPF (WEPF) is utilized in this study to reduce the computational complexity and improve the robustness of the method. WEPF simply lies in defining a window beyond which the edge points are ignored. Thus, the definition of the WEPF is written as

$$\text{EPF}(x,y) = \frac{Q}{4\pi\varepsilon_{eq}} \sum_{(x_i,y_i)\in w} \frac{1}{\sqrt{(x-x_i)^2 + (y-y_i)^2}} \quad (8)$$

where w is the window chosen; ε_{eq} is the constant permittivity of the image windowed; and Q is equal to the charge of

each edge points. In this way, the edge potential can be easily calculated from the edge map extracted from the original image. Moreover, the Sobel edge extractor is applied in our study.

In order to determine whether the target image contains the object whose shape is similar to the sketch for a given position, rotation, and scale factor, the matching function named EP Fenergyis is defined as

$$f(c_k) = \frac{1}{N^{(c_k)}} \sum_{n^{(c_k)}=1}^{N^{(c_k)}} \left\{ \text{EPF}\left(x_n^{c_k}, y_n^{c_k}\right) \right\} \quad (9)$$

Equation 9 defines the average value of the EPF energy, which represents the degree of the sketch attracted by the target image. Additionally, the EPF is maximized when the given sketch translates, reorients, and scales itself to obtain the accurate match. However, EPF is a multidimensional and complicated function, which is hard to optimize. Therefore, it is significant to select the appropriate optimization algorithm in this study.

In the model of the basic PIO, the landmark operator of the original PIO is conducted in the last few generations. However, the algorithm could be already convergent, thus the landmark operator could not work. In our model, a probability $P_{conduct}$ is exploited, which is defined as

$$P_{conduct} = \log sig \left(\frac{Nc_{max}/2 - t}{k} \right) \qquad (10)$$

where $\log sig()$ denotes a logarithmic sigmoid transfer function, Nc_{max} is the maximum iteration number, t is the current iteration number, and k is changing for the slope of $\log sig()$ function, which is set as 20 in this study. In each iteration, a random number within $[0, 1]$ is generated to be compared with the $P_{conduct}$. If the random number is less than $P_{conduct}$, the map and compass operator is conducted. Otherwise, the landmark operator is carried out. Moreover, while conducting the landmark operator, the number of pigeons is reduced by 90% in the SAPIO. Since the value of the $P_{conduct}$ decreases from 1 to 0 nonlinearly, there is a higher probability to perform the map and compass operator at the initial moment. Then, the value of the $P_{conduct}$ gets smaller as the iteration goes, leading to a higher probability to conduct the landmark operator.

As the search behavior of pigeons only use the best position of all the pigeons, although performed with a fast convergence, it seems easy to get into local optima. In order to avoid the local optima with higher efficiency, the Gaussian factor and the simulated annealing (SA) mechanism are exploited to improve the performance of the basic PIO.

SA was first proposed in the 1980s to solve the combination optimization problems (Kirkpatrick, 1984). In our model, the Gaussian disturbance is added to the newly generated pigeons, enhancing its local optimization ability. In addition, as iteration proceeds, the Gaussian factor is supposed to apply a small disturbance with a higher probability. Thus, individuals added Gaussian disturbance can be expressed as

$$X_{ig}(t) = X_i(t) + \log sig \left(\frac{Nc_{max}/2 - t}{k} \right) N(\mu, \sigma) \qquad (11)$$

where $\log sig()$ denotes a logarithmic sigmoid transfer function, which decreases the amplitude of the Gaussian disturbance as the iteration proceeds.

In the model of the SA, the worse individual is reserved with the probability P_r. In this study, suppose that the difference between the fitness of the individual-added Gaussian disturbance X_{ig} and the old one X_i is Δf, and the EPF-based matching is the maxim optimization problem, then the probability P_r is defined as

$$P_r = \exp (\Delta f / T) \qquad (12)$$

where T is the annealing temperature, which decreases as the iteration goes. In case that the initial temperature is high enough and the annealing rate is low, the improved algorithm contributes to the jumping out of the local optima.

3.2 The implementation procedure of SAPIO-optimized EPF

Step 1: Image preprocessing. Filter the noise of the matched image and then extract the edge map using the Sobel operator.

Step 2: Calculate the edge potential distribution of the matched image according to Equation 8.

Step 3: Initialize parameters. Initialize parameters of SAPIO algorithm, such as the number of pigeons n, the solution dimension space D, the maxim number of iteration Nc_{max} and the initial annealing temperature, and so on.

Step 4: Evaluate the fitness of pigeons. The rotation, translation, and scaling of the given sketch are initialized in step 3. Subsequently, the transformed sketch is fitted within the potential field according to Equation 9 to compute the matching index, namely, the fitness value of the pigeon.

Step 5: Select the operator to be conducted. Compare with the given probability $P_{conduct}$; if a random value between 0 and 1 is smaller, then perform the map and compass operator. Otherwise, conduct the landmark operator.

Step 6: Update the pigeons. If the map and compass operator is selected, the velocity and position of each pigeon are updated by Equations 1 and 2, respectively. Otherwise, utilize Equations 3 and 4 to update the individual.

Step 7: Add Gaussian disturbance. Add the Gaussian disturbance to the newly generated individuals according to Equation 11.

Step 8: Compare the fitness of pigeons before and after Gaussian disturbance.

Table 2. Set of parameters for SAPIO algorithm.

Parameter	Description	Value
n	Number of pigeons	200
Nc_{max}	Maximum times of iteration	200
R	Map and compass operator	0.2
D	Dimension of the search problem	2
μ	Mean of the Gaussian function	0
σ	Variance of the Gaussian function	1
k	Slope of the log sig() functions	20
T	Initial temperature	1000
C	Temperature decay coefficient	0.5

(a) If the individual-added Gaussian disturbance is better, reserve the better individual.

(b) If the individual-added Gaussian disturbance is worse, then calculate the difference between them, and determine the probability P_r according to Equation 12. Subsequently, a random value between 0 and 1 is generated to be compared with P_r. If rand $< P_r$, reserve the worse individual.

Step 9: Conduct the temperature annealing operation.

Step 10: If n pigeons have been generated, go to step 11. Otherwise, go to step 5.

Step 11: Terminate whether the current number of iterations Nc reaches the Nc_{max}. Otherwise, go to step 5.

3.3 Experimental results

In order to investigate the feasibility and effectiveness of the hybrid model in this chapter, a series of experiments are conducted for four different cases. As presented in the Dao's model, they compared the GA-EPF model with the optimized Hausdorff distance matching (Huttenlocher, Klanderman, and Rucklidge, 1993) and Charmfer matching (Borgefors, 1988), and WEPF is the only method to achieve a correct matching in some cases. In addition, SAPIO algorithm proposed in this chapter was further compared with other algorithms, that is, the basic GA, PSO, ABC, and PIO. In this study, the control parameters of PIO and SAPIO are shown in Table 2, and the parameters for GA and ABC are the same as those in Xu's model (Xu and Duan, 2010). For SAPIO, the initial temperature is set as 1000 experimentally. All the pictures used in this chapter originate from the Google earth.

The task of this case is to find the wharf with the specified shape in the original image (Figure 4a). The comparative results of GA, PSO, ABC, PIO, and SAPIO for case 1 in 10 runs are presented in Table 3. Additionally, all the following

Table 3. The comparative results of GA, PSO, ABC, PIO, and SAPIO for case 1 in 10 runs.

Algorithm	Best fitness	Best parameter	Mean
GA	13.5656	(132,224,267,1.2)	10.9862
PSO	16.8679	(119,225,286,1.3)	13.9812
ABC	16.8679	(119,225,286,1.3)	15.4111
PIO	16.8679	(119,225,286,1.3)	14.5277
SAPIO	16.8679	(119,225,286,1.3)	15.8120

experiments were implemented on a PC with 8GB RAM using Windows 7 and were encoded in MATLAB 2013b for 64 bit.

From Table 3, it is natural to conclude that the SAPIO outperforms GA, PSO, ABC, and PIO obviously, which has the highest mean value in 10 runs among all the algorithms. It should be further noted that GA runs into local optima easily and is not suitable for this case. Moreover, the best matching parameters for SAPIO, PIO, PSO, and ABC are the same.

To further compare the SAPIO with other algorithms, the evolution curves of the function's mean value in independently 10 runs for GA, PSO, ABC, PIO, and SAPIO are shown in Figure 4c and the best curves of those are compared in Figure 4d, respectively. As clearly indicated in Figure 4c, PIO has a faster convergence speed compared with the SAPIO algorithm. However, due to the simulated annealing mechanism, it is obvious that SAPIO is more stable compared to the basic PIO from the curve. Moreover, the SAPIO algorithm also converges quickly compared to the ABC algorithm, which indicates it can search the better solution with higher efficiency. Furthermore, it should be noted that GA has the poor performance on case 1. Finally, the satisfactory detection and matching result can be obtained using our method as shown in Figure 4e.

The experimental results show that the SAPIO clearly improves the performance of the basic PIO owing to the simulated annealing mechanism. Furthermore, the experimental comparison also shows the effectiveness and stability of our method over other algorithms, which provides a more effective way for more complicated UAV target detection tasks.

4 PIO FOR UAV PATH PLANNING

4.1 Path planning using PIO

4.1.1 *Modeling of path planning*

The threat sources modeling is the most important issue in UAVs optimal path planning. There are two kinds of threat

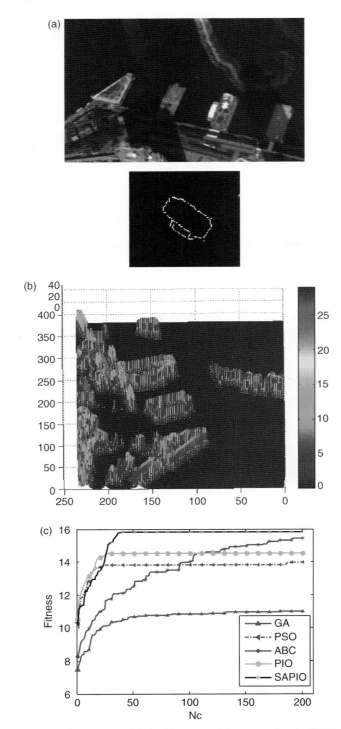

Figure 4. Experimental results. Experiments of case 1: (a) Original image and the target sketch. (b) The energy potential distribution of the original image. (c) Comparative evolution curves of the mean value of the objective function. (d) Comparative evolution curves of the best value of the objective function. (e) Comparative results of target detection and matching using GA, PSO, ABC, PIO, and SAPIO, respectively. (Reproduced with permission from C. Li and H. Duan. © Elsevier, 2014.)

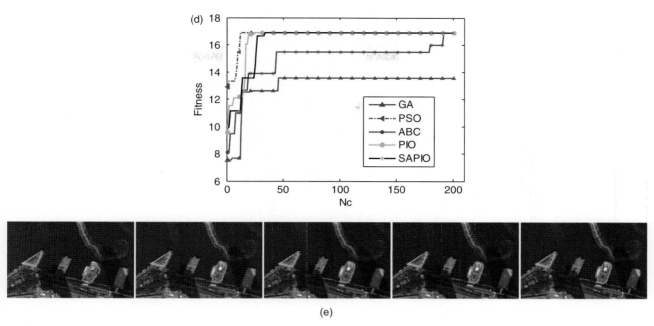

(e)

Figure 4. (*Continued*).

sources: artificial threats and natural threats. The artificial threats include the enemy's radar, missiles and artillery, and so on. There are appropriate models of them under different circumstances (Zhu *et al.*, 2008; Duan *et al.*, 2010; Zhu and Duan, 2014). The traditional optimization algorithms generally use circle models to describe these threats, and the radius of the circle is the range of threat source, and the treat level can also be defined to calculate the threat cost (Li and Duan, 2012; Zhou *et al.*, 2013). Mathematically, the problem of 3D path planning for UAVs can be described as follows.

Given the starting point A and target point B, $k$threats set$\{T_1, T_2, \ldots, T_k\}$, and the parameters of UAV's maneuvering performance constraints (such as the restrictions of turning angle α, climbing/diving angle β, and flying height h), our aim is to find a set of waypoints $\{W_0, W_1, \ldots, W_n, W_{n+1}\}$ with $W_0 = A$ and $W_{n+1} = B$ such that the resultant path is safe and flyable.

Suppose that the terrain of the environments and the information of threat regions are known, and the starting points and targets are also known. The cost function of UAV flight path can be defined as follows:

$$F = w_1 f(l) + w_2 f(h) + w_3 f(c) \tag{13}$$

where w_1, w_2, and w_3 are weight coefficients, which relate to length, height, and threat cost separately, and $w_1 + w_2 + w_3 = 1$.

For the given path, the length cost can be defined as

$$f(l) = \sum_{i=1}^{n} l_i^2 \tag{14}$$

where l_i is the length of the ith path segment.

The height cost $f(h)$ can be defined as

$$f(h) = \sum_{i=1}^{n} h_i \tag{15}$$

where h_i is the average altitude above the sea level of the ith route segment.

In order to simplify the calculations, more efficient approximation to the exact solution is adopted. In this chapter, threat cost of each edge connecting two discrete points was calculated at five points along it, as is shown in Figure 5.

Suppose the UAV fly in path $L_{i,j}$, we can divide the path $L_{i,j}$ into five sections in this case, and the treat cost $f_{\min}$ can be calculated by

$$f_{\min} = \begin{cases} 0, & R_{ij} > R_j \\ \dfrac{L_{ij}}{5} \sum_{k=1}^{N_t} t_k \left(\dfrac{1}{d_{0.1,k}^4} + \dfrac{1}{d_{0.3,k}^4} + \dfrac{1}{d_{0.5,k}^4} + \dfrac{1}{d_{0.7,k}^4} + \dfrac{1}{d_{0.9,k}^4} \right), & R_{ij} \leq R_j \end{cases} \tag{16}$$

where L_{ij} is the length of $L_{i,j}$, t_k is the kth treat level, R_j is the radius of the jth threat, N_t is the number of threat, R_{ij} denotes

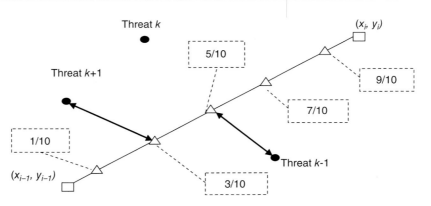

Figure 5. Computation of threat cost of UAV. (Reproduced with permission from H. Duan and P. Qiao. © Emerald Group Publishing Limited, 2014.)

the average distance between the ith path segment and the jth threat, and $d_{0.1,k}$ is the length of the 1/10 point and the kth threat center. By controlling the threat cost defined here, the survival probability of UAV can be increased accordingly.

4.1.2 Experimental results

The PIO procedure can be implemented in various ways by setting up PIO algorithm's parameters differently. In order to investigate the feasibility and effectiveness of our proposed PIO algorithm, a series of experiments are conducted, and further comparative experimental results with the standard DE algorithm are also given.

Set the coordinates of the starting point as $(0, 0, 30)$ and the target point as $(65, 100, 30)$, while the initial parameters of PIO algorithm were set as $NP = 300$, $D = 20$, $R = 0.2$, $\text{Nc}_{1\max} = 150$, and $\text{Nc}_{2\max} = 200$. We also set

$D = 20$. The comparative results with DE are shown in Figures 6 and 7.

From Figures 6 and 7, it is obvious that our proposed PIO algorithm can converge more quickly and are more stable compared to the standard DE algorithm, and the optimal path generated by using PIO is more smooth and satisfactory than with the standard DE algorithm. With the increasing of pigeon number, the convergence performance gets better. Generally, the experimental results also show that our PIO algorithm is much better in stability and superiority over the standard DE algorithm.

This chapter presents a novel swarm intelligence optimizer, named PIO. We also applied this new algorithm for solving the UAV path planning problem. Computational experiments are conducted to validate the performance of the proposed PIO algorithm. The comparative simulation results show that our proposed PIO algorithm is a feasible and effective algorithm for UAV path planning.

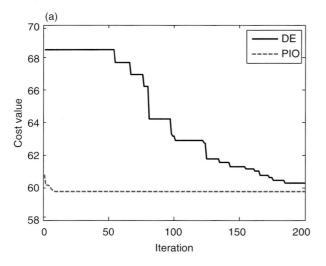
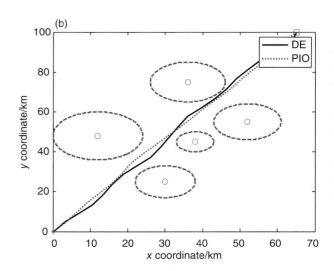

Figure 6. Comparative path planning results of PIO and DE ($NP = 300$) for case 1. (a) Evolutionary curves. (b) Path planning results. (Reproduced with permission from H. Duan and P. Qiao. © Emerald Group Publishing Limited, 2014.)

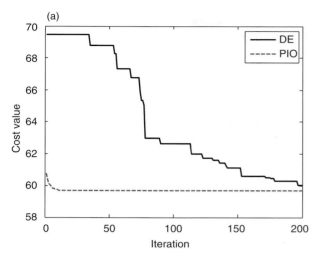

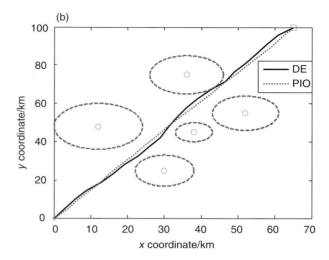

Figure 7. Comparative path planning results of PIO and DE (NP = 300) for case 2. (a) Evolutionary curves. (b) Path planning results. (Reproduced with permission from H. Duan and P. Qiao. © Emerald Group Publishing Limited, 2014.)

4.2 PP-PIO-based three-dimensional path planning

4.2.1 Predator–prey pigeon-inspired optimization

Three-dimensional path planner is an essential element of the UAV autonomous control module. It allows the UAV to compute the best path from a start point to an end point autonomously. Whereas commercial airlines fly constant prescribed trajectories, UAVs in operational areas have to travel constantly changing trajectories that depend on the particular terrain and conditions prevailing at the time of their flight.

Predatory behavior is one of the most common phenomena in nature, and many optimization algorithms are inspired by the predator–prey strategy from ecology. In nature, predators hunt prey to guarantee their own survival, while the preys need to be able to run away from predators. On the other hand, predators help to control the prey population while creating pressure in the prey population. In this model, an individual in predator population or prey population represents a solution, each prey in the population can expand or get killed by predators based on its fitness value, and a predator always tries to kill preys with least fitness in its neighborhood, which represents removing bad solutions in the population. In this chapter, the concept of predator–prey is used to increase the diversity of the population, and the predators are modeled based on the worst solutions that are demonstrated as follows:

$$P_{\text{predator}} = P_{\text{worst}} + \rho(1 - t/t_{\max}) \tag{17}$$

where P_{predator} is the predator (a possible solution), P_{worst} is the worst solution in the population, t is the current iteration, $t_{\max}$ is the maximum number of iterations, and ρ is the

hunting rate. To model the interactions between predator and prey, the solution to maintain a distance of the prey from the predator is shown as follows:

$$\begin{cases} P_{k+1} = P_k + \rho e^{-|d|}, & d > 0 \\ P_{k+1} = P_k + \rho e^{-|d|}, & d < 0 \end{cases} \tag{18}$$

where d is the distance between the solution and the predator and k is the current iteration.

On the basic algorithm of PIO, the landmark operation is used after several iterations of map and compass operation. For example, when the number of generations N_c is larger than the maximum number of generations of the map and compass operation $N_{c\max 1}$, the map and compass operator will stop and the landmark operation will start. During my experiment, we found that it was easy to fall into a local best solution before the number of generations got to $N_{c\max 1}$. Furthermore, half of the number of pigeons is decreased by N_p in every generation on the landmark operator. The population of pigeons is decreased too rapidly according to Equation 3, which would reach to zero after a small amount of iterations. The landmark operator would make only a small impact on the pigeons' position by this way. Therefore, we make a small modification on the basic PIO algorithm. The map and compass operation and the compass operation are used in parallel in each iteration. A parameter ω is used to define the impact of the landmark increase with a smoothly path. And a constant parameter c is used to define the number of pigeons that are in the landmark operator. Our new formula of landmark operator is as follows:

$$N_P(t) = c \cdot N_{P\max}, \quad c \in (0, 1) \tag{19}$$

$$X_c(t) = \frac{\sum X_i(t) \cdot \text{fitness}(X_i(t))}{N_P \sum \text{fitness}(X_i(t))} \qquad (20)$$

$$\omega = s + (1 - s) \cdot t/N_{c\,max}, \quad s \in (0, 1) \qquad (21)$$

$$X_i(t) = X_i(t-1) + \omega \cdot \text{rand} \cdot (X_c(t) - X_i(t-1)) \qquad (22)$$

where s is a constant experimentally defined.

4.2.2   Modeling of three-dimensional path planning

The first step of three-dimensional path planning is to discretize the world space into a representation that will be meaningful to the path planning algorithm. In our implementation, we use a formula to indicate the terrain environment. The mathematical function is of the following form (Zhang and Duan, 2014):

$$z(x, y) = \sin(x/5 + 1) + \sin(y/5) + \cos\left(a \cdot \sqrt{x^2 + y^2}\right) \\ + \sin\left(b \cdot \sqrt{x^2 + y^2}\right) \qquad (23)$$

where z indicates the altitude of a certain point, and a and b are constants experimentally defined. Cylindrical danger zones (or no-fly zones) are represented by a separate matrix, where each row represents the coordinates (x_i, y_i) and radius r_i of the ith cylinder, as shown in Equation 18. Complex no-fly zone can be built by partially juxtaposing multiple cylinders:

$$\text{danger zones} = \begin{pmatrix} x_1 & y_1 & r_1 \\ x_2 & y_2 & r_2 \\ \vdots & \vdots & \vdots \\ x_n & y_n & r_n \end{pmatrix} \qquad (24)$$

The three-dimensional trajectories generated by the algorithm are composed of line segments and (x_i, y_i, z_i) represents the coordinates of the ith way point. The trajectories are flown at constant speed.

In the situation of UAV path planning, the optimal path is complex and includes many different characteristics. To take into account these desired characteristics, a cost function is used and the path planning algorithm becomes a path that will minimize the cost function. We define our cost function as follows:

$$F_{cost} = C_{length} + C_{altitude} + C_{danger\ zones} + C_{power} \\ + C_{collision} + C_{fuel} \qquad (25)$$

In the cost function, the term associated with the length of a path is defined as follows:

$$C_{length} = 1 - \left(\frac{L_{p1p2}}{L_{traj}}\right) \qquad (26)$$

$$C_{length} \in [0, 1] \qquad (27)$$

where L_{p1p2} is the length of the straight line connecting the starting point P2 and the end point P2 and L_{traj} is the actual length of the trajectory.

The term associated with the altitude of the path is defined as follows:

$$C_{altitude} = \frac{A_{traj} - Z_{min}}{Z_{max} - Z_{min}} \qquad (28)$$

$$C_{altitude} \in [0, 1] \qquad (29)$$

where Z_{max} is the upper limit of the elevation in our search space, Z_{min} is the lower limit, and A_{traj} is the average altitude of the actual trajectory. Z_{max} and Z_{min} are respectively set to be slightly above the highest and lowest points of the terrain.

The term associated with the violation of the danger zones is defined as follows:

$$C_{danger\ zones} = \frac{L_{inside\ dz}}{\sum_{i=1}^{n} d_i} \qquad (30)$$

$$C_{danger\ zones} \in [0, 1] \qquad (31)$$

where n is the total number of danger zones, $L_{inside\ dz}$ is the total length of the subsections of the trajectory that go through danger zones, and d_i is the diameter of the danger zone i.

The term associated with a required power higher than the available power of the UAV is defined as follows:

$$C_{power} = \begin{cases} 0, & L_{not\ feasible} = 0 \\ P + \left(L_{not\ feasible}/L_{traj}\right), & L_{not\ feasible} > 0 \end{cases} \qquad (32)$$

$$C_{power} \in 0 \bigcup [P, P+1] \qquad (33)$$

where $L_{not\ feasible}$ is the sum of the lengths of the line segments forming the trajectory that require more power than the available power of the UAV, L_{traj} is the total length of the trajectory, and P is the penalty constant. This constant must be higher than the cost of the worst feasible trajectory, which would have, based on our cost function, a cost of 3. By adding this penalty P, we separate no feasible solutions from the feasible ones.

The term associated with ground collisions is defined as follows:

$$C_{collision} = \begin{cases} 0, & L_{under\ terrain} = 0 \\ P + (L_{under\ terrain}/L_{traj}), & L_{under\ terrain} > 0 \end{cases}$$
(34)

$$C_{collision} \in 0 \cup [P, P+1]$$
(35)

where $L_{under\ terrain}$ is the total length of the subsections of the trajectory that travels below the ground level and L_{traj} is the total length of the trajectory.

The term associated with an insufficient quantity of fuel available is defined as follows:

$$C_{fuel} = \begin{cases} 0, & F_{traj} \leq F_{init} \\ P + 1 - (F_{P1P2}/F_{traj}), & F_{traj} > F_{init} \end{cases}$$
(36)

$$C_{fuel} \in 0 \cup [P, P+1]$$
(37)

where F_{P1P2} is the quantity of fuel required to fly the imaginary straight segment connection from the starting point $P1$ to the end point $P2$, F_{traj} is the actual amount of fuel needed to fly the trajectory, and F_{init} is the initial quantity of fuel on board the UAV.

The search engine will be adopted to find a solution, which can minimize the cost function during the optimization phase of our path planner algorithm. This can also be explained as to find a trajectory that best satisfies all the qualities represented by this cost function. Our cost function demonstrates a specific scenario where the optimal path minimizes the distance traveled, the average altitude (to increase the stealthiest of the UAV), and avoids danger zones, while respecting the UAV performance characteristics. This cost function is highly complex and demonstrates the power of our path planning algorithm. However, this cost function could easily be modified and applied to a different scenario.

4.2.3   Experimental results

In order to evaluate the performance of our proposed PPPIO algorithm in this chapter, series of experiments are conducted in Matlab2012a programming environment. Coordinates of a starting point are set as (10, 16, 0) and the target point as (55, 100, 0). The initial parameters of PIO algorithm were set as NP = 150. The comparative results of PPPIO with PIO and PSO for 3D path planning are shown in Figures 8–10.

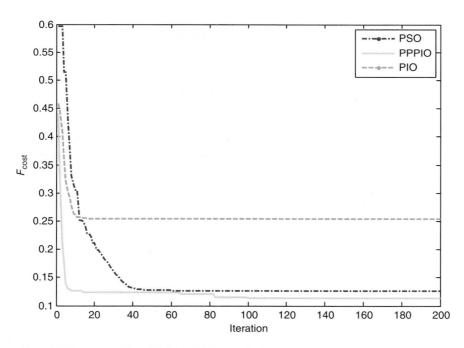

Figure 8. Comparative evolutionary curves of PPPIO, PIO, and PSO. (Reproduced with kind permission from Zhang and H Duan (2014). © Springer Science + Business Media.)

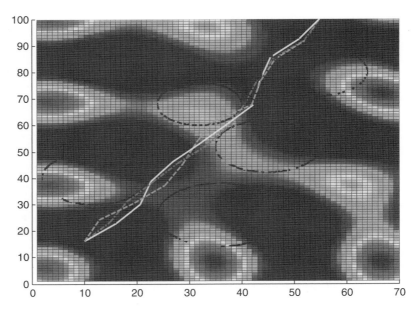

Figure 9. Comparative path planning results of PPPIO, PIO, and PSO. (Reproduced with kind permission from Zhang and H Duan (2014). © Springer Science + Business Media.)

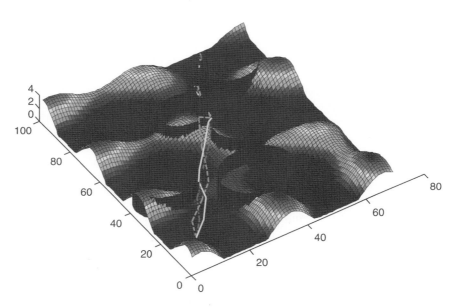

Figure 10. Comparative path planning results of PPPIO, PIO, and PSO on 3D version. (Reproduced with kind permission from Zhang and H Duan (2014). © Springer Science + Business Media.)

This chapter proposed a novel PPPIO algorithm for solving the UAV three-dimensional path planning problem in complex environments. We introduce the concept of predator–prey to the PIO algorithm and optimize the PIO algorithm. Series of comparative simulation results were given to show that our proposed PPPIO algorithm is more efficient than basic PIO and PSO in solving UAV three-dimensional path planning problems.

5 MISSION ASSIGNMENT BASED ON PIO

5.1 Mission assignment problem formulation

Using multiple UAVs cooperatively to accomplish a complicated task is a very challenging problem. In this, the mission assignment for multiple UAVs is a key issue that attracts great attention and it has different objective functions according to concrete applications (Luo *et al.*, 2006; Pohl and Lamont, 2008). The specific scenario is described as follows. $S = [0, L_1] \times [0, L_2]$ can be seen as the scale of the place where we assume all the UAVs performing their tasks. There exist M targets, denoted as $T = \{T_1, T_2, T_3, \ldots, T_M\}$ and N UAVs, denoted as $I = I\{_1, I_2, I_3, \ldots, I_N\}$. To make our solution have more practical and strategic significance, we assume that in this situation the number of our UAVs is less than the number of targets. In order to avoid the request confliction between adjacent UAVs, each UAV is located in a certain position in advance and forms a formation with others. Here the necessary assumptions for our model (Hao *et al.*, 2014) are presented:

1. All the UAVs are homogeneous, which means all UAVs have the ability to perform the task on all targets and the difference $N < M$ of internal operational ability of UAVs can be ignored.
2. A UAV can be allocated to execute another task after it accomplished the current one.
3. In this case, all the targets have the same priority. The sequence for task execution will be ignored, and we will consider this situation in future works.

This section presents a new formulation using weighted aggregation approach as the criterion for the rationality of assignment. Focusing on the least energy consumption, an evaluation function $f(d, a, v)$ is proposed according to the properties of UAVs and targets. We mainly consider the principal elements in order to simplify the space dimensions. The establishment of evaluation function is given as follows.

The ammunition or other equipment of each UAV can be effective only in a certain range. We use this range as the evaluation criterion for the weapon performance of each UAV. This can be related into the distance evaluation presented as follows:

$$\text{fd} = \begin{cases} 1 - \dfrac{D}{\text{range}}, & D \leq \text{range} \\ 0, & D > \text{range} \end{cases} \quad (38)$$

where D is the Euclidean distance between I_i and T_i and range denotes the properties of weapons of I_i. Distance function f_d

describes the performance of UAV with the changing distance from targets. Its value declines as the distance increases, which means the decrease of the prosecuting possibility for T_i. Once the distance is beyond the capture range of the UAV, there is no advantage so far for I_i to destroy T_i, and we set $f_d = 0$.

When the moving distance of UAV is $\Delta D \to 0$, the energy consumption can be seen as a constant C. We define f_c as the energy consumption function for UAV I_i, which denotes the least energy costs for I_i prosecuting T_i. According to the memory property of fractional derivative, it has

$$\frac{d^q C}{dx^q} = \frac{1}{\tau(1-q)} \frac{d}{dx} \int_0^x (x-t)^{-q} C dt = C \frac{x^{-q}}{\tau(1-q)} \quad (39)$$

where q is the fractional order, $q \in [0, 1)$, and C is the energy consumption constant, which can be obtained by the following synthetic effectiveness of UAV, when $q = 1/2$:

$$\frac{d^q C}{dx^q} = \frac{C}{\sqrt{\pi x}}. \quad (40)$$

In this case, x denotes the distance D mentioned above. And the energy consumption function f_c can be described as

$$f_c = \frac{C}{\sqrt{\pi D}} \quad (41)$$

This function denotes the accumulation of energy consumption. It shows that the value of f_c decreases as D increases.

The direction shifts of I_i and T_i will decide if the next move is appropriate or not, which can be evaluated here by angle evaluation function:

$$f_a = e^{-\frac{|\theta|}{\pi}} \quad (42)$$

where θ is the relative angle between I_i and T_i, and $\theta \in (-(\pi/2), (\pi/2)]$.

When target T_i comes close to our UAV I_i, θ will be negative. On the other hand, when T_i gets far away from I_i, we set $\theta > 0$, as is shown in Figure 11.

The angle evaluation function f_a evaluates the effect of relative angle on the assignment. As the angle increases, the

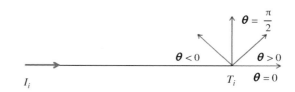

Figure 11. Definition of θ.

probability of prosecuting T_i for I_i will decline and the situation is the same for f_a.

Finally, we build a velocity function to describe the UAV velocity performance to assignment.

The velocity function can be expressed as follows:

$$f_v = \begin{cases} e^{-|V_I - V_T|}, & |V_I - V_T| \le 1\ \mathrm{km\,s^{-1}} \\ 0, & |V_I - V_T| > 1\ \mathrm{km\,s^{-1}} \end{cases} \quad (43)$$

where V_I denotes the velocity of our UAV I_i and V_T denotes the velocity of the target T_i.

When the velocity differences between two sides get closer, the chance of I_i to accomplish the mission becomes higher, and the value of velocity function f_v will augment.

According to these four functions, we build a comprehensive function:

$$f(d, a, v) = k_d f_d + k_a f_a + k_v f_v + k_c f_c \quad (44)$$

where $f(d, a, v)$ is the evaluation function; k_d, k_c, k_a, and k_v are the weights of distance function, energy consumption function, angle function, and velocity function, respectively. Here

$$k_d + k_a + k_v + k_c = 1 \quad (45)$$

The values taken for these parameters depend on the real situation and experiments.

Table 4. Parameters for mission assignment.

Values of range and C	UAV number			
	1	2	3	4
Range/km	50	50	40	60
C	10	10	10	10

5.2   Experimental results

To verify the validity and feasibility of the modified PIO-based multiple UAVs mission assignment approach, simulation experiments are carried out and the simulation results are compared with those obtained by using modified PIO, PIO, DE, and PSO-based algorithm.

The properties of UAVs, including range of weapons and energy consumption constant, are given in Table 4.

The weight parameters are set with $k_d = 0.3$, $k_c = 0.4$, $k_a = 0.2$, $k_v = 0.1$, and PIO parameters $N_p = 200$, $D = 4$, $R = 0.3$, Nc1 = 250, and Nc2 = 300. Some comparative patterns with PSO parameters $w = 0.5$, $c1 = c2 = 1.2$, and DE parameters CR = 0.6, F = 0.6, after 300 iterations at different target T_i are shown in Figures 12–14.

In case 1, parameters for the modified PIO algorithm are $N_p = 200$, $D = 4$, $P_R = 0.2$, and $P_C = 0.2$, at $T_1 = [0, 40]$. The maximum values of the evaluation function of modified PIO,

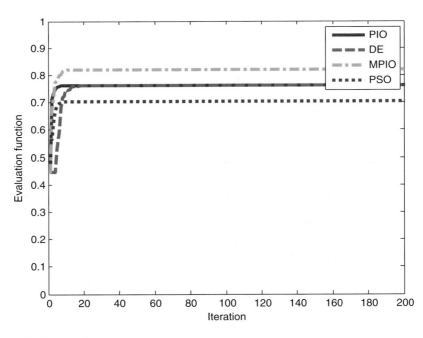

Figure 12. Experimental results for case 1.

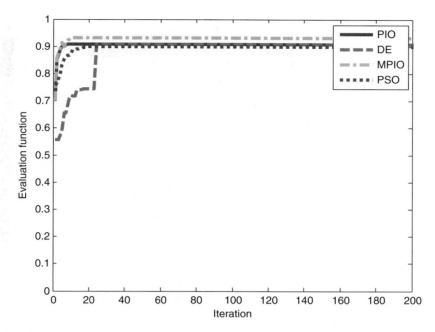

Figure 13. Experimental results for case 2.

PIO, DE, and PSO are 0.821651, 0.766442, 0.766440, and 0.705326, respectively, as shown in Figure 12.

In case 2, parameters are $N_p = 200$, $D = 4$, $P_R = 0.2$, and $P_C = 0.2$, at $T_2 = [10,30]$. The maximum values of the evaluation function of the modified PIO, PIO, DE, and PSO are 0.928657, 0.902653, 0.901448, and 0.898263, respectively, as shown in Figure 13.

In case 3, parameters are $N_p = 200$, $D = 4$, $P_R = 0.2$, and $P_C = 0.2$, at $T_7 = [40,40]$ and the maximum values of the evaluation function of the modified PIO, PIO, DE, and PSO are 0.806950, 0.762412, 0.762412, and 0.745253, respectively, as shown in Figure 14.

By comparing these different curves, it is easily observed that the modified PIO curves can get better evaluation

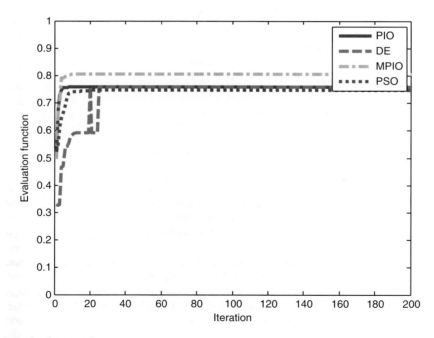

Figure 14. Experimental results for case 3.

Table 5. Assignment results using PIO.

The PIO assignment result	Targets						
	T_1	T_2	T_3	T_4	T_5	T_6	T_7
UAV number	1	2	2	3	3	4	4

Table 6. Assignment results using modified PIO.

The PIO assignment result	Targets						
	T_1	T_2	T_3	T_4	T_5	T_6	T_7
UAV number	1	2	1	3	3	4	4

Table 7. Assignment results using PSO.

The PSO assignment result	Targets						
	T_1	T_2	T_3	T_4	T_5	T_6	T_7
UAV number	1	1	2	2	3	3	4

Table 8. Assignment results using DE.

The DE assignment result	Targets						
	T_1	T_2	T_3	T_4	T_5	T_6	T_7
UAV number	1	1	2	2	3	4	4

functions than PIO, PSO, and DE curves. The convergence speed of modified PIO curve is faster than the other algorithms. Besides, classic PIO, PSO, and DE curves are easy to fall into local optimum. The simulation results show that the modified PIO procedure is more effective in handling the multiple UAVs mission assignment problem than using other algorithms.

After 100 runs, we can find the best solution for every target. The assignment solutions of four algorithms are presented in Tables 5–8.

In this chapter, one type of mathematical model of multiple UAVs mission assignment is established. An evaluation function is also constructed by taking energy consumption into account. A modified PIO algorithm is applied in optimizing the evaluation function to obtain the best assignment solution. The comparative simulation results show that the modified PIO approach is more effective in solving multiple UAVs mission assignment problem than using the classic PIO, DE, and PSO algorithms.

6 SUMMARY

As with manned aircraft flights, mission planning is a critical element in successful mission performance. In the simplest case, a vehicle is supposed to fly over a disaster area such as wood fire or a large traffic accident and deliver high-quality sensor data such as images or videos. Using multiple UAVs cooperatively to accomplish a complicated task is also a very challenging problem. This chapter deals with target detection, path planning, and mission assignment for UAVs using a newly proposed optimization method named PIO. In Section 2, biological behaviors of pigeons are introduced and the principle of PIO is described in detail. The hybrid model of EPF and SAPIO algorithm is proposed to accomplish the target detection task for UAVs at low altitude in Section 3. We deal with 2D and 3D path planning problems to search out an optimal or near-optimal flight path between an initial location and the desired destination in Section 4. In Section 5, PIO is applied to mission assignment problem for multiple UAVs. By incorporating the PIO algorithms into UAVs, it is possible to enhance the ability to understand and adapt to the environments, and also promote the performance of UAVs in cooperating with other autonomous systems.

REFERENCES

Borgefors, G. (1988) Hierarchical chamfer matching: a parametric edge matching algorithm. *IEEE Trans. Pattern Anal. Mach. Intell.*, **10**(6), 849–865.

Dao, M.S., Natale, F.G. De, and Massa, A. (2007) Edge potential functions (EPF) and genetic algorithms (GA) for edge-based matching of visual objects. *IEEE Trans. Multimed.*, **9**(1), 120–135.

Deng, Y. and Duan, H. (2014) Biological edge detection for UCAV via improved artificial bee colony and visual attention. *Aircr. Eng. Aerosp. Technol.*, **86**(2), 138–146.

Duan, H. and Li, P. (2014) *Bio-Inspired Computation in Unmanned Aerial Vehicles*, Springer, Berlin.

Duan, H. and Qiao, P. (2014) Pigeon-inspired optimization: a new swarm intelligence optimizer for air robot path planning. *Int. J. Intell. Comput. Cybern.*, **7**(1), 24–37.

Duan, H., Yu, Y., Zhang, X., and Shao, S. (2010) Three-dimension path planning for UCAV using hybrid meta-heuristic ACO-DE algorithm. *Simul. Model. Pract. Theory*, **18**(8), 1104–1115.

Ergezer, H. and Leblebicioglu, K. (2013) Path planning for UAVs for maximum information collection. *IEEE Trans. Aerosp. Electron. Syst.*, **49**(1), 502–520.

Guilford, T., Roberts, S., and Biro, D. (2004) Positional entropy during pigeon homing II: navigational interpretation of Bayesian latent state models. *J. Theor. Biol.*, **227**(1), 25–38.

Hao, R., Luo, D., and Duan, H. (2014) Multiple UAVs mission assignment based on modified pigeon-inspired optimization algorithm. *Proceeding of IEEE Chinese Guidance, Navigation and Control Conference*, Yantai, pp. 2692–2697.

Huttenlocher, D.P., Klanderman, G.A., and Rucklidge, W.J. (1993) Comparing images using the Hausdorff distance. *IEEE Trans. Pattern Anal. Mach. Intell.*, **15**(9), 850–863.

Kennedy, J. and Eberhart, R. (1995) Particle swarm optimization. *Proceedings of the 1995 IEEE International Conference on Neural Networks*, Perth, Australia, vol. 4, pp. 1942–1948.

Kirkpatrick, S. (1984) Optimization by simulated annealing: quantitative studies. *J. Stat. Phys.*, **34**(5), 975–986.

Li, P. and Duan., H. (2012) Path planning of unmanned aerial vehicle based on improved gravitational search algorithm. *Sci. China Technol. Sci.*, **55**(10), 2712–2719.

Li, C. and Duan, H. (2014) Target detection approach for UAVs via improved Pigeon-Inspired Optimization and edge potential function. *Aerosp. Sci. Technol.*, **39**(8), 352–360.

Luo, D., Duan, H., Wu, S., and Li, M. (2006) Research on air combat decision-making for cooperative multiple target attack using heuristic ant colony algorithm. *Acta Aeronaut. Astronaut. Sin.*, **27**(6), 1166–1170.

Pohl, A. and Lamont, G. (2008) Multi-objective UAV mission planning using evolutionary computation. *Proceedings of the 40th Conference on Winter Simulation*, pp. 1268–1279.

Williamson, W.R., Glenn, G.J., Dang, V.T., Stecko, S.M., and Takacs, J.M. (2009) Sensor fusion applied to autonomous aerial refueling. *J. Guid. Control Dyn.*, **32**(1), 262–275.

Xu, C. and Duan, H. (2010) Artificial bee colony (ABC) optimized edge potential function (EPF) approach to target recognition for low-altitude aircraft. *Pattern Recognit. Lett.*, **31**(13), 1759–1772.

Zhang, B. and Duan, H. (2014) Predator–prey Pigeon-Inspired Optimization for UAV three-dimensional path planning, in *Advances in Swarm Intelligence*, Springer International Publishing, pp. 96–105.

Zhou, Z., Duan, H., Li, P., and Di, B. (2013) Chaotic differential evolution approach for 3D trajectory planning of unmanned aerial vehicle. *Proceedings of the 10th IEEE International Conference on Control & Automation*, Hangzhou, pp. 368–372.

Zhu, W. and Duan, H. (2014) Chaotic predator–prey biogeography-based optimization approach for UCAV path planning. *Aerosp. Sci. Technol.*, **32**(1), 153–161.

Zhu, H., Zheng, C., Hu, X., and Li, X. (2008) Path planner for unmanned aerial vehicles based on modified PSO algorithm. *Proceeding of the IEEE International Conference on Information and Automation*, pp. 541–544.

Chapter 16
Autonomy Architectures

Antidio Viguria

Center for Advanced Aerospace Technologies (CATEC), Seville, Spain

1 INTRODUCTION TO AUTONOMY ARCHITECTURES FOR UAS

The use of UAS (unmanned aircraft system) has grown rapidly in the last decade, with the largest potential for the civil sector (European Commission, 2013). It is foreseen that in the next years the use of UAS in civil applications, such as inspection of infrastructures, environmental monitoring, and surveillance of critical infrastructures, will rise exponentially. For these types of applications, especially in complex and dangerous environments, the use of UAS that incorporate a high level of autonomy will make the system more effective, having a lower downside risk and higher confidence in success.

The use of high autonomy levels in UAS gains importance in operations when the ground pilot does not have a good situational awareness level or the UAS is flying beyond the visual line of sight (BVLOS scenario). The former arises, for example, when the UAS navigates in a cluttered environment

Unmanned Aircraft Systems. Edited by Ella Atkins, Aníbal Ollero,
Antonios Tsourdos, Richard Blockley and Wei Shyy.
© 2016 John Wiley & Sons, Ltd. ISBN: 978-1-118-86645-0.

or close to obstacles, while in the latter autonomy level could help to decrease the number of people that forms the crew (automatic detection and tracking, environment perception, etc.) and increase the safety level of the system in critical situations (avoiding collisions, landing in emergency situations, etc.).

Clough (2002) and Merz (2004) have commented about the differences among automatic, autonomous, and intelligent systems. An automatic system will do exactly as programmed because it has no capability of reasoning, decision-making, or planning. An autonomous system has the capability to make decisions and to plan its tasks and path in order to achieve its assigned mission, while an intelligent system has the capabilities of an autonomous system plus the ability to generate its goals by its own motivations and without any instruction or influence from outside. This chapter will be focused on automatic and autonomous systems architectures since due to the current regulation framework, where it is required to have a pilot-in-command at all time during a UAS operation, it will be difficult to obtain from the national civil aviation authorities (such as FAA in the United States or EASA in Europe) the required flying authorizations for complete intelligent systems.

1.1 Autonomy levels for UAS

There are many works that elaborate a comprehensive framework to distinguish levels of autonomy. One of the first ones is Sheridan's work (Sheridan, 1992) where he proposed a 10-level scale of degrees of autonomy based on who makes the decision (autonomous system or human) and on how those decisions are executed.

On the other hand, in Kendoul (2012) is presented a short overview of works that elaborates metrics and autonomy levels specifically for UAS. In fact, in this work, is presented the ALFURS framework that characterizes the autonomy levels from a research-oriented perspective that better suits UAS operating at low altitudes and in cluttered environments (from remote control to fully autonomous). The ALFURS framework is based on the NIST ALFUS (Huang *et al.*, 2005) and the AFRL ACL (Clough, 2002) studies. As the author states in this chapter: *Although the ALFURS framework was originally proposed for Rotorcraft UAS, it can be also used for UAS in general.*

This description of autonomy levels is important in order to easily and correctly determine the autonomy level of an existing system from an objective perspective. Also, it is helpful in order to identify the functionalities required to achieve a certain autonomy level during the design of a new system.

1.2   Overview of architectures for autonomous systems

In a general way, the architecture of an autonomous system is a methodology to structure algorithms, that is, the combination of languages, tools, and a philosophy to create and group functionalities. One of the key aspects to decide when designing the architecture for an autonomous system is its reactive and deliberative role and how to combine them. In a summarized manner, in reactive architectures the actions are obtained directly from the information acquired by the sensors from which a noncomplete view of the environment is conformed. However, the deliberative architectures are based on obtaining a model of the environment, and using it to plan a number of actions that the system should execute in order to fulfill the defined objectives. There are also architectures that combine both approaches, the so-called hybrid architectures.

Within the large and extensive different architectures, the most common ones are three-layer architectures, behavior-based architectures, and agent-oriented architectures.

1.2.1   *Three-layer architectures*

The three-layer architecture is a hybrid architecture widely used in the last 20 years (see, e.g., Gat (1998) and Alami *et al.* (1998)). In this type of architecture, there are functionalities in the highest levels of the architecture that require more time for their execution and are more deliberative (thinking about the future); and functionalities that are more reactive in the lowest part of the architecture.

This architecture is usually composed of three levels: planning and scheduling, executive, and functional. The

planning and scheduling (deliberative) layer receives one or more mission objectives, finds tasks to fulfill the defined objectives, and organizes them for their execution along an extended period of time. In order to understand the environment and make decisions, it is necessary that the planning system creates a model of the reality that gets updated at the same time that the environment evolves.

The intermediate level is an executive layer that synchronizes the execution of the different tasks planned by the planning and scheduling level and handles possible failures during the execution of the tasks.

The lowest level is the functional layer that is the interface between the software functionality and the hardware of the autonomous system. The complexity of this level could be very diverse from systems that execute simple lineal sequences of commands to complex systems that can organize themselves to failures and nonplanned changes that can put under risk the success of the mission.

1.2.2   *Behavior-based architectures*

These are reactive architectures, without a deliberative system, and based on behaviors instead of models of the environment. This type of architecture offers a way to combine real-time control with behaviors generated from the information coming from sensors (reactive control). In this type of architecture, it is defined behaviors that connect sensors and actuators. The behaviors are organized in layers and the higher level behaviors can inhibit other behaviors that are located at lower levels (Rosenblatt *et al.*, 2002; Flanagan, Toal, and Leyden, 2003). Another approach is the use of arbiters that weight the different votes sent by each behavior in order fuse the different received commands (Rosenblatt, 1997). This fusion allows considering multiple objectives and restrictions at the same time.

One of the advantages of this architecture is its simplicity and low computational requirements.

1.2.3   *Agent-oriented architectures*

In this type of architecture, the execution and deliberative parts are combined for specific tasks. Each of these tasks are taken care by a specific agent that implements both the executive and deliberative part of the task. Some examples of these architectures are T-REX (McGann *et al.*, 2007) and IDEA (Muscettola *et al.*, 2002).

The agents are executed in a concurrent way and they communicate among them. Therefore there is not an inherent architecture, although in most of the implementations there is some type of division in layers.

Each agent can exhibit a deliberative and a reactive behavior. The implementation of the deliberative behavior is performed in the agent itself making use of a planning system making use of models of the environment and reasoning about its own resources and objectives.

Once the most important concepts have been introduced, the rest of the chapter is organized as follows. In the next section, the particularities of autonomy architectures for UAS are commented and its decomposition into two major levels is presented. Therefore, these two levels will be commented in detailed explaining the different functionalities that could be implemented in each of them. Next, an example of an autonomy architecture implemented on a highly autonomous UAS and developed within the ARCAS project is presented. Finally, the main conclusions of this chapter are exposed.

2   AUTONOMY ARCHITECTURE FOR UAS

The implementation of autonomy architectures on UAS has its own particularities. First, the required level of safety and the number of critical systems are much higher than in other autonomous systems, for example ground robots. Another important factor to take into account is that UAS work in a complete 3D environment, which is important when implementing high computational functionalities, such as environment perception and trajectory planning.

In the last years, there have been several implementations of high-level autonomy architectures in UAS that have been successfully validated with flight experiments, especially with rotorcraft UAS, such as TAL (Doherty, Kvarnström,

and Heintz, 2009), APEX (Whalley *et al.*, 2005), ReSSAC (Fabiani *et al.*, 2007), ARTIS (Adolf and Andert, 2010), and AWARE (Maza *et al.*, 2011). Most of these implementations follow a layered architecture that, up to now, has been proved the most successful one for the UAS field.

Taking into account the particularities of UAS already mentioned (safety level, critical systems, and high computational capacity for certain functionalities), in a general way an autonomous architecture in UAS could be decomposed in two levels (see Figure 1):

1. *Low-level architecture:* This part of the architecture implements the required basic guidance, navigation, and control algorithms that allow an UAS to remain stable on the air and follow the commands transmitted by the modules implemented in the higher level. It is desirable that this level also implements behaviors in case of the most common failures, such as, for example, loss of GPS and low battery level. This level follows a standard GNC (guidance, navigation, and control) architecture, usually requires hard-real time constraints, and it is implemented in a specific hardware such as a microcontroller or a FPGA.

2. *High-level architecture:* This part of the architecture implements one of the following types of architectures: three-layer, behavior-based, and agent-oriented. As commented before, the most successful one so far in the UAS field has been the three-layer architecture. Therefore, this one is chosen as an example for the high-level architecture.

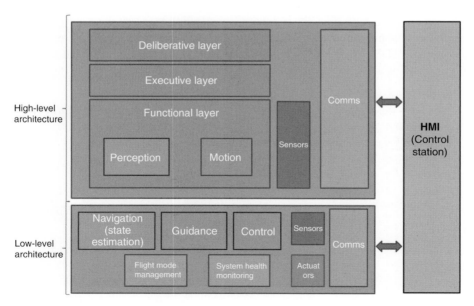

Figure 1. UAS general autonomy architecture.

In this type of architecture, the deliberative layer is in charge of decomposing the mission into tasks and organizes them for their execution along an extended period of time, while the executive level synchronizes the execution of the different tasks and handles possible failures during the execution of the tasks. Finally, the functional level implements the different functionalities that the UAS is able to execute (usually divided into perception and motion types of modules). This high-level architecture is usually implemented in an onboard computer with important computational capacity and an operating system that ensures at least soft real-time characteristics.

One of the key issues when designing an UAS architecture following this paradigm is controlling the time delays of the data flows, especially for the functional layer modules that require a high frequency of update, and also, require information from the low-level architecture. Therefore, it is important to define the frequency of update and the input data of each module before deciding to implement it in the functional layer of the high-level architecture or directly in the low-level architecture.

2.1   Low-level architecture

As commented before, this part of the architecture takes care of the implementation of the required guidance, navigation, and control algorithms to maintain stability of the UAS on the air (see Figure 2). Then, this level of the architecture should implement at least the set of sensors required to fly in a controlled way, which is usually formed by GPS (Global Positioning System), IMU (inertial measurement unit), magnetometer, and air data system (ADS). All this sensor information it is usually combined using a filtering technique such

as extended Kalman filter. Also, it is a good practice to use a specific for landing manoeuvre and increment the position accuracy in this operation (e.g., using an altimeter sensor).

Apart from this, this level also usually implements the following functionalities:

1. *Flight mode management:* This module handles the different flight modes of the aircraft (stabilized, fully autonomous, etc.) and the different phases of the flight (taking-off, cruise, landing, etc.).
2. *System health monitoring:* This module monitors the state of the different sensors, actuators, and the rest of the software modules and handles the alarms and possible failures at this level of the architecture. This module could just trigger a signal to alert the ground pilot or even implement sophisticated fault identification and failure techniques (Heredia and Ollero, 2010).
3. *Communications manager:* Handles the communication with the higher level and the ground control station.

For the implementation of this level of architecture, it has been successfully demonstrated the important benefits when using a model-based methodology (Santamaría *et al.*, 2012). This methodology has grown in popularity in the field of software design in the last 15 years (Bhatt *et al.*, 2005; Kirby and Kang, 2008; Ahmadian *et al.*, 2005; Orehek, 2001). This working methodology centralizes the classical software development process by the use of mathematical models of the system and the environment as the backbone of the whole product creation. This architecture will allow an easy rapid modeling and testing of the algorithms, and avoid many implementation problems by the use of automatic code generation tools, giving a complete flexibility to the process.

Unlike regular software modeling standards (such as UML 2.0), in the embedded critical software community the term model-based design implies a different paradigm. In

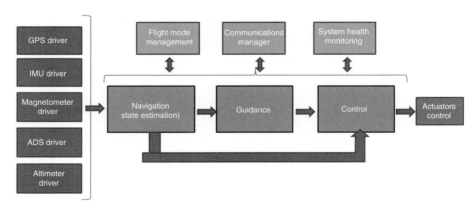

Figure 2. Low-level architecture for UAS.

this field, it refers to development of control models and simulation capabilities with tools such as Simulink or SCADE. Also, it is related to the ability to rapidly transfer the controller portion of the models to embedded target processors via automated code generation.

An important factor in any design and development of a critical process is the ability to verify performance at all stages of the design cycle. Traditional methods would require study of the requirements, develop the design specification documentation from which the design implementation and design testing plans are derived, and so on. However, this is a time-consuming process with a large degree of interpretation and a considerable risk for human error. This process can be improved by the use of a model-based design, development, and validation approach. Moreover potential errors, coming from manual coding, are eliminated by automated code generation. In recent years, the quality of autogenerated code has obtained a reasonable level, while the latest progress in microcontrollers makes automatically implemented systems very effective.

2.1.1 Model-based approach framework

Model-based design has proved to be a very reliable and convenient methodology for systems design and simulation in the last decade. Furthermore, the availability of code generation tools has improved the productivity of engineering teams, by the means of automating tedious and error-prone tasks, as it is hand coding of algorithms and model to code traceability. The use of this kind of tool also allows faster design iterations, easy reuse of the testing infrastructure, and rapid prototyping of the systems to promote a continuous verification and validation process.

The integration of code generation into the modeling and simulation tools results in a strong relationship between the models and the software, resulting in an unambiguous specification of the source code. This does not only allow establishing a common language for the different members of the development team, but also it eases the adoption of a component-based approach for the implementation and reuse of the models.

Using automatic code generation tools also has an impact on the roles of control and software engineers, freeing them from coding algorithms by hand, moving their focus to the software integration with the rest of system and designing the software architecture.

The goals of the model-based approach framework are as follows:

- Ease the reuse of the models in different hardware platforms, without requiring any changes in the models.
- Allow the reuse of the test harness models used during the design to validate the target system.

- Make the integration of models into the target hardware completely transparent to the control engineers.
- Establish a systematic and fast procedure to integrate automatic generated code into the target hardware, so the control engineers can test the model quickly and without the intervention of software engineers.

To achieve these goals, a component-oriented framework is used. This framework tightly fits the model-based development process, allowing a direct implementation of the models, following the componentization and modularity principles used in model-based software such as Simulink or SCADE (more information about this framework can be found in Santamaría *et al.*, 2012).

2.2 High-level architecture

Following the three-layer architecture, this high-level architecture could be divided in the three mentioned layers: deliberative (planning and scheduling), executive, and functional. As commented before, the deliberative layer is in charge of decomposing the mission into tasks and organizes them for their execution along an extended period of time, while the executive level synchronizes the execution of the different tasks and handles possible failures during the execution of the tasks. Finally, the functional level implements the different functionalities that the UAS is able to execute.

Usually, the deliberative layer makes use of planning frameworks. Regarding UAS, it is important that the planning framework takes into account concurrent operations and temporal dependencies. One of the most popular planning frameworks used in robotics is EUROPA 2 developed by NASA (Bedrax-Weiss *et al.*, 2004). The deliberative layer can involve, depending on the complexity of the UAS and the environment, symbolic planning, task planning, and even advanced motion planning taking into account the kinematic and even the dynamic constraints of the aerial system.

On the other hand and regarding the functional layer, it is common to makes use of software frameworks to facilitate the implementation of the communication mechanisms between the deliberative layer and the functional layer. One of the most popular software frameworks is ROS (Robot Operating System), that originally was designed for ground robotics (mainly indoor), but that due to the exponentially growth of the aerial robotics domain in the last 5 years it is implementing more functionalities especially designed for aerial robots. Another interesting framework that could help developing the executive layer is GenoM developed by LAAS.

Regarding the functional layer, and apart from the task specifics to the mission and related to the payload of the UAS, the most common functionalities required in an autonomous UAS can be grouped in two domains: perception and motion. Perception functionalities are focused on getting information from the environment increasing the situational awareness of the UAS, while motion functionalities are used to get the UAS moved in an intelligent and efficient way, optimizing specific metrics, and avoiding obstacles. Therefore, the most common functionalities implemented in an autonomous UAS are the following:

1. Localization and map management (perception)
2. Detection and tracking of obstacles (perception)
3. Global trajectory planning (motion)
4. Local navigation (motion)

There are different ways to combine these functionalities. In Figure 3 can be seen an example of the integration of the commented most common functionalities that are usually in an autonomous UAS. All these functionalities will be briefly commented in the next sections.

2.2.1   *Localization and map management*

One of the first functionalities that must be implemented in any autonomous UAS is the capability to localize itself in a specific coordinate framework. Most of the UAS make use of augmented INS systems using GPS, combining measurement of both sensors through an extended Kalman filter technique. However, if it is required to improve the precision on the localization solution or decrease the dependency with GPS, it is possible to use techniques that combine the IMU and ADS data with visual sensors range sensors such as LIDAR. One interesting technique, especially for indoor flights, is SLAM (simultaneous localization and mapping). This technique was originally developed for ground robots but in recent years it has been successfully implemented in small and light UAS (Weiss, Scaramuzza, and Siegwart, 2011; Caballero *et al.*, 2009). This technique allows creating in real time a map of the environment and localizing the UAS in it at the same time. Nevertheless, all these techniques require a large computational capacity onboard the UAS.

On the other hand and regarding the map representation, there are different alternatives:

1. *Cell division:*   This method consists in dividing the world into a series of representative areas, using grid cells, and describing the characteristics of the world in each one. Among the typical characteristics that are represented it can be found: ease of transit, safety level, etc. The use of hierarchical data structures, such as "octree" structures, allows managing map data in an efficient way getting high resolution only on the areas necessary for the mission.

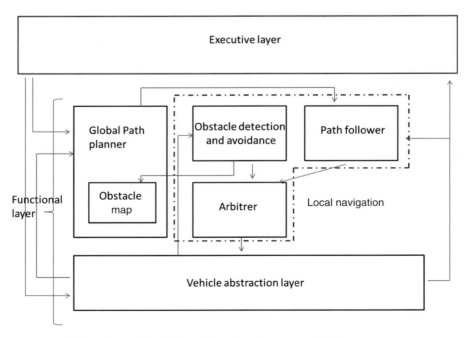

Figure 3. Detail of an example of integration of the different functionalities on an UAS.

2. *Roadmap:* This method consists in describing the environment in terms of how to reach from one to position to another and its associated costs. The most commonly used methods are the visibility graphs and the Voronoi diagrams. The most important drawback of these methods is that they require high computational power to create the map and update it; however, once the map is created, the path-planning module requires less time than using a cell-based map.

Usually, cell division maps are considered in dynamic environments where the UAS does not have a large a priori knowledge of the environment. However, in static environments where most of the obstacles are known a priori, the use of roadmaps could be an interesting option to consider.

2.2.2 Detection and tracking of obstacles

For UAS, this is related with the so-called "sense and avoid" functionality (Ramasamy and Sabatini, 2015), especially when considering noncooperative entities. "Sense and avoid" is nowadays one of the most important research topics in UAS, and it is one of the most promising technological challenges for the insertion of UAS in nonsegregated airspace.

The lack of small and light sensors with the appropriate range and precision is one of the main obstacles to solve the "sense and avoid" challenge, especially for fixed-wing UAS. Therefore, in recent years, most of the experimental research works that have dealt with detection and tracking of obstacles with UAS and aerial robotics are focused on short ranges with aerial vehicles moving at low speeds (mainly with rotary-wing configurations). In these works, the most popular sensors used in noncooperative sense and avoid are visual sensors (Geyer, Dey, and Singh, 2009), thermographic cameras, and LIDAR (Sabatini, Gardi, and Ramasamy, 2014).

Finally, for cooperative "sense and avoid," the most promising technology is the use of ADS-B (Ramasamy, Sabatini, 2015), a modern transponder technology that broadcasts the position and velocity of the aircraft based on GPS information.

2.2.3 Global trajectory planning

This functionality has an important relation with the automation of ATM considered in SESAR and NextGen programmes, and the insertion of UAS in the nonsegregated airspace. In fact global trajectory planning methods can be considered within the tactical type of operations in the ATM system. These methods can also be applied in noncontrolled airspace in order to calculate safe and obstacle-free routes. Therefore, this functionality takes care of planning a route or a motion that gets the UAS to the objective position without producing a collision with obstacles or other mobile systems.

Usually, global trajectory planning techniques requires high computational capacity and cannot be executed at a high rate.

The global trajectory planning methods usually includes two phases: compile information in an appropriate configuration space and use a search algorithm to find the best path in this space using predefined criteria. There is a large variety of motion planning techniques for autonomous systems coming for the robotics field (see a complete overview in Latombe, 2012). One important type of technique is based on graph search, where the A^* algorithm (Hart, Nilsson, and Raphael, 1968) is one of the best well-known algorithm of this type, and it has been successfully used in different applications. However, the A^* presents some drawbacks where applying in UAS. For this reason different variants have been developed in the last two decades: D^* (Stentz, 1994) can handle dynamic environments with noncomplete information, and Theta* (Nash et al., 2007) calculates trajectories that can be better followed by an UAS, minimizing changes in headings in comparison with the solutions obtained by A^* and D^* (originally developed for ground robots). Other variants of these algorithms not only take into account the minimization of changes in heading but also in altitude, in order to calculate efficient trajectories taking into consideration energy constraints (Garcia, Viguria, and Ollero, 2013).

2.2.4 Local navigation

Local navigation is usually in charge of following the trajectory provided by the global trajectory planning and avoiding obstacles (using information of the environment acquired by the onboard sensors). In principle, these methods are more reactive than the global trajectory planning and are executed in soft real time. It is important to point out that the velocity that the UAS can fly should be limited to the frequency than the local navigation module can be executed.

There are different strategies to implement this local navigation. One is the use of a fast trajectory planning, with a higher execution rate than the global trajectory planning, but only for the area covered by the UAS in the next seconds. Another option is to use some type of method that combines the different behaviors (trajectory following and obstacle avoidance). For example, it could use the DAMN architecture with arbiters (Rosenblatt, 1997).

3 EXAMPLE OF AUTONOMY ARCHITECTURE: THE ARCAS PROJECT

The European project ARCAS (Baizid et al., 2015) has as one of its main objectives the development of a highly

Figure 4. Multicopter aerial robot used in ARCAS project.

autonomous UAS that can not only acquire information from the environment but also interact with it by means of an onboard manipulator (see Figure 4).

As commented before, the complete UAS architecture is divided in two levels: low level and high level.

3.1 Low-level ARCAS architecture

The low-level architecture is composed of four main modules (Figure 5). There is a specific one used for the robotic arm and the rest are traditional in a multirotor controller that has been heavily modified to adapt it to the ARCAS system.

The main functionalities of each of these four modules are

- *Estimator and data processing:* This module is in charge of estimating and processing the state of the complete system (position, attitude, angular velocity, servos data, sensors, and safety operator radio references). Its input is all the sensor data onboard the platform.

- The position controller takes care of the platform stabilization in space whose output is the reference for the attitude controller. For this purpose, the controller needs the state of the platform, the position reference given by the high-level controller, and a compensation value coming from the attitude controller.
- The attitude controller block runs the lowest level controller of the platform. It receives references from the position controller and stabilizes the platform sending a control signal to each of the eight motors. It also runs a special compensator module that helps the controllers taking into account the movements of the robotic arm.
- The robotic arm controller module is the one in charge of the final checks of the references given to the arm, its deployment, retraction, and parsing the values to servo control signals.

The three controller modules are described with complete details in Ruggiero *et al.* (2015).

3.2 High-level ARCAS architecture

The high-level architecture follows a three-layer architecture and integrates the high-level modules that perform the navigation and perception tasks, such as the visual servoing or the object detection, making use of the lower control layer, and the different available sensors in the scenario. Furthermore, complete missions, like assembly, can be carried out by providing communication with the higher layer, which has visibility of the whole environment. The communication between layers and the concrete modules of the functional layer is represented in Figure 6, and it is implemented using the ROS middleware through the topics and machine states.

The different modules that are integrated in the high-level architecture are explained in the following sections.

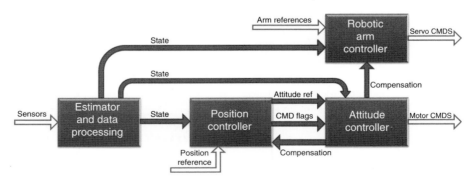

Figure 5. ARCAS low-level architecture.

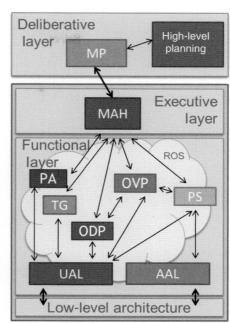

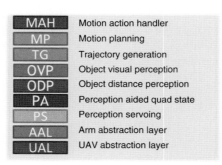

Figure 6. ARCAS high-level architecture scheme.

3.2.1 Motion planning (TP)

This module forms part of the deliberative layer. From the high-level planning modules (that implement assembly sequence planning and task planning functionalities), it generates optimal trajectories that reach the input waypoints avoiding the static obstacles taking into account the information about the environment.

3.2.2 Motion action handler (MAH)

This module implements the executive layer and acts like a supervisor of the whole mission. It is responsible to activate and to deactivate functional modules depending on the phase of the mission. In this sense, this block manages the transitions between the tasks that compose the mission.

On the other hand, the module communicates with the higher layer to request the trajectory planning for a specific configuration, or waypoints, of the aerial robotic manipulator.

The rest of the modules are part of the functional layer of the high-level architecture.

3.2.3 Perception servoing (PS)

It is responsible for the control of the aerial robotic manipulator only using visual information to accomplish the object grabbing and insertion tasks. Therefore, the outputs of the module are the control references for the aerial vehicle and the robotic arm. On one hand, the perception servoing module requires the UAV and arm states and the object, to be manipulated, pose estimation on the other.

3.2.4 Object visual perception (OVP)

This module has the function of localizing the objects, like for example bars or the landing and takeoff zones, analyzing the visual information from the available cameras in the scenario. It is devoted to the detection and identification of objects in the scene and the recognition and tracking of the different objects such as the bars of the structure. For this goal, robust techniques based on detecting visual markers and contours have been proposed for aerial robots involved in assembly tasks. Furthermore, its output is the object pose estimation that is the input to the perception servoing module.

In addition to the visual information, the module uses the UAS state and the objects pose estimation from the RO-SLAM perception module (see ODP module description). This block also estimates the object positions but presents less accuracy than the visual estimation based on markers. For that reason, the object perception modules work in conjunction with the RO-SLAM perception module. First, the RO-SLAM perception module estimates the object's position with 1 m of accuracy. Then, when the aerial vehicle is close to the object (around 1 m), the object visual perception can be launched.

3.2.5   Object distance perception (ODP)

This module localizes the objects of the scene, but in contrast to the previous module, this block is based on RO-SLAM algorithms that use the measurements of range sensors located in the scenario. As described above, the module works in conjunction with the object visual perception.

3.2.6   Perception-aided aerial vehicle state (PA)

Like the previous module, it is based on the RO-SLAM algorithms. It implements the vehicle location using only radio distance sensors and the vehicle altimeter (input data). Thus, the output of the module is only an estimation of the aerial vehicle position.

3.2.7   Trajectory generation (TG)

This module participates in the navigation tasks. The module receives the flight plan of an UAV as a waypoint list. A waypoint is defined by the desired position, heading, and cruise velocity.

This flight plan is a discretization of a continuous flight plan with very low sample time (usually 0.01 s). For this reason, a continuous path tracker has been implemented in order to follow the desired flight plan. The output of this module is the control reference of the aerial robot in order to follow the input flight plan, as well as, the takeoff and land commands (through ROS actions). Also, it needs the UAV state to control the aerial vehicle.

3.2.8   UAV Abstraction Layer (UAL)

This block acts as an interface between the ROS nodes and the autopilot onboard of the UAS (low-level architecture). It provides the UAV state using the perception aided aerial platform state module and the navigation sensors from the autopilot. Also, it receives the UAV control references and transmits them to the autopilot. The communication between the functional layer and the control layer is implemented through the UDP protocol.

3.2.9   Arm abstraction layer (AAL)

This module together with the UAL module forms the complete interface between high-level and low-level architectures. This block is responsible for providing the arm state to the functional layer nodes and to forward the arm control references to the low-level architecture. In an equivalent manner to the UAL module, the communication with the control layer is performed using the UDP protocol.

3.3   Example of ARCAS complex mission: assembly operations

In this section, the tasks sequence and messages flow of complex operations that show a high level of autonomy such as assembly operations are defined in order to show the relationships between the modules implemented in the high-level architecture.

The assembly mission objective is to find a bar in the environment and to insert it in the structure using the perception features and one aerial robotic manipulator, integrating all the necessary perception, and control and planning functionalities.

This mission can be split in the proposed tasks shown in Figure 7, and the necessary task sequence to carry out the mission is represented in Figure 8.

Taking into account the description of modules and interfaces, the inspection task can be performed with the messages flow represented in Figure 9. First, the MAH module requests to TP module a trajectory or path with certain points of interest to inspect for the UAS. Then, the response is traduced to a *Translation* action for the TG module who executes the flight plan using the PA and

Tasks	Activated specific modules	Always activated modules		
Inspection = Navigation + Object Detecction	TP, ROSB, TG	MAH, PA, ODP, UAL, AAL	ROSB	ROS bridge
			MAH	Motion action handler
Bar grabbing	PS, OVP		TP	Trajectory planning
			TG	Trajectory generation
Navigation	TP, ROSB, TG		OVP	Object visual perception
			ODP	Object distance perception
Bar insertion	PS, OVP		PA	Perception aided aerial vehicle state
			PS	Perception servoing
			AAL	Arm abstraction layer
			UAL	UAV abstraction layer

Figure 7. Tasks involved in the assembly mission.

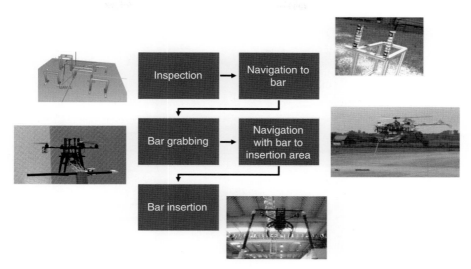

Figure 8. Task sequence for the assembly mission.

UAL modules. Furthermore, the ODP module is always active providing the object's estimations of the scene.

When the inspection finishes, the MAH has received the objects positions and can start other navigation task toward the place where the objects of interest are.

When the aerial vehicle is close the object to be manipulated, the MAH module can start, for this particular mission, a bar grabbing task. The bar grabbing and insertion tasks can be performed with the messages flow represented in Figure 10. The manipulation tasks are started using the *Close Interaction* action. This action activates the

perception servoing module that takes control of the aerial robotic manipulator to carry out the bar grabbing or insertion. For this action, the perception servoing uses the PA, OVP, UAL, and AAL modules. When the PS module starts to move the bar, it sends a *Manipulation* action to indicate to the ODP module to delete this specific bar from the RO-SLAM filters.

When the bar grabbing finishes, the MAH module can start a navigation task toward the insertion place. Finally, after reaching the place, the MAH module can start to execute a bar insertion task finishing the complete mission.

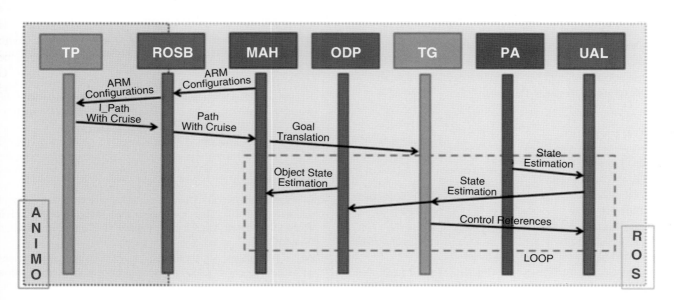

Figure 9. Messages flow in the inspection task.

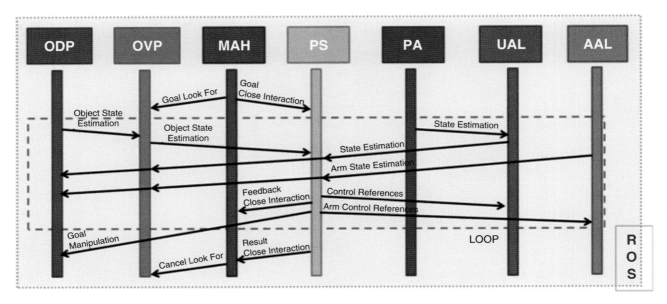

Figure 10. Messages flow in the bar grabbing/insertion tasks.

4 CONCLUSIONS

This chapter has briefly summarized the mains aspects related to autonomy architectures for UAS, including a case study that illustrates a real implementation of the presented concepts.

Although in the last 15 years it has been successfully proved that the implementation of these types of architectures in UAS, and their potential to execute complete missions with minimum intervention from the pilot on the ground, there is still a long way to work into autonomous UAS in order to make the technology robust and cheap enough for mass implementation in commercial products. However, the advances in computing and sensing technologies and the forecast exponential growth of civil application with UAS make us optimistic in the potential capacity of such autonomous systems in the near future.

REFERENCES

Adolf, F. and Andert, F. (2010) Onboard mission management for a VTOL UAV using sequence and supervisory control, in *Cutting Edge Robotics* (ed. V. Kordic), InTech, Croatia, pp. 301–316.

Ahmadian, M., Nazari, Z.J., Nakhaee, N., and Kostic, Z. (2005) Model based design and SDR, in *Proceedings of the lEEE/EURASIP Conference on DSPenbaleRadio*.

Alami, R., Chatila, R., Fleury, S., Ghallab, M., and Ingrand, F. (1998) An architecture for autonomy. *Int. J. Robot. Res.*, **17**(4), 315–337.

Baizid, K., Giglio, G., Pierri, F., Trujillo, M.A., Antonelli, G., Caccavale, F., Viguria, A., Chiaverini, S., and Ollero, A. (2015) Experiments on behavioral coordinated control of an Unmanned Aerial Vehicle manipulator system, in *IEEE International Conference on Robotics and Automation (ICRA)*, IEEE, pp. 4680–4685.

Bedrax-Weiss, T., Frank, J., Jónsson, A., and McGann, C. (2004) EUROPA2: Plan database services for planning and scheduling applications.

Bhatt, D., Hall, B., Dajani-Brown, S., Hickman, S., and Paulitsch, M. (2005) Model-based development and the implications to design assurance and certification, in *The 24th Digital Avionics Systems Conference (DASC)*, IEEE. vol. 2, p. 13.

Caballero, F., Merino, L., Ferruz, J., and Ollero, A. (2009) Vision-based odometry and SLAM for medium and high altitude flying UAVs. *J. Intell. Robot. Syst.*, **54**(1–3), 137–161.

Clough, B.T. (2002) Metrics, schmetrics! How the heck do you determine a UAV's autonomy anyway? in *Proceedings of the Performance Metrics for Intelligent Systems (PerMIS) Conference*, Gaithersburg, MD.

Doherty, P., Kvarnström, J., and Heintz, F. (2009) A temporal logic-based planning and execution monitoring framework for unmanned aircraft systems. *Auton. Agent Multi Agent Syst.*, **19**(3), 332–377.

European Commission. Directorate-General for Enterprise and Industry. (2013) Roadmap for the integration of civil Remotely-Piloted Aircraft Systems into the European Aviation System. OIB.

Fabiani, P., Fuertes, V., Piquereau, A., Mampey, R., and Teichteil-Königsbuch, F. (2007) Autonomous flight and navigation of VTOL UAVs: from autonomy demonstrations to out-of-sight flights. *Aerosp. Sci. Technol.*, **11**(2), 183–193.

Flanagan, C., Toal, D., and Leyden, M. (2003) Subsumption and fuzzy-logic, experiments in behavior-based control of mobile robots. *Int. J. Smart Eng. Syst. Des.*, **5**(3), 161–175.

Garcia, M., Viguria, A., and Ollero, A. (2013) Dynamic graph-search algorithm for global path planning in presence of hazardous weather. *J. Intell. Robot. Syst.*, **69** (1–4), 285–295.

Gat, E. (1998) On three-layer architectures, *Artificial Intelligence and Mobile Robots*, **195**, 210.

Geyer, C., Dey, D., and Singh, S. (2009) Prototype sense-and-avoid system for UAVs. Robotics Institute, Carnegie Mellon University, Technical Report CMU-RI-TR-09-09.

Hart, P. E., Nilsson, N.J., and Raphael, B. (1968) A formal basis for the heuristic determination of minimum cost paths. *IEEE Trans. Syst. Sci. Cybern.* **4**(2), 100–107.

Heredia, G. and Ollero, A. (2010) Virtual sensor for failure detection, identification and recovery in the transition phase of a morphing aircraft. *Sensors*, **10**(3), 2188–2201.

Huang, H.M., Pavek, K., Novak, B., Albus, J., and Messin, E. (2005) A framework for autonomy levels for unmanned systems (ALFUS). *Proceedings of AUVSI Unmanned Systems 2005*.

Kendoul, F. (2012) Survey of advances in guidance, navigation, and control of unmanned rotorcraft systems. *J. Field Robot.*, **29**(2), 315–378.

Kirby, B. and Kang, H. (2008) Model based design for power systems protection relays, using matlab & simulink, in *the Proceedings of the IET 9th International Conference on Developments in Power System Protectio*.

Latombe, J.C. (2012) *Robot Motion Planning*, vol. 124, Springer Science + Business Media.

Maza, I., Caballero, F., Capitan, J., Martinez-de-Dios, J.R., and Ollero, A. (2011) A distributed architecture for a robotic platform with aerial sensor transportation and self-deployment capabilities. *J. Field Robot.*, **28**(3), 303–328.

McGann, C., Py, F., Rajan, K., Thomas, H., Henthorn, R., and McEwen, R. (2007) T-rex: A model-based architecture for auv control, in *3rd Workshop on Planning and Plan Execution for Real-World Systems, vol. 2007*.

Merz, T. (2004) Building a system for autonomous aerial robotics research, *in 5th IFAC/EURONSymposiumon Intelligent Autonomous Vehicles*, Lisbon, Portugal.

Muscettola, N., Dorais, G.A., Fry, C., Levinson, R., and Plaunt, C. (2002) IDEA: Planning at the core of autonomous reactive agents.

Nash, A., Daniel, K., Koenig, S., and Felner, A. (2007) Theta*: Any-Angle Path Planning on Grids, in *Proceedings of the National Conference on Artificial Intelligence (vol. 22,*

No. 2, p. 1177). Menlo Park, CA; Cambridge, MA; London; AAAI Press; MIT Press; 1999.

Orehek, M. (2001) Model-based design of an ECU with data- and event-driven parts using auto code generation, in *Proceedings 2001 ICRA. IEEE International Conference on Robotics and Automation, 2001*, vol. 2, IEEE, pp. 1346–1351.

Ramasamy, S. and Sabatini, R. (2015) A unified approach to cooperative and non-cooperative sense-and-avoid, in International Conference on Unmanned Aircraft Systems (ICUAS), IEEE, pp. 765–773.

Rosenblatt, J. (1997) A distributed architecture for mobile navigation. *J. Exp. Theor. Artif. Intell.*, **9**(2/3), 339–360.

Rosenblatt, J., Williams, S., and Durrant-Whyte, H. (2002) A behavior-based architecture for autonomous underwater exploration. *Inf. Sci.*, **145**(1), 69–87.

Ruggiero, F., Trujillo, M., Cano, R., Ascorbe, H., Viguria, A., Peréz, C., Lippiello, V., Ollero, A., and Siciliano B., and Siciliano, B. (2015) A multilayer control for multirotor UAVs equipped with a servo robot arm, in *IEEE International Conference on Robotics and Automation*, Seattle, WA.

Sabatini, R., Gardi, A., and Ramasamy, S. (2014) A laser obstacle warning and avoidance system for unmanned aircraft sense-and-avoid, in *Applied Mechanics and Materials*, vol. 629, pp. 355–360.

Santamaría, D., Alarcón, F., Jiménez, A., Viguria, A., Béjar, M., and Ollero, A. (2012) Model-based design, development and validation for UAS critical software. *J. Intell. Robot. Syst.*, **65** (1–4), 103–114.

Sheridan, T.B. (1992) *Telerobotics, Automation, and Human Supervisory Control*, Massachusetts Institute of Technology, Cambridge, MA.

Stentz, A. (1994) Optimal and efficient path planning for partially-known environments, in *Proceedings of IEEE International Conference on Robotics and Automation, 1994. (pp. 3310–3317)*. IEEE, pp. 3310–3317.

Weiss, S., Scaramuzza, D., and Siegwart, R. (2011) Monocular-SLAM–based navigation for autonomous micro helicopters in GPS-denied environments. *J. Field Robot.*, **28**(6), 854–874.

Whalley, M., Freed, M., Harris, R., Takahashi, M., Schulein, G., and Howlett, J. (2005) Design, integration, and flight test results for an autonomous surveillance helicopter, in *Proceedings of the AHS International Specialists' Meeting on Unmanned Rotorcraft*.

Chapter 17

Obstacle Avoidance: Static Obstacles

Madhavan Shanmugavel,[1] Antonios Tsourdos,[2] and Brian White[3]

[1]*School of Engineering, Monash University Malaysia, Selangor, Malaysia*
[2]*School of Aerospace, Transport & Manufacturing and Centre for Autonomous and Cyber-Physical Systems, Cranfield University, Cranfield, UK*
[3]*Centre for Autonomous Systems and Cyber-Physical Systems, Cranfield University, Cranfield, UK*

1 INTRODUCTION

The obstacle avoidance is one of the prime objectives of path planning. The obstacle avoidance is in general applicable to anything the autonomous system wants to avoid, for example, threats, no-fly zones, restricted zones, or other autonomous systems in its group. Here, the term 'vehicle' is used to refer to the autonomous system. The obstacle avoidance provides the safety both to the obstacle and vehicle. The obstacle avoidance would be a small algorithm under the path planner. Hence, the discussion of obstacle avoidance is closely associated with the path planning. Also, the outcome of the obstacle avoidance would be collision-free path between the places of interest. Consider an environment E filled with obstacles whose space is represented by C_{obst},

while the remaining space will be $C_{free} = E/C_{obst}$. Finding a connectivity between two desired points P_s, and P_f avoiding the obstacles in a given space will be given by a path r in C_{free} is a continuous map

$$r : [0\ 1] \rightarrow C_{free} \qquad (1)$$

and

$$C_{free} \bigcap C_{obst} = \varnothing \qquad (2)$$

where $r(0) = P_s$ and $r(1) = P_f$, where P_s is the initial position and P_f is the final position.

The avoidance is required to avoid collision and hence safe navigation of the vehicle. A collision between two objects occurs when they try to occupy the same point at the same time. The method of avoidance would be based on the available information about the obstacles. The knowledge of size, shape, and mobility are important to handle the obstacles. Also, it is important to know the these parameters are known a priori or not. For the static obstacles, the first two characteristics are play a major roles in the path planning. The formulation of configuration space significantly simplifies the path planner by reducing the size of the vehicle to a point by enlarging the obstacle proportionately. Handling polygonal obstacles are widely studied by the robotics community for obstacle avoidance. Path planning among convex polygonal obstacles is is an ongoing research in the area of computational geometry. Producing curvature constrained path among the polygonal obstacles is NP hard (Reif and Wang, 1998; Agarwal *et al.*, 2002). The problem becomes complex following the evolution of cooperative/coordinated path planning where multiple vehicles are used.

Unmanned Aircraft Systems. Edited by Ella Atkins, Aníbal Ollero, Antonios Tsourdos, Richard Blockley and Wei Shyy.
© 2016 John Wiley & Sons, Ltd. ISBN: 978-1-118-86645-0.

The popular approaches for handling static obstacles are roadmap methods and potential field method. A comprehensive treatment of these methods can be seen in Latombe (1991) and LaValle (2006). Though these methods are proposed as path planning methods, they generate the path either by forming free configuration space C_{free} or by isolating the obstacles space C_{obst}. First, we discuss a brief review of these methods.

2   AVOIDING STATIC OBSTACLES

The case of static known obstacles is normally handled by a global path planner like Voronoi diagram, cell decomposition, and potential field, or global optimization techniques. This problem is relatively easy to handle as the obstacles are fixed with respect to time, whereas in the case of dynamic obstacles the trajectories are to be estimated prior to planning. However, in both cases, the presence of obstacles not only causes replanning of paths, but also communications, task assignment, resource allocation, and other functions depending on the mission. Furthermore, planning among static obstacles is a hard task as the number of vehicles are increased.

2.1   Voronoi diagram

The Voronoi diagram isolates the obstacles by forming polygonal cells around them such that any two obstacles are at equal distance from each other. As shown in Figure 1, the Voronoi diagram forms cells of isolated places. The cells are polygons sharing edges with other cells. The resulting

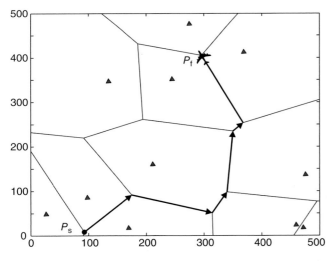

Figure 1. Path planning using Voronoi diagram: P_s is the start position, P_f is the goal position. Triangular dots are obstacles/threats.

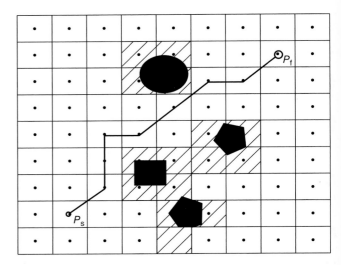

Figure 2. Path planning using cell decomposition: P_s is the start position, P_f is the goal position. Eliminating cells of obstacles and connecting the empty cells form the path/route.

diagram forms a network of nodes connected by straight lines. A search algorithm such as A* can be used to find a route/path between the start point P_s and goal point P_f.

2.2   Cell decomposition

The cell decomposition approach divides the given map into a number of cells. As shown in Figure 2, a polygonal path is formed by connecting the cells which are not occupied by the obstacles, while the cells occupied by the obstacles are eliminated.

2.3   Visibility graph

The visibility graph produces connectivity between the desired points by forming a network connecting the edges of polygonal obstacles. This provides visibility in an environment of obstacles. Again, a path is found by a search algorithm. A schematic sketch of this approach is shown in Figure 3.

2.4   Potential field and sampling-based methods

The potential field method is first proposed in Khatib (1985). This method treats the goal position as an attractive potential while the obstacles/threats as the repulsive potential. An unmanned aerial vehicle (UAV) with attractive potential will naturally be dragged toward the goal position. Representing the position of the vehicle $p(x, y)$, the field where the vehicle would be moving will be represented by a scalar

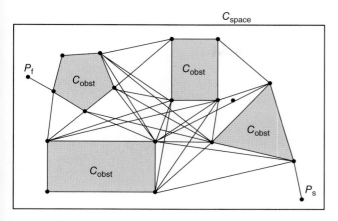

Figure 3. Visibility graph: network of all possible visible route.

function $U(p) = U_a(p) + U_r(p)$, where U_a is the attractive potential, and U_b is the repulsive potential. This approach initially suffers from the local minima, however, later modified approach have been proposed. The rapidly exploring random trees (RRT) method generates a tree of trajectories by randomly selecting a point in the given space. Each branch in the tree is tested for collision avoidance before adding into the tree until a connectivity is formed between the start and goal points. The probabilistic road map (PRM) method (Hsu, Latombe and Kurniawati, 2006) forms a connectivity by expanding the database graph/tree. The growth of the graph is done by adding a collision-free neighbor sampled from the given space.

3   RESEARCH ON OBSTACLE AVOIDANCE

As mentioned in the previous section, the obstacle avoidance algorithm is closely associated with the path planning because the former results in replanning of paths. This is applied in a different context in the literature, for example, minimizing risk of aircraft detection by radars, sensors, or surface air missiles (SAMs) (Chan and Foddy, 1985; Hebert *et al.*, 2001; Vian and More, 1989; Zabarankin, Uryasev and Pardalos, 2002), minimizing risk of submarine detection by sensors (Washburn, 1990), maximizing the probability of target detecting by a searcher (Assaf and Sharlin-Bilitzky, 1994; Benkoski, Monticino and Weisinger, 1991; Eagle and Yee, 1990; Koopman, 1980; Mangel, 1984; Stone, 1975; Thomas and Eagle, 1995; Washburn, 1983), airborne collision avoidance system (Williams, 2004).

The principle of configuration space (Lozano-Pérez, 1983) in robotics shrinks the size of the robots to point masses in proportion to enlarging the size of the obstacles. This approach enables efficient use of algorithms such as Voronoi diagram (Li *et al.*, 2002; Beard *et al.*, 2002; McLain,

2000; Bortoff, 2000), visibility diagram, and probabilistic methods to model the given map of landmarks and obstacle locations into a searchable database. The databases are graphs and trees forming a network of nodes and edges isolating the obstacles. A search algorithm is used to connect the start and finish points through the nodes of the network.

Obstacle avoidance methods are studied using potential field where attractive and repulsive forces are generated, respectively, by the goal and obstacle (Kim and Khosla, 1992). The potential field is used to avoid the obstacles in path planning of multiple UAVs, where the obstacles are assumed to be circular in shape with lethality described by its radius (Eun and Bang, 2006).

Most of the cases for aerial vehicle collision avoidance is done by assuming flight in a constant altitude. An analytical and discrete optimization approach is used in Zabarankin, Uryasev and Pardalos (2002) for optimal risk path generation in two-dimensional (2D) space, with an arbitrary number of sensors and a constraint on path length. Mixed integer linear programming (MILP) is used in Richards and How (2002) for the collision avoidance. The resulting path from these approaches does not provide a flyable path. Taking this point as initiation and in contrast to these approaches, flyable paths are used here to solve the problem of collision avoidance. Avoidance circles are used for obstacle avoidance of multiple mobile robots in Fujimori, Ogawa and Nikiforuk (2002), where two robots in avoidance circle cooperate to occupy the same place at the same time, while others have to wait outside the area. However, it is feasible in mobile robots, it is not feasible for aerial robotics applications.

Optimization and search algorithms are used to find the path in the cluttered environment. A branch and bound optimization used to avoid obstacles in 2D space is discussed in Eele and Richards (2009). Evolutionary optimization applied to find a route among the obstacles is studied in Zheng *et al.* (2005). A^* algorithm is used in Yang and Zhao (2004) to find the path in the midst of known obstacles and conflicts. The obstacles are modeled as solid objects such as ellipsoid, cuboid, cube, or pyramid. Optimal control would be a good choice for path finding problems. However, the computational load and complexity will be enormous with the number of obstacles and vehicles and constraints (Sasiadek and Duleba, 2000).

Path planning with obstacle avoidance in 3D is more complicated. The complication arises because there are more directions for maneuvers for both the UAV and the obstacle. Also, other constraints especially shortest length criterion is complicated in 3D space. Hence, most of the obstacle avoidance algorithms seek simplicity in maneuvering away from the obstacles. One more factor is preparation of the 3D space for the path planner. An octree representation is proposed in Kitamura *et al.* (1995, 1996) where the space is divided into searchable regions or cells and a collision-free path is generated by applying potential field to each cell of the octree.

Another direction in this research is use of UAVs in civilian airspace, which requires collision avoidance algorithm equivalent to those available in commercial aircrafts. The existing system uses traffic alert and collision avoidance system (TCAS), which issues resolution advisory (RA) based on the the position and altitude data from the transponder-equipped aircraft and from the ground if available. The safety aspects of using the UAVs in civilian airspace is discussed in Zeitlin and McLaughlin (1985). Generation of cooperative trajectories for air traffic with collision avoidance is discussed in Bicchi and Pallottino (2000).

4   AVOIDANCE OF STATIC OBSTACLES

Before avoiding the threats, it is necessary to detect it. In practice the threat is detected by sensor. There are several types of sensors available for obstacle detection, for example, ultrasonic, infrared, vision, and laser. The sensor is modeled with range. It is the distance at which a sensor can detect a threat. Once a threat is detected for possible collision, the path has to be replanned. If the vehicle is surrounded by an imaginary safety region R_{safe}, the intersection between R_{safe} and C_{obst} should be empty for collision avoidance. If it is not empty, the replanning has to be done by the path planner.

$$C_{\text{obst}} \bigcap R_{\text{safe}} = \varnothing \qquad (3)$$

Besides knowing the mobility of the obstacles (static or dynamic), it would be advantageous to know their locations a priori. For a known map, it is normally given in latitude and longitude. The position information will help to choose the direction of maneuver with respect to obstacle location. Irrespective of the size, shape of the static obstacles and the possible direction of collision avoidance in 2D is (i) either to left or (ii) right of the obstacle as shown in Figure 4. The number of choices are more in higher dimensions. For example, if we extend the circle into a sphere in three dimension, the number of choices are infinite. However, only a finite number of possibilities are available considering the kinematic constraints of the vehicle.

5   REACTIVE PLANNING

The collision avoidance for known static obstacles is best handled by the offline methods. Suppose the obstacles are

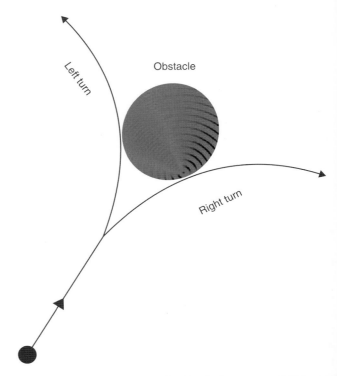

Figure 4. Obstacle avoidance in 2D: Only two possibilities are available for obstacle avoidance in 2D – fly either on left side or on right side.

not known a priori, then a reactive algorithm has to used to avoid the collision. Potential field method is a type of reactive algorithm where the vehicle is repelled from the obstacle once in its proximity. Another approach is the "sense-and-avoid" concept, which is one of the high priority research subjects for the use of unmanned air vehicles in civilian airspace. As mentioned in the previous section, the simple case of collision avoidance involves either turning left or right. An implementation of reactive collision avoidance is presented in Shanmugavel *et al.* (2009). The the avoidance is achieved by producing a way-point M midway between the current position and next waypoint, as shown in Figure 5, where the point M is assumed to lie on a safety circle around the obstacle. The replanned trajectory passes through M to reach the next waypoint. The selection of the point M is based on the minimum area of a triangle formed by lines connecting the current position, intersection point X_1 or X_2, and the center of the obstacle C. The points M and N on the safety circle are generated by intersection of the normal to the line segment $X_1 - X_2$. If the center C is left to the line $X_1 - X_2$, M is selected on right to the obstacle region and vice versa. The initial path is $r(t)$ and the replanned path is made of $r_1(t)$ and $r_2(t)$.

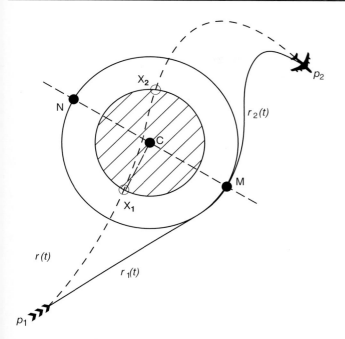

Figure 5. Threat handling by intermediate pose.

6   SUMMARY

This chapter discusses the path planning in an environment of threats. Two methods are explained: detection by intersection and by measuring the distance. The threats are avoided by replanning the flyable path. The replanning is done either by increasing the curvature of the path or by creating an intermediate waypoint. Both these approaches are illustrated with simulation results.

REFERENCES

Agarwal, K., Biedl, T., Lazard, S., Robbins, S., Suri, S. and Whitesides, S. (2002) Curvature-constrained shortest paths in a convex polygon. *SIAM J. Comput.,* **31**(6), 1814–1851.

Assaf, D. and Sharlin-Bilitzky, A. (1994) Dynamic search for a moving target. *J. Appl. Prob.,* **31**(2), 438–457.

Beard, R., McLain, T., Goodrich, M. and Anderson, E. (2002) Coordinated target assignment and intercept for for unmanned air vehicles. *IEEE Trans. Robo. Automat.,* **18**(6), 911–922.

Benkoski, S.J., Monticino, M.G. and Weisinger, J.R. (1991) A Survey of the search theory literature. *Naval Res. Logist.* **38**(4), 468–494.

Bicchi, A. and Pallottin, L. (2000) An optimal cooperative conflict resolution for air traffic management systems. *IEEE Trans. Intell. Transport. Syst.,* **1**(4), 221–232.

Bortoff, S. (2000) Path-planning for unmanned air vehicles. *Proc. Am. Control Conf.,* **1**, 364–368.

Chan, Y.K. and Foddy, M. (1985) Real time optimal flight path generation by storage of massive data bases. *Proceedings of the IEEE NEACON 1985,* Institute of Electrical and Electronics Engineers, New York, 516–521.

Eagle, J.N. and Yee, J.R. (1990) An optimal branch-and-bound procedure for the constrained path, moving target search problem. *Oper. Res.* **38**(1), 110–114.

Eele, A. and Richards, A. (2009) Path-planning with avoidance using nonlinear branch-and-bound optimization. *J. Guidance, Control Dyn.* **32**(2), 384–394.

Eun, Y. and Bang, H. (2006) Cooperative control of multiple Unmanned aerial vehicles using the potential field theory. *J. Aircr.* **43**(6), 1805–1814.

Fujimori, A., Ogawa, Y. and Nikiforuk, P.N. (2002) A modifiction of cooperative collision avoidance for multiple robots using the avoidance circle. *Proc. Instn. Mech. Engrs., Part – I, J. Syst. Control Eng.,* **216**, 291–299.

Hebert, J., Jacques, D., Novy, M. and Pachter, M. (2001) *Cooperative Control of UAVs.* Institute of Aeronautics and Astronautics.

Hsu, D. Latombe, J.C. and Kurniawati, H. (2006) On the probabilistic foundations of probabilistic roadmap planning. *Int. J. Rob. Res.,* **25**(7), 627–643.

Kim, J.O., and Khosla, P.K. (1992) Real-time obstacle avoidance using harmonic potential functions. *IEEE Trans. Rob. Autom.* **8**, 338–349.

Khatib, O. (1985) Real time obstacle avoidance for manipulators and mobile robots. *IEEE Int. Conf. Robot Autom.* **2**, 500–505.

Kitamura, Y., Tanaka, T., Kishino, F. and Yachida, M. (1995) 3D path planning in a dynamic environment using an octree and an artificial potential field. *Proc. IEEE/RSJ Int. Conf. Intell. Rob. Syst.* **2**, 474–481.

Kitamura, Y., Tanaka, T., Kishino, F. and Yachida, M. (1996) Real-time path planning in a dynamic 3D environment. *Proc. IEEE/RSJ Int. Conf. Intell. Robot Syst.* **2**, 925–931.

Koopman, B.O. (1980) *Search and Screening: General Principles with Historical Applications.* Elmsford, Pergamon Press, New York.

Latombe, J.C. (1991) *Robot Motion Planning,* Kluwer Academic Publishers, Boston, MA.

LaValle, S.M. (2006) *Planning Algorithm.* Cambridge University Press, New York.

Li, S.M., Boskovic, J.D., Seereeram, S., Prasanth, R., Amin, J., Mehra, R.K., Beard, R. and McLain, T.W. (2002) Autonomous hierarchical control of multiple unmanned combat air vehicles (UCAVs). *Am. Control Conf. Anchorage AK* (**1**), 274–279.

Lozano-Pérez, T. (1983) Spatial planning: A configuration space approach. *IEEE Trans. Comput.,* **C-32** (2), 108–120.

Mangel, M. (1984) Search Theory, *Lecture Notes.* Springer, Berlin.

McLain, T. (2000) Cooperative rendezvous of multiple unmanned air vehicles. *AIAA Guidance, and Control Conference,* August 2000, NC, Denver, AIAA-2000-3269.

Reif, J. and Wang, H. (1998) The complexity of the two dimensional curvature-constrained shortestpath problem. *Proceedings of the Third Workshop on the Algorthimic Foundations of Robotics on Robotics, WAFR'98*, MA.

Richards, A. and How, J.P. (2002) Aircraft trajectory planning with collision avoidance using mixed integer linear programming. *Am. Control Conf.* 1936–1941.

Sasiadek, J. and Duleba, I. (2000) 3D Local trajectory planner for UAV. *J. Intell. Rob. Syst.: Theor. Appl.*, **29**, 191–210.

Shanmugavel, M., Tsourdos, A., Zbikowski, R. and White, B.A. (2009) Co-operative path planning of multiple UAVs using dubins paths with clothoid arcs. *Control Engineering Practice*.

Stone, L.D. (1975) *Theory of Optimal Search*, Academic Press, New York.

Thomas, L.C. and Eagle, J.N. (1995) Criteria and approximate methods for path-constrained moving-target search problems. *Naval Res. Logist.* **42**, 27–38.

Vian, J.L. and More, J.R. (1989) Trajectory optimization with risk minimization for military aircraft. *AIAA J. Guid. Control Dyn.* **12**(3), 311–317.

Washburn, A.R. (1990) Continuous autorouters, with an application to submarines. Research Report, NPSOR-91-05, Naval Postgraduate School, Monterey, CA.

Washburn, A.R. (1983) Search for a moving target: the FAB algorithm. *Oper. Res.* **31**, 739–751.

Williams E.D. (2004) Airborne collision avoidance system. *9th Australian Workshop on Safety Related Programmable Systems*, Australia, 97–110.

Yang, H.I. and Zhao, Y.J. (2004) Trajectory planning for autonomous aerospace vehicles amid obstacles and conflicts. *J. Guid. Control Dyn.* **27**, 997–1008.

Zabarankin, M., Uryasev, S. and Pardalos, P. (2002) *Optimal Risk Path Algorithms, Cooperative Control and Optimization* (eds. R. Murphey and P. Pardalos), pp. 271–303.

Zeitlin, A.D. and McLaughlin, M.P. (2007) Safety of cooperative collision avoidance for unmanned aircraft. *IEEE Aerosp. Electron. Mag.*, **22**(4), 9–13.

Zheng, C., Li, L., Xu, F., Sun, F. and Ding, M. (2005) Evolutionary route planner for unmanned air vehicles. *IEEE Trans. Rob.*, **21**(4), 609–620.

Chapter 18

Guided Weapon and UAV Navigation and Path-Planning

Farhan A. Faruqi

Information Processing and Human Sciences Group, Combat and Mission Systems, WCSD, Defence Science and Technology Organisation, Edinburgh, South Australia

1 PROBLEMS OF GPS AND INS FOR MISSILES AND UAVs

1.1 Global Positioning System (GPS) Navigation

In recent years, the global positioning system (GPS) has become an important navigation aid for both military and civil applications. GPS is part of a satellite-based navigation system developed by the US Department of Defense under its NAVSTAR satellite program. It includes 28 active satellites approximately uniformly distributed around six circular orbits with four or more satellites each. The orbits are inclined at an angle of 55° relative to the Equator and are separated from each other by multiples of 60° right ascension. These non-geostationary orbits have a radius of 26,560 km

Unmanned Aircraft Systems. Edited by Ella Atkins, Aníbal Ollero, Antonios Tsourdos, Richard Blockley and Wei Shyy.

and an orbital period of one-half sidereal day (approx. 11.967 hours). Theoretically, three or more satellites will be visible from most points on Earth 24 hours per day and four or more of these satellites are sufficient to determine an observer's position on Earth. The GPS signals includes navigation information on the ephemeris of the transmitting satellite and an almanac for all GPS satellites, with parameters to correct for ionospheric signal propagation delays and an offset time between satellite clock time and true GPS time. Using the information from the satellites, it is possible to obtain vehicle-satellite relative range (known as the pseudo-range) and relative velocity (known as the pseudo-range rate) which can then be used to derive vehicle's position and velocity with respect to a fixed axes system.

In order to set up the kinematics equation for the navigation solution of a navigating vehicle I, we consider the geometry shown in Figure 1. Here (X, Y, Z) define the fixed navigation axes, and $\mathbf{S}_i$ is the ith GPS satellite ($i = 1, 2, 3, \ldots$).

Now

$$\|\boldsymbol{R}_i\| = \left[\left(x_{S_i} - x_I\right)^2 + \left(y_{S_i} - y_I\right)^2 + \left(z_{S_i} - z_I\right)^2 \right]^{\frac{1}{2}} \quad (1)$$

Now, $\|R_i\| = c\Delta T_i$ may be computed since ΔT_i the time taken for the radio waves to travel from the ith satellite transmitter to the vehicle is known, and $c = 0.299792458$ m ns^{-1} is the speed of light. In order to compute $R_I = \begin{pmatrix} x_I & y_I & z_I \end{pmatrix}^{\mathrm{T}}$, we need to solve Equation (1) using the data from at least three satellites.

This is not a trivial problem since the equations are nonlinear and require some kind of an iterative technique to solve. One method for solving this problem is given in

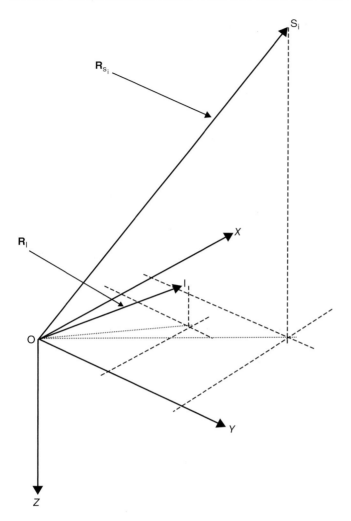

Figure 1. GPS navigation kinematics geometry.

Grewal, Weill and Andrews (2001). In current GPS systems least-square estimation (LSE) or Kalman filtering (KF) and extended Kalman filtering (EKF) techniques are commonly used. Theoretical foundations of these techniques are considered in Jazwinski (1998). The latter technique is usually the preferred option, as it allows the vehicle motion to be taken into account. Better accuracy is obtained if the satellites are well separated in space (i.e., significantly different angular separation of satellites with respect to the vehicle). Note that position calculations involve range differences, and where these are nearly equal and the angular separation between satellites is small, any small relative errors are greatly magnified. This effect brought about by the satellite geometry is known as *dilution of precision*. Any errors caused by other sources such as the clock bias gets magnified by this effect. For the case where clock bias is present, four satellites suitably located are required (e.g., three satellites

equally spaced near the horizon and the fourth overhead is regarded as a favorable satellite configuration). The GPS signals are referenced to a clock (the GPS time) that is derived from a combination of clocks at all monitoring stations and the satellite atomic clock. Over a long period, it is synchronized to within 1 ns of the Coordinated Universal Time (UTC) as maintained by the master clock at the US Naval Observatory, ignoring leap seconds.

In a manner similar to the computation of range from pseudo-range, the vehicle velocity can be computed from pseudo-range rate which is also available; thus we may write the relevant equation as

$$\|V_i\| = \left\{ \left(u_{S_i} - u_I\right)^2 + \left(v_{S_i} - v_I\right)^2 + \left(w_{S_i} - w_I\right)^2 \right\}^{\frac{1}{2}} \quad (2)$$

This also requires solving a set of nonlinear equations with data from four satellites, and application of techniques such as the LSE or KF is used. Typical navigation accuracy quoted for GPS is position accuracy under 20.0 m in range for stand-alone C/A (coarse) code receiver and under 5.0 m for (precision) P-code. For a differential GPS (DGPS) utilizing a secondary reference receiver station the range accuracy can be reduced to below 1.0 m. Groves (2008) gives a comprehensive exposition of GPS/INS-based navigation systems.

1.2 Inertial Navigation System (INS)

Inertial navigation is concerned with determining the current position and velocity of a vehicle with respect to a given frame of reference using measurements derived from sensors that measure motion utilizing the effect of inertia. There are a number of different frames of reference that may be used depending on particular mission; these are illustrated in Figures 2a–d.

1. The Earth-centered inertial (ECI) axes system has its origin at the Earth's center of mass with its *z*-axis along the Earth's spin axis and its *x*-axis in the direction of the vernal equinox; see Figure 2a. This axis system is generally used for space navigation.
2. The satellite orbital coordinate (SOC) system is illustrated in Figure 2 b and used in GPS ephemerides.
3. A local Earth (NED) axis (Geodetic) system is illustrated in Figure 2c with the *x*-axis pointing north, the *y*-axis pointing east and the *z*-axis pointing vertically down (towards the Earth center for a spherical Earth). This axes system is commonly used in Earth-based navigation (e.g., ships, aircraft, and land vehicles). Note that as a vehicle moves, the latitude and longitude change results in

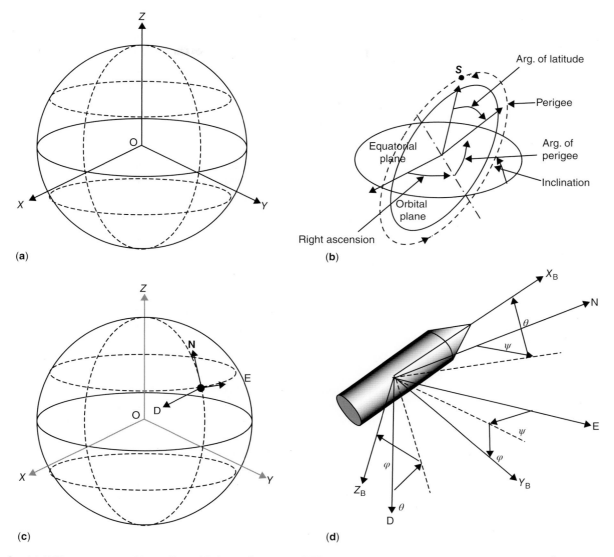

Figure 2. (a) ECI axes system; (b) satellite orbital coordinates; (c) NED axes system; (d) body axes system (X_B, Y_B, Z_B).

a rotational motion of the local axes, as the vehicle follows the Earth's curvature.

4. A vehicle body (VB) axis system, as illustrated in Figure 2d, is used to define the motion of a vehicle body with respect to a fixed-axis system; (ψ, θ, ϕ) are, respectively, the yaw, pitch, and roll angles also referred to as the *Euler angles*.

1.2.1 Inertial sensors

Commonly used inertial sensors in most navigation applications are combinations of gyros and accelerometers. Other sensors that are used in navigation include GPS, visual sensors, and magnetometers; however, these are not inertial

sensors. In this section we shall consider the characteristics of gyros and accelerometers and how measurements obtained from these are utilized to implement an inertial navigation system.

1.2.2 Gyro characteristics

Over the years a vast variety of gyroscopes (gyros) have been used for navigation; depending on the particular application, these devices have been categorised in terms of their accuracy and cost. Current gyros give angular rate information with respect to their output axes which is used to construct vehicle attitude angle; attitude or position sensors (that give vehicle body attitude directly) are not widely used in modern navigation systems. Examples of various types of gyros in use today include: laser gyros (high accuracy and high cost), fiber optics gyros (FOG) (medium to low accuracy), and micro-electro-mechanical systems (MEMS) gyros which are available in medium and low accuracy versions. The MEMS gyros are sturdy and that enables them to withstand very high g-forces. A rule of thumb that indicates the accuracy of a gyro package (cluster) is the value of the angular drift; for example, a highly accurate (inertial grade) gyro cluster may have a drift of between $0.01–0.1° h^{-1}$; for a medium-grade device this number would be in the range of $0.1–1° h^{-1}$, and a low grade (tactical grade) could be anywhere from $1.0–20° h^{-1}$ drift. From the perspective of the navigation systems designer the key performance parameters that define the gyro accuracy (or the error sources) are the following

1. Bias, which is any non-zero sensor (rate) output when the input rate is zero
2. Scale factor error that is due to manufacturing tolerances or ageing
3. Nonlinearity between input and output
4. Asymmetry of the scale factor that is generally a result of the electronics component mismatch
5. Thermal and other spurious noise including "random walk" noise effects
6. Scale factor dead-zone due to mechanical friction or lock-in for laser gyros
7. Quantization error that is present in all devices with digital output

A gyro package for measuring 3D angular motion of a vehicle usually comes as "cluster" of three mutually orthogonal gyros; non-orthogonal clusters have been used for some applications (e.g., high spinning vehicles) but it is not very common. A package of three orthogonal gyros combined with three mutually orthogonal accelerometers form an inertial measurement unit (IMU).

1.2.3   Accelerometers

Accelerometers used for inertial navigation utilize Newton's second law $F = ma$ to measure acceleration a by measuring the force F, with a scaling constant m called "proof mass." There are a wide variety of accelerometers available and their

selection depends on the particular application. The error sources for accelerometers are similar to those for gyros although the cause for these errors is quite different. These error sources are: (i) acceleration bias, (ii) scale factor error, and (iii) the (g^2) accelerometer sensitivity.

1.3   Inertial Navigation Algorithm

The navigation system for a vehicle may be implemented in two ways – stabilized platform system (SPS) and the strap-down system (SDS). In the SPS, the IMU is mounted on a stabilized platform containing a gimbaled arrangement that compensates for body rotation such that the platform (on which the IMU is mounted) always retains a fixed orientation with respect to the navigation axis. This type of arrangement was used in the past, particularly in ballistic missiles, mainly because the IMUs were not of sufficient accuracy to cope with large and rapid changes in body attitude. Current IMUs perform significantly better under large and rapid changes in body attitude, and navigation systems of choice today use SDS configuration (see Titterton and Weston, 2004). In the SDS implementation, the IMU measures the vehicle rotation rates and accelerations in the body axis; a transformation is required to convert the vehicle states (position, velocity, and acceleration) from the body axes to the navigation (or a fixed) axes. In this section we shall consider navigation of a vehicle with respect to a fixed frame in order to demonstrate the basic principles of navigation. We shall assume a flat, non-rotating Earth and that the vehicle body attitude, velocity, and acceleration change with time as the vehicle moves. Accordingly, the motion of the vehicle, in a fixed coordinate system may be written as

$$\frac{d}{dt} R_I = V_I \tag{3}$$

$$\frac{d}{dt} V_I = A_I + g_I \tag{4}$$

In order to use the information (measurements) from body mounted IMU we need to transform A_I from fixed to the body axes. This transformation is given by

$$A_I = \left[T_B^F \right] A_B \tag{5}$$

where

$$\left[T_B^F \right] = \begin{bmatrix} (\cos\psi\cos\theta) & (\cos\psi\sin\theta\sin\phi - \sin\psi\cos\phi) & (\cos\psi\sin\theta\cos\phi + \sin\psi\sin\phi) \\ (\sin\psi\cos\theta) & (\sin\psi\sin\theta\sin\phi + \cos\psi\cos\phi) & (\sin\psi\sin\theta\cos\phi - \sin\psi\sin\phi) \\ (-\sin\theta) & (\cos\theta\sin\phi) & (\cos\theta\cos\phi) \end{bmatrix} \tag{6}$$

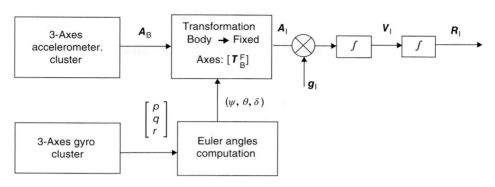

Figure 3. Inertial navigation system mechanisation in fixed axes.

The Euler angles (ψ, θ, ϕ) are computed from the body rates (p, q, r) (i.e., gyro outputs/measurements obtained from the vehicle strap-down IMU) using the following relations

$$\frac{\mathrm{d}}{\mathrm{d}t}\begin{bmatrix} \phi \\ \theta \\ \psi \end{bmatrix} = \begin{bmatrix} 1 & \sin\phi\tan\theta & \cos\phi\tan\theta \\ 0 & \cos\phi & -\sin\phi \\ 0 & \sin\phi\,sec\,\theta & \cos\phi\,sec\,\theta \end{bmatrix}\begin{bmatrix} p \\ q \\ r \end{bmatrix} \quad (7)$$

Equations (3), (4), (5), and (7) can be integrated at each time step to update the position, velocity, and acceleration values of the vehicle $(R_\mathrm{I}, V_\mathrm{I}, A_\mathrm{I})$ and navigate it in the fixed axes. A block diagram of the basic navigation algorithm for the case considered above is given in Figure 3.

1.3.1   Remarks

1. Computing Euler angles using Equation (7) may cause instabilities due to the presence of a discontinuity at $\theta = 90°$; for more efficient and stable methods of computing the transformation matrix, also known as the *direction cosine matrix* may be used (see Titterton and Weston, 2004).
2. For vehicle navigation in the local Earth axes (NED) for an oblate Earth, further transformation is required to convert the position, velocity, and acceleration values to the appropriate navigation frame; also, the Earth's rotation needs to be taken into account.
3. The complete navigation solution, where gyro and accelerometer errors are present, will require the use of a KF.
4. To start the navigation algorithm the initial conditions have to be specified. This is known as *initialization* and is a complex problem that requires either the use of a *static initialization* method known as gyro-compassing or *transfer alignment* where the vehicle INS is "slaved"

to a "master" navigation system that has accurate starting navigation information.

1.4   GPS/INS integration

The need for reliable and accurate navigation, for long-range flights as well as for precision approach and landing systems for aircraft or for achieving the desired aim-point for weapons, has focused attention on the use of coupled or integrated GPS/INS systems. There are a number of ways in which this can be achieved. In one approach, the GPS outputs (vehicle position and velocity) are used to update the outputs of an INS utilizing KF; this type of integration will be referred to as the *loosely coupled* GPS/INS. In the other approach, raw outputs from the GPS (pseudo-range, pseudo-range rate) are used in conjunction with the INS error model to estimate the navigation errors and the corrected states (i.e., estimates of position, velocity, acceleration, attitude, etc.) of the vehicle being navigated. This type of integration will be referred as a *tightly coupled* system. Here again KF or EKF is utilized to implement integration. The second of these techniques is the preferred option as it avoids undesirable coupling with the GPS internal filter and is more robust in case of either GPS or INS failures. GPS/INS integration using this so-called *navigation error-model* allows the estimation and correction of errors caused by the IMU and assists navigation in cases of GPS outages, data degradation, and jamming. The design of a tightly coupled system was considered by Faruqi and Turner (2000); it is complex and requires parameter tuning to obtain an optimal design. One possible manifestation of a tightly coupled GPS/INS is illustrated in Figure 4, where $\omega_\mathrm{B} = (p \quad q \quad r)^\mathrm{T}$, $\Phi_\mathrm{L} = (\psi \quad \theta \quad \phi)^\mathrm{T}$, (B, F) are the GPS clock biases, δ signifies state errors, $(\bar{X}, \hat{X})$ represent nominal and "optimally estimated" values, respectively, for a generic variable X.

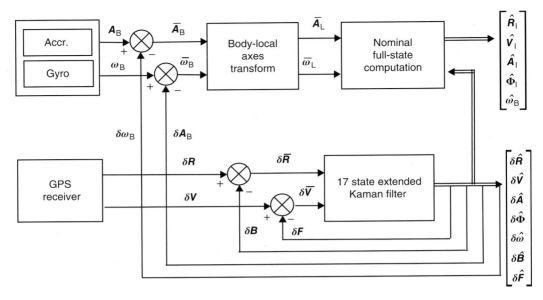

Figure 4. Integrated GPS/INS mechanization.

2   PRINCIPLES AND PRACTICE OF TERPROM AND TERCOM

The INS provides navigation data (body angular rates and accelerations) at a high rate (>50 Hz) that has relatively low short-term noise; however, its accuracy degrades with time due to drifts and biases. GPS on the other hand gives good long-term navigation accuracy, for example, for position (~5 m) and velocity (~0.1 m s^{-1}); however, the measurements are noisy and the attitude information is not available. Adversary jamming or the use of "selective availability" by the provider can deny usable navigation information. In order to circumvent these drawbacks the use of terrain-referenced navigation (TRN) has been proposed that provides yet another independent navigation source to augment GPS and INS. Prominent amongst the TRN systems are the terrain profile matching (TERPROM) system and the terrain contour matching (TERCOM) system.

The TRN system uses the (barometric) height data of the aircraft (or other navigating vehicle) above a given datum (e.g., sea level); its height above the ground (ground clearance), obtained through the use of a radio-altimeter (radalt) is used to construct the ground height above the same datum. The terrain height information is then used to obtain the aircraft position by comparing it with a prestored spatial-height map of the area. There are two ways in which this information can be processed. The first method is to use *batch processing* where a large number of height measurements are collected and then compared with the pre-stored terrain height information (digital map of the terrain); a spatial-height correlation technique is used to find the point

at which the best match is obtained. This is the principle of the TERCOM (Golden, 1980). In TERPROM (Robins, 1988), the matching between the data obtained from the aircraft radalt and the digital map is carried out on individual data points as these become available. In addition, the terrain gradients are utilized to further refine the matching process. In this way small corrections are continually made to aircraft position as it flies over the terrain. The navigation information required, that is, the aircraft position, velocity and attitude, is obtained using a KF (see Figure 5).

TRN offers a number of significant advantages over other forms of navigation systems and is commonly used for navigating strike-attack aircrafts and low-flying cruise missiles. These are

- The relative accuracy of the radalt allows terrain features to be detected with relative ease and these are then used to provide accurate and drift-free navigation. Actual flight tests undertaken confirm this fact. TERPROM has been shown to be instrumental in target acquisition under low-level flight conditions (Robins, 1988), a task which would otherwise not be possible to execute.
- TERPROM in conjunction with forward-looking, terrain-following radar can greatly enhance the terrain-following capability of a strike aircraft since the former provides the ability to "look forward" (using the digital map) and hence optimize the aircraft path and avoid obstacles. This enables the operator to plan the flight path so as to avoid detection by the enemy by utilizing terrain-following and terrain-screening strategies.

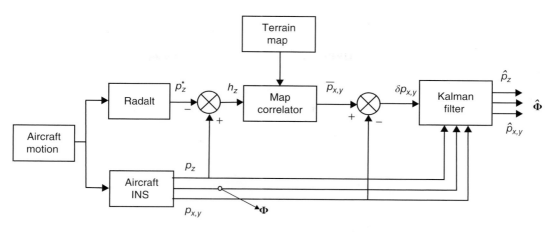

Figure 5. TERPROM-KF implementation using the radalt.

One of the key requirements for a TRN system to operate satisfactorily is that the terrain should be of sufficient definition (roughness) to allow the radalt to detect changes in height profile. Neither TERCOM nor TERPROM will operate satisfactorily over a flat terrain unless certain additional instrumentation is provided. Methods proposed for overcoming this disadvantage and assist in navigation over a flat terrain are the following

1. Augmentation of the radalt data with velocity measurements from a Doppler radar.
2. Use of a visual sensor to obtain imagery of the scene (e.g., roads, buildings and other landmarks) being overflown and match it with a pre-stored scene of the area. This method known as the terrain scene matching (TERSMAC) system gives accurate navigation information and is commonly used for navigating surveillance or combat UAVs as it is particularly suited for urban applications.

The accuracy of TRN can be greatly improved using an active laser line scanner (LLN) system (Groves, 2008; Handley *et al.*, 2003), that not only gives improved height data but also offers the possibility of building up 3D images of the terrain or scene of interest, thus offering a single sensor that can be used for both TERPROM and TERSMAC.

2.1 Aircraft and UAV Path Planning

One of the key elements of design, development and operation of UAVs is that of path planning. Algorithms used for path planning must satisfy a number of competing objectives, for example

- The path must be such as to avoid detection of the UAV by enemy radars, that is, the path to the target must be stealthy. Since the aircraft radar signature is not uniform in all directions, the path planning algorithm should steer the aircraft along a path that avoids signature "spikes" to be detected by enemy radars.
- In addition to stealth, the trajectories should be such as to minimize the path length to the target.
- The path planning algorithm must cater for operation of multiple UAVs working cooperatively against a single or multiple targets.
- The algorithm needs to be computationally efficient and must have the capability to re-plan and generate a new path in case of unforeseen threats or opportunity targets.

A number of different UAV path planning techniques and algorithms have been proposed, one such algorithm (Bortoff, 2000) is summarized here. In this algorithm a stealthy path is generated assuming known positions of enemy radar sites and target positions. Using a constrained optimization technique trade off is achieved between stealth and path length. Initially a suboptimal path is constructed taking into account known radar sites utilizing a graph based on Voronoi polygons. The vertices of the polygon define discrete points in space and the edges define possible paths that an aircraft or the UAV can take provided certain constraints with respect to aircraft orientation (chosen to minimize radar reflection) are satisfied. In order to avoid computational dimensionality problems, the algorithm starts off with a coarse graph with relatively few vertices and edges and searches for an optimum solution. Once the first optimal solution is obtained, a new graph is constructed in the neighborhood of the previous one, and a new search for an optimal solution is started. The process is repeated until an optimum path solution is obtained. Constraints with respect to minimizing radar signature exposure to enemy radars can be embedded in the optimization problem. Simulation results indicate that an optimum trajectory can be generated using this technique which can be implemented in real-time.

3 TACTICAL MISSILE GUIDANCE STRATEGIES

The first missile to have used the proportional navigation (PN) guidance was the US missile, the Lark, tested in December 1950. PN has been adapted for most RF guided and IR guided missiles since 1950 because of its simplicity, effectiveness, and ease of implementation. PN guidance has been studied by a number of authors and its validity has been established using the kinematics of engagement and the optimum control theory; the theoretical foundation of this is given by Sage (1968). Guidance laws such as the pursuit guidance (PG) may be regarded as a subset of the PN guidance. Other variations of the PN, such as the augmented proportional navigation (APN) and the optimal guidance law (OGL), based on the estimated target lateral acceleration have been implemented that allow the interceptor to engage highly maneuvering targets; in the OGL time-varying gains are employed to achieve a significantly reduced miss distance.

Other forms of interceptor guidance include command to line-of-sight (CLOS) and its offshoot, the beam rider (BR). The CLOS and its variants are derived from a slightly different underlying principle than the PN. The ultimate objective of all guidance strategies is to achieve successful intercept with the target. Finally, all the guidance laws mentioned above may be implemented using either the active, semiactive, or passive methods; however, their choice depends on operational requirements, cost, and engagement scenarios. Zarchan (1994) and Garnell and East (1977) have considered in detail the principles and performance characteristics of both PN and CLOS guidance systems. Techniques typically used during the design and analysis phase of guidance systems include the adjoint method, the covariance analysis method, and the Monte Carlo method (Zarhan, 1994; Bucco and Weiss, 2007).

3.1 CLOS Guidance and Variations

The command to line-of-sight guidance is so called because the objective of the guidance command is to force the interceptor to follow the line joining the target tracker (usually located at the interceptor launch point) and the current position of the target. An example of the interceptor trajectory for a typical CLOS guidance engagement is shown

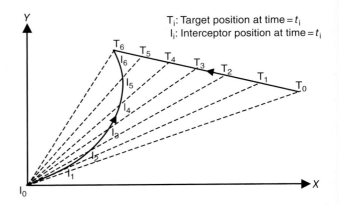

Figure 6. Typical CLOS guidance engagement trajectory.

in Figure 6; corresponding positions of the target and interceptor are indicated by subscript "i" as the target–interceptor engagement evolves until impact.

In order to derive the CLOS guidance equations, we consider the target–interceptor kinematics shown in Figure 7.

It follows directly from consideration of the target-interceptor kinematics in Figure 7 that in order for a successful intercept to occur: (i) the distance off beam (DOB) or the deviation of the interceptor from the target LOS must be nulled and (ii) the interceptor must successfully follow the target LOS; that is, once having nulled the DOB, the interceptor must apply additional acceleration perpendicular to the target LOS to keep up with it at a point P, where $OP = R_I$; $\delta\psi$ is small, and $\psi_I \approx \psi_T$. This acceleration is the Coriolis acceleration (A_P) required by the interceptor to keep up with the target LOS and can be derived (for the interceptor located at point P) as follows

$$R_I \dot{\psi}_T = V_I \sin(\psi_T + \theta_I) \tag{8}$$

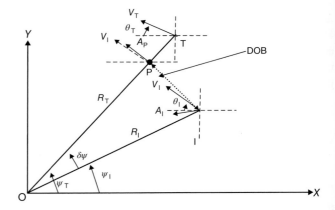

Figure 7. CLOS guidance engagement kinematics.

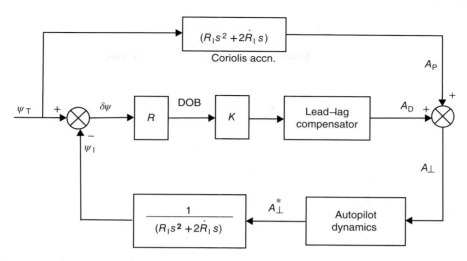

Figure 8. CLOS guidance block diagram.

$$\dot{R}_I = -V_I \cos(\psi_T + \theta_I) \qquad (9)$$

Here, Equations (8) and (9) define, respectively, the velocities perpendicular and parallel to the target LOS. Differentiating these equations with respect to time, and after some algebraic manipulation, we get

$$R_I \ddot{\psi}_T + 2\dot{R}_I \dot{\psi}_T = V_p \dot{\theta}_T \cos(\psi_T + \theta_I) + \dot{V}_p \sin(\psi_T + \theta_I) = A_P \qquad (10)$$

In terms of the Laplace operator (s), this equation may be written as

$$A_P = \left(R_I s^2 + 2\dot{R}_I s \right) \psi_T \qquad (11)$$

If we now define $A_\perp$ as the total interceptor acceleration demanded perpendicular to the target LOS required to null the DOB and to follow the LOS then

$$A_\perp = A_D + A_P \qquad (12)$$

Assuming that the interceptor achieves the necessary acceleration $A_\perp^*$, that enables it to stay on or near to the target LOS then rotation (ψ_I) of the interceptor LOS, is given by

$$A_\perp^* = R_I \ddot{\psi}_I + 2\dot{R}_I \dot{\psi}_I \qquad (13)$$

$$\rightarrow \psi_I = \frac{A_\perp}{\left(R_I s^2 + 2\dot{R}_I s \right)} \qquad (14)$$

Note that the demanded acceleration $A_\perp$ is the acceleration demand into the autopilot while achieved acceleration $A_\perp^*$ is the output of the autopilot, there being a dynamic time-delay

between the two, for example a first order approximation autopilot dynamics could be characterized by a transfer function of the form

$$\frac{A_\perp^*}{A_\perp} = \frac{1}{1 + sT_A} \qquad (15)$$

Using Equations (8)–(15), a block diagram of the CLOS guidance may be constructed; this is given in Figure 8. The object of the lead–lag compensator is to increase the guidance loop bandwidth (i.e., speed up the guidance response), K is the "beam-stiffness" gain, and the Coriolis operator $\left(R_I s^2 + 2\dot{R}_I s \right)$ provides the feed forward acceleration term that enables the interceptor to stay on the target LOS beam.

In the CLOS guidance implementation usually two beams (either RF or laser) are employed; one to track the target and the other to track the interceptor. The guidance commands are then passed on to the interceptor to enable it to catch up with the target LOS and follow it until intercept. A variant of the CLOS is the BR guidance that utilizes only one beam, the target LOS. The interceptor once having been gathered on to this beam then detects its DOB utilizing passive detectors mounted on the interceptor and corrects for this in the same way as the CLOS. In the BR, however, since no guidance commands are issued from the ground to the interceptor, the correction for the Coriolis acceleration is not generally possible. The BR trajectory is similar to the CLOS except for the fact that the interceptor now "sits" slightly below (behind) the target LOS due to its inability to correct for the Coriolis acceleration. For both CLOS and BR the miss distance (MD) performance is highly dependent on the target velocity, with MD increasing as the target velocity is increased. CLOS performs somewhat better than BR. In both CLOS and BR the interceptor acceleration demand increases as the engagement proceeds and may even exceed the interceptor limit at the end-game. Improvement in the MD performance can be achieved by appropriately tuning the guidance loop through the lead–lag network and the beam-stiffness gain K. However, for a highly responsive (fast) guidance loop CLOS guidance becomes noise at the end-game. The MD performance of both the CLOS and the BR with target maneuvers is highly dependent on the engagement geometry and the interceptor acceleration limit. If the acceleration limit of the interceptor is significantly higher than the target maneuver accelerations, then a successful intercept can be achieved.

3.2 Proportional Navigation (PN) Guidance

The proportional navigation (PN) guidance design is based on the principle that if the interceptor acts to null the target-interceptor LOS rate, and in fact achieves it, then a collision (intercept) will result.

A typical target-interceptor engagement geometry utilised in the derivation and analysis of the PN guidance is depicted in Figure 9. Here, I, T define the interceptor and the target positions, respectively, at time t, Note that

$$u_\mathrm{T} = V_\mathrm{T}\cos\theta_\mathrm{T} \quad v_\mathrm{T} = V_\mathrm{T}\sin\theta_\mathrm{T} \tag{16}$$

$$u_\mathrm{I} = V_\mathrm{I}\cos\theta_\mathrm{I} \quad v_i = V_\mathrm{I}\sin\theta_\mathrm{I} \tag{17}$$

$$a_{Tx} = -A_T\cos\theta_T \quad a_{Ty} = A_T\sin\theta_T \tag{18}$$

$$a_{Ix} = -A_\mathrm{I}\cos\theta_\mathrm{I} \quad a_{\mathrm{I}y} = A_\mathrm{I}\sin\theta_\mathrm{I} \tag{19}$$

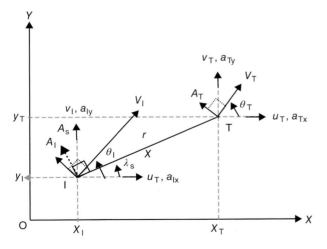

Figure 9. PN guidance engagement kinematics.

Mathematical expressions for the LOS angle (λ_s), LOS rate $(\dot\lambda_\mathrm{s})$, interceptor–target separation range (r), and the closing velocity $(V_\mathrm{c} = -\dot r)$ are given by

$$\lambda_\mathrm{s} = \tan^{-1}\left(\frac{y_\mathrm{T} - y_\mathrm{I}}{x_\mathrm{T} - x_\mathrm{I}}\right) = \tan^{-1}\left(\frac{y}{x}\right) \tag{20}$$

$$\Longrightarrow \dot\lambda_\mathrm{s} = \left[\frac{(v_\mathrm{T} - v_\mathrm{I})(x_\mathrm{T} - x_\mathrm{I}) - (u_\mathrm{T} - u_\mathrm{I})(y_\mathrm{T} - y_\mathrm{I})}{r^2}\right]$$
$$= \left\{\frac{vx - uy}{r^2}\right\} \tag{21}$$

$$r = \left[(y_\mathrm{T} - y_\mathrm{I})^2 + (x_\mathrm{T} - x_\mathrm{I})^2\right]^{\frac{1}{2}} = \left\{y^2 + x^2\right\}^{\frac{1}{2}} \tag{22}$$

$$\Longrightarrow$$
$$V_\mathrm{c} = -\dot r$$
$$= -\left[\frac{(v_\mathrm{T} - v_\mathrm{I})(y_\mathrm{T} - y_\mathrm{I}) + (u_\mathrm{T} - u_\mathrm{I})(x_\mathrm{T} - x_\mathrm{I})}{r}\right]$$
$$= -\left[\frac{vy + ux}{r}\right] \tag{23}$$

The PN guidance is so called because the steering commands, in terms of the acceleration (A_s) perpendicular to the LOS, applied to the interceptor is *proportional* to line-of-sight rate $(\dot\lambda_\mathrm{s})$ and the missile-target closing velocity (V_c). Mathematically, the guidance law may be written as

$$A_\mathrm{s} = NV_\mathrm{c}\dot\lambda_\mathrm{s} \tag{24}$$

The PN guidance law Equation (24) in terms of the interceptor lateral acceleration, may be written as

$$A_\mathrm{I} = A_\mathrm{s}\cos(\theta_\mathrm{I} - \lambda_\mathrm{s}) = NV_\mathrm{c}\dot\lambda_\mathrm{s}\cos(\theta_\mathrm{I} - \lambda_\mathrm{s}) = N'V_\mathrm{c}\dot\lambda_\mathrm{s} \tag{25}$$

Here, $N' = N\cos(\theta_\mathrm{I} - \lambda_\mathrm{s})$. For a constant velocity interceptor Equation (25) is equivalent to

$$\dot\theta_\mathrm{I} = N''\dot\lambda_\mathrm{s} \tag{26}$$

Here, $N'' = \frac{NV_\mathrm{c}}{V_\mathrm{I}}\cos(\theta_\mathrm{I} - \lambda_\mathrm{s})$. An expression of the form of Equation 26 has been quoted as the PN guidance by a number of authors. From Figure 9, it is seen that the interceptor heading with respect to the LOS is given by the angle $\theta_\mathrm{L} = (\theta_\mathrm{I} - \lambda_\mathrm{s})$ (also referred to as the look angle or the lead angle). For a collision course the components of the velocities normal to the LOS, of interceptor and target must be equal, that is:

$$V_I \sin(\theta_I - \lambda_s) = V_T \sin(\theta_T + \lambda_s) \qquad (27)$$

$$\rightarrow \theta_L = (\theta_I - \lambda_s) = \sin^{-1}\left\{\frac{V_T \sin(\theta_T + \lambda_s)}{V_I}\right\} \qquad (28)$$

An interceptor following the PN guidance is fired along the collision course heading (θ_L), however due to errors in locating the target, interceptor seeker errors, and target manoeuvres, heading errors are introduced; the PN guidance then acts to null the effect of these errors.

3.3 Miss Distance (MD)

The target–interceptor miss distance (MD) is caused by a number of error sources noted above as well as dynamic lag in the interceptor autopilot. It will be defined as the minimum target–interceptor separation (r_{min}) that occurs during the engagement. *It can be shown that this condition occurs when this relative range vector is perpendicular to the relative velocity vector.* The time (t_f) at which the minimum target–interceptor separation occurs is the total flight time, and hence in the general case the MD = $r_{min} = r(t_f)$.

The analysis of PN guidance is conducted using a computer simulation model using a 2D interceptor-target model defined by the set of Equations (16)–(28). The simulation inputs may be taken as initial target and interceptor position and velocities, flight time and navigation constants; and the effect on these variables of target maneuver and heading error can be studied.

The relationships between the variables of interest such as $(r, \dot{r}, \lambda, \dot{\lambda})$ and (x, y, u, v) in Equations (16)–(28) are non-linear in nature. It turns out, however, that in a number of practical engagements of interest certain simplifying assumptions can be made that not only adhere closely to the non-linear model but also renders the PN guidance easier to understand and analyze. For the linearized analysis it is assumed that the target maneuver is small and that the target-interceptor velocities remain constant ($u = -V_c$ = constant) and that ($\dot{\lambda}$) is sufficiently small such that: (λ) = constant. We may therefore choose the x-axis of our axes system of Figure 9 to coincide with the initial LOS. Thus we may assume that ($r \approx x, \lambda \approx 0$) and the target–missile separation is equal to (y) and MD = $y(t_f)$. Under these assumptions, the following relationships may be derived

$$r = x \rightarrow V_c = -\dot{r} = -u \qquad (29)$$

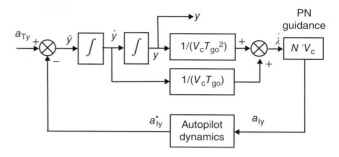

Figure 10. PN guidance system block diagram.

$$\lambda_s = \frac{y}{r} \quad \rightarrow \quad \dot{\lambda}_s = \left\{\frac{vr + V_c y}{r^2}\right\} = \frac{1}{V_c}\left[\frac{y}{T_{go}^2} + \frac{\dot{y}}{T_{go}}\right] \qquad (30)$$

where

$$r(t) = \int_{t_f}^{t} \dot{r}\, dt = V_c(t_f - t) = V_c T_{go} \qquad (31)$$

(Note that $\dot{y} = v$; $\ddot{y} = a_y$)

3.3.1 Performance characteristics of PN guidance system

A block diagram for the linearized analysis of the PN guidance is given in Figure 10. It allows the designer to undertake relatively straightforward simulations to study the effect, on the MD, of parameters such as the navigation constant, and target maneuver discussed earlier. A parameter that has a profound effect on the interceptor MD performance is the autopilot dynamics as defined by the autopilot time-constant and the order (i.e., first, second, third order, etc.) of the autopilot transfer function. Other interceptor parameters that affect MD performance include interceptor acceleration limit (or saturation effect), and the interceptor seeker radome aberration (also known as the parasitic effect).

Before we leave this section we mention two well-known variants of the PN guidance law particularly suited for engaging highly maneuvering targets. These are the augmented PN (APN) law and the optimum guidance law (OGL) (see, e.g., Zarchan (1994)). These guidance laws may be stated as

$$\text{(a) APN} \rightarrow A_I = N' V_c \dot{\lambda}_s + k_1 \hat{A}_T \qquad (32)$$

where k_1 is the *APN guidance constant*; and $\hat{A}_T$ is the estimated value of the target acceleration (usually obtained through the use of the Kalman filter.

(b)   OGL → A_I

$$= N' \left[V_c \dot{\lambda}_s + K_2 (T_{go}) \hat{A}_T + K_3 (t, T_{go}) A_I^* \right] \qquad (33)$$

3.3.2   Active, semiactive and passive homing

The guidance mode or method of implementing the guidance refers to the manner in which the interceptor acquires the target in order to generate the guidance commands for interception. The target location in terms of the angle or the angular rate of the LOS and range and range rate (closing velocity) are generated by the seeker mounted on-board the interceptor or via the ground tracker. There are a number of different ways in which target acquisition and tracking can be implemented and the information generated by the seeker utilized for computing the interceptor guidance laws.

Active mode guidance. In the active mode the seeker mounted on board the interceptor is used to illuminate the target and the return signals from the target contain information of the target location, speed, and other target states that allow a guidance law to be implemented. Active guidance mode includes RF or active laser seekers; an example of an active RF-guided missile is the AMRAAM.

Semiactive mode guidance. In the semiactive mode the seeker mounted on board the interceptor receives return signals from the target that is illuminated by a radar or a laser source that reside elsewhere; such as on the ground or a ship or a third-party aircraft. Information on the target location, target speed, and other target states are derived from target returns for the guidance law to be implemented. Examples of semiactive homing guidance include the SM-2 ship area defense missile and the Paveway laser-guided bomb.

Passive mode guidance. In the passive homing guidance no artificial illumination of the target occurs but rather the natural radiation (such as in visual or IR band) reflected or emanating from the target is utilized to acquire the target. The target acquisition is similar to the way a human eye views an objects as a result of the Sun's reflection from that object. The most common applications of this mode are the infra-red (IR) and imaging infra-red (I^2R) seekers. These seekers have the ability to detect radiations from objects at different wavelengths; and since different objects at different temperatures have different characteristic spectra, objects of interest can be isolated from their background by sophisticated signal processing methods. Information from passive target tracking can directly or after processing provide sufficient information to implement guidance laws. An example of a passive guided missile is the ASRAAM.

Multi-mode guidance. Current research indicates that for complex scenarios a mixed mode, say a combination of an active RF seeker and an imaging IR seeker can be used to achieve highly accurate guidance outcomes. Different combinations of modes are possible to satisfy mission requirements.

4   CONCLUSIONS

In this chapter, techniques and technologies used for weapon, UAV, and aircraft navigation and guidance were presented including method for analysis and synthesis of these systems. Vehicle navigation techniques such as INS and GPS for missiles and UAV were discussed including methods for deriving navigation information from GPS and INS and integrated GPS/INS. The application and principles of TRN systems such as TERPROM, TERCOM, and TERSMAC were described. Aircraft and UAV path planning was considered and a method for generating an optimum path to achieve a stealthy mission through an enemy territory was described. Guided-weapon autonomous guidance strategies commonly used today such as the CLOS and PN guidance were described and methods for analysis and synthesis of these guidance techniques were derived and their performance discussed. Different guidance phases of weapons during target engagements were discussed and the various guidance modes (active, semiactive, and passive) were described. This chapter should assists the reader in understanding the basic principle involved in techniques and technologies commonly used in analysis and design of guided weapons and missiles.

NOTATION

$A_I = \begin{pmatrix} a_{Ix} & a_{Iy} & a_{Iz} \end{pmatrix}^T$ acceleration vector of the vehicle with respect to the fixed axes

A_B acceleration vector if the vehicle in body axes (i.e., the output/measurement obtained from the vehicle strap-down IMU)

A_D interceptor acceleration (perpendicular to line-of-sight [LOS]) required to null DOB

$A_I^* = \left[\dfrac{A_I}{1 + sT} \right]$ Achieved interceptor acceleration taking into account the autopilot time-constant(T)

A_s — interceptor acceleration demanded perpendicular to LOS

(a_{Ix}, a_{Iy}) — x-acceleration and y-acceleration of the interceptor

$a_x = a_{Tx} - a_{Ix}$ — target–interceptor relative x-acceleration

$a_y = a_{Ty} - a_{Iy}$ — target–interceptor relative y-acceleration

(a_{Tx}, a_{Ty}) — x-acceleration and y-acceleration of the target

(B, F) — GPS clock biases

$DOB = R_I \delta\psi$ — distance off-beam for small $\delta\psi$

g_I — gravity vector in fixed axes

$\left[K_2(T_{go}); \; K_3(T_{go}) \right]$ — Time-varying gains functions of time and time-to-go (t, T_{go})

h_z — terrain height above datum

N — Uni-less designer chosen gain (usually in the range of 3–5), known as the navigation constant (or the effective navigation ratio)

$\hat{p}_{x,y}$ — KF estimates of the x, y positions of the aircraft

$\bar{p}_{x,y}$ — map-generated x, y positions of the aircraft

$p_{x,y}$ — INS-generated x, y positions of the aircraft

$\hat{p}_z$ — KF estimate of the z positions (height) of the aircraft above datum

p_z — INS-generated z positions (height) of the aircraft above datum

p_z^* — Radalt generated z positions (height) of the aircraft above the terrain

$\delta p_{x,y}$ — error between the map generated and the INS-generated x,y position of the aircraft

$R_{Si} = \begin{pmatrix} x_{Si} & y_{Si} & z_{Si} \end{pmatrix}^T$ — position vector of the ith satellite with respect to the fixed axes

(R_T, R_I) — range to the target and the interceptor

$R_i = R_{Si} - R_I$ — relative position of the vehicle with respect to the ith satellite

$R_I = \begin{pmatrix} x_I & y_I & z_I \end{pmatrix}^T$ — position vector of the vehicle with respect to the fixed axes

$r \left(t_f = \dfrac{r(0)}{V_c} \right)$ — interceptor-target separation range flight time

$T_{go} = (t_f - t)$ — time-to-go

$\left[T_B^F \right]$ — 3×3 transformation matrix that transforms a given vector from body axes to fixed axes

$U_i = U_{Si} - U_I = \dfrac{R_i}{\|R_i\|}$ — unit-pointing vector from the vehicle to the ith satellite

(u_I, v_I) — x-velocity and the y-velocity of the interceptor

(u_T, v_T) — x-velocity and the y-velocity of the target

$u = u_T - u_I$ — target-interceptor relative x-velocity

$v = v_T - v_I$ — target-interceptor relative y-velocity

$V_I = \begin{pmatrix} u_I & v_I & w_I \end{pmatrix}^T$ — velocity vector of the vehicle with respect to the fixed axes

$V_i = V_{Si} - V_I$ — Relative velocity of the vehicle with respect to the ith satellite

$V_{Si} = \begin{pmatrix} u_{Si} & v_{Si} & w_{Si} \end{pmatrix}^T$ — Velocity vector of the ith satellite with respect to the fixed axes

(V_T, V_I) — target and the interceptor velocities

(V_I, A_I) — longitudinal velocity and lateral acceleration of the interceptor

(V_T, A_T) — longitudinal velocity and acceleration of the target

(x_I, y_I) — x-position and the y-position of the interceptor

(x_T, y_T) — x-position and the y-position of the target

$x = x_T - x_I$ — target–interceptor relative x-position

$y = y_T - y_I$ — target–interceptor relative y-position

$\Phi = \begin{pmatrix} \psi & \theta & \phi \end{pmatrix}^T$ — Euler angles

$\omega_B = \begin{pmatrix} p & q & r \end{pmatrix}^T$ — gyro (rates) outputs

(θ_T, θ_I) — target and interceptor velocity angles with respect to the x-axis

(ψ_T, ψ_I) — LOS angle to the target and the interceptor

$\delta\psi = \psi_T - \psi_I$ — differential LOS angle

(θ_I, θ_T) — interceptor and target heading

λ_s — interceptor–target LOS

NOMENCLATURE

APN	augmented proportional navigation
BR	beam rider
CLOS	command to line-of-sight
CM	countermeasures
ECI	Earth-centered inertial
EKF	extended Kalman filter
FOG	Fiber optics gyro
GPS	global positioning system
IMU	inertial measurement unit
INS	inertial navigation system
KF	kalman filter
LLS	laser line scanner
LOS	line-of-sight
LSE	least-square estimation
MD	miss distance
MEMS	micro-electro-mechanical system
NED	(North, East down), local Earth axes
OGL	optimal guidance law
PN	proportional navigation
SDS	strap-down system
SOC	satellite orbit coordinates
SPS	stabilized platform system
TERPROM	terrain profile matching
TERCOM	terrain contour matching
TERSMAC	terrain scene matching
TRN	terrain referenced navigation
UTC	coordinated Universal Time

REFERENCES

Bortoff, S.A. (2000) Path Planning for UAV. *Proceedings of the American Control Conference, Chicago, IL*, pp. 364–368.

Bucco, D. and Weiss, M. (2007) Development of a Matlab/Simulink Tool to Facilitate System Analysis and Simulation via the Adjoint and Covariance Methods. *AIAA MST Conference*.

Faruqi, F.A. and Turner, K.J. (2000) Extended Kalman filter synthesis for integrated global positioning/inertial navigation systems. *J. Appl. Math. Comput.*, **115**, 213–227.

Garnell, P. and East, D.J. (1977) *Guided Weapon Control Systems*, Pergamon Press, Oxford.

Golden, P.J. (1980) Terrain contour matching (TERCOM): a cruise missile guidance aid. *Proc. SPIE Conf.*, **238**, 10–17.

Grewal, M.S., Weill, L.R. and Andrews, A.P. (2001) *Global Positioning Systems, Inertial Navigation and Integration*, John Wiley & Sons, New York.

Groves, P.D. (2008) *Principles of GNSS, Inertial, and Multisensor Integrated Navigation Systems*, Artech House, London.

Handley, R.J., Grover, P.D., McNeil, and Dack, L., (2003) Future terrain reference navigation techniques exploiting sensor synergy. *European Navigation Conference*, GNSS 2003, Graz, April 22–25, 2003.

Jazwinski, A.H. (1998) *Stochastic Processes and Filtering Theory*, Dover Publications, Mineola, NY.

Robins, A.J. (1988) Recent developments in the "TERPROM" integrated navigation system. *Proc. Inst. Navigat.*, **44**, 58–66.

Sage, A.P. (1968) *Optimum Systems Control*, Prentice Hall, Englewood NJ.

Titterton, D.H. and Weston, J.L. (2004) Strapdown inertial navigation technology, in *IEE Radar, Sonar, Navigation Series* (eds N. Stewart and H. Griffiths), IEE and AIAA, p. 17.

Zarchan, P. (1994) Tactical and strategic missile guidance, in *Progress in Astronautics and Aeronautics* (ed. A.R. Seebass), AIAA, Washington, DC, p. 157.

Chapter 19

Embedded UAS Autopilot and Sensor Systems

Randal W. Beard

Department of Electrical and Computer Engineering, Brigham Young University, Provo, UT, USA

1 INTRODUCTION

Unmanned air systems (UAS) are playing an increasingly prominent role in military operations. Technology advancements have enabled the development of large UAS like the Northrop Grumman Global Hawk and the General Atomics Predator, and also smaller UAS like the AeroVironment Raven and the *in situ* ScanEagle. As recent conflicts have demonstrated, there are numerous military applications for small UAS' including reconnaissance, surveillance, battle damage assessment, and communications relays. Civil and commercial applications are not as well developed, although potential applications are extremely broad in scope. Possible applications for UAS technology include environmental monitoring, forest fire monitoring, homeland security, border patrol, drug interdiction, aerial surveillance and mapping, traffic monitoring, precision agriculture, disaster relief, ad-hoc communications networks, and rural search and rescue.

The design of the autopilot system for a small UAS differs from manned aircraft in several ways. The first obvious difference is that a human pilot is not in the loop. This has several important implications. In particular, the pilot is not able to set trim conditions, which must therefore be determined automatically by the autopilot. Additional functions that are typically performed by the autopilot include take-off and landing and waypoint navigation. The second difference between small UAS and larger aircraft is the type of sensors that are available. Large aircraft typically have high quality inertial measurement unit (IMU), angle-of-attack sensors, and GPS. For a small UAS, the sensor suite is limited to global position system (GPS), microelectromechanical systems (MEMS) gyros and accelerometers, and differential and absolute pressure sensors. The implication is that estimates of the states of the aircraft will be of much poorer quality with small UAS. Another difference is that the airspeed and the wind speed are more closely aligned for small UAS than they are for larger aircraft. The implication is that navigation algorithms must be explicitly designed for high wind situations. Finally, small UAS have much less payload capacity than larger aircraft and can therefore not carry significant computational resources on-board. The consequence is that the autopilot algorithms must have relatively modest computational and memory requirements.

There are several commercially available autopilots for small UAS. These include the Kestrel Autopilot produced by Procerus Technologies (http://procerusuav.com), the Piccolo Autopilot produced by Cloud Cap Technologies (http://

Unmanned Aircraft Systems. Edited by Ella Atkins, Aníbal Ollero,
Antonios Tsourdos, Richard Blockley and Wei Shyy.
© 2016 John Wiley & Sons, Ltd. ISBN: 978-1-118-86645-0.

Figure 1. (a) TheKestrel autopilot. (Image courtesy of Procerus Technologies.) (b) The Piccolo II. (Image courtesy of Cloud Cap Technology, Inc.) (c) The MP2128. (Image courtesy of Micropilot, Inc.)

Table 1. Size, weight, and power specifications for the Kestrel, Piccolo II, and MP2128 autopilots.

	Kestrel	Piccolo II	MP2128
Size (in)	$2 \times 1.37 \times 0.47$	(includes datalink) $5.6 \times 1.8 \times 0.24$	$1.5 \times 4 \times 0.5$
Weight (g)	17	226	28
Power (W)	2	4	1

www.cloudcaptech.com), and the MP2128 Autopilot produced by MicroPilot Inc (http://www.micropilot.com). These autopilots are shown in Figure 1 and some of their specifications are shown in Table 1.

The objective of this chapter is to describe autopilot technologies for small UAS. The principles will be common to all autopilots in this size and weight class. In Section 2 we will describe a general autopilot architecture. In Section 3 we will describe low-level control design using successive loop closure. Section describes sensors that are typically on-board small UAS and sensor processing necessary to extract state information, and Section 5 discusses GPS navigation.

2   AUTOPILOT ARCHITECTURE

A typical system architecture for a small UAS is shown in Figure 2. The UAS is composed of a ground component and an air component. The air system includes the autopilot and a camera, which is often gimbaled. The autopilot communicates to the ground component via two communication links: one for telemetry and command and control, and the other for video. Typically, a 900 MHz transmitter is used for the command and control link, and a 2.4 GHz transmitter is

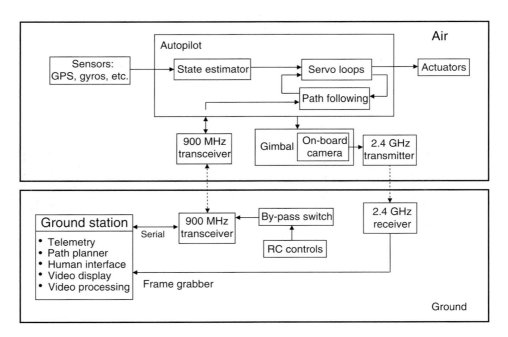

Figure 2. Autopilot architecture for small UAS. The architecture includes a ground component and an air component.

used for the video link. The autopilot software consists of a state estimator, servo loops, and a path following module.

The ground system includes a ground station, the corresponding communication hardware, and a mechanism to allow radio control (RC) override. The RC override is for safety and regulatory requirements. The ground station typically includes telemetry recording, video display, a human interface, and possibly a waypoint path planning module.

This chapter focuses primarily on the autopilot software. In Section 3 we discuss the servo loops. The input to the servo loops is the estimated state $\hat{x}$ as well as the commanded altitude h^c, airspeed V^c, pitch angle θ^c, and heading angle ψ^c. The servo loops send commands to the actuators which include a rudder δ_r, aileron δ_a, elevator δ_e, and throttle δ_t. The path following module receives commands from the ground station to go to a specified waypoint, or to orbit a specific point on the ground at a certain radius and airspeed. Section 5 discusses the path following module. The autopilot and the path following blocks rely on accurate state estimates that are obtained by filtering the on-board sensors that include accelerometers, rate gyros, pressure sensors, and GPS. The state estimator is discussed in Section 4.

3   INNER-LOOP CONTROL STRUCTURE

The equations of motion for fixed-wing aircraft are typically decomposed into lateral and longitudinal dynamics (Blakelock, 1991; Nelson, 1998; Etkin and Reid, 1996; Roskam, 1998). The lateral dynamics include the roll angle, the inertial heading direction (measured from North), and the inertial North–East position of the aircraft. The longitudinal dynamics include the pitch angle, the airspeed, and the altitude of the aircraft. Feedback loops are designed separately for the lateral and longitudinal dynamics. While there is coupling between lateral and longitudinal motion, the feedback loops are designed to reject the disturbances that are due to this coupling. Design of the feedback control loops for lateral motion is discussed in Section 3.1 and for longitudinal motion in Section 3.2. The autopilot design discussed in this chapter would not be appropriate for highly aggressive maneuvers where the coupling between lateral and longitudinal motion must be directly addressed.

Figure 3 depicts the definitions of the variables used throughout the rest of the chapter. The body frame x-axis is out the nose of the aircraft, the y-axis is out the right wing, and the z-axis is out the bottom. The rotation rates about the body frame axes are denoted by p, q, and r, and are called the roll rate, pitch rate, and yaw rate. The yaw angle ψ is defined by aligning the body axis with the inertial North–East-Down (NED) axes and rotating about the body z-axis (Down axis)

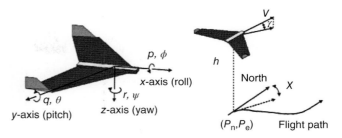

Figure 3. Definition of variables used throughout this chapter. The angular rates are given by p, q, and r. The Euler angles are denoted as ϕ, θ, ψ. The airspeed is V, the heading angle is χ and the flight path angle is γ. The inertial position is (p_n, p_e), and the altitude is h.

by ψ. The pitch angle is defined by rotating the resulting coordinate system about the body y-axis by θ. The roll angle is defined by rotating the yawed and pitched coordinate system about the body x-axis by ϕ. The North–East-Down position of the aircraft is denoted by (p_n, p_e, p_d). The altitude is measured along the negative Down axis and is denoted by $h = -p_d$. The airspeed of the airframe, which is defined as the speed of the aircraft relative to the surrounding air mass, is V. The flight path angle, which is the angle that the velocity vector makes with the North–East plane is given by γ and is equal to the pitch angle minus the aircraft angle of attack: $\gamma = \theta - \alpha$. The course angle χ is the angle of the velocity vector from North and is equal to the yaw angle plus the crab angle: $\chi = \psi + \chi_c$, where the crab angle is the angle between the inertial velocity vector of the aircraft, and the body frame x-axis which points out the noise of the aircraft. Note that the crab angle is different than the sideslip angle, which is the difference between the relative wind vector and the body frame x-axis. The lateral dynamics involve the variables p, ϕ, r, ψ, χ, p_n, and p_e. The longitudinal dynamics involve the variables q, θ, χ, V, and h.

3.1   Lateral autopilot

Figure 4 shows the block diagram for a lateral–directional autopilot using successive loop closure. There are five gains associated with the lateral–directional autopilot. The derivative gain k_{d_ϕ} provides roll rate damping at the innermost loop. The roll attitude is regulated with the proportional and integral gains k_{p_ϕ} and k_{i_ϕ}. The heading is regulated with the proportional and integral gains k_{p_χ} and k_{i_χ}. The idea with successive loop closure, is that the gains are successively chosen one at a time beginning with the inner loop and working outward. In particular, k_{d_ϕ} is usually selected first, k_{p_ϕ} second, k_{i_ϕ} third, k_{p_χ} fourth, and k_{i_χ} last.

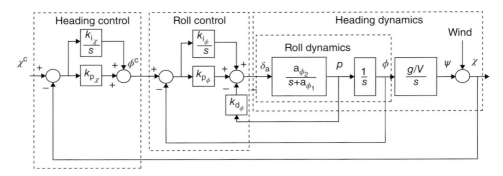

Figure 4. Control loops for the lateral autopilot using successive loop closure.

If a model of the dynamics is available, then the gains can be selected analytically. From Figure 4, note that if the integrator gain $k_{i_\phi} = 0$, then the transfer function from the commanded roll angle to the roll angle is given by

$$H_{\phi/\phi^c}(s) = \frac{k_{p_\phi} a_{\phi_2}}{s^2 + (a_{\phi_1} + a_{\phi_2} k_{d_\phi})s + k_{p_\phi} a_{\phi_2}} \quad (1)$$

If the desired response is given by the canonical second-order transfer function

$$\frac{\omega_{n_\phi}^2}{s^2 + 2\zeta_\phi \omega_{n_\phi} s + \omega_{n_\phi}^2} \quad (2)$$

then equating denominator polynomial coefficients, we obtain

$$\omega_{n_\phi}^2 = k_{p_\phi} a_{\phi_2} \quad (3)$$

$$2\zeta_\phi \omega_{n_\phi} = a_{\phi_1} + a_{\phi_2} k_{d_\phi} \quad (4)$$

Solving these expressions for k_{p_ϕ} and k_{d_ϕ} gives

$$k_{p_\phi} = \frac{\omega_{n_\phi}^2}{a_{\phi_2}} \quad (5)$$

$$k_{d_\phi} = \frac{2\zeta_\phi \omega_{n_\phi} - a_{\phi_1}}{a_{\phi_2}} \quad (6)$$

Therefore, selecting the desired damping ratio and natural frequency fix the values for k_{p_ϕ} and k_{d_ϕ}.

Since the roll dynamics is a type one system, zero steady-state tracking error in roll should be achievable without an integrator (Franklin, Powell, and Emami-Naeini, 2002). However, coupling from the longitudinal dynamics introduces a disturbance that enters at the summing junction before δ_a in Figure 4. Disturbances are also introduced into the roll

dynamics by gusts or turbulence. To reject these disturbances, an integrator must be included. If a_{ϕ_1} and a_{ϕ_2} are known, then k_{i_ϕ} can be effectively selected using root locus techniques. The closed-loop poles of the system are given by

$$s^3 + (a_{\phi_1} + a_{\phi_2} k_{d_\phi})s^2 + a_{\phi_2} k_{d_\phi} s + a_{\phi_2} k_{i_\phi} = 0 \quad (7)$$

which can be placed in Evans form as

$$1 + k_{i_\phi} \left[\frac{a_{\phi_2}}{s(s^2 + (a_{\phi_1} + a_{\phi_2} k_{d_\phi})s + a_{\phi_2} k_{d_\phi})} \right] = 0 \quad (8)$$

Figure 5 shows the root locus of the characteristic equation plotted as a function of k_{i_ϕ}. For small values of gain, the system remains stable.

The next step in the successive-loop-closure design of the lateral–directional autopilot is to design the heading hold outer loop. If the inner loop from ϕ^c to ϕ has been adequately tuned, then $H_{\phi/\phi^c} \approx 1$ over the range of frequencies from 0 to

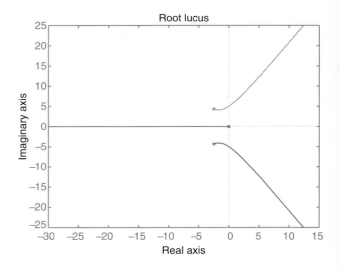

Figure 5. Roll loop root locus as a function of the integral gain k_{i_ϕ}.

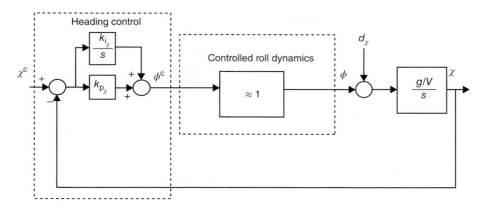

Figure 6. Heading hold outer feedback loop.

ω_{n_ϕ}. Under this condition, the block diagram of Figure 4 can be simplified to the block diagram in Figure 6 for the purposes of designing the outer loop, where the wind disturbance has been transformed to the input of the plant.

The objective of the heading hold design is to select k_{p_χ} and k_{i_χ} in Figure 6 so that the heading angle χ asymptotically tracks steps in the commanded heading angle χ^c. From the simplified block diagram, the transfer functions from the inputs χ^c and the input disturbance d_χ to the output χ are given by

$$\chi = \frac{(g/V)s}{s^2 + k_{p_\chi}(g/V)s + k_{i_\chi}(g/V)}d_\chi$$
$$+ \frac{k_{p_\chi}(g/V)s + k_{i_\chi}(g/V)}{s^2 + k_{p_\chi}(g/V)s + k_{i_\chi}(g/V)}\chi^c \tag{9}$$

Note that if d_χ and χ^c are constants, then the final value theorem implies that $\chi \to \chi^c$. The transfer function from χ^c to χ has the form

$$H_\chi = \frac{2\zeta_\chi\omega_{n_\chi}s + \omega_{n_\chi}^2}{s^2 + 2\zeta_\chi\omega_{n_\chi}s + \omega_{n_\chi}^2} \tag{10}$$

As with the inner feedback loops, we can choose the natural frequency and damping of the outer loop and from those values calculate the feedback gains k_{p_χ} and k_{i_χ}. Figure 7 shows the frequency response and the step response for H_χ. Note that because of the zero the standard intuition for the selection of ζ does not hold for this transfer function. Larger ζ results in larger bandwidth and smaller overshoot.

Comparing coefficients in Equations 9 and 10, we find

$$\omega_{n_\chi}^2 = \frac{g}{V}k_{i_\chi} \tag{11}$$

$$2\zeta_\chi\omega_{n_\chi} = \frac{g}{V}k_{p_\chi} \tag{12}$$

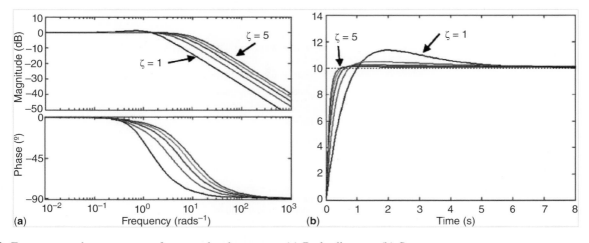

Figure 7. Frequency and step response for second-order system. (a) Body diagram. (b) Step response.

Solving these expressions for k_{p_χ} and k_{i_χ} gives

$$k_{p_\chi} = 2\zeta_\chi \omega_{n_\chi} \frac{V}{g} \qquad (13)$$

$$k_{i_\chi} = \omega_{n_\chi}^2 \frac{V}{g} \qquad (14)$$

To ensure proper function of this successive loop-closure design, it is essential that the inner (roll) and outer (heading) feedback loops are sufficiently separated. Adequate separation can be achieved using the rule of thumb

$$\omega_{n_\phi} > 5\omega_{n_\chi} \qquad (15)$$

Generally, more bandwidth separation is better. More bandwidth separation requires either slower response in the χ loop (lower ω_{n_χ}), or faster response in the ϕ loop (higher ω_{n_ϕ}). Faster response usually comes at the cost of requiring more actuator control authority, which may not be possible given the physical constraints of the actuators.

The rudder is typically used to regulate the sideslip angle β to zero in order to maintain coordinated turn conditions. Because of space limitations we will not give the equations for rudder control.

3.2   Longitudinal autopilot

The longitudinal autopilot is more complicated than the lateral autopilot because airspeed plays a significant role in the longitudinal dynamics. Our objective in designing the longitudinal autopilot will be to regulate airspeed and altitude using the throttle and the elevator as actuators. The method used to regulate altitude and airspeed depends on the altitude error. The flight regimes are shown in Figure 8.

In the take-off zone, full throttle is commanded and the pitch attitude is regulated to a fixed pitch angle θ^c using the elevator.

In the climb zone, the objective is to maximize the climb rate given the current atmospheric conditions. To maximize the climb rate, full throttle is commanded and the airspeed is regulated using the pitch angle. If the airspeed increases above its nominal value, then the aircraft is caused to pitch up which results in an increased climb rate and a decrease in airspeed. Similarly, if the airspeed drops below the nominal value, the airframe is pitched down thereby increasing the airspeed but also decreasing the climb rate. Regulating the airspeed using pitch attitude effectively keeps the airframe away from stall conditions. Note however, that we would not want to regulate airspeed with pitch attitude immediately after take-off because after take-off the airframe is always trying to gain airspeed but pitching down will drive the aircraft into the ground.

The descend zone is similar to the climb zone except that the throttle is commanded to zero. Again, stall conditions are avoided by regulating airspeed using the pitch angle thus maximizing the descent rate at a given airspeed. On some airframes zero throttle may not be possible or desirable. In these cases, the throttle is set to a minimum value.

In the altitude hold zone, the airspeed is regulated by adjusting the throttle, and the altitude is regulated by commanding the pitch attitude.

To implement the longitudinal autopilot shown in Figure 8 we need the following feedback loops: (i) pitch attitude hold using elevator, (ii) airspeed hold using throttle, (iii) airspeed hold using pitch attitude, and (iv) altitude hold using pitch attitude. The design of each of these loops will be discussed in the next four sections. Finally, the complete longitudinal autopilot will be presented in Section 3.2.5.

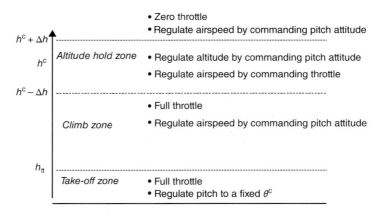

Figure 8. Flight regimes for the longitudinal autopilot.

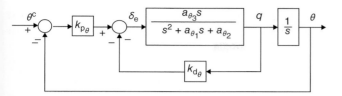

Figure 9. Pitch attitude hold feedback loop.

3.2.1 Pitch attitude hold

The pitch attitude hold loop is similar to the roll attitude hold loop and we will follow a similar line of reasoning in its development. From Figure 9, the transfer function from θ^c to θ is given by

$$H_{\theta/\theta^c}(s) = \frac{k_{p_\theta} a_{\theta_3}}{s^2 + (a_{\theta_1} + k_{d_\theta} a_{\theta_3})s + (a_{\theta_2} + k_{p_\theta} a_{\theta_3})} \quad (16)$$

Note that in this case, the DC gain is not equal to one.

If the desired response is given by the canonical second order transfer function

$$\frac{K_{\theta_{DC}} \omega_{n_\theta}^2}{s^2 + 2\zeta_\theta \omega_{n_\theta} s + \omega_{n_\theta}^2} \quad (17)$$

then equating denominator coefficients we obtain

$$\omega_{n_\theta}^2 = a_{\theta_2} + k_{p_\theta} a_{\theta_3} \quad (18)$$

$$2\zeta_\theta \omega_{n_\theta} = a_{\theta_1} + k_{d_\theta} a_{\theta_3} \quad (19)$$

Solving these expressions for k_{p_θ} and k_{d_θ} gives

$$k_{p_\theta} = \frac{\omega_{n_\theta}^2 - a_{\theta_2}}{a_{\theta_3}} \quad (20)$$

$$k_{d_\theta} = \frac{2\zeta_\theta \omega_{n_\theta} - a_{\theta_1}}{a_{\theta_3}} \quad (21)$$

Therefore, selecting the desired damping ratio and natural frequency fixes the value for k_{p_θ} and k_{d_θ}.

The DC gain of this inner-loop transfer function approaches one as the $k_{p_\theta} \to \infty$. The DC gain is given by

$$K_{\theta_{DC}} = \frac{k_{p_\theta} a_{\theta_3}}{(a_{\theta_2} + k_{p_\theta} a_{\theta_3})} \quad (22)$$

which for typical gain values is significantly less than one. For the design of the outer loop, we will use this DC gain to represent the gain of the inner loop over its full bandwidth. An integral feedback term could be employed to ensure unity DC gain on the inner loop. The addition of an integral term, however, can severely limit the bandwidth of the inner loop.

3.2.2 Altitude hold using commanded pitch

The altitude hold autopilot again utilizes a successive loop closure strategy with the pitch attitude hold autopilot as an inner loop. Assuming that the pitch loops function as designed and $\theta \approx K_{\theta_{DC}} \theta^c$, the altitude hold loop using the commanded pitch is shown in Figure 10.

In the Laplace domain we have

$$h(s) = \left[\frac{K_{\theta_{DC}} V k_{p_h} \left(s + \frac{k_{i_h}}{k_{p_h}} \right)}{s^2 + K_{\theta_{DC}} V k_{p_h} s + K_{\theta_{DC}} V k_{i_h}} \right] h^c(s) + \left(\frac{s}{s^2 + K_{\theta_{DC}} V k_{p_h} s + K_{\theta_{DC}} V k_{i_h}} \right) d_h(s) \quad (23)$$

where again we see that the DC gain is equal to one, and input disturbances d_h are rejected for low frequencies. The closed-loop transfer function is again independent of airframe parameters and is only dependent on the known airspeed. The gains k_{p_h} and k_{i_h} should be chosen such that the bandwidth of the altitude-using-pitch loop is approximately five times less than the bandwidth of the pitch attitude hold loop.

If the desired response of the altitude-hold loop is given by the canonical second order transfer function

$$\frac{\omega_{n_h}^2}{s^2 + 2\zeta_h \omega_{n_h} s + \omega_{n_h}^2} \quad (24)$$

then equating denominator coefficients and solving for k_{i_h} and k_{p_h} we obtain

$$k_{i_h} = \frac{\omega_{n_h}^2}{K_{\theta_{DC}} V} \quad (25)$$

$$k_{p_h} = \frac{2\zeta_h \omega_{n_h}}{K_{\theta_{DC}} V} \quad (26)$$

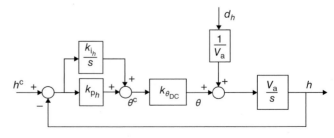

Figure 10. The altitude hold loop using the commanded pitch angle.

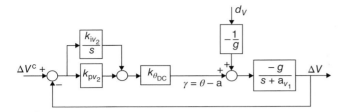

Figure 11. PI Controller to regulate airspeed using the pitch angle.

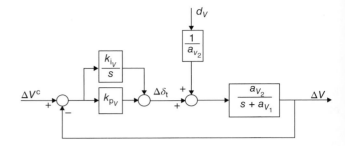

Figure 12. Airspeed hold using throttle.

3.2.3 Airspeed hold using commanded pitch

The airspeed can also be regulated using the pitch angle as a control variable since in level flight pushing the nose down increases airspeed and pulling the nose up decreases airspeed. The block diagram for airspeed hold using pitch attitude hold as an inner loop is shown in Figure 11. Disturbance rejection requires a proportional-integral (PI) controller.

In the Laplace domain we have

$$
\Delta V(s) = \left[\frac{(-K_{\theta_{DC}} g k_{p_{V_2}})\left(s + \frac{k_{i_{V_2}}}{k_{p_{V_2}}}\right)}{s^+(a_{V_1} - K_{\theta_{DC}} g k_{p_{V_2}})s - K_{\theta_{DC}} g k_{i_{V_2}}} \right] \Delta V^c(s)
$$

$$
+ \left[\frac{s}{s^2 + (a_{V_1} - K_{\theta_{DC}} g k_{p_{V_2}})s - K_{\theta_{DC}} g k_{i_{V_2}}} \right] d_V(s)
$$

(27)

Note that the DC gain is equal to one and that step disturbances are rejected. The angle of attack α enters into the block diagram at the same location as d_V. If we assume that the angle of attack is constant, then the PI controller winds up the integrator to reject α. Therefore, the output of the PI controller can be used as the input to the pitch attitude hold loop.

The gains $k_{p_{V_2}}$ and $k_{i_{V_2}}$ should be chosen such that the bandwidth of the airspeed to pitch loop is less than the bandwidth of the pitch attitude hold loop. Following a procedure similar to previous sections, we can determine values for the feedback gains by matching denominator coefficients in Equation 27 with those of a canonical second-order transfer function. Denoting the desired natural frequency and damping ratio we seek to achieve with feedback as $\omega^2_{n_{V_2}}$ and ζ_{V_2}, matching coefficients and solving for

the control gains gives

$$
k_{i_{V_2}} = -\frac{\omega^2_{n_{V_2}}}{K_{\theta_{DC}} g}
$$

(28)

$$
k_{p_{V_2}} = \frac{a_{V_1} - 2\zeta_{V_2}\omega_{n_{V_2}}}{K_{\theta_{DC}} g}
$$

(29)

3.2.4 Airspeed hold using throttle

The closed-loop system for the airspeed using the throttle as an input is shown in Figure 12.

The transfer function is given by

$$
\Delta V = \left[\frac{a_{V_2}(k_{p_V} s + k_{i_V})}{s^2 + (a_{V_1} + a_{V_2} k_{p_V})s + a_{V_2} k_{i_V}} \right] \Delta V^c
$$

$$
+ \left[\frac{1}{s^2 + (a_{V_1} + a_{V_2} k_{p_V})s + a_{V_2} k_{i_V}} \right] d_V
$$

(30)

where ΔV is the deviation of the airspeed from trim, ΔV^c is the commanded deviation from trim, where it is clear that the DC gain is one and that input disturbances d_V are rejected for low frequencies. If a_{V_1} and a_{V_2} are known, then the gains k_{p_V} and k_{i_V} can be effectively determined using the same technique we have used previously. Equating the closed-loop transfer function denominator coefficients with those of a canonical second-order transfer function, gives

$$
\omega^2_{n_V} = a_{V_2} k_{i_V}
$$

(31)

$$
2\zeta_V \omega_{n_V} = a_{V_1} + a_{V_2} k_{p_V}
$$

(32)

Inverting these expressions gives the control gains

$$
k_{i_V} = \frac{\omega^2_{n_V}}{a_{V_2}}
$$

(33)

$$
k_{p_V} = \frac{2\zeta_V \omega_{n_V} - a_{V_1}}{a_{V_2}}
$$

(34)

Note that since $\Delta V \triangleq V - \hat{V}$ and $\Delta V^c \triangleq V^c - \hat{V}$, then

$$\Delta V^c - \Delta V = V^c - V \qquad (35)$$

Therefore, the control loop shown in Figure 12 can be implemented without knowledge of the trim velocity $\hat{V}$. Similarly, since the integrator will wind up to reject step disturbances, and a constant error in $\hat{\delta}_t$ can be thought of as a step disturbance, we can set $\delta_t = \Delta \delta_t$.

3.2.5 Altitude control state machine

The longitudinal autopilot modes can be combined to create the altitude control state machine shown in Figure 13.

In the climb zone, the throttle is set to its maximum value ($\delta_t = 1$) and the airspeed hold from commanded pitch mode is used to control the airspeed and thus ensure that the airframe avoids stall conditions. In simple terms, this causes the UAS to climb at its maximum possible climb rate until it is close to the altitude set point. Similarly, in the descend

$$\delta_t = 0$$
$$\theta^c = \left(k_{p_{V_2}} + \frac{k_{i_{V_2}}}{s} \right) (V^c - V)$$
$$\delta_e = \left(k_{p_\theta} + \frac{k_{i_\theta}}{s} \right) (\theta^c - \theta) - k_{d_\theta} q$$

Descend zone

$h < h^c + \overline{\Delta h}$ ▲ $h \geq h^c + \overline{\Delta h}$
▼

$$\delta_t = \left(k_{p_V} + \frac{k_{i_V}}{s} \right) (V^c - V)$$
$$\theta^c = \left(k_{p_h} + \frac{k_{i_h}}{s} \right) (h^c - h)$$
$$\delta_e = \left(k_{p_\theta} + \frac{k_{i_\theta}}{s} \right) (\theta^c - \theta) - k_{d_\theta} q$$

Altitude hold zone

$h < h^c + \overline{\Delta h}$ ▲ $h \geq h^c + \overline{\Delta h}$
▼

$$\delta_t = 1$$
$$\theta^c = \left(k_{p_{V_2}} + \frac{k_{i_{V_2}}}{s} \right) (V^c - V)$$
$$\delta_e = \left(k_{p_\theta} + \frac{k_{i_\theta}}{s} \right) (\theta^c - \theta) - k_{d_\theta} q$$

Climb zone

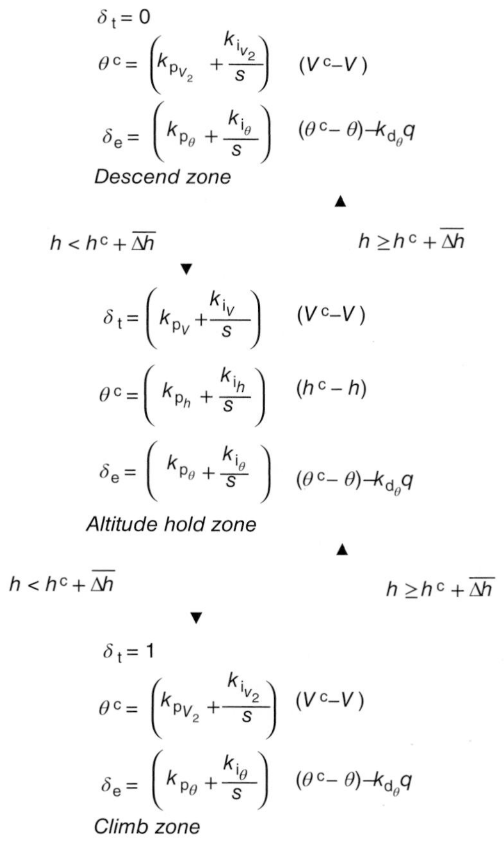

Figure 13. Altitude control state machine. The commanded altitude is h^c. δh defines a small altitude windows around h^c.

zone, the throttle is set to its minimum value ($\delta_t = 0$) and the airspeed hold from commanded pitch mode is again used to control airspeed. In this way, the UAS descends at a steady descent rate until it reaches the altitude hold zone. In the altitude hold zone, the airspeed from throttle mode is used to regulate the airspeed around V^c, and the altitude from pitch mode is used to regulate the altitude around h^c.

4 ON-BOARD SENSORS AND SENSOR PROCESSING

One of the biggest challenges with small UASs is to accurately estimate the dynamic states of the aircraft. Commercially available autopilots like the Kestrel, Piccolo II, and MP 2128 include the following sensor suite:

- Three-axis MEMS rate gyros
- Three-axis MEMS accelerometers
- Absolute and differential pressure sensors
- GPS.

Optional sensors that may be available include a two-axis magnetometer and an ultrasonic range finder (for landing). In Section 3 we saw that the low-level autopilot requires knowledge of the following states:

- Roll rate p, pitch rate q, yaw rate r
- Roll angle ϕ, pitch angle θ
- Airspeed V, altitude h
- Heading angle χ.

The navigation algorithm discussed in Section 5 will also make use of the North–East position of the aircraft, which will be denoted by p_n and p_e. The objective of this section is to describe how the available sensors are used to estimate the required states.

4.1 Angular rates, airspeed, and altitude

The angular rates, airspeed, and altitude of the aircraft can be estimated by directly inverting the sensor models. A brief discussion of this process is described below.

A MEMS rate gyro contains a small vibrating lever. When the lever undergoes an angular rotation, Coriolis effects change the frequency of the vibration, thus detecting the rotation. The output of the rate gyro is given by

$$y_{gyro} = k_{gyro}\Omega + \beta_{gyro} + \eta_{gyro} \qquad (36)$$

where y_{gyro} is in volts, k_{gyro} is a gain, Ω is the angular rate in radians per second, β_{gyro} is a bias, and η_{gyro} is zero mean white noise. The gain k_{gyro} and the bias β_{gyro} are strongly dependent on temperature and need to be carefully measured. The commercially available autopilots described in Section 1 have preprogrammed firmware routines that include look-up tables for the gain and bias as a function of temperature. Unfortunately, this process does not completely compensate for the bias term β_{gyro} which drifts with time. Therefore, during the preflight phase, it is necessary to calibrate the rate gyros.

If three rate gyros are aligned along the x, y, and z axes of the vehicle, then the rate gyros measure the angular body rates p, q, and r as follows:

$$y_{\text{gyro},x} = k_{\text{gyro},x}p + \beta_{\text{gyro},x} + \eta_{\text{gyro},x} \tag{37}$$

$$y_{\text{gyro},y} = k_{\text{gyro},y}q + \beta_{\text{gyro},y} + \eta_{\text{gyro},y} \tag{38}$$

$$y_{\text{gyro},z} = k_{\text{gyro},z}r + \beta_{\text{gyro},z} + \eta_{\text{gyro},z} \tag{39}$$

Therefore, assuming knowledge of the gain and bias, the angular rates p, q, and r can be estimated by low-pass filtering as

$$\hat{p} = \frac{[H_{\text{lpf}}(y_{\text{gyro},x}) - \beta_{\text{gyro},x}]}{k_{\text{gyro},x}} \tag{40}$$

$$\hat{q} = \frac{(H_{\text{lpf}}(y_{\text{gyro},y}) - \beta_{\text{gyro},x})}{k_{\text{gyro},x}} \tag{41}$$

$$\hat{r} = \frac{(H_{\text{lpf}}(y_{\text{gyro},z}) - \beta_{\text{gyro},x})}{k_{\text{gyro},x}} \tag{42}$$

where $H_{\text{lpf}}(\cdot)$ denotes the application of a low-pass filter, and where the "hat" notation denotes an estimated quantity.

The airspeed is measured using a pitot tube attached to a differential pressure sensor. If the pitot tube is oriented in the direction of motion, then the output of the differential pressure sensor is

$$y_{\text{diff pres}} = \frac{1}{2}\rho V^2 + \eta_{\text{diff pres}} \tag{43}$$

where V is the airspeed, ρ is the density of air, and η_{diffpres} is zero mean white noise. A simple estimation scheme for the airspeed is

$$\hat{V} = \sqrt{\frac{2}{\rho}H_{\text{lpf}}(y_{\text{diff pres}})} \tag{44}$$

The altitude of the UAS is measured using an static pressure sensor, where the output of the sensor is given by

$$y_{\text{static pres}} = \rho g h + \eta_{\text{static pres}} \tag{45}$$

where ρ is the density of air, g is the gravity constant, h is the altitude of the UAS, and $\eta_{\text{static pres}}$ is zero mean white noise. Therefore, the altitude of the UAS can be estimated as

$$\hat{h} = \frac{H_{\text{lpf}}(y_{\text{static pres}})}{\rho g} \tag{46}$$

4.2 Roll and pitch angles

The most difficult states to measure are the vehicle roll and pitch angles. The basic idea that we will discuss in this section is to estimate the roll and pitch angles by integrating the rate gyros and using the accelerometers to correct for integration constants and biases and drift in the rate gyros.

A MEMS accelerometer contains a small plate attached to torsion levers. The plate rotates under acceleration and changes the capacitance between the plate and the surrounding walls (http://www.silicondesigns.com/tech.html). The output of a single-axis accelerometer is given by

$$y_{\text{acc}} = k_{\text{acc}}A + \beta_{\text{acc}} + \eta_{\text{acc}} \tag{47}$$

where y_{acc} is in volts, k_{acc} is a gain, A is the acceleration in meters per second, β_{acc} is a bias term, and η_{acc} is zero mean white noise. The gain k_{acc} and the bias term β_{acc} are strongly dependent on temperature. Therefore, the accelerometer must be calibrated in a temperature chamber.

Accelerometers measure the specific force in the body frame of the vehicle. A physically intuitive explanation is given in Stevens and Lewis (2003, pp. 13–15). Additional explanation is given in Rauw (1998, p. 27). Mathematically, we have

$$[a_x\, a_y\, a_z]^T = \frac{1}{m}(\boldsymbol{F} - \boldsymbol{F}_{\text{gravity}}) \tag{48}$$

where $\boldsymbol{F}$ is the total force acting on the center of mass of the UAS, and $\boldsymbol{F}_{\text{gravity}}$ is the force due to gravity. For a flying vehicle, the total force is composed of four components:

$$\boldsymbol{F} = \boldsymbol{F}_{\text{thrust}} + \boldsymbol{F}_{\text{drag}} + \boldsymbol{F}_{\text{lift}} + \boldsymbol{F}_{\text{gravity}} \tag{49}$$

where $\boldsymbol{F}_{\text{thrust}}$ is the thrust force, $\boldsymbol{F}_{\text{drag}}$ is the drag force, and $\boldsymbol{F}_{\text{lift}}$ is the lift force. For a fixed-wing aircraft in unaccelerated flight, the total forces on the UAS sum to zero. Therefore, from Equation 48 we have that the accelerometers measure the direction of the gravity vector, which can be used to extract the roll and pitch angles.

We note here that while accelerometers can be used effectively to estimate roll and pitch for fixed-wing vehicles or for hovercraft in the forward motion regime, they are not useful for hovercraft in the hover regime. Near hover, the lift and drag forces are essentially zero. Therefore, from Equation 48 the accelerometers measure the thrust vector and not the gravity vector. Since the thrust vector is a body fixed quantity, the orientation of the aircraft cannot be extracted.

The acceleration of any rigid body is given by

$$\dot{v} + \omega_{b/i} \times v = \frac{1}{m}F \tag{50}$$

where $v \triangleq (u, v, \omega)^{\mathrm{T}}$ is the inertial velocity expressed in the body frame, and $\omega_{b/i} \triangleq (p, q, r)^{\mathrm{T}}$ is the angular velocity of the body with respect to the inertial frame, expressed in the body frame. Plugging Equation 50 into Equation 48, and expressing the gravity vector in terms of the roll angle ϕ and the pitch angle θ give

$$a_x = \dot{u} + qw - rv + g \sin \theta \tag{51}$$

$$a_y = \dot{v} + ru - pw - g \cos \theta \sin \phi \tag{52}$$

$$a_z = \dot{w} + pv - qu - g \cos \theta \cos \phi \tag{53}$$

where a_* is the specific acceleration along the $*$-axis of the body. If the gains of the accelerometers are calibrated and normalized to $1/g$, and the biases calibrated and removed, then the outputs of the accelerometers are given by

$$y_{\mathrm{acc},x} = \frac{\dot{u} + qw - rv + g \sin \theta}{g} + \eta_{\mathrm{acc},x} \tag{54}$$

$$y_{\mathrm{acc},y} = \frac{\dot{v} + ru - pw - g \cos \theta \sin \phi}{g} + \eta_{\mathrm{acc},y} \tag{55}$$

$$y_{\mathrm{acc},z} = \frac{\dot{w} + pv - qu - g \cos \theta \cos \phi}{g} + \eta_{\mathrm{acc},z} \tag{56}$$

The body frame velocity vector can be expressed as

$$\begin{bmatrix} u \\ v \\ w \end{bmatrix} = V \begin{bmatrix} \cos \alpha \cos \beta \\ \sin \beta \\ \sin \alpha \cos \beta \end{bmatrix} \tag{57}$$

where α is the angle of attack and β is the sideslip angle (Stevens and Lewis, 2003). Since α and β are typically not

measured on small UAS, and since for small fixed-wing vehicles α and β are typically small, we assume that $\alpha \approx \theta$ and $\beta \approx 0$ to obtain

$$\begin{bmatrix} u \\ v \\ w \end{bmatrix} \approx V \begin{bmatrix} \cos \theta \\ 0 \\ \sin \theta \end{bmatrix} \tag{58}$$

Approximating $\dot{u} = \dot{v} = \dot{w} = 0$, Equation 55 becomes

$$y_{\mathrm{acc},x} \approx \frac{qV \sin \theta}{g} + \sin \theta + \eta_{\mathrm{acc},x} \tag{59}$$

$$y_{\mathrm{acc},y} \approx \frac{rV \cos \theta - pV \sin \theta}{g} - \cos \theta \sin \phi + \eta_{\mathrm{acc},y} \tag{60}$$

$$y_{\mathrm{acc},z} \approx \frac{-qV \cos \theta}{g} - \cos \theta \cos \phi + \eta_{\mathrm{acc},z} \tag{61}$$

Therefore, a simple inversion scheme that does not include the integration of rate gyros is given by

$$\hat{\theta}_{\mathrm{accel}} = \sin^{-1} \left[\frac{H_{\mathrm{1pf}}(y_{\mathrm{acc},x})}{\frac{\hat{q}\hat{V}}{g} + 1} \right] \tag{62}$$

$$\hat{\phi}_{\mathrm{accel}} = \tan^{-1} \left[\frac{H_{\mathrm{1pf}}(y_{\mathrm{acc},y}) - \left(\frac{\hat{r}\hat{V}\cos \hat{\theta} - \hat{p}\hat{V}\sin \hat{\theta}}{g} \right)}{H_{\mathrm{1pf}}(y_{\mathrm{acc},z}) + \left(\frac{\hat{q}\hat{V}\cos \hat{\theta}}{g} \right)} \right] \tag{63}$$

A fixed gain Kalman filter can be used to fuse the accelerometer and rate gyro information. The relationship between the angular rates $\dot{\phi}$ and $\dot{\theta}$ and the body fixed angular velocities p and q is given by Stevens and Lewis (2003):

$$\dot{\phi} = p + q \sin \phi \tan \theta + r \cos \phi \tan \theta \tag{64}$$

$$\dot{\theta} = q \cos \phi - r \sin \phi \tag{65}$$

$$= q + [q(\cos \phi - 1) - r \sin \phi] \tag{66}$$

If we lump the nonlinearities into noise variables and make the (admittedly bad) assumption that the noise will be zero mean and Gaussian, then we have

$$\dot{\phi} = p + \xi_\phi \tag{67}$$

$$\dot{\theta} = q + \xi_\theta \tag{68}$$

where we assume that $\xi_\phi \sim N;(0, Q_\phi)$ and $\xi_\theta \sim N;(0, Q_\theta)$.

The accelerometers are used for measurement correction. Therefore

$$y_\phi = \hat{\phi}_{\text{accel}} = \phi + \eta_\phi \tag{69}$$

$$y_\theta = \hat{\phi}_{\text{accel}} = \theta + \eta_\theta \tag{70}$$

where we again assume that $\eta_\phi \sim \mathcal{N}(0, R_\phi)$ and $\eta_\theta \sim \mathcal{N}(0, R_\theta)$. The steady-state Kalman filter equations are (Dorato, Abdallah, and Cerone, 1995)

$$\dot{\hat{x}} = A\hat{x} + Bu + L(y - C\hat{x}) \tag{71}$$

$$L = PC^{\text{T}}R^{-1} \tag{72}$$

$$0 = AP + PA^{\text{T}} + GQG^{\text{T}} - PC^{\text{T}}R^{-1}CP \tag{73}$$

The Riccati equation for the roll angle simplifies to

$$Q_\phi - \frac{P_\phi^2}{R_\phi} = 0 \tag{74}$$

Therefore, we have

$$P_\phi = \sqrt{Q_\phi R_\phi} \tag{75}$$

and the Kalman gain becomes

$$L_\phi = P_\phi R_\phi^{-1} = \sqrt{\frac{Q_\phi}{R_\phi}} \tag{76}$$

We can calculate L_θ using similar reasoning. Therefore, the steady state Kalman filters for roll and pitch angles, which fuse rate gyro and accelerometer data, are given by

$$\dot{\hat{\phi}} = p + \sqrt{\frac{Q_\phi}{R_\phi}}(\hat{\phi}_{\text{accel}} - \hat{\phi}) \tag{77}$$

$$\dot{\hat{\theta}} = q + \sqrt{\frac{Q_\theta}{R_\theta}}(\hat{\phi}_{\text{accel}} - \hat{\phi}) \tag{78}$$

4.3 Inertial position and heading

This section describes an estimation scheme for the inertial position and heading using GPS as an input. GPS signals have an inertial position bias due to timing and other errors that will not not be corrected by the techniques discussed in this section. We assume that position, heading, and groundspeed are directly measured by the GPS unit.

The simplest technique for estimating position and heading is to simply low-pass filter the GPS signals. However, since the GPS update rate is on the order of one second, we desire to estimate the states in between GPS updates.

The kinematic equations of motion for inertial position and heading are given by

$$\begin{bmatrix} \dot{p}_{\text{n}} \\ \dot{p}_{\text{n}} \\ \dot{\chi} \end{bmatrix} = \begin{bmatrix} V_{\text{g}} \cos \chi \\ V_{\text{g}} \sin \chi \\ q\dfrac{\sin \phi}{\cos \theta} + r\dfrac{\cos \phi}{\cos \theta} \end{bmatrix} + \xi \tag{79}$$

where V_{g} is the ground speed of the UAS and ξ is a zero mean Gaussian process with covariance Q. Obviously, Equation 79 has problems as θ approaches $\pm \pi/2$, and as such the scheme described in this section assumes relatively small pitch angles. If we let $x = (p_{\text{n}}, p_{\text{e}}, \chi)^{\text{T}}$ and $u = (V_{\text{g}}, q, r, \phi, \theta)^{\text{T}}$, then we have

$$\dot{x} = f(x, u) + \xi \tag{80}$$

GPS returns measurements of p_{n}, p_{e}, and χ directly. Therefore, the output model is

$$y_{\text{gps}} = \begin{bmatrix} p_{\text{n}} \\ p_{\text{e}} \\ \chi \end{bmatrix} + \eta \tag{81}$$

where $\eta \sim \mathcal{N}(0, R)$ and $C = I$. To implement the Kalman filter, we need the Jacobian of f that can be calculated as

$$\frac{\partial f}{\partial x} = \begin{bmatrix} 0 & 0 & -V_{\text{g}} \sin \chi \\ 0 & 0 & V_{\text{g}} \cos \chi \\ 0 & 0 & 0 \end{bmatrix} \tag{82}$$

The resulting Kalman filter is listed in Algorithm 1.

Algorithm 1 Kalman filter for p_{n}, p_{e}, and χ

1:   Initialize: $\hat{x} = 0$.
2:   Pick an output sample rate T_{out} which is much less than the sample rates of the sensors.
3:   At each sample time T_{out}:
4:   **for** $i = 1$ to N **do** {Propagate the equations.}
5:      $\hat{x} = \hat{x} + \left(\frac{T_{\text{out}}}{N}\right)f(\hat{x}, u)$
6:      $A = \frac{\partial f}{\partial x}(\hat{x}, u)$
7:      $P = P + \left(\frac{T_{\text{out}}}{N}\right)(AP + PA^{\text{T}} + Q)$
8:   **end for**
9:   if GPS measurement received, **then** {Measurement Update}
10:     $L = P(R + P)^{-1}$
11:     $P = (I - L)P$
12:     $\hat{x} = \hat{x} + L(y_{\text{gps}} - \hat{x})$
13: **end if**

---

5 GPS NAVIGATION

In this section we will describe guidance laws for tracking straight-line paths and orbits. The primary challenge in tracking straight line segments and circular orbits is the constant winds, which are almost always present. For small UAS, wind speeds are commonly 20–60% of the desired airspeed. Effective path tracking strategies must overcome the effect of this ever present disturbance. For most fixed-wing small UAS, the minimum turn radius is in the range of 10–50 m, which places a fundamental limit on the spatial frequency of paths that can be tracked.

Implicit in the notion of trajectory tracking is that the vehicle is commanded to be at a particular location at a specific time. For fixed-wing UAS, the desired position is constantly moving (at the desired ground speed), which can result in significant problems if wind disturbances are not accounted for properly. If the UAS is flying into a strong wind (relative to its commanded groundspeed), the progression of the trajectory point must be slowed accordingly. Similarly, if the UAS is flying down wind, the speed of the tracking point must be increased to keep it from overrunning the desired position. Given that wind disturbances vary and are often not easily predicted, trajectory tracking can be challenging in anything other than calm conditions.

Rather than using a trajectory tracking approach, this section focuses on path following where the objective is to be *on the path* rather than at a certain point at a particular time. With path following, the time dependence of the problem is removed. A key insight is to use groundspeed and heading as opposed to airspeed and yaw angle. By using groundspeed instead of airspeed and heading instead of yaw, asymptotic path following can be guaranteed, even in the presence of constant wind disturbances. The path following approach is discussed in more detail in Nelson *et al.* (2007), Aguiar, Hespanha, and Kokotovic (2005), Encarnação and Pascoal (2001), Skjetne and Kokotovíc (2004), and Rysdyk (2003).

This section is limited to constant altitude motion. The corresponding equations of motion are given by

$$\dot{p}_n = V\cos\chi + w_n \tag{83}$$

$$\dot{p}_e = V\sin\chi + w_e \tag{84}$$

where we will assume that the input to the guidance law is the heading angle χ, and where (w_n, w_e) are the magnitude of the wind vector in North–East coordinates. The objective is to develop a method for accurate path following in the presence of wind. For a given airframe, there is an optimal airspeed for which the airframe is the most aerodynamically efficient.

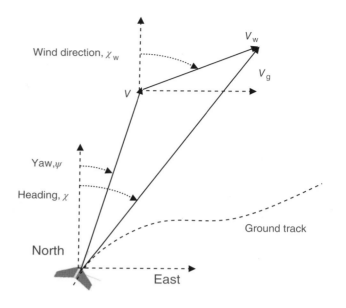

Figure 14. This figure shows the relationship between the airspeed V, the windspeed V_w, and ground speed V_g, as well as the relationship between yaw ψ, heading χ, and wind direction χ_w. The crab angle is defined as heading minus yaw.

Therefore, to conserve fuel, it is desirable that the UAS maintain a constant airspeed. Accordingly, in this section we will assume a constant airspeed V. The relationship between airspeed, windspeed, and groundspeed are shown in Figure 14.

The equations of motion for the North and the East directions can be written in terms of the heading and groundspeed as

$$\dot{p}_n = V\cos\psi + V_w\cos\chi_w = V_g\cos\chi \tag{85}$$

$$\dot{p}_e = V\sin\psi + V_w\sin\chi_w = V_g\sin\chi \tag{86}$$

where the groundspeed V_g is given by

$$V_g = \sqrt{V^2 + V_w^2 + 2VV_w\cos(\psi - \chi_w)} \tag{87}$$

and the heading angle χ is given by

$$\chi = \tan^{-1}\left(\frac{V\sin\psi + V_w\sin\chi_w}{V\cos\psi + V_w\cos\chi_w}\right) \tag{88}$$

Section 5.1 will develop a guidance strategy for following straight line paths, and Section 5.2 will develop a guidance strategy for following constant altitude orbits.

5.1 Straight-line path following

A straight line path is described by two vectors in $\mathbb{R}^2$, namely,

$$\mathcal{P}_{\text{straight}}(\boldsymbol{r}, \boldsymbol{q}) = \{ \boldsymbol{x} \in \mathbb{R}^2 : \boldsymbol{x} = \boldsymbol{r} + \alpha\boldsymbol{q}, \alpha \in \mathbb{R} \} \quad (89)$$

where $\boldsymbol{r} \in \mathbb{R}^2$ is the origin of the path, and $\boldsymbol{q} \in \mathbb{R}^2$ is a unit vector whose direction indicates the desired direction of travel. Figure 15 shows a top-down view of $\mathcal{P}_{\text{straight}}(\boldsymbol{r}, \boldsymbol{q})$.

The heading of $\mathcal{P}_{\text{straight}}(\boldsymbol{r}, \hat{\boldsymbol{q}})$, as measured from North is given by

$$\chi_q \triangleq \tan^{-1} \frac{q_e}{q_n} \quad (90)$$

The path following problem is most easily solved in a frame relative to the straight-line path. Let

$$R_{\text{i}}^{\mathcal{P}} \triangleq \begin{bmatrix} \cos\chi_q & \sin\chi_q \\ -\sin\chi_q & \cos\chi_q \end{bmatrix} \quad (91)$$

be the transformation from the inertial frame to the path frame, and let

$$\tilde{\mathbf{p}} = \begin{bmatrix} \tilde{p}_x \\ \tilde{p}_y \end{bmatrix} \triangleq R_{\text{i}}^{\mathbf{P}}(\boldsymbol{p} - \boldsymbol{r}) \quad (92)$$

be the relative path error in the path frame. Therefore, the relative dynamics in the path frame are given by

$$\begin{bmatrix} \dot{\tilde{p}}_x \\ \dot{\tilde{p}}_y \end{bmatrix} = R_{\text{i}}^{\mathcal{P}}(\mathbf{p}^i - \mathbf{r}^i) \quad (93)$$

$$= \begin{bmatrix} \cos\chi_q & \sin\chi_q \\ -\sin\chi_q & \cos\chi_q \end{bmatrix} \begin{bmatrix} V_g \cos\chi \\ V_g \sin\chi \end{bmatrix} \quad (94)$$

$$= \begin{bmatrix} V_g \cos(\chi - \chi_q) \\ V_g \sin(\chi - \chi_q) \end{bmatrix} \quad (95)$$

Our strategy is to construct a desired heading angle at every point relative to the straight-line path that results in the UAS moving toward the path. The set of desired heading angles at every point will be called a vector field because the desired heading angle specifies a vector (relative to the straight line) with a magnitude of unity. Figure 16 is an example of a vector field for straight-line path following. The objective is to construct the vector field so that when $\tilde{p}_y$ is large the UAS is directed to approach the path with heading angle χ^∞, and so that as $\tilde{p}_y$ approaches zero, the heading $\tilde{\chi}_q \chi - \chi_q$ also approaches zero. Toward that end, define the commanded heading of the UAS as

$$\chi^c(\tilde{p}_y) = \chi_q - \chi^\infty \frac{2}{\pi} \tan^{-1}(k\tilde{p}_y) \quad (96)$$

where k is a positive constant that influences the rate of the transition from χ^∞ to zero. Figure 17 shows how the choice of k affects the rate of transition. Large values of k result in short, abrupt transitions, while small values of k cause long, smooth transitions in the desired heading.

If χ^∞ is restricted to be in the range $\chi^\infty \in (0, \frac{\pi}{2}]$ then clearly

$$-\frac{\pi}{2} < \chi^\infty \frac{2}{\pi} \tan^{-1}(k\tilde{p}_y) < \frac{\pi}{2} \quad (97)$$

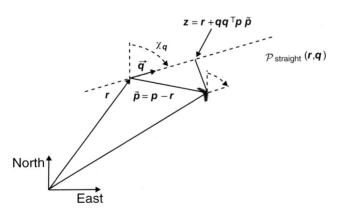

Figure 15. This figure shows the configuration of the UAS indicated by $(\boldsymbol{p}, \chi)$, and the configuration of the UAS relative to $\mathcal{P}_{\text{straight}}$ indicated by $(\tilde{\boldsymbol{p}}, \tilde{\chi})$.

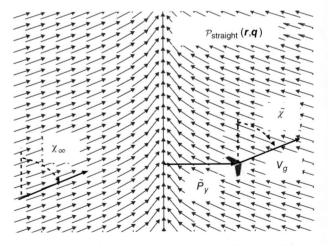

Figure 16. Vector field for straight-line path following. Far away from the waypoint path, the vector field is directed with an angle χ^∞ from the perpendicular to the path.

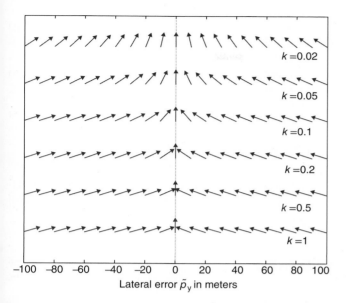

Figure 17. Vector fields for various values of k. Large values of k yield abrupt transitions from χ^∞ to zero, while small values of k give smooth transitions.

for all values of $\tilde{p}_y$. Therefore, since $\tan^{-1}(\cdot)$ is an odd function and $\sin(\cdot)$ is odd over $(-\frac{\pi}{2}, \frac{\pi}{2})$, we can use the Lyapunov function $W_1(\tilde{p}_y) = \frac{1}{2}\tilde{p}_y^2$ to argue that if $\chi = \chi^c(\tilde{p}_y)$, then $\tilde{p}_y \to 0$ asymptotically. Evaluating the Lie derivative of W_1 under the assumption that $\chi = \chi^c(\tilde{p}_y)$ gives

$$\dot{W}_1 = -V_g \tilde{p}_y \sin\left[\tilde{\chi}^\infty \frac{2}{\pi}\tan^{-1}(k\tilde{p}_y)\right] \qquad (98)$$

which is less than zero for $\tilde{p}_y \neq 0$. Asymptotic convergence of this scheme in the presence of yaw and roll dynamics is shown in Nelson *et al.* (2007).

5.2 Orbit following

An orbit is described by a center $\mathbf{c} \in \mathbb{R}^2$, a radius $R \in \mathbb{R}$, and a direction $\lambda \in \{-1, 1\}$, as

$$\mathcal{P}_{orbit}(\mathbf{c}, R, \lambda) = \{r \in \mathbb{R}^2 : r = c + \lambda R(\cos\varphi, \sin\varphi)^T,$$
$$\varphi \in [0, 2\pi)\} \qquad (99)$$

where $\lambda = 1$ signifies a clockwise orbit and $\lambda = -1$ signifies a counterclockwise orbit. Figure 18 shows a top-down view of an orbital path.

The guidance strategy for orbit following is best derived in polar coordinates. Let

$$d \triangleq \sqrt{(p_n - c_n)^2 + (p_e - c_e)^2} \qquad (100)$$

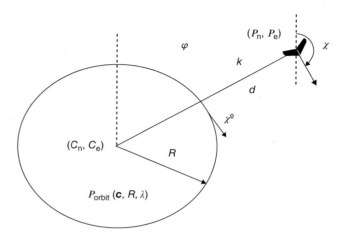

Figure 18. Orbital path with center (c_n, c_e), and radius R. The distance from the orbit center to the UAS is d, and the angular position of the UAS relative to the orbit is φ.

be the lateral distance from the desired center of the orbit to the UAS, and let

$$\varphi \triangleq \tan^{-1}\left(\frac{p_e - c_e}{p_n - c_n}\right) \qquad (101)$$

be the phase angle of the relative position, as shown in Figure 18. Differentiating d and using Equations 85 and 86 gives

$$\dot{d} = \frac{(p_n - c_n)\dot{p}_n + (p_e - c_e)\dot{p}_e}{d} \qquad (102)$$

$$= \frac{(p_n - c_n)V_g \cos\chi + (p_e - c_e)V_g \sin\chi}{d} \qquad (103)$$

Using Equation 101 gives

$$\dot{d} = V_g \frac{(p_n - c_n)\cos\chi + (p_e - c_e)\sin\chi}{d} \qquad (104)$$

$$= V_g\left(\frac{p_n - c_n}{d}\right)(\cos\chi + \sin\chi\tan\varphi) \qquad (105)$$

$$= V_g\cos\varphi(\cos\chi + \sin\chi\tan\varphi) \qquad (106)$$

$$= V_g(\cos\chi\cos\varphi + \sin\chi\sin\varphi) \qquad (107)$$

$$= V_g\cos(\chi - \varphi) \qquad (108)$$

As shown in Figure 18, for a clockwise orbit, the desired heading angle when the UAS is located on the orbit is given by $\chi^o = \varphi + \pi/2$. Similarly, for a counterclockwise orbit, the desired angle is given by $\chi^o = \varphi - \pi/2$. Therefore, in general we have

$$\chi^\circ = \varphi + \lambda \frac{\pi}{2} \qquad (109)$$

The control objective is to drive $d(t)$ to the orbit radius R. Defining the error variable $\tilde{d} \triangleq d - R$, the orbital kinematics can be restated as

$$\dot{\tilde{d}} = -\lambda V_g \sin(\chi - \chi^\circ) \qquad (110)$$

Our approach to orbit following is similar to the ideas developed in Section 5.1. The strategy is to construct a desired heading field that moves the UAS onto the orbit $\mathcal{P}_{orbit}(\mathbf{c}, R, \lambda)$. When the distance between the UAS and the center of the orbit is large, it is desirable for the UAS to fly toward the orbit center. In other words, when $\tilde{d} \gg R$, the desired heading is

$$\tilde{\chi}^d \approx \chi^\circ + \lambda \frac{\pi}{2} \qquad (111)$$

and when $\tilde{d} = 0$ the desired heading is χ°. The proposed guidance law is given by

$$\chi^c(\tilde{d}) = \chi^\circ + \lambda \tan^{-1}(k\tilde{d}) \qquad (112)$$

where $k > 0$ is a constant that specifies the rate of transition from $\lambda\pi/2$ to zero. This expression for $\chi^c(\tilde{d})$ is valid for all values of $\tilde{d} \geq -R$.

Consider the Lyapunov function $W_1 = \frac{1}{2}\tilde{d}^2$. When $\chi = \chi^c(\tilde{d})$, the Lie derivative of W_1 is

$$\dot{W}_1 = -\lambda V_g \tilde{d} \sin[\lambda \tan^{-1}(k\tilde{d})] \qquad (113)$$

$$= -V_g \tilde{d} \sin[\tan^{-1}(k\tilde{d})] \qquad (114)$$

which is negative definite since the argument of sin is in the set $(-\pi/2, \pi/2)$ for all $\tilde{d}$, implying that $\tilde{d} \to 0$ asymptotically. Asymptotic convergence of this scheme in the presence of yaw and roll dynamics is shown in Nelson et al., (2007).

6 SUMMARY

The objective of this chapter was to present basic principles in the design of autopilots for small UAS. We presented a basic architecture that is common to commercially available autopilots for small UAS. We discussed in detail the design of the feedback loops and how those loops are structured to control the lateral and longitudinal modes of the aircraft. Small UAS have a limited sensor suite. Common sensors were discussed and simple techniques for extracting the state of the system were described. We also presented a method for tracking waypoint paths and circular orbits.

ACKNOWLEDGMENTS

The author gratefully acknowledges many fruitful discussions about the topics of this chapter with his colleague Tim McLain at BYU.

END NOTES

1. In this chapter, we will use the term *small UAS* to refer to fixed wing aircraft with wingspan between 1 and 10 f, and payloads that may range from 0.25 to 10 kg. Operation times may range from 15 min to 10 h.

REFERENCES

Aguiar, A.P. Hespanha, J.P., and Kokotovic, P.V. (2005) Path-following for nonminimum phase systems removes performance limitations, *IEEE Trans. Autom. Control*, **50**(2), 234–238.

Blakelock, J.H. (1991) *Automatic Control of Aircraft and Missiles*, 2nd edn, John Wiley & Sons, Inc.

Dorato, P., Abdallah, C., and Cerone, V. (1995) *Linear-Quadratic Control: An Introduction*, Prentice Hall.

Etkin, B. and Reid, L.D. (1996) *Dynamics of Flight: Stability and Control*, John Wiley & Sons, Inc.

Encarnação, P. and Pascoal, A. (2001) Combined trajectory tracking and path following: an application to the coordinated control of marine craft, *Proceedings of the IEEE Conference on Decision and Control*, Orlando, FL. IEEE, pp. 964–969.

Franklin, G.F., Powell, J.D., and Emami-Naeini, A. (2002) *Feedback Control of Dynamic Systems*, 4th edn, Addison-Wesley.

Nelson, R.C. (1998) *Flight Stability and Automatic Control*, 2nd edn, McGraw-Hill, Boston, MA.

Nelson, D.R. Barber, D.B. McLain, T.W., and Beard, R.W. (2007) Vector field path following for miniature air vehicles, *IEEE Trans. Rob.*, **37**(3), 519–529.

Roskam, J. (1998) *Airplane Flight Dynamics and Automatic Flight Controls*, Parts I & II. DAR Corporation, Lawrence, KS.

Rauw, M. (1998) *FDC 1.2 - A SIMULINK Toolbox for Flight Dynamics and Control Analysis*. Available at http://www.docstoc.com/docs/7988762/FDC-12-%E2%80%93-A-Simulink-Toolbox-for-Flight-Dynamics-and-Control-Analysis (February 1998).

Rysdyk, R. (2003) UAV path following for constant line-of-sight. *Proceedings of the AIAA 2nd Unmanned Unlimited Conference*, AIAA, Paper No. 2003-6626.

Stevens, B.L. and Lewis, F.L. (2003) *Aircraft Control and Simulation*, 2nd edn, John Wiley & Sons, Inc., Hoboken, NJ.

Skjetne, R. Fossen, T. and Kokotović, P. (2004) Robust output maneuvering for a class of nonlinear systems, *Automatica*, **40**, 373–383.

PART 6
Control

Chapter 20

Modeling and Frequency-Domain Parameter Identification of a Small-Scale Flybarless Unmanned Helicopter[*]

Jessica Alvarenga, Nikolaos I. Vitzilaios, Matthew J. Rutherford, and Kimon P. Valavanis

Ritchie School of Engineering and Computer Science, DU Unmanned Research Institute, University of Denver, Denver, CO, USA

[*] This work was supported in part by NSF CNS-1229236.

1 INTRODUCTION

The complexity and high nonlinearity of rotorcraft dynamics require a high-fidelity model and robust methods of identifying system dynamics in order to design flight controllers. There are two main approaches to deriving rotorcraft dynamic models: first-principles modeling and system identification. First-principles modeling uses physical understanding of the system to derive mathematical equations describing dynamics. The result of using the first-principles approach is a highly complex set of nonlinear differential equations. These equations may be linearized about some operating point or set of initial conditions of interest to arrive at simplified models. In the case of rotorcraft, this is most commonly done through analysis of the forces and moments acting upon the rotorcraft followed by linearization around hover (Budiyono and Lesmana, 2007). System identification, on the other hand, relies on the use of experimental data to derive a model structure that closely reflects the behavior of the observed experimental data. Methods of system identification include neural networks, artificial intelligence, and time-domain and frequency-domain analysis techniques (Shim and Sastry, 2000). For rotorcraft, the use of the Comprehensive Identification from FrEquency Responses (CIFER) software has proven to produce accurate linear models for full-scale rotorcraft (Tischler and Remple, 2006) and has been used successfully in identification of small-scale

Figure 1. DU²SRI Bergen Industrial Turbine Helicopter.

helicopters (Mettler, 2003; La Civita, 2002; Raptis and Valavanis, 2011). This chapter summarizes the process of system identification using frequency-domain techniques and the CIFER software to develop a linear model around hover for the DU²SRI Bergen Industrial Turbine helicopter, as shown in Figure 1.

1.1   Testbed

The DU²SRI helicopter is equipped with an autopilot consisting of a flight computer, a NovAtel Global Positioning System (GPS) receiver with an update rate of 20 Hz, a Microstrain Attitude-Heading Reference System (AHRS) capable of updating at 100 Hz, and a Microbotics Servo Safety-Switch Controller for pilot takeover capability with data rates of up to 50 Hz, as shown in Figure 2. To increase the accuracy of GPS measurements, differential GPS capabilities are used with the addition of a ground-based GPS antenna connected to a Ground Control Station (GCS). The GCS, as shown in Figure 3, consists of a NovAtel GPS receiver, a wireless antenna, and a Panasonic Toughbook running on a Linux OS and QGroundControl (QGC) opensource software, as shown in Figure 4.

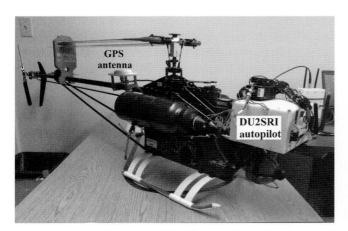

Figure 2. DU²SRI Bergen Industrial Turbine Helicopter with autopilot module.

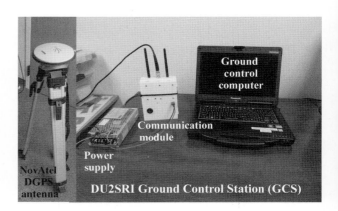

Figure 3. DU²SRI Ground Control Station.

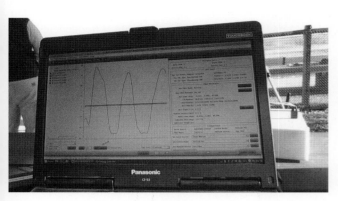

Figure 4. Toughbook with QGroundControl software showing flight data during frequency sweep data collection.

2   SYSTEM IDENTIFICATION OVERVIEW

The overall frequency-domain system identification process, as shown in Figure 5, consists of (i) collection of pilot input (main/tail rotor collectives and lateral/longitudinal cyclic inputs) and rigid-body state output (position, velocity, attitude, and angular rates) measurements, (ii) conditioning data, (iii) constructing a mathematical model, and (iv) verifying the model against experimental data. Collection of measurement data requires following experiments designed to ensure sufficiently rich excitation of system dynamics and modes of interest. Data are conditioned by reconstructing any lost measurements, removing faulty data, filtering out noise, removing biases, and resampling data to ensure even sampling times across all data points to be used during the analysis process. Once the data are ready to be processed, frequency-domain analysis is performed to ensure that collected data meet specific requirements with minimal correlation between input channels and a sufficient frequency bandwidth over the channels of interest. Next, frequency-response results are used to obtain model stability and control derivatives by an optimization process that starts with a specified model structure and uses optimization techniques to fit predicted model parameters to the given frequency responses. Finally, the predicted model is validated against a separate set of flight data to ensure model accuracy.

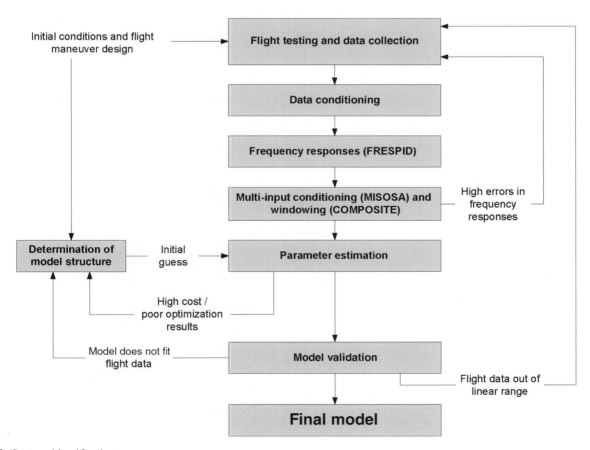

Figure 5. System identification process.

3   COLLECTION OF TIME HISTORY DATA

Data may be collected either experimentally (flight tests) or offline (simulation), see Section 4.3. In order to properly identify the model, it is first necessary to excite the model dynamics of interest. In the case of frequency-response analysis, this involves exciting the dynamics over a sufficient range of frequencies such that the system is persistently excited. This requires telemetry instrumentation for accurate dynamic measurements of both system inputs and states at proper sampling intervals. Additionally, the considerations that must be taken into account when selecting necessary instruments depend on the type of model to be identified. For example, in identifying handling qualities, the only measurements needed are those of the pilot stick deflection inputs and the angular rates and vertical accelerations of the rotorcraft. However, for 6-DOF bare-frame dynamics, it is necessary to record control surface deflections as well as translational velocities, accelerations, angular rates, and attitudes. In the case of rotorcraft, the main and tail rotor collective pitch deflections and cyclic flapping angles are not measured directly; instead the signals sent to the actuators are used (Hamel and Kaletka, 1997).

Collection of flight data requires experiments designed to ensure persistent excitation of the dynamics of interest. There are a number of flight inputs that can be used to design flight experiments. One such signal is known as DLR3211 (Hamel and Kaletka, 1997; Tischler and Remple, 2006), a multistep signal consisting of singlets whose time intervals vary between control reversals in the pattern 3–2–1–1, hence the name. This type of input is able to excite a wide range of frequencies and is good for highly unstable systems. DLR3211 has been used in time-domain system identification.

A second type of input, which is used in this work, is known as a frequency sweep. As the name implies, this type of input is a sinusoidal signal that sweeps through the range of frequencies of interest, beginning at low frequencies, in a safe manner. Neither of these approaches is optimal, but frequency sweeps have been used in related research (Mettler, 2003; Raptis and Valavanis, 2011; Tischler and Remple, 2006) for frequency-domain system identification due to the uniform distribution, symmetry in response time histories, and ability to control the frequency range of excitation. The flight experiment process consists of applying the frequency sweep signal to one input signal at a time, as shown in Figure 6. However, because of the highly coupled nature of rotorcraft dynamics, it is also necessary to apply a sufficient input to the remaining off-axis control inputs to keep the rotorcraft near the operating condition in a manner that minimizes any correlation between channels.

3.1   Criteria and Guidelines

For rotorcraft, the input signal should cover a large enough range of frequencies that will excite all aerodynamic modes. Aerodynamic modes include the long-period modes, such as phugoid, which occurs at the minimum frequency, and short-period dynamics that occur between approximately 1 and

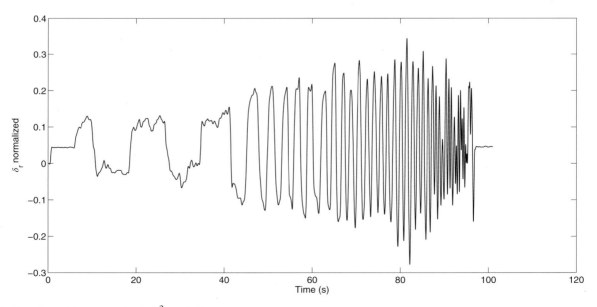

Figure 6. Sample rudder sweep for DU^{2}SRI helicopter.

$10\,\text{rad s}^{-1}$. For rotorcraft, a reasonable range of frequencies for applying the frequency sweep is between 0.3 and 12 rad s^{-1}, corresponding to a maximum period of approximately 20 s and a minimum period of 0.5 s. When collecting input–output measurement data, it is important that the sampling period is sufficiently small, following the Nyquist sampling criterion, and consistent across all the measurement channels. Noise in sensor measurements is usually filtered either online in the flight hardware or later when processing the data. The processing of measurement data to ensure equal sampling periods and elimination of noise and biases may greatly affect the frequency analysis results. Differences in filter frequencies may result in phase errors in frequency responses.

In order to avoid these errors, identical filters should be used on all input and output signals. In addition to proper filtering and sampling frequencies, flight tests should provide sufficiently long record lengths to capture enough frequencies. A general guideline is to have a record length two to five times the longest period of interest. The guidelines in Table 1 (Tischler and Remple, 2006) summarize these criteria.

Conducting piloted frequency sweeps requires multiple flights and careful planning to ensure consistent flight conditions and proper execution of pilot inputs. The following is a list of guidelines (Tischler and Remple, 2006) to take into consideration during flight tests:

(i) Each sweep should begin and end at the trim flight condition, in this case a steady hover, for about 3 s.

(ii) Begin the sweep with two long-period inputs of T_{max} period each.

(iii) Exact frequency progression is not necessary. This, in fact, may cause the pilot some confusion. Rather, smooth progression of frequencies is ideal with emphasis not to rush through the mid-frequencies.

(iv) It is not necessary to maintain constant amplitude or exact sinusoidal shape during sweeps. Although rotorcraft have a lower response at higher frequencies, it is

not necessary to increase the input amplitude to obtain a larger response.

(v) The pilot should apply off-axis controls to maintain aircraft responses symmetric around the trim condition without introducing correlation between channels. This can be done by applying low frequency or pulsed inputs.

(vi) Drift in aircraft response can be eliminated, or minimized, by adjusting the center of the sweep on the pilot stick.

(vii) Telemetry data should be monitored during flight tests, specifically the channels of interest.

(viii) The flight test engineer and pilot should establish a set of timing and flight cues. Timing indicators may prove useful in assisting pilots during the frequency sweeps, especially during the long-period sweeps. However, they may add confusion to both the pilot and flight test engineer. It is important that the pilot and flight test engineer practice to find the best method of providing timing indicators to assist the pilot in the frequency sweeps. A consistent and agreed upon vocabulary should also be established prior to flight testing.

While these guidelines provide a sound basis for performing piloted sweeps, they are developed with full-scale rotorcraft in mind. During flight testing of the DU^2SRI Bergen helicopter, the following were observed:

(i) Flight testing must be performed such that the pilot maintains line of sight and keeps the helicopter within range of ground station instrumentation and the pilot transmitter. This limits the amount of stick deflection that is applied during the long-period sweeps.

(ii) Since small-scale rotorcraft have faster dynamics than full-scale rotorcraft, large stick deflections can cause the rotorcraft to exhibit too large an attitude change, especially at lower frequencies.

(iii) In order to keep the rotorcraft from exhibiting unsafe responses during the long-period sweeps, the pilot was instructed to limit stick deflection amplitude. As the sweep progressed, the pilot increased the stick deflection amplitudes.

(iv) During the DU^2SRI flight testing, keeping the ground control station next to the pilot allowed the team to speak directly to the pilot. However, the use of headsets or other communication devices can be used.

(v) Small rotorcraft use feedback gyros that stabilize the yaw dynamics and, in the case of flybarless rotorcraft, pitch and roll dynamics. To identify the behavior of the embedded gyro controller, both the pilot commands to the gyro and the signals from the gyro to the servos are recorded.

Table 1. System identification measurement criteria.

Parameter	Description	Units	Criteria
ω_{min}	Minimum frequency of interest	rad s^{-1}	–
ω_{max}	Maximum frequency of interest	rad s^{-1}	–
T_{max}	Period of minimum frequency of interest	rad s^{-1}	$2\pi/\omega_{min}$
ω_f	Filter cutoff frequency	rad s^{-1}	$\omega_f \geq 5\omega_{max}$
ω_s	Sampling frequency	rad s^{-1}	$\omega_s \geq 5\omega_f$
T_{rec}	Record length	s	$T_{rec} \geq 5T_{max}$

(vi) Flight tests were performed in both feedback gyro operational modes: (i) heading-hold mode and (ii) rate mode.

(vii) The DU^{2}SRI helicopter uses a turbine engine that switches off when the collective stick deflection goes to 0 on the transmitter. To avoid accidental engine shutoff, the collective sweep amplitudes are kept small.

3.2   Computer-Generated Sweeps

Computer-generated sweeps have been used in full-scale rotorcraft for system identification. This is especially useful in wind tunnel testing. However, computer-generated sweeps are necessary in computer simulations, where a piloted frequency sweep is not possible. Using flight software, such as X-Plane, as shown in Figure 7, or a set of accurate equations in simulated environments, such as MATLAB/Simulink, as shown in Figure 8, computer-generated sweeps may be used to collect input–output measurements for system identification. There are a number of ways to model the frequency progression, including Schroeder-phase signals (optimal), linear progressions, and logarithmic or exponential progressions. The method developed by Tischler (Tischler and Remple, 2006) uses exponential progression as follows:

$$\delta_{sw} = A \sin\left(\theta(t)\right) \tag{1}$$

$$\theta(t) = \int_0^{T_{rec}} \left[\omega_{min} + K(\omega_{max} - \omega_{min})\right] dt \tag{2}$$

$$K = C_2[e^{C_1 t / T_{rec}} - 1] \tag{3}$$

where δ_{sw} is the sweep signal, A is the sweep amplitude, ω_{min} and ω_{max} are the starting and ending sweep frequencies, respectively, T_{rec} is the total desired record length, and the constants $C_1 = 4.0$ and $C_2 = 0.0187$ have been found to give satisfactory results for rotorcraft. White noise, δ_n, is then added to ensure a sufficient spectral richness necessary for persistent excitation. The total sweep signal is given as

$$\delta = \delta_{sw} + \delta_n \tag{4}$$

During real flights, the pilot keeps the rotorcraft response symmetric and near trim while performing the frequency sweep on the primary, or on-axis, input by applying enough control to the secondary, or off-axis, inputs in a way that does not introduce cross-correlation between channels and reduces feedback. For example, in a roll sweep, the pilot will apply the sweep to the aileron input and provide secondary controls to the remaining three channels. For computer-generated sweeps, this must be done by introducing feedback in the pitch, roll, and yaw channels. This can be done with simple proportional–integral–derivative (PID) controller feedback loops. It is important that these gains remain small enough to minimize cross-correlation, while still allowing the

Figure 7. The DU^{2}SRI Bergen helicopter modeled and simulated in X-Plane.

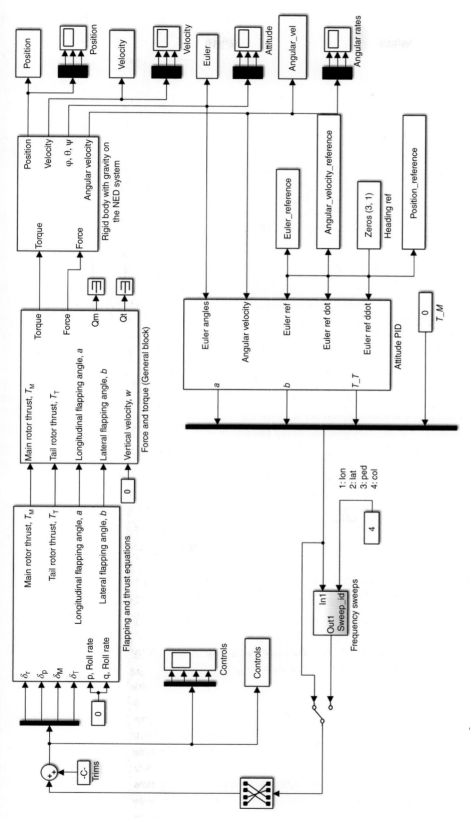

Figure 8. The DU²SRI Bergen helicopter system identification simulation in MATLAB/Simulink.

dynamics to remain near the trim conditions. An additional outer loop may be implemented for the heave channel and lateral-longitudinal position using small gains to maintain trim. One problem that arises from the addition of feedback to the loop is correlation between the on-axis and off-axis inputs. To reduce the effect of feedback, distinct white noise is added to each of the off-axis controls. Typical noise RMS levels are 10% of the sweep amplitude for the on-axis control and 10–50% of the sweep amplitude for the off-axis controls. These noise signals should pass through a filter to eliminate high-frequency content (Tischler and Cauffman, 1992).

3.3   Data Collection

To determine a 6-DOF dynamic rotorcraft model of the bare-frame dynamics, it is necessary to obtain specific measurements. In fixed wing aircraft it is possible to measure control surface deflection through the use of sensors on the control surfaces themselves or through servo feedback sensors. In rotorcraft, the control surfaces are the blades that are controlled through the use of servos and a complex mechanism (swash plate) that transfers the movement of one or more servos to the corresponding change in the collective or cyclic pitch angles. It is not feasible to measure these pitch angles directly. As a result, rather than obtaining control surface deflection measurements, the pilot stick deflections and/or servo pulse width signals are recorded. While estimation algorithms may be implemented during flight in order to estimate these values, doing so will require a reasonably reliable model of the blade flapping dynamics. In addition to the control measurements, it is important to record both translational and angular dynamics, including linear velocities, accelerations, angular rates, and attitude and heading angles, all with respect to the vehicle's body-fixed frame, as shown in Figure 9. For the DU²SRI Bergen helicopter, the following collected measurements were used:

- Body frame accelerations $\dot{u}$ and $\dot{v}$, and linear velocity w.
- Angular rates p, q, and r.
- Euler angles ϕ, θ, and ψ.

4   FREQUENCY-RESPONSE ANALYSIS

The frequency response of any system is the Fourier transform of the system's impulse response or the ratio of the input and output transforms. The frequency response represents the mapping of input to output expressed in the frequency

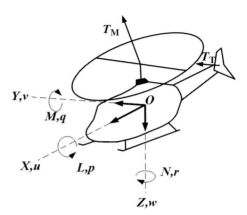

Figure 9. Helicopter body-fixed frame coordinate system (Raptis and Valavanis, 2011).

domain and can be expressed by its real and imaginary parts as

$$H(j\omega) = \frac{Y(j\omega)}{X(j\omega)} = H_{\mathrm{r}}(j\omega) + jH_{\mathrm{i}}(j\omega) \tag{5}$$

where

$$H_{\mathrm{r}}(j\omega) = |H(j\omega)| = \sqrt{H_{\mathrm{r}}^2(j\omega) + H_{\mathrm{i}}^2(j\omega)} \tag{6}$$

$$H_{\mathrm{i}}(j\omega) = \mathrm{e}^{j\angle H(j\omega)} \tag{7}$$

$$\angle H(j\omega) = \tan^{-1}\left(\frac{H_{\mathrm{i}}(j\omega)}{H_{\mathrm{r}}(j\omega)}\right) \tag{8}$$

The Fourier transform of a signal $p(t)$ is given as

$$P(j\omega) = \int_{-\infty}^{\omega} p(t)\mathrm{e}^{-j\omega t}\mathrm{d}t \tag{9}$$

A second method to obtain the frequency response of a system is through analysis of the spectral densities. The autospectral densities are given by S_{xx} and S_{yy}, and the *cross-spectral* density is given by S_{xy}.

In single-input single-output (SISO) systems, the coherence function is given as

$$\gamma_{xy}^2(j\omega) = \frac{|S_{xy}(j\omega)|^2}{|S_{xx}(j\omega)||S_{yy}(j\omega)|} \tag{10}$$

The coherence gives insight into the part of the output spectrum that is linearly correlated to the input spectrum. In multi-input multi-output (MIMO) systems, it gives a good indication of the correlation of each input–output pair and can be used to determine the frequency responses that will be eliminated before the identification process. In general, responses must have coherence values of $\gamma_{xy}^2 \geq 0.6$. Below this value, the correlation between input–output pairs is not reliable. An additional criterion requires that there must be a valid frequency range of interest whose decade span meets the condition $\mathrm{dec_{span}} = \log\left(\omega_{\max}/\omega_{\min}\right) \geq 0.3$ for which $\gamma_{xy}^2 \geq 0.5$. Including responses with decade spans below this guideline has been shown to degrade the identification (Tischler and Cauffman, 1992).

4.1   Frequency-Response Selection

The previously discussed guidelines are used to select the frequency responses and ranges of frequencies that are used for the system identification process. Frequency responses are computed for each input–output pair and examined to determine the frequency ranges for which $\gamma_{xy}^2 \geq 0.6$. This is done using the *CIFER* functions *FRESPID* for the computation of individual frequency responses, *MISOSA* that performs multi-input conditioning, and *COMPOSITE* that performs window averaging to smooth the data. For each input, three to four sweeps were performed. Of these, the best two were selected based on progression of frequencies in the sweep, record lengths, symmetry in responses, coherences, and magnitude and phase plots. These selected sweeps are concatenated in *FRESPID*. By providing more than one sweep, more frequency information is included. Frequency-response analysis results, using flight data from the DU^{2}SRI Bergen helicopter, are shown in Figures 10–12 for the on-axis input–output pairs and in Figures 13 and 14 for the off-axis input–output pairs. Improper performance of the frequency sweep experiments and collection of data will be reflected in the frequency responses. Improper selection of sampling frequency or rushed frequency sweep will give poor results in the coherence calculation for those input–output pairs.

Table 2 shows the selected frequency responses and ranges according to the coherence value criterion for each input–output pair for the DU^{2}SRI Bergen helicopter. The bold entries correspond to the on-axis responses. The frequency sweeps for the main collective and cyclics are performed with the feedback gyro in heading-hold mode. The sweeps for the tail collective are performed in both heading-hold and rate modes to examine the effects of the feedback gyro on the system dynamics.

5   MODEL LINEARIZATION

In order to perform system identification, a linearized state-space model structure must be determined. Linearization is achieved either by applying Taylor series expansion about some initial condition (Hald and Hesselbaek, 2005) or by applying small perturbation theory to the vehicle dynamic equations of motion. Small perturbation theory is widely used to linearize the nonlinear rotorcraft dynamics about a trim flight condition, usually hover. Examples may be found in Etkin and Reid (1996), Prouty (1995), Raptis and Valavanis (2011), and Ren *et al.* (2012). In the case of the rotorcraft dynamics, the total force is made up of forces and torques contributed by the various rotorcraft subsystems. These forces are referred to as either controlled, a result of the pilot input, or uncontrolled, a result of the dynamic parameters (Alvarenga *et al.*, 2015).

Small perturbation analysis involves applying a small incremental force Δf, resulting in small perturbations to the dynamics. For a rotorcraft, the dynamic parameters that make up the state vector are given in (11), while the control inputs are given in (12):

$$\vec{x} = \begin{bmatrix} u & v & p & q & \phi & \theta & w & r & a & b & r_{\mathrm{fb}} & c_{\mathrm{fb}} & d_{\mathrm{fb}} \end{bmatrix}^{\mathrm{T}} \tag{11}$$

$$\vec{u}_{\mathrm{c}} = \begin{bmatrix} \delta_{\mathrm{lat}} & \delta_{\mathrm{lon}} & \delta_{\mathrm{ped}} & \delta_{\mathrm{col}} \end{bmatrix}^{\mathrm{T}} \tag{12}$$

Table 2. Frequency (Hz) responses and corresponding frequency ranges.

	δ_{lat}	δ_{lon}	δ_{col}	δ_{ped} (Rate mode)	δ_{ped} (Heading-hold mode)
u	0.06–4.0	**0.5–5.7**	–	–	3.27–11.4
v	**0.55–7.8**	–	–	–	0.62–11.4
w	–	2.7–10	**0.3–12**	–	–
p	**0.4–12**	–	–	0.6–0.48	0.62–7.39
q	0.06–12	0.4–12	–	2.2–5.11	3.48–6.87
r	–	–	–	**0.3–6.5**	**0.4–12**
ϕ	**0.5–12**	2.4–10	–	–	–
θ	1.6–12	**0.52–12**	–	–	–
ψ	–	–	–	**0.6–3.5**	**0.4–12**

Bold entries correspond to on-axis responses.

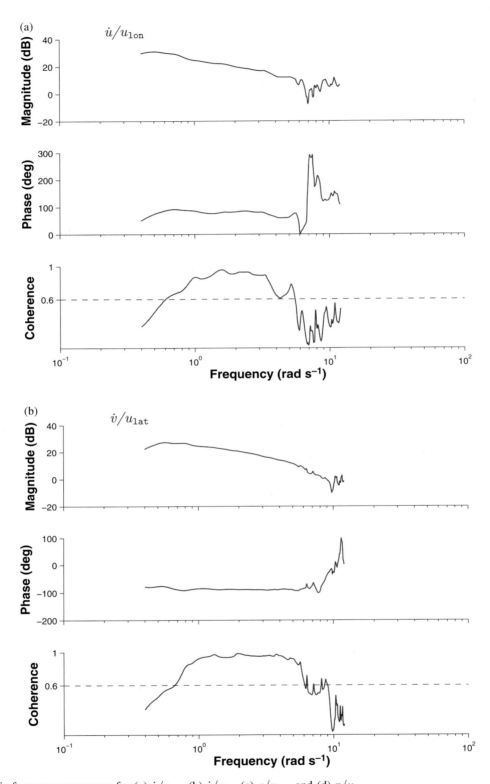

Figure 10. On-axis frequency responses for (a) $\dot{u}/u_{\mathrm{lon}}$, (b) $\dot{v}/u_{\mathrm{lat}}$, (c) q/u_{lon}, and (d) p/u_{lat}.

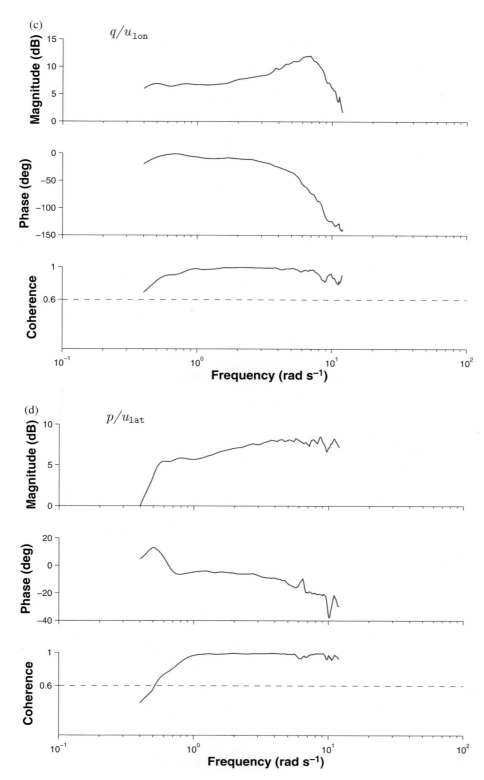

Figure 10. Continued

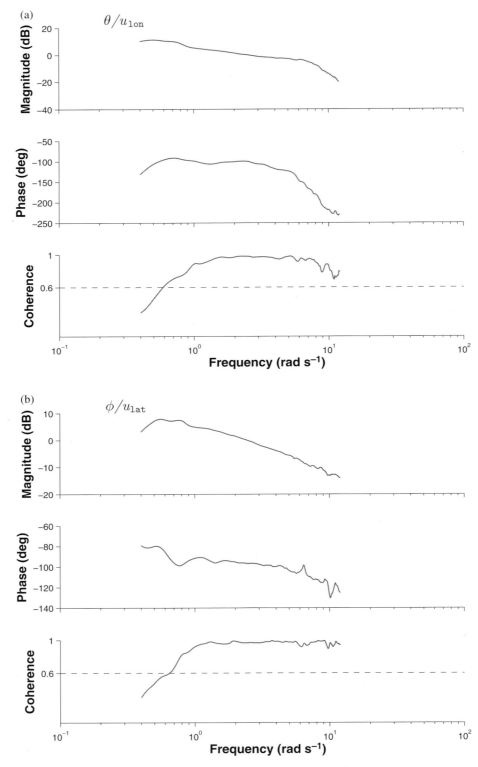

Figure 11. On-axis frequency responses for (a) θ/u_{lon}, (b) ϕ/u_{lat}, (c) $\dot{w}/u_{\mathrm{col}}$, and (d) r/u_{ped}.

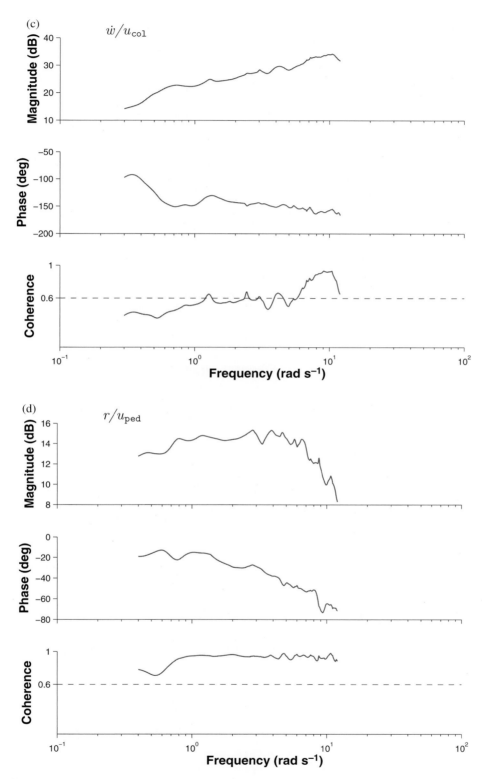

Figure 11. Continued

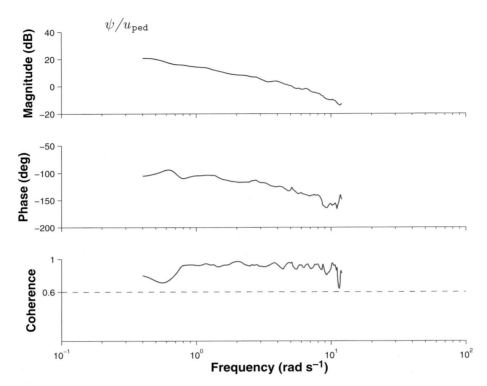

Figure 12. On-axis frequency responses for ψ/u_{ped}.

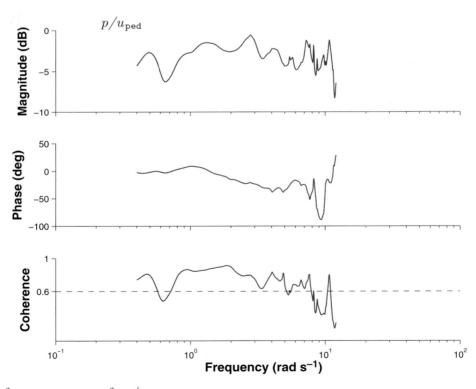

Figure 13. Off-axis frequency responses for p/u_{ped}.

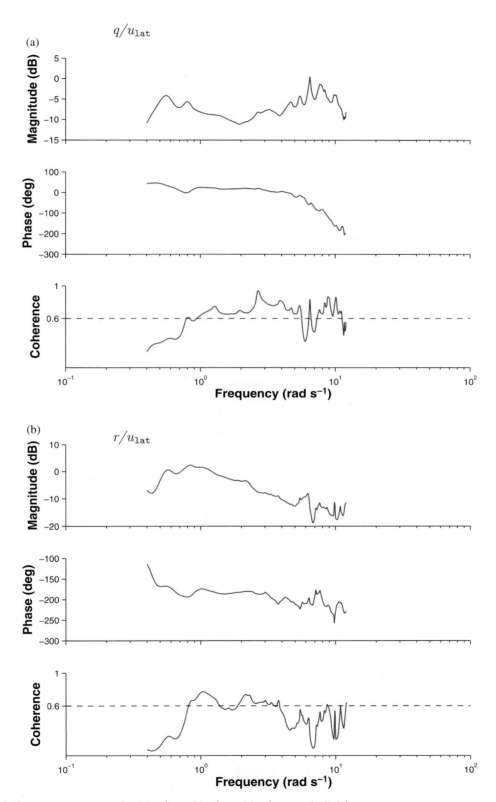

Figure 14. Off-axis frequency responses for (a) q/u_{lat}, (b) r/u_{lat}, (c) p/u_{lon}, and (d) $\dot{v}/u_{\text{ped}}$.

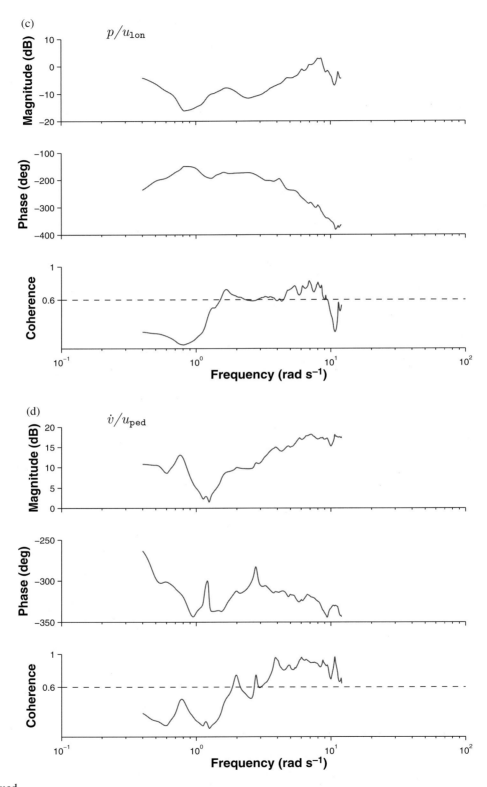

Figure 14. Continued

In Etkin and Reid (1996), the forces and moments are defined to be strictly functions of the state and input variables. A linear approximation of the forces can be found using a first-order Taylor approximation about the initial condition $x_i(t_0) = a$, given by

$$f_{x_i} \approx f(a) + f'(a)(x_i - a) \quad (13)$$

Although it may be desired to retain terms of higher order derivatives or nonlinear terms for the sake of accuracy and completeness, as in La Civita (2002), often only the first-order terms are considered. For small-enough motion, the effects of the nonlinear terms (e.g., $\partial^2 F/\partial x^2$), and derivatives of dynamic parameters (e.g., $\dot{u}, \dot{q}$) are insignificant (Padfield, 2007). For small-scaled unmanned helicopters, this corresponds to low speeds, below 3 m s^{-1}. As flight speed exceeds this limit, the validity of the linearized model will degrade. Separate models may be identified for low-speed forward flight, as seen in Mettler (2003).

The derivative of the forces, in terms of the disturbed variables Δx_i and $\Delta \delta_i$, and the disturbed force $\Delta F_{x_i} = f(x + \Delta x) - f(x)$, is given by

$$\frac{\partial f}{\partial x} = \frac{\Delta F x_i}{\Delta x} \quad (14)$$

This allows for the perturbed forces and moments to be defined as linear functions of the perturbed variables and the force derivatives ($F_{x_i} = \partial f/\partial x_i$). This combination is given by

$$\Delta F = \sum_{x_i \in \bar{x}} F_{x_i} \cdot \Delta x_i + \sum_{\delta_i \in \bar{u}} F_{\delta_i} \cdot \Delta \delta_i \quad (15)$$

The derivatives with respect to the controlled inputs are referred to as the *control derivatives*, while those with respect to the uncontrolled states are known as the *stability derivatives*. The notation is simplified to $\partial f/\partial a = F_a$. The derivatives are listed in Table 3.

The forces ($F = \begin{bmatrix} X & Y & Z \end{bmatrix}^T$) and moments ($\tau = \begin{bmatrix} L & M & N \end{bmatrix}^T$) that drive the rigid-body dynamics consist

Table 3. Control and stability derivatives.

Derivative type	Notation	Description
Control derivatives	$F_{\delta_{col}}$	Collective input
	$F_{\delta_{ped}}$	Tail rotor collective
	$F_{\delta_{lat}}$	Lateral cyclic
	$F_{\delta_{lon}}$	Longitudinal cyclic
Stability derivatives	F_u, F_v, F_w	Translational velocities
	F_p, F_q, F_r	Angular rates
	F_ϕ, F_θ, F_ψ	Orientation angles
	F_a, F_b	Main rotor cyclic angles
	F_{cfb}, F_{dfb}	Main rotor cyclic feedback gyro gains
	F_{rfb}	Tail feedback gyro gain

of the X, Y, and Z body forces and L, M, and N moments about the body axes acting on the rotorcraft, as shown in Figure 9. A small increment of each of these forces and moments is a sum of the derivatives and perturbations, as shown in Table 3 and (15), and are given by

$$\begin{aligned}
\Delta X &= X_u \Delta u + X_v \Delta v + \cdots + X_{\delta_{col}} \Delta \delta_{col} + \cdots \\
\Delta Y &= Y_u \Delta u + Y_v \Delta v + \cdots + Y_{\delta_{col}} \Delta \delta_{col} + \cdots \\
\Delta Z &= Z_u \Delta u + Z_v \Delta v + \cdots + Z_{\delta_{col}} \Delta \delta_{col} + \cdots \\
\Delta L &= L_u \Delta u + L_v \Delta v + \cdots + L_{\delta_{col}} \Delta \delta_{col} + \cdots \\
\Delta M &= M_u \Delta u + M_v \Delta v + \cdots + M_{\delta_{col}} \Delta \delta_{col} + \cdots \\
\Delta N &= N_u \Delta u + N_v \Delta v + \cdots + N_{\delta_{col}} \Delta \delta_{col} + \cdots
\end{aligned} \quad (16)$$

Next, small perturbations ($\delta = \Delta \delta + \delta_0$) are applied to the rigid-body dynamics. Applying the perturbed variables to the forward velocity component of the rigid-body dynamics reduces to (17):

$$\dot{u}_0 + \Delta \dot{u} = (r_0 + \Delta r)(v_0 + \Delta v) - (q_0 + \Delta q)(w_0 + \Delta w) \\ -\sin(\theta_0 + \Delta\theta)g + \frac{X_0 + \Delta X}{m} \quad (17)$$

It is assumed that the perturbations and any derivative have very small values. As a result, the product of perturbations is subsequently negligible (Raptis and Valavanis, 2011). These assumptions result in the following properties: (i) $\Delta x \Delta y = 0$, (ii) $\cos(\Delta\theta) = 1$, and (iii) $\sin(\Delta\theta) = \Delta\theta$. Applying these properties to the forward velocity component in (17) gives

$$\dot{u}_0 + \Delta\dot{u} = r_0 v_0 + r_0 \Delta v + v_0 \Delta r - q_0 w_0 - q_0 \Delta w - w_0 \Delta q \\ -\sin\theta_0 g - \Delta\theta \cos\theta_0 g + \frac{X_0}{m} + \frac{\Delta X}{m} \quad (18)$$

In hover flight, the rotorcraft is operating in conditions where $u_0 = v_0 = w_0 = p_0 = q_0 = r_0 = \dot{u}_0 = 0$. This reduces the forward velocity equation to

$$\Delta\dot{u} = -\sin\theta_0 g - \Delta\theta \cos\theta_0 g + \frac{X_0}{m} + \frac{\Delta X}{m} \quad (19)$$

At hover, the rotorcraft flies at nearly level flight such that the attitude angles are very small. For the forward velocity dynamics, this reduces (19) to (20)

$$\Delta\dot{u} = -\theta_0 g - \Delta\theta g + \frac{X_0}{m} + \frac{\Delta X}{m} \quad (20)$$

5.1 Trim

Aircraft in steady flight must operate at some conditions where the forces and moments are in equilibrium about the center of gravity, known as trim flight (Prouty, 1995). Trim conditions correspond to certain trim values of the state and input variables, given by x_0 and δ_0, respectively. These trim values can be found both analytically, as seen in Prouty (1995), and numerically, as seen in Ren *et al.* (2012). In trim flight, it is assumed that there are no disturbances so that $\Delta \dot{u} = \Delta \theta = \Delta X = 0$. Given these assumptions, the forward velocity dynamics in (17) becomes (21). This equilibrium condition is combined with the forward velocity dynamics at hover given in (20) to form the trimmed linear forward velocity dynamics at hover. The complete set of equilibrium equations at hover are derived and given as

$$X_0 = mg \sin \theta_0 \tag{21}$$

$$Y_0 = -mg \sin \phi_0 \cos \theta_0 \tag{22}$$

$$Z_0 = -mg \cos \theta_0 \cos \phi_0 \tag{23}$$

$$L_0 = M_0 = N_0 = 0 \tag{24}$$

$$\dot{x}^{\mathrm{I}} = \dot{y}^{\mathrm{I}} = \dot{z}^{\mathrm{I}} = 0 \tag{25}$$

At level cruise, the trim conditions mimic those of hover except that the initial condition for translational velocity, usually forward, is set to a nonzero value. In the case of level forward flight, $u_0 \neq 0$. In Oktay (2012), the trim condition for a level-banked turn is given using a constant forward velocity, constant yaw angle, and no sideslip.

5.2 Comprehensive Linear State-Space Model

Following this same procedure with the entire set of dynamic equations, a complete set of linear dynamic equations is derived and is given in state-space form as

$$\vec{x} = A\vec{x} + B\vec{u} \tag{26}$$

The state matrix A contains the stability derivatives, the input matrix B contains the control derivatives, $\vec{x}$ contains the state variables, and $\vec{u}$ contains the input variables. Following procedures from Etkin and Reid (1996), Mettler (2003), Prouty (1995), Padfield (2007), and Raptis and Valavanis (2011), a comprehensive linear model is derived. From this generalized structure, the rotorcraft model may be further simplified according to the assumed flight condition (i.e., hover, cruise, turn, etc.). For example, at hover, all of the terms containing initial translational and angular velocity parameters $(u_0, v_0, \ldots)$ are eliminated, and terms with initial condition attitude angles reduce due to trigonometric identities. An example of this type of structure can be found in Mettler (2003). The remaining derivative terms must be identified by the linearization of the forces under the identified trim condition. However, in straight-level cruise or a banked turn, the initial velocities and angular rates must be taken into consideration and included in the model structure. Using the proposed linear model structure, a bank of linear models may be developed by considering various flight conditions. In Joelianto *et al.* (2011), the linear model derivatives are determined at hover in addition to various forward speeds in straight-level flight. The state-space structure for the DU^{2}SRI Bergen helicopter in hover flight is given in (27)–(30):

$$x = \begin{bmatrix} u & v & p & q & \phi & \theta & a & b & w & r & r_{\mathrm{fb}} & c_{\mathrm{fb}} & d_{\mathrm{fb}} \end{bmatrix}^{\mathrm{T}} \tag{27}$$

$$u = \begin{bmatrix} \delta_{\mathrm{lat}} & \delta_{\mathrm{lon}} & \delta_{\mathrm{ped}} & \delta_{\mathrm{col}} \end{bmatrix}^{\mathrm{T}} \tag{28}$$

$$B = \begin{bmatrix} 0 & 0 & 0 & 0 & 0 & 0 & A_{\mathrm{lat}} & B_{\mathrm{lat}} & 0 & 0 & 0 & 0 & I_{\mathrm{lat}} \\ 0 & 0 & 0 & 0 & 0 & 0 & A_{\mathrm{lon}} & B_{\mathrm{lon}} & 0 & 0 & 0 & H_{\mathrm{lon}} & 0 \\ 0 & Y_{\mathrm{ped}} & 0 & 0 & 0 & 0 & 0 & 0 & 0 & N_{\mathrm{ped}} & 0 & 0 & 0 \\ 0 & 0 & 0 & M_{\mathrm{col}} & 0 & 0 & 0 & 0 & Z_{\mathrm{col}} & N_{\mathrm{col}} & 0 & 0 & 0 \end{bmatrix}^{\mathrm{T}} \tag{29}$$

$$A = \begin{bmatrix} X_u & 0 & 0 & 0 & 0 & -g & X_a & 0 & 0 & 0 & 0 & 0 & 0 \\ 0 & Y_v & 0 & 0 & g & 0 & 0 & Y_b & 0 & 0 & 0 & 0 & 0 \\ L_u & L_v & 0 & 0 & 0 & 0 & 0 & L_b & L_w & 0 & 0 & 0 & L_{d\mathrm{fb}} \\ M_u & M_v & 0 & 0 & 0 & 0 & M_a & 0 & M_w & 0 & 0 & M_{c\mathrm{fb}} & 0 \\ 0 & 0 & 1 & 0 & 0 & 0 & 0 & 0 & 0 & 0 & 0 & 0 & 0 \\ 0 & 0 & 0 & 1 & 0 & 0 & 0 & 0 & 0 & 0 & 0 & 0 & 0 \\ 0 & 0 & 0 & -1 & 0 & 0 & -1/\tau_{\mathrm{f}} & A_b & 0 & 0 & 0 & A_{c\mathrm{fb}} & 0 \\ 0 & 0 & -1 & 0 & 0 & 0 & B_a & -1/\tau_{\mathrm{f}} & 0 & 0 & 0 & 0 & B_{d\mathrm{fb}} \\ 0 & 0 & 0 & 0 & 0 & 0 & Z_a & Z_b & Z_w & Z_r & 0 & 0 & 0 \\ 0 & N_v & N_p & 0 & 0 & 0 & 0 & 0 & N_w & N_r & N_{r\mathrm{fb}} & 0 & 0 \\ 0 & 0 & 0 & 0 & 0 & 0 & 0 & 0 & 0 & K_r & K_{r\mathrm{fb}} & 0 & 0 \\ 0 & 0 & 0 & H_q & 0 & 0 & 0 & 0 & 0 & 0 & 0 & H_{c\mathrm{fb}} & 0 \\ 0 & 0 & I_p & 0 & 0 & 0 & 0 & 0 & 0 & 0 & 0 & 0 & I_{d\mathrm{fb}} \end{bmatrix} \tag{30}$$

6 STABILITY AND CONTROL DERIVATIVES

The stability and control derivatives can first be estimated using the linearization of the 6-DOF dynamic equations around the operating condition of interest. The stability and control derivatives are summarized as follows:

- X_u, Y_v, Z_w: Linear velocity stability derivatives resulting from parasitic drag forces and rotor thrust generation
- X_a, Y_b, Z_a, Z_b: Linear velocity stability derivatives resulting from the main rotor force component
- X_θ, Y_ϕ: Linear velocity stability derivatives resulting from the gravitational force component
- $L_u, L_v, L_w, M_u, M_v, M_w, N_v, N_w$: Moment stability derivatives resulting from main rotor thrust and horizontal and vertical tail fin drag forces
- L_b, M_a: Moment stability derivatives resulting from main rotor thrust and flapping forces
- A_b, B_a: Main rotor flapping stability derivatives directly from flapping dynamics
- $K_{rfb}, H_{rfb}, I_{rfb}, K_r, H_q, I_p$: Three-dimensional gyro feedback gains
- $N_{rfb}, M_{cfb}, L_{dfb}$: Angular rate responses to 3D gyro feedback
- Y_{ped}: Lateral velocity control derivative due to tail input force
- M_{col}: Moment control derivative due to main rotor input force
- A_{lon}, B_{lat}: On-axis flapping control derivatives due to cyclic inputs
- A_{lat}, B_{lon}: Off-axis flapping control derivatives due to cyclic inputs
- Z_{col}: Vertical velocity control derivative due to main rotor input force
- N_{ped}: Moment control derivative due to tail rotor input force
- H_{lon}, I_{lat}: Cyclic direct feed-forward gains
- Z_r, N_{col}: Heave–yaw coupling derivatives
- N_p: Roll–yaw coupling derivative
- N_r: Yaw damping

7 FREQUENCY RESPONSE

The stability and control derivatives can first be estimated using the linearization of the 6-DOF dynamic equations around the operating condition of interest. At hover, the main rotor thrust magnitude is assumed to be equal to the entire weight of the rotorcraft. These parameter estimates

may be further tuned by using the associated frequency responses. In CIFER, the transfer identification function (NAVFIT) may be used to specify the order of the transfer function, add time delays, and make corrections by powers of s.

For the heave dynamics, the equation of motion linearized about hover is given as

$$\dot{w} = Z_w w + Z_{col}\delta_{col} \tag{31}$$

where Z_{col} is the control derivative associated with the collective pitch input δ_{col} and Z_w is the stability derivative associated with the helicopter's vertical speed, rotor damping, and drag (Mettler, 2003).

The equation of motion can be rewritten in transfer function form as

$$\frac{d\dot{w}}{d\delta_{col}} = \frac{z_{col}}{s + z_w} \tag{32}$$

Using the NAVFIT function in CIFER, the stability and control derivatives can be estimated. Figure 15 shows the comparison between the transfer function model and the flight data frequency-response results.

For the yaw dynamics, the linearized equation of motion about hover is given as

$$\dot{r} = N_v v + N_r r + N_{ped}\delta_{ped} \tag{33}$$

where N_v is the stability derivative associated with the lateral speed v, N_r is the stability derivative associate with the yaw rate r, and N_{ped} is the control derivative associated with the tail rotor collective pitch.

The equation can be rewritten in transfer function form as

$$\frac{dr}{d\delta_{ped}} = \frac{N_{ped}}{s + N_r} \tag{34}$$

In small-scale helicopters, the addition of a yaw feedback gyro introduces additional dynamics into the yaw rate response. Following Mettler (2003), we assume that the yaw feedback dynamics can be modeled as a first-order low-pass filter:

$$\frac{d_{rfb}}{dr} = \frac{K_r}{s + K_{rfb}} \tag{35}$$

resulting in the closed-loop dynamics:

$$\frac{dr}{d\delta_{ped}} = \frac{N_{ped}(s + K_{rfb})}{s^2 + (K_{rfb} - N_r)s + (K_r N_{ped} - N_r K_{rfb})} \tag{36}$$

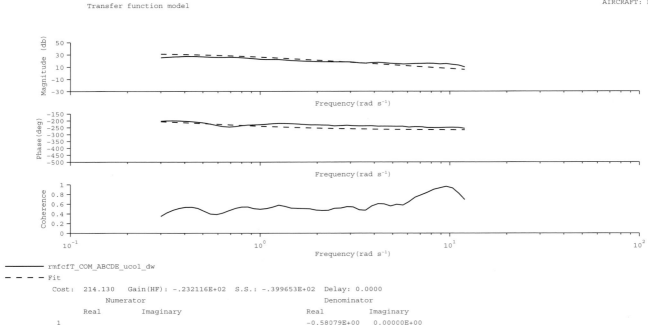

Figure 15. Heave frequency response ($\dot{w}/\delta_{\mathrm{col}}$) generated by CIFER of flight data (solid line) and transfer function fit (dashed line).

As with the heave response, NAVFIT is used to find a transfer function fit according to this structure. Figure 16 shows the comparison between the transfer function model and the flight data frequency-response results. In order to ensure good identification results, a constraint is placed on the derivatives. The pole of the low-pass filter is designed to be faster than that of the bare-frame dynamic model:

$$K_{rfb} > -2N_r \tag{37}$$

This will ensure proper low-pass filter performance. A similar approach is taken to determine initial model guesses for the cyclic dynamics.

8   THREE-DIMENSIONAL RATE GYRO

Unlike their full-scale counterparts, small-scale rotorcraft use feedback gyros and flybars, also known as stabilizer bars, to reduce the effort on the pilot to maintain the rotorcraft in stable flight. Flybars act as mechanical dampeners in the pitch and roll axis. Their design results in flapping dynamics that cannot be measured during flight. As a result, the flybar flapping dynamics must be estimated through state estimation algorithms, such as state observers, in order to

implement full-state feedback systems. Additionally, a feedback gyro works to aid the pilot in maintaining the commanded yaw rate. These feedback gyros usually work in two modes: (i) rate mode and (ii) heading-hold mode. For flybarless helicopters, the stabilizer bar is replaced with a 3D feedback gyro. The DU^{2}SRI Bergen helicopter is equipped with a Beast-X Microbeast 3D rate gyro. These gyros act as individual gyros for each of the yaw, pitch, and roll axes. The tail (yaw) channel operates as it would with a regular single-axis tail feedback gyro, as shown in Figure 17. The cyclic channels, according to the Beast-X manual, operate slightly differently. The cyclic gyros not only provide a dampening effect, but the pilot can also adjust the direct feed-forward gain. This gain effectively controls how much of the pilot command is mixed with the feedback loop, as shown in Figure 18. In this work, the cyclic feed-forward gain is set to about 75% of the total range. This gives the pilot more control over the helicopter.

As shown in Figures 17 and 18, the addition of the feedback gyro and the actuators augment the bare-frame dynamics. In order to properly identify the bare-frame dynamics, the feedback gyro and actuators must either be included in the first-principles model and linearization process, where the bare-frame dynamics are measured directly, or the dynamics may be extracted through frequency-

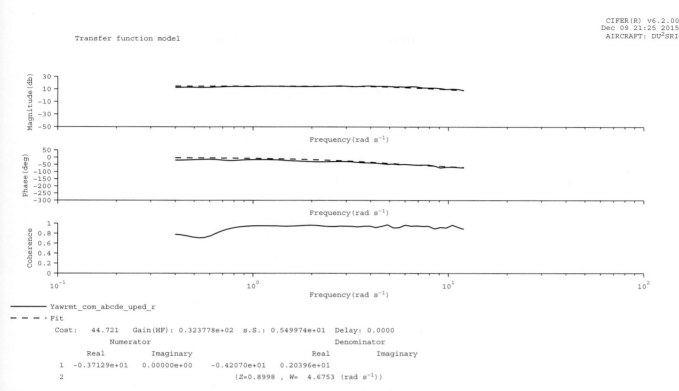

CIFER(R) v6.2.00
Dec 09 21:25 2015
AIRCRAFT: DU²SRI

Transfer function model

Figure 16. Yaw rate frequency response (r/δ_{ped}) generated by CIFER of flight data (solid line) and transfer function fit (dashed line).

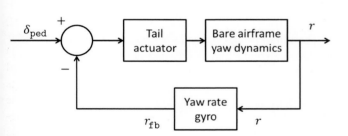

Figure 17. Tail feedback gyro yaw dynamic block diagram.

response arithmetic, as seen in Mettler (2003) for a single-axis yaw feedback gyro. While there have been a number of groups that have tackled the problem of identification of yaw feedback gyros and flybar systems, there have been few examples with flybarless helicopters and 3D feedback gyros. In the DU²SRI helicopter, multiple flight experiments were performed in order to identify the behavior of the 3D feedback gyro. To do so, the frequency sweeps were performed while measuring both the pilot stick commands going to the MicroBeast and those from the MicroBeast to the servos. The measured sweeps for the tail and lateral channels are shown

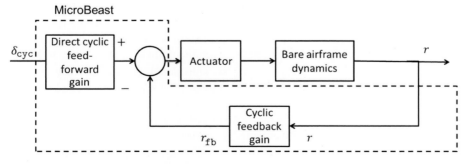

Figure 18. Proposed cyclic feedback dynamics block diagram.

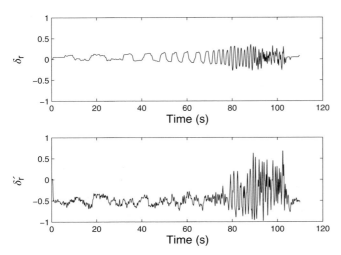

Figure 19. Normalized pilot rudder command δ_r and normalized augmented rudder command from feedback gyro δ'_r.

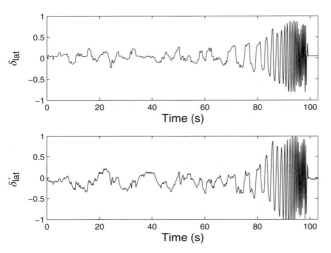

Figure 20. Normalized pilot lateral cyclic command δ_{lat} and normalized augmented lateral cyclic command from feedback gyro δ'_{lat}.

in Figures 19 and 20, respectively. The cyclic feedback gyro dynamics are modeled similar to the yaw feedback gyro dynamics.

9   STATE-SPACE MODEL IDENTIFICATION

For the DU²SRI helicopter, a minimum of three flights were performed for each type of sweep. Frequency analysis results are used to determine the best two sweeps to concatenate. These concatenated sweeps are then analyzed again. Next, an initial model structure is established. The results are then processed using *DEREVID* in *CIFER*. This function uses the frequency-response results and given model structure and finds the best fit to the frequency-response data. This is done through an optimization algorithm that iteratively seeks to reduce the magnitude and phase errors between the flight data and estimated model while minimizing the cost of doing so for each input–output pair. This algorithm iterates until the average cost across all pairs is minimized. Ultimately, the cost function gives a good measure of how accurately the estimated model fits the flight data. Two additional metrics are provided as a result of the *DEREVID* process. The first is the Cramer–Rao bound. This provides a measure of the reliability of each derivative. The second metric is the insensitivity which occurs when a parameter has little to no effect on the cost function. The Cramer–Rao bound represents the smallest expected standard deviation. A high Cramer–Rao bound means the parameter could either be eliminated or fixed in the model structure. High insensitivities mean that the parameter does not have much importance in the model structure and should be eliminated from the model. These guidelines are summarized as follows:

- Cost function, $J_a \leq 100$
- Cramer–Rao bound, CR (%) ≤ 20
- Insensitivity, I (%) ≤ 10

The initial model structure is obtained using the DU²SRI Bergen helicopter parameters and equations of motion linearized around hover. The resulting initial model and final model derivatives, including Cramer–Rao bound and insensitivity percentages, are shown in Table 4.

10   FREQUENCY RESPONSE OF THE IDENTIFIED MODEL

Figure 21 shows the frequency responses of the identified model against the frequency responses generated from flight data. The identified model responses show good agreement with the flight data. According to Mettler (2003), a cost of 70 is considered an exceptionally good fit. Table 5 shows the transfer function costs for each of the input–output pairs that were obtained during the identification process. Most of the responses had costs well below the threshold of $J = 100$. The iteration process was stopped to eliminate overestimation, which may result in undesirable high-frequency modes in the estimated model.

Table 4. Identified linear state-space model parameters.

	A Matrix		
Parameter	Final value	CR (%)	Insensitivity (%)
X_u	−0.07225	0.08124	45.36
X_a	−0.9498	20.02	121.3
Y_v	0.0728	0.08691	40.82
Y_b	17.22	55.59	35.9
L_v	−0.02969	0.1757	58.38
L_p	−0.2367	2.497	55.62
L_b	−383.2	3.812	0.226
$L_{d_{fb}}$	−0.3562	143.1	59.86
M_u	0.1687	0.1573	332.6
M_q	−0.7982	7.374	10.12
M_a	−12.29	134	10.62
M_{cfb}	−3.595	72.89	5.128
A_p	1.355	20.41	3.16
A_a	12.86	127.1	3.355
A_b	29.24	468.6	10.89
B_q	0.176	2.736	8.641
B_a	17.64	128.7	1.386
B_b	15.12	81.17	5.812
Z_w	−0.0259	0.2802	444
N_r	−0.5349	0.6665	29.13
N_{rfb}	330	43720	10.69
K_r	−0.03157	1.393	3.564
K_{rfb}	0.678	1.281	33.9
H_q	−5.787	82.92	3.557
H_{cfb}	0.9504	17.58	47.81
I_p	−0.1062	0.2516	0.7222
I_{dfb}	−0.07811	0.6809	154.7
g	9.81		
τ_f	0.01968		

Table 5. Transfer function costs for input–output pairs.

Response	Transfer function cost
v/δ_{lat}	77.87
p/δ_{lat}	84.29
q/δ_{lat}	105.2
ϕ/δ_{lat}	101.22
u/δ_{lon}	20.13
q/δ_{lon}	45.76
θ/δ_{lon}	45.00
r/δ_{ped}	43.72
w/δ_{col}	80.25

11　TIME-DOMAIN VALIDATION

Once the linear stability and control derivatives are identified, the model must be validated to ensure robustness to other control inputs. This is done with a completely separate set of flight data than that used for the system identification process. For this, a doublet input is applied to each control channel individually while maintaining the off-axis channels at trim. The maneuver starts and ends in hover. This input is characterized by two unit steps, or singlets, of opposite sign performed consecutively for a short period, 1–3 s. For the DU^{2}SRI Bergen helicopter, the doublets were performed using 10–40% pilot stick deflection from trim conditions with a duration of 2 s per singlet. This was done to prevent large translational speeds that would take the rotorcraft out of the pilot's line of sight as well as keep the rotorcraft from exhibiting too large of an

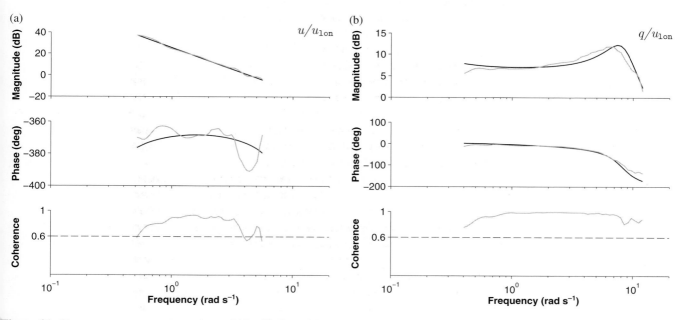

Figure 21. Frequency-response comparison of identified model responses (black line) and flight data (gray line).

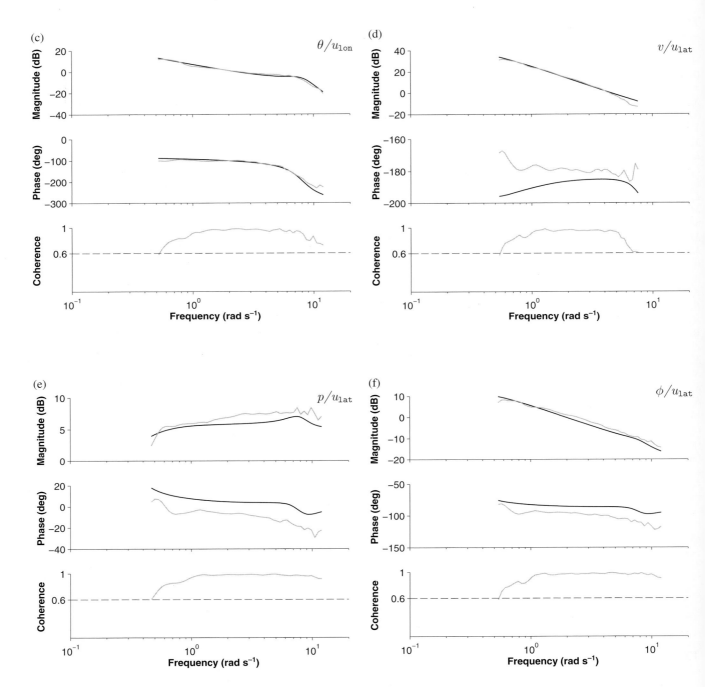

Figure 21. Continued

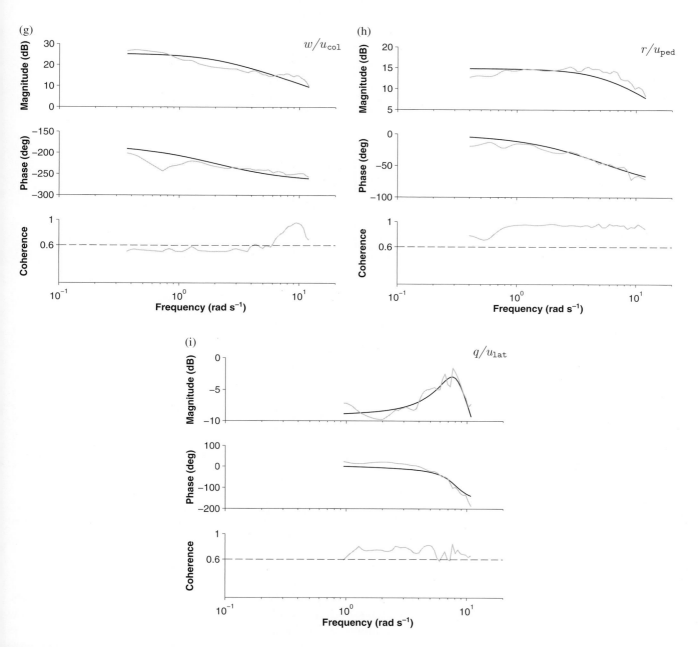

Figure 21. Continued

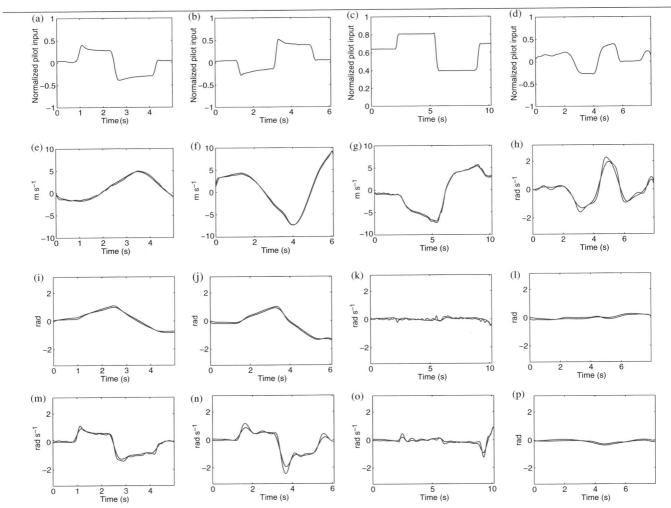

Figure 22. Time history response of identified model (black) versus flight data (gray). The pilot commands (top row) are given in normalized values: (a) u_{lat}, (b) u_{lon}, (c) u_{col}, (d) u_{ped}, (e) v, (f) u, (g) w, (h) r, (i) ϕ, (j) θ, (k) p, (l) ϕ, (m) p, (n) q, (o) q, and (p) θ.

attitude change. The *CIFER* function *VERIFY* is used to compare the flight results to the derived model by applying the pilot input to the model and observing the resulting output. These results give insight into the robustness of the identified linear model to different inputs. Figure 22 shows on-axis response results of the verification of the estimated model using CIFER.

12   CONCLUSION

In this chapter, the system identification process, as it pertains to unmanned rotorcraft, has been presented. First, the requirements for flight testing, maneuvers, and data collection were shown, including considerations that must be taken when

dealing with small-scale, unmanned rotorcraft. Next, the process of linearizing the model dynamics is given, resulting in a linear model structure for a small-scale, flybarless rotorcraft. Next, frequency-domain analysis is presented using the CIFER software. Then, the resulting model derivatives are presented. Finally, model validation is discussed.

REFERENCES

Alvarenga, J., Vitzilaios, N.I., Valavanis, K.P., and Rutherford, M.J. (2015) Survey of unmanned helicopter model-based navigation and control techniques. *J. Intell. Robot. Syst.*, **80**(1), 87–138.

Budiyono, A. and Lesmana, H. (2007) First principle approach to modeling of small scale helicopter, in *International*

Conference on Intelligent Unmanned Systems, Bali, Indonesia, pp. 100–110.

Etkin, B. and Reid, L. (1996) *Dynamics of Flight: Stability and Control*, 3rd ed., John Wiley & Sons, Inc., New York.

Hald, U.B. and Hesselbaek, M.V. (2005) *Autonomous Helicopter: Modeling and Control*, Technical Report, Aalborg University, Aalborg, Denmark.

Hamel, P.G. and Kaletka, J. (1997) Advances in rotorcraft system identification. *Prog. Aerosp. Sci.*, **33** (3-4), 259–284.

Joelianto, E., Sumarjono, E.M., Budiyono, A., and Penggalih, D.R. (2011) Model predictive control for autonomous unmanned helicopters. *Aircr. Eng. Aerosp. Technol.*, **83**(6), 375–387.

La Civita, M. (2002) *Integrated modeling and robust control for full-envelope flight of robotic helicopters*. PhD thesis, Carnegie Mellon University.

Mettler, B. (2003) *Identification Modeling and Characteristics of Miniature Rotorcraft*, Springer, Boston, MA.

Oktay, T. (2012) *Constrained Control of Complex Helicopter Models*. PhD thesis, Virginia Polytechnic Institute and State University.

Padfield, G.D. (2007) *Helicopter Flight Dynamics*, Blackwell Publishing Ltd, Oxford, UK.

Prouty, R. (1995) *Helicopter Performance, Stability and Control*, Krieger Publishing Company.

Raptis, I. and Valavanis, K. (2011) *Linear and Nonlinear Control of Small-Scale Unmanned Helicopters, Intelligent Systems, Control and Automation: Science and Engineering*, Springer, The Netherlands.

Ren, B., Ge, S.S., Chen, C., Fua, C.-H., and Lee, T.H. (2012) *Modeling, Control and Coordination of Helicopter Systems*, Springer, New York.

Shim, H. and Sastry, S. (2000) Control system design for rotor-craft-based unmanned aerial vehicles using time-domain system identification, C in *2000 IEEE International Conference on Control Applications*, pp. 808–813.

Tischler, M. and Cauffman, M. (1992) Frequency-response method for rotorcraft system identification: flight applications to BO 105 coupled rotor/fuselage dynamics. *J. Am. Helicopter Soc.*, **7**(3), 3–17.

Tischler, M. and Remple, R. (2006) *Aircraft and Rotorcraft System Identification: Engineering Methods with Flight-Test Examples*, AIAA Education Series, American Institute of Aeronautics and Astronautics.

Chapter 21
Trajectory Planning and Guidance

Pantelis Isaiah and Tal Shima

Faculty of Aerospace Engineering, The Technion—Israel Institute of Technology, Haifa, Israel

1 INTRODUCTION

The theme of this chapter revolves around trajectory planning and guidance of vehicles. Each topic has garnered independent, long-standing interest, and has been expounded in dedicated monographs (LaValle, 2006; Zarchan, 2012); the presentation herein brings them together toward a unifying goal: the tracking of a trajectory—be it that of a target or a precomputed one—by an autonomous vehicle. Within the present context, *trajectory planning* refers to the methods used to generate a trajectory in physical space that satisfies a set of requirements and constraints. *Guidance* refers to the implementation of a geometric rule, by means of suitable controls, enforcing a desired behavior on the part of a vehicle. For example, it can be a rule that leads to the interception or tracking of a target.

In a nutshell, the rationale behind decomposing a problem of interception or path following into separate problems of trajectory planning, guidance, and, ultimately, control, is the need to cope with analytic and computational complexity. To wit, if a desired objective is modeled mathematically as an optimal control problem with constraints and the underlying vehicle models tend to be realistic—which typically means that the models will be nonlinear and high-dimensional—a numerical solution is likely to be computationally prohibitive for real-time implementation. In adversarial scenarios where the theory of differential games (Isaacs, 1999) may offer a more natural framework than optimal control, for example, the interception of a maneuvring target or trajectory planning in the presence of uncertainty, the limitations that apply to optimal control formulations are only accentuated, since the numerical solution of differential games is computationally more intensive than that of optimal control problems (Patsko and Turova, 2011).

It is, therefore, a blend of theoretical and practical considerations that has led to a hierarchical paradigm for the solution of problems related to navigation, interception, motion control, coordination of vehicles, and, in general, of problems where the ultimate goal is to control one or more vehicles along the desired trajectories. Figure 1 shows a conceptual diagram that summarizes the steps involved in progressing from a tractable analysis to a practical implementation. Given a task to be accomplished, an admissible trajectory is computed for a model that is simple enough to be amenable to analytic and computational tools, but also captures some of the essential features of the problem to be ultimately solved. Then, a guidance law for tracking is devised, or chosen, that is either independent of the vehicle model or applicable to a more granular, realistic model. The last step is to derive the control command that implements the guidance law and leads to the tracking of the computed trajectory by a vehicle whose model is possibly more

Unmanned Aircraft Systems. Edited by Ella Atkins, Aníbal Ollero,
Antonios Tsourdos, Richard Blockley and Wei Shyy.
© 2016 John Wiley & Sons, Ltd. ISBN: 978-1-118-86645-0.

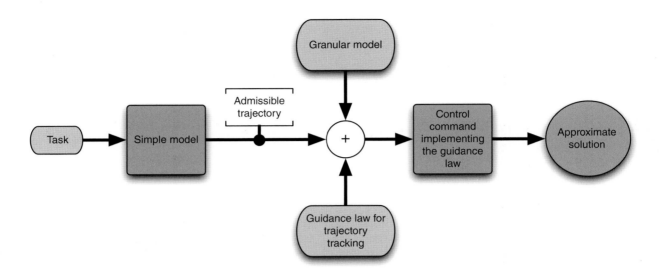

Figure 1. A conceptual diagram that summarizes the steps involved in the solution of a typical problem of trajectory planning.

complex than the one used to generate the trajectory in the first place. It is, perhaps, worth drawing the reader's attention to the analogy between the approach just described and the standard control-theoretic idea of tracking an open-loop trajectory via feedback control. The control of articulated robotic arms, for example, often follows a similar pattern. An idealized model of the robot is constructed using rigid body dynamics and the resulting inverse dynamics model is used to map the time series of desired position, velocity, and acceleration of the end effector to controls—that is, forces and torques—at the joints of the robotic arm. However, because the links of the arm are not perfectly rigid and the actuators introduce further modeling uncertainties, a feedback control is employed to minimize the tracking error (Murray, Li, and Sastry, 1994). An important difference, however, between this example and the use of guidance laws for path following, in the present context at least, is that the feedback control that implements the guidance law is often designed for a vehicle model that is different from the model used to generate the open-loop trajectories. The power of the methods described in the following sections is partly due to this flexibility in choosing models of appropriate complexity.

The problems addressed in this chapter are the generation of admissible trajectories for simplified models of vehicles that undergo planar motion and the application of guidance laws that allow the tracking of these trajectories. The literature on trajectory planning and guidance is vast enough to warrant a dedicated chapter to review it. In what follows, the choice of topics is necessarily biased by the authors' research experience and, as mentioned above, tilts toward a coherent presentation of a general method rather than a comprehensive

review of all the significant contributions that have appeared to date. The bibliography, however, should provide the reader who is interested in exploring further the topics of this chapter with sufficiently many pointers to the literature.

2 TRAJECTORY PLANNING

2.1 A basic kinematic model for planar motion under curvature constraints

With Figure 1 in mind, the focus in this section is on a particularly popular kinematic model for vehicles that undergo planar motion. The model is a control system, and at least two reasons contribute to its popularity: (i) There exist complete characterizations[1] of its minimum–time trajectories that can be computed efficiently (it is even possible to find the type of the optimal trajectory that connects two given boundary conditions by first constructing a look-up table (Shkel and Lumesky, 2001)), and (ii) it incorporates an essential characteristic that the trajectories of many real vehicles have: curvature that is bounded from above by a fixed constant. Extensions of the model to describe three-dimensional motion have also been considered in the literature and relevant references are given at the end of the section. However, the associated results are considerably

---

[1]The plural refers to the fact that the structure of the optimal trajectories depends on the choice of the set of control values and, therefore, different choices lead to different families of optimal trajectories. Section 2.2 elaborates on this observation.

more scarce than in the two-dimensional case and, depending on the way the extension is performed, analytic complications also arise.

Consider the control system

$$
\begin{aligned}
\dot{x} &= u \cos \theta \\
\dot{y} &= u \sin \theta \\
\dot{\theta} &= v
\end{aligned}
\tag{1}
$$

where $(x, y, \theta) \in \mathbb{R}^2 \times \,] - \pi, \pi[$ are coordinates of a point in the state space $M = \mathbb{R}^2 \times \mathbb{S}^1, \mathbb{S}^1$ denotes the unit circle, and (u, v) is the (vector-valued) control that takes values in a compact subset $U \times V$ of $\mathbb{R}^2$. Equation 1 can be thought of as describing the motion of a point in $\mathbb{R}^2$ along continuously differentiable curves whose curvature κ satisfies

$$
\kappa = \left| \frac{v}{u} \right|
\tag{2}
$$

when $u \neq 0$. We refer to such curves as (admissible) *paths*. A *trajectory*[2] is a solution to Equation 1. A trajectory $\gamma : I \subset \mathbb{R} \to M$ from an interval I into the state space M gives rise to a unique path $c : I \to \mathbb{R}^2$, which is the projection of γ on $\mathbb{R}^2$. It follows from Equation 2 that bounds on the controls u and v yield bounds on the curvature of the paths. To avoid pedantry that is irrelevant to the purpose of this chapter, and for the economy of language, mathematical objects are occasionally identified with the physical objects they model, when there is no risk of confusion. A typical instance of this convention being in effect is the identification of a control system and its trajectories with the vehicle it models and the actual trajectories of the vehicle, respectively.

Variations on the problem of finding shortest paths of bounded curvature between two given points in the plane have been studied since the end of the nineteenth century (Markov, 1887; Dubins, 1957; Reeds and Shepp, 1990; Boissonnat, Cérézo, and Leblond, 1991; Sussmann and Tang, 1991), beginning with the work of Markov (1887) on laying railroad track connecting existing sections of track. The continued interest in Equation 1 today stems mostly from its usefulness as a model for autonomous vehicles that are kinematically constrained and cannot undergo arbitrarily sharp turns. The quintessential type of vehicle to which such a constraint applies is a fixed-wing airborne vehicle.

2.2   Variations on the basic model

We deliberately refrained from explicitly specifying the exact nature of the set of control values $U \times V$ because it deserves elaboration: different choices for U lead to models[3] with different qualitative characteristics. As for the set V, it is almost always taken to be of the form $[-1/\rho, 1/\rho] \subset \mathbb{R}$ where ρ is a fixed, positive, real number, referred to as the *minimum turn-radius*. One exception to this choice of is the work by Bakolas and Tsiotras (2011) where the optimal trajectories of Equation 1 are analyzed when is not symmetric around the origin. It follows from Equation 2 that

$$
\kappa \leq \left| \frac{1}{\rho \, \overline{u}} \right|
\tag{3}
$$

where $\overline{u}$ is the maximum allowable value for the control u, that is, the maximum speed of the vehicle. Some standard terminology related to specific choices for U is as follows.

When U contains a single positive value, that is, $U = \{ \overline{u} \}$, for some fixed positive $\overline{u} \in \mathbb{R}$, the control u in (1) is essentially eliminated and Equation 1 is referred to as the *Dubins vehicle*. When $U = \{ -\overline{u}, \overline{u} \}$, Equation 1 is called the *Reeds–Shepp vehicle*. From a practical viewpoint, the difference between the two cases is that a Dubins vehicle can only move forward at maximum speed $\overline{u}$, whereas a Reeds–Shepp vehicle can move both forward and backward at maximum speed $\overline{u}$. Without loss of generality, and because the exact value of $\overline{u}$ is immaterial for the analysis of Equation 1, it is assumed hereafter that $\overline{u} = 1$. Alternatively, since it is the bound Equation 3 that is of interest, $\overline{u}$ can be absorbed into ρ.

One of the reasons that make Equation 1 a popular mathematical model for autonomous vehicles is that relatively simple sufficient families of minimum-time[4] trajectories are known for both the Dubins and the Reeds–Shepp vehicles. A *sufficient family* $\mathcal{F}$ of optimal trajectories is a collection of trajectories such that, given any two states $p, q \in M$, there exists an optimal trajectory $\gamma : [0, T] \to$ M *in* $\mathcal{F}$ satisfying $\gamma(0) = p$ and $\gamma(T) = q$. The projections of the trajectories $\gamma \in \mathcal{F}$ form a sufficient family of optimal paths. To recall the two main results that describe the sufficient families of minimum-time trajectories of Equation 1 when $U = \{1\}$ and when $U = \{-1, 1\}$, we first introduce some notation.

Let C_ϕ denote a circular arc of ϕ radians and of radius ρ, and S_d a straight line segment of length d. A C^1 planar concatenation

---

[2]An admissible path is a C^1 curve in $\mathbb{R}^2$, whereas a trajectory is, in general, an absolutely continuous curve in the state space M. Absolute continuity is an assumption that guarantees almost everywhere differentiability.

[3]At this point, we can be more precise about the meaning of the word "model": It is a control system together with a set of control values.

[4]Since |u| is always equal to 1, minimum-time and minimum-length trajectories coincide.

of such arcs and straight line segments is denoted by juxtaposition of the corresponding symbols. For example, $C_\alpha S_d C_\beta$ denotes a C^1 curve that consists of an arc of α radians, followed by a straight line segment of length d, followed by an arc of β radians. If I is an interval in $\mathbb{R}$, the space of locally absolutely continuous curves in M defined on I is denoted by $W^{1,1}_{\mathrm{loc}}(\mathrm{I}; \mathrm{M})$. The following theorem is proven in Boissonnat, Cérézo, and Leblond (1991); Dubins (1957); Sussmann and Tang (1991).[5]

Theorem 1. Let $U \times V = \{1\} \times [-1/\rho, 1/\rho]$. Given two points $p, q \in$ M, there exists a minimum-time trajectory $\gamma \in W^{1,1}_{\mathrm{loc}}([0, \mathrm{T}]; \mathrm{M})$ of (1) such that $\gamma(0) = p$ and $\gamma(\mathrm{T}) = q$. The projection on $\mathbb{R}^2$ of a minimum-time trajectory γ is either of the form $C_\alpha C_\beta C_\delta$ or of the form $C_\alpha S_d C_\delta$, where $0 \le \alpha, \delta < 2\pi$, $\pi < \beta < 2\pi$, and $d \ge 0$.

Moreover, it is shown in Sussmann and Tang (1991) that if $C_\alpha C_\beta C_\delta$ is an optimal path, that is, the projection of an optimal trajectory, then

$$\min\{\alpha, \delta\} < \beta - \pi \quad \text{and} \quad \max\{\alpha, \delta\} < \beta$$

In words, a Dubins vehicle starting from a given initial configuration[6]—which is a position and orientation in the plane—has to move along straight line segments and circular arcs of maximum curvature in order to reach in minimum time a given final configuration (Figure 2). Moreover, the value of the control v (see Equation 1) can change at most twice. Therefore, if L denotes a left turn, R denotes a right turn, and S denotes a straight line segment (equivalently, if L, R and S correspond to the control values $\nu = 1/\rho$, $\nu = -1/\rho$, and $v = 0$, respectively), then a sufficient family of optimal paths for the Dubins vehicle, that is also minimal, is

$$\mathcal{F}_\mathrm{D} = \{\mathrm{LSL}, \mathrm{LSR}, \mathrm{RSR}, \mathrm{RSL}, \mathrm{RLR}, \mathrm{LRL}\}$$

The elements of $\mathcal{F}_\mathrm{D}$ are called *Dubins paths*. A sufficient family $\mathcal{F}_\mathrm{RS}$ is also known for the Reeds–Shepp vehicle and it has a structure similar to that of $\mathcal{F}_\mathrm{D}$ in the sense that it is finite and consists of concatenations of circular arcs of maximum curvature and straight line segments. Not surprisingly,

---

[5]The proof in Dubins (1957) consists of a hands-on analysis of C^1 curves of bounded curvature between two given points in the plane and with prescribed slopes at each point, without any reference to control theory. The proofs in Boissonnat, Cérézo, and Leblond (1991); Sussmann and Tang (1991), on the other hand, rely on the Maximum Principle (Pontryagin and Boltyanskii (1962)).

[6]A configuration and a state refer to the same information: a position and an orientation. We use the term "configuration" in the context of planar motion and the term "state" when referring to points in the state space of Equation 1.

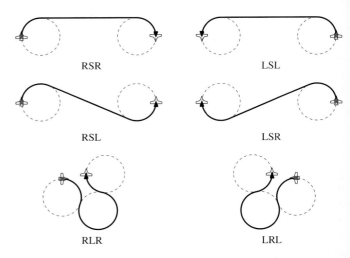

Figure 2. Examples of Dubins paths connecting pairs of configurations in the plane.

however, $\mathcal{F}_\mathrm{RS}$ is larger than $\mathcal{F}_\mathrm{D}$ since the increased maneuverability of the Reeds–Shepp vehicle allows for more admissible paths between two given positions and orientations. Let us call the paths that are concatenations of circular arcs of maximum curvature and of straight line segments "simple." The following theorem establishes the existence of a sufficient family $\mathcal{F}_\mathrm{RS}$ of optimal paths whose elements are called *Reeds–Shepp paths* (Reeds and Shepp, 1990; Boissonnat, Cérézo, and Leblond, 1991; Sussmann and Tang, 1991).

Theorem 2. Let $U \times V = \{-1, 1\} \times [-1/\rho, 1/\rho]$. Given two points $p, q \in$ M, there exists a minimum-time trajectory $\gamma \in W^{1,1}_{\mathrm{loc}}([0, \mathrm{T}]; \mathrm{M})$ of (1) such that $\gamma(0) = p$ and $\gamma(\mathrm{T}) = q$, and such that the projection of γ on $\mathbb{R}^2$ is a simple path consisting of at most five circular arcs of maximum curvature and straight line segments. Moreover, along such an optimal trajectory, the value of the control u can change at most twice.

There is a subtlety related to the minimum-time trajectories of the Reeds–Shepp vehicle that makes the formulation of Theorem 2 slightly different from that of Theorem 1. Not every optimal path for the Reeds–Shepp vehicle is simple; however, given any optimal path, there exists a simple one with the same length (Sussmann and Tang, 1991). This is one of the motivations for introducing the concept of sufficient families of optimal paths.

Assuming that $L^\pm$, $R^\pm$, and $S^\pm$ denote left turns, right turns, and motion on a straight line, respectively, and the signs correspond to forward (plus sign) and backward (minus sign) motion, then the main result in Reeds and Shepp (1990) states that a sufficient family of shortest paths between two configurations of a vehicle consists of 48 words constructed from the alphabet $\{L^-, L^+, R^-, R^+, S^-, S^+\}$, each word contains at most five letters, and there can be at most two points of direction reversal, that is, points where the vehicle changes from forward to backward motion or vice versa. In Sussmann and Tang (1991), the classification of the minimum-time paths for the Reeds–Shepp vehicle is made systematic[7] within the framework of geometric and optimal control theory, and further improved upon by eliminating two words ($L^- L^+ L^-$ and $R^- R^+ R^-$) from the sufficient family of optimal paths. The resulting 46 words represent the elements of $\mathcal{F}_{RS}$. The construction of the family $\mathcal{F}_{RS}$ reduces the basic motion planning problem of finding a shortest path between two given configurations of the Reeds–Shepp vehicle to the computation of 46 paths and the selection of the shortest among them. This procedure is further refined in Souères and Laumond (1996) where a minimum-time "synthesis" is constructed in the sense that the state space $\mathbb{R}^2 \times \mathbb{S}^2$ of the Reeds–Shepp vehicle is partitioned into sets of initial conditions according to the type of minimum-time Reeds–Shepp path that connects each initial condition to the configuration (0, 0, 0).

The optimal paths for the Dubins and the Reeds–Shepp vehicles assume a simpler form when the terminal condition that has to be satisfied by the optimal trajectories is relaxed to allow for a free orientation. That is, when the vehicle is required to reach some final position in the plane without a constraint on its orientation. In the case of the Dubins vehicle, a minimal sufficient family $\mathcal{F}_D^r$ of optimal paths is given by

$$F_D^r = \{LS, RS, LR, RL\}$$

The family $\mathcal{F}_D$ is reduced to the family $\mathcal{F}_D$ by supplementing the necessary conditions of optimality with a transversality condition at the final time $t = T$ (Pontryagin and Boltyanskii, 1962, Chapter II, Section 16; Boissonnat and Bui, 1994). The elements of $\mathcal{F}_D^r$ are called *relaxed Dubins paths* and Figure 3 shows one example for each type of path in $\mathcal{F}_D^r$. A similar simplification occurs for the Reeds–Shepp vehicle (Figure 4) (Souères, Fourquet, and Laumond, 1994, Theorem 1):

Theorem 3. A sufficient family $\mathcal{F}_{RS}^r$ of relaxed Reeds–Shepp paths consists of paths of the following form:

1. $C_a | C_{\pi/2} S_d, a < \pi/2, d \geq 0$
2. $C_a | C_b, a < b < \pi/2$
3. $C_a S_d, a \leq \pi/2, d \geq 0$

---

[7]Similar to the case of the Dubins vehicle, the first results on the Reeds–Shepp vehicle that appeared in Reeds and Shepp (1990) made no reference to control theory. Rather, the proof in Reeds and Shepp (1990) consists of laborious trigonometric arguments and relies also on computer simulation. The proofs in Sussmann and Tang (1991) and Boissonnat, Cérézo, and Leblond (1991) on the other hand, center on the Maximum Principle.

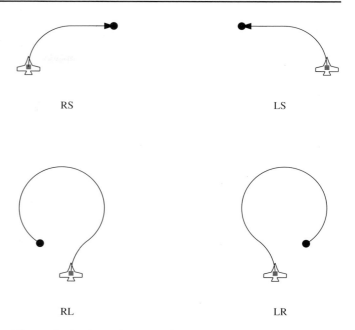

Figure 3. Typical relaxed Dubins paths.

where the symbol | between the segments of a path indicates the existence of a cusp, that is, a change in the sign of the control u.

2.3   An incrementally more complicated model

The goal of this section is to partially justify the widespread use of Equation 1 as a model for autonomous vehicles. The point to be made is that, although it is appealing to consider more realistic models, Equation 1 strikes a satisfactory balance between tractability and efficacy. Suppose, for example, that (1) is modified so that v controls the angular acceleration, instead of the angular velocity

$$\begin{aligned} \dot{x} &= u \cos \theta \\ \dot{y} &= u \sin \theta \\ \dot{\theta} &= w \\ \dot{w} &= v. \end{aligned} \qquad (4)$$

One reason for considering Equation 4, as opposed to Equation 1, is that, in the case of both the Dubins and the Reeds–Shepp vehicles, the angular velocity can have jump discontinuities as a function of time along an optimal path, that is, its values can switch instantly between any two values in $[-1/\rho, 1/\rho]$. Therefore, the angular acceleration is not even defined in a classical sense, let alone being physically realizable. In (4), however, if we confine ourselves to standard differentiability notions, it is necessary to posit stronger regularity assumptions (e.g., absolute continuity of trajectories) on the angular

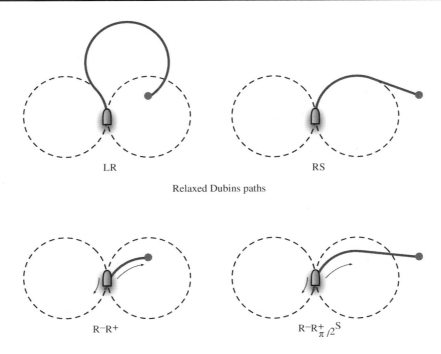

LR   RS

Relaxed Dubins paths

R⁻R⁺   R⁻R⁺_{π/2}S

Relaxed Reeds–Shepp paths

Figure 4. A comparison between relaxed Dubins paths and relaxed Reeds–Shepp paths.

velocity $\dot{\theta}$ so that the system of equations makes sense. This innocuous-looking extension has the following consequence as far as the analysis of the optimal trajectories is concerned. It can be shown that there exist extremals (i.e., trajectories that satisfy the *necessary* conditions for optimality) associated with controls that undergo infinitely many switchings, a phenomenon known as "chattering." Moreover, it is not known whether these extremals are in fact minimizers (Sussmann, 1997). Of course, this is not meant to imply that Equation 4 is not a useful model. Rather, that the wealth of analytic results about Equation 1 offers a more complete and clear picture of the structure of optimal trajectories that can be directly exploited in trajectory planning and guidance for autonomous vehicles. The minimum-time trajectories of (4) and the computation of approximate paths for trajectory planning have been studied in Boissonnat *et al.* (1999); Fraichard and Scheuer (2004) and the references therein.

2.4   Further model variations

The basic model Equation 1 has been extended in several directions in the literature in an effort to realistically model autonomous vehicles early on in the analysis of trajectory planning and guidance problems. Put differently, it is desirable

to narrow the gap between the simple and the granular model in Figure 1. From a mathematical point of view, many of these extensions—for example, all the kinematic models mentioned thus far—amount to simply changing the set of control values or a boundary condition. However, the ensuing changes in the structure of optimal solutions can be drastic. One example is the significant difference in complexity between the sufficient families of optimal paths for the Dubins and the Reeds–Shepp vehicle. In this line of research, variations on the theme of characterizing the minimum-time trajectories of Equation 1 have appeared in Bakolas and Tsiotras (2011, 2013); Dolinskaya and Maggiar (2012); Isaiah, Weiss, and Shima (2014). In Bakolas and Tsiotras (2011) an optimal synthesis (in the sense of Souères and Laumond (1996), explained above) is constructed for a Dubins vehicle whose minimum turn radii for left and for right turns are different. That is, the vehicle can perform sharper turns in one direction than in the other. An optimal synthesis that accounts for the presence of a constant drift field is presented in Bakolas and Tsiotras (2013), whereas time–optimal trajectories when the maximum speed and the minimum turn radius are functions of the orientation of the vehicle are analyzed in Dolinskaya and Maggiar (2012). A sufficient family of optimal paths for a vehicle that moves faster forward than backward is described in Isaiah, Weiss, and Shima (2014). This last case is used as an example in the remainder of this section to explain the interest in variations on the basic model Equation 1.

There are types of vehicles, such as marine vehicles, that are heavily optimized for forward motion and are not adequately described by the Reeds–Shepp kinematic model. The reason is that, although these vehicles may be able to move forward and backward, their maximum backward speed is smaller than their maximum forward speed. Another example is ground vehicles with mounted sensors that are not omnidirectional and, hence induce a preferred direction of motion for the vehicle. Regardless of the actual maximum speed of the vehicle while moving backward, backward motion can be modelled as being slower so as to penalize it more heavily when searching for minimum-time trajectories.

With this motivation in mind, the set of control values for Equation 1 can be taken to be $U \times V = \{-c, 1\} \times [-1/\rho, 1/\rho] \subset \mathbb{R}^2$, where c is a positive real number such that $c \in]0, 1[$, and the problem considered is that of finding the minimum-time paths between a given initial configuration and a final position. Because of the asymmetry of the set $U \times V$ and because the terminal constraint is relaxed to allow for a free choice of orientation for the vehicle, the problem is called the relaxed asymmetric Reeds–Shepp (RARS) problem. An application of the Maximum Principle to this model leads to the conclusion that the sufficient family $\mathcal{F}_{RS}^r$ is also sufficient for the RARS problem (Isaiah, Weiss, and Shima, 2014).

The fact that $\mathcal{F}_{RS}^r$ is a sufficient family for the RARS problem does not imply that a given instance of the relaxed asymmetric Reeds–Shepp problem admits the same solutions as the relaxed Reeds–Shepp problem, when the boundary conditions are the same for the two problems. This is a point where the implications of the different maximum forward and backward speeds emerge. Consider, for example, the scenario depicted in Figure 5. The initial state of the vehicle is $(0, 0, 0)$ and the final position is $(x_f, 0) = (-d, 0)$, $d > 0$. It can be shown that there exists a positive real number d^* such that

1. if $d < d^*$, the shortest path is of the form S_d^-, that is, the vehicle moves backwards on a straight line segment for time $T = d/c$, and

2. if $d > d^*$, a shortest path is of the form $R_\alpha^- R_{n/2}^+ S_\delta^+$, $\alpha > 0$, $\delta > 0$. An illustration of this path is given in Figure 5.

Intuitively, it is to be expected that, when seeking a minimum-time path to the final destination, the vehicle should exploit its maneuvrability to travel longer distance at higher speed, as shown in Figure 5. If the value of c is significantly smaller than 1 (the maximum forward speed), this example shows why the original Reeds–Shepp model would have led to unsatisfactory paths.

2.5   Extensions to three dimensions

The fact that Equation 1 describes planar motion is less restrictive than what might appear initially because, even if a vehicle has an additional spatial degree of freedom, there is a large number of motion planning and guidance scenarios that take place in a plane in three-dimensional space. Nevertheless, models with three spatial degrees of freedom, analogous to Equation 1, have appeared in the literature and it is, of course, of interest to understand and classify their optimal trajectories. Representative works in this direction are Chitsaz and LaValle (2007) and Sussmann (1995). In Chitsaz and LaValle (2007) the minimum-time trajectories of the system,

$$\begin{aligned}
\dot{x} &= \cos\theta \\
\dot{y} &= \sin\theta \\
\dot{z} &= u \\
\dot{\theta} &= v
\end{aligned}$$

are studied when an initial and a final state are given, and $(u, v) \in [-1, 1]^2$. In Sussmann (1995) the minimum-time trajectories between the given initial and final states are classified for the control system:

$$\begin{aligned}
\dot{x} &= y \\
\dot{y} &= y \times w
\end{aligned} \tag{5}$$

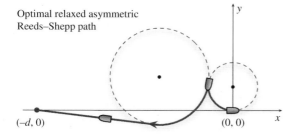

Optimal relaxed asymmetric Reeds–Shepp path

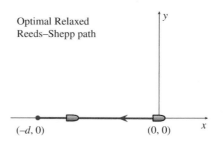

Optimal Relaxed Reeds–Shepp path

Figure 5. Comparison between RRS and RARS paths.

where $(x, y) \in \mathbb{R}^3 \times \mathbb{S}^2$, $\mathbb{S}^2$ is the unit sphere in $\mathbb{R}^3$, and w is the control taking values in the closed-unit ball in $\mathbb{R}^3$. The operation between y and w in the second equation in Equation 5 is the usual cross-product of two vectors. Unlike the optimal trajectories of Equation 1, not every minimum-time trajectory of Equation 5 is a concatenation of circular arcs and straight line segments; there are also minimizers that are helicoidal arcs. Of more theoretical interest is the generalization in Jurdjevic (2014) of the problem of Dubins (1957) to any finite-dimensional, simply connected space of constant curvature.

2.6   Trajectory planning with interior point constraints

The previous sections reviewed several aspects of the problem of finding minimum-time trajectories of the control system Equation 1 for a given set of control values and given boundary conditions. In this section we pick up the thread that connects the analysis of a simple kinematic model such as Equation 1 to practical trajectory planning. A natural next step is to consider the problem of finding admissible, preferably optimal, paths for a vehicle that not only connect an initial configuration to a final configuration or position, but also satisfy certain constraints. The introduction of state constraints into an optimal control problem can raise serious analytic (Vinter, 2000, Chapter 9) and computational (Polak, 1997, Chapter) challenges. Here, we purposely focus on a type of constraint known in the control literature as interior point or intermediate constraints and for which conclusions about the optimal trajectories can be drawn by applying the classic technique of Lagrange multipliers (Bryson and Ho, 1975, Chapter 3; Dmitruk and Kaganovich, 2011). An interior point constraint is a condition of the form

$$N[x(\tau), \tau] = 0, \quad t_i < \tau < t_f,$$

where x represents the state of a control system such as (1), which has to be satisfied by an optimal trajectory defined on a time interval $[t_i, t_f]$. From a mathematical point of view, the main consequence of introducing interior point constraints is that the components of the costate vector may exhibit discontinuities of the first kind, also known as jump discontinuities, along an optimal trajectory.

The rationale for considering such constraints is twofold. First, the very formulation of a trajectory planning problem may posit such a requirement. That is, it could be the case that a vehicle has to pass through a sequence of predefined configurations or positions. Second, there are trajectory planning problems whose solution is facilitated by introducing auxiliary interior point constraints. For example, if obstacles have to be taken into account, it is a common approach to consider the vertices of the obstacles as candidate waypoints through which a vehicle may pass.

In the motion planning[8] literature, the problem of computing minimum-time trajectories of (1) that satisfy interior point constraints has been given particular attention in the context of the so-called Dubins traveling salesperson problem (DTSP) (Isaiah and Shima, 2015; Edison and Shima, 2011; Savla, Frazzoli, and Bullo, 2008; Ma and Castañón, 2006; Tang and Özgüner, 2005). Given an initial configuration, the DTSP consists in finding an admissible path for the Dubins vehicle (Section 2.2) that passes through a finite set of fixed locations in the plane such that the vehicle eventually assumes its initial configuration again. The requirement that the vehicle return to its initial configuration is not essential and methods that yield admissible paths for the DTSP can be used in trajectory planning with interior point constraints. In other words, the requirement that the path be closed can be simply dropped. Approaches to the DTSP include heuristic algorithms (Savla, Frazzoli, and Bullo, 2008; Tang and Özgüner, 2005), receding horizon control (Ma and Castañón, 2006), discretization (Edison and Shima, 2011; Le Ny, Feron, and Frazzoli, 2012), a combination of receding horizon control with graph algorithms (Isaiah and Shima, 2015), and randomized algorithms (Isaiah and Shima, 2015). Finding *optimal* solutions to trajectory planning problems with interior point constraints becomes impractical as the number of constraints increases and, therefore, the goal is, typically, to find satisfactory admissible paths. Figure 6 shows the output of five different algorithms from the literature when applied to an instance of the Dubins travelling salesperson problem in which a Dubins vehicle has to pass through 10 a priori given locations in the plane.

3   GUIDANCE FOR PATH FOLLOWING

3.1   Introduction

As explained in Section 1, once a method to generate paths for a simple kinematic model such as (1) has been established, it is necessary to devise a feedback control that allows an actual vehicle to track these paths. A plethora of path following techniques have appeared in the literature over the years and in various contexts (Zhao and Tsiotras, 2012; Tsourdos, White, and Shanmugavel, 2011; Lawrence, Frew, and Pisano, 2008). In what follows, the focus is on

---

[8] A synonym for trajectory planning, more commonly used in the robotics community.

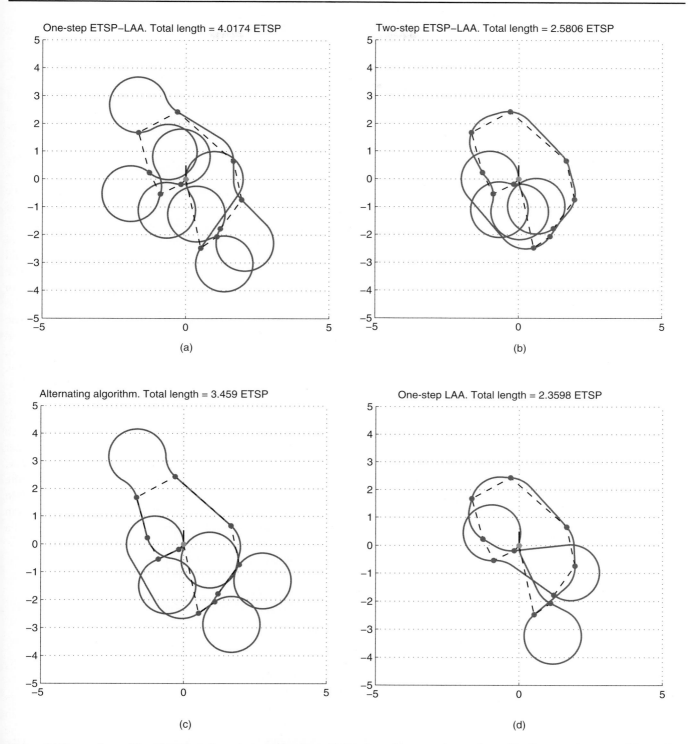

Figure 6. Comparison of five algorithms for trajectory planning with interior point constraints and, specifically, on an instance of the Dubins traveling salesperson problem with 10 cities. The minimum turning radius is $\rho = 1$, the initial condition of the Dubins vehicle is $p = (0, 0, \pi/2)$, and the positions of the targets are randomly and uniformly distributed in $[-2.5, \ 2.5]^2 \subset \mathbb{R}^2$. Parts (a) and (b) correspond to the two-point and the look-ahead algorithms in Ma and Castañón (2006), part (c) corresponds to the alternating algorithm in Savla, Frazzoli, and Bullo (2008), and parts (d) and (e) correspond to algorithms in Isaiah and Shima (2015).

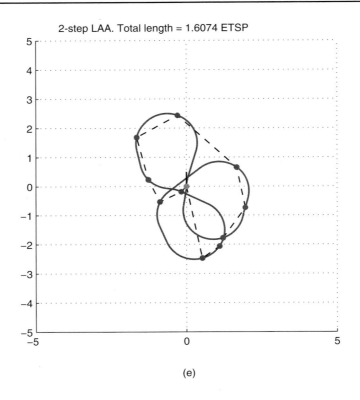

2-step LAA. Total length = 1.6074 ETSP

(e)

Figure 6. (*Continued*)

the use of guidance laws for path following. There are several reasons that contribute to the appeal of applying guidance techniques to the path following problem. The implementation of guidance laws leads to *feedback* controls; guidance laws can be simple, intuitive, and computationally inexpensive; there is accumulated experience and a vast array of results that can serve as the basis for the development of new path following algorithms.

Guidance laws dictate a relative geometry between a vehicle and a target, and can be independent of the mathematical model for the vehicle. On the other hand, the form of the controls that implement a guidance law depends on the equations that describe the motion of the vehicle and it is partly for this reason that we refrain from delving into the problem of describing explicitly controls for guidance.

Problems of guidance have been the subject of rigorous analysis since the first half of the eighteenth century (Shneydor, 1998, Chapter 3), if not earlier, and they continue to be the subject of active research carried out by various scientific communities (e.g., in robotics, aerospace engineering, and control engineering). Particularly prominent is the place that guidance holds in aerospace engineering where missile guidance, in particular, is essentially a subdiscipline. However, as is often the case in applied mathematics, the modern abstract formalism used to describe and solve problems of guidance—optimal control theory and differential games are standard frameworks in which missile guidance is studied—renders the relevant results transferable to other application areas and to path following specifically. There are at least three avenues for deducing path following controls and algorithms from guidance laws:

(i)   Guidance of a vehicle to a succession of stationary fictitious targets (waypoints) along a given path (Section 3.2).
(ii) Tracking of a virtual target that traces out the path to be followed (Section 3.3).
(iii) Cross-track guidance that relies on the idea of minimizing the deviation of a vehicle's trajectory from a given path (Section 3.4).

In principle, virtually any guidance law—suitably modified, if necessary—can be employed in each one of the cases (i)–(iii), as long as it accomplishes the guidance objective. The *performance*, however, of the overall path following scheme depends crucially on the choice of the guidance law. In the following sections, where the three approaches (i)–(iii) are outlined, the dependence of the tracking error on the guidance law is illustrated by means of representative results from the literature.

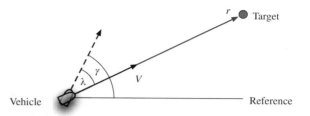

Vehicle   Reference

Figure 8. Illustration of the geometric rule for pure pursuit. The velocity V has to be collinear and in the same direction with r. Therefore, a guidance law for pure pursuit is designed to minimize the error λ.

vectors. One guidance law that results in pure pursuit against a static target is given by

$$\dot{\gamma} = -\frac{k|V|\sin\lambda}{|r|} \tag{7}$$

where γ is the heading of the vehicle, $k > 2$ is a constant,[9] λ is the angle between the velocity V of the vehicle and the line of sight r, and $|\cdot|$ denotes the ℓ^2 (or standard euclidean) norm of a vector. An early occurrence of the guidance law (7) in the context of path following can be found in Ollero, García-Cerezo, and Martinez (1994).

Because of its simplicity and its efficacy in a multitude of guidance scenarios, another natural candidate guidance law for path following is proportional navigation (PN) (Shneydor, 1998, Chapter 5). It can be shown (Cons, Shima, and Domshlak, 2013) that there is a trade-off between the number of waypoints located on the path to be tracked and the magnitude of the navigation constant. The goal of PN is to bring a pursuer to a collision course with a target. This is achieved by keeping the angle of the line of sight (LOS) between the pursuer and the target constant. With reference to Figure 9, the kinematics of the engagement are as follows. As before, let V and γ denote the velocity and the heading of the vehicle, respectively, $R = |r|$ be the distance between the vehicle and the target/waypoint, and $\theta \triangleq \gamma - \lambda$ be the angle contained between the line of sight and the horizontal axis. Then, the following relations hold:

$$\dot{R} = -|V|\cos(\gamma - \theta) = -|V|\cos\lambda$$

$$\dot{\theta} = \frac{-|V|\sin(\gamma-\theta)}{R} = \frac{-|V|\sin\lambda}{R}$$

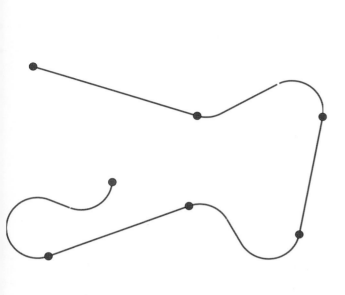

Figure 7. A guidance law that leads consecutively a vehicle to a sequence of fictitious targets along a path can be used to track the path.

3.2 Waypoint guidance

Suppose that an admissible path for the Dubins vehicle has been generated using one of the methods described in Section 2, or any other method for that matter, and the goal is to have a vehicle track that path. One way to tackle this problem is to disperse fictitious static targets, also referred to as waypoints, along the computed path and use a guidance law to lead the vehicle to each target consecutively (Figure 7). Pure pursuit (Shneydor, 1998, Chapter 3) is one of the simplest guidance laws to consider for this task. The geometric rule for pure pursuit is to always keep the vehicle oriented toward the target. In terms of vector algebra, pure pursuit imposes the conditions

$$\begin{aligned} V \times r &= 0 \\ V \cdot r &> 0 \end{aligned} \tag{6}$$

where V is the velocity of the vehicle and r is the vector connecting the vehicle to the target (Figure 8). The operation on the left-hand side of the first equation in (6) is the cross-product of two vectors, whereas the operation on the left-hand side of the inequality is the dot product between two

---

[9]The condition $k > 2$ implies that $\lambda \to 0$ and $\dot{\gamma} \to 0$ as $|r| \to 0$ (Shneydor, 1998, p. 73).

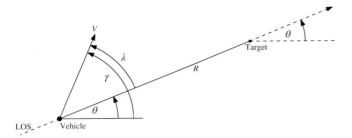

Figure 9. Relative geometry between a vehicle and a target and the quantities that appear in the definition of proportional navigation.

and the acceleration command that corresponds to proportional navigation is

$$u = N|V|\dot{\theta} = -\frac{N|V|^2\sin\lambda}{R} = -\frac{N|V|^2\sin\lambda_0}{R_0}\left(\frac{R}{R_0}\right)^{N-2} \quad (8)$$

where N is the navigation constant and the naught subscript implies the initial value of a variable. It follows from (8) that the inequality $N \geq 2$ has to hold for the acceleration command to remain bounded as the distance between the vehicle and target decreases.

A problem that arises at this point is that, when a vehicle that undergoes planar motion has minimum turn radius ρ bounded away from zero, then there are two disks D_L and D_R of radius ρ associated with the configuration of the vehicle at any given moment, as shown in Figure 10, and the vehicle cannot reach a target inside D_L or D_R by implementing the proportional navigation law with a finite $N \geq 2$. One way to overcome this issue is to introduce an additional waypoint on the relaxed Dubins path that connects the initial configuration of the vehicle to the final waypoint, at the point on the relaxed Dubins path where the control v (see (1)) changes value. Figure 11 illustrates the idea. A similar approach can be taken for Dubins paths as well. That is, Dubins paths between given

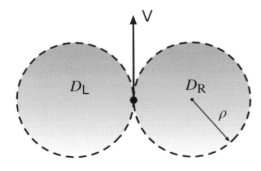

Figure 10. The disks associated with the configuration of a Dubins vehicle whose radii are equal to the minimum turn radius of the Dubins vehicle.

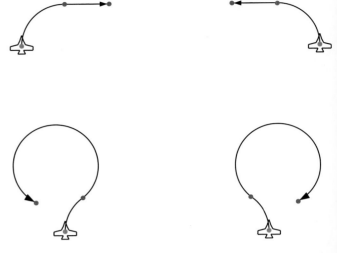

Figure 11. Targets inside the disks of minimum turn radius tangent to the configuration of a Dubins vehicle can be reached using proportional navigation with a finite gain $N \geq 2$, if additional waypoints are introduced at the points where the control changes value.

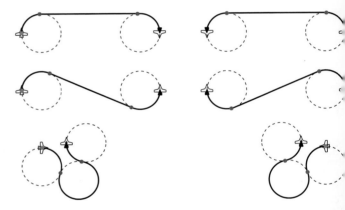

Figure 12. A Dubins path that connects two configurations can be tracked using proportional navigation with a finite navigation constant $N \geq 2$, if waypoints are introduced at the points on the path where the control changes value.

configurations can be tracked using proportional navigation with a finite navigation constant $N \geq 2$ by introducing additional waypoints at the points on the Dubins paths where the control changes value (Figure 12). Therefore, if an admissible path for the Dubins vehicle is a concatenation of Dubins paths, then the path can be tracked by a vehicle that implements the PN law with $N \geq 2$, if sufficiently many waypoints are introduced along the path (Cons, Shima, and Domshlak, 2013).

Proposition 1. A concatenation of Dubins paths can be tracked by a vehicle that implements proportional navigation

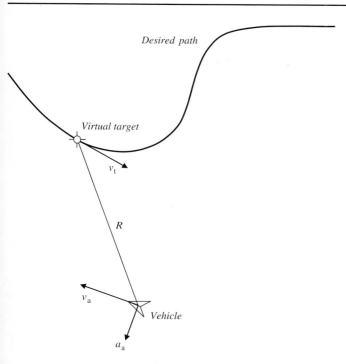

Desired path

Virtual target

v_t

R

v_a

Vehicle

a_a

Figure 13. A schematic representation of virtual target following.

with a finite gain $N \geq 2$, if waypoints are introduced at every point along the overall path where the control changes value and at the final position of the vehicle.

3.3 Virtual target guidance

In the previous section, a guidance law was used to "intercept" a sequence of fictitious static targets along a given path. Another way to link path following with guidance laws is to use a guidance law to track a virtual target that traces out the path to be followed. To this end, consider the setup depicted in Figure 13, where R is the distance between the controlled vehicle and the virtual target, v_a and a_a are the velocity and the lateral acceleration of the vehicle, respectively, and v_t is the velocity of the target. The speed of the target varies as a function of the speed of the vehicle according to the rule

$$|v_\mathrm{t}| = |v_\mathrm{a}|\frac{R_\mathrm{m}}{R}, \text{ when } R < R_\mathrm{m} \tag{9}$$

$$|v_\mathrm{t}| = |v_\mathrm{a}|, \quad \text{ when } R \leq R_\mathrm{m} \tag{10}$$

In Equation (9), R_m is a positive constant that represents a minimum distance between the vehicle and the target, and can be considered as a design parameter. Once the minimum distance has been achieved, the target moves at the speed of

the tracking vehicle. In this scenario, the target is an artifact and, hence its speed profile can be chosen freely. The idea of tracking a virtual target opens up the possibility of using existing guidance laws, or modifications thereof, for path following. For example, pure pursuit (PP) is used in Medagoda and Gibbens (2010) to follow a path that consists of straight line segments. Because in pure pursuit the heading of the vehicle points toward the target, whereas in path following it is preferable to have the heading of the vehicle aligned with the tangent to the path at the location of the target, guidance laws that impose a constraint on the angle of interception lead to lower tracking errors. The guidance law in Ryoo, Cho, and Tahk (2005) belongs to this category and is used for path following in Ratnoo *et al.* (2014), where its performance is compared with pure pursuit guidance and with the guidance law from Park, Deyst, and How (2004). The acceleration command for the guidance law in Ratnoo *et al.* (2014) is given by

$$a_\mathrm{a} = \frac{|v_\mathrm{a}|}{t_\mathrm{go}}(-6\theta + 4\gamma_\mathrm{a} + 2\gamma_\mathrm{t}) \tag{11}$$

where a_a, v_a, θ, γ_a, and γ_t are the vehicle's lateral acceleration,[10] the vehicle's velocity, the line-of-sight angle, the heading of the vehicle, and the heading of the virtual target, respectively (Figures 9 and 13) (Ryoo, Cho, and Tahk, 2005, Equation 26). The time-to-go in Equation (11) is defined to be $t_\mathrm{go} \triangleq R/|v_\mathrm{a}|$, unlike the case of guidance laws for scenarios of actual interception (as opposed to path following) that are often posited on the assumption that the time-to-go equals the ratio $-R/\dot{R}$. The analysis in Ratnoo *et al.* (2014) shows that the use of the guidance law Equation (11), which controls the angle of interception, leads to fast and global convergence, at the cost of higher (lateral) acceleration values. Representative simulations are shown in Figure 14.

3.4 Cross-track guidance

Cross-track guidance for path following can be viewed as a hybrid of waypoint and virtual target guidance. The idea behind cross-track guidance is to place waypoints along a path and connect them with straight line segments (tracks) to obtain a piecewise linear approximation of the path. Each waypoint is considered visited when the distance between the vehicle that is following the path and the waypoint becomes

---

[10]A scalar quantity that represents the acceleration applied in the direction of $\frac{\mathrm{d}}{\mathrm{d}t}\frac{v_\mathrm{a}}{|v_\mathrm{a}|}$.

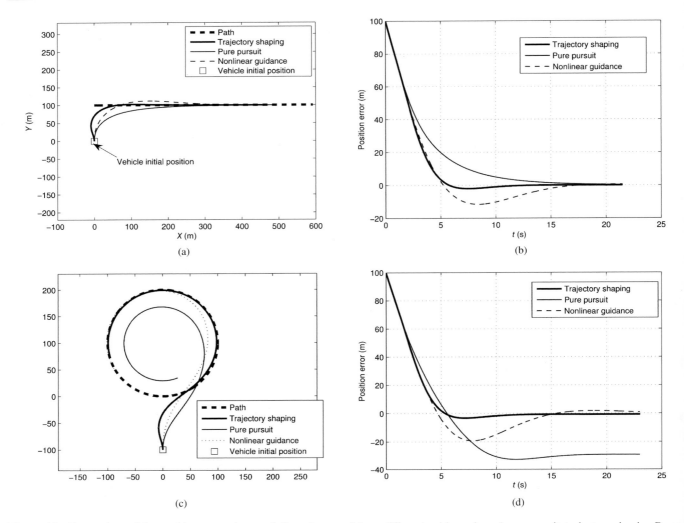

Figure 14. Comparison of the tracking errors that result from the use of three different guidance laws (pure pursuit, trajectory shaping Ryoo, Cho, and Tahk, 2005; Ratnoo *et al.* (2014), and the nonlinear guidance law of Park, Deyst, and How (2004)) for path following. The path to be followed, shown with a thick dashed line in parts (a) and (c), is a straight line segment in parts (a) and (b), and a circle in parts (c) and (d). For the simulation, it is assumed that $v_a = 50 \text{ m s}^{-1}$, $|a_a| \le 15g$, and $R_m = 100$ m. (Taken from Ratnoo *et al.* (2014).

less than a given threshold r_{min} (Figure 15); the *current track* is the track that connects the last visited waypoint to the next one. The goal of a guidance law in this setting is to minimize at any given moment the distance d of the vehicle from the current track (Shima and Rasmussen, 2009, Appendix B). Particularly pertinent to viewing cross-track guidance as a fusion of waypoint and virtual target guidance is the approach taken in Medagoda and Gibbens (2010), where a virtual target is assumed to travel from one waypoint to the next along the track that connects them. The vehicle whose goal is to follow the overall path implements a pure pursuit guidance law against the virtual target.

By viewing the current track as a line of sight, cross-track guidance can be implemented using LOS guidance. Line-of-sight guidance means guiding a vehicle toward a target by

keeping the vehicle on the line of sight between a reference point and the target (Shneydor, 1998, Chapter 2). With the notation being as in Figure 16a, where, in the case of cross-track guidance, the point O represents the last visited waypoint and the target represents the next waypoint, the lateral

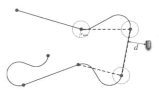

Figure 15. A schematic representation of cross-track guidance for path following.

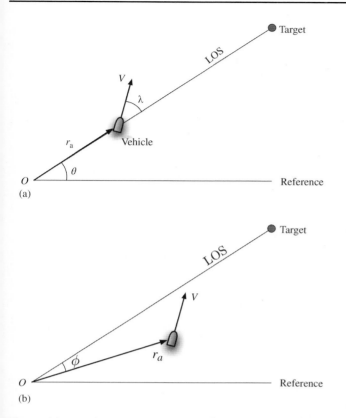

Figure 16. Relative geometry between a vehicle and a target in LOS guidance.

acceleration for maintaining planar LOS guidance is given, in general, by (Shneydor, 1998, p. 16)

$$a_{\mathrm{a}} = 2|V|\dot{\theta} + \frac{|r_{\mathrm{a}}|\ddot{\theta}}{\cos\lambda} - \frac{\mathrm{d}}{\mathrm{d}t}|V| \cdot \tan\lambda$$

A guidance law that implements LOS guidance can be devised by considering a deviation of the relative geometry between the vehicle and the target from the geometric rule dictated by LOS guidance. That is, the vehicle is assumed to lie off the LOS between the reference point O and the target so that there is a second LOS, r_{a}, between O and the vehicle (Figure 16b). Then, if ϕ is the angle between the two lines of sight, a guidance law results from assuming a lateral acceleration a_{a} that is proportional to the error ϕ:

$$a_{\mathrm{a}} = A\phi, \tag{12}$$

for some constant A. It turns out that the guidance law Equation 12 leads to instability, which, however, can be prevented by the addition of derivative control (Shneydor, 1998, p. 34). Another alternative for the implementation of cross-track guidance is to measure directly the distance of the vehicle from the LOS, as opposed to measuring the angle ϕ

and the range $|r_{\mathrm{a}}|$. This method is known as beam riding (Shneydor, 1998, pp. 35, 39).

4 CONCLUSIONS

Many problems of trajectory planning and guidance can be elegantly and concisely described, and occasionally solved, within general mathematical frameworks. Most notably, as optimal control problems or as differential games. However, practical considerations, such as the need for robust real-time implementation of control algorithms on autonomous vehicles, prohibit the transition from theory to practice of solutions that put excessive demands on computational time and infrastructure, or of solutions that can only be obtained by means of a trial-and-error process that requires manual tuning and intervention. By adopting a hierarchical approach that starts with the generation of trajectories for simple kinematic models and culminates with the implementation of guidance laws on complex realistic models of vehicles, it is possible to obtain solutions that remain within the scope of current technology and are, therefore, relevant to engineering practice.

REFERENCES

Bakolas, E. and Tsiotras, P. (2011) Optimal synthesis of the asymmetric sinistral/dextral Markov–Dubins problem. *J. Optim. Theory Appl.*, **150**(2), 233–250.

Bakolas, E. and Tsiotras, P. (2013) Optimal synthesis of the Zermelo–Markov–Dubins problem in a constant drift field. *J. Optim. Theory Appl.*, **156**(2), 469–492.

Boissonnat, J.-D. and Bui, X.-N. (1994) Accessibility region for a car that only moves forwards along optimal paths. Rapport de recherche 2181, INRIA Sophia Antipolis.

Boissonnat, J.-D., Cérézo, A., and Leblond, J. (1991) Shortest paths of bounded curvature in the plane. Robotique, Image et Vision 1503, INRIA Sophia Antipolis.

Boissonnat, J.-D., Cérézo, A., Degtiariova-Kostova, E.V., Kostov, V.P., Leblond, J. (1999) Shortest plane paths with bounded derivative of the curvature. *C. R. Acad. Sci. Paris* **329**, 613–618.

Bryson, A.E. and Ho, Y.-C. (1975) *Applied Optimal Control: Optimization, Estimation, and Control*, Taylor & Francis.

Chitsaz, H. and LaValle, S.M. (2007) Time-optimal paths for a Dubins airplane, in *Proceedings of the 46th IEEE Conference on Decision and Control*, IEEE, pp. 2379–2384.

Cons, M.S., Shima, T., and Domshlak, C. (2013) Integrating task and motion planning for unmanned aerial vehicles. *Unmanned Syst.*, **2**(1), 1–20.

Dmitruk, A.V. and Kaganovich, A.M. (2011) Maximum principle for optimal control problems with intermediate constraints. *Comput. Math. Model.*, **22**(2), 180–215.

Dolinskaya, I.S. and Maggiar, A. (2012) Time-optimal trajectories with bounded curvature in anisotropic media. *Int. J. Robotics Res.*, **31**(14), 1761–1793.

Dubins, L.E. (1957) On curves of minimal length with a constraint on average curvature, and with prescribed initial and terminal positions and tangents *Am. J. Math.*, **79**(3), 497–516.

Edison, E. and Shima, T. (2011) Integrated task assignment and path optimization for cooperating uninhabited aerial vehicles using genetic algorithms. *Comput. Oper. Res.*, **38**(1), 340–356.

Fraichard, T. and Scheuer, A. (2004) From Reeds and Shepp's to continuous-curvature paths. *IEEE Trans. Robotics* **20**(6), 1025–1035.

Isaacs, R. (1999) *Differential Games: A Mathematical Theory with Applications to Warfare and Pursuit, Control and Optimization*, Dover Books on Mathematics Series, Dover Publications.

Isaiah, P. and Shima, T. (2015) Motion planning algorithms for the Dubins travelling salesperson problem. *Automatica* **53**, 247–255.

Isaiah, P., Weiss, M., and Shima, T. (2014) The relaxed asymmetric Reeds–Shepp problem. *Proceedings of the 19th IFAC World Congress*, The International Federation of Automatic Control, pp. 2646–2651.

Jurdjevic, V. (2014) The Delauney–Dubins problem, in *Geometric Control Theory and Sub-Riemannian Geometry* (eds G. Stefani, U. Boscain, J.-P. Gauthier, A. Sarychev, and M. Sigalotti), vol. **5**, INdAM, Springer, pp. 219–240.

LaValle, S.M. (2006) *Planning Algorithms*, Cambridge University Press.

Lawrence, D.A., Frew, E.W., and Pisano, W.J. (2008) Lyapunov vector fields for autonomous unmanned aircraft flight control. *AIAA J. Guid. Control Dyn.*, **31**(5), 1220–1229.

Le Ny, J., Feron, E., and Frazzoli, E. (2012) On the Dubins travelling salesman problem. *IEEE Trans. Automat. Control*, **57**(1), 265–270.

Ma, X. and Castañóon, D.A. (2006) Receding horizon planning for Dubins travelling salesman problems, in *Proceedings of the 45th IEEE Conference on Decision and Control*, San Diego, CA, pp. 5453–5458.

Markov, A.A. (1887) Some examples of the solution of a special kind of problem on greatest and least quantities. *Soobshch. Karkovsk. Mat Obshch.* **1**, 250–276 (in Russian).

Medagoda, E. and Gibbens, P. (2010) Synthetic-waypoint guidance algorithm for following a desired flight trajectory. *AIAA J. Guid. Control Dyn.*, **33**(2), 601–606.

Murray, R.M., Li, Z., and Sastry, S.S. (1994) *A Mathematical Introduction to Robotic Manipulation*, CRC Press.

Ollero, A., García-Cerezo, A., and Martinez, J.L. (1994) Fuzzy supervisory path tracking of mobile robots. *Control Eng. Pract.* **2**(2), 313–319.

Park, S., Deyst, J., and How, J.P. (2004) A new nonlinear guidance logic for trajectory tracking. *AIAA Guidance, Navigation, and Control Conference*, Providence, RI.

Patsko, V.S. and Turova, V.L. (2011) Homicidal chauffeur game: history and modern studies. *Advances in Dynamic Games*, Springer, pp. 227–251.

Polak, E. *Optimization: Algorithms and Consistent Approximations*, vol. **124**, Applied Mathematical Sciences, Springer, 1997.

Pontryagin, L.S. and Boltyanskii, V.G. (1962) *The Mathematical Theory of Optimal Processes*. John Wiley & Sons, Inc.

Ratnoo, A., Hayoun, S.Y., Granot, A., and Shima, T. (2014) Path following using trajectory shaping guidance. *AIAA J. Guid. Control Dyn.*, **38**(1), 106–115.

Reeds, J.A. and Shepp, L.A. (1990) Optimal paths for a car that goes both forwards and backwards. *Pac. J. Math.*, **145**(2), 367–393.

Ryoo, C.K., Cho, H., and Tahk, M.J. (2005) Optimal guidance laws with terminal impact angle constraint. *AIAA J. Guid. Control Dyn.*, **28**(4), 724–732.

Savla, K., Frazzoli, E., and Bullo, F. (2008) Travelling salesperson problems for the Dubins vehicle. *IEEE Trans. Automat. Control*, **53**(6), 1378–1391.

Shima, T. and Rasmussen, S. (eds.) (2009) UAV Cooperative Decision and Control: Challenges and Practical Approaches, *Advances in Design and Control*, vol. **18**, SIAM.

Shkel, A.M. and Lumesky, V. (2001) Classification of the Dubins set. *Robotics Auton. Syst.*, **34**, 179–202.

Shneydor, N.A. (1998) *Missile Guidance and Pursuit: Kinematics, Dynamics and Control*, Horwood.

Souères, P., Fourquet, J.-Y., and Laumond, J.-P. (1994) Set of reachable positions for a car *IEEE Trans. Automat. Control*, **39**(8), 1626–1630.

Souères, P. and Laumond, J.-P. (1996) Shortest paths synthesis for a car-like robot. *IEEE Trans. Automat. Control*, **41**(5), 672–688.

Sussmann, H.J. (1995) Shortest 3-dimensional paths with a prescribed curvature bound, in *Proceedings of the 34th Conference on Decision and Control* IEEE, pp. 3306–3312.

Sussmann, H.J. (1997) The Markov–Dubins problem with angular acceleration control, in *Proceedings of the 36th Conference on Decision and Control*, vol. **3**, IEEE, pp. 2639–2643.

Sussmann, H. and Tang, G. (1991) *Shortest paths for the Reeds–Shepp car: a worked out example of the use of geometric techniques in nonlinear optimal control. Rutgers Center for Systems and Control Technical Report, No. 10*, pp. 1–71.

Tang, Z. and Özgüner, Ü. (2005) Motion planning for multitarget surveillance with mobile sensor agents. *IEEE Trans. Robotics*, **21**(5), 898–908.

Tsourdos, A., White, B., and Shanmugavel, M. (2011) *Cooperative Path Planning of Unmanned Aerial Vehicles*, vol. **32**, Aerospace Series, John Wiley & Sons, Inc., New York.

Vinter, R.B. (2000) Optimal Control, *Modern Birkhäuser Classics*. Birkhäuser.

Zarchan, P. (2012) *Tactical and Strategic Missile Guidance*, American Institute of Aeronautics & Astronautics.

Zhao, Y. and Tsiotras, P. (2012) Time-optimal path following for fixed-wing aircraft. *AIAA J. Guid. Control Dyn.*, **36**(1), 83–95.

Chapter 22
Sensor Fusion

Giancarmine Fasano, Domenico Accardo, and Antonio Moccia

University of Naples "Federico II", Napoli, Italy

1 INTRODUCTION

The general concept of sensor fusion is actually a biologically inspired concept, as both humans and animals typically perform decision making by fusing information from different senses, previous experiences, and communication with other living creatures. Sensor fusion, data fusion, and information fusion are typically used as synonyms or with slight semantic differences relevant to the type of data that are combined (Castanedo, 2013).

Within the military community, a well-known definition of data fusion is the one provided by the Joint Directors of Laboratories (JDL) workshop (JDL, 1991): "A multi-level process dealing with the association, correlation, combination of data and information from single and multiple sources to achieve refined position, identify estimates and complete and timely assessments of situations, threats and their significance." Another common definition is the one by Hall and Llinas (1997): "Data fusion techniques combine data from multiple sensors and related information from associated databases to achieve improved accuracy and more specific inferences than could be achieved by the use of a single sensor alone."

Given these general definitions, the domain of data fusion is very wide, comprising ground, airborne, spaceborne, fixed and/or moving sensors and databases providing context information such as maps and digital elevation models. Data fusion applications are widespread, including guidance navigation and control of aerospace vehicles, remote sensing, military and civilian surveillance, monitoring of manufacturing processes, medical applications, and robotics. Moreover, fusion approaches are typically multidisciplinary as they are based on a set of various disciplines such as digital signal processing, statistical estimation, control theory, artificial intelligence, and classic numerical methods.

Data fusion has a strong link with Unmanned Aircraft Systems (UAS), representing indeed a key enabling technology for several missions and unmanned platforms. In fact, in many cases, only the proper combination of different sources can provide the integrated information needed for safe flight and/or mission success. This chapter will briefly highlight architectures, algorithms, practical implementation issues, and two application examples relevant to the adoption of sensor fusion in UAS.

2 SENSOR FUSION IN UAS

The need for data fusion in UAS is related to different applications and requirements:

Unmanned Aircraft Systems. Edited by Ella Atkins, Aníbal Ollero, Antonios Tsourdos, Richard Blockley and Wei Shyy.
© 2016 John Wiley & Sons, Ltd. ISBN: 978-1-118-86645-0.

- *Autonomous navigation:* The absence of the human pilot onboard, and consequently of a direct feedback from the pilot, makes reliable and accurate autonomous navigation even more important than in the case of manned flight (Gross *et al.*, 2012). In fact, the remote pilot has no physical feeling of the aircraft, which makes all the onboard feedbacks (e.g., weight on wheel, surface deflections, and flaps extension) extremely important for flight safety and amplifies the need for (multisensor-based) failure detection and identification (Heredia *et al.*, 2009).
- *Situational awareness:* A proper combination of data from multiple sensors is usually required to provide to the remote pilot the best possible situational awareness, for example, in terms of surrounding traffic (in civil scenarios) or identification–friend–foe–neutral systems (in military applications) (FAA, 2013; Liggins *et. al.*, 2009). Moreover, data link limitations and the necessity to limit the remote pilot workload lead to the necessity of generating and transmitting synthetic estimates of more immediate usage.
- *Enhanced surveillance from a single UAS:* Medium/large UAS are usually equipped with multisensor systems such as optical cameras (visible to far-infrared) and synthetic aperture radars (SAR) (Gundlach, 2012). The combination of different sensors information typically guarantees an enhanced observability and thus significant advantages in terms of ground object detection, tracking, and classification. Moreover, even in a single sensor perspective, effective air or ground target tracking requires proper integration of raw detections with navigation measurements, and all the relevant uncertainties must be taken into account to have a reliable performance. This is especially true for small UAVs that cannot rely on high-quality inertial navigation systems.
- *Collaborative sensing:* Typically, a UAV is smaller than a manned aircraft used in the same role (Austin, 2010). This paves the way for multi-UAV applications where payload information generated by the single platforms can be properly combined to get a complete picture of the observed area (Merino *et al.*, 2006). Application advantages are related to enhanced observability in the case of heterogeneous sensors exploiting different physical phenomena, statistical advantages when overlapping information is acquired with the same type of sensors, and/or enhanced coverage due to the combination of single sensor fields of view (FOVs).
- *Simultaneous localization and mapping (SLAM):* In GPS-denied environments, navigation and mapping tasks usually have to be carried out at the same time. In these applications, fusing measurements from different sensors guarantee an invaluable advantage (as an example,

monocular SLAM suffers from scale ambiguity, which can be tackled by proper integration with inertial sensors/laser scanners/stereocameras (Weiss *et al.*, 2013; Alpen *et al.*, 2010).
- *Photogrammetric mapping:* Small and micro-UAS represent a disruptive technology in the field of high-accuracy 3d mapping. In these applications, building an accurate digital elevation model usually results from a combination of a stream of optical data with highly accurate navigation estimates obtained, for example, from carrier-phase differential GPS (Nex and Remondino, 2014).

In general, in all these cases, sensor fusion allows estimating variables of interest that are unobservable using single sensors, increasing accuracy and/or integrity, achieving a better trade-off between false alarms and missed detections, improving autonomous health management, and so on.

It is interesting to note that in several cases, different sensor technologies are complementary in view of the application requirements:

- Considering autonomous navigation, global navigation satellite systems (GNSS) and inertial navigation systems (INS) are complementary in terms of measurement rate, error drift, and measurement continuity.
- Within obstacle detection applications, radar and optical systems can be complementary in terms of detection range, sensitivity to weather and illumination conditions, angular resolution, and availability of direct range and range rate information.

The trend toward sensor miniaturization, the continuous performance improvements even for commercial low-cost sensors, and the availability of computational power even on tiny CPU boards have been further drivers for an increasing adoption of data fusion onboard small and micro-UAS. As an example, GPS/INS fusion, together with real-time stabilization and control, is the key technology adopted in commercial multicopters that gained a dramatic diffusion in recent times.

As an example of the different information content related to typical data processed onboard a UAS, Figure 1 reports visible (panchromatic) and thermal infrared images of a close approaching ultralight aircraft taken in low-altitude flight from FLARE, an optically piloted aircraft of the Italian Aerospace Research Center (Fasano *et al.*, 2008a). Table 1 lists the main characteristics of the panchromatic and thermal infrared cameras adopted in the encounter test (Fasano *et al.*, 2008b). While the daylight sensor has a better angular resolution that allows discriminating a higher level of details of the aircraft, the thermal image highlights the engine

(a)   (b)

Figure 1. Airborne visible and thermal images of an approaching ultralight aircraft.

Table 1. Main characteristics of the optical sensors adopted in the considered encounter test.

Sensor	Wavelength	Detector technology	Picture size	Field of view (FOV)	Instantaneous FOV (IFOV)
Panchromatic camera	400–700 nm	Charge-coupled device (CCD)	1280 × 960	52.9° × 40.8°	0.041°
Infrared camera	7.5–13 μm	Microbolometer	320 × 240	24° × 18°	0.075°

emission and the different temperatures of ground objects. It should also be noted that since infrared cameras detect radiation emitted by the target, they offer better performance under poor sun illumination conditions, thus complementing reduced panchromatic data quality.

A similar example can be shown comparing low-altitude images taken from a panchromatic camera and a pulsed forward-looking radar (Figure 2). Table 2 lists the main characteristics of the adopted radar (Fasano *et al.*, 2008b). Actually, pulse radar outputs range (i.e., distance) and azimuth and elevation angle measurements, along with echo amplitude. Figure 2a has been obtained by projecting echoes from a 3D radar scan on a horizontal plane. In this case, in spite of a much coarser angular resolution, radar range and echo amplitude measurements provide an information content that overcomes the limitations deriving from the

unfavorable observation geometry of the optical sensor, thus improving awareness of ground obstacles.

3   ARCHITECTURES

In general, the high-level objective of a data fusion system is to integrate in the best possible way the different information sources while ensuring high reliability and possibly limiting the computational burden, especially in the cases of real-time applications.

Given this objective, various architectures can be used from the point of view of the type of information that is combined and that results from the fusion process, the location of the fusion processes, and the eventual sensors' hierarchy. In practical cases, the selected fusion architecture

Table 2. Main characteristics of the considered pulsed radar.

Wavelength	Scan rate	Range resolution	Azimuth FOV	Elevation FOV	Angular resolution
0.9 cm	≤1.3 Hz (depends on FOV)	≥4.5 m	30–180°, pointable	21–30°, pointable	1.7°

Figure 2. Low-altitude images in similar geometries acquired with a pulsed forward-looking radar (a) and a daylight panchromatic camera (b).

will depend on the application, sensor resolution, processing resources, and communication constraints.

Some of these classifications are briefly recalled here, and the reader also is advised to refer to JDL (1991), Hall and Llinas (1997), Durrant-Whyte (1988), Dasarathy (1997), Luo, Yih, and Su (2002), and Klein (2004) for a more detailed and general treatment.

Regarding the type of information that is combined, it is possible to define the following:

- *Low-level fusion:* Sensors' raw or preprocessed data are combined. This is related to the concept of pixel-level fusion found in Klein (2004).
- *Medium-level fusion:* It refers to the combination of features or characteristics in the framework of classification problems, and is equivalent to the concept of feature-level fusion (Klein, 2004).
- *High-level fusion:* It is also known as decision fusion, as it is relevant to decision-making problems.
- *Multiple level fusion:* It refers to a combination of the previously defined levels.

This classification is closely related to the five fusion levels (0–4) defined by JDL (1991). Also recalling the descriptions of Section 2, it is clear that most UAS applications (e.g., detection, tracking, classification, navigation, and/or mapping) are mainly low- or medium-level fusion tasks and are classified at level 1 within the JDL classification.

This discussion is also relevant to the location of the fusion processes. In fact, following Klein (2004), it is possible to define the following:

- *Sensor-level fusion:* A significant amount of processing (detection, tracking, state estimation, and classification) is carried out before data entry into the fusion processor. The fusion processor then combines the estimates from the sensors to improve the information content. In UAS applications, this approach is useful to combine "smart sensors" and allows one to "plug and play" information sources, thus improving configuration flexibility. It is also appealing because of the communication constraints that usually hinder multi-UAV applications. However, this is a suboptimal architecture from the theoretical point of view.
- *Central-level fusion:* Each sensor provides minimally processed data to a single-fusion processor. While this architecture is theoretically optimal, possible main disadvantages consist in less flexibility and in the communication bandwidth required to transfer large quantities of data.
- *Hybrid fusion:* A combination that tries to gain the benefits of both the previous approaches, but can pose

challenges in terms of both computational and communication resources.

The type of information that is fused and the location of fusion processes are also related to the interaction among the sensors. In particular, the concept of cross-sensor cueing refers to an exchange of information at sensor level that can be used to improve performance. Implementation of the concept usually brings advantages, but also an additional complexity, besides requiring a proper architecture customization and the access to raw data. In integrated navigation applications, the concept resembles the definition of deep integration versus loose/tight coupling (Schmidt, 2010).

Another possible choice in multisensor architectures is the implementation of a hierarchy among sensors, for example, distinguishing main and auxiliary sensors on the basis of their role in the fusion framework.

Thinking about data fusion in UAS, it is also possible to define other architectural choices:

- *Onboard versus on ground fusion:* The choice depends on the constraints related to onboard processing resources and communication bandwidth and reliability.
- *Real-time versus off-line fusion:* Navigation, SLAM, and sense and avoid are examples of the first category, while the second comprises, among other applications, 3d photogrammetric mapping, and (off-line) high-accuracy trajectory reconstruction to be used as reference for specific applications.
- *In multi-UAV systems, decentralized versus centralized fusion:* The first term refers to a system that works without a central node to process the information. An important point in these architectures is to take into account the statistical relation among estimates formulated on different vehicles (Indelman *et al.*, 2012). Moreover, homogeneous/heterogeneous UAS can be combined with homogeneous/heterogeneous payloads.

4   ALGORITHMS

Data fusion systems usually make use of different types of algorithms. In the following state estimation techniques, which are adopted in both navigation and tracking problems and are actually focused on how to combine uncertainties from different sources, will be introduced. Tracking systems also require proper methods for data association. The reader is referred to Blackman and Popoli (1999) and Bar-Shalom, Willett, and Tian (2011) for a detailed mathematical presentation of these approaches.

A basic need in sensor fusion systems is the capability to derive the best estimate from the output of multiple sources of information in terms of one or more parameters that represent the state for a system of interest. First of all, each source of information must be analyzed in order to assess the following issues:

- If it can be adequately modeled as instantaneous or dynamic system, that is, if its current output depends only on current input or on the whole temporal history of past inputs.
- If it has a nonnegligible random component or it is a fully deterministic system.

Moreover, the concept of best estimate must be exploited by introducing a proper cost function J that must be minimized in order to attain the estimate of the system state condition at a specific time instant that is closest to the effective one that the system state assumed at that time. In general, a least-square concept is adopted to derive J (Gelb, 1974).

Once the above-described analysis is completed, each source of information can be identified as belonging to one of the types reported in the following list:

(a) Dynamic source of information with nonnegligible random component.
(b) Dynamic source of information with negligible random component.
(c) Instantaneous source of information with nonnegligible random component.
(d) Instantaneous source of information with negligible random component.

In order to assess the dynamical behavior of the system of interest, sources belonging to types (a) and (b) are defined *process* sources, whereas sources belonging to types (c) and (d) are defined *measurement* sources. In order to assess the noise behavior of the system of interest, sources belonging to types (a) and (c) are defined *stochastic* sources, whereas sources belonging to types (b) and (d) are defined *deterministic* sources. Analytically, process and measurement models are represented by a set of equations as reported in Equation 1:

$$x_k = f(x_{k-1}, \ k, u_k) + v_k$$
$$y_k = h_k(x_k) + w_k \tag{1}$$

where x_k is a state vector at discrete time k, u_k is the input to the source of information, that is, the effective physical term that is estimated, v_k is a random vector representing the noise for process source of information, $f()$ is the state update function for process source of information, y_k is the measurement vector at discrete time k, $h()$ is the state-measurement transfer function, and w_k is a random vector representing the noise for measurement source of information. In general, x_k, u_k, and y_k have different dimensions. A discrete formulation is presented because this is the most common form adopted for computer-coded sensor fusion algorithms.

Sensor fusion algorithms are developed to generate a best estimate of system state each time an updated information is made available from information sources. They account for both dynamical and noise terms to fulfill the above-reported task. In order to handle each type of information source, sensor fusion algorithms need to have a common framework that includes the following processing components:

(i) A *state process update segment* that handles system state information from process sources by exploiting Equation 1.1. This segment is often referred to *a priori* state update because it is performed before a measurement is made available.

(ii) A *state measurement update segment* that handles system state information from measurement sources by exploiting Equation 1.2. This segment is often referred to *a posteriori* state update because it is performed after a measurement is made available.

(iii) A *probability process update segment* that handles system noise information from process sources by estimating proper stochastic terms derived from Equation 1.1. This segment is often referred to *a priori* probability update because it is performed before a measurement is made available.

(iv) A *probability measurement update segment* that handles system noise information from measurement sources by estimating proper stochastic terms derived from Equation 1.2. This segment is often referred to *a posteriori* probability update because it is performed after a measurement is made available.

In general, probability update is performed by applying a specific formulation of the Bayes theorem (Gelb, 1974). For this reason, all the sensor fusion algorithms described in this section are addressed as Bayesian filters. The main techniques reported can be summarized in the following list:

1. The linear Kalman filter (LKF) that can be applied if all sources of information are conveniently modeled by linear dynamical and measurement equations and the noise is modeled as Gaussian white noise. In this case, both $f()$ and $h()$ shall be linear and v_k and w_k are Gaussian random variables (GRVs). It can be demonstrated (Gelb, 1974) that under the Gaussian assumption, the Kalman filter is the minimum mean-square error state estimator.

2. The extended Kalman filter (EKF), a suboptimal technique that can be applied if some sources of information have a nonlinear dynamical model, but their behavior can be adequately modeled through the linear term of Taylor series of process and measurement equations in the vicinity of a known solution. The noise must be modeled as Gaussian white noise. In this case, $f()$ and/or $h()$ can be nonlinear and v_k and w_k are GRVs,

3. The unscented Kalman filter (UKF) that is also suboptimal and is used when the sole linear term in the Taylor series is not enough accurate to model process propagation. In this case, a proper set of *sigma points* is propagated by nonlinear $f()$ so that mean and covariance are preserved through the transformation. In this case, $f()$ and $h()$ can be nonlinear, but v_k and w_k shall be GRVs.

4. The Monte Carlo sequential sampling (MSS) techniques that exploit Monte Carlo integration by removing any type of hypothesis on process model, measurement model, and the type of noise. The most common MSS method is represented by particle filters (PFs). In this case, a large set of samples of the probability density function (PDF), that is, the particles, are propagated by nonlinear update functions. The effect of measurement is accounted by a processing step that updates the weights associated with the particles. An additional processing step is usually adopted, which is called *resampling*. The value of the state at a specific time step is the weighted mean of the values of each particle at the same time step. No hypothesis is made on the type of probability distribution function adopted. However, theoretical optimality is ensured for a sufficiently large number of particles.

Indeed, several other techniques have been developed to perform sensor fusion. The initial ones were essentially least-square/maximum-likelihood methods or applications of Bayes theorem, such as probabilistic grids (Stone, Barlow, and Corwin, 1999) that do not account for process dynamical model (Gelb, 1974). In some cases, they were suboptimal techniques based on heuristic parameters. Subsequently, other filters were developed that accounted for process dynamics, but they did not account for noise stochastic models, such as α–β complementary filters (Brookner, 1998). However, all these techniques were mainly intended for application in computers with reduced processing resources. Indeed, current computers are largely capable of handling the computational effort requested by the Kalman filter and the other techniques reported in the previous list. Moreover, they can accomplish this task in real time with negligible cost for both acquisition and power consumption. For this reason, only the techniques reported in the list will be described in detail in the following sections.

4.1 Linear Kalman Filters

LKF is based on a linear model of Equation 1:

$$x_k = \mathbf{\Phi}_k\, x_{k-1} + B_k u_k + v_k$$
$$y_k = H_k\, x_k + w_k \tag{2}$$

where $\mathbf{\Phi}_k$ is a time-varying matrix reporting state update, B_k is a time-varying matrix reporting direct input-state transfer, and H_k is a time-varying matrix reporting state-measurement transfer. Process noise v_k and measurement noise w_k are assumed as temporally uncorrelated Gaussian white noises with zero mean and covariance, respectively, given by the two time-varying matrices $Q_k = E\left[v_k v_k^{\mathrm{T}}\right]$, that is, the process noise covariance matrix, and $R_k = E\left[w_k w_k^{\mathrm{T}}\right]$, that is, the measurement noise covariance matrix. The mutual covariance matrix $E\left[v_k w_k^{\mathrm{T}}\right]$ is assumed as null matrix. Standing system linearity, the effect of deterministic input signal u_k on process can be addressed by standard linear systems analysis techniques. As a consequence, the problem discussed in Equation 2 can be reduced as reported in Equation 3:

$$x_k = \mathbf{\Phi}_k\, x_{k-1} + v_k$$
$$y_k = H_k\, x_k + w_k \tag{3}$$

Since the value of noise cannot be estimated, the best estimate of state update $\tilde{x}_k$ is given by Equation 4:

$$\tilde{x}_k(-) = \mathbf{\Phi}_k\, \tilde{x}_{k-1} \tag{4}$$

The symbol $(-)$ is referred to *a priori* quantity, whereas the symbol $(+)$ is referred to *a posteriori* quantity. The expression of the *a priori* state covariance $P_k(-) = E\left[\tilde{x}_k(-)\tilde{x}_k(-)^{\mathrm{T}}\right]$ can be derived by applying the definition of expected value to the square of state derived from Equation 3.1. It is given in Equation 5:

$$P_k(-) = \mathbf{\Phi}_k\, P_{k-1}\mathbf{\Phi}_k^{\mathrm{T}} + Q_k \tag{5}$$

Each time a new measurement $\hat{y}_k$ is acquired, the Kalman filter assumes that the *a posteriori* state estimate $x_k(+)$ can be derived as a linear combination of *a priori* state estimate $x_k(-)$ and the "weighted error" between the effective new measurement value $\hat{y}_k$ and the process estimated measurement value $\tilde{y}_k = H_k x_k$. Analytically, $\tilde{x}_k(+)$ is given by Equation 6:

$$\tilde{x}_k(+) = \tilde{x}_k(-) + K_k(\hat{y}_k - H_k\tilde{x}_k(-)) \tag{6}$$

where the linear time-varying matrix K_k is called "Kalman gain." The main issue to complete the development of the Kalman filter was to select K_k so that the trace of the *a posteriori* error covariance matrix $P_k(+) = E\left[\tilde{x}_k(+)\tilde{x}_k(+)^{\mathrm{T}}\right]$ is minimized. Indeed, the trace of $P_k(+)$ matrix has been selected as cost function, since it is usually a matrix that have diagonal terms greater than off-diagonal ones. $P_k(+)$ can be estimated by applying the expected value operator to the square of Equation 5. The relevant expression is given by Equation 7:

$$P_k(+) = (I - K_k H_k)P(-)(I - K_k H_k)^{\mathrm{T}} + K_k Q_k K_k^{\mathrm{T}} \tag{7}$$

where $P_k(-)$ is given by Equation 5. Equation 5 is known as "Joseph form" of the covariance update equation. The Jacobi matrix of the trace of $P_k(+)$ as a function of K_k can be derived and it can be set equal to zero in order to find the minimum. The resulting value of the Kalman gain is given as reported in Equation 8:

$$K_k = P_k(-)H_k^{\mathrm{T}}\left(H_k P_k(-)H_k^{\mathrm{T}}\right)^{-1} \tag{8}$$

This is the searched value of the Kalman gain that closes the loop of the Kalman filter. Indeed, the expression of the Kalman gain can be read as a division between a term that depends on process noise Q_k and a term that depends on the measurement noise R_k. As expected, large values of the Kalman gain are determined for large process noise or small measurement noise.

The application of the Kalman filter as sensor fusion algorithm can be summarized in the procedure described in the following section.

4.1.1 Kalman filter algorithm

Initialization: The initial state x_0 is determined by external reference, the initial value of process covariance matrix Q_k, and the measurement covariance matrix R_k is determined by noise specifications reported by system manufacturers in system data sheets or by other considerations, the initial value of P_k is often set so that $P_0 = Q_0$. Indeed, the steady-state value of P_k does not depend on initial condition because it is a linear system.

Step 1: As soon as information is output from process information sources, the *a priori* state update (Equation 4) and covariance update (Equation 5) can be used to determine updated values of $x_k()$ and $P_k()$. In all applications of the Kalman filter, the process update rate is greater than or equal to the measurement update rate. As an example, when inertial systems and GPS are integrated, the following condition is determined:

(a) The inertial system is the process source of information and its data rate can be up to 1000 samples per second.

(b) The GPS receiver is the measurement source of information and its data rate is 1 sample per second.

If no new data $\hat{y}_k$ are output by measurement source of information, then no *a posteriori* estimate can be derived and the procedure shall loop on step 1 in order to get a new *a priori* estimate at time $k + 1$. Differently, if new $\hat{y}_k$ is output by measurement source of information, then the algorithm must process step 2.

Step 2: The Kalman gain K_k is estimated by exploiting Equation 8.

Step 3: The *a posteriori* state estimate $x_k(+)$ is estimated by processing Equation 6.

Step 4: The *a posteriori* state covariance estimate $P_k(+)$ is given by Equation 4.

Step 5: The algorithm can return to step 1 to process new information at time $k + 1$. The *a posteriori* terms derived at steps 3 and 4 are assumed as *a priori* values in Equations 4 and 5 to initiate a new process update sequence.

In brief, the algorithm performs two separate update processes, that is, the state update and the measurement update, for both *a priori* and *a posteriori* conditions. The only link that is present between the two processes is the Kalman gain determination by Equation 8.

4.2 Extended Kalman Filters

The application of EKF is similar to the application of LKF. The main difference is that the state update function $f()$ and/or the state-measurement transfer functions are not linear. To overcome this issue, it is assumed that a known approximate value of the state $\hat{x}_k$ is available at time k. The state update function can be expressed as Taylor series in the vicinity of $\hat{x}_k$, as reported in Equation 9:

$$x_k = \hat{x}_k + \nabla f_x(x_k - \hat{x}_k) + v_k + \text{hots} \qquad (9)$$

where ∇f_x is the Jacobi matrix of $f()$ with respect to x and hots stands for "higher order terms." By setting $\delta x_k = x_k - \hat{x}_k$, Equation 4 can be expressed as reported in Equation 10:

$$\tilde{\delta x}_k(-) = \nabla f_x \tilde{\delta x}_{k-1} \qquad (10)$$

At the same time, the measurement Equation 3.2 can be expressed as reported in Equation 11:

$$\tilde{\delta y}_k = \nabla h_x x_k \qquad (11)$$

where ∇h_x is the Jacobi matrix of $h()$ with respect to x.

Standing Equations 10 and 11, the Kalman filter can be applied to $\tilde{\delta x}_k$ process and $\tilde{\delta y}_k$ measurement by assuming $\Phi_k = \nabla f_x$ and $H_k = \nabla h_x$. As a result, the nonlinear model of Equation 1 can be used to generate the *a priori* state and measurement estimates in the initial *predict* phase. Subsequently, the linear model of Equations 5–8 can be derived with the above-reported positions and the best estimate can be derived in the final *update* phase. The algorithm is identical to LKF in terms of implementation.

4.3 Unscented Kalman Filters

The unscented Kalman filter represents the state distribution by a GRV in the same manner as the EKF, but now this distribution is specified using a minimal set of chosen sample points that completely defines the mean and covariance of the GRVs. These samples can express the posterior mean and covariance accurately to the third order for any nonlinearity when propagated through the true nonlinear system. The UKF is an extension of the Unscented Transformation to the recursive estimation of the state space redefined as the concatenation of the original state x_k and noise variables: $x_k^a = \begin{bmatrix} x_k^T & v_k^T & w_k^T \end{bmatrix}^T$.

A complete description of the Unscented Transformation and of the UKF algorithm is reported in Van der Merwe and Wan (2000). It is worth noting that the number of samples needed to model the unscented Kalman filter is very small with respect to the one needed for sequential Monte Carlo sampling techniques. This result is determined by the fact that process and measurement noises are still assumed as GRVs. This means that unscented Kalman filter still has the limitation of its applicability to solve problems where only Gaussian noise is involved.

4.4 Particle Filters

Particle filter is a Monte Carlo method; in particular, it is a technique for implementing a recursive Bayesian filter by Monte Carlo simulations. Different algorithm variants exist, among which one of the most commonly adopted is called sequential importance resampling (SIR). The key idea is to represent the required posterior density function by a set of random samples with associated weights and to compute estimates based on these samples and weights. As the number of particles becomes very large, the SIR filter approaches the optimal Bayesian estimates since the Monte Carlo characteristics are very close to the posterior PDF.

In particular, the particle filter is based on the recursive generation of random measures that approximate the distribution of unknowns (Monte Carlo integration). The random measures are composed of particles (samples) drawn from relevant distributions and of importance weights of the particles. These random measures allow for the computation of all sorts of estimates of the unknown. As new observation become available, the particles and the weights are propagated following the concept of sequential importance resampling (Ristic, Arulampalam, and Gordon, 2004; Gustafsson *et al.*, 2002; Hwang and Huber, 2007; Doucet, Godsill, and Andrieu, 2000).

The resampling step is implemented in order to prevent the sample degeneracy phenomenon. Indeed, a common framework for PF is given by the fact that at each update, the weights of all but one particle tend to zero (Doucet, Godsill, and Andrieu, 2000). As a consequence, most particles are propagated, but have very small effect on the determination of the state estimate. This is a waste of computational resources that can be prevented through the resampling step. This step is essentially a contest among particles. In the case of uniform distribution, each particle shall have a weight equal to $1/N$ where N is the total number of particles. However, the sum of weight shall also be equal to 1 in the case of nonuniform distribution as it happens in the case of sample degeneracy. Particles with a weight much smaller than $1/N$ are deleted and particles with a weight larger than $1/N$ are splitted into two or more new particles, in order to get a more uniform particle distribution. New particles, generated from the same initial one, are separated from each other by adding a proper amount of noise to their state.

4.4.1   Particle filter SIR algorithm

Initialization: The initial state of the particles is derived by extracting N_p random samples $\mathbf{x}_k^0$ from the process PDF $p(x)$. The number N_p shall be large enough to validate the Monte Carlo hypothesis within a stated percentage of error. The initial weight of each sample is set as $\mathbf{w}_0 = 1/N_p$.

Step 1 – a priori estimate: Each particle state $\mathbf{x}_k^i$, $i = 1$, . . . , N_p is propagated from time $k - 1$ to time k by means of a nonlinear expression derived from Equation 1 that is reported in Equation 12:

$$\mathbf{x}_k^i = \mathbf{f}\left(\mathbf{x}_{k-1}^i, k, \mathbf{u}_k\right) \qquad (12)$$

At each step, the value assumed for the current state $\mathbf{x}_k$ is the mean of the values of all $\mathbf{x}_k^i$, such as

$$\mathbf{x}_k = \frac{1}{N_p} \sum_{i=1}^{N_p} w_i \mathbf{x}_k^i \qquad (13)$$

If no new data $\hat{\mathbf{y}}_k$ are output by measurement source of information, then no *a posteriori* estimate can be derived and the procedure shall loop on step 1 in order to get a new *a priori* estimate at time $k + 1$. Differently, if new $\hat{\mathbf{y}}_k$ is output by measurement source of information, then the algorithm must process step 2.

Step 2—a posteriori estimate: The weight w_i of each particle is updated by means of the expression reported in Equation 14:

$$w_i = \frac{p\left(\hat{\mathbf{y}}_k | \mathbf{x}_k^i\right) p\left(\mathbf{x}_k^i | \mathbf{x}_{k-1}^i\right)}{q\left(\mathbf{x}_k^i | \mathbf{x}_{k-1}^i, \hat{\mathbf{y}}_k\right)} \qquad (14)$$

where $q\left(\mathbf{x}_k^i | \mathbf{x}_{k-1}^i, \hat{\mathbf{y}}_k\right)$ is an importance density, $p\left(\mathbf{x}_k^i | \mathbf{x}_{k-1}^i\right)$ is the state transition probability, and $p\left(\hat{\mathbf{y}}_k | \mathbf{x}_k^i\right)$ is the joint posterior density. The total weight sum is computed as $S = \sum_{i=1}^{N_p} w_i$ and each weight is normalized by dividing per S as $\tilde{W}_k^i = w_i/S$.

Step 3—Resampling:

- For each $i = 1, \ldots, N_p$, retain $k_i = \left\lfloor N_p \tilde{W}_k^i \right\rfloor$ copies of $\mathbf{x}_n^i$
- Let $N_r = N_p - k_1 - \cdots - k_{N_p}$, obtain N_r independent identically distributed draws from $\{\mathbf{x}_n^i\}$ with probabilities proportional to $N_p \tilde{W}_k^i - k_i (i = 1, \ldots, N_p)$
- Reset $w_k^i = 1/N_p$
- Add noise to $\mathbf{x}_n^i$, estimated considering $p\left(\mathbf{x}_k^i | \mathbf{x}_{k-1}^i\right)$, in order to separate identical samples.

Step 4: The algorithm can return to step 1 to process new information at time $k + 1$. The *a posteriori* terms derived at steps 2 and 3 are assumed as *a priori* values in Equations 12 and 13 to initiate a new process update sequence.

A key issue for the correct adoption of PF is the selection of the number of particles. Indeed, a proper guess can be derived by capping process PDF with a Gaussian and applying Chebishev inequality to the variance of the derived Gaussian PDF. Usually, large values N_p are needed to obtain satisfactory performance. As a consequence, the overall computational load of PF can be up to two orders of magnitude worse than EKF and UKF.

4.5   Summary of Sensor Fusion Algorithms

Table 3 summarizes pros and cons of each reported sensor fusion algorithm.

Table 3. Comparative chart of Sensor Fusion algorithms.

	LKF	EKF	UKF	PF
Linear model required for process and measurement	Yes	No, but must be reduced to linear by Taylor series	No	No
Gaussian noise required for process and measurement	Yes	Yes	Yes	No
Computational effort	Standard	Standard	Slightly worse than standard	Considerably worse than any form of KF
Settling time	Slow	Slow	Faster than EKF	Very fast
Optimality ensured by theorem	Yes	No, suboptimal	No, suboptimal	Yes for large N_p

5   IMPLEMENTATION ISSUES

Typical sensor fusion applications on UAS have to tackle several implementation issues that are not dependent on the selected architectures and the theoretical optimality of the adopted estimation algorithms, such as spatial and temporal registration of the information provided by the different sensors, algorithm tuning, and reliable filtering of spurious measurements.

Spatial registration is related in most single platform cases with sensors alignment, comprising for example, relative alignment among multiple optical sensors, alignment of single or multiple cameras with respect to the onboard attitude and heading reference system (AHRS), and relative and absolute (i.e., with respect to the AHRS) alignment of optical, radar, and/or laser sensors.

While relative alignment of multiple cameras can be carried out with well-assessed extrinsic calibration procedures based on image processing and least-squares approaches (Bouguet, 2013), in the other cases problem solution can be much less trivial, and it is interesting to underline that its importance is often overlooked especially in small UAS applications.

In general, spatial registration approaches can be classified as off-line and online techniques, depending on how the calibration solution is provided. Online techniques (see, for example, Weiss *et al.* (2013)) estimate alignment parameters including them within the dynamic state vector that is estimated in flight. Off-line techniques, instead, foresee an alignment procedure that is carried out on ground without the need for real-time processing. An example is the cameras–AHRS calibration procedure based on carrier-phase differential GPS, image analysis, and AHRS attitude measurements in static conditions, which is described in Fasano *et al.* (2010). While the main rationale of off-line approaches is in the much better accuracy that can be achieved in static conditions and in the fact that alignment parameters can

usually be considered as constant even in a multimission perspective, online approaches have the advantage of performing estimation without requiring additional instrumentation and in the real operation environment experimented on the vehicle, which can differ from ground calibration conditions. It is also possible to adopt hybrid approaches that combine the different benefits. In multi-UAS applications, online registration approaches (including translation and rotation between different sensors) are always required because of dynamically changing and/or *a priori* unknown conditions.

Time registration is another key issue to be tackled for effective data fusion. In fact, different sensors usually provide their measurements at different times, with different rates, and sometimes with nonnegligible latencies. All these factors have to be taken into account and compensated to avoid the introduction of additional unmodeled errors in the fusion framework (Weiss *et al.*, 2013; Fasano *et al.*, 2010). As in other multisensor applications, another common issue to be considered is represented by out of sequence measurements, that is, measurements that are received before data relevant to previous instants. Approaches selected to compensate these effects mainly depend on the fusion computational load (Blackman and Popoli, 1999). From a practical point of view, especially when sensors are installed in strapdown configuration, unmanned aircraft dynamics usually requires that all the data to be fused are properly time referenced, which is also dependent on the capability of processing units, operating systems, and developed software components to satisfy real-time requirements.

Algorithm tuning is usually aimed at performance optimality and robustness. While it is very important to accurately model sensor performance in order to avoid solution inconsistency, other parameters may have to be chosen on the basis of a compromise between contradicting requirements, thus involving practical considerations about the set of operating conditions for the considered UAS. A possible example is the process noise in multitarget tracking applications. In general, tuning depends on the unmanned aircraft dynamics, maneuverability, and so on. Indeed, some sensor fusion algorithms comprise the aircraft dynamic model, although parameter uncertainties can hinder this estimation approach. Tuning is also related to the fact that theoretical assumptions on which common estimation algorithms such as LKF are based are not usually satisfied in practical applications.

Inclusion of spurious measurements within a fusion algorithm can lead to dangerously inaccurate estimates; thus, proper techniques have to be used to reduce this risk. These techniques can work at preprocessing level or within the fusion algorithm itself. Some examples are as follows:

- Removal of ground clutter echoes in radar/EO airborne tracking for sense and avoid (Accardo *et al.*, 2013).
- Basic or advanced data association algorithms in multisensor multitarget tracking systems (Blackman and Popoli, 1999; Bar-Shalom, Willett, and Tian, 2011).
- Randomized sample consensus (RANSAC) methodologies adopted to remove outliers in vision-aided navigation applications (Hartley and Zisserman, 2000).

6 SENSOR FUSION EXAMPLES

6.1 Multisensor-Based Sense and Avoid

Sense and avoid is one of the key challenges for UAS integration within the civil airspace. In general, it consists of two functions such as separation assurance, which ensures that the aircraft remain "well clear" of each other, and collision avoidance, which is related to extreme maneuvers to prevent collision in cases where safe separation is lost. Information sources can be cooperative systems, wherein an aircraft is equipped with a transponder to interrogate and/or broadcast information, for example, TCAS (Kuchar and Drumm, 2007) and ADS-B (Helfrick and Buckwalter, 2013), or noncooperative sensors, which are able to detect targets autonomously (mainly, radar and

optical sensors). Separation assurance is mainly based on cooperative systems that guarantee very high detection ranges, while noncooperative systems are aimed at collision avoidance for unequipped intruders or in case of data link losses/integrity problems. While safe UAS integration requires cooperative and noncooperative information sources to be smoothly integrated, the great potential of multisensor systems and data fusion within the noncooperative framework derives from the fact that the different sensors have complementary features, as it is shown qualitatively in Table 4.

Looking at today's technology (research efforts are under way that can rapidly change this situation), radars are demanding in terms of cost, size, weight, and required electric power, while EO cameras offer the potential of highly accurate angular measurements at fast update rates and with limited budgets. Thus, while stand-alone EO sensors are a viable solution for small platforms, integrated radar/EO systems can be a good sensing option to provide the required situational awareness in the case of medium/large UAS platforms that have to attain a reliable full autonomy from ground.

An example of data fusion implementation for sense and avoid is the multisensor-based system developed within project TECVOL by the Italian Aerospace Research Center and the University of Naples "Federico II" (Fasano *et al.*, 2008a, 2008b, 2011; Accardo *et al.* 2013), which reached flight demonstration of closed-loop autonomous avoidance. TECVOL sensor fusion architecture was designed based on the following concepts:

- *Hierarchy among sensors:* The radar is used as the main sensor that determines track status; the EO/IR sensors are auxiliary information sources.
- *Central-level fusion:* Raw sensor data are combined on the basis of their estimated uncertainty in a unique tracking algorithm using an EKF for state estimation.
- *Cross-sensor cueing:* EO/IR obstacle detection is cued by the tracking module (initially fed by the radar) providing information about intruder range and (initially coarse) line of sight.

Table 4. Characteristics of noncooperative systems.

Sensor	Detection range	Range/range rate measurements	Typical angular accuracy	Typical refresh rate (Hz)	Sensitivity to false measurements	Sensitivity to weather/ illumination
RADAR	High	Yes	Coarse: 1°	≤ 1	Low	None/low
EO/IR	Low/moderate	No	Fine: 0.1°	≥ 10	High/moderate	High

A synopsis of TECVOL project from the point of view of the sensing system is shown in Figure 3:

- Figure 3a shows the experimental aircraft with the sensors mounted in central position on the top of the aircraft wing.
- Figure 3b reports the whole sense and avoid hardware architecture (including the flight control computer with sensors and actuators).

- Figure 3c shows the logical architecture with the central fusion processor and the scheme for cross-sensor cueing.
- Figure 3d describes the EO obstacle detection process that takes advantage from cross-sensor cueing and is carried out only in a relatively small search window.
- Figure 3e and f are flight images showing the highly variable operating conditions that have to be dealt with by the optical detection system.

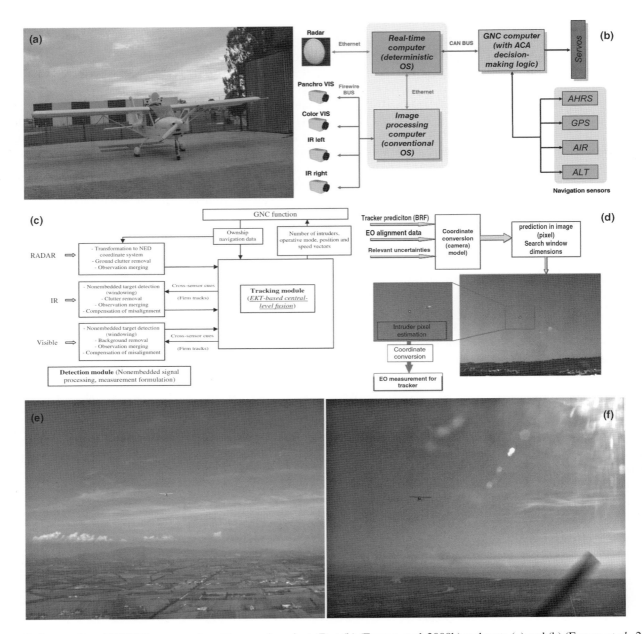

Figure 3. Images from TECVOL sense and avoid research project. (Part (b) (Fasano *et al.* 2008b) and parts (g) and (h) (Fasano *et al.*, 2011) have been reprinted with permission from the American Institute of Aeronautics and Astronautics, Inc.)

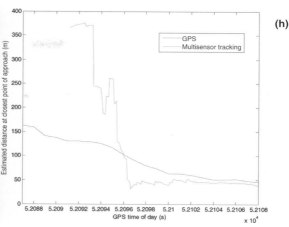

Figure 3. (*Continued*)

- Figure 3g is an example of search window binarization and obstacle line of sight extraction.
- Figure 3h reports the distance at closest point of approach as estimated by the multisensor system compared with the estimate based on GPS data (reference used in off-line analysis). In the second part of the diagram, the estimation error decreases significantly, which is due to the integration of optical and radar measurements. This demonstrates quantitatively the improvement of situational awareness.

Among other experiences, a data fusion-based sense and avoid system was also demonstrated within the MIAA (Multiple Intruders Autonomous Avoidance) project by the US Air Force Research Lab and Northrop Grumman (Chen *et al.*, 2011; Graham *et al.*, 2011). The adopted data fusion architecture shows similarities and differences with TECVOL. In fact, MIAA project also used EKF as basic fusion algorithm, but tracks were generated at sensor level, no cross-sensor cueing logic or sensors' hierarchy was implemented, and the fusion architecture comprised both noncooperative and cooperative sensors. Moreover, multiple intruders were considered.

6.2   Multisensor Integrated Navigation

Inertial Navigation Units (INUs) are systems composed of triaxial accelerometers and gyros able to determine, by integration, the navigation status of a platform, that is, its attitude, position, and velocity. This process is described in detail in several references: Rogers (2007), Groves (2013),

Savage (2002), and Chatfield (2007). The navigation state vector is formed by nine independent scalar terms as follows:

$$\underline{r}^{\mathrm{n}} = \left\{ \begin{array}{c} \varphi \\ \lambda \\ h \end{array} \right\} = \left\{ \begin{array}{c} \text{geodetic latitude} \\ \text{longitude} \\ \text{geodetic altitude} \end{array} \right\}$$

$$v_e^{\mathrm{n}} = \left\{ \begin{array}{c} v_{\mathrm{N}} \\ v_{\mathrm{E}} \\ v_{\mathrm{D}} \end{array} \right\} = \left\{ \begin{array}{c} \text{North component of ground speed} \\ \text{East component of ground speed} \\ \text{Down component of ground speed} \end{array} \right\}$$

C_b^{n} = Direct cosine matrix between platform body reference frame and locally level reference frame. The matrix has nine terms, but just three are independent.

Standard-level inertial units cannot be adopted as standalone source of information for the navigation system installed onboard an aircraft since they are limited by the following drawbacks (Rogers, 2007; Groves, 2013; Savage, 2002; Chatfield, 2007; Grewal, 2013; Titterton, 2005; Farrell, 2008):

- Their error has a drift with time due to the fact that navigation state terms are estimated by integrating accelerometers and gyro measurement that are affected by random noise and bias drift.
- They are not capable to derive autonomously their initial status.

As a consequence of the above issues, aiding sensors, such as GPS, magnetometers, odometers, and imaging cameras, are needed to compensate drift and provide information about initial conditions. Nevertheless, the adoption of inertial

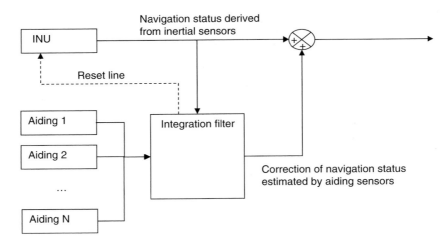

Figure 4. Architecture of an integrated navigation system.

sensor is mandatory standing the reasons reported in the following list:

- They are the main source of attitude information.
- They are the only source that is capable to attain a measurement rate up to 100 Hz and higher. This data rate is needed to preserve an adequate bandwidth for platform control.
- They are a very reliable and integrous sources of information since their principle of operation is based on the basic Laws of Newtonian Mechanics rather than on measuring some parameters of a force field that can be affected by nuisance and intentional jamming.

Standing the above considerations, the typical layout of an onboard navigation system is composed by the integration of an INU and one or more aiding sensors in a data fusion framework. The general data fusion scheme is based on the following integration architecture:

(i) Inertial sensors determine an update of the navigation status at each measurement update by integrating navigation equations that will be presented in Section 6.2.1.

(ii) Each time a new estimate is available from aiding systems, that is, GPS or EO, a Kalman filter will be run to perform optimal correction of status and keep inertial estimate drift (Chatfield, 2007) under control (Sections 6.2.2 and 6.2.3).

Two types of integrated navigation schemes can be adopted, such as uninterrupted integrated navigation and integrated navigation with reset.

Following the scheme reported in Figure 4, the only difference between the above reported forms is the presence of the reset link reported by the dashed line.

Reset operation is composed of the following processing steps:

1. The integration filter block detects that the value of correction is over a stated threshold that is the limit of EKF linear approximation. As a consequence, it activates the reset link. In some applications, the reset line is periodically activated to support hard real-time configurations.
2. The navigation state in the INU is set equal to the sum of the current INU state plus correction estimated from integration filter.
3. The current value of integration filter correction, that is, the opposite of error estimate, is reset to zero in the integration filter.

Reset operation is performed to keep the value of the correction within a stated threshold so that the linear approximation of EKF is verified. Indeed, EKF is the typical data fusion technique used as integration filter.

In order to perform the error containment process described above by exploiting a data fusion filter, such as an EKF, a proper measurement model is needed for each aiding sensor. Following Figure 4 and sensor data fusion filter models reported in Section 4, three analytical frameworks must be defined to develop an integrated navigation system:

(a) An inertial navigation framework that is used to derive estimates of the navigation status by exploiting initial conditions and inertial sensor measurements.

(b) A process model for the data fusion filter that accounts for a linear representation of the error for navigation state terms as a function of the error in inertial sensor measurements.

(c) A measurement model for the data fusion filter that accounts for a linear representation of the error for aiding sensor measurements as a function of navigation status error.

These three topics are briefly discussed in Sections 6.2.1, 6.2.2, and 6.2.3.1 for the GPS-aided system. Section 6.2.4 discusses the determination of initial conditions. Finally, Section 6.2.5 shows an application example.

6.2.1 Inertial navigation mechanization in local geographic navigation frame

In this section, the following definitions of reference frames are adopted:

- The body reference frame (BRF) is a frame with origin in the moving platform, that is, the aircraft, center of mass, and axis x, y, and z mutually orthogonal.
- The navigation or locally level reference frame (NRF) is a frame with origin in the aircraft center of mass and axes directed parallel to local, that is, referred to platform nadir, geographic directions. A typical sequence of axis is north, east, and down (NED).
- The earth reference frame (ERF) is a frame that has origin in the center of Earth and axes directed following the sequence: X pointing at intersection of Greenwich meridian and Equator, Z pointing at geographic north, and Y to form a left-rotating reference frame. This reference frame is commonly defined as earth-centered earth-fixed (ECEF).
- The inertial reference frame (IRF) is a frame that has origin in the center of Earth and axes constantly directed toward the fixed stars. In particular, X_I axis is pointing toward the First Pont of Aries, the Z_I is pointing the geographic north, and Y_I is selected to form a left-rotating reference frame. This reference frame is commonly defined as earth-centered inertial (ECI).

The inertial navigation equations can be described in NRF as reported in Equation 15 (Rogers, 2007; Groves, 2013; Savage, 2002; Farrell, 2008):

$$\left\{ \begin{array}{c} \dot{\underline{r}}^n \\ \dot{\underline{v}}^n_e \\ \dot{C}^n_b \end{array} \right\} = \left\{ \begin{array}{c} [D]^{-1}\underline{v}^n_e \\ C^n_b\underline{f}^b - \left[2\,\underline{\omega}^n_{ie} + \underline{\omega}^n_{en}\right] \times \underline{v}^n_e + \underline{g}^n_l \\ C^n_b\left(\Omega^b_{ib} - \Omega^n_{in}\right) \end{array} \right\} \quad (15)$$

where

n superscript refers to NRF components,
b superscript refers to BRF components,
b subscript refers to quantities defined in BRF;
e subscript refers to quantities defined in ECEF,
$\underline{f}^b$ is the specific force sensed by accelerometers installed in strapdown configuration, that is, with sensor input axes that have fixed orientation with respect to BRF.
$\underline{\omega}^b_{ib} = \left[\omega_x\,\omega_y\,\omega_z\right]$ is the inertial rate of body reference system sensed by rate gyros installed in strapdown configuration. The letters x, y, and z refer to the axes of BRF.
Ω^b_{ib} is the skew-symmetric matrix form of $\underline{\omega}^b_{ib}$ as reported in Equation 16:

$$\Omega^b_{ib} = \begin{bmatrix} 0 & -\omega_z & \omega_y \\ \omega_z & 0 & -\omega_x \\ -\omega_y & \omega_x & 0 \end{bmatrix} \quad (16)$$

$\Omega_\oplus$ is the module of Earth rate that is equal to $7.292115 \times 10{-}5\,\text{rad s}{-}1$ for WGS84 Earth model.
$\underline{\omega}^n_{ie}$ is the Earth rate expressed in NRF as reported in Equation 17:

$$\underline{\omega}^n_{ie} = \left[\omega^n_{ie_N}\,\omega^n_{ie_E}\,\omega^n_{ie_D}\right] = \left[\Omega_\oplus\,\sin\phi\,0 - \Omega_\oplus\,\cos\phi\right] \quad (17)$$

$\underline{\omega}^n_{en}$ is the transport rate, that is, the rate acting on the navigation frame because of the nonzero $\underline{v}^n_e$ velocity of the platform with respect to ECEF, expressed in NRF.
$\underline{\omega}^n_{in} = \underline{\omega}^n_{ie} + \underline{\omega}^n_{en}$ is the overall inertial rate of NRF with respect to ECI expressed in the NRF.
Ω^n_{in} is the skew-symmetric matrix form of $\underline{\omega}^n_{in}$.
$\underline{g}^n_l$ is the local gravity expressed in the NRF. In general, it is a function of altitude h and geodetic latitude φ as reported in Savage (2002).
The matrix $[D]$ is given by Equation 18:

$$[D]^{-1} = \begin{bmatrix} 1/(M+h) & 0 & 0 \\ 0 & 1/((N+h)\cos\,\varphi) & 0 \\ 0 & 0 & -1 \end{bmatrix} \quad (18)$$

M and N are Earth meridian and transverse radii that can be given as reported in Equation 19:

$$M = \frac{R\left(1 - e^2\right)}{\left(1 - e^2\sin^2\varphi\right)^{3/2}}$$
$$N = \frac{R}{\left(1 - e^2\sin^2\varphi\right)^{1/2}} \quad (19)$$

R is the Earth semimajor axis, that is, the radius at Equator, that is equal to 6378137.0 m for WGS84 Earth Model (Rogers, 2007).

e is the Earth major eccentricity of the ellipsoid that is equal to 0.0818191908426 for WGS84 Earth Model (Rogers, 2007).

Standing Equations 18 and 19, the term $\underline{\omega}^{n}_{en}$ can be estimated, as reported in Equation 20:

$$\underline{\omega}^{n}_{en} = \left[\frac{v_E}{N+h} \quad -\frac{v_N}{M+h} \quad -\frac{v_E \tan \varphi}{N+h} \right]^{T} \quad (20)$$

The INU block of Figure 4 integrates the discrete form of Equation 15 each time a new update of inertial sensor measurements is output. For typical systems, this process can be executed as rated from 100 up to 500 Hz. Initial conditions are assumed as known by external reference.

6.2.2   Process model for the integration filter

Following Equation 9, by deriving the first-order Taylor series model of Equation 15, it is possible to build a nine-state linear process model, similar to the one reported in Equation 10, to be used in an extended Kalman filter with the nine-dimensional state vector given in Equation 21:

$$\underline{x}_k = \left\{ \begin{array}{c} \delta \boldsymbol{r}^{n} \\ \delta \underline{v}^{n} \\ \underline{\varepsilon}^{n} \end{array} \right\}_k \quad (21)$$

The expression of the relevant state transition matrix $\boldsymbol{\Phi}_k = \nabla f_x$ is reported in Groves (2013). The process noise covariance matrix $\boldsymbol{Q}_k$ can be derived from the same reference by taking into account error terms of inertial sensors reported by sensor manufacturer in system documentation.

The integration filter block in Figure 4 updates the process model equation at each update of INU.

An augmented model of the filter is capable of estimating the bias of accelerometers and gyro sensors. These are added as pure random terms in the state represented by Equation 21. In this form, their derivative is expressed by random walk process with zero mean and a variance determined as a function of their bias instability and random walk term (Savage, 2002).

6.2.3   Measurement model for the integration filter

A single measurement model will be described in the following, that is, GPS/INS integration. Indeed, several others can also be found:

- Magnetometer and inertial integration (Groves, 2013)
- Air data and inertial integration (Merhav, 1996)
- Odometer and inertial integration (Rogers, 2007)

- VOR/DME and inertial integration (Rogers, 2007)
- Stellar and inertial integration (Chatfield, 2007)
- Optical flow and inertial integration (Bhanu, 1990)
- Simultaneous localization and mapping integration (Lupton and Sukkarieh, 2012)

In any case, the procedure to derive the measurement model is based on a linear transformation between the navigation state and sensor observables. In general, when the typical transformation is not linear, a Taylor series approach is adopted. The measurement error matrix can be carried out by performing a covariance analysis of measurement sensor output based on technical data from sensor manufacturers.

6.2.4   Measurement model for GPS/INS integration

The Global Positioning System (GPS) is a space-based satellite navigation system that provides location and time information in all weather, anywhere on or near the Earth, where there is an unobstructed line of sight to four or more GPS satellites (Axelrad, 2010). It is maintained by the US government and is freely accessible by anyone with a GPS receiver. The position and velocity from GPS can be considered as measurements. So the straightforward formulation of measurement equation can be expressed by Equation 22:

$$\underline{z}_k = \left\{ \begin{array}{c} \boldsymbol{r}^{n}_{INS} - \boldsymbol{r}^{n}_{GPS} \\ \underline{v}^{n}_{INS} - \underline{v}^{n}_{GPS} \end{array} \right\}_k = \left\{ \begin{array}{c} \varphi_{INS} - \varphi_{GPS} \\ \lambda_{INS} - \lambda_{GPS} \\ h_{INS} - h_{GPS} \\ \underline{v}^{n}_{INS} - \underline{v}^{n}_{GPS} \end{array} \right\}_k \quad (22)$$

where subscripts INS and GPS refer to corresponding techniques, and the state-measurement transfer matrix is expressed as reported in Equation 23:

$$\boldsymbol{H}_k = \begin{bmatrix} \boldsymbol{I}_{3\times3} & \boldsymbol{0}_{3\times3} & \boldsymbol{0}_{3\times3} \\ \boldsymbol{0}_{3\times3} & \boldsymbol{I}_{3\times3} & \boldsymbol{0}_{3\times3} \end{bmatrix}_k \quad (23)$$

where $\boldsymbol{I}$ and 0 are the identity and null matrix and subscript 3×3 refers to matrix dimensions. The error measurement matrix reported in Equation 24 will be used:

$$\boldsymbol{R}_k = \text{diag}\left(\sigma^2_{\varphi} \; \sigma^2_{\lambda} \; \sigma^2_{h} \; \sigma^2_{v_n} \; \sigma^2_{v_e} \; \sigma^2_{v_d} \right) \quad (24)$$

which can be obtained by GPS processing. However, this approach causes numerical instabilities when calculating $\left(\boldsymbol{H}_k \boldsymbol{P}^{-}_k \boldsymbol{H}^{T}_k + \boldsymbol{R}_k \right)^{-1}$ for the Kalman gain $\boldsymbol{K}_k$, because φ and λ are in radians, so their values are very small. To solve this problem, it is necessary to multiply the first and the second row by $(M+h)$ and $(N+h)\cos\varphi$, respectively, so that $\underline{z}_k$

and $\boldsymbol{H}_k$ are given as reported in Equations 25 and 26:

$$\underline{z}_k = \left\{ \begin{array}{c} (M+h)\,(\varphi_{\mathrm{INS}} - \varphi_{\mathrm{GPS}}) \\ (N+h)\cos\varphi(\lambda_{\mathrm{INS}} - \lambda_{\mathrm{GPS}}) \\ h_{\mathrm{INS}} - h_{\mathrm{GPS}} \\ \underline{v}^{\mathrm{n}}_{\mathrm{INS}} - \underline{v}^{\mathrm{n}}_{\mathrm{GPS}} \end{array} \right\}_k \quad (25)$$

$$\boldsymbol{H}_k = \begin{bmatrix} (M+h) & 0 & 0 & & \\ 0 & (N+h)\cos\varphi & 0 & \boldsymbol{0}_{3\times3} & \boldsymbol{0}_{3\times3} \\ 0 & 0 & 1 & & \\ & \boldsymbol{0}_{3\times3} & & \boldsymbol{I}_{3\times3} & \boldsymbol{0}_{3\times3} \end{bmatrix}_k \quad (26)$$

It is worth nothing that the above-reported scheme requires that an updated position and velocity fix is available from the GPS receiver to perform aiding. Since signals from at least four satellites must be received to allow for a position fix, no aiding can be performed when less than four satellites are observed. For this reason, this type of aiding is called "loosely coupled" GPS aiding. The advanced mode for GPS aiding is the "tightly coupled" GPS aiding (Farrell, 2008). It adopts pseudorange measurements from the GPS receiver as source of information for the measurement model instead of position and velocity fixes. Indeed, pseudoranges to satellites are the original information estimated by GPS receiver by evaluating the difference between the time of arrival of a signal with respect to the relevant time of start from the satellite. "Tightly coupled" GPS aiding has several advantages with respect to "loosely coupled" model:

1. Aiding can be performed even if just one satellite is observed.
2. The linear approximation and Gaussian noise assumption of EKF are more accurate if they are referred to pseudorange aiding rather than to position and velocity aiding.

It is worth noting that the application of "tightly coupled" pseudorange integration can be performed only if the adopted GPS receiver supports the output of pseudoranges and GPS ephemeris at each measurement update.

6.2.5   System initialization

In general, system initialization is performed by assuming the following procedure:

1. Initial velocity is assumed equal to zero because, in general, the initialization process is performed when the hosting platform is at rest. In case of hosting platform that is carried by another moving platform, such as a plane on a carrier deck or an airborne missile, the initial velocity is assumed equal to the one of the main platforms. A proper transmission link must be provided to get this information.
2. Initial position must be given by an external source, such as a GPS receiver that can be used as aiding after initialization. Some systems allow for manual input of initial position.
3. Initial tilt, that is, pitch and roll, can be deduced autonomously by means of a stationary processing of accelerometer output called "leveling" (Savage, 2002). Indeed, in stationary condition, accelerometers measure the gravity vector that is directed toward local vertical. As a consequence, its component is equal to last column in the attitude matrix $\boldsymbol{C}^{\mathrm{n}}_b$.
4. In general, initial heading is determined by exploiting magnetic field information by a triaxial fluxgate magnetometer. If installed gyros are in the tactical grade, that is, they are capable of accurately estimating the horizontal component of Earth rate, autonomous initialization of heading is feasible by means of a procedure called "gyrocompassing" (Savage, 2002).

6.2.6   Example of system application: A MEMS-based AHRS

This example of application is related to an AHRS developed for airborne application and based on microelectro-mechanical sensors (MEMS). The unit is AX1-GNS™ manufactured by Italian producer Axitude™. The AX1-GNS testing activities included a flight test campaign to be performed at an airfield in Castel Volturno, Italy. The scope of these activities was to evaluate the Axitude AX1-GNS in-flight performances through a comparison between attitude, velocity, and position provided by the Axitude system and the same quantities measured in the same dynamic conditions by nominally more accurate systems, taken as reference. Testing activities were carried out on a TECNAM P92-S™ aircraft equipped with advanced flight instrumentation, as shown in Figure 5.

The INS/GPS iNAV-FMS™ of the iMAR GmbH Company has been selected as the reference inertial system during the flight test campaign; its attitude measurement performance is on the order of 0.1° rms for each axis with GPS aiding.

Figure 6 shows the flight path obtained through the geodetic positions given by the AX1-GNS. Heading measurement accuracy is a key issue to estimate the performance of AHRS. Indeed, the standard attitude measurement accuracy FAA requirements, that is, 1° rms for pitch and roll and 3° rms for heading (FAA, 2012), are intended for typical mission profiles that were adopted in the past. Currently, aeronautical platforms, such as Unmanned Aircraft Systems

Figure 5. Testing setup installed onboard the Tecnam P92-S™ aircraft.

and Personal Aircraft, require proper augmentation systems and thus an increased performance in terms of attitude determination that must be realized by exploiting low-cost hardware. This performance will support the installation of directional strapdown observation sensors, such as the one needed for sense and avoid, and the development of advanced integrated navigation systems. As a matter of fact, using accurate inertial units with aiding sensor increases the length of inertial coasting and improves the overall accuracy. Moreover, this type of sensors and architectures support increased length of pure inertial coasting in case of extended GPS outages and, in the near future, will be the base

Figure 6. AX1-GNS flight test ground path.

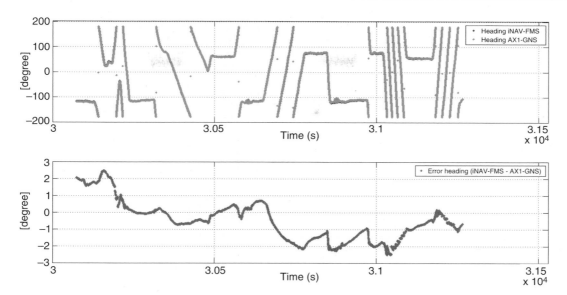

Figure 7. AX1-GNS heading error performance.

Table 5. AX1-GNS heading error main statistics.

Parameter	Standard deviation (1σ)	Mean error	RMS error	Max error	Reference system
ΔHeading (°)	<1.2	<0.5	<1.3	<2.6	iNAV-FMS

to investigate the feasibility of magnetometer-less configurations. The heading performance of AX1-GNS™, as resulting from the flight tests, is evaluated in Figure 7 and Table 5.

7 FUTURE DEVELOPMENTS

Based on the current technology scenario and main research trends, it can be foreseen that the importance and the role of sensor fusion in UAS will continue to increase in the future.

Advanced fusion capabilities will be required by various missions and operating environments such as indoor navigation, planetary exploration, and safe UAS integration into the civil airspace, due to several possible reasons:

- The necessity to fly for long time in GPS-denied environments.
- The demand for onboard autonomy as a way to cope with the large communication latency or the outages of the command and control link, which can derive, for example, from security problems or from frequency congestion with many UAS operating in the field.
- The increasing integration of payload and the guidance, navigation and control (GNC) system that can foster mission effectiveness overcoming traditional boundaries relevant to the air vehicle operator and the mission payload operator in common medium/large UAS. Cognitive-based guidance systems are being actively studied in the unmanned systems community.
- The installation of a large number of low-cost sensors that must be handled at central level to attain high-quality information, that is, large data management.
- In general, an ever-increasing need for performance, reliability, safety, integrity, and cost-effectiveness.

This evolution will be fueled by research and technology progress at different levels.

At hardware level, miniaturization and performance improvement will likely be the main trends for both sensors and processing units. Miniaturization of optical sensors will follow the path of recent technology progress, while a significant effort in this direction can be also foreseen for radars, which represent a disruptive technology on small UAS due to their capability to work in extreme environments. Looking at inertial sensors, the fast progress of MEMS technology will likely be a key factor in the evolution of GNC systems for small flight platforms.

This technology progress by itself will not fulfill the requirements posed by future UAS applications, and will be complemented by the evolution of sensor fusion algorithms. Besides incremental development of existing algorithms, innovative fusion techniques are likely to be derived from biomimetic or bioinspired strategies in order to handle multiple sources of information and to support temporary or permanent failure of one or more sensors. Moreover, while until today fusion algorithms have been used within specific onboard subsystems, in the future they will be increasingly aimed at creating tight logical links between these subsystems, such as payload, GNC, and communications. An example is the combination of cognitive-based and communication-aware guidance in multi-UAV systems.

At architectural level, progress in data fusion should lead to changes in the design approach, enabling a more efficient sensor selection carried out in view of the multisensor-based system.

Finally, sensor fusion importance as a basic building block of unmanned systems will be further amplified by the diffusion of distributed systems (UAS swarms) based on formation flying. Within an evolution that will follow the two main paths of cooperation and autonomy, effective, reliable, and yet affordable sensor fusion will be the key to unlock the full potential of these two concepts.

ACKNOWLEDGMENTS

The authors acknowledge the Italian Aerospace Research Center, responsible for the research project TECVOL, funded by the Italian Government in the frame of National Aerospace Research Program (PRORA) and aimed at development of enabling technologies for autonomous flight.

REFERENCES

Accardo, D., Fasano, G., Forlenza, L., Moccia, A., and Rispoli, A. (2013) Flight test of a radar-based tracking system for UAS sense and avoid. *IEEE Trans. Aerosp. Electron. Syst.*, **49**(2), 1139–1160.

Alpen, M., Willrodt, C., Frick, K., and Horn, J. (2010) On-board SLAM for indoor UAV using a laser range finder. *Proc. SPIE*, 7692. doi: 10.1117/12.849984.

Austin, R. (2010) *Unmanned Aircraft Systems: UAVS Design, Development and Deployment*, John Wiley & Sons, Inc., New York.

Axelrad, P. (2010) Global navigation satellite systems, in *Encyclopedia of Aerospace Engineering* (eds R. Blockley and W. Shyy), John Wiley & Sons, Ltd, Chichester.

Bar-Shalom, Y., Willett, P.K., and Tian, X. (2011) *Tracking and Data Fusion*, YBS Publishing.

Bhanu, B. (1990) Inertial navigation sensor integrated motion analysis for obstacle detection. *Proceedings of the IEEE International Conference on Robotics and Automation,* Cincinnati, OH, May 13–18, 1990.

Blackman, S.S. and Popoli, R.F. (1999) *Design and Analysis of Modern Tracking Systems*, Artech House, Norwood, MA.

Bouguet, J.Y. (2013) *Camera Calibration Toolbox for Matlab.* Available at http://www.vision.caltech.edu/bouguetj/calib_doc/index.html.

Brookner, E. (1998) *Tracking and Kalman Filtering Made Easy*, John Wiley & Sons, Inc., New York, NY.

Castanedo, F. (2013) A review of data fusion techniques. *Sci. World J.* 2013, doi: 10.1155/2013/704504.

Chatfield, A. (2007) *Fundamentals of High Accuracy Inertial Navigation*, American Institute of Aeronautics and Astronautics, Arlington, VA.

Chen, R.H., Gevorkian, A., Fung, A., and Chen, W.Z. (2011) Multi-sensor data integration for autonomous sense and avoid. *AIAA Infotech@Aerospace Technical Conference 2011,* St. Louis, MO, March 2011.

Dasarathy, B.V. (1997) Sensor fusion potential exploitation-innovative architectures and illustrative applications. *Proc. IEEE*, **85**(1), 24–38.

Doucet, A., Godsill, S., and Andrieu, C. (2000) On sequential Monte Carlo sampling methods for Bayesian filtering. *Stat. Comput.*, **10**(3), 197–208.

Durrant-Whyte, H.F. (1988) Sensor models and multisensor integration. *Int. J. Robot. Res.*, **7**(6), 97–113.

Federal Aviation Administration (FAA) (2012) *Technical Standard Order C-201: Attitude and Heading Reference Systems (AHRS)*, Washington, DC.

Federal Aviation Administration (2013) *Integration of Civil Unmanned Aircraft Systems (UAS) in the National Airspace System (NAS) Roadmap*, 1st edn. Available at http://www.faa.gov/about/initiatives/uas/media/UAS_Roadmap_2013.pdf.

Farrell, J.A. (2008) *Aided Navigation: GPS with High Rate Sensors*, McGraw-Hill, New York, NY.

Fasano, G., Accardo, D., Moccia, A., Carbone, C., Ciniglio, U., Corraro, F., and Luongo, S. (2008a) Multi-sensor-based fully autonomous non-cooperative collision avoidance system for unmanned air vehicles. *AIAA J. Aerosp. Comput. Inform. Commun.*, **5**, 338–360.

Fasano, G., Accardo, D., Moccia, A., and Rispoli, A. (2010) An innovative procedure for calibration of strapdown electro-optical sensors onboard unmanned air vehicles. *Sensors*, **10**(1), 639–654.

Fasano, G., Forlenza, L., Accardo, D., Moccia, A., and Rispoli, A. (2011) Data fusion for UAS collision avoidance: results from flight testing. *Proceedings of the AIAA Infotech@Aerospace 2011, Reston VA*, American Institute of Aeronautics and Astronautics (AIAA), St. Louis, MO, March 29–31, 2011, pp. 1–16.

Fasano, G., Moccia, A., Accardo, D., and Rispoli, A. (2008b) Development and test of an integrated sensor system for autonomous collision avoidance. *Proceedings of the 26th*

International Congress of the Aeronautical Sciences, Anchorage, AK, September 14–19, 2008.

Gelb, A. (1974) *Applied Optimal Estimation,* MIT Press, Cambridge, MA.

Graham, S., Chen, W.-Z., De Luca, J., Kay, J., Deschenes, M., Weingarten, N., Raska, V., and Lee, X. (2011) Multiple intruder autonomous avoidance flight test. *AIAA Infotech@Aerospace Technical Conference 2011,* St. Louis, MO, March 2011.

Grewal, M.S. (2013) *Global Navigation Satellite Systems, Inertial Navigation, and Integration,* 3rd ed., Wiley-Interscience, New York NY.

Gross, J.N., Gu, Y., Rhudy, M.B., Gururajan, S., and Napolitano, M.R. (2012) Flight-test evaluation of sensor fusion algorithms for attitude estimation. *IEEE Trans. Aerosp. Electron. Syst.,* **48**(3), 2128–2139.

Groves, P.D. (2013) *Principles of GNSS, Inertial, and Multisensor Integrated Navigation Systems,* Artech House, London, UK.

Gundlach, J. (2012) *Designing Unmanned Aircraft Systems: A Comprehensive Approach (AIAA Education Series)* American Institute of Aeronautics & Astronautics, Arlington, VA.

Gustafsson, F., Gunnarsson, F., Bergman, N., Forssell, U., Jansson, J., Karlsson, R., and Nordlund, P.-J. (2002) Particle filter for positioning, navigation and tracking. *IEEE Trans. Signal Process.,* **50**(2), 425–437.

Hall, D.L. and Llinas, J. (1997) An introduction to multisensor data fusion. *Proc. IEEE,* **85**(1), 6–23.

Hartley, R.I. and Zisserman, A. (2000) *Multiple View Geometry in Computer Vision,* Cambridge University Press, Cambridge, UK.

Helfrick, A. Buckwalter, L. (2013) *Principles of Avionics,* 8th ed., Avionics Communications Inc.

Heredia, G., Caballero, F., Maza, I., Merino, L., Viguria, A., and Ollero, A. (2009) Multi-Unmanned Aerial Vehicle (UAV) Cooperative Fault Detection Employing Differential Global Positioning (DGPS): inertial and vision sensors. *Sensors,* **9,** 7566–7579.

Hwang, R.R. and Huber, M. (2007) A particle filter approach for multi-target tracking. *IEEE/RSJ International Conference on Intelligent Robots and Systems, 2007 (IROS'07),* San Diego, CA, pp. 2753–2760.

Indelman, V., Gurfil, P., Rivlin, E., and Rotstein, H. (2012) Graph-based distributed cooperative navigation for a general multi-robot measurement model. *Int. J. Robot. Res.,* **31**(9), 1057–1080.

JDL (1991) *Data fusion Lexicon.* Technical Panel for C3, F.E. White, San Diego, CA, Code 420.

Klein, L.A. (2004) *Sensor and Data Fusion: A Tool for Information Assessment and Decision Making,* SPIE Press, Bellingham, WA.

Kuchar, J.K. and Drumm, A.C. (2007) The traffic alert and collision avoidance system. *Lincoln Lab. J.,* **16**(2) 277–295.

Liggins, II, M., Hall, D., and Llinas, J., *Handbook of Multisensor Data Fusion: Theory and Practice,* 2nd ed., CRC Press.

Luo, R.C., Yih, C.-C., and Su, K.L. (2002) Multisensor fusion and integration: approaches, applications, and future research directions. *IEEE Sens. J.,* **2**(2), 107–119.

Lupton, T. and Sukkarieh, S. (2012) Visual-inertial-aided navigation for high-dynamic motion in built environments without initial conditions. *IEEE Trans. Robot.,* **28**(1), 61–76.

Merhav, S. (1996) *Aerospace Sensor Systems and Applications,* Springer, New York, NY.

Merino, L., Caballero, F., Martínez-de Dios, J.R., Ferruz, J., and Ollero, A. (2006) A cooperative perception system for multiple UAVs: application to automatic detection of forest fires. *J. Field Robot.,* **23** (3–4), 165–184.

Nex, F. and Remondino, F. (2014) UAV for 3D mapping applications: a review. *Appl. Geomatics,* **6**(1), 1–15.

Ristic, B., Arulampalam, S., and Gordon, N. (2004) *Beyond the Kalman Filter: Particle Filters for Tracking Applications,* Artech House, New York.

Rogers, R.M. (2007) *Applied Mathematics in Integrated Navigation Systems,* AIAA Education Series, American Institute of Aeronautics and Astronautics, Arlington, VA.

Savage, P.G. (2002) *Strapdown Analytics,* Strapdown Associates Inc., Minneapolis, MN.

Schmidt, G. (2010) INS/GPS Technology Trends, NATO Report RTO-EN-SET-116.

Stone, L.D., Barlow, C.A., and Corwin, T.L. (1999) *Bayesian Multiple Target Tracking,* Artech House, Norwood, MA.

Titterton, D.H. (2005) *Strapdown Inertial Navigation Technology,* 2nd ed., Peter Peregrinus Ltd., New York, NY.

Van der Merwe, R. and Wan, E.A. (2000) The unscented Kalman filter for nonlinear estimation, *Proceedings of the IEEE Adaptive Systems for Signal Processing, Communications, and Control Symposium,* IEEE, Lake Louise, Alberta, Canada, October 1–4, 2000, pp. 153–158.

Weiss, S., Achtelik, M.W., Lynen, S., Achtelik, M.C., Kneip, L., Chli, M., and Siegwart, R. (2013) Monocular vision for long-term micro aerial vehicle state estimation: a compendium. *J. Field Robot.,* **30**(5), 803–831.

PART 7
Human Oversight

Chapter 23

Function Allocation between Human and Automation and between Air and Ground

Karen Feigh and Amy Pritchett

Cognitive Engineering Center, Georgia Tech, Atlanta, GA, USA

1 INTRODUCTION

Function allocation is the design decision determining which functions should be executed by humans and which by automated agents on the air and on the ground – more colloquially, it specifies "who does what." Function allocation in unmanned aerial systems (UAS) has its own unique considerations given that, unlike manned aircraft, the humans are removed to remote ground control stations. To help address these considerations, this chapter frames several aspects of effective function allocation. First, it must consider all the activities required to complete the UAS operations, that is, the taskwork functions. Furthermore, allocating functions between different agents then creates the need for additional teamwork functions as required to coordinate between agents with that specific function allocation. All of these functions need to be performed in a complex operational environment that has dynamics and timing requirements, and thus can constrain allowable function allocations to those feasible with given communication links.

Unmanned Aircraft Systems. Edited by Ella Atkins, Aníbal Ollero, Antonios Tsourdos, Richard Blockley and Wei Shyy.
© 2016 John Wiley & Sons, Ltd. ISBN: 978-1-118-86645-0.

Throughout this chapter, we review conceptual debates about the relative roles of human and automation, and the purposes that should be ascribed to machines. However, we also note the importance of including perspectives spanning the dynamics of the work, the interactions within the team, and the agents' collective ability to perform the work that is required of the team. We specifically discuss a particular concern with UAS operations, that is, how human–machine function allocations in UAS operations are often particularly sensitive to reliable communication linkages between ground control station and vehicle. Thus, different types of UAS, and different types of UAS operations, will require different function allocations.

Throughout, our intention in identifying key requirements for function allocation is practical: We wish to guide design such that key trade-offs can be considered and fundamental concerns with human factors addressed before UAS technologies and ground operator interfaces are created. Indeed, from a human factors perspective, function allocation is the earliest possible design decision, from which training, procedures, technological functions and interface specifications should all flow. We first framed this human factors perspective in a series of papers from which this chapter is derived (Feigh and Pritchett, 2014; Pritchett, Kim, and Feigh, 2014a, 2014b). This perspective is described in terms of five key requirements for effective function allocation, starting with the most basic requirement that each agent must be allocated functions that it is capable of performing, then additionally requiring that each agent can handle its collective set of allocation functions, and the teamwork that corresponds to the allocation, and requiring that the function allocation supports the dynamics of the work; the last requirement defines how function allocation should be the result of deliberate design decisions.

1.1   Requirement 1: Each agent must be allocated functions that it is capable of performing

The first most basic requirement for an effective function allocation is that every agent in the team must be able to perform each of the functions assigned to him/her/it. A common method of addressing this requirement follows what may be termed a comparison strategy for function allocation: Each function is individually compared with the capabilities of each agent, and assigned to the most capable. Such a comparison has been fostered by the "Fitts List" that lists types of functions that Men Are Better At versus Machines Are Better At (thus, also called the MABA–MABA list) (Fitts, 1951).

Sometimes one agent's capabilities are articulated relative to those of others. For example, distinctions have been made between automation that is capable of being a tool for a human versus a prosthesis, depending on whether it enables a human's normal capabilities or extends a human's capabilities beyond that normally achievable (Reason, 1987). As automation has become capable of initiating or conducting its own actions, a further distinction has been made of automation as an agent (Lee, 2006). Likewise, it is common to see one agent – human or automated – described as a backup or monitor for another agent who is presumed to be fallible (Bainbridge, 1983; Wiener and Curry, 1980); this role has been especially common in aviation, and UAS in particular, where manual control by a remote pilot or pilot intervention is the assumed reversionary mode.

Constructs such as the MABA–MABA list have been criticized for the extent to which human capabilities are framed in machine terms (see, for example, Sheridan, 2000). Indeed, the general definition of automation as "machines capable of taking over functions normally conducted by humans" has perhaps been viewed as an engineering challenge to increase autonomy through developing machines capable of more human functions.

Function allocations established around machine capabilities can implicitly apply an unfortunate function allocation strategy: automate as many functions as technology will permit, and assume the human will pick up whichever functions are left over. Commonly automated functions usually can accomplish standard processes within conditions that can be clearly defined, such as nominal operations or established emergency shutdown procedures. However, this makes for brittle automation that can fail in an unexpected manner in off-nominal conditions, which is generally when the human needs the most support (Norman, 1990). Operational studies have noted that function allocations based on this approach often result in designs in which humans are assigned to monitor automation – or monitor the environment for conditions beyond which the automation can operate (Bainbridge, 1983; Wiener and Curry, 1980), despite consistent findings that humans are poor monitors (Molloy and Parasuraman, 1996). Indeed, the report authoring the MABA–MABA list suggested instead that "machines should monitor men" (Fitts, 1951).

Unfortunately, categorizations of levels of automation as defined by Sheridan and Verplank (1978), or on a more multidimensional scale by Parasuraman, Sheridan, and Wickens (2000), are only defined by what the machine is capable of doing. Thus, they do not properly consider what the UAS operator's corresponding functions will need to be, and whether those functions will be feasible for the human to perform consistently and reliably (notably, the heavy emphasis on monitoring described in the previous paragraph).

In particular, a facile implementation of the so-called higher levels of automation can result in two problems. First, if the automation is conducting the operation with minimal human intervention, then the human operator is left with an infeasible task of passive monitoring. Second, implementing the machine capabilities does not imply they will actually be used: counterexamples in manned aircraft have noted where the human pilot may consistently override or second-guess high levels of automation for a number of reasons, including concerns that the automation cannot "account for everything" or difficulties in configuring or programming the automation if the mission profile needs to change quickly (Pritchett, 2009).

A related concern notes that human–automation function allocation often considers only authority for a function without also examining responsibility. Specifically, whereas *authority* is generally used to describe who is assigned the execution of a function in operational sense, *responsibility* is used here to identify who will be held accountable in an organizational and legal sense for the function's outcome. In general, human operators remain vested with the responsibility for the outcome of the UAS operation, particularly where the safety of others must be maintained. However, if the responsible human operator cannot knowledgeably oversee the automation, s/he is forced to trust the automation or to not rely on it at all. Unfortunately, without a concrete basis for assessing whether the automation is correct, humans often over- and undertrust the automation: Either way, incorrect trust is viewed as human error, despite its basis in the function allocation (Parasuraman and Riley, 1997).

1.2   Requirement 2: Each agent must be capable of performing its collective set of functions

Assuming that the previous requirement is met, this requirement now examines whether each agent can perform his/her/its collective set of functions under realistic operating conditions. This emphasis on each agent's collective set of functions underlies methods for global function allocation, and for understanding the agents (human and automation) as limited resources (see Dearden, Harrison, and Wright, 2000 for a review). One potential obstruction may be that the set of functions is too large, that is, its task load overwhelms an agent's ability to process and execute. This has been a concern with overloading automated systems, leading to technical developments such as real-time embedded systems capable of prioritizing demands on processors and communication devices.

Likewise, the set of functions assigned to a human agent may create a task load that corresponds to excessive workload. Task load that is consistently excessive is usually readily apparent – and quickly remedied. However, workload spikes are not as easily predicted, as they represent brief durations where the task load is too high. Such spikes are endemic to emergencies and unexpected situations such as unusual system malfunctions that must be diagnosed and isolated even as the vehicle must be controlled in a basic reversionary mode.

Extrapolating to UAS operations, concerns with human workload in human–automation teams found in related domains, particularly manned aircraft, find that the autoflight and control technologies supporting UAS operations are generally good at reducing the average workload of their human operators. However, such automation can also create a workload spike for the human operator (Bainbridge, 1983; Wiener and Curry, 1980), particularly during off-nominal or emergency situations outside automation's boundary conditions. For example, many autoflight systems (for both manned aircraft and UAS) are not rated to control the aircraft in unusual attitudes or when there may be ice accretion on vital sensors. Such situations can also be created – or compounded – by automation that demands significant interaction from the human operator at brief points in time, as also discussed next in Requirement 3. For example, highly automated autoflight systems can suddenly require a significant amount of programming from their human operator in response to a reroute commanded by an air traffic controller or revised mission profile, even as the s/he must also respond to the controller, and potentially execute other tasks such as finding charts or reconfiguring ground control station displays as appropriate to the new routing.

The corollary of workload spikes is the problem of excessively low workload during normal operations in between the spikes. Low workload by itself corresponds to concerns with task engagement, boredom, and the human becoming "out of the loop" (Bainbridge, 1983; Endsley and Kiris, 1995). Furthermore, this low workload typifies passive activities, such as monitoring, discussed earlier as being not well suited to the human.

Estimates of the workload of any human agent in the team must also ensure to weigh properly the cognitive workload allocated to the human even as the manual workload is allocated to the machine (Bainbridge, 1983; Wiener and Curry, 1980). These cognitive functions can demand significant information gathering, judgment, detection, and decision-making activities, sometimes even up to the point of reverse engineering or second-guessing the output of the automation. Estimates of the workload incurred by such cognitive functions need to account for myriad effects. One is the human's expertise at the task, and thus their ability to apply less effortful strategies, rules, and skills while maintaining performance (Rasmussen, 1983).

Furthermore, nonlinear effects arise when several cognitive functions rely on awareness of the same information or on interdependent judgments of similar phenomena. In such cases, adding or removing one cognitive function may seem to have little impact because it does not change what information the human operator needs to gather and interpret. In contrast, adding just one other function may implicitly require a significant amount of new information gathering and interpretation, and thus appear to require a disproportionate amount of effort.

Interdependence between tasks suggests that Requirement 2 can be better achieved when the function allocations establish coherent roles for the human operator and for each of the technical components (on the ground and on the vehicle). One attribute of a coherent function allocation can be viewed from the bottom-up: Its functions share (and build upon) obvious, common constructs underlying all their activities, such as a shared information and knowledge basis within each agent, and the distribution of functions prevents conflicts over low-level actions and resources between agents. Another attribute can be viewed from the top-down: The functions collectively contribute toward work goals in a manner that is not only apparent to the human operator but can be purposefully coordinated and adapted in response to context – for example, the functions here may correspond with succinct statements of intent such as "loiter" or "survey this area" that make sense to the human operator within a mission plan.

In contrast, with an incoherent division of functions, the human operator can neither coordinate underlying information search and judgment activities nor purposefully adapt their behavior in support of performance (Dekker and Woods, 2002). For example, an incoherent function allocation in UAS operations divide up a high-level task such as "survey this area" such that the vehicle's onboard automation and the human operator need to tightly coordinate, for example, the human operator needing to command each turn for an autoflight system to execute: Such an allocation may frustrate the human operator with the simplicity of her/his task even as s/he needs to attend to it too frequently to allow her/him to perform other tasks. Other problems may arise when it is not clear which small task elements the human must execute to achieve a particular intent, and when. While such problems may be partly addressed by an interface that can illustrate the allocation, clearly coordinate human and automation tasks, and enable better monitoring of automation (Palmer *et al.*, 1995), the general concern of an incoherent function allocation is that it cannot be abstracted into clearly defined roles and intents.

1.3 Requirement 3: The function allocation must be realizable with reasonable teamwork

The allocation of taskwork functions within any team – whether it includes automation or not – demands additional teamwork functions. These teamwork functions serve to coordinate the taskwork across agents. They include human-to-human communication, human–automation interaction, and coordination of the timing and conduct of taskwork. Continuing the example of an incoherent allocation of functions within the intent to "survey this area" as an exemplar, if the human operator be made procedurally responsible for commanding (or approving) turns in the flight path, then s/he would need to be allocated the activities of commanding these turns through an operator interface and communicating them to the onboard autoflight system. Conversely, a more coherent function allocation is typical in many present-day UAS, where the vehicle can automatically follow its own flight path. Even so, other functions such as terrain avoidance are typically left to the human operator, generating teamwork activities in the form of monitoring terrain clearance, which can be a significant tasking during low-altitude operations.

Thus, while the underlying taskwork is the same, each different function allocation of the same taskwork adds a unique set of teamwork functions. Basic considerations here include whether the combined taskwork and teamwork allocated to any agent is feasible. For example, with a clumsy ground control interface, it is conceivable that the time required to command a change in trajectory may be sufficiently long to create a limit on how quickly and precisely the UAS can be expected to maneuver in response to, for example, terrain or collision avoidance – or it may limit when changes in trajectory can be commanded given other competing demands on the human operator's time. Likewise, the examples given above of incoherent function allocations can create extensive teamwork for the human operator, particularly in the form of monitoring.

Furthermore, more subtle considerations apply a team perspective to function allocation: the extent to which automation acts properly as a team member must be examined. For example, models and measures of trust, social judgment, and roles have been extended from the social sciences and studies of teams to human–automation interaction (see, for example, Muir (1994), Muir and Moray (1996), and Woods and Hollnagel (2006)). However, this perspective usually highlights where automation does not have same teamwork skills that humans naturally possess. Automation has no motivation to live up to its obligations, does not experience shame or embarrassment, and cannot be assessed for attributes such as loyalty, benevolence, and agreement in values (Lee and See, 2004). When placed outside its boundary conditions, automation often cannot function properly, unlike a human team member who will generally strive for effective performance in unfamiliar circumstances. Thus, rather than implying the human operator should "trust" the automation as s/he would trust a human, it is more appropriate to instead apply a rational decision of whether to employ automated functions based on perceived cost and benefit.

Especially in UAS operations, communication within the team must also be addressed. Literature within human factors studies of effective human teams finds that team members should be able to anticipate each other's information needs and provide information at useful, noninterruptive times (e.g., Hutchins, 1995). Timing must be considered as well: Inter-leaved tasks can leave one agent waiting on another, and poorly timed communication can disrupt performance. An interruption may be demanded by circumstances, in which case it can spur knowledge acquisition and facilitate decision-making performance.

Unfortunately, too often automation is "clumsy": It cannot anticipate the information needs of its human teammates and unduly interrupts (Christoffersen and Woods, 2002; see Dorneich *et al.* (2012), for an example of efforts to improve this aspect of automation). With human–automated teams, predefined sets of function allocations may serve as more explicit coordination strategies, such as the playbook proposed by Miller and Parasuraman (2007) and such as the function allocations that the pilot of a modern aircraft or UAS may invoke. Overall, the implication here is the need for caution in assuming that automation can be swapped in as a team member without also accounting for the cost of its clumsiness within team dynamics, compared to humans' general ability to interact effectively (a skill that is further refined in manned aircraft through training in Crew Resource Management (CRM), particularly the given recent findings that poor communication is an important contributing factor to accidents (FAA, 2013)).

1.4   Requirement 4: The function allocation must support the dynamics of the work

In dynamic, complex work environments, many work activities are interdependent and heavily coupled. However, these couplings may be hidden. For example, in a function allocation with the operator remotely controlling pitch and an onboard autothrottle regulating speed, the operator controls elevator and the automation controls throttle; however, pitch and speed are intrinsically coupled such that the actions of one will interfere with the actions of the other when, for example, a change in speed by the autothrottle requires the operator to compensate for a resulting change in pitch. Thus, dynamic analysis is often required to identify situations where the interleaving of functions assigned to disparate agents requires significant coordination or idling as one waits on another, or where workload may accumulate, or where one agent will be unduly interrupting another.

These dynamics can become particularly important in UAS operations when the function allocation requires significant communication between air and ground systems. This communication may have nontrivial limitations in its transmission rate, latency, update rate, and dropout rate, to the extent that it constrains the effective bandwidth of the air–ground interaction. In the example given above, for example, ground control of pitch with on-aircraft control of throttle may introduce a significant time delay into one channel of the autopilot that is simultaneously controlling other channels without delay.

Thus, communication delays may drive the need for onboard automation to control fast dynamics, or to respond to fast-developing hazards. Such onboard automation does not need to be complex, but may instead enact some simple, basic functions that serve to stabilize the vehicle; these simple functions then can perhaps interact with the UAS operator for more detailed instructions or for parameterization of their functions. At an extreme, such capabilities that are already established in the "lost-link" behavior automatically invokes when all communication is lost. Many of these dynamic aspects are amenable to classic dynamic systems analysis; for example, function allocations in spacecraft

operations have been driven by communication time delays (e.g., Ferrell and Sheridan, 1967).

Furthermore, more qualitative human factors analysis should also examine whether the function allocation supports the dynamics of the human operator. For example, the sequence and timing of human activity are dictated by established procedures in some domains such as manned aircraft and, in other domains, by less formally defined work practices. Less prescribed constructs for managing complex, dynamic environments have proposed the construct of resilience, which describes a team's ability to work in unexpected situations as maintaining control of or regulating those situations (Hollnagel 2004; Hollnagel, Woods, and Leveson, 2006). However, the brittleness of automation (i.e., degradation in its performance when the environment exceeds its boundary conditions) reflects how allocation of key functions to automation may degrade the team's resilience in unexpected situations. Likewise, resilience is fostered when a human operator may select strategies (courses of action) appropriate to the state of the environment and their own capabilities. For example, Hollnagel's concept of *cognitive control* describes how humans adapt their activities (and sequence them) in response to their competency and their perception of resources available to them (such as information availability) and demands on them (such as subjective available time) (Feigh and Pritchett, 2006; Hollnagel, 1993).

Unfortunately, the dynamics of the human operator (and thus their ability to plan and adapt their activities) can be overly constrained or obstructed by a function allocation, particularly where the automation demands a specific sequence of activities from its human operator. The adverse effects of such overly prescribed function allocations can manifest in work-arounds or disuse of automation (Feigh and Pritchett, 2010; Kirlik, 1993; Parasuraman and Riley, 1997). Indeed, a trade-off exists when designing function allocations between allowing the human operator the flexibility to actively manage and control their activities (with the potential cost of simplifying the contribution of automation) versus applying complex automated capabilities and dynamically allocating functions (with the potential cost of limiting the human operator's authority to adapt to the demands of stressful or unexpected situations) (Miller and Parasuraman, 2007).

1.5   Requirement 5: The function allocation should be the result of deliberate design decisions

We continue to argue that function allocation decisions should be integrated into the earliest stages of the broader engineering design process. Many "engineering" models commonly exist even at the earliest design stages (e.g., for UAS, models of the flight dynamics of the aircraft and for atmospheric effects such as turbulence and wind, as well as of key technologies such as anticipated communication links). The corresponding need is to model how the operation will be conducted, such as the functions required for systems management, procedures for interacting with air traffic control (or other vehicles or command and control, as appropriate to the mission), communications, and human–automation interactions. Such models need not be complex or attempt to model the human operator – instead, at the earliest stages, there are advantages in modeling how the functions ought to be performed such that these normative models can then drive the design of training, procedures, the operator interface, and automated functions (Pritchett *et al.*, 2014).

The value of examining function allocations early in the design of UAS operations, that is, before technologies and interfaces have been created, may also be argued in terms of cost and development risk. Function allocation decisions made early in design can be used as a cost-effective method for defining the specifications for technologies and for identifying the information and interactivity specifications for interfaces. This minimizes the risk of significant problems being identified later. In contrast, once significant effort has been placed in creating these technologies – sufficient for human-in-the-loop evaluations, for example – fundamental problems with function allocation pose a significant risk to the project, as the cost of redesigning the technologies can be large. In some cases, the cost of redesigning the technologies is considered prohibitive, and a cost-benefit analysis decides that, given the sunk costs in technology design and the notional value of a redesign, problems inherent to the system design at the function allocation level can only be mitigated operationally through procedures and training.

2 CONCLUSIONS

UAS are enabled by sophisticated autonomous machine capabilities. Even so – or perhaps especially so – the allocation of function between a UAS human operator and autonomous systems (on the vehicle and on the ground) remains a driving factor in overall UAS performance and safety. This chapter reviewed, from first-principles, the requirements for an effective function allocation between human and machine and between air and ground. To summarize, these requirements are as follows:

1. Each agent must be allocated functions that it is capable of performing.

2. Each agent must be capable of performing its collective set of functions.
3. The function allocation must be realizable with reasonable teamwork.
4. The function allocation must support the dynamics of the work (including communication).
5. The function allocation should be the result of deliberate design decisions.

Throughout this review of these requirements, several themes have emerged. The first concerns the degree to which machine capability should dominate the function allocation. Several cautions were noted throughout with such a technology-driven approach: that the functions assigned to the human – implicitly or explicitly – will be those "left-over" as an incoherent set; that systems that function properly only in certain operating conditions (or systems that might fail) leave significant passive monitoring to the human operator, working only to a human weakness, and that leaving the human operator with responsibility for the outcome often effectively requires them to take over, or override, machine functions when s/he has a reasonable suspicion about what the machine is doing relative to the demands of the immediate situation.

The second theme is the need to balance the advantages of sophisticated automated capabilities with the complexity it may create in its interactions with the human operator. A poor function allocation creates a complex task environment for the human operator, such as one demanding an incoherent set of low-level tasks, perhaps where these tasks need to wait upon the automation and then respond quickly. Furthermore, an allocation of authority to sophisticated autonomy but responsibility to the human operator can leave her/him second-guessing what the autonomy is doing and why, thereby creating a complex decision task of "Should I intervene or not?" Thus, a challenge to UAS designers may be framed as "creating complex autonomy that is simple to monitor, understand and override"; a potential solution is to ensure that the machine's functions enable it to be allocated functions that can accomplish a clear, sensible intent within the mission.

A third theme is that we cannot simply say that the operator should "trust" an autonomous system to perform the functions it has been allocated. Several drivers usually require the human operator to be integral to the operation: partial automation that requires human tasks in nominal conditions; the fallibility of the autonomous system in the face of system failures or off-nominal conditions; and the requirement for human responsibility for safety and policy concerns. At best, the human operator is required to "trust, but verify," where this verification is a rational process driven by an immediate assessment based on immediate information. Thus, function allocation decisions should recognize – welcome – this component of the human operator's task, and highlight what information and knowledge should then be provided to support it.

Finally, the function allocation is not a simple static concept in isolation, but instead part of a broader dynamic in which communication may be lost even as the vehicle is maneuvering quickly. Thus, the function allocation should be evaluated using a model with sufficient detail to capture specific interactions between agents and the coordinated dynamics of the work and the environment. In many cases, such analysis can identify clear cases where fast dynamics must be controlled onboard the vehicle, and where interaction between the human operator and onboard systems is limited by both operator communication links to require high-level intent specifications that can be easily inputted and shared.

Taken together, there is no simple solution to the problem of function allocation – each system and mission may need to find its own best answer to the concerns and trade-offs noted here. However, the difficult design decision of allocating functions to automation can be addressed systematically, guided by the key requirements here – to account for the demands placed on both autonomous systems and the human operator by each function alone and all the functions collectively, recognizing the additional demands created by responsibility and teamwork, and predicting how the function allocation will work within the dynamics of the operation.

REFERENCES

Bainbridge, L. (1983) Ironies of automation. *Automatica*, **19**, 775–779.

Christoffersen, K. and Woods, D.D. (2002) How to make automated systems team players, in *Advances in Human Performance and Cognitive Engineering Research* (ed. E. Salas), vol. 2, Elsevier, pp. 1–12.

Dearden, A., Harrison, M., and Wright, P. (2000) Allocation of function: scenarios, context and the economics of effort. *Int. J. Hum. Comput. Stud.*, **52**, 289–318.

Dekker, S. and Woods, D. (2002) MABA–MABA or abracadabra? Progress on human–automation co-ordination. *Cogn. Technol. Work*, **4**, 240–244.

Dorneich, M., Ververs, P., Mathan, S., Whitlow, S., and Hayes, C. (2012) Considering etiquette in the design of an adaptive system. *J. Cogn. Eng. Decis. Mak.*, **6**, 243–265.

Endsley, M. and Kiris, E. (1995) The out-of-the-loop performance problem and level of control in automation. *Hum. Factors*, **37**, 381–394.

Federal Aviation Administration (FAA) (2013) Operational Use of Flight Path Management Systems. Final Report of the

Performance-Based Operations Aviation Rulemaking Committee (PARC)/Commercial Aviation Safety Team (CAST) Flight Deck Automation Working Group. Retrieved from http://www.faa.gov/about/office_org/headquarters_offices/avs/offices/afs/afs400/parc/parc_reco/media/2013/130908_parc_fltdawg_final_report_recommendations.pdf.

Feigh, K. and Pritchett, A. (2006) Contextual control modes during an airline rescheduling task. *Proceedings of the Annual Meeting of the Human Factors and Ergonomics Society*, Human Factors and Ergonomics Society, San Francisco, CA.

Feigh, K. and Pritchett, A. (2010) Modeling work for cognitive work support system design in operational control centers. *J. Cogn. Eng. Decis. Mak.*, **4**, 1–26.

Feigh, K. and Pritchett, A. (2014) Requirements for effective function allocation: a critical review. *J. Cogn. Eng. Decis. Mak.*, **8**, 23–32.

Ferrell, W. and Sheridan, T. (1967) Supervisory control of remote manipulation. *IEEE Spectr.*, **4**, 81–88.

Fitts, P. (ed.), (1951) *Human Engineering for an Effective Air-Navigation and Traffic Control System (DTIC Accession No. ADB815893)*, National Research Council, Committee on Aviation Psychology, Washington, DC.

Hollnagel, E. (1993) *Human Reliability Analysis: Context and Control*, Academic Press, London, UK.

Hollnagel, E. (2004) *Barriers and Accident Prevention*, Ashgate Publishing Ltd., London, UK.

Hollnagel, E., Woods, D.D., and Leveson, N. (2006) *Resilience Engineering: Concepts and Precepts*, Ashgate Publishing Ltd., Aldershot, UK.

Hutchins, E. (1995) *Cognition in the Wild*, The MIT Press, Cambridge, MA.

Kirlik, A. (1993) Modeling strategic behavior in human–automation interaction: why an "aid" can (and should) go unused. *Hum. Factors*, **35**, 221–242.

Lee, J. (2006) Human factors and ergonomics in automation design, in *Handbook of Hum. Factors and Ergonomics*, (ed. G. Salvendy), John Wiley & Sons, Inc., New York, pp. 1570–1596.

Lee, J. and See, K. (2004) Trust in automation: designing for appropriate reliance. *Hum. Factors*, **46**, 50–58.

Miller, C. and Parasuraman, R. (2007) Designing for flexible interaction between humans and automation: delegation interfaces for supervisory control. *Hum. Factors*, **49**, 57–75.

Molloy, R. and Parasuraman, R. (1996) Monitoring an automated system for a single failure: vigilance and task complexity effects. *Hum. Factors*, **38**, 311–322.

Muir, B. (1994) Trust in automation: Part I. Theoretical issues in the study of trust and human intervention in automated systems. *Ergonomics*, **37**, 1905–1922.

Muir, B. and Moray, N. (1996) Trust in automation. Part II. Experimental studies of trust and human intervention in a process control simulation. *Ergonomics*, **39**, 429–460.

Norman, D.A. (1990) The "problem" of automation: inappropriate feedback and interaction, not "over-automation". *Philos. Trans. R. Soc. Lond. B*, **327**, 585–593.

Palmer, M., Rogers, W., Press, H., Latorella, K., and Abbott, T. (1995) *A Crew-Centered Flight Deck Design Philosophy for High-Speed Civil Transport (HSCT) Aircraft*. NASA Technical Memorandum 109171, NASA Langley Research Center, Hampton, VA. Retrieved from http://citeseerx.ist.psu.edu/viewdoc/summary?doi=10.1.1.51.1426.

Parasuraman, R., Sheridan, T.B., and Wickens, C.D. (2000) A model for types and levels of human interaction with automation. *IEEE Trans. Syst. Man Cybern. A Syst. Hum.*, **14**, 286–297.

Parasuraman, R. and Riley, V. (1997) Humans and automation: use, misuse, disuse, abuse. *Hum. Factors*, **39**, 230–253.

Pritchett, A. (2009) Aviation automation: general perspectives and specific guidance for the design of modes and alerts. *Reviews of Human Factors and Ergonomics*, vol. 5, Human Factors and Ergonomics Society, Santa Monica, CA, pp. 82–113.

Pritchett, A., Feigh, K., Kim, S.Y., and Kannan, S.K. (2014) Work models that compute to describe multiagent concepts of operation: Part 1. *J. Aerosp. Inform. Syst.*, **11**(10), 610–622.

Pritchett, A., Kim, S.Y., and Feigh, K. (2014a) Modeling human–automation function allocation. *J. Cogn. Eng. Decis. Mak.*, **8**, 33–51.

Pritchett, A., Kim, S.Y., and Feigh, K. (2014b) Measuring human–automation function allocation. *J. Cogn. Eng. Decis. Mak.*, **8**, 52–77.

Rasmussen, J. (1983) Skills, rules, and knowledge; signals, signs, and symbols, and other distinctions in human performance models. *IEEE Trans. Syst. Man Cybern.*, **13**, 257–266.

Reason, J. (1987) Cognitive aids in process environments: prostheses or tools? *Int. J. Man–Machine Stud.*, **27**(5–6), 463–470.

Sheridan, T. (2000) Function allocation: algorithm, alchemy or apostasy? *Int. J. Hum. Comput. Stud.*, **52**, 203–216.

Sheridan, T. and Verplank, W. (1978) Human and computer control of undersea teleoperators. Technical Report 780815025, Office of Naval Research, Arlington, VA. Retrieved from http://www.dtic.mil/cgi-bin/GetTRDoc?AD=ADA057655%26Location=U2%26doc=GetTRDoc.pdf.

Wiener, E. and Curry, R. (1980) Flight-deck automation: promises and problems. *Ergonomics*, **23**, 995–1011.

Woods, D. and Hollnagel, E. (2006) *Joint Cognitive Systems: Patterns in Cognitive Systems Engineering*, CRC Press, Boca Raton, FL.

Chapter 24

Coordination with Manned Aircraft and Air Traffic Control

Douglas M. Marshall

TrueNorth Consulting LLC, Grand Forks, ND, USA
De Paul University College of Law, Chicago, IL, USA

1 INTRODUCTION

Any discussion of the strategies for coordination or integration of unmanned aircraft (commonly referred to as "drones") with manned aircraft and air traffic control providers must begin with a description of the environment in which this coordination must take place. The unique flight, weight, speed, kinetic energy, and maneuverability characteristics of small, commercial-grade unmanned systems that have proliferated in the last decade have introduced an additional layer of complexity to the air traffic control system. To fully understand the challenges presented to manned aviation and the air traffic management organizations, it is useful to examine each of the elements of the system to see how each component is affected by the introduction of a new, potentially disruptive technology.

1.1 The airspace system

Controlled airspace is a generic term that covers the different classifications of airspace (Class A, B, C, D, and E airspace) and defined dimensions within which air traffic control service is provided to IFR (instrument flight rules) flights and to VFR (visual flight rules) flights in accordance with the airspace classification. Figure 1 depicts the five categories of controlled airspace in the United States and many other parts of the world, as well as uncontrolled Class G airspace. A seventh category of uncontrolled airspace, Class F, may be found in other parts of the world. These classifications were created by the International Civil Aviation Organization (ICAO), and are adhered to by all ICAO member states (numbering 191 at this writing) (International Civil Aviation Organization, n.d.).

Air traffic control clearances are required for all aircraft operating in Class A and B airspaces. A mode C transponder with altitude reporting capability is required within 30 nmi of a Class B airport from the surface up to 10 000 ft MSL (mean sea level).

Class C airspace is generally that airspace from the surface to 4000 ft above the airport elevation surrounding those airports that have an operational control tower, are serviced by a radar approach control, and have a certain number of IFR operations or passenger enplanements. Although the configuration of each Class C airspace area is individually

Unmanned Aircraft Systems. Edited by Ella Atkins, Aníbal Ollero, Antonios Tsourdos, Richard Blockley and Wei Shyy.
© 2016 John Wiley & Sons, Ltd. ISBN: 978-1-118-86645-0.

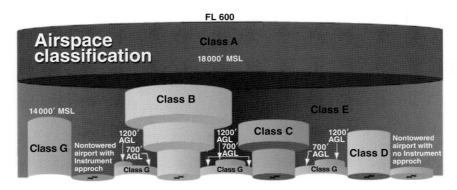

Figure 1. Airspace classifications. (Reproduced from www.faasafety.gov.)

tailored, the airspace usually consists of a 5 NM radius core surface area that extends from the surface up to 4000 ft above the airport elevation, and a 10 NM radius shelf area that extends no lower than 1200 ft up to 4000 ft above airport elevation. No specific pilot certification is required. Unless otherwise authorized by ATC, an operable two-way radio is required.

Class D airspace is usually that airspace from the surface to 2500 above the airport elevation surrounding those airports that have an operational control tower. The configuration of each Class D airspace area is individually tailored and when instrument procedures are published, the airspace will normally be designated to contain the procedures. Two-way radio communication must be established with the ATC facility providing ATC services prior to entry and thereafter maintain those communications while in Class D airspace. Pilots of arriving aircraft should contact the control tower on the publicized frequency and give their position, altitude, destination, and any request(s). Radio contact should be initiated far enough from Class D airspace boundary to preclude entering the Class D airspace boundary before two-way radio communications are established.

Generally, if the airspace is not Class A, B, C, or D, and is controlled airspace, it is Class E airspace. There are no specific pilot certification or equipment requirements to operate in Class E airspace. Special VFR operations are permitted but clearance must be obtained from the controlling facility.

Class G airspace is uncontrolled and goes from the surface to varying altitudes, dependent upon the floor of surrounding airspace (Federal Aviation Administration webpage, n.d.).

1.2   Unmanned aircraft systems

Unmanned aircraft systems, or "drones," have been in use in mostly military applications for nearly a century (Fahlstrom

and Gleason, 2012). Civilian uses include aerial photography, precision agriculture, infrastructure inspection, construction, insurance investigation, film and entertainment, wildlife and resource management, climate monitoring and observation, search and rescue, newsgathering, security, and geographical surveys, to name just a few (Austin, 2010). UAS are now being produced and deployed in many countries. The vast majority is in the "small" category (weighing between 5 and 12 lbs). One analysis estimated that over 300,000 nonmilitary UAVs would be sold worldwide in 2015, and an aerospace consulting firm predicted that UAVs would be "the most dynamic growth sector of the world aerospace industry" (Congressional Research Service, 2015). Worldwide drone sales in 2015 exceeded the predictions and reached 4.3 million, with the United States leading, but Europe and China are not far behind (Washington Examiner, 2016).

The introduction of these easily deployable and small aircraft into the national airspace has created an immediate need for reexamination and revision of air traffic management policies and procedures worldwide.

1.3   Safety management

The FAA (Federal Aviation Administration) and other national CAAs (Civil Aviation Authorities) charged with the responsibility for safely managing the airspace falling within their respective jurisdictions must develop procedures and operational restrictions that mitigate the chances of a drone crashing into a building, another aircraft, or people and objects below a drone flying in that airspace. The challenges to this mandate derive from the small sizes, relatively slow operating speeds, and short endurance of most commercial drones. Figure 2 depicts an eight-rotor drone equipped with a camera and presumably with navigation, control data links, and telemetry technology that allow it to be operated

Figure 2. Generic eight-rotor drone. (Reproduced from Privacy Research Papers, "Drones in Canada" webpage www.priv.gc.ca.)

remotely by a pilot and a ground control station. These devices are often deployed at low altitudes, near or over people and structures, and in the case of multirotor aircraft, are extremely maneuverable, which allows them to hover over a point, move in any direction or climb and descend at high rates. These characteristics are more akin to birds than traditional aircraft, and present unique operational demands for air traffic controllers charged with separating drones from manned aircraft or other drones.

1.4   Role of controllers in air traffic management

Depending upon the requirements of the civil aviation authority as dictated by the regulations in their jurisdiction, air traffic controllers may be "in-the-loop" during UAS operations, communicating with the drone operator via aviation radio frequencies or by telephone. The primary function of the air traffic controller is surveillance of the airspace and aircraft, with the goal of ensuring that aircraft follow their assigned clearances and separating them from other traffic within the boundaries of the national airspace (NAS). Air traffic control facilities are equipped with radar systems that update the positions of aircraft in their sector. In some airspace classes (Classes A and B), the aircraft must report its position and altitude with a Mode C transponder, and in others (Classes C and D and some E) the pilot must establish and maintain two-way radio communication with the controlling authority before entering the airspace. Class G, uncontrolled airspace (generally from the surface to 1200 ft AGL, but as high as 14 500 ft in some areas), does not have similar requirements, although there may still be some operating restrictions that do not involve communication with air traffic controllers. Classes A and B are positive

control airspaces, and aircraft may not enter them without the required operating transponder equipment, two-way radio communication with the controlling authority, and permission or clearance from that authority to proceed.

One of the key safety elements of manned aviation is the pilot's duty to "see and avoid" other aircraft when operating in visual flight rules conditions, and even when flying in Class A positive control airspace, where air traffic controllers are providing separation services, pilots are still obligated to look out the windscreen of their aircraft when possible (Federal Aviation Regulations, n.d.). Although ATC provides separation services in various classes of airspace, it is still ultimately the pilot's responsibility to see and avoid other aircraft, to give right of way when appropriate, and to not operate an aircraft so close to another aircraft so as to create a collision hazard (Federal Aviation Regulations, n.d.).

1.5   Communication between ATC and unmanned aircraft

For the drone operation, the pilot is not in the aircraft to provide "see-and-avoid" hazard mitigation, and air traffic controllers can only communicate approximate location, altitude, and airspeed of intruding aircraft to the drone operator via two-way radio or telephone communication, if at all. And that capability is dependent upon the class of airspace involved and the reporting and communication equipment aboard the manned aircraft according to the regulatory requirements that apply to that sector of airspace. The drone operator/pilot may or may not have the drone within visual line of sight, and may be severely limited in its ability to identify and respond to a collision hazard. In such circumstances, particularly where the drone may be too small to be positively identified on a radar display, the controller's ability to provide separation advice or potential collision warning to the manned aircraft or the drone pilot may be severely limited, if not nullified.

Accordingly, in most jurisdictions that allow commercial drone activities, flights are only allowed under strict operating rules or exemptions that are intended to keep them a safe distance and altitude away from normal manned aircraft operations. Generally, these restrictions involve altitude limitations, prohibit night operations, and require the drone to remain within visual line-of-sight, among others. If operating within a certain distance from an airport (5 nm for towered airports and 3 nm for nontowered), permission from the airport operator or ATC facility may be required. The certificates of waiver or authorization or special flight operations certificates issued by different countries typically set forth an extensive list of requirements and prohibitions designed to protect the airspace and minimize the workload impact on air traffic control facilities. The drone may be equipped with a very high-frequency (VHF) radio relay that allows for direct communication between the drone pilot and the air traffic control.

However, the proliferation of affordable and highly capable multirotor drones has led to hundreds of reports in the United States of near misses between commercial airliners and small drones that are being flown illegally by individuals who are either ignorant of or indifferent to the risks of operating a drone near large aircraft in airport operational environments (*Los Angeles Times*, 2016). These reports come from the pilots of the passenger aircraft, as Terminal Approach Radar Control Facility (TRACON) radars cannot see these drones. The potential for disaster is very much on the minds of regulators, controllers, pilots, responsible drone operators, and law enforcement agencies. In these circumstances, coordination between air traffic control and manned aircraft to avoid collision with an unmanned aircraft is essentially impossible with the current state of the technology, since these rogue operators are not communicating with anyone, and are extremely difficult to track down after a reported airspace violation.

2 EVOLVING UAS TECHNOLOGY AND CAPABILITY

As noted in previous sections, the evolution of small unmanned aircraft, particularly the multirotors, has led to the wholesale democratization of the drone operator community. With over 4 million drones being sold worldwide in 2015 alone, the opportunities for inexperienced and untrained aircraft operators to contaminate the airspace with unregistered and unidentifiable drones are virtually unlimited.

2.1 Fixed wing versus multirotor

The great majority of commercial off-the-shelf systems (COTS) are multirotors. The typical quadcopters or hex copters are electric motor powered with GPS navigation, command and control data links, and telemetry down links for video and photo camera payloads. They are typically programmable for semiautonomous flight, and have safety features such as return-to-home software to deal with lost link or loss of control events and geofencing software to prevent fly aways and intrusion into restricted or unprotected airspace. They use three types of data links: flight control link,

system monitoring (telemetry) link, and a payload or task link used to control, manage, or monitor onboard sensors or similar equipment.

Communications technology solutions for UAS are in a dynamic state of development and no single architecture fits all cases. Issues of communication modes and frequency allocations, bandwidth, security, integrity, and interoperability with legacy airborne and ground-based communication systems continue to challenge manufacturers and end users.

Until relatively recently, fixed wing drones were the favored system, and their mature technology was largely derived from military applications. These systems have the advantage of longer range and endurance, and arguably a less hazardous recovery capability in the event of a loss of control or propulsion because they can either glide to a recovery point or fly into a "deep stall" and settle to the ground with very low kinetic energy.

Multirotors have the advantage of greater maneuverability, and the ability to station over a set position to enhance the quality of imaging from onboard systems. Their disadvantage is shorter endurance (due to limitations of battery weight and life) and fewer alternatives for recovery in the event of a loss of power or propulsion.

2.2 Commercial applications

In the United States alone, over 5200 petitions for exemptions to operate commercial drones have been granted as of May 31, 2016 (FAA webpage Section 333, n.d.). These exemptions are granted to aircraft registered with the FAA, operated by a pilot with an FAA airman certificate, and pursuant to a Certificate of Waiver or Authorization (COA). The COA typically restricts operations to remain below 200 ft AGL and within visual line of sight of the pilot, daytime operations only, no closer than 500 ft to persons, and with specific requirements on communication with air traffic control and nearby airports. The exempted operations run the gamut of uses, from aerial inspections to real estate photography to motion picture and television production to news gathering, and so on. All employ multirotor (dual helicopter type or 4, 6, 8, or 12 rotor) platforms and require skilled pilots, visual observers, and video technicians to operate. None are exempted from communicating with air traffic control facilities.

The FAA Air Traffic Organization (ATO) reviews all proposed UAS operations and evaluates the safety of these operations relative to the requested airspace through the existing COA process. The majority of current UAS operations occurring in the NAS are coordinated through air traffic control by the issuance of a COA. This process not only makes local ATC facilities aware of UAS operations, but also provides ATC the ability to consider airspace issues that are unique to UAS operations. A typical exemption granting the commercial use of a UAS to perform aerial surveying contains approximately 30 conditions and limitations. Among those conditions is the requirement for the operator to request a NOTAM (Notice to Airmen) to alert other users of the NAS to the UAS activities being conducted. The COA may authorize flights within Class C, D, E, and G airspaces, as well as within 5 miles of an airport (in which case a letter of agreement with airport management must be obtained).

2.3   Communications architecture

A systems approach is favored when considering communications architecture for unmanned systems. The UAS ground control station, air traffic control demands, and the presence of manned aircraft must all be considered equally in the equation. Long-term solutions to UAS communications must be designed, frequency allocation negotiated, and implemented. Requirements for communications must be defined before solutions are achieved. On a global scale, and in the interest of harmonization, that definition should neither identify a particular technology or system nor be exclusive to a particular country or region. "Requirements should be flexible enough to allow for innovative ideas and applications to be explored. The proper venue for negotiating these requirements should be through the required communications performance (RCP) being developed by ICAO and civil aviation authorities. The field of wireless and networked communications is changing rapidly with the advent of new concepts, technologies, and capabilities; it would be wrong to limit future UAV communication systems to what is known and proven today" (DeGarmo, 2004).

3   SYSTEM REQUIREMENTS FOR UAS AIRSPACE INTEGRATION AND COORDINATION

The expanding need (or demand) for both military and civil unmanned aircraft systems to operate in the same unrestricted or unsegregated airspace as manned aircraft faces one significant barrier, which is that the UAS pilot is not onboard the aircraft to provide "see-and-avoid" capability. The traditional communications link in manned aircraft is from pilot-to-ground or pilot-to-pilot. With UAS, the "cockpit" remains on the ground with the pilot, so the radio communications link is now between the pilot and the flight control system on the

aircraft. The additional link in the chain of communications presents new challenges in the manner in which UAS operations integrate with manned aircraft.

3.1   Sense and avoid

The term "sense and avoid" has evolved to describe the technology intended to mitigate the "see-and-avoid" requirements of civil aviation regulations. In the United States, the FAA endorses the concept that *sense and avoid* "is the capability of an unmanned aircraft to remain well clear from and avoid collisions with other airborne traffic" (FAA SAA workshop, 2009). The two components of this requirement are self-separation and collision avoidance. Self-separation requirements are intended to reduce the probability of a collision by remaining "well clear" of other aircraft. The term "well clear" has yet to be defined to anyone's satisfaction. Collision avoidance refers to extreme maneuvers taken just prior to the closest point of separation to avoid collision where safe separation is lost (Lacher *et al.*, 2010).

Sense-and-avoid technology has yet to achieve the same level of confidence or accuracy as human perception, which is a prerequisite to unlimited integration of UAS into unsegregated airspace. The command and control (C2) link between the aircraft and the pilot on the ground is vulnerable to interruptions due to radio frequency interference, radio or satellite system failure, and electromagnetic interference (EMI). And there are latencies in the transmission of data that could influence maneuverability and response-to-threat times. As a result, air traffic management handles unmanned aircraft differently than manned aircraft. Safe separation is achieved by not allowing UAS to operate in areas likely to be occupied by manned aircraft (above 500 ft, AGL, near airport operations areas, and in congested or densely populated areas). Visual observers are required and telephone connection between the ground control station and the air traffic controllers in the event of a lost link is mandated. Unmanned aircraft operating in positively controlled airspace must be equipped with Mode C transponders, and be in two-way radio communication with ATC. There may even be creation of temporary flight restrictions to prevent manned aircraft from entering the sector of airspace (such as over a forest fire or similar emergency where the UAS is the preferred aircraft due to various hazards inherent in flying in that environment).

3.2   Ground-based sense and avoid

Where airborne sense-and-avoid capability is not available, and operations beyond visual line of sight are desired, there is a need for some other technology to sense or detect airborne targets in the area where the UAS is flying. A ground-based sense and avoid (GBSAA) system has been suggested to utilize air surveillance radars with a three-dimensional position information display to support UAS flight teams. These systems may be cost-prohibitive for the typical commercial UAS operator, but could provide an acceptable means of compliance for see-and-avoid requirements in complex operational environments (multiple UAS in a confined airspace, for example) (Lacher *et al.*, 2010).

3.3 Airborne sense and avoid

Airborne sense-and-avoid (ABSAA) technology is still evolving, as noted in Section 3.1. A true ABSAA system has yet to be fully certified by any civil aviation authority, and thus no technology solution available today can satisfy the see-and-avoid requirement of most national and international aviation regulations.

ABSAA solutions may be divided into two categories: cooperative and noncooperative. Cooperative systems receive information from other aircraft regarding their absolute and relative location. The information is communicated to the UAS pilot on the ground or is used by the UAS directly to autonomously sense-and-avoid other aircraft. For this system of systems to be viable, all aircraft operating in the desired airspace would have to be equipped with position identification capability. Currently available technology could include Mode C transponders and Automatic Dependent Surveillance – Broadcast (ADS-B) reporting devices. Very few aircraft are equipped with ADS-B, although in the United States the FAA intends to mandate ADS-B OUT, and other nations are mandating ADS-B OUT as well (Lacher *et al.*, 2010).

Transport category aircraft worldwide are equipped with airborne collision avoidance systems (ACAS), which alerts pilots to potential collision threats and suggests specific resolution maneuvers. A similar system, based upon cooperative sensor capabilities, could offer a solution to airborne sense-and-avoid requirements for UAS. The specific technical requirements for a certifiable ABSAA system are beyond the scope of this chapter, but the concept is very pertinent to the discussion of cooperation between UAS, manned aircraft, and the ATC system.

The UAS pilot's ability to perform self-separation and collision voidance maneuvers relying upon onboard ABSAA equipment is a function of whether the pilot is "in-the-loop" or if the system is autonomous. If the pilot is in-the-loop, traffic information is communicated to the pilot, who then makes the decision on how to respond. The type of notice or information provided to the pilot will be dependent upon the architecture of the C2 link, operational requirements and limitations, and procedural and separation criteria. The viability of this arrangement in turn depends upon the surveillance accuracy of the position information of the other aircraft.

An autonomous ABSAA system will make the same decisions that would be made by a pilot on the ground, based upon information collected by the onboard sensors, but no pilot action is required. The pilot on the ground would be provided with the same information in the GCS, and could override the automation if necessary. Both the autonomous and pilot-in-the-loop alternatives present their own challenges with respect to vulnerability, latencies in the C2 link, and the integrity of the link.

Noncooperative ABSAA is similar to cooperative systems, except that the sensor package must deal with other traffic that is not similarly equipped. The technical solution to the challenge of accurately identifying other aircraft, their relative and absolute positions, and their speed is years away due to a variety of factors, not the least of which is cost. Noncooperative ABSAA presents additional complexities over cooperative technologies due to difficulties detecting traffic, dealing with dynamic meteorological conditions, ground clutter, aircraft of varying sizes, speeds, and configurations, fusion of data with multiple onboard sensors, and range limitations, among others. Small commercial UAS have size, weight, and power limitations that would preclude implementation of this technology for the immediate and foreseeable future (Lacher *et al.*, 2010).

4 COMMAND AND CONTROL INTEGRATION

The challenge of coordinated communication between air traffic control, manned aircraft pilots, and UAS operators has compelled aviation authorities to create a variety of approaches to reduce operational risk to all users of the airspace.

One relatively painless alternative is to only allow UAS to operate in segregated airspace, which is airspace that is blocked from all other traffic. This can be implemented where there is an immediate threat from manned aircraft or activities taking place on the ground, such as a natural or man-made disaster. This airspace may lie within a restricted area (such as around a military installation or sensitive infrastructure complex) or in a warning area where military operations are in effect.

Other strategies could integrate standardized lost link, emergency, and contingency procedures, such as onboard technology that automatically returns the UAS to home or a predesignated location when there is an interruption of the C2 or data link. Air traffic controllers would necessarily need to be in the communications loop so that they would know to clear other traffic out of the area or to otherwise secure the airspace of any potential conflicts. The requirement in most operating authorizations for UAS require that the pilot establish and maintain two-way telephone or radio communication with ATC if operating under a Certificate of Authorization or Waiver (or similar permissions available in other countries).

4.1   New ATM procedures

As there is greater demand for access to unsegregated airspace by commercial and noncommercial UAS users, modified or new procedures for air traffic management professionals and agencies will need to be developed. These changes may include operational and technical responses to lost link and other flight contingencies; methodologies to incorporate into the ATM system differences on flight performance capabilities between manned and unmanned aircraft; uniform procedures for providing routine separation services; and modifications of emergency procedures (such as a dedicated emergency transponder code just for UAS) to accommodate the unique nature of UAS emergencies (they have no person onboard, so the aircraft may be "sacrificed" to avoid collisions with aircraft or persons or property on the ground).

Some civil aviation authorities have created "no-drone zones" around sensitive areas, where UAS are prohibited from operating under any circumstances (FAA No Drone Zone, 2016). While it may prove difficult to enforce these restrictions because the agency must rely upon interested citizens and local law enforcement officials to report violations, work is underway to develop defensive surveillance and detection systems to place around airport areas and other locations where drones are prohibited. One such system uses a radio beam to disable drones, causing them to crash (not necessarily a desirable result, depending upon where it crashes), and also as an optical disruptor that can disrupt the autofocus on the drone's camera, rendering it useless (Szondy, 2015). This may not be useful to air traffic controllers, but this or similar devices have the potential for disabling drones that create an immediate hazard to navigation around airports or sensitive areas.

Perhaps the greatest challenge, however, to the safe integration of unmanned systems into unsegregated airspace is the reality that the great majority of the millions of small rotorcraft UAS that have been sold over the last 2 years are in the hands of noncertificated operators who have had little or no exposure to the intricacies of flying an airplane in any nation's airspace. Without proper training and experience in the skills required to be safe and proficient in a manned aircraft, these amateur (and sometimes professional) drone pilots have regrettably created unnecessary risks and hazards to other manned aircraft and to persons or property on the ground. They typically are neither familiar with airspace designations and restrictions, meteorology, aerodynamics, data link functions, air traffic control procedures and communication requirements, and emergency procedures, nor any of the knowledge, skills, and aptitudes that characterize the licensed aviator. In short, they do not know the rules, and in many cases, apparently do not care. The rise in drone hazards not only irritates pilots, but has also created an enormous burden on the regulators as they attempt to rein in the rogue operators and establish a regulatory environment that enables the growth of the responsible commercial sector of the industry (Huerta, 2015)

5   CONCLUSIONS

The challenges presented to the manned aviation community and the world's air traffic control system by the introduction of unmanned aircraft into national and international airspace are well known. The lack of airborne see-and-avoid capability in today's unmanned systems precludes full airspace integration, a milestone, when achieved, which will allow properly equipped UAS to follow many of the same procedures as manned aircraft. Filing a flight plan, or launching in class E or G airspace without one, establishing and maintaining contact with air traffic control, assuring separation from other aircraft, avoiding collisions when imminent, and recovering safely in the event of an emergency or unplanned descent will all become routine operations. The role of air traffic controllers will continue to be to coordinate manned and unmanned aircraft flights so that they do not conflict with one another, to provide separation and flight following services, to clear airspace in an emergency, to monitor restricted or segregated airspace limitations, and to offer a reliable method for drone operators to remain in contact with the air traffic management system as required to maintain safe operations.

The roles of the participants in this system of system will be continually refined to accommodate advances in technology and the capabilities of both manned and unmanned aircraft. As reliable onboard sense-and-avoid systems become available, scalable and affordable, the ATM system will also have to adapt and upgrade its technology to meet the growing demand for access to the airspace. Controllers will require additional training on new systems and procedures, and manufacturers and developers will continue to evolve their systems to maintain their place in the marketplace. Technology that does not evolve with demand is doomed to failure.

REFERENCES

Austin, R. (2010) *Introduction to Unmanned Aircraft Systems (UAS)*, John Wiley & Sons, Inc., New York, pp. 273–279.

Congressional Research Service (2015) Unmanned Aircraft Systems (UAS): Commercial Outlook for a New Industry, 7–5700 www.crs.gov, R44192.

DeGarmo, M.T. (2004) *Issues Concerning Integration of Unmanned Aerial Vehicles in Civil Airspace*, The MITRE Corporation.

Fahlstrom, P.G. and Gleason, T.J. (2012) *Introduction to UAV Systems*, John Wiley & Sons, New York, pp. 3–7.

Federal Aviation Administration webpage: www.faa.gov

Federal Aviation Administration webpage (Section 333): www.faa.gov

Federal Aviation Administration (2009) Sense and Avoid (SAA) for Unmanned Aircraft Systems (UAS), October 2009.

Federal Aviation Administration No Drone Zone webpage www.faa.gov/unmanned aircraft systems/no drone zone.

Federal Aviation Regulations, General Operating and Flight Rules, 14 CFR Part 91, §91.113(b).

Huerta, M. (2015) Rise in drone hazards irritates pilots, concerns officials, interview with CBS News reporter Kris Van Cleave, Aug. 12, 2015, CBS News Webpage: www.cbsnews.com

International Civil Aviation Organization webpage: www.icao.int

Lacher, A., Zeitlin, A. Maroney, D., Markin, K., Ludwig, D., and Boyd, J. (2010) Airspace Integration Alternatives for Unmanned Aircraft, Presented at AUVSI's Unmanned Systems Asia-Pacific 2010.

Los Angeles Times (2016), Lufthansa jet and drone nearly collide near LAX, LA Times Webpage: www.latimes.com

Szondy, D. (2015) Anti-UAV Defense System uses radio beam to disable drones, Gizmag, Oct. 10, 2015.

Washington Examiner (2016) Drone sales surge 167% to 4.3 million, U.S. leads but China catching up, citing a report by marketing and investment firm Kleiner Perkins Caufield & Byers.

Chapter 25

Aircraft Pilot and Operator Interfaces

Mary L. Cummings[1] and Greg L. Zacharias[2]

[1]*Humans and Autonomy Laboratory, Duke University, Durham, NC, USA*
[2]*Charles River Analytics, Cambridge, MA, USA*

1 INTRODUCTION

In 1907, the first instruments in a cockpit were introduced in Wright aircraft because of the high numbers of crashes due to stalls (the first documented cases of pilot error!). These crude instruments included a vane-type incidence indicator for attitude control, as well as manometric tubes used to indicate speed (Coombs, 2005). As exemplified in Figure 1, a 1910-era Boeing and Westervelt (B&W) floatplane cockpit illustrates these early cockpit designs with just a crude airspeed representation. Design of cockpits at this time was relatively *ad hoc* in that gauges were added to cockpits through more of a trial-and-error process than through principled design. Cockpit design choices were made out of convenience and often only added after critical events occurred, such as pilot errors that lead to crashes. Even with the formalization of human factors research by the Army Air Force in the 1940s,

which occurred primarily due to aviation mishaps in WWII, as new missions and technologies were added (e.g., additional engines, radars, weapons, etc.), new instruments and displays were also added. This increasing complexity caused a significant increase in workload for pilots, which was often managed by increasing the number of crew. In military and commercial transport planes, this meant not just an increase in the number of pilots, but also navigators and flight engineers. Figure 2 shows a Boeing 377 1940s-era Stratocruiser with five people on the flight deck.

The introduction of digital processors across a spectrum of displays, controls, and subsystems, first with the Space Shuttle design of the early 1970s, and later with military and commercial aircraft (in the late 1970s and early 1980s) revolutionized cockpit design, leading to the introduction of glass cockpits (Wiener and Nagel, 1988). Figure 3 shows a 1980s-era Boeing 737, which incorporates a few "glass" displays, but still retaining many more traditional, dedicated, and hard-wired instruments that represent individual pieces of information. The increased use of both digital processors and glass displays allowed airplane designers to reduce the physical space needed for individual instruments, by integrating selected pieces of information into more "holistic" views of aircraft/subsystem state and status, so that individual instruments were no longer needed.

Another significant change that dramatically influenced cockpit design was the increasing use of control augmentation. This trend started in the 1940s with the introduction of hydraulic force augmentation of direct linkages between cockpit controls and aerodynamic control surfaces (e.g., between the pitch control "yoke" of early cockpits and aircraft elevators), and was accelerated with the introduction of "fly-by-wire" controls. These electronic controls

Unmanned Aircraft Systems. Edited by Ella Atkins, Aníbal Ollero, Antonios Tsourdos, Richard Blockley and Wei Shyy.

Figure 1. A Boeing and Westervelt 1910-era floatplane cockpit. (Reproduced with permission from The Boeing Co.)

Figure 2. Boeing 1940s-era 377 stratocruiser. (Reproduced with permission from The Boeing Co.)

Figure 3. 1980s-era Boeing 737. (Reproduced with permission from The Boeing Co.)

hydraulic actuators controlling the aerodynamic surfaces. This "wire" is the pilot's only (virtual) linkage to the control surface, and the signal transmitted across that wire is mediated by a flight control system that can change the stability and handling qualities of the entire aircraft. The ultimate result for both military and commercial transport aircraft has been a dramatic reduction in manning. What once required two pilots, a flight engineer, and a navigator to effectively do the job as seen in Figure 2 was streamlined into something that could be handled by just two pilots, such as most modern-day commercial airliners.

Accompanying the transition from mechanical gauges to glass cockpit displays and fly-by-wire controls, aircraft control generally began to shift from manual or "pilot-in-the-loop" control to human supervisory control (often called "pilot-on-the-loop"). Human supervisory control (HSC) is the process by which a human operator intermittently interacts with an on-board computer/processor, receiving feedback from and providing commands to a controlled process or task environment, which is connected to that computer through actuators and sensors (Figure 4). Pilots of modern

essentially eliminated the earlier mechanical linkages, and replaced them with controllers that transduce pilot forces/ motions into electrical signals that drive digital processor-based flight control systems. These, in turn, drive the

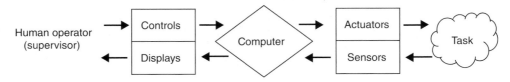

Figure 4. Human supervisory control. (Reproduced with permission from Sheridan, 1992. © The MIT Press.)

"fly-by-wire" aircraft, which constitute the majority of commercial and military aircraft, are consequently supervisors of highly complex systems. The introduction of modern autopilots and flight management systems as was also a major contributor to the manning reductions seen in the cockpit.

For pilots, the move towards HSC means that controls, displays, and computers now separate the crew from the aircraft and subsystems they are meant to manage. The implication is that the design of these intermediary systems is critical for safe and efficient operation. This is not a trivial point. Fly-by-wire HSC systems must have virtual forces programmed into them since pilots can no longer "feel" the forces on the flight surfaces. This is particularly the case for those aircraft with side-stick controllers like those found in Airbus aircraft and military aircraft like the US Air Force F-16. The classic yoke controller can be replaced with a much smaller side stick controller, since control has evolved from actuator-augmented mechanical linkages to ones that are fully electronic, computer-mediated, and hydraulically effected.

Complementing these dramatic advances in control and actuation design are equally compelling advances in display and sensor design, typically manifested through "glass cockpit" displays (Figure 5). All new commercial and military aircraft today are built with some type of glass cockpit display. Figure 5 illustrates the most recent advances in cockpit display technology, which include larger screens, dual heads-up displays, and full-route vertical situation displays. Glass cockpits are also found in spacecraft, railway cabs, ship command centers, and even every-day consumer goods like automobiles and the latest generation of smart phones.

While significant cost-reductions in design and maintenance can be achieved in these domains as a result of glass cockpits, such displays are not always guaranteed to promote optimal human performance. Moreover, because there are many more sources of information, often hidden in many layers of hierarchical software menus, it can be difficult to design displays to incorporate all the needed information in a dynamic fashion that follows the changing needs of the operator, while avoiding the "clutter" of unneeded information. Lastly, the transition from manual to supervisory control means that significantly more automation is introduced, which directly impacts the information displayed to the pilots. As will be discussed in the next section, embedded automation and the associated displays that go with that automation can add significant benefit in terms of workload reduction and information "decluttering," but there are also potential negative consequences to be considered.

2 BASIC COCKPIT DESIGN

In current commercial and military aircraft, and increasingly in the general aviation community, almost every aspect of flying is controlled via a human computer interface, so the need for seamless human–computer interaction (HCI) is paramount. Given the primarily visual nature of piloting tasks, and the human innate ability for dealing with exquisitely complex, dynamic, and detailed visual cues, most of the information presented to a pilot is via the visual channel (as opposed to the auditory and tactile channels). Given a pilot's heavy reliance on visual cues, there are a number of basic visual display considerations that must be taken into account in the design process, some of which are listed in Table 1. For a more extensive discussion of these parameters, see Boff and Lincoln (1988), Tsang and Vidulich (2003), Wickens and Hollands (2000), and Wiener and Nagel (1988).

While some of the visual design parameters listed in Table 1 seem obvious (such as the need to reduce clutter),

Figure 5. The cockpit of the Boeing 787 dreamliner. (Reproduced with permission from The Boeing Co.)

Table 1. Selected visual design parameters.

Density	Display size
Clutter	Screen resolution
Color	Contrast
Highlighting	Brightness
Grouping	Font size
Graphics vs. text	Field of view
Analog vs. digital	Vibrations
Dual coding	Dark adaptation

not all are so intuitive. Previous research has shown that aviation displays tend to use too much color (Xing, 2006). If personnel in supervisory settings are required to remember more than approximately six color categories, reaction time and the number of operational errors can increase, in comparison to performance seen under simpler coding schemes (Cummings and Tsonis, 2006). Often, selecting values for many of the different visual design parameters given in Table 1 requires that the designer address competing objectives; as a consequence, the design engineer must often perform some type of trade space analysis to specify a final design. For example, dual coding, the use of more than one visual cue for symbol representation such as shape and color, is critical since approximately 8% of the male population is color-blind (Boff and Lincoln, 1988). However, this design choice could compete with the just-mentioned design guideline that only about six colors should be used. This conflict needs to be dealt with via trade-off studies that are typical in any design effort, cockpit, or otherwise.

In addition to basic visual design variables that should be considered in any cockpit design, there are a few basic design principles that are core to supervisory control applications, particularly for aviation systems that are complex, conducted under time-pressure and characterized by many uncertainties (such as weather, other aircraft position, etc.) These include the principle of pictorial realism, the compatibility-proximity principle, and the principle of the moving part (Wickens and Hollands, 2000), which will be illustrated in turn using Figure 6, a Primary Flight Display (PFD), as an illustration.

For the principle of pictorial realism (Roscoe, 1968), the idea is to design the display so that it is a relatively accurate pictorial representation of the information it represents. An example of this can be seen in the center of Figure 6, where

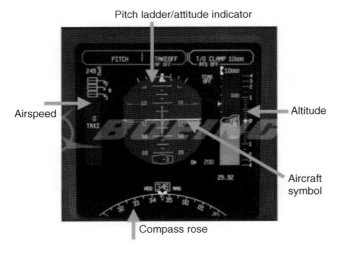

Figure 6. A typical primary flight display. (Reproduced with permission from The Boeing Co.)

even non-pilots can easily determine that the blue, upper area represents sky and the lower, brown area represents the earth. This allows the pilot to quickly determine the orientation of the aircraft with respect to the horizon line that serves as the blue-brown boundary separator. The pilot can also quickly determine whether the plane is climbing or descending, depending on the relative ratio of the blue and brown areas in the display. By designing displays that can graphically and accurately represent real-world relationships in this manner, designers can provide pilots with affordances (Norman, 1988) that enable them, without any conscious thought or interpretation, to immediately understand what is presented to them.

The Compatibility-Proximity Principle (Wickens and Carswell, 1995), which essentially states that correlated information across both task and time should be in close proximity, is also critical to effective interface design, particularly as the number of information sources continues to grow. An example is given in the PFD in Figure 6, which essentially integrates heading (lower section), airspeed (left side of display), altitude (right side of display), and attitude (center), as well as a few status indicators which in older aircraft would have taken more than four separate displays. Software-driven glass cockpit displays like these make implementation of the Compatibility-Proximity Principle easier than in older aircraft, and information integration has lead to the development of displays, which help reduce pilot workload. As a consequence, the PFD has become a central instrument for flying, with similar information often repeated on a Head-Up Display (HUD), both of which reduce pilot workload due to the information's close proximity.

The last basic display design principle is the Principle of the Moving Part (Tsang and Vidulich, 2003), which states that those elements of the display which move, should move in accordance with a pilot's mental model of the associated dynamics. So for example, in Figure 6 the altitude tape on the right is vertically oriented because in our earth-bound reference frame, we think of altitude as up and down. One common violation of this principle is the use of the egocentric aircraft symbol on the pitch ladder in the center of Figure 6 (as depicted, wings level). In western aircraft, the aircraft symbol is static, while the "world" (the blue (top) and brown (bottom) circular earth representation in Figure 6) moves, in violation of the principle of the moving part. In many eastern aircraft, particularly those from the former Soviet Union, the representation is in keeping with the principle of the moving part, called an exocentric view because the aircraft symbol moves in the PFD and the world stays static. Both displays are effective (Johnson and Roscoe, 1972), which demonstrates that training is a critical aspect of piloting, and this point further underscores that while design principles are effective tools, they are not absolute, and some variations are acceptable.

While the focus of this section thus far has been on visual display of information, the other primary mode of information "display" deserves mentioning, the auditory channel. Audio displays in cockpits enable pilots to successfully divide their attention so that they can listen for relevant audio signals, while still attending to visual information. Audio displays work best when the inputs are short, simple, do not have to be remembered for long periods, and relate to an action required immediately (McCormick, 1976), so they are primarily used for alerts and warnings in the cockpit. Pilots have been shown to respond faster to auditory alerts than visual ones by up to 30–40 ms. However, auditory alerts have problems in that they can be distracting, and if not integrated into the overall cockpit in a holistic manner, they can cause confusion. One example is the Helios Boeing 737 crash in 2005. At 10 000 ft, an alarm sounded, indicating a pressurization problem. The pilots confused this alarm with a similar sounding alarm meant to alert pilots of an incorrect takeoff configuration, which only sounded on the ground. This alert confusion caused them to ignore the alarm, causing the pilots to succumb to hypoxia and ending with the crash of the plane (Air Accident Investigation & Aviation Safety Board (AAIASB), 2006). Since it is difficult for humans to either remember or clearly distinguish alarms when multiple alarms are present, it is critical that aircraft cockpit designers take a systems engineering approach to all alerts and warnings.

The least common information presentation modality is through the haptic (i.e., touch) information-processing channel. However, one effective use of haptic alerts in many cockpits is the use of "stick shakers". As an aircraft detects the onset of a stall, the stick/yoke column will shake, which simulates the buffet that a plane would experience if approaching a fully developed stall. The shaking of the stick alerts the pilot through vibrations and is an extremely salient cue, which immediately informs the pilot that the attitude must be decreased and the airspeed increased or a full stall could develop. In everyday life, similar haptic alerts can be seen in actual (or virtual) rumble strips, grooved or raised pavement patterns on highway shoulders, which alert drivers to possible departures from the roadway.

3 THE IMPACT OF INCREASING AUTOMATION IN THE COCKPIT

As previously mentioned, automation has revolutionized the cockpit design of all modern aircraft. However, while automation brings with it many advances in terms of pilot workload, it does not always reduce workload, and in most cases, merely changes the nature of the work (Billings, 1991). In any system that is both highly automated but also allows operator intervention during various phases of operation, there is a significant risk of operator "mode confusion" (Billings, 1997; Sarter and Woods, 1994). As evidenced by numerous airline crashes in the 1990s during the transition to highly automated flight control and navigation systems, mode confusion occurs when an operator attempts to take control of a highly automated system, but does not understand the current mode of automation (i.e., the goals or objectives the automation is attempting to achieve). This lack of understanding can cause catastrophic human-system failure due to confusion over who is in control (the human or system), especially when the desired goal state of the operator differs from that of the automation.

An example of mode confusion caused by poor interface design is the fatal Airbus A320 crash near the Strasburg-Entzheim Airport in France in 1992. The crew inserted 3.3 into the flight management computer (FMC) intending to fly a 3.3° glide slope to touchdown. However, there was confusion about the current operating mode of the FMC, and the pilots inadvertently inserted the command while the FMC was in the vertical descent mode instead of the intended flight path control mode, which placed the aircraft in a 3300 ft min^{-1} descent, causing it to crash in the surrounding foothills (Hughes, 2005). The single button that toggled between these modes was not well designed, hard to read, and had no independent mode presentation cue. The pilots were highly trained with years of experience, yet they failed to correctly interpret, or effectively communicate their intent to the FMC, a situation which was exacerbated by poor interface design.

Another critical consideration in the design of cockpit automation and interfaces is recognizing problems with automation bias, or overtrust in automation, which occurs when a human decision maker unconditionally accepts a computer-generated solution, and disregards or does not search for information that could potentially contradict the "accepted" automation solution (Mosier and Skitka, 1996; Parasuraman and Riley, 1997). Human errors that result from automation bias can be further decomposed into errors of commission or omission. Errors of commission occur when humans erroneously follow automated directives or recommendations, while errors of omission occur when humans fail to notice problems because the automation does not alert them (Mosier and Skitka, 1996). Many studies in aviation have demonstrated the human tendency towards automation bias, causing both errors of commission and omission.

One study that looked more closely at *errors of commission* during enroute flight planning found that commercial pilots, when given a computer-generated plan, exhibited significant automation over-reliance and complacency, leading them to accept flight plans that were significantly sub-optimal. When presented with an automated solution characterized by some uncertainty in the problem space, 40% of pilots reasoned little or none at all about the solution and deferred to erroneous automation recommendations, even though they were provided with tools with which to explore the solution space (Layton, Smith, and McCoy, 1994). In a similar experiment looking at automated assistance in a military in-flight replanning task, pilots tended to accept the computer's solution without question, despite having the ability to change and improve on the computer's solutions (Johnson *et al.*, 2002).

One example of an automation-induced *error of omission* occurred in 1972, with an Eastern L-1011 crash into the Florida Everglades. Upon execution of the landing checklist, the nose gear indicator signaled a potentially unlocked nose gear, an unsafe condition. After engaging the autopilot, the crew intently focused on the (potentially) unsafe nose gear, failing to notice several minutes later that the aircraft had begun gradually losing altitude, a situation, which was likely caused by one of the pilots inadvertently bumping the control stick and disengaging the autopilot. The crew mistakenly relied on the (now non functioning) automation to both keep the plane at the correct altitude and to warn them if the autopilot failed, leading to a gradual descent into the Everglades, and the subsequent deaths of 101 crew and passengers (NTSB, 1973). An ironic postscript to this event is that the nose gear indicator, which triggered the entire sequence of crew errors, was faulty, and that the crew could have landed safely without any intervention to "fix" the nose gear.

Neither the A320 nor L-1011 designers ever anticipated that operators would make such mistakes, and as is often the case in engineering, it is not until accidents occur that engineers make important design changes (Petroski, 1992). These accidents highlight a design consideration that is not a formal design principle *per se*, but is a well-established usability HCI design consideration: the critical importance of providing adequate feedback to the human operators interacting with the automation system (Nielsen, 1993). In both cases cited above, pilots had some indication of the status of the aircraft, but neither cue (the small glide slope FMC mode indicator light on the A320 or the C-chord chime altitude alert on the L-1011) was salient enough, and not designed with an understanding of how the cue would have been interpreted in noisy and cluttered visual and auditory environments. So while proper design of each individual display element is important, it is also critical to understand the context and environment in which the displays will be used, so that critical pieces of information get transmitted to the crew at the critical times needed. The need for proper feedback about both aircraft status and external information has become even more critical for pilots of "remote" aircraft, which is the focus of the next section.

4 UNMANNED AERIAL VEHICLE GROUND CONTROL INTERFACES

Unmanned aerial vehicles (UAVs) are quickly becoming ubiquitous, not just in military intelligence, surveillance, and reconnaissance (ISR) operations, but also in civilian applications like crop monitoring and border patrol. With reduced radar signatures, increased endurance, and the removal of humans from immediate threat, unmanned (also known as uninhabited) aerial vehicles have become indispensable assets to militarized forces and governmental agencies around the world. Their future growth in all sectors is projected to be nearly exponential (Bharat, 2008). Despite the absence of any crew onboard UAVs, UAVs still operate as manned systems (as illustrated earlier in Figure 4), and human operators are still needed for supervisory control (Erickson and Zacharias, 2004).

A very common misconception in the design of a UAV ground control station interface is that it should be very similar, if not nearly identical to, a traditional manned aircraft cockpit. This naïve assumption is problematic because a UAV "pilot" does not have the ability to rely on an external, rich set of physical cues, which a pilot of a manned aircraft has. Such cues that UAV operators do not have include a significantly larger visual field of view, ambient sounds, and a rich array of vestibular and tactile cues signaling aircraft attitude, motion, and system failures. However, because of the supervisory control nature of the UAV flight task, there is often *no need* for sensory augmentation or complicated cockpit replicas that is a critical point often overlooked in the design of UAV ground control interfaces.

Automated flight control and subsystems management have advanced to the point that UAV operators can now offload many or all of their "locomotion" tasks of flying the aircraft, and associated "housekeeping" tasks of managing the subsystems. As will be discussed in detail below, UAV operators need to concentrate on higher level mission management tasks (like waypoint planning and route modification) and critical payload management tasks (e.g., sensor management, weapons use, etc), which are tasks that typically are very complex and cognitively demanding. As a result, increasing the number of displays a remote pilot has in order to replicate the "feel" of a real cockpit, or even adding advanced sensory displays like spatial audio (i.e., the location of an aural cues in three dimensions), is not only unnecessary, but potentially could lead to operator error. A study

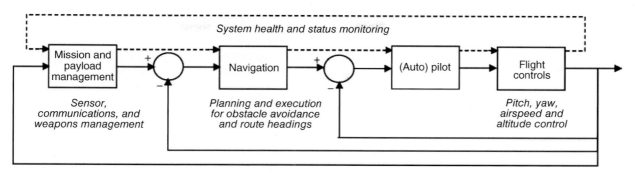

Figure 7. Nested control loops for supervision of UAVs.

examining human error rates for UAV operators found that pilots using a traditional pilot-centric cockpit representation to "fly" their vehicles made more mistakes than their lesser-trained point-and-click counterparts who were not "flying" their UAVs, but rather were managing them (Williams, 2004).

As illustrated in Figure 7, the concept of a human "piloting" a UAV is fundamentally flawed because of the embedded control loops that occur in supervisory control of a UAV. The inner loop shown represents the basic flight control loop, which is the most critical loop that must obey the physical laws of nature such as aerodynamic stability and control surface effectiveness. In this loop, operator actions are focused only on the short term and local control (keeping the aircraft in stable flight). The second loop, the navigation loop, represents the actions that some agent, whether human or computer-driven, must execute to meet local geo-spatial constraints such as routes to waypoints or points of interest, time on target, and avoidance of obstacles. The navigation loop corresponds to the control of the path necessary to accomplish the mission. The outermost loop represents the highest levels of control, that of mission management. It is this loop where the actual mission requirements are typically met. For example, for ISR UAV missions, sensors must be monitored and decisions made enroute based on the incoming information to meet overall mission requirements. Finally, the system health and status monitoring loop represents the continual supervision that must occur, either by a human or automation or both, to ensure all systems are operating within normal limits.

From the human-in-the-loop and UAV mission perspective, if the inner loops fail, then the higher or outer loops will also fail. The dependency of higher loop control on the successful control of the lower loops drives the need for high inner-loop reliability. If UAV operators must interact in the guidance and motion control loop (i.e., fly a UAV through traditional displays to control roll, pitch, and yaw), the cost is high because this effort requires significant

perceptual, cognitive, and manual control resources. Consequently, the chances for pilot error increase significantly, especially because the UAV pilot is on the ground, deprived of the many "in cockpit" cues he or she would normally rely on for situation assessment. Additionally, UAV operators are subject to potentially destabilizing time delays due to remote operations, which could be 4 or more seconds, making inner-loop flight control impossible to achieve when dealing with a responsive airframe. At best, such delays could lead to severe pilot-induced oscillations. Given these issues, if a pilot has to fly a UAV in the innermost loop, little spare mental capacity is available for accomplishing objectives demanded by the outer navigation and mission management control loops.

The insistence that pilots hand-fly UAVs violates the priority scheme represented in Figure 7, in that human operators are required to use most of their cognitive resources to fly the plane. This manual control approach for a highly automated system in combination with poor ground control station interface design have led to numerous Predator crashes (the US Air Force's primary UAV) (Carrigan et al., 2008; Williams, 2004). When operators become saturated at either or both the perceptual-motor or cognitive levels, or do not correctly allocate their cognitive resources to the appropriate control loops in the correct priorities, they fail to properly satisfy the competing demands of the multiple control loop, potentially causing catastrophic failure. As a result, the key to successful UAV operations is twofold: (i) leveraging automation appropriately, so that operators can concentrate on the "outer" mission management loop, and (ii) designing effective ground control interfaces that provide low cognitive overhead, appropriate feedback on mission, platform and payload status, a means for preventing mode confusion, and some way to ensure appropriate trust in the system.

While how much automation is appropriate for the supervision of UAVs is an ongoing debate (Cummings et al., 2007; Dixon and Wickens, 2003; Ruff, Narayanan and Draper, 2002), discussion typically revolves around how much automation should be used in the navigation and mission management loops depicted in Figure 7. The inner-most flight control loop in all UAVs can be completely automated, including automated takeoff and landing. Indeed this technology has existed for some time and has been proven in actual operations, for example, by the Shadow 200, which is an Army UAV flown through a point and click interface by enlisted personnel who are *not* trained pilots. For example, if an operator is told to go to a heading of 150, he can simply command the vehicle through a point and click mechanism (i.e., click on a compass rose heading of 150) instead of actually turning the aircraft through a joystick to the correct heading. By relieving the operator from these low-level joystick actions, the operator's cognitive resources are freed to concentrate on the mission tasks, such as searching for enemy combatants or clearing a convoy route.

The need to shift from the pilot-centric manual control paradigm to the supervisory control architecture as depicted in Figure 4 is not only critical for improvement of current UAV operations where one UAV is typically controlled by a team of personnel, but also to achieve the future vision of one person controlling multiple UAVs. Such architecture will allow agencies more flexibility in manning and missions, as well as overall lower system costs (Johnson, 2003). This move towards a future of one operator controlling many UAVs will mean that the design of displays, particularly the integration of information, and advanced inter- and intra-vehicle autonomy, will be even more critical.

Given then that the operator of a UAV is not faced with the same tasks required of a traditional aircraft pilot (including maintaining stick and rudder skills for "inner loop" control), the question should then be asked, how and to what degree do the interface design variables, principles, and guidelines change? The short answer is, not very much at all! All of the basic principles and guidelines remain the same, since these are applicable to all supervisory control systems. What changes, however, in going from a set of traditional cockpit displays to UAV ground control displays is what constitutes the primary and secondary displays.

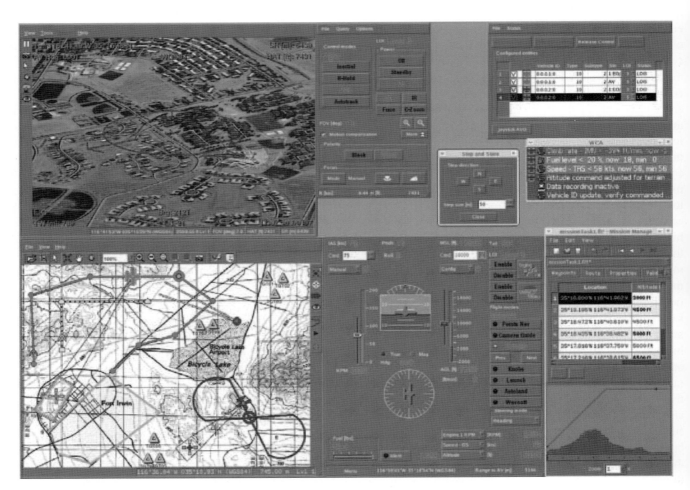

Figure 8. A UAV ground control station interface. (Reproduced with permission from CDL Systems.)

In the cockpits of Figures 3 and 5, the primary displays were those displays that aided the pilot in maintaining stable flight and navigating from point to point. Because a significant amount of this work can be offloaded to automation in UAVs, the conventional instrument panel display is relegated to a secondary display, where the primary display becomes the map and/or the camera view as can been seen in Figure 8. The primary interface in Figure 8, used to fly such UAVs as the Army's Shadow 200, is the map display for geo-spatial awareness, and the camera view, which supports the surveillance mission typical of the majority of current UAV operations.

The most striking feature of the display for a UAV operator in Figure 8 is the secondary priority of the primary flight display. Instead of the integrated PFD seen in Figure 6 that is the central instrument used by pilots, the same information is relegated to a minor panel (lower middle of Figure 8). While the same attitude display is used, it is significantly smaller, and the heading compass now represents an actual compass, which the operator can click on to command new headings. This same point-and-click approach is also used for the altitude and airspeed "sliders," which are similar to the vertical tape representations in Figure 6, but in a UAV display, they are interactive, (i.e., to fly an altitude of 1000 ft, it is simply clicked on with a traditional computer mouse.) This interactive point-and-click approach allows UAV operators to command the onboard autopilot quickly and easily, so that they can allocate their attention to the most important (and largest) windows for the camera image and the geo-spatial location. Thus, displays such as the one in Figure 8 supports the outer two navigation and mission management loops depicted in Figure 7, while offloading the basic inner loop flight control tasks to the automation.

One critical display that is important in both UAV and manned aviation in Figure 8 is the vertical display in the lower right corner, which allows operators the ability to visualize possible vertical obstacles. While the use of three dimensional (3D) displays may seem like a desirable option (i.e., the use of one 3D display instead of two two-dimensional (2D) displays for horizontal and vertical situation awareness as shown in Figure 8), research has shown that 3D displays do not improve UAV operator performance, and actually can cause problems due to the difficulties in supporting detailed judgments (Smallman and St. John, 2005).

5 CONCLUSION

This chapter has been an introductory overview for both manned and unmanned aircraft cockpit and ground station interface design. While the purpose was not to provide a comprehensive "how-to" checklist for design, this chapter demonstrated that regardless of the supervisory control system of interest, there are a set of variables, principles, and guidelines that should be taken into account as early as possible in the systems engineering design phases. Aircraft systems, both manned and unmanned, are becoming increasingly complex, with new cutting-edge display technologies on the horizon such as tunnel-in-the-sky, embedded synthetic views, helmet-mounted displays, retinal displays, and so on. While these technologies show great promise, the responsibility is that of the cockpit/interface design engineer to ensure that such systems actually support operators' needs in the context of overall missions, and to be aware of the complex interactions that can occur when humans interact with these advanced automation technologies.

ACKNOWLEDGMENTS

The Boeing Company and CDL Systems, Ltd. graciously provided the cockpit pictures and interface screenshots.

REFERENCES

Air Accident Investigation & Aviation Safety Board (AAIASB) (2006) *Aircraft Accident Report: Helios Airways Flight HCY522 Boeing 737-31S at Grammatiko, Hellas on 14 August 2005*, Hellenic Republic Ministry of Transport & Communications, Glyfada, Greece.

Bharat (2008) Emerging UUV&UGV Markets: forecasts and business opportunities to 2020, Mumbai India, June 2008.

Billings, C.E. (1991) *Human-Centered Aircraft Automation Philosophy: A Concept and Guidelines*, NASA, Moffett Field, CA.

Billings, C.E. (1997) *Aviation Automation: The Search for a Human-Centred Approach*, Lawrence Erlbaum Associates, Mahwah, NJ.

Boff, K. and Lincoln, J. (1988) *Engineering Data Compendium: Human Perception and Performance*, U.S. Air Force, Wright-Patterson AFB, OH.

Carrigan, G., Long, D., Cummings, M.L., and Duffner, J. (2008) Human factors analysis of predator B crash. *AUVSI 2008: Unmanned Systems North America*. San Diego, CA.

Coombs, L.F.E. (2005) *Control in the Sky: The Evolution and the History of the Aircraft Cockpit*, Pen and Sword Aviation, South Yorkshire.

Cummings, M.L., Bruni, S., Mercier, S., and Mitchell, P.J. (2007) Automation architecture for single operator-multiple UAV command and control. *Int. Command Control J.*, **1**(2), 1–24.

Cummings, M.L. and Tsonis, C. (2006) Partitioning complexity in air traffic management tasks. *Int. J. Aviat.Psychol.*, **16**(3), 277–296.

Dixon, S.R. and Wickens, C.D. (2003) *Imperfect Automation in Unmanned Aerial Vehicle Flight Control*, Institute of Aviation, Savoy, IL.

Erickson, J. and Zacharias, G. (eds) (2004) *Report on Human-System Integration in Air Force Weapon Systems Development and Acquisition: SAB-TR-04-04*.

Hughes, D. (2005) Incidents Reveal Mode Confusion. *Aviation Week & Space Technology*, January 30, 2005.

Johnson, K., Ren, L., Kuchar, J. and Oman, C. (2002) Interaction of automation and time pressure in a route replanning task, in *International Conference on Human-Computer Interaction in Aeronautics* (eds S. Chatty, R.J. Hansman, and G. Boy), AAAI Press, Cambridge, MA.

Johnson, R. (ed) (2003) *Report on Unmanned Aerial Vehicles in Perspective: Effects, Capabilities, and Technologies: SAB-TR-03-01*.

Johnson, S.L. and Roscoe, S.N. (1972) What moves, the airplane or the world. *Hum. Factors*, **14**(2), 107–129.

Layton, C., Smith, P. J. and McCoy, C.E. (1994) Design of a cooperative problem-solving system for en-route flight planning – an empirical evaluation. *Hum. Factors*, **36**(1), 94–116.

McCormick, E.J. (1976) *Human Factors in Engineering and Design*, 4th edn, McGraw-Hill Book Co, New York.

Mosier, K.L. and Skitka, L.J. (1996) Human decision makers and automated decision aids: Made for each other? in *Automation and Human Performance: Theory and Applications* (eds R. Parasuraman and M. Mouloua), Lawrence Erlbaum Associates, Inc, Mahwah, New Jersey.

Nielsen, J. (1993) *Usability Engineering*, Academic Press, Cambridge, MA.

Norman, D.A. (1988) The psychopathology of everyday things. *The Design of Everyday Things*, Doubleday, New York.

NTSB (1973) *Aircraft Accident Report, Eastern Airlines Inc, L-1011 N310EA*, Miami, FL, December 19, 1972. Washington, DC, National Transportation Safety Board.

Parasuraman, R. and Riley, V. (1997) Humans and automation: use, misuse, disuse, abuse. *Hum. Factors*, **39**(2), 230–253.

Petroski, H. (1992) *To Engineer Is Human*, Vintage Books, New York.

Roscoe, S.N. (1968) Airborne displays for flight and navigation. *Hum. Factors*, **10**(4), 321–332.

Ruff, H., Narayanan, S. and Draper, M.H. (2002) Human interaction with levels of automation and decision-aid fidelity in the supervisory control of multiple simulated unmanned air vehicles. *Presence*, **11**(4), 335–351.

Sarter, N.B. and Woods, D.D. (1994) Decomposing automation: autonomy, authority, observability and perceived animacy, in *Human Performance in Automated Systems: Recent Research and Trends* (eds M. Mouloua and R. Parasuraman), Lawrence Erlbaum Associates, Hillsdale, NJ, pp. 22–27.

Sheridan, T.B. (1992) *Telerobotics, Automation and Human Supervisory Control*, The MIT Press, Cambridge, MA.

Smallman, H.S. and St. John, M. (2005) Naïve realism: misplaced faith in the utility of realistic displays. *Ergon. Des.*, **13**(3), 6–13.

Tsang, P. and Vidulich, M.A. (2003) *Principles and Practice of Aviation Psychology*, Lawrence Erlbaum Associates, Mahwah, NJ.

Wickens, C.D. and Carswell, C.M. (1995) The proximity compatability principle: its psychological foundation and relevance to display design. *Hum. Factors*, **37**(3), 473–494.

Wickens, C.D. and Hollands, J.G. (2000) *Engineering Psychology and Human Performance*, 3rd edn, Prentice-Hall Inc., Upper Saddle River, NJ.

Wiener, E.L. and Nagel, D.C. (1988) *Human Factors in Aviation*, Academic Press, San Diego, CA.

Williams, K.W. (2004) *A Summary of Unmanned Aircraft Accident/Incident Data: Human Factors Implications*, Civil Aerospace Medical Institute, Federal Aviation Administration, Washington DC.

Xing, J. (2006) *Color and Visual Factors in ATC Displays*, FAA Civil Aerospace Medical Institute, Washington DC.

PART 8
Multi-Vehicle Cooperation and Coordination

Chapter 26
Multi-UAV Cooperation

Iván Maza,[1] **Jesús Capitán,**[1] **Luis Merino,**[2] **and Aníbal Ollero**[3]

[1]*Grupo de Robótica, Visión y Control, Universidad de Sevilla, Seville, Spain*
[2]*Grupo de Robótica, Visión y Control, Universidad Pablo de Olavide, Seville, Spain*
[3]*Universidad de Sevilla and Scientific Advisory Department of the Center for Advanced Aerospace Technologies, Seville, Spain*

1 INTRODUCTION

The range of applications of unmanned aerial vehicles (UAVs) can be widened if teams of multiple UAVs are considered. The cooperation between the vehicles of the team allows them to accomplish some tasks that they could not perform alone, reduce the time and/or space required to achieve some missions, or enhance the robustness of the system as a whole. For instance, one of the limitations of small and micro aerial vehicles (MAVs) is their reduced payload. However, a team of MAVs may be used instead to overcome these limitations (Kushleyev *et al.*, 2013). In the last years, several demonstrations have shown the possibilities in the civilian arena (see Figure 1).

Cooperation entails many aspects, from perception to decision making, including coordination and explicit cooperation between the vehicles of the team. This chapter considers the case of small and medium size teams in which each vehicle has a high level of autonomy and explicitly cooperate with its teammates. Swarm techniques (De Nardi *et al.*, 2006), in which cooperative emergent behaviors can arise from simple single-vehicle behaviors, are not considered here.

In the chapter, different approaches employed in the literature are discussed. First, multi-UAV architectures for cooperation are presented. Then, different relevant issues for multi-UAV cooperation are considered: in particular, perception and decision making. The chapter also shows an example of the application of a multi-UAV team for cooperative tracking.

2 MULTI-UAV ARCHITECTURES FOR COOPERATION

In the first part of this section, the concepts of coordination and cooperation are briefly presented due to their relevance in any system with multiple autonomous vehicles. Then, a classification based on the coupling between the vehicles is outlined.

2.1 Coordination and cooperation

In platforms involving multiple vehicles, the concepts of coordination and cooperation play an important role. In general, coordination deals with the sharing of resources, and both temporal and spatial coordination should be considered. The temporal coordination relies on synchronization

Unmanned Aircraft Systems. Edited by Ella Atkins, Aníbal Ollero,
Antonios Tsourdos, Richard Blockley and Wei Shyy.

Figure 1. A team of helicopters and an airship were demonstrated in forest fire-fighting activities in the EU-funded COMETS Project (Ollero *et al.*, 2005).

among the different vehicles and it is required in a wide spectrum of applications. For instance, for objects monitoring, several synchronized perceptions of the objects could be required. In addition, spatial coordination of UAVs deals with the sharing of the space among them to ensure that each UAV will be able to perform safely and coherently regarding the plans of the other UAVs, and the potential dynamic and/ or static obstacles.

Regarding cooperation, according to Cao *et al.*, (1997), given some task specified by a designer, a multiple-robot system displays cooperative behavior if, due to some underlying mechanism (i.e., the "mechanism of cooperation"), there is an increase in the total utility of the system. The cooperation of heterogeneous vehicles requires the integration of sensing, control, and planning in an appropriated decisional architecture. These architectures can be either centralized or decentralized depending of the assumptions on the knowledge's scope and accessibility of the individual vehicles, their computational power, and the required scalability. A centralized approach will be relevant if the computational capabilities are compatible with the amount of information to process, and the exchange of data meets both the requirements of speed (up-to-date data) and expressivity (quality of information enabling well-informed decision-taking). On the other hand, a distributed approach will be possible if the available knowledge within each distributed vehicle is sufficient to perform "coherent" decisions, and this required amount of knowledge does not endow the distributed components with the inconveniences of a centralized system (in terms of computation power and communication bandwidth requirements). One way to ensure that a minimal global coherence will be satisfied within the whole system is to enable communication between the vehicles of the system, up to a level that will guarantee that the decision is globally coherent. One of main advantages of the distributed approach

relies on its superior suitability to deal with the scalability of the system.

2.2   Classification of multi-UAV architectures

Multi-UAV systems can be classified from different points of view. One possible classification is based on the coupling between the UAVs:

1. *Physical coupling.* In this case, the UAVs are connected by physical links and then their motions are constrained by forces that depend on the motion of other UAVs. The lifting and transportation of loads by several UAVs lies in this category (Bernard *et al.*, 2011). The main problem is the motion coordinated control taking into account the forces constraints. From the point of view of motion planning and collision avoidance, all the members of the team and the load can be considered as a whole. As the number of vehicles is usually low, both centralized and decentralized control architectures can be applied.

2. *Formations.* The vehicles are not physically coupled but their relative motions are strongly constrained to keep the formation. Then, the motion planning problem can be also formulated considering the formation as a whole. Regarding the collision avoidance problem within the team, it is possible to embed it in the formation control strategy. Scalability properties to deal with formations of many individuals are relevant and then, decentralized control architectures are usually preferred (Turpin *et al.*, 2012). Techniques in this area can be applied to control coordinated motions of vehicles even if they are not in formation.

3. *Swarms.* They are homogeneous teams of many vehicles in which interactions generate emerging collective behaviors. The resulting motion of the vehicles does not

lead necessarily to formations. Scalability is a main issue due to the large number of vehicles involved and then pure decentralized control architectures are mandatory.

4. *Intentional cooperation.* The UAVs of the team move according to trajectories defined by individual tasks that should be allocated to perform a global mission (Viguria *et al.*, 2010). These UAV trajectories typically are not geometrically related as in the case of the formations. In this case, problems such as multi-UAV task allocation, high-level planning, plan decomposition, and conflict resolution should be solved taking into account the global mission to be executed and the different UAVs involved. Both centralized and decentralized decisional architectures can be applied.

The rest of this section focuses on multi-UAV architectures for intentional cooperation, showing different examples.

2.3 Intentional cooperation multi-UAV architectures

In the intentional cooperation approaches, each individual executes a set of tasks (subgoals that are necessary for achieving the overall goal of the system, and that can be achieved independently of other subgoals) explicitly allocated to perform a given mission in an optimal manner according to planning strategies (Viguria *et al.*, 2010). The UAVs cooperate explicitly and with purpose, but also has the limitation of independent subgoals: if the order of task completion is mandatory, additional explicit knowledge has to be provided to state ordering dependencies in the preconditions of those tasks. It is also possible to follow a design based on "collective" interaction, in which entities are not aware of other entities in the team, yet they do share goals, and their actions are beneficial to their teammates (Parker, 2008).

Teams composed by heterogeneous members involve challenging aspects, even for the intentional cooperation approach. Ollero and Maza (2007) study the existing problems and potentialities of platforms with multiple UAVs (with emphasis on systems composed by heterogeneous UAVs). This heterogeneity is twofold: firstly in the UAV platforms looking to exploit the complementarities of the aerial vehicles, such as helicopters and airships, and secondly in the information processing capabilities on-board, ranging from pure remotely teleoperated vehicles to fully autonomous aerial robots.

The multi-UAV coordination and control architecture developed in the EU-funded COMETS Project was demonstrated for the autonomous detection and monitoring of fires (Merino *et al.*, 2006) by using two helicopters and one

airship. Regarding teams involving aerial and ground vehicles, the CROMAT architecture (Viguria *et al.*, 2010) also implemented cooperative perception and task allocation techniques that have been demonstrated in fire detection, monitoring and extinguishing. Multi agent (combined ground and air) tasking and cooperative target localization has been also demonstrated (Hsieh *et al.*, 2007).

A distributed architecture for the autonomous coordination and cooperation of multiple UAVs for civil applications was developed in the framework of the EU-funded AWARE Project (Maza *et al.*, 2011). The architecture is endowed with different modules that solve the usual problems that arise during the execution of multi purpose missions, such as task allocation, conflict resolution, complex task decomposition, and so on. One of the main objectives in the design of the architecture was to impose few requirements to the execution capabilities of the UAVs to be integrated in the platform. Basically, those vehicles should be able to move to a given location and activate their payload when required. Thus, heterogeneous autonomous vehicles from different manufacturers and research groups can be integrated in the architecture developed, making it easily usable in many multi-UAV applications. The software implementation of the architecture was tested in simulation and finally validated in field experiments with four autonomous helicopters (see Figure 2). The

Figure 2. Autonomous helicopters (Technische Universität Berlin) used for the demonstration of the multi-UAV architecture developed in the framework of the EU funded AWARE Project.

validation process included several multi-UAV missions for civil applications in a simulated urban setting: surveillance; fire confirmation, monitoring, and extinguishing; load transportation and deployment with single and multiple UAVs; and people tracking.

In the above-mentioned architectures, two key elements can be identified: perception capabilities and autonomous decision-making in a multi-UAV context. Cooperative perception can be considered as an important tool in many applications based on multi-UAV schemes. It can be defined as the task of creating and maintaining a consistent view of a world containing dynamic objects by a group of UAVs each equipped with one or more sensors. Thus, a team of vehicles can simultaneously collect information from multiple locations and exploit the information derived from multiple disparate points to build models that can be used to take decisions. Section 3 details cooperative perception in further detail. Finally, the "decision-making" process, regarding intelligent systems, deals with different mechanisms focusing on the autonomous and coherent processing of a mission (ranging from simple requests, to complex sequences of high-level tasks), either within a centralized or decentralized system. Key aspects related to decision making are discussed in Section 4.

3 COOPERATIVE PERCEPTION

One of the main advantages of a multi-UAV team is the possibility to leverage a data-fusion framework that combines information provided by the different sensors on board the vehicles of the team. This way, it is possible to enhance the final perception of the team, reducing the uncertainties due to sensor noise, occlusions, and so on. In the last years, there have been several works by different researchers on cooperative perception architectures for teams of UAVs.

These architectures can be classified according to the underlying representations employed, and there are different approaches for that. For instance, Yanli et al. (2005) present a multi-UAV map-building approach based on Dempster-Shafer theory of evidence for information fusion. The objective is that a group of UAVs builds a certainty grid about potential targets on an area.

However, most works on data fusion using teams of aerial vehicles are based on Bayesian approaches, whereby the sensors are modeled as uncertain sources. Different representations and filters within a Bayesian framework are considered, depending on the application. Evidence grids are employed in multi-UAV search and exploration missions (Flint et al., 2004). In Bertuccelli and How (2006), Beta probabilities are used to encode uncertain information about prior knowledge provided by mission designers.

Gaussian representations and their associated Kalman and Information filters are, however, the most used for cooperative tracking and Simultaneous Localization and Mapping (SLAM) applications, as, for instance, in Sukkarieh et al. (2003), Merino et al. (2006), Maza et al. (2011), and Stachura and W. Frew (2011).

As second characterization of the different architectures is their level of centralization. A first potential solution would be a centralized scheme, in which each sensor just sends all its measurements to a central node where the data fusion is performed (Merino et al., 2006; Yanli et al., 2005). However, this architecture presents some drawbacks. These drawbacks include: (i) high bandwidth requirements, especially for transmission of high-frequency motion data; (ii) limited range, since each sensor should be within communication range of the central node; and (iii) robustness issues, because a failure in the central node entails that the whole system fails.

Therefore, an alternative are decentralized architectures. Within these decentralized approaches, each node of the cooperative perception system only uses information from local sensors and shares its estimations with others nodes of the fleet, without any knowledge of the topology of the whole UAV fleet or of the broadcast facilities. Thus, the need for a central node is eliminated and, since only local communications take place, scalability is achieved. Moreover, the different members of the multi-UAV platform can work more independently without the need to keep continuously communication range with a central node.

For instance, Sukkarieh et al. (2003) present a decentralized information fusion system for multi-UAV mapping; Beard et al. (2006) employ decentralized consensus techniques for cooperative aerial surveillance; while Capitan et al. (2011) present an scalable ·cooperative perception system involving multiple UAVs and sensor networks for object tracking in the context of disaster scenarios Maza et al., (2011).

4 DECISION MAKING IN A MULTI-UAV CONTEXT

In Section 2, it has been explained how an architecture for cooperation with multiple UAVs is made of several components. Apart from the components in charge of the cooperative perception (Section 3), the component for planning or making decisions is also quite relevant. If cooperation is sought during a mission, UAVs should implement decision-making algorithms in order to follow behaviors, allocate tasks, perform actions, and so on, in a collaborative manner.

The techniques employed for multi-UAV cooperation stem in many cases from decision making and control algorithms developed in the field of multi-robot cooperation. This section intends to give an overview of this state-of-art classifying works according to several criteria.

4.1   Different formulations

A first criterion to classify planning approaches could be the formulation that is given to the problem. According to this, a decision-making problem with multiple UAVs could be formally described as a discounted infinite-horizon dynamic programming problem with deterministic models, and solved by means of conventional optimization techniques such as gradient methods, Simplex or convex optimization theory (Derenick et al., 2007; Beard et al., 2006).

However, for any relatively complex scenario, these deterministic techniques become intractable and alternative techniques, such as heuristic (Parker, 2002; Pirjanian and Mataric, 2000) ones can be used. These heuristic methods exploit additional information from the particular environment or application in order to enhance the behavior of the team (e.g., information about the capabilities of the vehicles participating in the mission). There are also many probabilistic approaches for decision-theoretic planning under uncertainties (Wong et al., 2005). These methods model uncertainties in the scenario and reason about them, being usually more flexible than heuristic methods.

4.2   Centralized versus decentralized

As already discussed in Sections 2 and 2.1, another relevant feature to classify works on decision making with multiple UAVs could be the distribution of the algorithm. The final goal is always to solve a control/planning problem in some optimal manner. However, this control problem can be solved in a centralized or decentralized scheme. On the one hand, in a fully centralized approach (Derenick et al., 2007), a single control problem is solved for the whole team. There is a central node that is able to access all the information from the vehicles. On the other hand, fully decentralized approaches (Waslander et al., 2003; Wong et al., 2005; Beard et al., 2006; Choi et al., 2009) solve a different controller for each UAV, so the algorithm is totally distributed. In the middle, semi decentralized approaches (Gancet et al., 2005) could also be considered. They exploit the distributed computational capacity that multiple aerial platforms offer but they still require a central node to fuse information or resolve global constraints.

In the decentralized case, further classifications can be made with regard to the information that the vehicles share.

Despite being decentralized, some algorithms require that some global information is available for all the UAVs; whereas in others, UAVs are just required to access local information. The need to maintain global sensor and vehicle models or any other global information can make a difference. In Grocholsky et al. (2003), for instance, a distinction is made between cooperative and coordinated solutions. In the former, predicted optimal negotiated group decisions are taken. In the latter, however, there is no negotiation or need to share global information or models. Even though decision makers act locally and do not have knowledge about the rest of UAVs' models or control actions, they exchange some information that may influence implicitly other UAVs' subsequent decisions.

4.3   Abstraction level

Decision-making algorithms can also be distinguished by the level of abstraction they consider. Thus, many of them act directly over control inputs of the UAVs, whereas others work in a higher level of abstraction where the commands to the UAVs are tasks or behaviors. This second branch includes all the multitask allocation algorithms, which distribute a set of tasks among a set of UAVs in some optimal manner (Bethke et al., 2008). Of course, tasks are no longer basic control actions but higher-level commands, such as visiting certain way-point points or following some mission-related behaviors.

Marked-based techniques (Viguria et al., 2010) are very well-known approaches in the literature for multi-UAV task allocation. The system performs a negotiation based on market rules where each task has an associated cost and a reward (which can depend on the type of UAV). Finally, tasks are allocated in such a way that a goal function is optimized. Furthermore, these techniques are quite suitable for distributed platforms since tasks and resources can be allocated and then, performed by different teammates. Despite this, a centralized entity with global knowledge about all the vehicles and their features is usually required in order to distribute the tasks optimally. Due to this, many of these algorithms can be seen as semi-decentralized approaches.

As seen above, some methods work directly over control inputs and others over a higher-level workspace. Nevertheless, some techniques could be placed in a middle position, since they exploit features from both branches. This is the case of POMDPs (*Partially Observable Markov Decision Processes*) (Capitan et al., 2014), which provide an elegant way to model the interaction of a UAV with its environment. Based on prior knowledge of the sensor models and the environment dynamics, policies which indicate the UAV how to act optimally can be computed.

Therefore, the abstraction level of the actions which POMDPs deal with can vary. A POMDP could be solved to determine a set of low-level control inputs over a UAV (e.g., translational and rotational speed), or on the contrary, the actions could be modeled as higher-level tasks (e.g., patrol a specific area, track a specific target, etc.). In fact, within the same framework, both low-level and high-level actions, can be modeled without changing the formulation. Furthermore, POMDPs allow the system to combine different kinds of objectives at the same time, just by optimizing a merged goal function.

4.4 Time horizon

The time horizon is another interesting consideration for decision-making algorithms. Control algorithms usually try to calculate an optimal action to take in the next time step. However, the optimization problem could consider predicted future steps to evaluate the goal function along a time horizon. Otherwise, when the optimization is carried out just over the current time step, the algorithm is called *greedy*.

Many of the previous works apart from POMDPs perform optimization in a greedy manner, that is, they do not plan forward, so the time horizon is reduced to the current step. However, POMDP techniques can work with a variable time horizon and long-term goals in order to learn a policy. This is more than just calculating the current optimal action, since a policy describes the behavior of a UAV along a wide variety of situations.

4.5 Classification

Throughout this section, several criteria have been used in order to differentiate among a wide range of approaches for cooperative perception and decision making with UAVs. Table 1 depicts the features used here to establish a possible classification. The *formulation* of the algorithm indicates the nature of the mathematical framework used by each approach. Thus, conventional optimization techniques with deterministic models have been compared to heuristic and probabilistic techniques. Moreover, the analysis can also be focused on the degree of *distribution* that the algorithms present. In this sense, centralized, semidecentralized, and decentralized approaches have been cited. The *abstraction level* is highly relevant as well. Algorithms can be aimed at dealing with control actions at different levels, from low-level control commands to high-level task definitions. Finally, according to their *time horizon*, greedy and variable-horizon algorithms have been described.

5 APPLICATION: PEOPLE TRACKING WITH MULTIPLE UAVS

In this section, the issues discussed above are illustrated through a particular example of cooperation of multiple UAVs for people tracking. This application was one of the objectives of the previously cited AWARE project (EU-funded), and consisted of a team of UAVs trying to track firemen in a rescue scenario. In particular, an architecture for multi-UAV cooperation developed within the framework of the project (Maza *et al.*, 2011) was used during the experiments shown.

In this tracking application, two firemen were located in an area in front of a building assisting injured people and moving equipment. The objective of the system was to have an estimation of the location of the firemen on the map at all times, and also to provide images of their operations. Two UAVs were available and ready on the landing pads for this mission, both equipped with fixed visual cameras aligned with the fuselage and pointing downwards 45°.

Table 1. Summary of approaches for decision making with multiple UAVs and some related references.

Formulation	Distribution	Abstraction	Horizon	Examples
Deterministic	Centralized	Low-level control	Greedy	Derenick *et al.* (2007)
Heuristic	Semi-decentralized	Low-level control	Greedy	Gancet *et al.* (2005)
Heuristic	Cooperative	Low-level control	Greedy	Parker (2002)
Heuristic	Cooperative	Task level	Greedy	Pirjanian and Mataric (2000)
Probabilistic	Cooperative	Low-level control	Greedy	Wong *et al.* (2005)
Deterministic	Cooperative	Low-level control	Greedy	Waslander *et al.* (2003)
Deterministic	Cooperative	Low-level control	Variable	Beard *et al.* (2006)
Probabilistic	Coordinated	Low-level control	Variable	Grocholsky *et al.* (2003)
Heuristic	Semi-decentralized	Task level	Greedy	Viguria *et al.* (2010)
Probabilistic	Variable	Variable	Variable	POMDPs (Capitan *et al.*, 2014)

The firemen were equipped with sensor nodes, and a cooperative perception approach (Capitan *et al.*, 2011) was used to obtain an estimation of their location based on the measurements from a wireless sensor network (WSN) deployed in front of the building and the visual images of a ground camera network. Later, when the UAVs took off, this information was also fused online with the estimations computed from the visual images gathered by the helicopters in order to decrease the uncertainty in the location of the firemen.

In this mission, two tasks were sent to the UAVs at different times in order to track the operations of the fireman labeled as fireman2. The task are allocated automatically to the vehicles by the decision-making modules using a distributed negotiation process based on the SIT algorithm described by Viguria *et al.*, (2007).

- The first task was announced and allocated to UAV 2 due to its lowest bid (lowest insertion cost) in the negotiation process. In order to compute the insertion cost for the tracking task, the required associated waypoint and heading were computed. Once the waypoint was reached, the UAV captured images from the fireman and processed them in order to contribute to the estimation of his position. UAV 2 broadcast its position relative to the tracked object location; if more tracking tasks for the same object were commanded, the next UAVs will need this information to compute their positions around the object accordingly.
- Later, the second task was announced and allocated to UAV 1 (UAV 2 bid with infinite cost because it was already tracking the same object). In general, if more than one UAV is commanded to monitor the same object, the second and next UAVs will consider the places already occupied around that object.

The trajectories followed by the UAVs during the execution of the mission are shown in Figure 3. The waypoints labelled as wp1 and wp4 are initial locations defined by the user for each UAV after take off. The WSN and ground cameras of the platform started to provide estimations from the fireman labelled as fireman2. Based on those estimations, the tracking tasks can be sent to the UAVs. In the case of UAV 1, the computed waypoint wp2 for the observation was very close to the first waypoint, but the heading was different.

The decentralized cooperative perception system is fed with all the sensor information provided by the WSN, camera network, and UAV cameras. This system is based on Information Filters that estimate the full trajectory of the persons being tracked. The decentralized filters launched a prediction/updating step every 0.2 s and included the local and

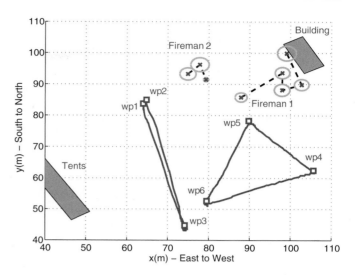

Figure 3. Paths followed by the two helicopters during the people tracking mission. The trajectories correspond to UAVs 1 and 2. The different waypoints are represented by squares. Some samples of the firemen trajectories during the experiment are also shown: crosses denote the estimated position of the firemen and ellipses stand for their uncertainty.

neighbor measurements when available. Ground cameras and UAVs provided local measurements at 1–3 Hz approximately, while the WSN updated local measurement at 10 Hz. All this information generated the estimation presented in Figure 4 for firemen 1 and 2.

A centralized filter processing offline all the data was used to provide the ground truth of the fireman tracking. This centralized estimation is not affected by communications delays, double counting of information, asynchronous data, or partial information. The comparison with the position of the firemen estimated by means of the centralized filter is also shown in the figure. Notice that the estimated standard deviation from the filter is coherent with the errors and is always within the 3σ confident interval. The estimation in the Z-axis is not shown because throughout the experiment the firemen stayed on the ground, so the analysis of the results do not provide significant information about the performance of the perception system. It can also be seen that the average localization error for fireman 1 (top figure) is about 1 m while for fireman 2 (bottom figure) it is about 1.5 m during the entire trajectory. This error can be considered small if all the error sources are taken into account, that is, errors in the camera calibration, position, orientation, and communications issues.

As an example, the standard deviation computed by the decentralized approach and the estimation with the centralized filter for fireman 2 are shown in Figure 5. As expected, the decentralized approach presents more conservative

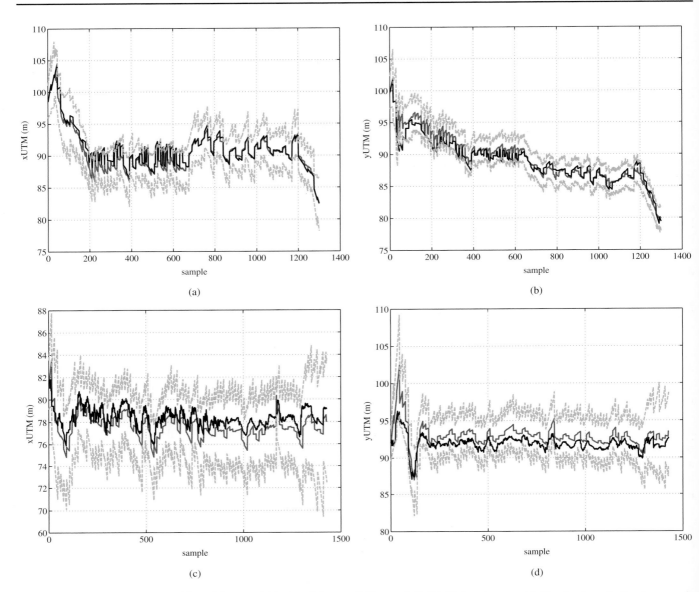

Figure 4. Estimated X and Y position of fireman 1 (Top) and fireman 2 (Bottom). Solid: Estimation provided by the decentralized approach. Dashed: Estimated 3σ confidence interval of the decentralized approach. The centralized estimation used as ground-truth is also shown.

estimations than the centralized filter because of communications issues, the Covariance Intersection algorithm, and the fact that the decentralized approach cannot access all the information at the same time as the centralized filter does. However, it is worth mentioning the closeness of the two estimations, which differ by 1 m approximately. In general, the standard deviation grows during the prediction steps and decreases after updating with local or external information showing a high-frequency component. Since the centralized filter is accessing all of the information at every time step, it is updated more often, presenting a lower variation over the

mean standard deviation. In the decentralized case, the effect of the longer prediction periods can be seen.

6   CONCLUSIONS

This chapter has discussed architectures for multi-UAV cooperation. Perception and decision making are two of the main issues that such architectures have to consider. The chapter offers a description of the main approaches employed in the literature, and presents as an example an

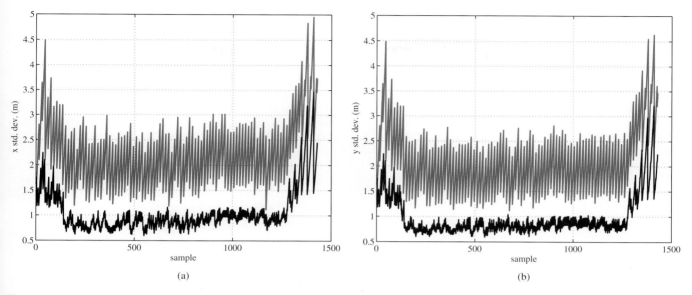

Figure 5. Estimated standard deviation in X and Y for fireman 2. Gray: Estimation using the decentralized approach. Black: Centralized estimation

application in which a team of UAVs cooperatively track firemen in a disaster management scenario, and which involves all these elements.

The chapter shows how teams of cooperative UAVs can offer an added value for certain applications. Cooperative perception techniques allow the fleet to enhance its situation awareness, acquiring a global knowledge that is more informative than the sum of the individual separate perceptions of the UAVs. Intelligent decision making can reduce time, distance, and enhance the robustness of the system for many applications. The chapter will allow the reader to explore the alternative solutions for the issues mentioned that can be found in the literature.

ACKNOWLEDGMENTS

This work has been partially supported by the FP7 EC-SAFEMOBIL Project (FP7-ICT-288082) funded by the European Commission, and the PAIS-MultiRobot Project (TIC-7390) funded by Regional Government of Andalucia

REFERENCES

Beard, R., McLain, T., Nelson, D., Kingston, D., and Johanson, D. (2006) Decentralized cooperative aerial surveillance using fixed-wing miniature UAVs. *Proc. IEEE*, **94**(7):1306–1324.

Bernard, M., Kondak, K., Maza, I., and Ollero, A. (2011) Autonomous transportation and deployment with aerial robots for search and rescue missions. *J. Field Robot.*, **28**(6):914–931.

Bertuccelli, L. and How, J. (2006) Search for dynamic targets with uncertain probability maps, in *Proceedings of the American Control Conference*, 737–742.

Bethke, B., Valenti, M., and How, J. (2008) UAV task assignment. *IEEE Robot. Autom. Mag.*, **15**(1):39–44.

Cao, Y.U., Fukunaga, A.S., and Kahng, A. (1997) Cooperative mobile robotics: antecedents and directions. *Auton. Robots*, **4**(1):7–27.

Capitan, J., Merino, L., Caballero, F., and Ollero, A. (2011) Decentralized delayed-state information filter (DDSIF): a new approach for cooperative decentralized tracking. *Robot. Auton. Syst.*, **59**:376–388.

Capitan, J., Merino, L., and Ollero, A. (2014) Decentralized cooperation of multiple uas for multi-target surveillance under uncertainties, in *2014 International Conference on Unmanned Aircraft Systems (ICUAS)*, 1196–1202.

Choi, H.-L., Brunet, L., and How, J. (2009) Consensus-based decentralised auctions for robust task allocation. *IEEE Trans. Robot.*, **25**:912–926.

De Nardi, R., Holland, O., Woods, J., and Clark, A. (2006) SwarMAV: a swarm of miniature aerial vehicles, in *Proceedings of the 21st Bristol International UAV Systems Conference*.

Derenick, J., Spletzer, J., and Hsieh, M. (2007) A graph theoretic approach to optimal target tracking for mobile robot teams, in *Proceedings of International Conference on Intelligent Robots and Systems*, pp. 3422–3428.

Flint, M., Fernandez, E., and Polycarpou, M. (2004) Efficient bayesian methods for updating and storing uncertain search information for UAVs, in *Proceedings of the IEEE Conference on Decision and Control*, pp. 1093–1098.

Gancet, J., Hattenberger, G., Alami, R., and Lacroix, S. (2005) Task planning and control for a multi-UAV system:

architecture and algorithms, in *Proceedings of the IEEE/RSJ International Conference on Intelligent Robots and Systems*, pp. 1017–1022.

Grocholsky, B., Makarenko, A., Kaupp, T., and Durrant-Whyte, H. (2003) Scalable control of decentralised sensor platforms. *Information Processing in Sensor Networks* (eds. F. Zhao and L. Guibas), *Vol. 2634 of Lecture Notes in Computer Science*, Springer, Berlin, pp. 551–551.

Hsieh, M.A., Chaimowicz, L., Cowley, A., Grocholsky, B., Keller, J.F., Kumar, V., Taylor, C.J., Endo, Y., Arkin, R.C., Jung, B., Wolf, D.F., Sukhatme, G., and MacKenzie, D.C. (2007) Adaptive teams of autonomous aerial and ground robots for situational awareness. *J. Field Robot.*, **24**(11):991–1014.

Kushleyev, A., Mellinger, D., Powers, C., and Kumar, V. (2013) Towards a swarm of agile micro quadrotors. *Auton. Robots*, **35**(4):287–300.

Maza, I., Caballero, F., Capitan, J., de Dios, J.M., and Ollero, A. (2011) A distributed architecture for a robotic platform with aerial sensor transportation and self-deployment capabilities. *J. Field Robot.*, **28**(3):303–328.

Merino, L., Caballero, F., de Dios, J.M., Ferruz, J., and Ollero, A. (2006) A cooperative perception system for multiple UAVs: application to automatic detection of forest fires. *J. Field Robot.*, **23** (3–4):165–184.

Ollero, A., Lacroix, S., Merino, L., Gancet, J., Wiklund, J., Remuss, V., Perez, I.V., Gutierrez, L.G., Viegas, D.X., Gonzalez, M.A., Mallet, A., Alami, R., Chatila, R., Hommel, G., Colmenero, F.J., Arrue, B.C., Ferruz, J., Martinez-de Dios, J., and Caballero, F. (2005) Multiple eyes in the skies: architecture and perception issues in the COMETS unmanned air vehicles project. *IEEE Robot. Autom. Mag.*, **12**(2):46–57.

Ollero, A. and Maza, I., editors (2007) Multiple heterogeneous unmanned aerial vehicles. *Springer Tracts on Advanced Robotics*. Springer.

Parker, L. (2002) Distributed algorithms for multi-robot observation of multiple moving targets. *Auton. Robots*, (12):231–255.

Parker, L. (2008) Distributed intelligence: overview of the field and its application in multi-robot systems. *J. Phys. Agents*, **2**(1), 5–14.

Pirjanian, P. and Mataric, M. (2000) Multi-robot target acquisition using multiple objective behavior coordination, in *Proceedings of the IEEE International Conference on Robotics and Automation*, Vol. **3**, pp. 2696–2702.

Stachura, M. and W. Frew, E. (2011) Cooperative target localization with a communication-aware unmanned aircraft system. *J. Guid. Control Dynam.*, **34**:1352–1362.

Sukkarieh, S., Nettleton, E., Kim, J.-H., Ridley, M., Goktogan, A., and Durrant-Whyte, H. (2003) The ANSER project: data fusion across multiple uninhabited air vehicles. *Int. J. Robot. Res.*, **22** (7–8):505–539.

Turpin, M., Michael, N., and Kumar, V. (2012) Decentralized formation control with variable shapes for aerial robots, in *2012 IEEE International Conference on Robotics and Automation (ICRA)*, pp. 23–30.

Viguria, A., Maza, I., and Ollero, A. (2007) SET: an algorithm for distributed multirobot task allocation with dynamic negotiation based on task subsets, in *Proceedings of the IEEE International Conference on Robotics and Automation*, Rome, Italy, pp. 3339–3344.

Viguria, A., Maza, I., and Ollero, A. (2010) Distributed service-based cooperation in aerial/ground robot teams applied to fire detection and extinguishing missions. *Adv. Robot.*, **24** (1–2):1–23.

Waslander, S.L., Inalhan, G., and Tomlin, C.J. (2003) Decentralized Optimization via Nash Bargaining, in *Theory and Algorithms for Cooperative Systems*, pp. 565–583.

Wong, E.-M., Bourgault, F., and Furukawa, T. (2005) Multi-vehicle Bayesian search for multiple lost targets, in *Proceedings of the IEEE International Conference on Robotics and Automation*, pp. 3169–3174.

Yanli, Y., Minai, A., and Polycarpou, M. (2005) Evidential map-building approaches for multi-UAV cooperative search, in *Proceedings of the American Control Conference*, pp. 116–121.

Chapter 27

Coordinated Standoff Tracking of Moving Ground Targets Using Multiple UAVs

Hyondong Oh,[1] Seungkeun Kim,[2] and Antonios Tsourdos[3]

[1]*Department of Aeronautical and Automotive Engineering, Loughborough University, Loughborough, UK*
[2]*Department of Aerospace Engineering, Chungnam National University, Daejeon, Republic of Korea*
[3]*School of Aerospace, Transport & Manufacturing and Centre for Autonomous and Cyber-Physical Systems, Cranfield University, Cranfield, UK*

1 INTRODUCTION

In the last two decades, using a team of UAVs has received increasing attention to accomplish a higher level of missions in both military and civil applications. Searching and subsequent tracking of moving ground target of interest is one of the important capabilities of UAVs required to increase an overall knowledge of target's intent and to take proactive measures. Persistent tracking of a ground target is not an easy task due to unknown movements of the target as well as kinematic constraints of the UAV. For this, a standoff tracking concept is introduced for the UAV, which is to keep a certain distance (termed standoff distance) from the

moving target, resulting in orbiting around the target due to the possible speed superiority of the UAV over the ground targets. Coordinated standoff tracking is also proposed by distributing a team of UAVs on the same standoff orbit in order to obtain better estimation accuracy as well as more robust performance against sensing failure of individual sensors or obscuration of the target.

Lawrence (2003) first proposed the application of Lyapunov vector fields for standoff coordination of multiple UAVs, which was further investigated by Frew, Lawrence, and Morris (2008), Summers, Akella, and Mears (2009), and Lim *et al.* (2013) to include phase-keeping between UAVs as well as standoff distance tracking. They invented a decoupled control structure in which speed and rate of heading change are separately controlled for standoff distance and phase angle keeping, respectively. Quigley *et al.* (2005) applied the concept of Hopf bifurcation to standoff tracking that has faster convergence to the standoff orbit than the Lyapunov vector field approach. Chen (2013) proposed the use of a tangent-plus-Lyapunov vector field that includes a simple switching logic between tangent and Lyapunov vector fields to make convergence to the standoff orbit faster. Yoon, Park, and Kim (2013) used the concept of spherical pendulum stabilization to obtain the standoff tracking guidance commands. Wise and Rysdyk (2006) well surveyed and compared the different methodologies for standoff tracking: Helmsman behavior, Lyapunov vector field, controlled collective motion, and model predictive control. Prevost *et al.* (2009) also applied the model-based predictive control for standoff tracking. Although the aforementioned two works

Unmanned Aircraft Systems. Edited by Ella Atkins, Aníbal Ollero, Antonios Tsourdos, Richard Blockley and Wei Shyy.
© 2016 John Wiley & Sons, Ltd. ISBN: 978-1-118-86645-0.

tried to apply the model predictive control (also known as receding horizon control) to the standoff tracking problem, Wise and Rysdyk (2006) used a simple relative kinematic variation between UAV and a target while still decoupling speed and heading change, and Prevost *et al.* (2009) focused on a single UAV application for only the standoff distance keeping with simple UAV/target trajectory prediction.

Recently, another model predictive standoff tracking approach was proposed by Kim, Oh, and Tsourdos (2013) to achieve an optimal performance in terms of tracking accuracy as well as control efforts for a pair of UAVs by combining speed and heading control. To ensure the safety under realistic situations, inequality constraints were also considered for collision avoidance between UAV members and control input saturations in the model predictive control scheme. Oh *et al.* (2014b) proposed a modified tangent vector field guidance using a decentralized adaptive sliding mode control to tackle the effect of unmodeled dynamics and disturbances in the heading-hold autopilot for coordinated standoff tracking. As slightly different point of views for coordinated standoff tracking, Oh *et al.* (2013c) proposed a two-phase orbit approach based on path shaping, and Oh *et al.* (2013b) proposed a differential geometric guidance law using the change of geometry between a ground target and the UAV. To deal with multiple ground targets, coordinated standoff group tracking guidance using the Lyapunov vector field was proposed incorporating online local replanning for target hand-off and discard so that UAVs can simultaneously monitor multiple ground targets in an efficient way (Oh *et al.*, 2013a).

The purpose of this chapter is to introduce above recent algorithms on coordinated standoff tracking with realistic numerical examples along with relevant required techniques such as target tracking filter design and also discuss their advantages and disadvantages/limitations. The overall structure of this chapter is as follows: Section 2 gives the problem formulation for UAV dynamics, ground target and sensor modeling, target tracking filter design, and the review on vector field guidance strategies. Section 3 discusses recent techniques on coordinated standoff tracking for a single target: decentralized adaptive sliding mode control approach, nonlinear model predictive approach, and other relevant approaches. Section 4 introduces coordinated standoff group tracking using multiple UAVs. Finally, conclusions are given in Section 5.

2   PROBLEM FORMULATION

In this study, a two-dimensional (2D) space is considered for the UAV flying at a constant altitude. It is also assumed that initial target information is given by other sources such as a search-and-monitoring UAV, and an on-board sensor can

point at the target or group center position using a gimbal system. Note that data association for multiple targets and communications between UAVs are not the scope of this study. UAV team members share a known global coordinate system such as GPS (global positioning system) for their own and target's position.

2.1   UAV dynamic model

Assuming each UAV has a low-level flight controller for heading and velocity hold functions, this study aims to design guidance command inputs to this low-level controller for standoff target tracking. Consider a 2D UAV kinematic model (Kim, Oh, and Tsourdos, 2013) as

$$
\begin{pmatrix} \dot{x} \\ \dot{y} \\ \dot{\psi} \\ \dot{v} \\ \dot{\omega} \end{pmatrix} = f(\mathbf{x}, \mathbf{u}) = \begin{pmatrix} v\cos\psi \\ v\sin\psi \\ \omega \\ -\dfrac{1}{\tau_v}v + \dfrac{1}{\tau_v}u_v \\ -\dfrac{1}{\tau_\omega}\omega + \dfrac{1}{\tau_\omega}u_\omega \end{pmatrix} \quad (1)
$$

where $\mathbf{x} = (x, y, \psi, v, \omega)^{\mathrm{T}}$ are the inertial position, heading, speed, and yaw rate of the UAV, respectively. τ_v and τ_ω are time constants for considering actuator delay. $\mathbf{u} = (u_v, u_\omega)^{\mathrm{T}}$ are the commanded speed and turning rate constrained by the following dynamic limits of fixed-wing UAV: $|u_v - v_0| \leq \Delta v_{\max}$ and $|u_\omega| \leq \omega_{\max}$, where v_0 is a nominal speed of the UAV. The continuous UAV model in Equation 1 can be discretized by Euler integration into

$$
\mathbf{x}_{k+1} = f_{\mathrm{d}}(\mathbf{x}_k, \mathbf{u}_k) = \mathbf{x}_k + T_{\mathrm{s}}f(\mathbf{x}_k, \mathbf{u}_k) \quad (2)
$$

where $\mathbf{x}_k = (x_k, y_k, \psi_k, v_k, \omega_k)^{\mathrm{T}}$, $\mathbf{u}_k = (u_{vk}, u_{\omega k})^{\mathrm{T}}$, and T_{s} is a sampling time.

2.2   Ground target and sensor model

After analyzing the car trajectory data acquired by running a traffic simulation software (SIAS, 2011) and considering general behavior of ground vehicles, it is observed that the jerk is not negligible, with the acceleration best modeled using a piecewise constant profile over a specific duration of time. Therefore, a good model to apply to the tracking of ground targets considers acceleration dynamics (Mehrotra and Mahapatra, 1997). This acceleration model defines the target acceleration as a correlated process with a decaying exponential autocorrelation function, which means if there is a certain acceleration at a time t, then it is likely to be correlated via the exponential at a time instant $t + \tau$. A discretized system equation for the acceleration model for a ground vehicle is thus expressed in the form

$$\mathbf{x}_{k+1}^t = F_k \mathbf{x}_k^t + \eta_k \tag{3}$$

where the state vector is $\mathbf{x}_k^t = (x_k^t, \dot{x}_k^t, \ddot{x}_k^t, y_k^t, \dot{y}_k^t, \ddot{y}_k^t)^T$ and η_k is a process noise that represents the acceleration characteristics of the target. The details of acceleration dynamics can be found in Mehrotra and Mahapatra (1997) and Kim, Oh, and Tsourdos (2013).

In addition, this study assumes that UAVs are equipped with a GMTI (ground moving target indicator) sensor to localize the position of a target. To produce appropriate surveillance data for multiple targets, a GMTI is a well-suited sensor due to its wide-coverage and real-time capabilities (Koch, Koller, and Ulmke, 2006). The GMTI radar measurement $\mathbf{z} = (r, \theta)^T$ can be defined as the following nonlinear relation using the target position $(x_k^t, y_k^t)^T$ and the UAV position $(x_k, y_k)^T$ as

$$
\begin{aligned}
\mathbf{z}_k = \begin{pmatrix} r_k \\ \theta_k \end{pmatrix} &= h(\mathbf{x}_k^t) + \boldsymbol{\nu}_k \\
&= \begin{pmatrix} \sqrt{(x_k^t - x_k)^2 + (y_k^t - y_k)^2} \\ \tan^{-1} \dfrac{y_k^t - y_k}{x_k^t - x_k} \end{pmatrix} + \boldsymbol{\nu}_k
\end{aligned} \tag{4}
$$

where $\boldsymbol{\nu}_k$ is a measurement noise vector, and its noise covariance matrix is defined as

$$V_n[\nu_k] = R_k = \begin{bmatrix} \sigma_r^2 & 0 \\ 0 & \sigma_\theta^2 \end{bmatrix} \tag{5}$$

2.3 Ground target tracking filter

Considering the nonlinear measurement equation in Equation 4 and the advantage of using information form in multisensor systems, target localization is performed by using extended information filter (EIF) (Mutambara, 1998) as

Prediction

$$\mathbf{y}_{k|k-1}^t = Y_{k|k-1} F_k Y_{k-1|k-1}^{-1} \mathbf{y}_{k-1|k-1}^t \tag{6}$$

$$Y_{k|k-1} = (F_k Y_{k-1|k-1}^{-1} F_k^T + Q_k)^{-1} \tag{7}$$

Update

$$\mathbf{y}_{k|k}^t = \mathbf{y}_{k|k-1}^t + H_k^T (R_k)^{-1} [\mathbf{z}_k - h(\mathbf{x}_{k|k-1}^t) + H_k \mathbf{x}_{k|k-1}^t] \tag{8}$$

$$Y_{k|k} = Y_{k|k-1} + H_k^T (R_k)^{-1} H_k \tag{9}$$

where $Y_k = (P_k)^{-1}$ and $\mathbf{y}_k^t = Y_k \mathbf{x}_k^t$ represent the information matrix and information state vector, respectively. The output matrix H_k is a Jacobian of Equation 4 with respect to the time-update state $\mathbf{x}_{k|k-1}^t$. Given that multiple UAVs carry out the coordinated standoff tracking of groups of targets, each UAV's GMTI sensor obtains its own measurement and executes the tracking filter algorithm separately. After each UAV receives the other's estimation via communication, a decentralized EIF or state-vector sensor fusion technique can be applied to enhance the estimation accuracy (Mutambara, 1998; Oh *et al.*, 2012a).

2.4 Review on vector field guidance

This section reviews broadly used vector field approaches for a standoff tracking problem that will be used throughout this chapter. Let us begin with the Lyapunov vector field, which was initially proposed by Lawrence (2003) and further developed by Frew, Lawrence, and Morris (2008) as

$$V(x, y) = (r^2 - r_d^2)^2 \tag{10}$$

where $r = \sqrt{\delta x^2 + \delta y^2} = \sqrt{(x - x_t)^2 + (y - y_t)^2}$ is the distance between the UAV and the ground vehicle. Here (x_t, y_t) is the position of the ground vehicle that can be estimated from the tracking filter, and r_d is a desired standoff distance from the UAV to the ground vehicle. Differentiating Equation 10 gives the total time derivative of V:

$$\dot{V}(x, y) = \nabla V [\dot{x}, \dot{y}]^T \tag{11}$$

The Lyapunov vector field uses the following desired velocity $[\dot{x}_d, \dot{y}_d]^T$ that makes above to be nonpositive:

$$\begin{bmatrix} \dot{x}_d \\ \dot{y}_d \end{bmatrix} = \frac{-v_d}{k_l r(r^2 + r_d^2)} \begin{bmatrix} \delta x (r^2 - r_d^2) + \delta y (2r r_d) \\ \delta y (r^2 - r_d^2) - \delta x (2r r_d) \end{bmatrix} \tag{12}$$

where k_l is positive constant and v_d is a desired UAV speed. Note that k_l can be used to adjust the converging speed of the generated field to the standoff orbit. The desired heading can be decided using the desired 2D velocity components in

Equation 12 as

$$\psi_d = \tan^{-1}\frac{\dot{y}_d}{\dot{x}_d} \qquad (13)$$

If the velocity of a ground target could be estimated, the guidance vector can be adjusted in order to take a target velocity into account as discussed in Frew, Lawrence, and Morris (2008). The guidance command u_ω for turn rate is selected as the sum of proportional feedback and feedforward terms as

$$u_\omega = -k_\psi(\psi - \psi_d) + \dot{\psi}_d \qquad (14)$$

where $\dot{\psi}_d = 4v_d[r_d r^2/(r^2 + r_d^2)^2]$ can be obtained by differentiating Equation 13.

Another vector field approach is the tangent vector field generated by a desired heading angle (Kingston and Beard, 2007):

$$\psi_d = \psi_p + \tan^{-1}(k_t d) \qquad (15)$$

where $d = r - r_d$ is the distance of the UAV from the desired standoff orbit and ψ_p is tangent to the standoff orbit along the ray connecting the UAV and the target position as

$$\psi_p = \theta + \frac{\pi}{2} \qquad (16)$$

where $\theta = \tan^{-1}(\delta y/\delta x)$. The geometry for the tangent vector field is illustrated in Figure 1. The convergence analysis of a desired heading angle to ψ_p can be found in Kingston and Beard (2007).

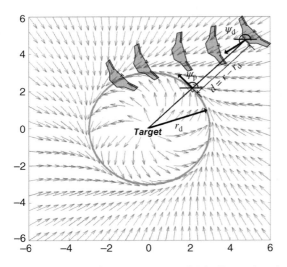

Figure 1. Geometry of tangent vector field. (Reproduced with permission from Oh *et al.*, 2014b. © Springer.)

There are also other variations of the vector field approach such as in Quigley *et al.* (2005), Lim *et al.* (2013), and Summers, Akella, and Mears (2009). Although these vector fields have different characteristics, they can generate a similar trajectory by adjusting the field gain such as k_l and k_t. It is worth noting that the field gain should be determined not to exceed the turning rate constraint of the UAV, $\omega_{\max}$, along with a control gain.

In case that a pair of UAVs are involved for the same target flying on the same orbit, the angular separation between UAVs could be additionally performed by controlling the speed of UAVs in order to obtain more accurate target information and to avoid collision between them as explained in Frew, Lawrence, and Morris (2008):

$$u_v = \pm k_v(\gamma - \theta_d)r_d + v_d \qquad (17)$$

where k_v is a control gain, θ_i is the azimuth angle of the ith UAV relative to the target position, $\gamma = \theta_2 - \theta_1$ is the angular phase separation of UAVs, and θ_d is a desired phase difference between the UAVs. Different approaches to this angular separation can be applied, such as controlling the orbit radius instead of the speed (Wise and Rysdyk, 2006; Kingston and Beard, 2007) and a decentralized approach to address more than two UAVs (Summers, Akella, and Mears, 2009; Oh *et al.*, 2012a, 2012b), as will be discussed in the following section.

3 RECENT TECHNIQUES ON COORDINATED STANDOFF TRACKING

This section introduces recent coordinated standoff tracking guidance algorithms, some of which is in conjunction with a roadmap-assisted filtering and sensor fusion technique. These are (i) decentralized vector field guidance using adaptive sliding mode control, (ii) nonlinear model predictive coordinated standoff tracking (NMPCST), (iii) a two-phase orbit approach using path shaping, and (iv) differential geometric guidance. Compared with existing methods, the proposed methods have various benefits such as combining decentralized estimation and guidance considering different information/communication networks or guaranteeing optimal performance in terms of tracking and control efforts.

3.1 Decentralized adaptive sliding mode control approach

To track an unknown moving ground target, a tangent vector field guidance algorithm using sliding mode control (SMC) is proposed in Kingston and Beard (2007). This section now introduces adaptive terms in the SMC approach to reduce the effect of unmodeled dynamics and disturbance in the heading-hold autopilot (Oh *et al.*, 2014b). Decentralized angular separation guidance along with decentralized estimation is also proposed using either velocity change or orbit radius change by different communication/information architectures.

3.1.1 Adaptive sliding mode control

The UAV is assumed to follow a first-order heading dynamics by the autopilot, as given by

$$\dot{\psi} = \alpha(\psi_c - \psi) + \nu \quad (18)$$

where ψ_c is the commanded heading, α is a positive constant that characterizes the speed response of heading-hold autopilot loop, and ν represents unmodeled dynamics or disturbances of the autopilot loop. In general, α is difficult to determine experimentally and inevitably contains error in its estimated value, and ν can have a considerable value with known bounds due to disturbances or faults. Since these factors could result in less precise target tracking, this study makes use of an adaptive sliding mode control method that can estimate and consequently compensate the effect of unknown parameters.

To this end, first of all, let us define a sliding surface as

$$S = e + k_I \int_0^t e \, d\tau \quad (19)$$

where $e = \psi - \psi_d$ is a tracking error and k_I is an integral gain. Differentiating the sliding surface with respect to time gives

$$\dot{S} = \dot{\psi} - \dot{\psi}_d + k_I e = \alpha(\psi_c - \psi) + \nu - \dot{\psi}_d + k_I e \quad (20)$$

Consider a Lyapunov function candidate $W_1 = (1/2)S^2$ and take the derivative to obtain

$$\dot{W}_1 = S\dot{S} = S(\alpha(\psi_c - \psi) + \nu - \dot{\psi}_d + k_I e) \quad (21)$$

Then, the control command can be selected as

$$\psi_c = \psi + \frac{1}{\hat{\alpha}}(-\hat{\nu} + \dot{\psi}_d - k_I e - c_1 S - c_2 \text{sign}(S)) \quad (22)$$

where $\dot{\psi}_d$ is a derivative of Equation 15, $\hat{\alpha}$ and $\hat{\nu}$ are the estimation of α and ν, respectively, and c_1 and c_2 are positive constants. Rearranging Equation 21 using $\psi_s = \psi_c - \psi$ gives

$$\dot{W}_1 = S(\alpha\psi_s + \nu - \dot{\psi}_d + k_I e)$$
$$= S\{\tilde{\alpha}\psi_s + \hat{\alpha}(\psi_c - \psi) + \nu - \dot{\psi}_d + k_I e\} \quad (23)$$

where $\tilde{\alpha} = \alpha - \hat{\alpha}$. Substituting ψ_c from Equation 22 into Equation 23 yields

$$\dot{W}_1 = S\{\tilde{\alpha}\psi_s - \hat{\nu} + \dot{\psi}_d - k_I e - c_1 S - c_2 \text{sign}(S) + \nu - \dot{\psi}_d + k_I e\}$$
$$= S\{\tilde{\nu} - c_1 S - c_2 \text{sign}(S) + \tilde{\alpha}\psi_s\} \quad (24)$$

where $\tilde{\nu} = \nu - \hat{\nu}$.

To obtain the adaptation rule for the parameter estimations of $\hat{\nu}$ and $\hat{\alpha}$, consider another Lyapunov function candidate W_2 as

$$W_2 = \frac{1}{2}S^2 + \frac{1}{2}\gamma_\nu^{-1}\tilde{\nu}^2 + \frac{1}{2}\gamma_\alpha^{-1}\tilde{\alpha}^2 \quad (25)$$

where γ_ν and γ_α are positive constants. Differentiating Equation 25 and using Equation 25 gives

$$\dot{W}_2 = S(\tilde{\nu} - c_1 S - c_2 \text{sign}(S) + \tilde{\alpha}\psi_s)$$
$$+ \gamma_\nu^{-1}\tilde{\nu}\dot{\tilde{\nu}} + \gamma_\alpha^{-1}\tilde{\alpha}\dot{\tilde{\alpha}} = -c_1 S^2 - c_2|S| \quad (26)$$
$$+ \tilde{\nu}(S - \gamma_\nu^{-1}\dot{\hat{\nu}}) + \tilde{\alpha}(\psi_s S - \gamma_\alpha^{-1}\dot{\hat{\alpha}})$$

where the relations $\dot{\tilde{\nu}} = -\dot{\hat{\nu}}$ and $\dot{\tilde{\alpha}} = -\dot{\hat{\alpha}}$ are used under the assumption that ν and α are constant. Then, the adaptation law for $\hat{\nu}$ and $\hat{\alpha}$ can be obtained as $\dot{\hat{\nu}} = \gamma_\nu S$ and $\dot{\hat{\alpha}} = \gamma_\alpha \psi_s S$. Then,

$$\dot{W}_2 = -c_1 S^2 - c_2|S| \leq 0 \quad (27)$$

from which it can be concluded that S goes to zero in finite time, and finally the error state e tends to zero by LaSalle–Yoshizawa theorem (Krstic, Kanellakopoulos, and Kokotovic, 1995). This means that the heading angle ψ of the UAV can follow the desired heading ψ_d provided from Equation 15 in spite of model uncertainties in α and ν.

3.1.2 Decentralized angular separation control

In performing a coordinated target tracking mission, it would be useful for UAVs to keep a certain intervehicle angular separation. This might be required to maximize sensor coverage, enhance the estimation accuracy of the target, or avoid collision between UAVs while maintaining a standoff distance from a target as mentioned earlier. To do so, this study introduces decentralized angular separation control of multiple UAVs using velocity or orbit radius change by different information architectures. It builds upon a rigid graph theory utilizing asymmetric minimally persistent leader–follower and symmetric nonminimally persistent architectures based on the previous study (Summers, Akella, and Mears, 2009).

In a minimally persistent leader–follower architecture, one of the UAVs (leader) flies on the standoff orbit around a target with desired airspeed and orbit radius using the TVFG with the adaptive SMC. The remaining vehicles (followers) try to remain on the same orbit; however, they keep a prescribed angular spacing with the neighboring vehicle ahead of it by adjusting airspeed or orbit radius. This architecture can be modeled by a directed graph and requires a minimum number of communication/sensing links to achieve the circular orbiting and angular spacing. For instance, while the leader remains on the desired orbit, followers adjust their orbit radius by the modified desired heading angle as

$$\psi_i^{\mathrm{d}} = \psi_i^{\mathrm{p}} + \tan^{-1}(k_{\mathrm{t}}(d_i - k_{\mathrm{o}}\delta\theta_i)) \qquad (28)$$

where k_{o} is a control gain weighting the convergence to a desired orbit, and $\delta\theta_i$ is ith angular spacing error given by $\delta\theta_i = \theta_{i+1} - \theta_i - \theta_{\mathrm{d}}$, where θ_{d} is the desired angular separation between UAVs. Note that $d_i = r_i - r_{\mathrm{d}}$ is a distance from the desired orbit to the UAV position in the normal TVFG as Equation 15; now it is modified by additional term $(k_{\mathrm{o}}\delta\theta_i)$ according to the angular spacing error, resulting in the temporary change of orbit radius, as illustrated in Figure 2. The difference of time spent on the different orbits allows for the control of the angular separation of UAVs. Similarly, the speed of the UAV can also be adjusted according to $\delta\theta_i$ while flying on the same standoff orbit to maintain the desired angular separation.

In a nonminimally persistent architecture, the airspeed or orbit radius is adjusted such that each vehicle moves toward the midpoint of its two nearest neighbours on the standoff orbit; thus, now $\delta\theta_i = (1/2)(\theta_{i-1} + \theta_{i+1}) - \theta_i$ for all UAVs to adjust orbit radius accordingly. This is modeled by an undirected graph. This architecture does not need to know the number of engaging vehicles in advance, compared with the minimally persistent case that requires a desired separation angle θ_{d}; hence, this control structure can be viewed as fully decentralized.

3.1.3 Numerical simulations

To verify the feasibility and benefits of the proposed approach, numerical simulations are performed by using realistic ground vehicle tracking scenario. The vehicle trajectory data acquired from the S-Paramics traffic simulator (SIAS, 2011) are used to generate the GMTI radar measurements composed of relative range and azimuth angle with respect to a position of the UAV. These true GMTI measurements were mixed with the white noise having the corresponding standard deviations. Detailed parameters for this simulation study can be found in Oh et al. (2014b).

Table 1 shows mean errors in the decentralized estimation and the decentralized TVFG using adaptive SMC with different information/communication structures for three UAVs. Here, the velocity change scheme with a fully connected communication network and nonminimally information architecture shows the best performance in terms of the estimation accuracy, standoff distance, and phase angle keeping. Although the velocity change scheme shows much better performance than that of orbit radius change, in case that frequent velocity change is undesirable or unattainable, the angular separation can be achieved without velocity control with a bounded error. Note that this is done by

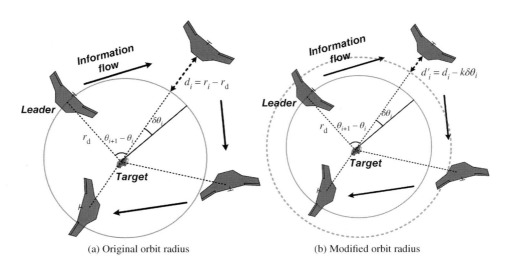

(a) Original orbit radius   (b) Modified orbit radius

Figure 2. Illustration of orbit radius change in minimally persistent case. (Reproduced with permission from Oh *et al.*, 2014b. © Springer.)

Table 1. Performance comparison of guidance algorithms with different information/communication structures.

Mean error	Single UAV (EIF)		Multiple UAVs (DEIF)			
			Orbit radius change		Velocity change	
	SMC	ASMC	Minimal	Nonminimal	Minimal	Nonminimal
Position (m)	14.3291		7.6990	4.9703	7.6790	4.7218
Velocity (ms^{-1})	3.1445		2.5216	2.1412	2.5082	2.0913
Standoff distance (m)	26.8211	14.4717	63.3150	70.5503	15.9357	14.0693
Angular separation (deg)	—	—	23.9987	16.7073	11.6899	2.7884

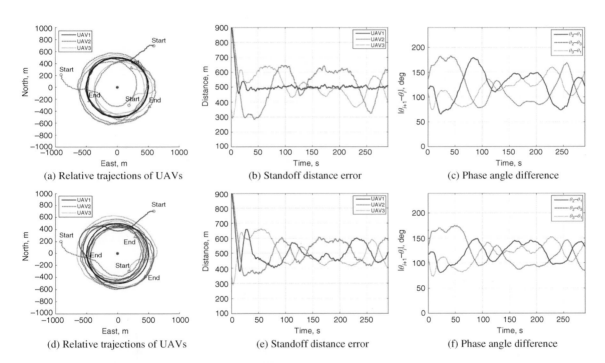

(a) Relative trajectories of UAVs   (b) Standoff distance error   (c) Phase angle difference

(d) Relative trajectories of UAVs   (e) Standoff distance error   (f) Phase angle difference

Figure 3. Tracking results using orbit radius change for angular separation and DEIF: (a–c) minimally persistent information architecture and minimum communication link; (d–f) nonminimally persistent and fully connected communication link. (Reproduced with permission from Oh *et al.*, 2014b. © Springer.)

adjusting orbit radius appropriately at the expense of performance in standoff distance as shown in Figure 3.

3.2 Nonlinear model predictive control approach

This section introduces the NMPCST framework for a pair of UAVs. Unlike a decoupled control structure (heading for standoff distance and speed for phase angle keeping) based

on vector field approaches, the NMPCST makes use of a coupled approach that can provide better performance in terms of tracking guidance accuracy as well as control efforts. Note that the performance of the tracking guidance algorithms heavily depends on the sensing and estimation capability to determine target's position and intent. Thus, sensor fusion using other UAV's measurements and estimates for the same target along with contextual information could enhance the performance of the guidance as well as the estimation. In many applications for the ground target tracking, the majority of ground vehicles are moving on road networks whose topographical coordinates could be known with certain accuracy. Such roadmap information can be used for improving the estimation accuracy, by constraining the state of the ground target in its position and velocity within the road geometry. In particular, for a prediction-based control scheme, since the prediction of the target movement plays a more important role in the tracking performance, this study presents the effect of improved estimation accuracy utilizing road information on the tracking guidance performance (Oh et al., 2012b).

3.2.1 NMPC formulation

The model predictive coordinated standoff tracking (Kim, Oh, and Tsourdos, 2013) computes a control input sequence for N sampling times: $U_k = \{\mathbf{u}_0, \mathbf{u}_1, \ldots, \mathbf{u}_{N-1}\}$ that minimizes the following performance index J for maintaining a standoff distance between a UAV and a ground target as well as a relative phase angle between UAVs:

$$
\begin{aligned}
J &= \phi(\tilde{r}_N, \tilde{d}_N) + \sum_{k=0}^{N-1} L(\tilde{r}_k, \tilde{d}_k, \mathbf{u}_k) \\
\phi(\tilde{r}_N, \tilde{d}_N) &= \frac{1}{2}(p_r \tilde{r}_N^2 + p_d \tilde{d}_N^2) \\
L(\tilde{r}_k, \tilde{d}_k, \mathbf{u}_k) &= \frac{1}{2}\left\{ q_r \tilde{r}_k + q_d \tilde{d}_k^2 + r_v \left(\frac{u_{vk} - v_0}{v_{\max}}\right)^2 \right. \\
&\quad \left. + r_\omega \left(\frac{u_{\omega k} - (v_0/r_d)}{\omega_{\max}}\right)^2 \right\}
\end{aligned}
\tag{29}
$$

where

$$
\tilde{r}_k = \frac{r_d^2 - |r_k|^2}{r_d^2}
\tag{30}
$$

$$
\tilde{d}_k = \frac{r_k^T r_k^p + |r_k| |k_p|}{r_d^2}
\tag{31}
$$

where r_k and r_k^p represent the relative vectors from the target position to the positions of the current UAV and its pair UAV, respectively, r_d is a desired standoff distance from the UAVs to the target position, v_0 is a nominal speed of UAVs, and (v_0/r_d) is a nominal angular velocity. p_r, p_d, q_r, q_d, r_v, and r_ω are constant weighting scalars. In Equation 31, $\tilde{d}_k$ is derived from the inner product of r_k and r_k^p as $\langle r_k, r_k^p \rangle = r_k^T r_k^p = |r_k| |r_k^p| \cos \Delta\theta_k$, where $\Delta\theta_k = |\theta_k^p - \theta_k|$ with the phase angles of UAV positions with respect to the current target location. If the desired phase difference $\Delta\theta_k$ is π, the

above equation can be rearranged as $r_k^T r_k^p + |r_k| |r_k^p| = 0$ since $\cos \pi = -1$. Therefore, minimizing $\tilde{d}_k$ results in keeping the phase angle between a pair of UAVs.

Furthermore, by incorporating the dynamics of the UAVs and admissible control input ranges as equality and inequality constraints, an augmented performance index can be derived as

$$
\begin{aligned}
J_a &= \phi(\tilde{r}_N, \tilde{d}_N) + \sum_{k=0}^{N-1} \left[L(\tilde{r}_k, \tilde{d}_k, \mathbf{u}_k) + \lambda_{k+1}^T \{ f_d(\mathbf{x}_k, \mathbf{u}_k) \right. \\
&\quad \left. - \mathbf{x}_{k+1} \} + \frac{1}{2} \mu_v l_{vk} S_v^2(\mathbf{u}_k) + \frac{1}{2} \mu_\omega l_{\omega k} S_\omega^2(\mathbf{u}_k) \right]
\end{aligned}
\tag{32}
$$

where $S_v(\mathbf{u}_k) = (|u_{vk} - v_o| - v_{\max})/v_{\max} \leq 0$, $S_\omega(\mathbf{u}_k) = (|u_{\omega k}| - \omega_{\max})/\omega_{\max} \leq 0$, λ_k is a Lagrange multiplier, and μ_v and μ_ω are penalty function parameters. l_{vk} and $l_{\omega k}$ are defined to avoid unnecessary computation for satisfying inequality constraints:

$$
l_{*k} = \begin{cases} 0, & S_* \leq 0 \\ 1, & S_* > 0 \end{cases}
\tag{33}
$$

Then, gradient descent-based online optimization can be performed to find the optimal solution for above performance index. Each UAV runs the optimization routine in a decentralized way at each sampling time. When the measurement for the target comes in, each UAV performs the target localization and then shares control commands/states of UAVs as well as their state/covariance estimation of the target via communication.

3.2.2 Numerical simulations

To verify the feasibility and benefits of the proposed approach, the same car tracking scenario used in the previous section was tested for the LVFG as well as the NMPCST. As can be seen in Table 2, the mean error of standoff distance keeping by the NMPCST is much lower than that of the LVFG (which was tuned at best by trial and error procedure) even though its control consumption on the turn rate is slightly higher than that of the LVFG. The NMPCST has a slightly lower mean error for phase angle keeping than the LVFG with a similar level of speed consumption. Intuitively, this merit results from the fact that part of turn rate control as well as speed change by the NMPCST contributes on keeping a phase difference to meet the inner product relation of position vectors in Equation 31 during optimization.

In addition, to evaluate effect of the improved estimation accuracy on the guidance performance, another numerical

Table 2. Performance comparison between LVFG and NMPCST.

Parameter	LVFG		NMPCST	
Standoff distance error (m)	Mean±std	Min/max	Mean±std	Min/max
	22.90±8.65	2.61/48.56	5.38± 4.14	0.09/25.90
Phase keeping error (deg)	Mean±std	Min/max	Mean±std	Min/max
	7.62±4.52	0.03/20.17	7.28± 4.21	0.01/17.97
u_v control consumption (m s^{-1})	18.79e2		17.33e2	
u_ω control consumption (rad s^{-1})	28.61		29.30	

Table 3. Performance comparison between estimation methods.

	Single UAV		Multiple UAVs (state-vector fusion)			
	Unconstrained		Unconstrained		Road-constrained	
Mean error	EKF$_{UAV,1}$	EKF$_{UAV,2}$	EKF	UKF	ECEKF	ECUKF
Position (m)	18.2612	18.9317	14.4422	14.1452	8.5729	8.1752
Velocity (m s^{-1})	3.3433	3.5795	3.1123	3.1166	2.1712	2.1565

simulation is performed by comparing tracking guidance performance using road-constrained filtering techniques given that the road information is available. To address nonlinearity of road constraints and provide good estimation performance, both extended Kalman filter (EKF) and unscented Kalman filter (UKF) are implemented along with state-vector fusion technique for multiple UAVs as detailed in Oh *et al.*, 2012b. The measurement noise is set to significantly larger values in this simulation in order to see the benefit of constrained filtering more clearly. Table 3 shows mean tracking error in position and velocity among different filtering methods. The EKF and the UKF using the decentralized sensor fusion based on the state-vector fusion with two UAVs show better performance than the one using only single UAV, and the equality (road)-constrained UKF

(ECUKF) with data fusion provides the best estimation accuracy. Table 4 compares tracking guidance performance for standoff distance and phase keeping between LVFG and NMPCST using the EKF and ECUKF. It is worthwhile noting that the performance improvement of NMPCST with changing estimation method from the EKF to the ECUKF is more remarkable than that of LVFG, since NMPCST uses predicted target's information to a certain future time explicitly.

3.3 Other approaches

There are also other notable approaches other than using vector field or optimization-based (i.e., NMPC) techniques.

Table 4. Tracking performance with different estimation methods.

	LVFG		NMPCST	
Mean error	EKF	ECUKF	EKF	ECUKF
Standoff distance (m)	15.8830	13.0514	14.8521	9.2001
Phase keeping (deg)	13.1350	13.0020	13.3673	11.3492

Among them, a two-phase orbit approach based on path shaping is of interest (Oh *et al.*, 2014c). Unlike the most existing approaches focusing on the guidance law design with turn rate and speed, this work describes a novel path shaping strategy taking kinematic constraints of multiple UAVs into account for a coordinated standoff target tracking. Since fixed-wing UAVs fly at a nominal airspeed for a longer duration with better fuel efficiency than a rotary-wing UAV with hovering capability, it is desirable that they keep angular separation between vehicles while holding a constant velocity. In order to produce flyable paths to fit to the fixed-wing UAV dynamics, the path shaping algorithm needs to consider the most critical constraint of fixed-wing UAVs: curvature of turning maneuver, which is dependent on the operating range of speed and bank angle. Moreover, it is advantageous for the shaped trajectories to have a shorter length as well as lower curvature differences at discontinuous points in order to minimize flight time or energy consumption. For this, two constant curvature segments are exploited for path shaping. As a measure to coordinated target tracking for multiple UAVs, a simultaneous arrival concept is first introduced, which initializes UAVs on a standoff orbit with a desired angular separation. In order to address arrival time delay or failure of the UAV, a two-orbit approach is newly proposed in which UAVs first arrive at the outer orbit and subsequently shrink to the desired inner orbit at different time, while adjusting the angular separation between constant-speed UAVs.

In Oh *et al.* (2013b), standoff tracking guidance to a moving ground target is proposed using differential geometry motivated by previous works (White, Zibkowski, and Tsourdos, 2007). Using the relative geometry, convergent, divergent, and parallel solutions can be obtained depending on their initial positions and the velocity ratio between them. Then, the convergent solution can be used to guide the UAV on the corresponding rendezvous geometry associated with the target movement. In a similar way, a novel guidance law for standoff tracking is derived by superimposing a standoff orbit circle around the target position. The proposed differential geometric guidance law has several advantages along with its inherent simplicity over the other standoff tracking guidance laws. First of all, while most of the literature has focused on the stability analysis limited to standoff tracking of a fixed target, the proposed approach can analyze stability for both rendezvous and standoff tracking of a moving target. Another benefit is that the guidance command can explicitly consider a target velocity for enhancing the tracking performance when its estimation by the localization filter is reasonably accurate. Finally, the proposed guidance law requires the reduced number of tuning variables, only a curvature command, unlike other approaches such as vector field guidance requiring more parameters for appropriate

vector field generation as well as guiding vehicle into the field as explained in Section 2.4.

3.4 Discussion

This section introduces four different guidance algorithms for coordinated standoff tracking of a moving ground target. The following summarizes the advantages and disadvantages/limitations of each method:

- *Two-phase orbit approach based on path shaping:* the path shaping approach with a constant speed would be desirable for a mission duration point of view; this requires less communication between UAVs to achieve simultaneous arrival to the standoff orbit, compared with the other guidance approaches in which information needs to be shared continuously; but this would be difficult to be applied to an unknown moving target tracking.

- *Standoff tracking guidance based on differential geometry:* this approach brings the following advantages along with its inherent simplicity: rigorous stability, explicit use of a target velocity, and tuning parameter reduction; in case that UAVs are inside the standoff orbit, an *ad hoc* method is used; and coordination strategy based on differential geometry needs to be developed.

- *Decentralized vector field guidance using adaptive sliding mode control:* adaptive sliding mode control concept can significantly reduce the effect of unmodeled dynamics and disturbance in the UAV autopilot; either velocity or orbit radius change can lead to successful standoff tracking; in case that frequent velocity change is undesirable or unattainable, the angular separation can be achieved without velocity control but with a bounded error; and decentralized approach can directly combine the estimation and guidance considering different information/communication structures.

- *Nonlinear model predictive coordinated standoff tracking:* NMPCST guidance can provide an optimal performance in terms of tracking as well as control efforts; this method depends heavily on the UAV model and target estimation accuracy to propagate commands; high-frequency control inputs are required to be followed by UAVs for both velocity and turning rate, which is hard to achieve in practice; and this would require significant computation power to perform a real-time optimization.

Note that this section has addressed the problem of a single-target tracking only. However, in reality, several distinct or a group of suspicious targets could be observed in geographically close areas. This requires a more complicated consideration on a standoff tracking problem involving a task allocation between UAVs and targets as well as tracking of clustered targets. In this regard, coordinated standoff tracking of groups of moving targets using multiple UAVs will be discussed in the next section.

4   COORDINATED TARGET GROUP TRACKING USING MULTIPLE UAVs

In case that multiple moving ground vehicles are identified as targets of interest from reconnaissance or surveillance systems within the ground road traffic (Oh *et al.*, 2013c, 2014a), it is required to develop strategies on how to deploy multiple UAVs to persistently follow them. Although various different methodologies have been developed for the multitarget tracking using multiple ground (Parker, 2002; Jung and Sukhatme, 2002) or aerial vehicles (Tang and Ozguner, 2005; Deghat *et al.*, 2014), there is relatively little research on multiple or group target tracking in the context of cooperative standoff tracking considering uncertain dynamic environments and UAV sensing capability. Thus, this section introduces a coordinated standoff tracking methodology of moving target groups using multiple UAVs. In order to track a group of targets using the sensor with the limited field of view (FOV), the vehicle should be positioned as close as possible to multiple targets to obtain better estimation accuracy and far enough to keep the group of targets within its FOV. The objective of this study is to develop an active sensing/guidance algorithm to maximize information or accuracy of estimation of the targets as well as persistently keep all of them (or as many as possible) within the view of multiple UAVs considering physical (turning radius and speed) and sensing (FOV and range) constraints.

4.1   LVFG with variable standoff distance

This study (Oh *et al.*, 2013a) utilizes a LVFG (Lyapunov vector field guidance) for standoff group tracking with the vector field function using the same Equations (10)–(12). Now, $\delta x = x - x^{tc}$, $\delta y = y - y^{tc}$, and $r = \sqrt{\delta x^2 + \delta y^2}$ becomes the distance of the UAV from the group enter. Herein (x^{tc}, y^{tc}) is the center position of a target group estimated from the tracking filter as shown in Figure 4. r_d is a desired standoff distance from the UAV to the center of a target group that can be computed considering the FOV α_f of the UAV as

$$r_d = \frac{d_{max} + d_m}{\sin((\alpha_f - \varepsilon_m)/2)} \qquad (34)$$

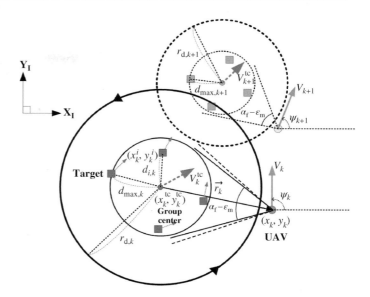

Figure 4. Geometric relation between UAV, ground target, and target group at time steps k and $k + 1$. (Reproduced with permission from Oh *et al.*, 2013a. © IEEE.)

where d_{max} is the distance between the group center and the target furthest from the center in the group, $d_m > 0$ is a distance margin for d_{max}, and $\varepsilon_m > 0$ is an angle margin for the FOV of the UAV. By using aforementioned standoff distance r_d, the UAV can keep all the target in the group within its FOV as shown in Figure 4.

The guidance command u_ω for turn rate takes the same form as in Equation 14 but with the different feedforward term considering the effect of the variable standoff distance as

$$\dot{\psi}_d = 4v_d \frac{r_d r^2}{(r^2 + r_d^2)^2} - \frac{2r\dot{r}_d}{r^2 + r_d^2} \qquad (35)$$

which can be obtained by differentiating Equation 13. As r approaches r_d, the left term of Equation 35 increases monotonically, and magnitude of the right term also increases. Then, the guidance vector field will be feasible as long as the loiter circle pattern itself is feasible considering variable $\dot{r}_d$, which satisfies the following when $r = r_d$:

$$\dot{\psi}_d = \frac{v_d}{r_d} - \frac{\dot{r}_d}{r_d} < \omega_{max} \qquad (36)$$

Using the above equation, feasible standoff distance can be determined as

$$r_d \geq \frac{v_d}{\omega_{max}} - \frac{\dot{r}_d}{\omega_{max}} = r_{d,min} \qquad (37)$$

Therefore, $r_{d,min}$ can be determined by both the maximum speed of a ground vehicle that determines $\dot{r}_d$ and the UAV kinematic constraints ω_{max}. Note that for the turning rate command to be feasible (i.e., within ω_{max}), the gain k_ω and standoff distance r_d need to be carefully determined.

4.2 Multiple target group tracking by multiple UAVs

This section presents a multitarget group surveillance strategy by cooperating multiple UAVs with its benefits such as better estimation accuracy with sensor/data fusion and more robust tracking performance. Since multitarget tracking using multiple UAVs is typically NP-hard in the number of both sensing agents and targets (Tang and Ozguner, 2005), this study uses a two-step heuristic approach: target clustering/resource allocation and cooperative standoff group tracking with local replanning.

4.2.1 Target clustering and resource allocation

Since this study uses a standoff tracking concept in which UAVs are continuously orbiting around moving targets, one of the suboptimal approaches to partition the targets would be treating geographically close targets as the same target group. This is done by K-means clustering algorithm to group objects based on attributes into predefined K number of groups (Bishop, 1995). After clustering, UAVs need to be assigned to the corresponding target group. An optimal assignment approach is used as the one that gathers the most information of targets defined by Fisher information matrix (FIM). The FIM describes the amount of information a set of measurements contains about the state variable in terms of sensitivity of the estimation process. The details of the FIM and the initial assignment process can be found in Frew (2008), Taylor (1979), and Oh et al. (2013a).

4.2.2 Online local replanning

Once initial assignment of UAVs to target groups is completed, online local replanning is followed either by handing over targets between groups or by discarding target out of the group against a sensing range or the convergence speed of the vector field to the desired orbit while UAVs are persistently following corresponding groups.

- *Target hand-off:* By running K-means clustering algorithm in a recursive manner, a target hand-off event between groups can be done inherently since clustering itself can regroup targets according to their proximity to the target group and UAVs. To avoid frequent change of

the group for the target on the boundary between two (passing/receiving) groups as well as make sure that the target passed to the receiving group is inside the FOV of UAVs of that group, hand-off occurs for a certain period of time T_{hd}:

$$T_{hd} \geq \left[\frac{|r - r_d|}{|\dot{r}_{new} - (r_d/r)\dot{r}_d|} \right]_{t=t^i_{hd}} \tag{38}$$

where t^i_{hd} represents the time when the target is first requested for the hand-off by the clustering algorithm. For T_{hd}, the hand-off target will be included in both passing and receiving groups.

- *Target discard:* If the standoff distance for the group tracking becomes larger than the sensing range (i.e., $r_d > r_{d,\,max}$), or radial velocity difference between the vector field and desired standoff orbit is larger than a certain value, the target furthest from the center in the group is removed from the group.

4.3 Numerical simulations

Figure 5 shows a numerical simulation result with seven ground target (six randomly moving and one maneuvering) and four UAVs using the proposed standoff tracking framework. First, targets are clustered into two groups, and UAVs are assigned to the appropriate group using the proposed assignment algorithm as shown in Figure 5a. Note that data association regarding which measurement comes from which target is assumed to be solved in this study. However, since our approach to track multiple targets is to exploit the center of the group and furthest target information from the center only, even some false data association at the beginning in a cluttered situation as shown in Figure 5a would not affect the guidance performance in terms of keeping all targets in the standoff orbit. At around 35 s, since a target hand-off event triggered, the target (moving toward north-east direction) is included in both target groups until UAVs of the receiving group (group 2) reach the desired standoff orbit for T_{hd} seconds as shown in Figure 5b. Figure 5c shows the situation after target hand-off (from group 1 to group 2) process is finished. As targets in the group get dispersed widely, the furthest target from the center is removed from the group depending on the sensing range or convergent limit.

5 CONCLUSIONS

This chapter has presented the coordinated standoff tracking of groups of moving targets using multiple UAVs, some of

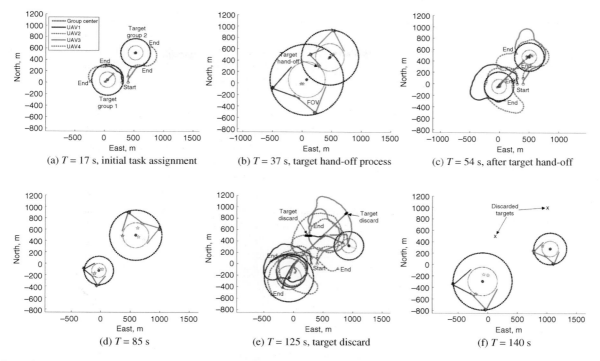

Figure 5. Trajectories of standoff tracking of seven ground targets (six randomly moving and one maneuvering) with four UAVs.

which is in conjunction with a roadmap-assisted filtering and sensor fusion technique. Four recent standoff tracking algorithms were introduced for a single ground target, and their properties and performance were compared with one another. This study particularly investigated the effect of improved estimation accuracy on the tracking guidance performance. In addition, based on the vector field guidance approach, an active guidance algorithm was developed to maximize information of the multiple targets and keep all of them inside the view of UAVs considering physical and sensing constraints. For multiple group target surveillance by multiple UAVs, a two-phase approach was presented consisting of target clustering/assignment and cooperative standoff tracking with online local replanning. Numerical simulation showed successful standoff tracking as well as local replanning while keeping all targets within the FOV of the UAV sensor at all times in a dynamic environment.

REFERENCES

Bishop, C. (1995) *Neural Networks for Pattern Recognition*, Oxford University Press.

Chen, H. (2013) UAV path planning with tangent-plus-Lyapunov vector field guidance and obstacle avoidance. *IEEE Trans. Aerospace Electron. Syst.*, **49**(2), 840–856.

Deghat, M., Xia, L., Anderson, B., and Hong, Y. (2014) Multi-target localization and circumnavigation by a single agent using bearing measurements. *Int. J. Robust Nonlinear Control*. doi: 10.1002/rnc.3208.

Frew, E. (2008) Sensitivity of cooperative target geolocalization to orbit coordination. *J. Guid. Control Dyn.*, **31**(4), 1028–1040.

Frew, E., Lawrence, D., and Morris, S. (2008) Coordinated standoff tracking of moving targets using Lyapunov guidance vector fields. *J. Guid. Control Dyn.*, **31**(2), 290–306.

Jung, B. and Sukhatme, G. (2002) A region-based approach for cooperative multi-target tracking in a structured environment. *IEEE/RSJ International Conference on Intelligent Robots and Systems*, pp. 2764–2769.

Kim, S., Oh, H., and Tsourdos, A. (2013) Nonlinear model predictive coordinated standoff tracking of moving ground vehicle. *J. Guid. Control Dyn.*, **36**(2), 557–566.

Kingston, D. and Beard, R. (2007) UAV splay state configuration for moving targets in wind. *Lect. Notes Control Inform.*, **369**, 109–128.

Koch, W., Koller, J., and Ulmke, M. (2006) Ground target tracking and road map extraction. *ISPRS J. Photogramm. Remote Sens.*, **61**, 197–208.

Krstic, M., Kanellakopoulos, I., and Kokotovic, P. (1995) *Nonlinear and Adaptive Control Design*, vol. 8, John Wiley & Sons, Inc., New York.

Lawrence, D. (2003) Lyapunov vector fields for UAV flock coordination. *2nd AIAA Unmanned Unlimited Conference, Workshop, and Exhibit, Reston, VA.*

Lim, S., Kim, Y., Lee, D., and Bang, H. (2013) Standoff target tracking using a vector field for multiple unmanned aircrafts. *J. Intell. Robot. Syst.*, **69** (1–4), 347–360.

Mehrotra, K. and Mahapatra, P.R. (1997) A jerk model for tracking highly maneuvering targets. *IEEE Trans. Aerospace Electron. Syst.*, **33**(4), 1094–1105.

Mutambara, A. (1998) *Decentralized Estimation and Control for Multisensor Systems*, CRC Press, Boca Raton, FL.

Oh, H., Kim, S., Shin, H., Tsourdos, A., and White, B. (2013a) Coordinated standoff tracking of groups of moving targets using multiple UAVs. *21st Mediterranean Conference on Control and Automation, Crete, Greece*.

Oh, H., Kim, S., Shin, H., Tsourdos, A., and White, B. (2014a) Behaviour recognition of ground vehicle using airborne monitoring of UAVs. *Int. J. Syst. Sci.*, **45**(12), 2499–2514.

Oh, H., Kim, S., Shin, H., White, B., Tsourdos, A., and Rabbath, C. (2013b) Rendezvous and standoff target tracking guidance using differential geometry. *J. Intell. Robot. Syst.*, **69**, 389–405.

Oh, H., Kim, S., Tsourdos, A., and White, B. (2012a) Decentralised road-map assisted ground target tracking using a team of UAVs. *9th IET Data Fusion & Target Tracking, London, UK*.

Oh, H., Kim, S., Tsourdos, A., and White, B. (2012b) Road-map assisted standoff tracking of moving ground vehicle using nonlinear model predictive control. *American Control Conference, Montreal, Canada*.

Oh, H., Kim, S., Tsourdos, A., and White, B. (2014b) Decentralised standoff tracking of moving targets using adaptive sliding mode control for UAVs. *J. Intell. Robot. Syst.*, **76**, 169–183.

Oh, H., Shin, H., Kim, S., Tsourdos, A., and White, B. (2013c) Airborne behaviour monitoring using Gaussian processes with map information. *IET Radar Sonar Navigation*, **7**(4), 393–400.

Oh, H., Turchi, D., Kim, S., Tsourdos, A., Pollini, L., and White, B. (2014c) Coordinated standoff tracking using path shaping for multiple UAVs. *IEEE Trans. Aerospace Electron. Syst.*, **50**(1), 348–363.

Parker, L. (2002) Distributed algorithms for multi-robot observation of multiple moving targets *Autonomous Robots*, **12**(3), 231–255.

Prevost, C., Theriault, O., Desbiens, A., and Poulin, E. (2009) Receding horizon model-based predictive control for dynamic target tracking: a comparative study. *AIAA Guidance, Navigation, and Control Conference, Chicago, IL*.

Quigley, M., Goodrich, M., Griffiths, S., Eldredge, A., and Beard, R. (2005) Target acquisition, localisation, and surveillance using a fixed-wing mini-UAV and gimbaled camera. *IEEE International Conference on Robotics and Automation, Barcelona, Spain*.

SIAS (2011) S-Paramics Software. Available at http://www.sias.com.

Summers, T., Akella, M., and Mears, M. (2009) Coordinated standoff tracking of moving targets: control laws and information architectures. *J. Guid. Control Dyn.*, **32**(1), 56–69.

Tang, Z. and Ozguner, U. (2005) Motion planning for multitarget surveillance with mobile sensor agents. *IEEE Trans. Robot.*, **21**, 898–908.

Taylor, J. (1979) Cramer–Rao estimation error lower bound computation for deterministic nonlinear systems. *IEEE Trans. Autom. Control*, **24**(2), 343–344.

White, B., Zibkowski, R., and Tsourdos, A. (2007) Direct intercept guidance using differential geometry concepts. *IEEE Trans. Aerospace Electron. Syst.*, **43**(3), 899–919.

Wise, R. and Rysdyk, R. (2006) UAV coordination for autonomous target tracking. *AIAA Guidance, Navigation and Control Conference, Keystone, CO*.

Yoon, S., Park, S., and Kim, Y. (2013) Circular motion guidance law for coordinated standoff tracking of a moving target. *IEEE Trans. Aerospace Electron. Syst.*, **49**(4), 2440–2462.

Chapter 28

Distributed Situational Awareness and Control

Seng Keat Gan, Zhe Xu, and Salah Sukkarieh

Australian Centre for Field Robotics, The University of Sydney, Sydney, Australia

1 INTRODUCTION

The coordination of distributed mobile sensors such as unmanned aerial vehicles (UAVs) has attracted significant interest in recent years. An important focus is its application to situational awareness tasks. Example applications include environmental monitoring and mapping (Kovacina *et al.*, 2002; Antonelli, Chiaverini, and Marino, 2012), area coverage and exploration (Renzaglia, 2012), search and rescue, (Hollinger *et al.*, 2009) and target tracking (Frew *et al.*, 2008).

The first challenge in many of these applications is the size and dynamic nature of the environment. In most situational awareness applications, the operational field is usually large compared to the sensing capabilities of each UAV. Further, the state of the underlying object of interest is usually nonstationary, both spatially and temporally. Representative examples include searching for a target in remote areas and the mapping of atmospheric weather. Figure 1 shows a motivating example of a team of UAVs in cooperation with a team of ground robots in a search and rescue mission.

While platform mobility is beneficial in maintaining situational awareness in large, time-varying environments, the ability to coordinate distributed mobile teams is essential in many problem instances for three key reasons:

- *Survivability*: A single platform is vulnerable to a range of faults such as mechanical faults, software errors, or communication failure. On the other hand, a distributed team can be robust to failures of individual platforms.
- *Scalability*: As the problem size increases, more platforms can be added to a distributed system.
- *Modularity*: Each platform is able to operate independently of others. Platforms may be heterogeneous and an example is shown in Figure 2; different sensor agents may carry different sensor payloads to suit a variety of missions.

In this chapter, we will focus on probabilistic approaches to the situational awareness problem. Probabilistic approaches are preferred as they model a probability distribution over the target state estimate, from which we can derive the target state estimate (the mean estimate) as well a measure of confidence (the estimate uncertainty). In the context of a probabilistic target state estimate, control for situational awareness is to minimize the uncertainty in the estimate, also known as information-theoretic control.

2 DISTRIBUTED SITUATIONAL AWARENESS

Before one can consider how to control a team of UAVs to improve situational awareness, one must first consider the decentralized data fusion (DDF) problem. Decentralized data

Unmanned Aircraft Systems. Edited by Ella Atkins, Aníbal Ollero, Antonios Tsourdos, Richard Blockley and Wei Shyy.

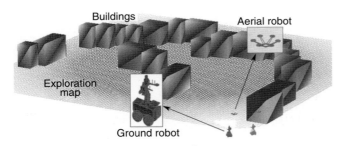

Figure 1. Example of a multiagent exploration mission. Two ground robots and two aerial robots exploring an unknown large-scale urban environment.

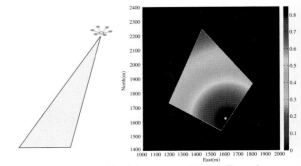

(a) Left: An aerial robot with vision sensor; Right: Example sensor observation model

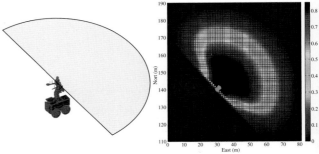

(b) Left: A ground robot with radar; Right: Example sensor observation model

Figure 2. A team of heterogeneous mobile sensor agents. The aerial robot has a vision sensor that classifies and localizes targets from the air. The ground robot is equipped with a radar that further provides high accuracy localization information. Aerial robots usually have higher mobility while ground robots have longer endurance, processing power, and could carry more sensor payloads. A team of distributed heterogeneous mobile sensors in cooperation provides higher flexibility and applicability to a larger range of missions.

fusion is a class of algorithms that probabilistically processes the sensor observations from a team of intelligent agents such as UAVs to collectively estimate the state of a target.

Consider the previous example of a team of ground robots and UAVs performing a search and rescue mission. Each robot, which could have different sensing capabilities such as vision and radar, continuously takes local observations about the environment. Images taken from the vision sensor help to identify the target and provide relative bearing information. Radar, on the other hand, could not distinguish target type by itself but it provides high-accuracy range information for objects within its field of view. These local observations contain partial information about the state of the target of interest which are less informative individually. However, a more informative understanding of the target state can be obtained if they were to be combined and processed collectively.

In this section, we describe three common ways of representing the target state estimate: Kalman and information filters (IF), certainty grids (CGs) (CG) and particle filters, and how they may be applied to a distributed team of UAVs in a situational awareness mission.

2.1   Kalman and Information Filters

The well-known Kalman filter and its variant the information filter (IF) have been widely used in decentralized data fusion problems. The Kalman and information filters represent the probability distribution over the target state as a Gaussian distribution. This representation means the estimate and its uncertainty can be represented compactly, thus imparting low communication overheads when UAVs share information. These filters are commonly used for target tracking scenarios. However, this representation limits its applicability in certain problem domains. In target search problems, for example, the distribution can be highly multimodal.

For brevity, we will describe the mathematical formulation of only the information filter. In the IF, the target state $\mathbf{x}$ and its covariance $\mathbf{P}$ are parameterized into an information vector $\mathbf{y} = \mathbf{P}^{-1}\mathbf{x}$ and matrix $\mathbf{Y} = \mathbf{P}^{-1}$. A {sensor model} $\mathbf{H}$ must be defined such that an observation $\mathbf{z}$ at time k is modeled by $\mathbf{z}_k = \mathbf{H}_k\mathbf{x}_k + \mathbf{v}$, where $\mathbf{v}$ represents sensor noise with covariance $\mathbf{R}$.

A process model that describes the dynamics of the target must also be defined. The process model is encapsulated in a state transition matrix $\mathbf{F}$ such that $\mathbf{x}_k = \mathbf{F}_{k-1}\mathbf{x}_{k-1} + \mathbf{w}$, where $\mathbf{w}$ represents process noise with covariance $\mathbf{Q}$ and $\mathbf{G}$ describes how process noise impacts the state.

The first stage of the IF, prediction, uses the process model to propagate the estimate forward in time. The propagation of the estimate from $k-1$ to k is shown below (Maybeck, 1982). Here, $\mathbf{M}_k = \mathbf{F}_k^{-T}\mathbf{Y}_{k-1\,|\,k-1}\mathbf{F}_k^{-1}$, $\mathbf{\Sigma}_k = \mathbf{G}_k^T\mathbf{M}_k\mathbf{G}_k + \mathbf{Q}_k^{-1}$, and $\boldsymbol{\Omega}_k = \mathbf{M}_k \mathbf{G}_k \boldsymbol{\Sigma}_k^{-1}$:

$$\mathbf{Y}_{k|k-1} = \mathbf{M}_k - \boldsymbol{\Omega}_k \boldsymbol{\Sigma}_k \boldsymbol{\Omega}_k \tag{1}$$

$$\mathbf{y}_{k|k-1} = [\mathbf{I} - \boldsymbol{\Omega}_k \mathbf{G}_k^T] \mathbf{F}_k^{-T} \mathbf{y}_{k-1} \tag{2}$$

As time increases, the process noise effectively increases the target state estimation uncertainty, while the process model shifts the mean of the distribution.

The second stage, update, fuses target observations:

$$\mathbf{I}_k = \mathbf{H}_k^T \mathbf{R}_k^{-1} \mathbf{H}_k \tag{3}$$

$$\mathbf{i}_k = \mathbf{H}_k^T \mathbf{R}_k^{-1} \mathbf{z}_k \tag{4}$$

$$\mathbf{Y}_{k|k} = \mathbf{Y}_{k|k-1} + \mathbf{I}_k \tag{5}$$

$$\mathbf{y}_{k|k} = \mathbf{y}_{k|k-1} + \mathbf{i}_k. \tag{6}$$

A sensor update reduces the state estimation uncertainty and performs a correction on the estimation mean. Figure 3 shows an example of a UAV tracking a target. It illustrates the changes in target IF estimate during prediction and after a sensor observation update.

In a decentralized UAV team, each UAV maintains its own local IF, updating it based on its own sensor observations. UAVs also share their target state estimates by communicating the information vector $\mathbf{y}$ and information matrix $\mathbf{Y}$. Established techniques for incorporating information communicated by other UAVs include channel filters (Grime, 1994) and covariance intersection (Julier et al., 2008). Below, we will describe covariance intersection, a technique that is applicable to general networks but is conservative, that is, can lead to some of the communicated information not to be used.

When one UAV a shares its target state estimate (by communicating $\mathbf{y}_a$ and $\mathbf{Y}_a$) with a second UAV b, the second UAV can update its existing estimate ($\mathbf{y}_b$, $\mathbf{Y}_b$) to an updated estimate ($\mathbf{y}'_b$, $\mathbf{Y}'_b$) as follows:

$$\mathbf{y}_{b'} = \omega \mathbf{y}_b + (1 - \omega)\mathbf{y}_a \tag{7}$$

$$\mathbf{Y}_{b'} = \omega \mathbf{Y}_b + (1 - \omega)\mathbf{Y}_a. \tag{8}$$

In the above equations, ω is a scaling factor between 0 and 1 and is often chosen to maximize the information content in the information matrix, as measured by its determinant $|\mathbf{Y}_{b'}|$.

2.2 Certainty Grid

The certainty grid approach models the target state estimate by discretizing the state space into a grid. The probability of the target being in a given state is recorded against each grid

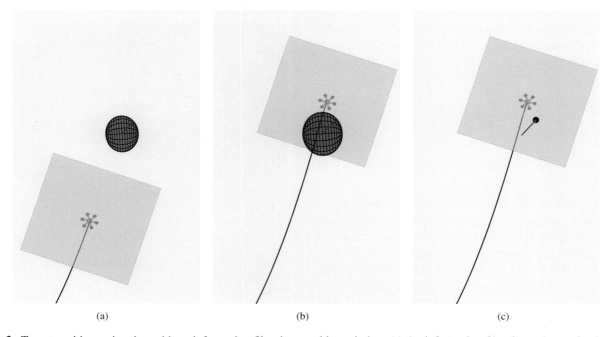

(a)   (b)   (c)

Figure 3. Target position estimation with an information filter in a tracking mission. (a) An information filter is used to maintain a target position estimate, represented as an ellipsoid in 3D space. Position uncertainty increases (size of ellipsoid increases) during the prediction stage due to process noise. (b) The target appears within the UAV's field of view. A relative range-bearing sensor observation of the target is taken. (c) The target estimation uncertainty reduces after fusing the sensor observation (a reduction in the size of ellipsoid).

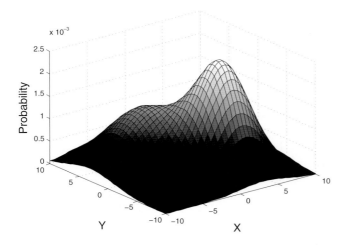

Figure 4. Example of a certainty grid for a two-dimensional state. The state space is discretized into grid cells. The probability of the target being in each state is stored against each grid cell.

cell. The advantage of a CG representation is its ability to represent multimodal distributions, making it applicable to problems such as target search. However, CG representations suffer from the "curse of dimensionality," and have high communication overheads. An example of a CG representation is shown in Figure 4.

As with the information filter, the certainty grid is maintained in two stages: {prediction} and {update}. The prediction stage convolves the current target state estimate $p(\mathbf{x}_k \mid \mathbf{z}_k, ..., \mathbf{z}_0)$ with the process model $p(\mathbf{x}_{k+1} \mid \mathbf{x}_k)$. The update stage corrects this distribution with the sensor model $p(\mathbf{z}_k \mid \mathbf{x}_k)$ and normalizes the result by a constant c such that it remains a valid distribution whose integral equals to one. The prediction and update equations are shown as follows:

$$p(\mathbf{x}_k \mid \mathbf{z}_{k-1}, ..., \mathbf{z}_0) = \int p(\mathbf{x}_{k-1} \mid \mathbf{z}_{k-1}, ..., \mathbf{z}_0) p(\mathbf{x}_k \mid \mathbf{x}_{k-1}) d\mathbf{x}$$

$$(9)$$

$$p(\mathbf{x}_k \mid \mathbf{z}_k, ..., \mathbf{z}_0) = \frac{1}{c} p(\mathbf{x}_k \mid \mathbf{z}_{k-1}, ..., \mathbf{z}_0) p(\mathbf{z}_k \mid \mathbf{x}_k). \quad (10)$$

In a decentralized team, each UAV maintains its own CG representation of its target state estimate. A UAV must communicate the CG to other team members to share its estimate, which can incur significant communication overheads. Each UAV can then follow a simplistic policy whereby it selects the most informative target state estimate among the CGs it receives from other team members and its own CG estimate. The measure of how informative an estimate is can be the entropy $H = \Sigma_{\mathbf{x}} p(\mathbf{x}) \log p(\mathbf{x})$ or an application-specific measure. Figure 5 shows an example of two UAVs performing a certainty grid Bayesian update in a target search mission.

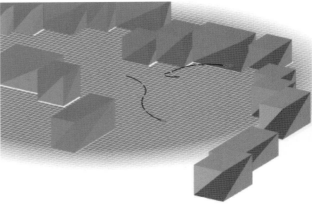

(a) Prior target probability distribution

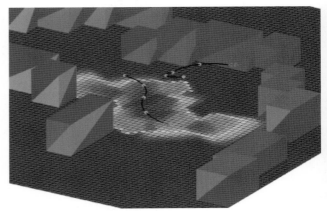

(b) Sensor likelihood

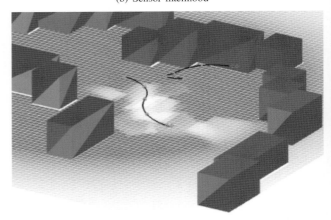

(c) Posterior target probability distribution

Figure 5. Demonstration of a certainty grid Bayesian update process in a target search mission. (a) Prior target probability distribution represented as a certainty grid. (b) Each UAV took five vision-based sensor observations and communicated these observations to the other. These observations collectively form a joint sensor observation likelihood. (c) Sensor likelihood is fused with prior distribution to form a posterior distribution. In and, the gray-scale intensity of each cell represents the probability of the target being in a particular cell. Black indicates that it is likely that the target could be in a cell, while white indicates cells where the target is unlikely to be.

2.3 Particle Filters

Particle filters address the "curse of dimensionality" in the CG representation by sampling the distribution instead of discretizing it. Particle filters approximate the true distribution using the sum of n-weighted samples, or particles $\mathbf{x}_i$, such that the weights w_i sum to 1. In this approximation, δ represents the Dirac delta function (Arulampalam, 2002):

$$p(\mathbf{x}_k \mid \mathbf{z}_k, ..., \mathbf{z}_0) \approx \Sigma_i^n w_i \delta(\mathbf{x}_k - \mathbf{x}_i). \tag{11}$$

In the prediction stage, each particle is moved according to the process model (including process noise) so as to approximate the distribution $p(\mathbf{x}_k \mid \mathbf{x}_{k-1})$. In the update stage, the particle weights are modified according to the sensor model $p(\mathbf{z}_k \mid \mathbf{x}_k)$ and normalized such that they sum to 1. This update often results in many particles with low weights which do not contribute significantly to the representation of the target state estimate. The particle filter is then {resampled} by removing particles with low weight and adding new particles near particles with high weight.

One approach to applying particle filters in a decentralized estimation problem is to have each UAV maintain a local particle filter. Each UAV then approximates distribution represented by the particle filter with a Gaussian distribution or a mixture of Gaussians (Ong, 2008). This Gaussian mixture is then shared with other team members, who fuse it with their local estimates using extensions of the techniques described for the Kalman filter/information filter (e.g., channel filters or covariance intersection) (Upcroft, 2005). Finally, the Gaussian mixture approximation can be sampled to return to a particle-based representation.

3 DISTRIBUTED CONTROL

Distributed control builds on top of the distributed situational awareness described in the previous section. In this section, we first introduce two main classes of distributed control strategies: coordination and cooperation strategies. We then give details to how the general complex distributed control problem can be reduced to a problem that can be solved online followed by a review on different types of online decision-making algorithms.

3.1 Coordination and Cooperation

In a multiagent team, coordination and cooperation are two types of distributed control strategies used for situational awareness missions. Coordination strategies are passive strategies based on the team having a shared estimate of the target state, often achieved through the decentralized data fusion techniques outlined in the previous section. Decisions are made locally without the need for negotiation among team members. A UAV's influence, if any, is implicitly realized over time.

On the other hand, members of a cooperative team actively negotiate among themselves to decide on a joint action that explicitly maximizes some future expected information gain. On top of the shared target state estimate, cooperative strategies require an additional communication layer for its team negotiation scheme, a process known as decentralized decision making (DDM).

3.2 Problem Definition

The DDM problem aims to find a control policy for a team of UAVs which minimizes uncertainty (or maximizes information) in the aforementioned target state estimates. In the context of distributed information-theoretic control, the target state estimate is also known as the belief as it is the basis upon which the team makes decisions.

A control policy π comprises a set of mapping functions over a discrete set of steps that define the mission length M, such that $\pi = [\mu_1, ..., \mu_M]$. Each mapping function μ_k defines a mapping from the space of beliefs β to the joint team control space U such that $\mu_k : B \rightarrow U$. In other words, given a belief, the mapping function asks each UAV in the team what control actions to take. The *goodness* of a policy is measured by a loss function $L : B \times U \rightarrow R$, which provides a scalar metric for how much loss in mission performance occurs if an action is taken in a belief state. In the context of information-theoretic control, this function is the information lost or negative information gain. The general team decision-making problem is to find the optimal policy π^*, which is defined by

$$\pi^* = \underset{\pi}{arg\,min}\ E\left\{ \sum_{k=0}^{M} L(b_k, \mu_k(b_k)) \right\}. \tag{12}$$

This optimal policy can be numerically solved for finite-state systems using dynamic programming (DP), first introduced by Bellman (Bellman, 1956) and further refined by Bertsekas (Bertsekas, 1995). The DP approach recursively solves for the optimal policy offline, stores it in memory onboard, and applies it online through a table lookup. Nevertheless, this approach is well known to be intractable for real-life operations, especially for problems with large belief space (Powell, 2007). DP requires a significant amount of computational resources to solve for the optimal policy.

Subsequently, a substantial amount of memory is needed to store the optimal policies for online operations.

3.3 Problem Simplifications

Instead of solving for the globally optimal policy offline, a more practical approach is to solve for a suboptimal policy online. The complex decision-making problem is reduced into a simpler subproblem using an open-loop receding horizon model-based forward planning architecture. The reduced problem allows tractable solution approaches to run online. The simplified problem preserves the continuous form of the system state and control, suitable for decision-making problems in continuous domains.

3.3.1 Open-loop planning

Instead of solving for an optimal policy that covers every possible situation, a simplified approach attempts to solve for an optimal open-loop plan based on the latest available belief. An open-loop plan is expected to be executed for the duration of the plan regardless of any observed future events. By planning open-loop, the optimal policy is simplified to the following equation, where $v_k \triangleq [u_k, u_{k+1}, ..., u_{k+M-1}]$ is a vector of actions over a planning horizon:

$$v_k^* = \underset{v_k}{arg\,min}\ E\{\sum_{l=k}^{M-1} L_B(b_l, u_l)\}. \qquad (13)$$

This approach avoids planning for the optimal policy for all possible scenarios described by the general team decision-making problem. Instead, the optimization process is performed online, thus removing the need to store the optimal policy in memory. However, since an optimal open-loop plan is obtained based on the expected future observations and models, the plan is no longer optimal if the actual observations and system behaviors are different from their expectations. The optimization process has to be repeated online to incorporate any new state information and changes to the models in order for this approach to be adaptive.

3.3.2 Receding horizon planning

In spite of the simplifications made in the open-loop planning model, the evaluation of each option v_k can still be computationally expensive if the mission length M is large. Receding horizon planning alleviates this computational burden by solving the same problem but for a shorter planning horizon.

At each replanning instants k, instead of solving for an optimal plan for the full mission length $M - k$, a receding

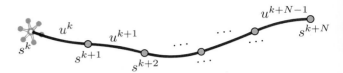

Figure 6. Parameterized action vector for the length of N-step look-ahead actions. The dots are the agent's expected sensing state. The segment between each consecutive sensing state is an action, which is constant within the segment.

horizon approach solves for a suboptimal plan for a shorter action horizon N, where $N \ll M$ (Chung et al., 2007). Once a suboptimal plan is obtained, a subset of the N steps is executed before the next replanning occurs. An open-loop plan with limited look-ahead actions is illustrated in Figure 6. The solution is a relatively myopic plan and only optimal over the action horizon N.

3.3.3 Model-based planning

The last simplification to the problem assumes that the UAVs' actions are deterministic and that there is a reliable closed-loop controller capable of executing a plan v_k. With this deterministic model, the planner is able to forward predict the future UAV states. Since the platforms' states and state transitions are deterministic, the belief now contains only the probability distribution of the target.

Under these assumptions, a planner is able to explicitly model the future sensor and belief states, given an action, the planner can model where this action will take the UAV. Given the new UAV state, the planner can then model what sensor observations will be made using the sensor model. This observation allows the planner to model how the belief will change had the UAV made that observation. This removes the need for the decision-making process to perform expectations over the belief states, thus reduces the team decision-making problem into a (tractable) optimization problem.

3.3.4 Forward planning

We have reduced a complex problem that could be solved suboptimally online. Nevertheless, solving such problem still requires certain amount of online computation time. Many autonomous systems, such as a fixed-wing UAV system, do not have the flexibility to stop and plan. One possible solution to this problem is to plan for future actions while executing the previously planned actions. Algorithm 1 outlines our proposed algorithmic framework for real-time decision making in a decentralized system. A diagram of this process is depicted in Figure 7.

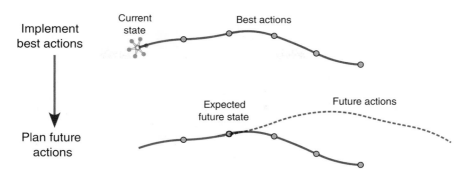

Figure 7. A fixed-time forward planning architecture. During each replanning, the UAV first transmits its latest plan to a controller for closed-loop execution of this plan. The UAV then forward propagates its state and target belief to the next replanning instant and performs real-time decision making from that future state and belief.

Forward planning of agent i 1.

```
1: repeat
2:   if Replanning required then
3:     Implement latest local plan v_k^i
4:     Update current state s_k^i
5:     Forward propagate state s_k^i and belief
         b_k^i to next replanning step
6:   end if
7:   Perform decentralized data fusion
8:   Perform decentralized decision
       making at future state and belief
9: until mission completes
```

Following on the example in Figure 7, the UAV has an action horizon of five steps and planning horizon of two steps. At replanning instant k, the UAV plans for its open-loop action vector v_k which is composed of five piecewise constant control actions. Given the initial state s_k and an action vector v_k, the UAV forward propagated its state and target belief for two steps through its process model and data fusion algorithm. Decentralized decision making is performed at this future state and belief. The information-theoretic measure of some future action vector can be evaluated in a similar fashion by forward propagating the states and belief along the future action vector.

3.4   Solution Approaches

This section gives an overview to the approaches used in solving the simplified optimization problem. The advantages and difficulties of these approaches are discussed, mainly with respect to their applicability to online trajectory planning for a team of decentralized agents.

Agents are usually involved in an iterative process during decentralized planning. There are mainly two categories of online decentralized algorithms: sequential and asynchronous. These can be depicted in Figure 8. Sequential algorithms perform update and message passing in a predefined agents' ordering. Such approaches are usually easier in terms of implementation, but they usually require longer planning time to reach consensus and suffer modularity issues for large team sizes.

Asynchronous algorithms, on the other hand, do not rely on an ordered sequence. Each planning agent is able to perform iterations continuously based on the latest knowledge communicated from its neighbors, fully utilizing the time slots allocated for team planning (Androulakis, 1999).

3.4.1   Brute force search

In this approach, the action space is discretized into a finite set. This discretization is usually uniform over the entire action space or a subset of the action space. The loss function is evaluated for each member of this action set. The action with minimum loss is selected. The result of a brute force search is a suboptimal solution. The approach is relatively simple to implement and it is a constant time operation. However, this approach scales exponentially with the action horizon and it is difficult to decentralize as each agent is required to solve the equivalent centralized problem.

Nevertheless, due to its simplicity, this decision-making approach is commonly applied to many information gathering demonstrations (Cole, 2009). To avoid exponential scaling with respect to the number of planning horizons, modified versions of this approach such as depth-first search and best-first search on a decision tree have been developed.

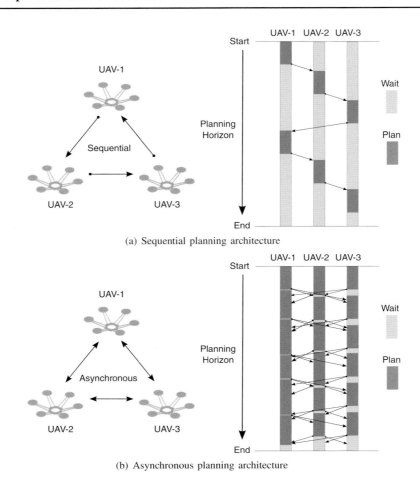

(a) Sequential planning architecture

(b) Asynchronous planning architecture

Figure 8. (a) In a sequential planning architecture, while one UAV is planning, other UAVs have to stop planning and wait for their turn to plan in the sequential link. (b) Asynchronous planning architecture allows UAVs to continuously plan and communicate their decision to their neighbors until team solution converges or until the end of the planning horizon.

3.4.2   Sampling-based planning

Instead of uniformly discretizing the action space, this approach generates samples in the action space using randomized or probabilistic sampling techniques. These samples build a graph in the action space in which a graph-based search algorithm can be used to obtain the optimal actions. Representative approaches include the rapidly exploring random tree (RRT), rapidly exploring random graph, and probabilistic road map. A good review for sampling-based algorithms can be found in (Karaman, 2001; Lavalle, 2006). Figure 9 shows a UAV plan for its action using a sampling-based algorithm. Within this set of plans, the UAV selects the one that minimizes the lost function and executes this plan.

Probabilistic approaches are often able to cover the action space more efficiently than brute force methods and can naturally incorporate constraints such as obstacles into the

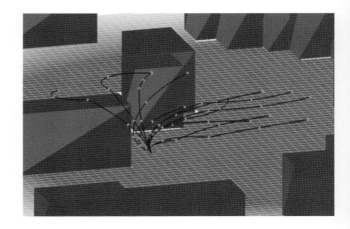

Figure 9. A UAV plans its trajectory using a sampling-based algorithm.

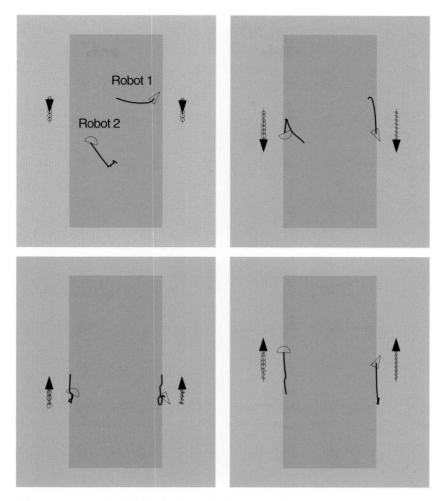

Figure 10. Two UAVs tracking two targets. The UAVs (hollow triangle and semicircle) are constrained to operating in the shaded area. The targets (solid triangles) traverse outside this area. The UAVs adopt a "divide-and-conquer" strategy, an emergent behavior from the decentralized optimization that aims to minimize uncertainty in the target states.

planning process. However, sampling-based approaches are difficult to decentralize for multi-UAV planning. A sequential ordering-type team planning is mandatory and no local convergence can be guaranteed.

3.4.3   Explicit solution

For some situational awareness problems, the analytical form to the optimal solution may be explicitly obtained. These problems usually involve relatively simple process and sensor models. Simplifications such as the assumptions of Gaussian noise and the availability of Gaussian belief representations are usually required. Even so, these problems are usually analytically solvable only for the case of a local zero look-ahead action, also known as a greedy solution. The

solution is essentially a control law that explicitly computes the optimal action based on the current system state. Example applications can be found in (Grocholsky, 2003; Zhou, 2010).

3.4.4   Gradient-based optimization

Another approach is to apply gradient-descent optimization to find actions that minimize the loss function. Starting with an initial action, the joint team action is perturbed in the direction defined by the derivative of the loss function. This process is applied iteratively to refine the action. With this approach, convergence to minima of the loss function can be guaranteed, either to a local or a global optimal solution, depending on the nature of the objective function, specifically its convexity. This approach can be run asynchronously across planning agents. It is easily decentralized through variations of standard gradient-based optimization (Mathews, 2008) and has been further extended to handle problems with various spatial-temporal constraints (Gan *et al.*, 2014).

In many situational awareness problems, the loss function is nonconvex. Thus, the selection of the initial action vector highly influences which local minima the gradient-descent algorithm can achieve. This can be obtained using primitive motion approaches or through sampling-based algorithms we have described earlier such as an RRT (Yang, Gan, and Sukkarieh, 2010). While selecting which algorithm to use, it is important to tune the algorithm such that it is computationally efficient. Their main purpose here is to rapidly explore the action space, generate some good initial solution estimate that has a higher chance for the optimization routine to yield a good local minimum.

While most gradient-based optimization formulations tackle problems that are continuous in nature, some multi-UAV coordination problems can involve both continuous and discrete variables. An example of such a problem is where multiple UAVs track multiple targets and must coordinate both which UAV is assigned to which target and how each UAV moves. This class of mixed discrete continuous optimization problems can be solved by relaxing the discrete constraints and applying gradient-descent optimization (Xu *et al.*, 2013).

4   SIMULATION EXAMPLE

In this section, we present a simulation example of distributed situational awareness and decentralized control for target tracking. In the target tracking problem, the target state distributions are well approximated by Gaussian distributions and thus we will use the information filter to fuse observations by the UAVs. To solve the distributed decision-making problem, we apply gradient-based optimization to minimize the uncertainty in the target states over a fixed planning horizon.

In this example, as illustrated in Figure 10, two UAVs track two targets. The UAVs are constrained to operating in the shaded area while the targets traverse outside. The constrained operating region is modeled as a constraint in the optimization. The UAVs adopt a "divide-and-conquer" strategy where the UAVs divide and track a target each. This strategy was not "hard-coded" a priori rather it is an emergent behavior from the decentralized optimization that minimizes uncertainty in the target states.

5   CONCLUSION

In this chapter, we described the benefits of applying distributed UAV teams to situational awareness problems. The situational awareness problem was decomposed into two components: decentralized data fusion and team decision making to maximize information gain. The decentralized data fusion problem aims to build and share a target state estimate (or belief) across the UAV team based on each UAV's observations. Three approaches to decentralized data fusion were presented: the information filter, certainty grids, and particle filters. The information filter is appropriate in situations where the state being estimated is well represented by Gaussian distributions. Certainty grids can represent non-Gaussian distributions but are more demanding with respect to storage and communication bandwidth due to the "curse of dimensionality." Particle filters address this curse of dimensionality by sampling the distribution being represented.

The team decision-making problem selects the best team actions to minimize uncertainty (or maximize information) in the team's belief. The general distributed decision-making problem is intractable; however, in this chapter, we introduce a number of commonly used simplifications that allow for tractable and real-time solutions. This simplified problem can be solved using a number of techniques such as brute-force search, sampling-based planning, optimal control, and numerical optimization.\

REFERENCES

Androulakis, I. and Reklaitis, G. (1999) Approaches to asynchronous decentralized decision making. *Comput. Chem. Eng.*, **23**(3), 339–354.

Antonelli, G., Chiaverini, S., and Marino, A. (2012) *A coordination strategy for multi-robot sampling of dynamic fields*. Proc. of IEEE ICRA.

Arulampalam, M. S., Maskell, S., Gordon, N., and Clapp, T. (2002) A tutorial on particle filters for online nonlinear/non-Gaussian Bayesian tracking. *IEEE Trans. Sig. Process.* **50**(2) 174–188.

Bellman, R. (1956) Dynamic programming and {Lagrange} multipliers. *Proc. Natl. Acad. Sci. U.S.A.*, **42**(10), 767.

Bertsekas, D. (1995) *Dynamic Programming and Optimal Control*, Athena Scientific.

Chung, T. and Burdick, J. (2007) *A decision-making framework for control strategies in probabilistic search*. Proc. IEEE ICRA.

Cole, D. (2009) A cooperative UAS architecture for information-theoretic search and track. Ph.D. thesis. The University of Sydney.

Frew, E., Lawrence, D., and Morris, S. (2008) Coordinated standoff tracking of moving targets using Lyapunov guidance vector fields. *J Guid. Control Dynam.*, **31**(2), 290–306.

Gan, S. K., Fitch, R., and Sukkarieh, S. (2014) Online decentralized information gathering with spatial–temporal constraints. *Auton. Robots*, **37**(1), 1–25.

Grime, S. and Durrant-Whyte, H. F. (1994) Data fusion in decentralized sensor networks. *Control Eng. Practice*, **2**(5), 849–863.

Grocholsky, B., Makarenko, A., Kaupp, T., and Durrant-Whyte, H. (2003) Scalable control of decentralised sensor platforms. *Infor. Process. Sensor Netw.*, **2634**, 96–112.

Hollinger, G., Singh, S., Djugash, J., and Kehagias, A. (2009) Efficient multi-robot search for a moving target. *Int. J. Robot. Res.*, **28**(2), 201–219.

Julier, S. J. and Uhlmann, J. K., *General Decentralized Data Fusion with Covariance Intersection (CI)*, CRC Press, 2001.

Karaman, S. and Frazzoli, E. (2011) Sampling-based algorithms for optimal motion planning. *Int. J. Robot. Res.*, **30**(7), 846–894.

Kovacina, M., Palmer, D., Yang, G., and Vaidyanathan, R. (2002) Multi-agent control algorithms for chemical cloud detection and mapping using unmanned air vehicles. *Proc. IEEE/RSJ IROS*.

LaValle, S. (2006) *Planning Algorithms*, Cambridge University Press.

Mathews, G. (2008) Asynchronous decision making for decentralised autonomous systems. Ph.D. thesis. The University of Sydney.

Maybeck, P. S. (1982) *Stochastic Models, Estimation, and Control*, vol.1, Academic Press.

Ong, L.-L., Bailey, T., Durrant-Whyte, H., and Upcroft, B. Decentralised particle filtering for multiple target tracking in wireless sensor networks. *Information Fusion, 2008 11th Internatioxnal Conference on, IEEE*, 2008, pp. 1–8.

Powell, W., (2007) *Approximate Dynamic Programming: Solving the Curses of Dimensionality*, Wiley-Blackwell.

Renzaglia, A., Doitsidis, L., Martinelli, A., and Kosmatopoulos, E., (2012) Multi-robot three-dimensional coverage of unknown areas. *Int. J. Robot. Res.*, **31**(6), 738–752.

Upcroft, B., Ong, L. L., Kumar, S., Ridley, M., Bailey, T., Sukkarieh, S., and Durrant-Whyte, H. Rich probabilistic representations for bearing only decentralised data fusion. *2005 8th International Conference on Information Fusion, vol. 2, IEEE, 2005*, pp. 8–pp.

Xu, Z., Fitch, R., Underwood, J., and Sukkarieh, S. (2013) Decentralized coordinated tracking with mixed discrete–continuous decisions. *J. Field Robot.*, **30**(5), 717–740.

Yang, K., Gan, S., and Sukkarieh, S. (2010) An efficient path planning and control algorithm for {RUAV}s in unknown and cluttered environments. *J. Intell. Robot. Syst.*, **57**(1), 101–122.

Zhou, K. and Roumeliotis, S. (2010) Multirobot active target tracking with combinations of relative observations. *IEEE Trans. Robot.*, **27**(4), 1–18.

Chapter 29

Cooperative Search, Reconnaissance, Surveillance

José Joaquin Acevedo,[1] **Begoña C. Arrue,**[1] **and Aníbal Ollero**[2]

[1]*Grupo de Robótica, Visión y Control, Universidad de Sevilla, Seville, Spain*
[2]*Universidad de Sevilla and Scientific Advisory Department of the Center for Advanced Aerospace Technologies, Seville, Spain*

1 INTRODUCTION

The surveillance and monitoring issues are related to a large set of problems that have been considered in the literature: from designing robots for surveillance operations Beainy and Commuri (2009) to high-level efficient frameworks to detect and avoid intruders Darbha *et al.* (2010) using a set of ground stations. Other authors, as in Zhang and Xiao (2013), address the cooperation between ground stations and unmanned aerial vehicles in border defense missions.

A complete surveillance system involves many activities, but they all may be summarized as three main tasks:

- Detecting new events, intruders, or information of interest, which is related to plan efficiently the motion of the mobile agents based on the known information and estimations about the problem, the environment, and the rest of the agents.

- Reporting the new detected information to the rest of the team of agents such that every element in the system can update its information about the mission.

- Handling the detected events, which is related to deciding a more suitable agent to go the event location to tackle it. Given a new detected event or intruder, it can be considered a new target with a set of features.

This chapter focuses on the first defined task: search or monitoring in a cooperative manner, but taking into account the second defined task to consider the information propagation even under communication constraints.

1.1 Cooperative Surveillance with Multiple Aerial Robots

Surveillance, search, or reconnaissance with aerial robots or unmanned aerial systems (UAS) has been a very relevant topic over the past few years due to the interest for applications, including border defense, search and rescue to aid people in disaster or dangerous situations, fire or pollution detection in natural, industrial, or civilian environment, intruder detection in private properties, coast control, traffic control, and so on. The problem is well known in the literature and different aerial robotic platforms have been developed in the recent years.

This chapter addresses the monitoring missions with multiple aerial robots or UAS. Cooperative multi robot

Unmanned Aircraft Systems. Edited by Ella Atkins, Aníbal Ollero, Antonios Tsourdos, Richard Blockley and Wei Shyy.
© 2016 John Wiley & Sons, Ltd. ISBN: 978-1-118-86645-0.

systems (or multi-UAS systems) offer many advantages in these applications, mainly robustness against failures, higher spatial coverage, and an efficient deployment Viguria, Maza and Ollero (2010). Also, increasing the number of robots results in better performance and robustness compared to single-robot coverage N. Hazon and Kaminka (2008). For instance, while a UAV leaves the mission temporally to recharge energy, the rest of the team can assume its tasks to maintain overall mission performance Acevedo *et al.* (2015).

Nevertheless, cooperation between multiple aerial robots requires a robust coordination method and it is a very difficult challenge to overcome, even more than a distributed manner. In order to improve the robustness of the whole system, a distributed coordination control is the most efficient option to approach the surveillance missions with multiple aerial robots. Robustness and scalability are very relevant topics that are directly related to the distributed and decentralized control. Moreover, taking into account that a totally open communication among all the robots cannot be guaranteed in many cases (too large areas, for instance), a distributed and decentralized control is required. Another challenge to overcome in multi-UAS systems is to address how to permanently maintain a mission accounting for the limited energy storage capability of the UAS. The use of thermal sources allows to extend the flight of autonomous gliders and, therefore, the mission duration Acevedo *et al.* (2014).

2   A PATH PLANNING PROBLEM

Regardless of their ultimate goal, the search, reconnaissance, or surveillance missions address a path-planning problem where the aerial robots have to decide how to cover a defined area based on a set of potential conditions or constraints: zones with different priorities, hostile environment, smart targets, and so on When there are no additional constraints or they are *a-priori* unknown by the system, a frequency-based approach is the most suitable way to address the problem. It tries to maximize the frequency of visit at any position into the area of interest. More details about this approach are provided in Section 3.

The problem dealt with here can be stated as follows. Given an area and a set of aerial robots, they have to decide their paths to monitor the area optimizing a cost function, which can be related to finish time, security, power consumption, and so on. Some authors consider the actual area shape and aerial robot capabilities to plan the paths, but others simplify the area before addressing the path-planning problem.

2.1   View Point Based Search

These approaches assume that only a set of fixed targets or positions is interesting to be monitored by the robots. So, the question is how the robots should visit the different view points or way points taking into account the objective function and the different constraints. In Lim and Bang (2009), a path-planning algorithm to monitor a set of targets with different priorities assuming a hostile environment is proposed based on a combinatorial approach. This solution is computationally too expensive for multi robot problems. A way point-based surveillance problem assuming different priorities is solved in Bao, Fu and Gao (2010) based on a *particle swarm optimization*.

The main advantage of this problem is that the path-planning algorithm does not need to consider the area shape and the covering capabilities of the robots. However, the robots do not always have information about these way points of interest.

2.2   Grid-Map Based Search

Another way to facilitate the area coverage of path-planning problem when there are no way points is re defining the space as a grid, dividing it into disjoint cells (defined as squares, hexagons, etc.) such that each cell center may be considered as a way-point of interest to be monitored by the robots Bertuccelli and How (2005). Yang, Polycarpou and Minai (2007) propose to create a *cognitive map*, where the cells increment their associated value periodically. The union of all the cells should include the whole area and each robot should have the covering capabilities to monitor each cell being in its center.

These approaches have the same advantages as those based on way points and may be used even without previous information about the targets of interest. However, they have some drawbacks related to their performance when they are used by a team of heterogeneous aerial robots (with different covering capabilities), because the cell shape and size have to be chosen such that all the robots can monitor the whole cell, but the robots covering range may be different. So, some robots will cover the area with a lower rate than that actually could (based on their capabilities).

2.3   Coverage Path-Planning Algorithms

Based on a cell decomposition or not, a coverage path-planning algorithm has to be proposed to solve the problem. The coverage path-planning algorithms try to build an efficient path that ensures that every point in the area can be monitored from at least one position in the path. Minimizing the path length will be the criteria to optimize. Different approaches have been developed in the literature to solve the area coverage path planning.

The area coverage problem with mobile robots can be solved from two different approaches: using on line or off line algorithms. In the on-line coverage algorithms, the area to be covered is unknown *a priori*, and step by step the robots have to discover obstacles, compute their paths, and avoid collisions. In Guruprasad, Wilson and Dasgupta (2012), an on line algorithm is proposed, in which both Voronoi spatial partitioning and coverage are handled in a distributed manner considering a multi robot system with minimal communication overhead. Based on a incrementally cognitive map (defined as a grid), an on line path planning is proposed in Yang, Polycarpou and Minai (2007) using the *opportunistic learning*.

On the other hand, in the off line algorithms, the robots have a map of the area and the obstacles, and can plan the path to cover the whole area. The most well-known off line coverage path planning is called *boustrophedon cellular decomposition* and was presented in Choset and Pignon (1997). It proposes to divide the whole area into smaller sub-areas that can be covered with a simple *back and forth* motion.

Hazon and Kaminka (2008) propose other off line algorithms where a spanning tree based on the approximated cellular decomposition is created and the coverage path is generated as the boundary around it. A neural network that creates a gradient field is proposed in Yang and Luo (2004), and the coverage path is obtained from the gradient field.

The coverage path-planning algorithm presented in Acevedo *et al.* (2014) generates a *pseudo symmetrical* coverage path. The pseudo symmetrical paths are defined in Acevedo *et al.* (2014) as closed paths where, given a set of positions (so-called *link positions*), the distance for each pair of consecutive links is always the same. The proposed method is based on the *back and forth* method but assume some modifications to obtain a closed coverage path and keep equal length between each pair of consecutive link positions.

3 CRITERIA FOR MONITORING MISSIONS

The patrolling problem assumes that the area or path is covered again and again and it is required to define a criterion to measure the performance of the mission. Agmon, Kaminka, and Kraus (2011) consider the patrolling problem

in adversarial settings, assuming some information about the probability of intruder appearing. They solve the problem using a probabilistic approach, maximizing the chances of detecting an intruder trying to penetrate through the patrolled path.

Fixed stations and mobile robots often have to cooperate to handle priorities in a surveillance mission. Zhang and Xiao (2013) address a patrolling mission where mobile robots and fixed sensors collaborate to detect and handle events with different priorities. They define and use the *digital pheromones* to solve the problem.

Elmaliach (2009) analyzes in detail the cooperative patrolling problem from a frequency-based approach, where the parameter to optimize is the frequency of visit to any position in the monitored area or path. However, a frequency-based approach usually leads to deterministic solutions.

3.1 Frequency-Based Approach

Assuming that there is no information about the intruders or events of interest, it may be considered that they could appear in any position with the same probability. So, all the positions into the region of interest should be monitored with the same frequency, leading to the so-called frequency-based approaches Elmaliach (2009). They prove that a solution where each position in the path is visited with the same period is the optimal solution to cover a perimeter with a set of agents. Therefore, the probability p of having an object of interest in a position s at time t without being detected is inversely related to the frequency f at which that position is monitored (the higher frequency of visit) (Figure 1). This probability can be defined as a Poisson counting process related to the inverse of the frequency of visits to the positions f, as given by the following expression:

$$p(s,t) = 1 - e^{-\gamma(1/f(r,t))} \tag{1}$$

where γ is related to the rate at which objects of interest can appear.

Then, the objective is to maximize the frequency of visits of any position into the area. Maximizing the frequency of visit of a position is equivalent to minimizing the elapsed time ET between two consecutive visits to that position or refresh time. Given an updating time dT, the elapsed time can be periodically updated as

$$\mathrm{ET}(s,t) = \begin{cases} 0, & \text{if } s \text{ is being monitored at time } t \\ \mathrm{ET}(s,t-dT) + dT, & \text{in other case} \end{cases} \tag{2}$$

where $\mathrm{ET}(r,0) = 0, \forall r \in S$

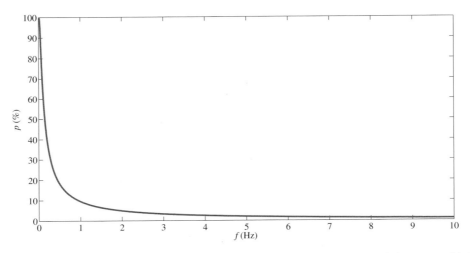

Figure 1. Probability p of having an object of interest in a position r at time t without being detected decreases with the frequency of visit.

Therefore, the objective is to decrease the maximum elapsed time through all the positions in the area or path. In this case, the objective function J to minimize can be formally defined as

$$J = \max_{s,t} \text{ET}(s, t) \qquad (3)$$

3.2 Urgency Criterion

In practical applications, such as security missions, it is usual to consider zones with different priorities. Based on the historical data and the updated information about the environment conditions and mission results, the nodes may predict the critical zones that, therefore, have to be more frequently monitored. For instance, in Saptharishi *et al.* (2002), the authors consider a surveillance problem with multiple autonomous vehicles, where the whole area is divided into zones with different priorities that have to be assigned to the vehicles.

The problem can not be solved based on the elapsed time and a new criterion has to be defined to solve the whole surveillance problem isolated from the priorities of different zones. In Pasqualetti, Franchi, and Bullo (2012), a team of homogeneous mobile robots monitor a set of viewpoints with different priorities. The authors consider the weighted elapsed time or urgency (based on the priorities) as parameter to minimize and propose a cyclic strategy with stops in the viewpoints. In these cases, there are some information about the intruders or event appearing (it means, the different priorities).

For instance, given a team of n heterogeneous aerial robots that patrol an area or perimeter, which is divided into m zones

with different priorities (P_i being the priority of the zone i). The priority of a path defines the probability of appearing an intruder or event of interest in this zone with respect to others. Therefore, the priorities can be related to the elapsed times. Given the ith zone whose priority is P^i, ET^i defines the maximum elapsed time computed along a mission. So, the jth zone whose priority is P^j should be patrolled with a maximum elapsed time $\text{ET}^j = (P^i/P^j)(\text{ET}^i)$. Then, this problem can be addressed from an urgency criterion, defining the urgency index in the position s at time t as a weighted elapsed time:

$$U(s, t) = P(s)\text{ET}(s, t) \qquad (4)$$

4 COOPERATIVE PATROLLING

Another related issue is how the aerial robots should behave to cooperate in the search or surveillance missions. A common patrolling strategy is required to allow the multiple aerial robots to cooperate in the surveillance and monitoring missions. Different patrolling strategies have been proposed to solve the problem of multi robot monitoring missions from a frequency-based approach.

4.1 Formation Strategy

A first straightforward solution would be that all the robots form a common structure and keep the formation while the leader (or virtual leader or mass center) is following a coverage path. Assuming an enough robust formation controller, the problem is similar to a single-aerial-robot coverage path planning, but with increased covering capabilities. In Sutton, Fidan, and van der Walle (2008) propose to use leader–follower-based formation to patrol a predefined path.

For instance, assuming that the maximum motion speed of all the robots is v, and that each aerial robot has a covering pattern defined by a circle whose radius is c_i, the rate at which each robot would monitor the area individually will be defined by $a_i = 2c_i v$. However, if they keep the a formation, this rate at which the team of robots would monitor the area will be defined by $a = 2v \sum_{i=1}^{n} c_i$, with n being the amount of aerial robots. Therefore, from a frequency-based approach, assuming that the team keeps the formation to patrol an area whose size is A, the maximum elapsed time will be upper-bounded by

$$ \text{ET}^{\text{max}} \leq \frac{A}{2v \sum_{i=1}^{n} c_i} \tag{5} $$

To get this lower bound, the robots should find a closed coverage path such that the motion of the aerial robots never overlaps. In this case, the length of the path will be defined by $L' = \frac{A}{2 \sum_{i=1}^{n} c_i}$. So, the maximum elapsed time may be redefined as follows:

$$ \text{ET}^{\text{max}} \leq \frac{L'}{v} \tag{6} $$

Assuming that all the aerial robots have the same maximum motion speed, this strategy could theoretically obtain the optimal performance. However, if any of the robots was able to move to a higher speed, it would have to adapt its motion to keep the formation, decreasing the performance compared to the theoretically optimal one.

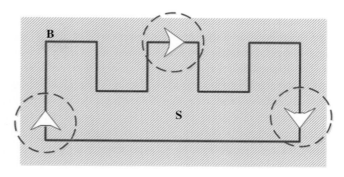

Figure 2. Implementation of the cyclic strategy with three aerial robots.

4.2 Cyclic Strategies

This strategy also implies to a closed path to cover the whole area, which is known to the whole team of robots. According to this strategy, all the robots move along the path at the same speed v, and in the same direction and equally spaced (see Figure 2).

Therefore, in a steady state and assuming a team of n aerial robots patrolling a path whose size is L, Figure 3 shows that the maximum elapsed time can be upper-bounded as follows:

$$ \text{ET}^{\text{max}} \leq \frac{L}{nv} \tag{7} $$

As all the robots have the same covering pattern (defined by a radius c) and defining the size of the area as A, in the best case scenario, the length of the path will be defined by $L = A/2c$. So, the maximum elapsed time could be redefined by

$$ \text{ET}^{\text{max}} \leq \frac{A}{2nvc} \tag{8} $$

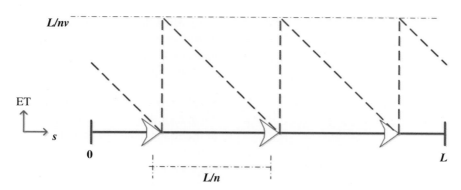

Figure 3. Elapsed time evolution along a closed path changes while a team of three robots executes a cyclic strategy.

It will be equal to the theoretically optimal one if and only if all the robots have the same covering capabilities and the same maximum motion speed as stated in Acevedo *et al.* (2015). Pasqualetti, Franchi, and Bullo (2012) propose a distributed algorithm to coordinate a team of identical robots to move along a closed road map following a cyclic strategy. They analyze this strategy in detail and probe that this strategy theoretically offers the optimal performance with homogeneous robots from an elapsed time criterion.

However, assuming robots with different motion speeds or non-closed paths, this strategy can not be applied directly. The motion speed of the robots or the path should be adapted obtaining a performance lower than the theoretically optimal one (based on the aerial robot capabilities). Also, assuming communication constraints with a communication range smaller than the total path length, the robots could not communicate between them to coordinate their motions and share informations.

4.3 Path Partitioning Strategies

The drawbacks of the cyclic strategy can be tackled using a partitioning strategy, obtaining near optimal solution from an elapsed time criterion. Carli, Cenedese, and Schenato (2011) describe the partitioning strategy to coordinate a set of video cameras in surveillance missions and assume asynchronous communications.

Given n aerial robots, a single coverage path whose size is L is generated to monitor the whole area and the path is divided into n non overlapping segments, as is shown in Figure 4. Each segment is allocated to a different aerial robot.

Now, each robot is in charge to patrol its assigned segment following a *back and forth* motion (from one endpoint to the other one, again and again). To minimize the elapsed time along the whole path, each robot Q_i moves along its segments at its maximum motion speed v_i^{max}. Also,

the maximum speed of each robot should determine the length $L_i = v_i^{max}(L/\sum_{j=1}^{n} v_j^{max})$ of its allocated segment.

From here, it is easy to deduce that all the robots in the sub team would take the same time T' to cover their assigned segments:

$$T' = \frac{L_i}{v_i^{max}} = \frac{L}{\sum_{j=1}^{n} v_j^{max}} \tag{9}$$

Figure 5 illustrates the elapsed time computed along a segment in a steady state and depending on the position of owner aerial robot, while it is performing a path partitioning strategy.

The theoretical maximum elapsed time can be computed as the time that a robot takes to cover its assigned segment twice:

$$ET^{max} \leq 2T' = 2\frac{L}{\sum_{j=1}^{n} v_j^{max}} \tag{10}$$

Assuming that all the robots have the same covering pattern (defined by a radius c) and that the size of the area to monitor is A, in the best case scenario the length of the path will be defined by $L = A/2c$. So, the maximum elapsed time could be redefined by

$$ET^{max} \leq \frac{A}{2c \sum_{j=1}^{n} v_j^{max}} \tag{11}$$

According to these results, the path partitioning strategy is worse than the cyclic strategy from an elapsed time criterion if all the robots are homogeneous and the path is closed. However, in general, the path partitioning strategy gets better results than the cyclic one when the robots have different maximum motion speeds, according to an elapsed time criterion. It is because the partitioning strategy allows the robots to exploit their different maximum speeds to patrol the whole path in a cooperative manner. Also, this strategy is useful even with non closed paths.

On the other hand, according to the partitioning strategy, all the robots take the same time T' to cover their assigned path. Therefore, it is possible to coordinate the robots such that each pair of neighbor robots meet periodically in their common segment endpoint, allowing a non-continuous, but periodic connectivity.

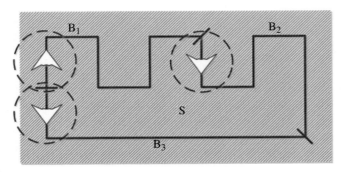

Figure 4. A team of three homogeneous mobile robots perform a path partitioning strategy to patrol cooperatively a path B

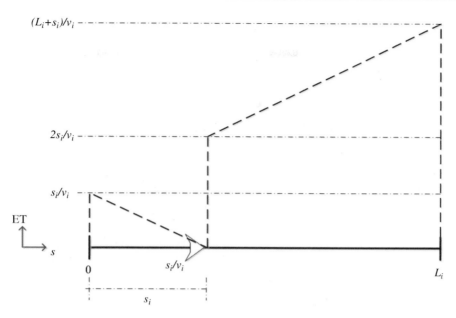

Figure 5. Elapsed time evolution along segment, while a team of aerial robots are performing a path partitioning strategy. The horizontal coordinate s represents the position in the path and the vertical coordinate shows the current elapsed time ET at each position s. The arrow shows the current position s_i of the aerial robots that are patrolling the segment moving at a constant speed v_i.

The periodic connectivity between neighbor robots ensures the information propagation between all the robots in a finite time. Then, the maximum time to share an information among all the aerial robots or latency time LT may be theoretically upper-bounded as follows:

$$LT \leq 2T' + T' + (n-3)T' = nT' = n\frac{L}{\displaystyle\sum_{j=1}^{n} v_j^{\max}} \qquad (12)$$

Assuming that all the robots can move with the same maximum motion speed v, this value would be $LT \leq L/v$, which is approximately half of the upper-bounds of the latency for the cyclic strategy.

The cyclic and partitioning patrolling strategies are defined, analyzed, and compared in Pasqualetti, Franchi, and Bullo (2012) from an elapsed time criterion.

4.4   Area Division Strategies

Given n aerial robots, this strategy proposes to divide the whole area into n non overlapping sub areas as shown in Figure 6. Each robot is in charge to patrol a different sub area whose size A_i depends on the robots capabilities $a_i^{\max} = 2c_i v_i^{\max}$, with $v_i^{\max}$ being its maximum motion speed

and c_i the radius of its circular covering pattern):

$$A_i = a_i^{\max} \frac{A}{\displaystyle\sum_{j=1}^{n} a_j^{\max}}, \forall i = 1, ..., n \qquad (13)$$

So, independent of their capabilities, theoretically all the aerial robots should take the same time T' to cover their assigned sub areas:

$$T' = \frac{A_i}{a_i^{\max}} = \frac{A}{\displaystyle\sum_{j=1}^{n} a_j^{\max}} \qquad (14)$$

Assuming that a robot can generate a closed path of minimum length $L_i = A_i/2c_i$ to continuously patrol its

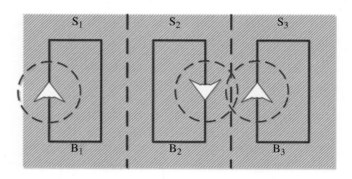

Figure 6. Implementation of an area partitioning strategy to monitor an area with three aerial robots.

assigned sub area using a cyclic motion, the maximum elapsed time would be theoretically upper-bounded by T'

$$\mathrm{ET}^{\max} \leq T' = \frac{A}{\sum_{j=1}^{n} a_j^{\max}} \qquad (15)$$

The path partitioning strategy ensured periodical communication and exploited different speeds, but it was not able to obtain the optimal performance from an elapsed time criterion because the robots move with a *back and forth* motion along their segments. In the area partitioning strategy, the robots implement a *cyclic* motion to patrol the area in order to improve the previous solution. Then, this strategy can theoretically obtain the optimal performance from a *frequency-based approach* exploiting the capabilities of different robots: maximum speed, coverage range, an so on.

Area monitoring is solved in Acevedo, *et al.* (2014); Acevedo *et al.* (2015) using an area partitioning strategy. Furthermore, in Acevedo *et al.* (2014), the required conditions to ensure multirobot synchronization and information propagation between the aerial robots even under communication constraints, based on a simplified model where the system is described as a visibility graph (see Figure 7), are provided:

- All the aerial robots travel in the same direction or each pair of neighbor aerial robots travel in opposite directions.
- All the aerial robots take the same time to patrol their assigned paths.
- The visibility graph defined for the problem is bipartite.
- The links are located in opposite positions in the paths of the neighbor robots, from an angular point of view.

A simple bipartite graph would be the grid-shape graph. Assuming a $r \times c$ grid-shape area division, it is possible to compute the maximum time to share an information between all of them. In a grid-shape division (graph), each path (node) has at most four link positions, and the distance between each pair of consecutive links is the same $T'/4$. The latency time LT depends on the number of links between the two farthest paths (nodes). The two farthest nodes will be located in two nonconsecutive corners. The amount of links between them will be $(r-1) + (c-1)$.

Therefore, following a monotone path, it is easy to compute an upper-bound time to share any information (see Figure 8):

$$\mathrm{LT} \leq T' + (r-2)T'/2 + T'/4 + (c-2)T'/2 = 5T'/4 + (r+c-4)T'/2 \qquad (16)$$

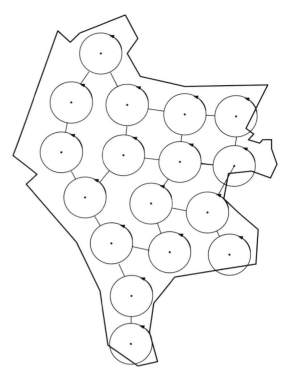

Figure 7. Visibility graph for a team of 17 aerial robots implementing an area division strategy. The circles represent the circular paths of the UAVs or the nodes in the graph. The lines represent the edges in the visibility graph and intersect the circular paths in the link positions.

4 x 5 Grid-shape division

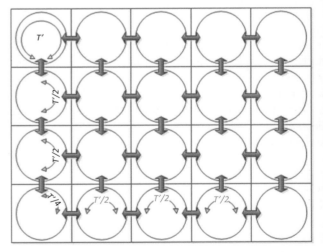

$$\mathrm{LT} = T' + 2\cdot T'/2 + T'/4 + 3\cdot T'/2 = 15\cdot T'/4$$

Figure 8. Maximum latency time between the two farthest paths in a 4×5 grid.

5 COORDINATION OF MULTIPLE AERIAL ROBOTS

After deciding a more appropriate cooperative strategy for the surveillance tasks, it is required to coordinate the robots to implement it in a more efficient manner. Different coordination schemes have been proposed in the literature: centralized, distributed, and mixed coordination methods.

In Frew (2009), the authors address the perimeter surveillance problem looking to find the optimal perimeter and coordinate a multiagent system in order to survey a defined target. They propose a distributed algorithm based on ordered upwind methods but using centralized decisions and assuming a continuously opened communication channel among all the agents. A centralized approach based on the *receding planning horizon* is proposed in Ahmadzadeh *et al.* (2008) to coordinate a set of UAVs to cooperate in surveillance missions. The *mixed integer lineal problem* is proposed in Bertuccelli and How (2005) to coordinate a set of UAVs to visit set of way points minimizing the finish time in a centralized manner.

However, surveillance scenarios usually involve large areas where the communications among the aerial robots and with the control stations can not be guaranteed. Therefore, a distributed and decentralized architecture offers higher robustness and dynamism in the solution.

5.1 Distributed Coordination

The distributed coordination techniques are useful to allow the multiple aerial robots to converge to a common and coherent objective, for instance, to a common patrolling strategy, from local information and asynchronous communications among pair of them. Moreover, a distributed and decentralized approach offers more scalability and dynamism to the solution.

Different distributed approaches have been proposed based on market-based algorithms, inhibitors, rewards, and so on. In Fazli *et al.* (2010), a distributed and decentralized system to survey a known area with multiple mobile ground robots in a cooperative way is presented. It assumes communication constraints and it is fault tolerant. On the other hand, in Pasqualetti, Franchi, and Bullo (2012), the authors propose a distributed algorithm to coordinate a team of cooperating mobile robots to execute cyclic and partitioning strategies. This strategy offers theoretically optimal results from an elapsed time-based approach with homogeneous robots.

A reward-based approach is proposed in Yang, Polycarpou and Minai (2007), where the authors propose to assign rewards to the dynamically potential waypoints related to the probability that this waypoint was visited by another aerial robot. The potential paths of the rest of the aerial robots are defined as "soft obstacles." A similar approach is presented in Lim and Bang (2009), but considering potential threats in solution. Other approach based on a particle filter and information-theoretic methods is proposed in Hoffmann, Waslander and Tomlin (2006) for cooperative search applications.

5.1.1 One-to-one coordination

The *one-to-one coordination* technique was introduced in Acevedo *et al.* (2004) to implement an area partitioning strategy in a distributed manner and considering communication constraint. The algorithms based on the one-to-one coordination assume a different problem to be solved by each pair of contacting robots. Each pair of contacting robots addresses a reduced version of the whole problem considering just their own information. For instance, a problem with just two robots and the part of the area (or the targets) that they are currently monitoring. The idea is to join the tasks that have been previously assigned to both robots and divide the sum between them according to their capabilities. They do not take into account the capabilities or tasks from the rest of the team.

In Carli, Cenedese, and Schenato (2011), a *one-to-one* (or peer-to-peer) coordination technique is proposed to coordinate a team of agents to patrol a path according to a path partitioning strategy in a distributed manner and assuming asynchronous communications among neighbors. This technique implies that each pair of robots solves a different coordination problem including only their own information. They share their assigned segments, join them, and divide them according to their capabilities.

These algorithms require low information storage capabilities for the aerial robots because they just have to store their own local information. Moreover, this technique has proved to converge to the desired solution, but its convergence time complexity is increased quadratically with the total number of robots Carli, Cenedese, and Schenato (2011). Moreover, it is a dynamic technique able to adapt to failures and changes in the initial conditions Acevedo *et al.* (2015) and it is totally scalable because each aerial robot just needs to know the information about the contacting robot Acevedo *et al.* (2014).

This method allows the system to obtain the whole coordination from local decision and information. The robots just needs information from nearby neighbors. Therefore, this technique is decentralized, scalable, dynamic, convergent, and fault-tolerant solution obtaining a suitable area partitioning strategy in a distributed manner.

5.1.2   *Coordination variables*

The coordination variables can be defined as the minimum information that all the robots should share in order to solve the problem in a coherent manner using the coordination function McLain and Beard (2005). Therefore, the algorithms based on the coordination variables assume that the problem can be totally described by a limited set of variables and that using these variables each aerial robot can independently solve the whole problem. The idea is that when a pair of robots meets, each one updates its own variables based on the information received from the other. Then, each aerial robot uses these updated coordination variables to solve in an independent manner, not a reduced version of the problem but the whole problem So, iteratively, the robots receive information in a direct or indirect manner from all the robots in the team such that in a finite time they have correctly updated their own coordination variables and can calculate the correct and coherent common solution. A very relevant challenge to overcome is deciding which information the aerial robots should interchange such that they can calculate the coordination variables.

In Kingston, Beard and Holt (2008), an algorithm based on coordination variables is proposed to coordinate a team of small homogeneous UAVs to cooperatively perform a path partitioning strategy through a perimeter in a distributed and decentralized manner and assuming limited communications. In Acevedo *et al.* (2015), an algorithm based on the coordination variables is proposed to coordinate a team of aerial robots to implement some types of area partitioning strategies. Finally, an approach based on the coordination variables is proposed in Acevedo *et al.* (2015) to solve a perimeter surveillance problem with multiple aerial robots and assuming prioritized zones.

They have proved to be dynamic, fault tolerant, and totally scalable. Although the robots have to store information about the whole problem, this information should be encoded as a limited set of variables and the size of this set should not depend on the total number of robots. Moreover, the use of the coordination variables has proved to get the desired solution for distributed coordination problems in few iterations Geng (2009), being a fast convergent solution to solve the problem in a coherent manner.

6   RELEVANT RESULTS

This section provides a set of relevant results related to the topics explained in this chapter.

6.1   Comparison Among Patrolling Strategies

A large set of simulations have been considered to compare the different patrolling strategies based on two different indices: the *detection time* defined as the time since an information appears until any robot detects it and the *reporting time* defined as the time since an information is detected by a robot until that information is reported to the control station.

The objective is to detect events in an irregular area and report to the ground station in a minimum time, assuming a limited communication range for the aerial robots. The simulations have been executed with a team of 20 homogeneous aerial robots in order to properly compare all the strategies, because the cyclic strategy can not exploit the advantages of a heterogeneous team.

Figure 9 presents the average time to detect the events (which is related to the elapsed time ET), to report to the control station (which is related to latency time LT), and the sum of both for different strategies.

Simulations show that the lowest detection times are obtained using the cyclic strategy. The area partitioning strategy offers slightly higher times, and the path partitioning one obtains the worst results for detection. However, the times to report to the control (or police) station using the area partitioning strategy are significantly lower than using the other ones. Adding both times, it is shown that the best global performance is achieved using the area partitioning strategy.

6.2   Convergence Analysis

The approaches based on the one-to-one coordination and the ones based on the coordination variables have been compared considering a large set of scenarios to coordinate the aerial robots to implement a path partitioning strategy assuming a limited communication range.

Each scenario has been executed using both approaches and they both algorithms converge to the same path division between the robots, but the convergence times are different. The convergence time has been calculated for both algorithms, assuming that a system has converged if the maximum difference between the segment actually assigned to each and the theoretically optimal one is less than 5%.

The results are summarized in Figure 10 depending on the number of aerial robots. It shows that by increasing the number of robots, the algorithm based on the coordination variables converges much faster than the one-to-one coordination algorithm.

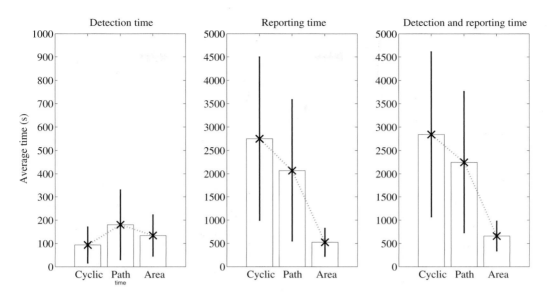

Figure 9. Detection time, time to report to the control station, and sum of both during simulations, considering area partitioning path partitioning, and cyclic strategies. It shows the average times (± its standard deviation).

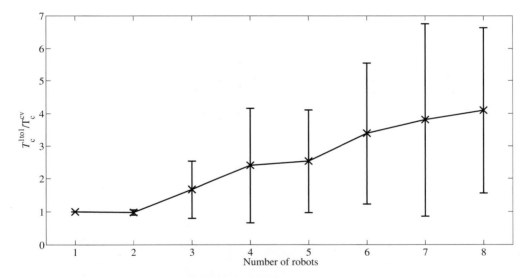

Figure 10. Rate between the average convergence times using the algorithms based on one-to-one coordination and the ones based on the coordination variables (± its standard deviation).

6.3 Experimental Results

This section presents experimental results, where the one-to-one coordination technique is used to implement an area partitioning strategy in a distributed manner. More details are provided in Acevedo *et al.* (2015). The robots were a team of hummingbird quadrotors by Ascending Technologies with 200 g payload and up to 20 min of flight autonomy.

The algorithm based on one-to-one coordination has been implemented in the experiments. In the first experiment, a team of two quadrotors cover the whole area. It is assumed that one

robot is faster than the other one. Figure 11 shows that each robot covers an area related to its own maximum speed, taking advantage of the different capabilities in a heterogeneous multirobot system. The first robot (solid lines on the right in Figure 11) has limited its maximum speed to 0.35 ms^{-1} and the other one (solid lines on the left) can move with a maximum speed of 0.45 ms^{-1}.

In the second experiment, a team of three quadrotors with a maximum speed of 0.3 ms^{-1} covers the area. Figure 12 shows the area division obtained using the one-to-one technique. One of the robots leaves the mission at time $t = 250$ s

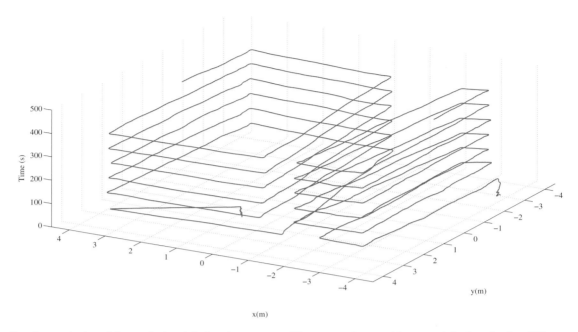

Figure 11. Quadrotors (x, y) positions calculated during the area surveillance experiment with two aerial robots having different capabilities.

and Figure 13 shows how the system can adapt to the area distribution in a decentralized manner dynamically.

7   SUMMARY

This chapter poses the cooperative patrolling problem with multiple aerial robots or UAS. It does not focus on low-level problems such as motion control, communication links, or vision-based event detection, but on the distributed and decentralized coordination of the aerial robots.

The use of multiple aerial robots offers many advantages for surveillance and monitoring application against a single aerial robot, but it implies a relevant challenge to overcome: the coordination of the aerial robots such that they can cooperate in the mission. However, using a centralized system, the solution is less robust to failures, less dynamic, and less scalable. Therefore, a distributed

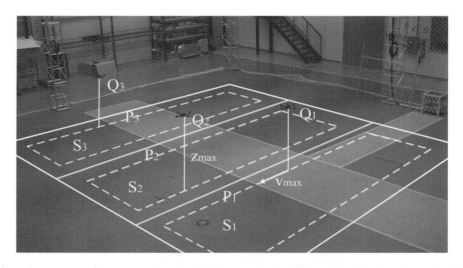

Figure 12. Snapshot from the area surveillance experiment with three quadrotors. The solid white lines determine the area division and the dashed lines indicate the coverage paths followed by the quadrotors.

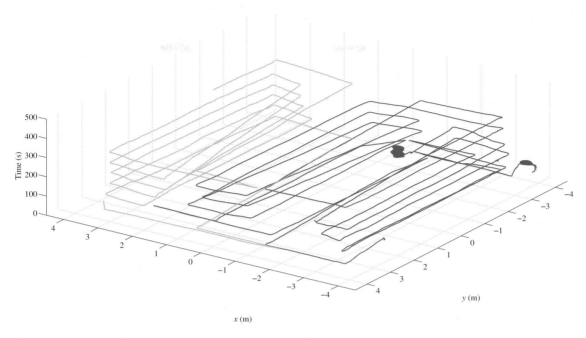

Figure 13. Quadrotors (x, y) positions calculated during the area surveillance experiment with three aerial robots and dynamic conditions.

coordination system is a more suitable solution to get this cooperation.

It is required to define a criterion to patrol the monitored area in order to maximize the amount of detected events. Assuming no information about when and where the events or intruders can appear, a more efficient solution is to maximize the frequency in which any position into the area is monitored by any aerial robots. It is equivalent to minimize the refresh time or elapsed time between each pair of consecutive visits to any position into the area. Four types of cooperative patrolling strategies allow addressing the problem from a refresh time criterion: formation control strategies, cyclic strategies, path partitioning strategies, and area partitioning strategies. The partitioning strategies (both the path and area partitioning) are more suitable for these applications because they ensure the information propagation among all the robots (reporting information goal) even under communication constraints and exploit different capabilities of the heterogeneous robots in the solution.

Many distributed coordination approaches have been proposed in the related literature. Two of them have been proposed to coordinate the multi-UAS system to converge to the desired partitioning strategies (both the path and area partitioning) in a distributed manner: the techniques based on the one-to-one coordination and those based on the coordination variables. The approaches based on the one-to-one coordination require less information storage capabilities for the aerial robots than those based on the coordination variables. Although both converge to the partitioning strategy,

the algorithms based on the one-to-one coordination quadratically converge with the number of aerial robots, while those based on the coordination variables converge linearly. On the other hand, both algorithms are totally scalable and robust to single robot-failures and changes in the initial conditions of the problem.

REFERENCES

Acevedo, J.J., Arrue, B., Maza, I. and Ollero A. (2015) Dynamic zone assignment under priorities for perimeter surveillance missions with aerial robots. *International Conference on Unmanned Aircraft Systems (ICUAS)*, 2015 pp. 871–878.

Acevedo, J.J., Arrue, B.C., Maza, I. and Ollero A. (2015) Distributed cooperation of multiple UAVs for area monitoring missions, in G. Carbone and F. Gomez-Bravo, ed., *Motion and Operation Planning of Robotic Systems*, Mechanisms and Machine Science, vol. 29, pp. 471–494. Springer International Publishing.

Acevedo, J.J., Lawrance, N.R.J., Arrue, B.C., Sukkarieh, S. and Ollero A. (2014) Persistent monitoring with a team of autonomous gliders using static soaring. *International Conference on Intelligent Robots and Systems*.

Acevedo, J.J., Arrue, B.C., Diaz-Banez, J.M., Ventura, I., Maza, I. and Ollero A. (2014) One-to-one coordination algorithm for decentralized area partition in surveillance missions with a team of aerial robots. *J. Intell. Robot. Syst.*, **74**, 1–17.

Acevedo, J.J., Arrue, B.C., Maza, I. and Ollero A. (2015) A distributed algorithm for area partitioning in grid-shape and vector-shape configurations with multiple aerial robots. *J. Intell. Robot. Syst.*, pp. 1–15.

Agmon Noa, Kaminka Gal A, and Kraus Sarit. (2011) Multi-robot adversarial patrolling: facing a full-knowledge opponent. *J. Artif. Int. Res.*, **42**(1), 887–916.

Ahmadzadeh Ali, Keller James, Pappas George, Jadbabaie Ali, and Kumar Vijay. (2008) An optimization-based approach to time-critical cooperative surveillance and coverage with UAVs, in O. Khatib, V. Kumar, and D. Rus, ed, *Experimental Robotics*, Springer Tracts in Advanced Robotics, vol. **39**, pp. 491–500. Springer, Berlin.

Beainy, F. and S. Commuri. (2009) Development of an autonomous atv for real-life surveillance operations. *17th Mediterranean Conference on Control and Automation, 2009 (MED'09)* pp. 904–909.

Bertuccelli, L.F. and J.P. How,. (2005) Robust UAV search for environments with imprecise probability maps. *44th IEEE Conference on Decision and Control, 2005 and 2005 European Control Conference (CDC-ECC '05.)*, F. pp. 5680–5685.

Bao, Y., Fu, X. and Gao, X. (2010) Path planning for reconnaissance UAV based on particle swarm optimization. *Computational Intelligence and Natural Computing Proceedings (CINC), 2010*, vol. 2, pp. 28–32.

Carli, R., Cenedese, A. and Schenato, L. (2011) Distributed partitioning strategies for perimeter patrolling. *American Control Conference (ACC)*, 2011, pp. 4026–4031.

Choset, Howie and Pignon, Philippe. (1997) Coverage path planning: the boustrophedon decomposition. *International Conference on Field and Service Robotics.*

Darbha, S., Krishnamoorthy, K., Pachter, M. and Chandler, P. (2010) State aggregation based linear programming approach to approximate dynamic programming. *49th IEEE Conference on Decision and Control (CDC 2010)*, pp. 935–941.

Elmaliach Yehuda. (2009) *Multi-Robot Frequency-Based Patrolling*. PhD thesis, Bar Ilan University.

Fazli, P., Davoodi, A., Pasquier, P. and Mackworth, A.K. (2010) Complete and robust cooperative robot area coverage with limited range. *IEEE/RSJ International Conference on Intelligent Robots and Systems (IROS 2010)*, pp. 5577–5582.

Frew, E.W. (2009) Combining area patrol, perimeter surveillance, and target tracking using ordered upwind methods. *IEEE International Conference on Robotics and Automation, 2009 (ICRA '09)*, pp. 3123–3128.

Geng, X. (2009) Consensus-reaching of multiple robots with fewer interactions. *2009 WRI World Congress on Computer Science and Information Engineering*, vol. **5**, pp. 249–253.

Guruprasad, K.R., Wilson, Z. and Dasgupta, P. (2012) Complete coverage of an initially unknown environment by multiple robots using voronoi partition. *International Conference on Advances in Control and Optimization in Dynamical Systems.*

Hazon, N. and Kaminka, G. A. (2008) On redundancy, efficiency, and robustness in coverage for multiple robots. *Robot. Auton. Syst.*, **56**(12), 1102–1114.

Hoffmann, G. M, Waslander, S. L and Tomlin, C. J. (2006) Distributed cooperative search using information-theoretic costs for particle filters, with quadrotor applications. *Proceedings of the AIAA Guidance, Navigation, and Control Conference and Exhibit*, pp. 21–24.

Kingston, D., Beard, R.W. and Holt, R.S. (2008) Decentralized perimeter surveillance using a team of UAVs. IEEE Trans. *Robot.*, **24**(6), 1394–1404.

Lim, S. and Bang, H. (2009) Waypoint guidance of cooperative UAVs for intelligence, surveillance, and reconnaissance. *IEEE \break International Conference on Control and Automation, 2009. ICCA 2009*, pp. 291–296.

McLain, T. W and Beard, R. W. (2005) Coordination variables, coordination functions, and cooperative timing missions. *J. Guidance Control Dyn.*, **28**(1), 150–161.

Pasqualetti, F., Franchi, A. and Bullo, F. (2012) On cooperative patrolling: optimal trajectories, complexity analysis, and approximation algorithms. IEEE Trans. *Robot.*, **28**(3), 592–606.

Saptharishi, M., Spence Oliver, C., Diehl, C.P., Bhat, K.S., Dolan, J.M., Trebi-Ollennu, A and Khosla, P.K. (2002) Distributed \break surveillance and reconnaissance using multiple autonomous atvs: Cyberscout. IEEE Trans. *Robot. Autom.*, **18**(5), 826–836.

Sutton, A., Fidan, B. and van der Walle, D. (2008) Hierarchical UAV formation control for cooperative surveillance. *Proc. 17th World Congress of International Federation of Automatic Control (IFAC'08)*, pp. 12087–12092.

Viguria, A., Maza, I. and Ollero, A. (2010) Distributed service-based cooperation in aerial/ground robot teams applied to fire detection and extinguishing missions. *Adv. Robot.*, **24** (1–2), 1–23.

Yanli, Y., Marios, M.P. and Ali, A.M. (2007) Multi-UAV cooperative search using an opportunistic learning method. *J. Dyn. Sys. Meas. Control.*, **129**(5), 716–728.

Yang, S. and Luo, C. (2004) A neural network approach to complete coveragepath planning. *IEEE Trans. Man Cybernet. Syst.*, **34**, 718–724.

Zhang, Y. and Xiao, Y. (2013) Digital pheromone based patrolling algorithm in wireless sensor and actuator networks. *Consumer Communications and Networking Conference (CCNC 2013) IEEE*, pp. 496–501.

Chapter 30
UAV Swarms: Decision-Making Paradigms

Hyo-Sang Shin and Pau Segui-Gasco
Centre for Autonomous and Cyber-Physical Systems, SATM, Cranfield University, Cranfield, UK

1 INTRODUCTION

The large scale of unmanned aerial vehicle (UAV) applications has proliferated vastly in the last few years with the fielding of Pointer, Pathfinder Raven, and Dragoneye SUAVs, among others. The operational experience of UAVs has proven that their technology can have a dramatic influence in the military and civilian arenas. Inexpensive UAVs have considerable potential for use in many different military and civilian operations. Moreover, small UAVs are cheaper and more versatile than manned vehicles, and are ideally suited for dangerous, long, and/or monotonous missions that would be inadvisable or impossible for a human pilot. Especially, swarms of UAVs are of special interest due to their ability to coordinate simultaneous coverage of large areas or cooperate to achieve common goals.

Despite many operational successes demonstrating the UAV capabilities in the military arena, there are certain challenging issues and capability gaps to operate swarms of UAVs. Currently, UAVs are always remotely piloted by humans operating from ground control stations. This requires several highly skilled pilots, working in shifts, which is expensive and simply unfeasible if a swarm of UAVs is to be deployed in numbers in the civilian airspace. Hence, for civilian use such as search and rescue, there is a strong need to provide reliable UAV technology so that some of the decision-making responsibility resides with the vehicle. Human supervision will still have to be retained, whether in the form of an operator planning mission of UAV swarms or a human air traffic controller (ATC), or both. From the viewpoint of a human supervisor, autonomous UAVs must verifiably behave as predictably as human pilots. The main technological challenge is to develop a rigorous approach for designing and analyzing cooperative, dynamic decision-making for autonomous UAVs in swarms so that they will complete the required tasks, recover safely from faults and emergencies, and respond predictably to operator instructions at all times.

Allowing the distributed systems to have capability of task management or allocation will enable a certain level of decision-making authority onboard. One of the main considerations for the task management will be the scalability of the architecture as the operation involves a very large number of distributed UAVs, that is, a swarm of them. In this case, it is impractical or unscalable to have a centralized decision-maker allocation tasks and distributing them among UAVs. Therefore, distributed or decentralized task management is most likely desirable in UAV swarm operations.

Another important issue in UAV swarm operations is guidance and control. Similar to the task management, it will be unscalable and unsustainable to have a centralized decision-maker computing guidance and control commands, and distributing them around. Therefore, guidance,

Unmanned Aircraft Systems. Edited by Ella Atkins, Aníbal Ollero, Antonios Tsourdos, Richard Blockley and Wei Shyy.

navigation, and control should be distributed as a swarm of the distributed systems will be most likely favored. In order to best apply this concept, it might be desirable to design guidance and control schemes not in a conventional way, but based on *a set of individual rules* leading to a certain behavior. Following these rules, guidance and control commands should be generated for each UAV and a global guidance and control goal of the distributed systems must be accomplished.

Meaningful cooperation and/or coordination of UAV swarms can only happen if the UAVs communicate among themselves and also with the operator by exchanging messages about their flight trajectories, mission status, and vehicle health. It is therefore crucial to be able to assess the effects of nonideal communications on UAV swarm cooperation and to develop group behaviors that will guarantee sufficient communication for meaningful UAV swarm cooperation.

Under the aforementioned background, this chapter will address the following challenges in UAV swarm operations:

- Task management, that is, allocation
- Communication network connectivity
- Guidance and control

2   TASK ALLOCATION IN UAV SWARMS

Efficient cooperation of a swarm of UAVs, termed as task allocation, is a vital aspect for UAV mission success. This is because the strength of UAV swarms hinges on the distributed nature of the resources available, making the successful assignment of these resources key to maximize its operational advantages.

Moreover, a real-world UAV swarm operation presents fundamental features that complicate the problem even more: dynamic and uncertain communication conditions on the one hand and the fact that the information and computational resources are distributed throughout the different agents complicating the reliable access to a central entity on the other. A direct consequence of these issues is that situational awareness is not common among agents because it is not shared instantaneously and homogeneously across the network. Therefore, in order to preserve resilience in the absence of reliable communications to a central entity, the decision-making must happen in a decentralized manner. This makes the already difficult allocation problem even more complex, posing trade-offs with deep structural implications.

A key trade-off that must be performed concerns the optimality of the solution versus the convergence time in a decentralized setup. Most of the task allocation problems are,

as we shall see, NP-Hard, requiring exponential time to be solved optimally, thus requiring the careful craft of approximation strategies. Two extreme cases succinctly describe the situation: on one side, we would have an algorithm that would guarantee an optimal solution but could take an unacceptably long time; on the other extreme, we would have an algorithm that would run arbitrarily fast but with arbitrary poor performance. The first case would yield optimal solutions however, this solution would take a very long time to be obtained and they would be only as optimal as the knowledge of each agent evaluating the cost functions. On the other hand, arbitrary fast algorithms, such as metaheuristics, can deliver some results with certain convergence speed; however, they offer no guarantee whatsoever that this result is close to an optimal solution.

Formally, the multi robot task allocation problem is defined in (Dias *et al.* 2006):

Given a set of tasks $\mathcal{T}$, a set of robots $\mathcal{R}$, and a function for each subset of robots $r \in \mathcal{R}$ specifying the cost/utility of completing each subset of tasks $c_r : 2^{\mathcal{T}} \to \mathbb{R}^+ \cup \{\infty\}$, find the allocation $\mathcal{A}^* \in \mathcal{R}^{\mathcal{T}}$ that minimizes/maximizes a global objective function $\mathcal{C} : \mathcal{R}^{\mathcal{T}} \to \mathbb{R}^+ \cup \{\infty\}$.

Ideally, the task allocation solution consists in quickly assigning each task to an agent of the team while optimizing some overall performance metric. In recent years, there has been a wealth of literature published studying different but closely related applications. The problem is that each application looks at its own particularization of the problem with little or no reference to a more general framework or its relations with other works that might be reduced to the same abstract problem. To resolve this situation, a very important contribution to this young field was made in the form of a domain-independent taxonomy to classify, interpret, and abstract the different versions of task allocation problems. In 2004, Gerkey and Mataric (2004) published what is now the standard taxonomy of the field, it provided a unifying theory to the task allocation family of problems, mapping instances of the task allocation problem to corresponding combinatorial optimization problems. This taxonomy proposes three axes to characterize an instance of the task allocation problem:

Single-task robots (ST) versus multi-task robots (MT).
Distinguishing whether the robots are able to execute a single or multiple tasks at the same time.
Single-robot tasks (SR) versus multi-robot tasks (MR).
Distinguishing whether the tasks require one or multiple robots to be executed.
Instantaneous assignment (IA) versus time-extended assignment (TA). Distinguishing whether the robots construct a plan to be executed imminently or they can construct a more elaborate plan to be executed over a given horizon of time.

A particular instance of the task allocation problem is, therefore, defined by the triple ({ST or MT},{SR or MR}, {IA or TA}), for a detailed explanation of each instance the reader is referred to (Gerkey and Mataric, 2004). The most common problems solved in the literature are the single-task robot, single-robot tasks, both instantaneous assignment and time-extended assignment, that is, (ST,SR,IA) and (ST,SR, TA). The simplest case (ST,SR,IA) is an instance of the optimal assignment problem that can be solved optimally in polynomial time (Burkard, Dell'Amico, and Martello, 2009). Unfortunately, in (ST,SR,TA), the addition of the time-extended assignment feature transforms the problem into an instance of the set packing problem (SPP) which happens to be both NP-Hard and impossible to approximate within a constant factor (Paschos, 1997).

This taxonomy by (Gerkey and Mataric, 2004) provided a common reference frame to describe task allocation problems; however, there was a fundamental limitation: it did not explicitly cover dependencies between the tasks. Very recently, this gap was filled by the work of Korsah, Stentz, and Dias (2013). Based on the taxonomy by Gerkey and Mataric, Korsah et al. provided the theoretical framework to extend the taxonomy to cover situations with task dependencies such as related utilities and task-coupling constraints, This provided a mapping of instances of the task allocation problem with dependencies to well-studied combinatorial optimization problems. The types of dependencies considered by this taxonomy are as follows:

No dependencies (ND). Simplest case. Occurs when all the tasks are fully decoupled and, therefore, the agent's utilities of the tasks are independent.

In-schedule dependencies (ID). The assessments of an agent depend on what other tasks are being executed by that same agent. This dependency could be in the utility function or constraints within the tasks schedule for each agent. Consequently, its valuations are independent of the allocations of other agents.

Cross-schedule dependencies (XD). Occurs when the task valuations depend not only on the executing agent's schedule but also on the other agents scheduled tasks. This can happen, for example, when multiple vehicles are needed to execute a given set of tasks or when temporal or precedence constraints are imposed in the tasks, and so on. Nevertheless, these dependencies are simple in the sense that they are known to the agents before the actual allocation.

Complex dependencies (CD). Occurs in the same case as cross-schedule dependencies but with the added compilation that these dependencies do not have a simple structure. They have a complex structure in the sense that the tasks that are being allocated have multiple decompositions into subtasks that are coupled to the allocation. Hence, the

dependencies can only be resolved simultaneously with the allocation, this coupling of the task decomposition problem and the task allocation problems create a more complicated set of dependencies than the cross-schedule.

With this classification of the dependencies, the problem instances are defined by combining a dependency type with an instance of the triple ({ST or MT},{SR or MR},{IA or TA}). This expands the type of situations that can be modeled significantly, for a detailed discussion of each of the cases the reader is referred to Korsah, Stentz, and Dias (2013). The most studied cases involve either no dependencies or in-schedule dependencies, with some works devoted to cross-schedule and complex dependencies.

2.1 Decentralized task allocation approaches

We now discuss representative examples of studies solving the decentralized task allocation problem.

When there are no dependencies between the tasks at all, the SR–ST task allocation problems are instances of the optimal assignment problem that is solvable in polynomial time. For the most detailed and current treatise on the centralized approaches to this problem, the reader is referred to the book by Burkard, Dell'Amico, and Martello (2009). Due to the tractability of these problems, there have been a number of algorithms proposed that do guarantee optimal performance in a decentralized setup. The first distributed task allocation strategy was that proposed by Bertsekas, David, and Castañon (1991), where an auction algorithm was proposed based on the idea of a shared memory model. However, the shared memory model required a topology of the networked system that is not always achievable in real scenarios. To address this issue in Zavlanos, Spesivtsev, and Pappas (2008), an algorithm is proposed to handle a networked system in which agents interact with its neighbors, rather than having access to a shared database. Similarly, in Choi, Brunet, and How (2009), the authors present the consensus-based auction algorithm (CBAA). This algorithm uses the concept of maximum consensus to distribute a series of single item auctions over the network, successfully achieving guarantees of both convergence if the network is connected and of an optimal assignment. Another approach based on task swapping also delivers a global optimal solution based on local task swaps (Liu and Shell, 2013). In the work by Moon, Oh, and Shim *et al.,* (2012), a recent application of a qualitatively similar algorithm for UAV task allocation in a dynamic environment is described alongside an account of its performance in real flight.

In-Schedule dependencies occur when the valuations of the tasks that each agent performs are coupled, that is, they cannot be valued as single task but rather they have a valuation as a bundle. This makes the problem hard, to illustrate why, for example, consider the intuitive scenario of two tasks far away from an agent but very close to each other. If executed independently, they may have very low performance indices for an agent due to the costly trip required from the agent's location. However, if they are executed together, they can synergize costs: once the agent has traveled to a location, the cost of the remaining nearby task is zero, the task is somewhat "free," because the agent has already incurred in the trip cost. Hence, it can generate a much larger payoff than if it was to be valued independently with the double computation of the trip cost. This kind of scenarios are paradigmatic of multi-UAV, task allocation problems.

This situation, unfortunately, leads to the fact that all in-schedule task allocation problems are very hard to solve. Indeed, most of these problems can be modeled as a combinatorial auction, a type of auction in which agents submit bids for bundles of item (task) rather than single items (tasks). Resolving the winners in a combinatorial auction and, hence finding the optimal allocation of items (tasks) to agents is known as the winner determination problem, which is NP-Hard. Even worse, they are an instance of the abstract problem known as the set packing problem (SPP) (Vries and Vohra, 2003), which happens to be proven not to have a constant factor approximation algorithm (Paschos, 1997). Consequently, it not only takes exponential time to find solutions with in-schedule dependencies, but it is also impossible to approximate it uniformly.

With such a bleak outlook, one of the approaches that is gaining more momentum in decentralized task allocation is to incorporate assumptions into the problem formulation in order to enforce performance guarantees. Currently, a popular decentralized task allocation employing this strategy is the consensus-based bundle algorithm (CBBA) developed by Choi, Brunet, and How (2009). It is based on a decentralization of a greedy heuristic to solve the winner determination problem. To guarantee the performance of the greedy heuristic, it imposes a diminishing marginal gains (DMG) condition on the bids of the agents. DMG is a well-studied property used to prove approximation bounds in greedy heuristics and is also related to the submodularity condition or triangle inequality. Thus, by restricting the bundle-bid space, CBBA gives guarantees on convergence and approximation ratio while at the same time it is distributed across the network with no central entity needed. Another example of an algorithm is the work by Luo *et al.* (2012) that guarantees performance bounds by leveraging the triangle inequality assumption and a greedy heuristic.

The CBBA algorithm has two phases: bundle construction and conflict resolution (maximum consensus). In the bundle construction phase, each agent creates a bundle by greedily adding tasks until there are no tasks left or its bids are inferior than the current highest bidder. The bid value that each agent places on each individual task is computed as the marginal gain as a result of adding that task to the bundle. Once each agent has built their own bundle of tasks, the consensus phase starts. In this phase, agents exchange with each other the bids that they have for each value and the agent with the highest bid is assigned a given task and the outbid agents drop their bids. This is accomplished by following a series of communication exchange rules, detailed in Choi, Brunet, and How (2009). In this process each agent maintains information regarding the current winners and its bids value. If the tasks reward functions satisfy the diminishing marginal gains property, and the network is connected, the algorithm converges to a solution that is guaranteed to be at least 50% of the optimal.

Several authors have extended CBBA so as to perform better in dynamic environments. For example, Ponda, Johnson, and How (2012b) propose a framework to handle stochastic environments through chance-constrained reward functions. Recently, the same group (Johnson *et al.*, 2012) tackles the limitation of DMG task scoring by using warping functions so that the bids appear (sic) as if they were submodular in the consensus space while they are handled as non-DMG in the agent own domain, consequently allowing improved synergies within the bundles. This lifts the requirement for the DMG condition, although surrendering all the performance guarantees.

CBBA assumes that all the agents know all the tasks, this means that all the agents must be informed when new tasks are dynamically introduced. However, in practical environments, only a small number of agents end up bidding for any given task because the reward functions depend on the fuel cost and distance and are competitive only for those agents nearby the specific task. This makes that much time is spent in synchronizing all the agents for every new task. Furthermore, agents may become unavailable or disconnected from the network, which can stop the whole consensus process. In order to overcome this problem, Johnson *et al.* (2011) propose a new set of interaction rules for CBBA so that it allows local agreement within asynchronous networks. Mercker and Casbeer (2010) propose another approach to overcome these problems in dynamic environments by the introduction of local interaction rules to handle "pop up" tasks within local agents, speeding up the convergence with a modest loss of optimality of about 2%.

Limiting the bid space, imposing, for example, the DMG assumption like CBBA does, provides guarantee to converge in polynomial time to a solution that is close to the optimal to certain extent. However, this is done at the expense of a large limitation on the types of situations that can be modeled within those assumptions. Furthermore, its trade-off between computation time and solution quality is a structural nature of the algorithm rather than something that can be tuned when setting up the framework for a specific mission: if a given mission satisfies the conditions, then a certain performance is guaranteed; otherwise nothing is guaranteed. This prevents its application to many real-world scenarios.

Recently, Zhang, Collins, and Shi (2012) and Zhang, Collins, and Barbu (2013) have presented a different approach, a series of stochastic clustering auction algorithms (SCA). The SCA algorithms that they propose are conceptually similar to the classical metaheuristic approach simulated annealing, they allow the designer to choose the rate at which the stochastic exploration of the solution space should take place. As with simulated annealing, a global optimal solution an be obtained if the cooling rate (in SA terms) or the rate of the proportion of stochastic steps in the search is reduced or is slow enough. However, the problem is NP-Hard, and hence if global optimal results are desired, the cooling rate will have to be very slow, and an exponential convergence time should be expected. Nevertheless, in practice, obtaining a "good" solution quickly is preferred to obtaining an optimal solution very slowly. The notion of what a "good" solution is depends on each specific problem, consequently allowing direct control of the speed optimality trade-off, empowering the designer with the tools to adjust the convergence speed to its specific notion of what a "good" solution is in each situation. In some cases a fastest convergence at the expense of optimality can be considered a "good" solution and in others optimality will be more important and some time can be spared to achieve it.

The works presented so far assume that the tasks' rewards are independent of the tasks allocated to other agents and only considered in-schedule dependencies within an agent's own schedule. Because of this, much work has been recently developed trying to include cross-schedule dependencies, such as constraints, coupling the tasks and enabling a richer representation for real world scenarios. In Choi, Whitten, and How (2012), a CBBA-based mechanism to allocate tasks involving two agents is presented. Later, the same group, Choi *et al.* (Whitten, 2010), introduced an extension to handle the following task dependencies: unilateral dependency, mutual dependency, mutual exclusion, and timing constraints, all involving possibly more than two agents. With the purpose of adequate human supervision of large UAV autonomous networks, Argyle, Casbeer, and Beard (2011) extended CBBA with the notion of teams, each team allocates tasks independently using CBBA and then, through an outer loop, teams exchange unassigned tasks. In another work Hunt, Meng, and Hinde (2012) introduced group-dependent tasks by modifying the score functions and the conflict resolution strategy of CBBA at the expense of higher communication overhead.

These works are a first step enabling cross-schedule dependencies in a decentralized environment; however, none of them provides any performance guarantee.

The task allocation problem is NP-Hard; hence, it is very unlikely that a polynomial time algorithm will ever be found. Consequently, researchers have had to apply two different approaches to try to devise practical decentralized task allocation frameworks. The first one is the approach followed by the designers of CBBA (Choi, Brunet, and How, 2009): leveraging assumptions about the nature of the problem to enforce tractability. The second one is the approach by the SCA designers (Zhang, Collins, and Shi, 2012): enabling algorithm whose convergence speed can be easily traded off against the optimality of the solution. Imposing assumptions is advantageous because it provides the means to enforce performance guarantees, but it is also limiting because it rules out all the other problem instances that fall outside the assumptions. Algorithms such as SCA are advantageous because they can be tailored to the scenario, but problematic because they aim to solve the most complicated instance of the problem without no problem-specific information to lower the computational burden. Bearing this in mind, it is clear that future successful decentralized task allocation algorithms will have to leverage both approaches. On one hand, they should allow the incorporation of problem-specific knowledge to try to make the solution space more computationally tractable. On the other hand, they should facilitate the tailoring of the speed-optimality trade-offs so that they can be adjusted to fit the requirements of real applications. Such a decentralized task allocation algorithmic framework is yet to be found.

3 COMMUNICATION NETWORK CONNECTIVITY

In order to make a proper decision in the UAV swarm operation, the decision maker requires to obtain appropriate information such as situational awareness and motion information of all the UAVs in the swarm. As some of the decision-making responsibility is desired to reside with the vehicles in UAV swarm operations, this information needs to be shared among UAVs. Therefore, communication plays an important role in the operation of UAV swarms. Especially, when the human operator is in the loop, maintaining communication connectivity between UAVs and the mission control station becomes essential. However, communication among UAVs or between the mission control station and the UAV is almost impossible to be guaranteed at all times due to limited bandwidth, communication range, transmission power, and physical obscuration or occlusion in the mission environment. Options to address this problem are (i) to have one UAV function (partly) as a communication hub and (ii) to operate them in loose formation all the time.

In the first approach, while small UAVs perform their tasks, UAVs with the required communication capability loiter at a designated position ensuring network connectivity between themselves and the mission control station via satellite links, as well as themselves and small UAVs via line-of-sight radio frequency (RF) modem. In the second approach, a decentralized setup can be exploited by local communication and coordination among small UAVs without using satellite communication resources. In this case, the UAVs might need to be flying in formation within all the other UAVs' communication ranges and LOS, including the mission control station while performing their tasks; if this is not achievable, some of UAS need to be used for a relay purpose temporarily, in order to ensure the network connectivity of the whole team.

3.1   Communication models

In order to plan effectively with communications constraints, one must be able to predict the communication performance at several positions in the problem domain in order to assess the feasibility of the trajectories. To make these predictions, a model of the communication environment is needed. In the following, a description of the current literature on models of communications-aware planning strategies for unmanned aircraft is reviewed. The literature is structured in two main parts: model-based approaches and measurement-based approaches. Both are fundamentally useful and complementary, because to produce a plan before deployment model-based representations are the only way to go, and to optimize the locations of already deployed vehicles measurement approaches could be best suited.

3.1.1   Model-based approaches

Range-only approaches Most of the work on communications-constrained UAV cooperative control has been carried out assuming a range-based communications model. Among the most prolific authors in range-based communications-constrained distributed control for robotic networks are F. Bullo, J. Cortés, and S. Martínez. Most of their approaches are described in their book (Bullo, Cortes, and Martinez, 2009). They propose several constraints to enforce range-only connectivity and algorithms to achieve a range-only connected deployment. This algorithm can be interesting as a simple model, probably as an initial solution

for the optimization algorithm in order to speed up convergence to a good solution.

In the UAS domain, there are many examples of range-only communications-aware coordination; here we will name just a couple of the most recent ones. Acevedo *et al.* (2013) propose an approach to coordinate a heterogeneous team of UAS subject to a number of constraints, including coverage and communications. Another approach to coordinate heterogeneous teams while ensuring connectivity is represented in Ponda *et al.* (2012a) and Kopeikin (2012), which is based around the consensus bundle-based algorithm. This algorithm allows distributed coordination for task allocation on dynamic environments; in this approach, in order to maintain connectivity, relay nodes are added where the network might break according to its topology.

Range-only and visibility Visibility and range-based approaches have been proposed by several authors. In the robotics domain, the group of Bullo has been again one of its major contributors (Obermeyer, Ganguli, and Bullo, 2011). In these works the authors propose a series of deployment algorithms for full visibility in polygonal environments with holes. If a map-based deployment were to be used, it could be fruitful to exploit this sort of algorithm to speed up convergence of the optimizer.

Other approaches that involve range limited and visibility are mainly those formulated in terms of mixed integer linear programming (MILP) programs. In this approach, most of problems are formulated in terms of discrete variables and commercial solvers such as CPLEX are employed to find a solution.

Channel propagation Han, Swindlehurst, and Liu (2009) propose a control strategy to guide one UAV over a ground MANET to improve its connectivity. Its communication model is based on a deterministic part of an exponential distance path loss model as described previously and on a stochastic part describing small-scale fading by a Rayleigh distribution. They derive an expression for the probability distribution of the SNR. Given an SNR threshold and a probability distribution for, they calculate the probability that the link is above this threshold. Then, to construct the adjacency matrix of the network, they establish a certain probability threshold, say 99%, such that if the probability that the channel is above the SNR threshold is above this reliability threshold (99%), then there is a link, else, there is no link. With the probability of the link being above certain threshold, they also define a weighted graph that describes deconnectivity. With this graph established, they define several connectivity metrics: global message connectivity, worst-case connectivity, network bisection connectivity, and k-connectivity, which they then use in the optimization to find the best motion for the UAV.

Simulation based Grotli and Johansen (2011) present a communications-aware simulation-based path planning strategy for UAS. They exemplify their approach using two UAV acting as relays for an offshore application, although this framework can be applied more broadly. The communications channel that they use is the simulation package SPLAT! Fundamentally, the approach is based on a MILP formulation in combination with the SPLAT! simulator, a series of constraints is established to be satisfied, among them is a capacity constraint that is translated through Shannon–Hartley theorem to an SNR threshold. Basically, they try to iteratively solve the problem until they satisfy the constraints. They set a set of constraints and solve the problem, then the wireless channel is simulated using SPLAT!, and if the signal-to-noise ratio (SNR) is above certain threshold, then the constraints are modified and iteration starts again until the path satisfies the communication constraints.

3.1.2 Measurement-based approaches

Several measurement-based approaches have been proposed in the last few years to control UA Systems. These approaches measure some variables relevant to the communication channel, normally the signal-to-noise ratio and then use some sort of stochastic gradient estimators to optimize the control policy. In contrast to model-based approaches, measurement-based approaches have an advantage over the model-based approaches in that they can accommodate dynamic changes in the environment, although their knowledge of the RF environment is usually restricted to a more local area.

SNR estimation

Extremum seeking In Dixon and Frew (2012) and Frew *et al.* (2009), Frew *et al.* introduce a control architecture to optimize end-to-end connectivity in a chain of UAS-based relays. In these works, each vehicle estimates its local SNR by means of received signal strength indicator (RSSI) measurements. Then, using the fact that the capacity of a communication channel is a function of SNR, they use an extremum seeker controller to optimize the capacity of the communication channel in a distributed manner. Each vehicle follows a Lyapunov vector field guidance whose central point is driven by the optimizer to maximize the capacity with its neighboring vehicles. This way each vehicle uses local measurements to drive its motion and in conjunction the whole team achieves an end-to-end capacity maximizing strategy.

The key point of the works (Dixon and Frew, 2012; Frew *et al.*, 2009) is the use of the combination of extremum seeking and LGVF or, in general, the orbiting motion guidance, which happen to be a great way to optimizing in a distributed manner for Dubins vehicles (in this case, fixed wing UAV). Extremum seeking is an approach that finds the extremum (max or min) of an objective function that is only known to have such an extremum, no model whatsoever is required of this function, just values (in this case, measurements) at given points. This is achieved by estimating the gradient of the functions through sinusoidal perturbations, precisely, the same sort of periodic perturbations that are obtained by an orbiting vehicle. Hence, just by sensing along the orbital motion ES allows to maximize this sensed variable with no model at all. This happens to be a key strength of Frew *et al.* approach because communications field are incredibly difficult to predict, usually modeled as stochastic phenomena; hence, by using this approach no communication field prediction needs to be made and hence it can take into consideration unforeseen circumstances such as jamming or shadowing by unknown obstacles.

Simultaneous perturbation stochastic approximation Another gradient-based approach is that by Le Ny, Ribeiro, and Pappas (2012) where the authors combine potential field tasks with communication constraints. In this work, the authors use the signal-to-noise and interference ratio (SNIR) analogous to the SNR with the addition of interference alongside Shannon–Hartley theorem to model the channel capacity. With this capacity, the authors then construct a constraint and add it to its optimization problem alongside other constraints, such as tasks and vehicle dynamics, in order to maximize some reward function. To obtain adaptation to changing communication environments, the authors use an online sampling scheme to get a realistic estimate of the capacities. This sampling scheme is based on timescale separation and simultaneous perturbation stochastic approximation (SPSA) and it gives an estimate of the channel capacity at nearby locations and its gradient to be used within the optimization. A drawback however is that the sampling positions are random and the vehicle must deviate to visit those points for the estimation.

RSSI potential fields Goddemeier, Daniel, and Wietfeld (2012) proposed another idea to maintain a connected swarm of UAS. They defined a network connectivity metric that simply required each UAV to maintain d connections, with $d = c \log(n)$ where c is a constant, in this case 4, and n is the total number of vehicles in the swarm. Then, they defined a series of potential fields based on the RSSI and the metric d to steer the swarm while maintaining connectivity.

3.2 Communications-aware task allocation

Currently, the only communications-aware task allocation strategy is an extension to CBBA developed by Ponda *et al.* (Ponda *et al.*, 2012a; Kopeikin and Ponda, 2011) tackling the problem of enforcing connectivity of disconnected tasks.

They introduce the idea of an outer loop on CBBA by means of which after each CBBA iteration, relays are created to connect or remove the disconnected tasks, although with a rather simplistic obstacle-free range-only communication scenario. Later, the same group (Kopeikin *et al.*, 2012) introduce advances to handle more realistic communication criteria, including bit error rate and data rate, however, without handling obstacles in the radio environment. It is interesting to note that in both of these works, the authors implement the algorithm in an actual team of quadrotors for real-time flight testing successfully. These studies fall in the category of complex dependencies because the relays must be placed at the allocation time and they are not known beforehand, and no polynomial time algorithm with performance guarantees is known for this instances of the task allocation problem. This is a young area and these seem to be the first and only studies considering fully distributed task allocation coupled with communication constraints.

4   DISTRIBUTED GUIDANCE AND CONTROL

The main purpose of guidance and control for UAV swarm operations would be to guide and maintain UAVs within a proximity for the groups of those UAVs to perform the tasks assigned from task allocation. Thus, many guidance and control schemes for UAV swarms can be found in *formation flying'* literatures. In the UAV swarm operations, formation flying could be loose or tight.

As stated in Section 1, because it is impractical or unscalable to have a centralized decision maker computing guidance and control commands and distributing them to the individual UAVs, distributed guidance and control would be preferable in UAV swarm operations. In order to best apply the distributed guidance and control concept, it would be desirable to design guidance and control schemes based on *a set of individual rules* leading to a certain behavior, not using conventional guidance and control approaches such as proportional navigation and/or PID control systems. Therefore, the discussion for distributed guidance and control is limited to those that are based on *a set of individual rules* resulting in a certain behavior, for example, a certain subtask.

Approaches for formation flying in UAV swarms are often referred to as *behavioral* since they are mainly derived from the observed behavior of social living beings (birds, fish, and insects). The most common behavioral techniques that can be found in literature are *flocking*, *swarming*, and *schooling*, where the difference is in the type of motion obtained for the agents. With the behavioral approach, motion of agents is based on three simple rules, based on local information:

Separation: avoiding collisions with neighboring agents
Alignment: matching velocity with neighbors
Cohesion: staying close to the neighboring agents

Behavioral techniques have been widely applied for the control of groups of autonomous vehicles, since they allow the intuitive definition of control algorithms and the self-organization of the agents without the need for a given structure. With reference to the formation flight in UAV swarms, a behavioral approach implies that a vehicle, for its control action, considers not only the trajectory of another UAV but also that of the entire UAVs in the swarm.

4.1   Pure behavioral approach

Bennet *et al.* (2011) introduced *swarming* approach for formation flying. Unlike many other flocking and swarming algorithms available in literature, the proposed work provides a formal mean of verification and is thus eligible for application in real-engineered systems. A velocity field is defined for each agent, which is composed of a steering term, used to control the formation, and a repulsive term, responsible for collision avoidance. Both the terms are bounded, avoiding saturation on actuators and thus guaranteeing for the correctness of the final behavior. The steering term is defined on the basis of an artificial bifurcating potential field, which is a potential field that can assume two different states according to a switching parameter: such a commutation change number and position of the equilibrium states for the field. Repulsive potential function is a simple pairwise exponential function that has its maximum, and so bounded, value when two vehicles are at the minimum separation distance. The potential fields previously introduced are then used for obtaining three-dimensional formation patterns.

A first simple simulation is performed given maximum velocity and minimum separation, assuming that the agents can communicate freely and can move instantaneously in all the degree of freedom. Results shows that agents form the desired formation and travel at the required speed once in equilibrium, respecting given constraints. It must be noticed that in the double ring formation, the disposition of the agents is uncontrolled, and depends mainly on the initial conditions. The application of the previously described potential field allow defining references for forward speed and heading and pitch angles, implementing thus a guidance law.

A simple block diagram for the control system is shown in Figure 1. The desired motion for the swarm is along *x*-axis at

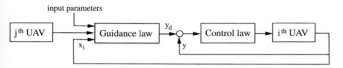

Figure 1. Guidance and control block diagram. (Adapted from Bennet *et al.*, 2011. Reprinted with permission from the American Institute of Aeronautics and Astronautics, Inc.)

a given constant speed, switching in time between the three shapes previously described (Figure 2). Numerical results show that with the proposed guidance law, the vehicles can safely form different patterns, operating within the limit of allowed control actions. Drawbacks of the proposed method are that external perturbations are not accounted for and that an ideal communication system is considered, where positions of all the other vehicles are always perfectly known. This issue could be easily solved considering sensing regions, since other vehicles position information is used only for the repulsive potential function, and is not of real interest when such vehicles are far away and thus do not represent a collision threat. It must be noticed, however, that if the sensing region would not be sufficiently large, accuracy in swarm shape could drastically reduce. One possible drawback important to notice is that the behavioral approach would not be able to strict specification of the desired swarm shape.

When there exist unhealthy, that is, faulty, UAVs in the swarm, a behavioral approach can also be used. As a possible solution, Giulietti, Pollini, and Innocenti (2001) considered a classical behavior of migrating birds: during a migration flight if one or more elements of the group lose their position in the formation, the others leave the migration trajectory and *wait* for the lost ones until the swarm shape is reconstituted. Aircraft dynamics is modeled as that of a point-mass moving in three-dimensional space: six first-order nonlinear differential equations are defined with reference to the following state variables–airspeed, three Earth frame position variables, and

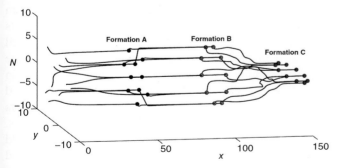

Figure 2. Flight trajectories with formation switching. (Adapted from Bennet *et al.*, 2011. Reprinted with permission from the American Institute of Aeronautics and Astronautics, Inc.)

flight path angle and heading angles. In order to maintain formation geometry, an imaginary point called *formation geometry centre* (FGC) is introduced and each aircraft must keep a prescribed relative position from this point. The kinematic equations for FGC are obtained averaging kinematics of all the vehicles in formation, and its position is the defined by integration. The considered controller is then defined with reference to a linearized dynamic for the aircraft about the steady flight condition. This type of behavioral inspired guidance shows interesting properties of robustness with respect to communication link fault events (control action is not based on information provided from a single vehicle).

4.2 Other behavioral approaches

Another important behavioral approach is formation using the concepts of *potential fields* and *virtual leaders* (VLs). Leonard and Fiorelli (2001) used artificial potential fields (APFs) for the coordinated and distributed control of multiple autonomous vehicles. An APF is a way for defining artificial attractive and repulsive forces that agents in the formation exert on the other agents, usually function of the relative distance between them. These interactions prevent collisions between the agents and allow formation keeping if proper equilibrium states are achieved. In this context, VLs are virtual vehicles that exert forces on the actual vehicles, but are free to move in the space, following arbitrary trajectories, so as to lead other UAVs in the swarm to achieve the given tasks. VLrs are thus used to "guide" the formation, exploiting the virtual interactions between vehicles and VLs. Assuming fully actuated vehicles, the actual control action on the latter is defined as the sum of the virtual forces exerted on the vehicle, which is given from

- sum of the interaction forces from the other vehicles in the formation,
- sum of the forces from the VLs, and
- a dissipative term, function of the vehicle speed that is designed to be zero when the desired speed is matched.

The absence of an actual leader furthermore makes the formation in a swarm robust to the vehicle loss event and quickly reconfigurable.

The problem of formation control using the virtual leaders technique is also addressed in Xi and Abed (2005b): agents are modeled as point particles, moving in a plane under two different actions–repulsive force exerted by other agents, responsible for maintaining separation, and attractive force from one virtual leader, which implements cohesion, alignment, and tracking of the desired trajectory. Two different cases are considered with reference to the established formation:

4.2.1   *Specified formation*

A specified formation in UAV swarms can be obtained by associating with each vehicle a different virtual leader. Constraints on feasible formations are that they must be symmetrical and distances between neighboring agents (i.e., agents that can communicate) are fixed and equal. Given a desired trajectory for the swarm center, and desired rotation in time, trajectories of the virtual leaders can be defined. The desired velocity for a vehicle, as mentioned before, is composed of one attractive term that drives the agent to track its VL, and is function of the distance from the latter, and a repulsive term that controls the distances between the vehicle and all its neighbors.

4.2.2   *Emergent formation*

With this design approach, the swarm structure cannot be directly defined, but a balanced formation will arise depending on initial condition and attraction/repulsion functions. In

this case, only one VL is introduced in the scenario and its trajectory will correspond to that of the swarm center. A proper formation must satisfy the following requirements:

1. There is no collision between any pair of agents.
2. Any agent has at least one neighbor, except possibly the one in the center of the formation.
3. It is a "connected" or "quasi-connected" formation, where the latter is composed of multiple connected subformations.

Figures 3 and 4 show the swarm behavior in both the considered cases, where "+" is the target position, "o" is the formation center, and the reference path is a straight line.

Xi and Abed extend the previous work (2005a): a feedback control law is defined for achieving flocking of a group of autonomous mobile agents in an obstacle-free scenario, considering the case of reduced communication. With reference to the "specified formation" approach depicted in Xi and Abed (2005b), the concept of "blind area" is introduced: it is a circular zone associated with a certain vehicle, with radius α, and centered on the desired

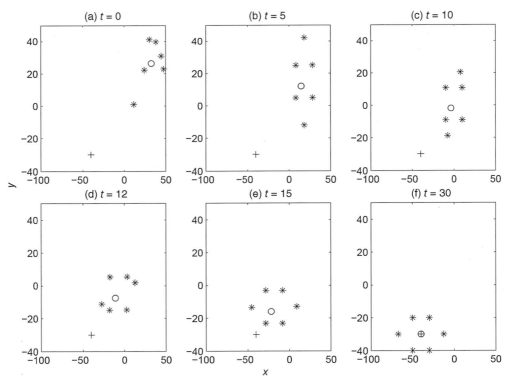

Figure 3. Simulation results for a specified formation case: the specified formation switch between two different topologies (obtained respectively in (b) and (e)) during the simulation. (Adapted from Xi and Abed, 2005b.)

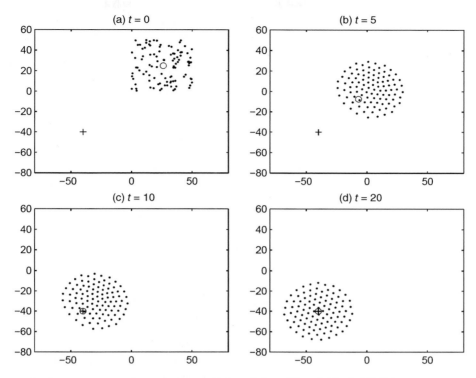

Figure 4. Simulation results for an emergent formation case. (Adapted from Xi and Abed, 2005b.)

position for the vehicle, that is, at a fixed distance from its virtual leader. When an agent is inside its blind zone, it will not communicate with its neighbors, and thus its velocity vector will depend only on the trajectory of its VL, and no repulsive action will be exerted from the other vehicles, regardless of their position. By design, blind zones are not overlapping, and this means that if two vehicles are inside one blind zone, one of them will be considering the repulsive force due to the presence of the other: separation principle (that guarantees collision avoidance) is thus preserved. It is important to notice that the desired formation can be achieved and maintained only when all the vehicles are inside or outside the relative blind area. The behavior and convergence properties of the vehicles when all the VLs are moving with constant, equal, velocities are analyzed. The case where all the agents are outside the respective blind areas is initially considered: a necessary condition is found for the agents to converge to their blind area, but not a sufficient[1] one. Convergence to an equilibrium for all the agents is anyway guaranteed, regardless that it is inside or outside the blind area. The behavior of the group in the time lapse between the first and the last agent entry to their respective blind zones is

very complicated, which results in difficulty in doing classical analysis. Analysis of the single agent when it enters its blind zone is then performed using the Lyapunov stability theory, proving the asymptotical convergence to a unique equilibrium state corresponding to the center of the area. Since the centers of all the agents' blind areas form the desired formation, it is concluded that by using this design, eventually, the agent group can achieve the desired formation and maintain it. Similar considerations are made for the case of time-dependent velocity vectors for the VLs, assuming such a quantity slowly varying and thus locally constant.

Simulations have been run for evaluating the blind zones approach: the results in Xi and Abed (2005a) showed that a group of agents, which succeed in entering their respective blind zones, move as a swarm toward the target position following a slowly varying trajectory. The implementation of blind areas allows energy consumption reduction, since a blind agent stops to consider other agents position, and can thus turn off its receiver or measurement sensor (depending on the how relative positions are obtained). A smaller, smooth, required control effort is also obtained using blind zones: the desired velocity vector for a blind agent will not depend on the relative positions with respect to other agents, "filtering" transitory trajectories of the latter. Some considerations about energy saving due to the reduced

[1] Simulations suggest that the necessary condition could also be sufficient.

communications are made. The latter result would however need further investigation, since it is not clearly defined how communications are implemented: in the case where agents broadcast their position in the neighborhood, for example, communications cannot be interrupted even when in the blind zone, since collision avoidance capability must be maintained.

5 SUMMARY

UAV technologies can have a dramatic influence in the military and civilian arenas. Especially, operations of UAV swarms are of special interest as they can coordinate simultaneous coverage of large areas or cooperate with increased efficiency and effectiveness of the operations. Unlike operations using single UAV, a small number of UAVs, or manned aircraft, there is a strong need for the onboard decision-making responsibility in UAV swarm operations. Therefore, this chapter addresses key challenging issues of key UAV technologies enabling onboard decision-making such as

- task allocation in UAV swarms,
- communication network connectivity, and
- guidance and control for UAV swarm operations.

Moreover, possible solutions to these issues are also discussed in this chapter. Note that these possible solutions are abstracted from existing literatures and thus they are based on the current paradigm to cope with the UAV swarm problem. However, as discussed in the previous sections, there might be a need to propose a new paradigm to more efficiently and effectively operate swarms of UAVs in military and civilian applications.

REFERENCES

Acevedo, J.J., Begona, C., Maza, I., and Ollero, A. (2013) Distributed approach for coverage and patrolling missions with a team of heterogeneous aerial robots under communication constraints. *Int. J. Adv. Robot. Syst.* **10**(28), 1.

Argyle, M., Casbeer, D.W., and Beard, R. (2011) A multi-team extension of the consensus-based bundle algorithm. *Proceedings of the American Control Conference*, pp. 5376–5381.

Bennet, D.J., MacInnes, C.R., Suzuki, M., and Uchiyama, K. (2011) Autonomous three-dimensional formation flight for a swarm of unmanned aerial vehicles. *J. Guid. Control Dyn.* **34**(6), 1899–1908.

Bertsekas, D.P., and Castañon, D.A. (1991) Parallel synchronous and asynchronous implementations of the auction algorithm. *Parallel Comput.* **17**(6–7), 707–732.

Bullo, F., Cortes, J., and Martinez, S. (2009) *Distributed Control of Robotic Networks*, Applied Mathematics Series, Princeton University Press.

Burkard, R.E., Dell'Amico, M., and Martello, S. (2009) *Assignment Problems*, SIAM e-books, Society for Industrial and Applied Mathematics Philadelphia, PA.

Choi, H.-L., Whitten, A.K., and How, J.P. (2012) Decentralized task allocation for heterogeneous teams with cooperation constraints. *Proc. Am. Control Conf.*. 3057–3062.

Choi, H.L., Brunet, L., and How, J.P. (2009) Consensus-based decentralized auctions for robust task allocation. *IEEE Trans. Robot.* **25**(4), 912–926.

Dias, M.B., Zlot, R., Kalra, N., and Stentz, A. (2006) Market-based multirobot coordination: a survey and analysis. *Proc. IEEE*, **94**(7), 1257–1270.

Dixon, C. and Frew, E.W. (2012) Optimizing cascaded chains of unmanned aircraft acting as communication relays. *IEEE J. Sel. Areas Commun.* **30**(5), 883–898.

Frew, E., Dixon, C., Elston, J., and Stachura, M. (2009) Active sensing by unmanned aircraft systems in realistic communication environments. *FAC Workshop on Networked Robotics*, pp. 62–67.

Gerkey, B.P., and Mataric, M.J. (2004) A formal analysis and taxonomy of task allocation in multi-robot systems. *Int. J. Robot. Res.* **23**(9), 939–954.

Giulietti, F., Pollini, L., Innocenti, M. (2001) Formation flight control: a behavioral approach. *AIAA paper*.

Goddemeier, N., Daniel, K., and Wietfeld, C. (2012) Role-based connectivity management with realistic air-to-ground channels for cooperative UAVs. *IEEE J. Sel. Areas Commun.* **30**(5), 951–963.

Grotli, E.I., and Johansen, T.A. (2011) Path planning for UAVs under communication constraints using SPLAT! and MILP. *J. Intell. Robot. Syst.* **65**(1-4), 265–282.

Han, Z., Swindlehurst, A.L., and Liu, K. (2009) Optimization of MANET connectivity via smart deployment/movement of unmanned air vehicles. *IEEE Trans. Vehicular Technol.* **58**(7), 3533–3546.

Hunt, S., Meng, Q., and Hinde, C.J. (2012) An extension of the consensus-based bundle algorithm for group dependent tasks with equipment dependencies. *Neural Information Processing*, Springer, Berlin, pp. 518–527.

Johnson, L.B., Ponda, S.S., Choi, H.L., and How, J.P. (2011) Asynchronous decentralized task allocation for dynamic environments. *AIAA Infotech at Aerospace Conference*, March, 2011.

Johnson, L., Choi, H.L., Ponda, S., and How, J.P. (2012) Allowing non-submodular score functions in distributed task allocation. *Proc. IEEE Conf. Decis. Control.* 1, 4702–4708.

Kopeikin, A.N., and Ponda, S.S. (2011) Real-time dynamic planning to maintain network connectivity in a team of unmanned air vehicles. *WiUAV 2011*, pp. 1303–1307.

Kopeikin, A.N., Ponda, S.S., Johnson, L.B., Toupet, O., and How, J.P. (2011) Real-time dynamic planning to maintain network connectivity in a team of unmanned air vehicles. *GLOBECOM Workshops 2011, IEEE*, 1303–1307.

Kopeikin, A., Ponda, S.S., Johnson, L.B., and How, J.P. (2012) Multi-UAV network control through dynamic task allocation: Ensuring data-rate and bit-error-rate support, *WiUAV 2012*.

Korsah, G.A., Stentz, A., and Dias, M.B. (2013) A comprehensive taxonomy for multi-robot task allocation. *Int. J. Robot. Res.*, **32**(12), 1495–1512.

Le Ny, J., Ribeiro, A., and Pappas, G.J. (2012) Adaptive communication-constrained deployment of unmanned vehicle systems. *IEEE J. Sel. Areas Commun*, **30**(5), 923–934.

Leonard, N.E., and Fiorelli, E. (2001) Virtual leaders, artificial potentials and coordinated control of groups. *Proceedings of the 40th IEEE Conference on Decision and Control, 2001*, vol. 3, pp. 2968–2973.

Liu, L., and Shell, D.A. (2013). An anytime assignment algorithm: from local task swapping to global optimality. *Auton. Robots*, **35**(4), 271–286.

Luo, C., Ward, P., Cameron, S., Parr, G., and Mcclean, S. (2012) Communication provision for a team of remotely searching UAVs: a mobile relay approach. *IEEE Globecom Workshops*, pp. 1544–1549.

Mercker, T., Casbeer, D.W., Millet, T., and Akella, M. (2010) An extension of consensus-based auction algorithms for decentralized, time-constrained task assignment. *American Control Conference*, pp. 6324–6329.

Moon, S., Oh, E., and Shim, D.H. (2012) An integral framework of task assignment and path planning for multiple unmanned aerial vehicles in dynamic environments. *J. Intell. Robot. Sys.* **70**(1–4), 303–313.

Obermeyer, K.J., Ganguli, A., and Bullo, F. (2011) Multi-agent deployment for visibility coverage in polygonal environments with holes. *Int. J. Robust Nonlinear Control*, 1–28.

Paschos, V.T. (1997) A survey of approximately optimal solutions to some covering and packing problems. *ACM Comput. Surveys*, **29**(2), 171–209.

Ponda, S.S., Johnson, L.B., Kopeikin, A.N., Choi, H.L., and How, J.P. (2012a) Distributed planning strategies to ensure network connectivity for dynamic heterogeneous teams. *IEEE J. Sel. Areas Commun*, **30**(5), 861–869.

Ponda, S.S., Johnson, L.B., and How, J.P. (2012b) Distributed chance-constrained task allocation for autonomous multi-agent teams. *ACC Control Conference*, pp. 4528–4533.

Vries, S.De., and Vohra, R.V. (2003) Combinatorial auctions: a survey. *INFORMS J. Comput.*, **15**, 284–309.

Whitten, A. (2010) Decentralized planning for autonomous agents cooperating in complex missions.

Xi, X., and Abed, E.H. (2005a) Formation control with virtual leaders and reduced communications. *44th IEEE Conference on Decision and Control, 2005 and 2005 European Control Conference (CDC-ECC '05)*, pp. 1854–1860.

Xi, X., and Abed, E.H. (2005b) New formation control designs with virtual leaders. *Proceedings of the 16th IFAC World Congress*.

Zavlanos, M.M., Spesivtsev, L., and Pappas, G.J. (2008). A distributed auction algorithm for the assignment problem. *47th IEEE Conference on Decision and Control*, pp. 1212–1217.

Zhang, K., Collins, E.G., and Adrian, B. (2013) An efficient stochastic clustering auction for heterogeneous robotic collaborative teams. *J. Intell. Robot. Syst.*, **72**, 541–558.

Zhang, K., Collins, E.G., and Shi, D. (2012) Centralized and distributed task allocation in multi-robot teams via a stochastic clustering auction. *ACM Trans. Auton. Adapt. Syst.* **7**(2), 1–22.

Chapter 31

Integrated Health Monitoring for Multiple Air Vehicles

Nicolas Léchevin, Camille A. Rabbath, and Chun-Yi Su

Department of Mechanical and Industrial Engineering, Concordia University, Montreal, Quéebec, Canada

1 INTRODUCTION

Despite fault detection, diagnosis, and fault-tolerant control software embedded onboard unmanned aerial vehicles (UAVs), the effectiveness of teaming UAV may be lost due to the occurrence of faults and failures on flight-critical components, such as actuator, sensor and communications devices, and due to body damage. An actuator fault reduces the achievable performance of the aircraft. A loss of effectiveness (LOE) in one UAV in turn may significantly deteriorate team performance. For example, a flight controller that is optimal under nominal operating conditions results in less-than-optimal closed-loop performance under degraded actuator operation. In formation flight, inefficient flight performance of one vehicle may adversely affect team cohesion and may in fact result in an increase in overall energy expenditure. To compensate for such deficiencies, a cooperative and distributed health monitoring system should be embedded onboard the UAVs to preserve multiaircraft mission objectives. A few approaches have been proposed to guarantee some level of performance of UAV formation flight. Formation reconfiguration under information flow faults has been studied in the studies by Giulietti, Pollini, and Innocenti (2000), Pollini, Giulietti, and Innocenti (2002), and Innocenti and Pollini (2004). In these works, graph theory and Dijkstra algorithm are utilized to command geometry changes and to optimally reconfigure communication channels once information flow faults have been detected. An interacting-multiple-model fault detection and isolation (FDI) approach for formations faced with communications failure is proposed in the study by Mehra, Boskovic, and Li (2000). A multiple aerial vehicles command-and-control architecture is proposed in Valenti *et al.* (2006), where centralized mission planning and health management is carried out on the ground in real time. Interestingly, decentralized model-based abrupt fault diagnosis of mobile robot formations is proposed in Daigle, Koutsoukos, and Biswas (2007) exploiting fault propagation dynamics among robots.

We focus in this chapter on team-level fault detection applied to two classes of faults, namely, abrupt and non-abrupt faults. The proposed detectors are utilized when the Component-Level Fault Detection and Recovery (CL-FDR) system fails to compensate for faulty behavior that may jeopardize the cohesion of the fleet of aerial vehicles. For instance, the concurrent occurrence of severe actuator faults and intervehicle communication loss may cause vehicles of

Unmanned Aircraft Systems. Edited by Ella Atkins, Aníbal Ollero, Antonios Tsourdos, Richard Blockley and Wei Shyy.
© 2016 John Wiley & Sons, Ltd. ISBN: 978-1-118-86645-0.

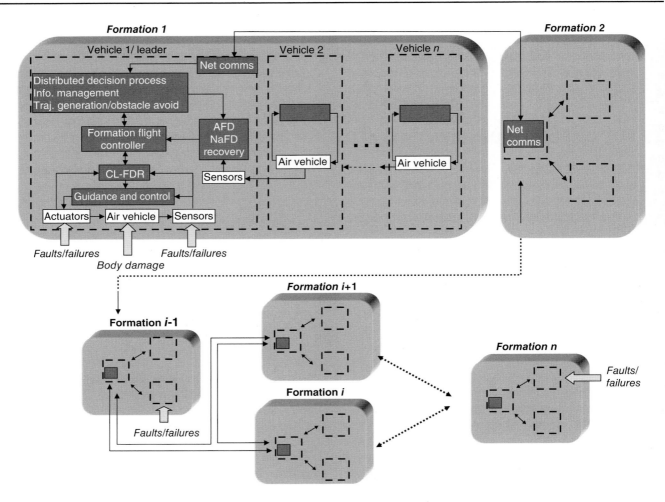

Figure 1. Integrated control and health monitoring system applied to fleets of UAVs.

the fleet to follow divergent paths and to adopt erratic behavior. Every vehicle of the formation is equipped with CL-FDR integrated with a team-level fault detector. The latter exploits information on neighboring vehicles and a set of data available through the communication network. The proposed integrated health monitoring is thus composed, as shown in Figure 1, of a system model- and a signal-based detectors dedicated to abrupt and nonabrupt faults, respectively, and of a command adaptation mechanism that is triggered whenever a transition occurs in the state of one of the detectors. It should be noted that a complete self-healing system should include, besides the proposed integrated health monitoring, a distributed decision-making capacity working in collaboration with a path planning and trajectory generation module, as suggested in Figure 1. At a higher level, the fleet should also be able to communicate with the command and control center to enable task replanning if necessary.

The implementation of fault detectors involves selecting thresholds so that the average detection time and the probability of missed detection are minimized while keeping the probability of false alarms within an acceptable bound. Threshold selection is discussed. The design of the abrupt fault detector (AFD) being based on models of vehicle dynamics allows to derive bounds on the thresholds. A stochastic search algorithm is, on the other hand, applied to select near-optimal thresholds of the nonabrupt fault detector (NaFD) since no particular assumptions can be leveraged to derive useful bounds of thresholds.

2   INTEGRATED CONTROL AND FAULT DETECTION

Each aerial vehicle of the fleet is equipped with autopilots and formation flight controllers. The outer-loop control law

(Léchevin, Rabbath, and Sicard, 2006), which stabilizes the formation $\mathcal{V} = \{1, \ldots, n\}$ of n UAVs, is based on local information; that is, only neighbors are sensed. Assuming that the autopilots are finely tuned, a simplified, linear, uncertain model of the formation flight dynamics is used to design the fault detector. It is assumed that a vehicle is not considered faulty by its neighbors as long as it can achieve position trajectory tracking based on the measured relative distances regardless of its attitude. The linear, parameter-dependent, uncertain model of vehicle $i \in \mathcal{V}$ in feedback with neighbors $j \in \mathcal{N}_i$, where $\mathcal{N}_i$ represents the set of neighboring vehicles in $\mathcal{V}$ that are sensed by i, is given by

$$q_i = A_i(\alpha_i)q_i + B_i(\alpha_i)\underbrace{\begin{bmatrix} h_i\sum_{j\in\mathcal{N}_i}k_i\left(x_j - x_{ij}^*\right) \\ h_i\sum_{j\in\mathcal{N}_i}k_i\left(y_j - y_{ij}^*\right) \\ z_t \end{bmatrix}}_{v_{i1}-v_{i2}} \quad (1)$$

$$\begin{bmatrix} x_i \\ y_i \\ z_i \end{bmatrix} = \begin{bmatrix} 1 & 0 & 0 & 0 & 0 & 0 \\ 0 & 0 & 1 & 0 & 0 & 0 \\ 0 & 0 & 0 & 0 & 0 & 1 \end{bmatrix} q_i \quad (2)$$

where

$$\begin{bmatrix} x_{ij}^* \\ y_{ij}^* \end{bmatrix} = \rho_{ij}^* \begin{bmatrix} \cos(\lambda_{ij}^* + \psi_i) \\ \sin(\lambda_{ij}^* + \psi_i) \end{bmatrix} \quad (3)$$

x_i, y_i, and z_i are the vehicle translations; ψ_i is the heading angle of vehicle i; z_t is the prescribed altitude of the formation, which is assumed to be known prior to mission; ρ_{ij}^* and λ_{ij}^* denote, respectively, the prescribed relative distance and line of sight (LOS) angle between i and j. State variable q_i is a vector in $\mathbb{R}^{n\times 1}$, with $n = 6$.

The model presented in (1) is specified by considering a polytopic-type flight envelope allowing the designer to account for possible parameter uncertainties, that typically arise with aerial vehicles due to time-varying operating conditions. Polytopic interpolation is a convenient approach to design controllers and observers that are robust to mathematical models characterized by uncertain, although bounded, parameters. Such approach has been used, for instance, to model a fighter aircraft and to design its fault-tolerant flight control systems (Liang, Wang, and Yang, 2002). More precisely, the state-space matrices vary within a polytope of matrices

$$A_i(\alpha) = \sum_{j=1}^s \alpha_{ij}A_{ij}$$
$$B_i(\alpha) = \sum_{j=1}^s \alpha_{ij}B_{ij} \quad (4)$$

with vertices $\{A_{i1}, \ldots, A_{is}\}$ and $\{B_{i1}, \ldots, B_{is}\}$, respectively. The uncertain parameter α_i is assumed to evolve in the unit simplex

$$\Gamma_i = \left\{ (\alpha_{i1}, \ldots, \alpha_{is}) \Big| \sum_{j=1}^s \alpha_{ij} = 1, \alpha_{ij} \geq 0 \right\} \quad (5)$$

Note that $A_i(\alpha)$ (resp. $B_i(\alpha)$) can be decomposed as the sum of a nominal matrix $A_i^* = A_i(\alpha^*)$ (resp. $B_i^* = B_i(\alpha^*)$) and a deviation matrix $\tilde{A}_i(\alpha)$ (resp. $\tilde{B}_i(\alpha)$) that evolves in the same polytope as that of $A_i(\alpha)$ (resp. $B_i(\alpha)$); that is,

$$\tilde{A}_i(\alpha) = \sum_{j=1}^s \alpha_{ij}\tilde{A}_{ij}$$
$$\tilde{B}_i(\alpha) = \sum_{j=1}^s \alpha_{ij}\tilde{B}_{ij} \quad (6)$$

Furthermore, it is important to note that the polytopic representation may be used, with some degree of conservatism however, to derive a single AFD function for the entire homogeneous fleet of UAVs. Indeed, the observer present with AFD depends on the simplified model of any vehicle i. Each vehicle is expected to utilize an observer that is tuned from the knowledge of $A_i(\alpha)$ and $B_i(\alpha)$. However, a single polytopic representation is used in Section 3.6 and is shown to provide satisfactory performance. Such polytopic representation thus has to include the polytopes that define $A_i(\alpha)$ and $B_i(\alpha)$ for all $i \in \mathcal{V}$.

Typical actuator faults are shown in Figure 2. Faults include lock in place (LIP), hard-over failure (HOF), float and loss of effectiveness (Boskovic, Bergstrom, and Mehra, 2005). Two types of abrupt faults can typically occur: HOF with $1 - u(t_f)/u_{HOF} \approx 1$ and $\theta_{HOF} \approx \pi/2$. Actuator faults of types LIP and float as well as HOF with $\theta_{HOF} \ll /2$ constitute the so-called class of nonabrupt faults considered in this chapter. Briefly, nonabrupt faults are characterized by slower dynamics than that found with abrupt faults. Techniques used to detect the occurrence of faults may be better suited for one type of fault than another.

3 DECENTRALIZED FAULT DETECTION

3.1 Abrupt fault detector

3.1.1 Design of robust observer

In this section, an observer-based decentralized AFD is presented to address the worst-case situation where a formation is faced with concurrent communications loss and a component-level fault in one follower vehicle, whose CL-FDR is unable to

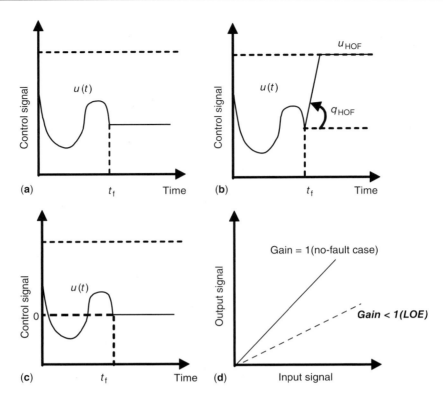

Figure 2. Actuator faults of type: (a) lock-in-place; (b) hard-over failure (HOF); (c) float; (d) loss of effectiveness.

compensate. The sequence of events that are subsequent to the appearance of a fault within the fleet of UAVs, as sketched in Figure 3, can be described as follows. First, the fault occurs in vehicle $i \in \mathcal{N}_k$ at t_f. Second, couplings between vehicle dynamics, which are caused by formation flight, impact vehicle i at $t_f + t_1$, resulting in a possible undesired trajectory. Third, AFD detects that k is faulty at $t_f + t_d$. The recovery mechanism is then immediately activated so that vehicle i catches up with the rest of the fleet.

The recovery is made possible by means of the abrupt fault detector depicted in Figure 4. The residue yield by the observer remains in the vicinity of zero when the system is exempt from fault or failure. Fluctuations of the residue are mainly caused by measurement noise and modeling uncertainties. The robust observer is designed to attenuate such phenomena. The observer should yield a residue that is sensitive to faulty behavior of i. It is important to notice that the formation flight controller of i feeds back the distance between i and j, which is not available to k in the case where j is not a neighbor of k ($j \notin \mathcal{N}_k$), as shown in Figure 3.

The specifications that the observer has to satisfy are twofold. First, it has to be robust to parametric uncertainties that are included in the polytopic models expressed in (1). Second, the observer is required to compensate for the fact that $j \notin \mathcal{N}_k$ so that the residue is sensitive to the occurrence of

faults and not to exogenous signals not related to faults. The latter is carried out by inverting the nominal model of i obtained by substituting A_i^* and B_i^* for A_i and B_i in (1). The former is achieved by applying techniques of robust filtering (Tuan, Apkarian, and Nguyen, 2003).

The structure of the observer $(A_{F,i}, B_{F,i}, C_{F,i})$ is shown in Figure 5, where the inverse nominal model of i is used to reconstruct j's Cartesian coordinates (x_j, y_j) that are used by the formation flight controller. The estimate $\hat{x}_j$ and $\hat{y}_j$ are corrupted by measurement noise, low-pass filtered derivative applied to $(x_k, y_k) + (x_{ki}, y_{ki})$ and used in the inversion of the model and by the fact that a nominal model is used instead of the actual, partially known, model of i. Furthermore, a disturbance δ^* is combined to $\hat{x}_j$ and $\hat{y}_j$ when a faulty/failure, modeled by signal δ, affects i. The low-pass filters in series with the derivatives are tuned so that the occurrence of δ^*, which is related to the abrupt faulty signal δ through the inverse dynamics, results in a high-frequency content of the residue signal r_i. In doing so, r_i can be discriminated from modeling errors, which are characterized by low frequency dynamics (Léchevin and Rabbath, 2007).

Robustness of the observer is thus required to attenuate the imprecise estimation of x_j and y_j and to take into account the polytopic representation of modeling uncertainties. To do so,

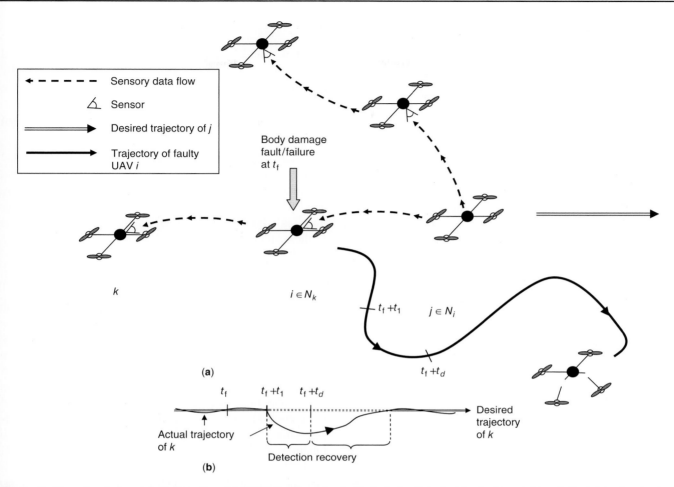

Figure 3. Team-level abrupt fault detection: (a) UAV k's dynamics is likely to be disturbed by faulty neighbor i. Fast detection by k of neighbor's faulty behavior is required to maintain the cohesion of the fleet; (b) time trajectory followed by k prior to and posterior to fault occurrence.

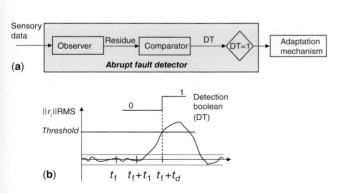

Figure 4. Abrupt fault detector: (a) AFD comprises a model-based observer and a comparator that triggers, if needed, an adaptation mechanism; (b) the comparator involves the observer residue and a threshold used by the comparator.

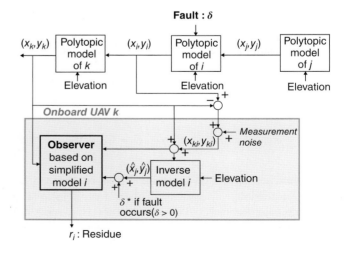

Figure 5. AFD observer onboard UAV k.

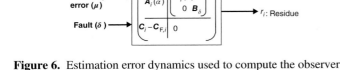

Figure 6. Estimation error dynamics used to compute the observer gain.

an estimation error dynamics is built from the observer, the inverse nominal model of i, and the polytopic model of i. The polytopic model thus obtained is excited, as shown in Figure 6, by two exogenous signals, namely, the faulty signal δ and the modeling error signal μ, which results from the use of A_i^* and B_i^* in the computation of the inverse model. The error dynamics can thus be represented by means of the structured matrix shown in Figure 6, where $A_i(\alpha)$ and $B_i(\alpha)$ are linearly dependent on the observer matrices $A_{F,i}$ and $B_{F,i}$. The objective of reducing the effect of disturbance μ on residue r_i leads us to the computation of observer matrices $A_{F,i}$, $B_{F,i}$, and $C_{F,i}$, obtained by solving the L_2-gain minimization problem defined as

$$\min_{A_{F,i}, B_{F,i}, C_{F,i}} \{\gamma; \|T_{\mu r_i}(\alpha)\|_\infty < \gamma, \alpha \in \Gamma_i\} \quad (7)$$

where

$$T_{\mu r_i}(\alpha) = (A_i(\alpha_i), B_i(\alpha), [\, C_i' - C_{F,i} \,], 0) \quad (8)$$

The objective defined in Equation 7 can be reached by solving, if feasible, an associated linear matrix inequality (Léchevin and Rabbath, 2007). The latter is carried out efficiently by software such as the linear matrix inequality (LMI) toolbox of Mathworks (Gahinet *et al.*, 2004), and Scilab 5.1.1 (2009), to name a few.

3.1.2   Threshold selection

It is shown in Léchevin and Rabbath (2007) that the threshold $J_{\text{th},i}$ of the AFD implemented onboard k to detect i's faulty behavior must satisfy

$$J_{\text{th},i} \geq (a_{i1} + \varphi_{i1}(\rho_{i0}^*))\|H_1(s)\tilde{A}_i(s)\|_\infty + \|H_2(s)\|_\tau \quad (9)$$

where $a_{i1} > 0$, φ_{i1} is an increasing function, and ρ_{i0}^* is the relative distance between the leader and follower i. $\|\|_\infty$ and $\|\|_\tau$ denote the H_∞ and the root mean square, respectively. H_1 and H_2 are transfer functions that depend on the observer matrices. The transfer functions enter the Laplace transform of the residue as follows

$$z_i(s) = H_1(s)\tilde{A}_i(s)Q_i(s) + H_2(s) + G_i(s)\delta(s) \quad (10)$$

where the vector Q_i depends on the state space vector of the observer and on the low-pass filtered faulty signal.

From inequality (9), the threshold $J_{\text{th},i}$ is an increasing function of ρ_{i0}^*. Therefore, vehicles whose distances from the leader are significant are bound to implement a detector with a higher threshold $J_{\text{th},i}$ than with vehicles that are close to the leader. On the other hand, it is shown in Léchevin and Rabbath (2007) that the magnitude of the minimum detectable fault is bounded from below by an increasing positive function of $J_{\text{th},i}$. Hence, the distance between a vehicle and the leader limits indirectly the magnitude of the abrupt fault that can be detected by AFD, thus impacting on the probability of false alarms.

3.2   Nonabrupt fault detector

The detector presented in the last section is unable to react to LIP and float faults in such a way that the health monitoring can be satisfactorily accomplished. Indeed, the observer is not sensitive to slow developing faulty behavior caused by nonabrupt faults, leading to residue r_i that remains below the threshold shown in Figure 4b. To address the issue of detecting nonabrupt faulty behavior that may arise in formation flight of UAVs, a distributed signal-based detector (Léchevin and Rabbath, 2009) is briefly presented. We then propose a near-optimal search algorithm that automates the selection of thresholds in order to minimize a cost function based on the detection time, the probability of false alarms and the missed detections.

NaFD is based on signal redundancy and on dynamic coupling between UAVs that are inherent to formation flight. To see this, consider the nominal case, where no fault occurs. As shown in Figure 7a, couplings between

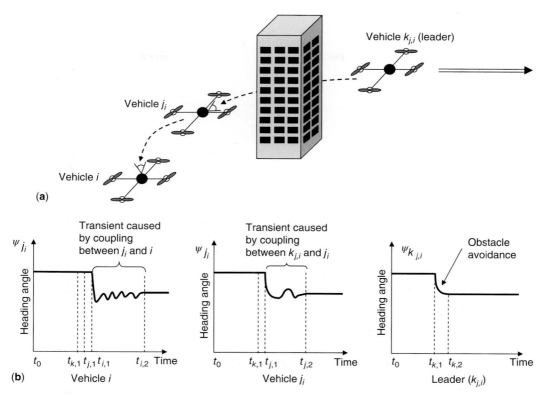

Figure 7. Trajectories of a (nominal) fault-free UAV formation maneuvering to avoid an obstacle. (a) The leader turns around the obstacle and then moves along a straight flight path; (b) the heading angle of the leader stabilizes at a new equilibrium after a transient occurring on $[t_{k,1}, t_{k,2}]$. The heading angles of followers j_i and i adopt similar patterns although over delayed time intervals $[t_{j,1}, t_{j,2}]$ and $[t_{i,1}, t_{i,2}]$, respectively.

UAVs caused by the distributed formation flight controller implemented in every vehicle (Léchevin, Rabbath, and Sicard, 2006) are such that changes in the behavior of the leader $k_{j,i} \in \mathcal{N}_{j_i}$, resulting for instance from a maneuver to avoid an obstacle, are propagated to that of the follower $j_i \in \mathcal{N}_i$, whose behavior in turn impacts that of i. Changes in vehicle motion are detected by monitoring the heading angle ψ. It should be noted that for the sake of focusing on the concept rather than on the mathematics, planar motions are assumed. The trajectory of the followers is similar to that of the leader, as illustrated by the heading angles trajectories in Figure 7b. The straight line piecewise trajectories of j_i and i include transients, which result from obstacle avoidance of $k_{j,i}$ and from couplings between $k_{j,i}$ and j_i and between j_i and i. The appearance of transients is delayed, that is, $t_{i,1} > t_{j,1} > t_{k,1}$.

Building on the notion of propagation of behavior change caused by formation control coupling, NaFD leverages the capability of vehicle i to detect the occurrence of j_i's faulty behavior. This is done by establishing a correlation between the time history of $\psi_{k_{j_i}}$ and the time history of ψ_i. We assume the worst-case situation, where information about j_i is either unavailable to i or deemed unreliable.

Faulty vehicle j_i yields a new steady-state heading angle trajectory, as shown in Figure 8b, or a drifting trajectory (not shown). The abnormal behavior of j_i caused by a fault at t_f propagates to i after a time delay of $\bar{t}_{i,1}$. The dynamics of i is thus impacted at $t_f + \bar{t}_{i,1}$, setting ψ_i to a value that is different from that of $\psi_{k_{j_i}}$ after $t_f + \bar{t}_{i,1}$. This situation is to be compared to that of the obstacle avoidance depicted in Figure 7, where $\psi_i(t) - \psi_{k_{j_i}}(t)$ is close to zero when $t > t_{i,2}$; that is, when the formation has completed its maneuver around the obstacle. Note that $t_{i,2}$ is associated with the transient caused by dynamic couplings among the formation.

Comparing ψ_i with $\psi_{k_{i,j}}$ allows obtaining the detection signal T at time instant t_k shown in Figure 9. Instrumental to deriving T is the classification of the heading angle trajectories in two types, namely, time-constant and time-varying heading angle trajectories, which corresponds, in Figure 9, to

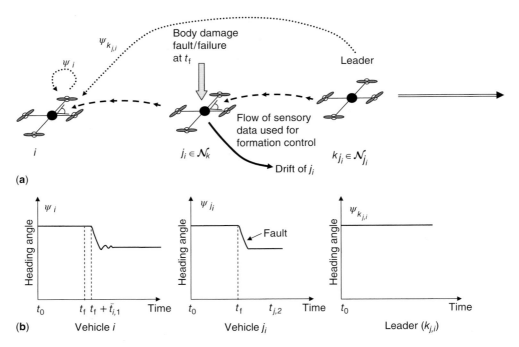

Figure 8. NaFD based on redundancy and coupling arising from the formation control system. (a) A fault causes vehicle j_i to drift from the rest of the formation; (b) signals $\psi_{k_{j_i}}$ and ψ_i estimated or transmitted through a communication network, if any. A fault occurs at t_f and yields a new steady-state heading angle trajectory.

$H'(i) = 1$ and to $H'(i) = 0$, respectively. Vehicle i achieved this step at t_k by collecting ψ_i and $\psi_{k_{i,j}}$ as the following vectors:

$$Y_{1, N_{k_{j,i}}}(k_{j,i}) = \{\psi_{k_{j,i}}(T_s), \dots, \psi_{k_{j,i}}(NT_s)\}$$
$$Y_{1, N_i}(i) = \{\psi_i(T_s), \dots, \psi_i(NT_s)\} \tag{11}$$

over a time interval $[t_k - NT_s, t_k]$, where N is an integer, and T_s stands for the sampling period at which the heading angle is estimated. Note that signals displayed in Figures 7 and 8 are exempt from noise, which is quite unrealistic. Actual signals are corrupted by measurement noises to which is combined the effect of wind turbulence on UAV dynamics. A statistical test (Basseville and Nikiforov, 1993; Poor, 1994) is thus required to decide whether $H'(i)$ is one or zero at each t_k. Once vehicle i determines, at t_k, $H'(i)$ and $H'(k_{j,i})$, T is obtained by comparing $H'(i)$ at t_k to $H'(k_{j,i})$ on $[t_k - t_{dc}, t_k]$. Intuitively, no fault is detected if $H'(i)$ and $H'(k_{j,i})$ are identical; a fault is likely to have occurred within $[t_k - t_{dc}, t_k]$, otherwise.

Letting

$$Y_i = \{Y_{1, N_{k_{j,i}}}(k_{j,i}), Y_{1, N_i}(i)\} \tag{12}$$

the preceding discussion can be summarized as follows

$$T(t_k, Y_i) = \begin{cases} 1 & \text{if } H'(i) \text{ at } t = t_k \text{ is not equal to } H'(k_{j,i}) \\ & \quad \text{for all } t \in [t_k - t_{dc}, t_k] \\ 0 & \text{if } H'(i) \text{ at } t = t_k \text{ is equal to } H'(k_{j,i}) \\ & \quad \text{for all } t \in [t_k - t_{dc}, t_k] \end{cases} \tag{13}$$

where t_k and t_{dc} denote the time at which the detection is performed and the length of the time window of the detector.

4 AUTOMATED THRESHOLD SELECTION FOR NAFD

The statistical test proposed in the study by Léchevin and Rabbath (2009), which is instrumental to (13), depends on two thresholds, labeled as v_i and $v_{k_{j,i}}$. Fixing values of v_i and $v_{k_{j,i}}$ is a critical step to the implementation of (13) and depends on such requirements as the level of false alarms and missed detections than can be tolerated. It should be noted that aggravating factors, as far as these two levels are concerned, include measurement noise and sensitivity of the vehicle dynamics to atmospheric turbulences. Hence, to

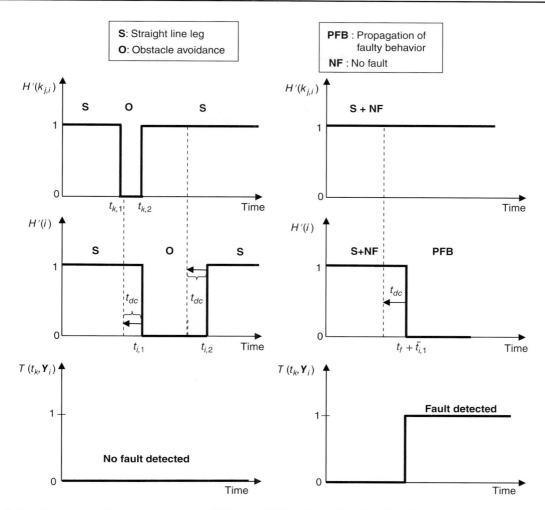

Figure 9. Fault detection process. Comparison between $H'(k_{j,i})$ and $H'(i)$ and resulting detection signal T: (a) the (nominal) non-faulty case; (b) the faulty case.

speed up the selection of thresholds v_i and $v_{k_{j,i}}$, automated computations can be carried out prior to mission, given a set of operating conditions, by means of near-optimal search techniques based on iterative simulations. Several randomized algorithms, such as simulated annealing, genetic algorithm, and cross-entropy method, to name a few, provide approximation of global extremum of complex optimization problems. We explicit in the sequel the application of the cross-entropy method (Rubinstein and Kroese, 2004) to the problem of automated search of the thresholds used in NaFD. The cross-entropy approach is adopted for its simplicity of implementation and its property of convergence. In the next section, the algorithm is applied to the fleet of vehicles equipped with NaFD.

The cost, which aims to minimize the detection time t_{detec}, the probability of false alarms, p_{F}, and the probability

of missed detections, p_{MD}, is expressed, for all $l = 1, \ldots, N$, as

$$d(t_{\text{detec}}, p_{\text{F}}, p_{\text{MD}}; v_l) = \alpha_d d(t_{\text{detec}}; v_l) + \alpha_{\text{F}} p_{\text{F}} + \alpha_{\text{MD}} p_{\text{MD}}$$

$$d(t_{\text{detec}}; v_l) = \frac{1}{N_l} \sum_{i=1}^{N_l} t_{\text{detec}}$$

$$t_{\text{detec}} = \begin{cases} (t_d - t_{\text{f}})/\Delta & \text{if } \bar{t}_d \in [0, \Delta] \\ 1 & \text{otherwise} \end{cases}$$

$$(14)$$

where $v_l = (v_{i,l}, v_{k_{j,i},l})$ is a pair of thresholds that belong to a sample drawn from a Gaussian distribution. This distribution is updated through the cross-entropy algorithm described in the sequel. α_d, α_{F}, and α_{MD} are positive weights satisfying $\alpha_d + \alpha_{\text{F}} + \alpha_{\text{MD}} = 1$. N_l is the number of simulation runs that allows to compute p_{F}, p_{MD}, and $d(t_{\text{detec}}; v_l)$ in response to

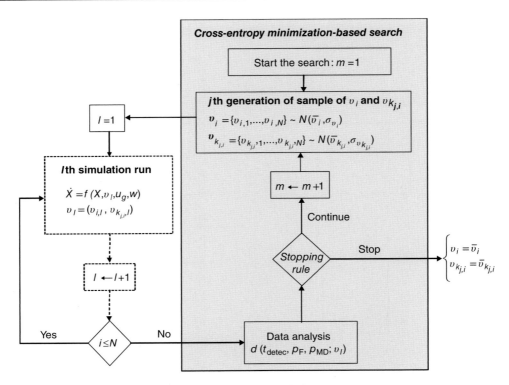

Figure 10. Automated search of NaFD thresholds based on iterative simulations.

measurement noise and wind turbulence processes. Δ is a tuning parameter, which is established so that a detection time greater than $t_f + \Delta$ is undesirable. t_d is the time instant at which the fault is detected.

The cost in (14) increases as the number of false alarms, the number of missed detections, or the detection time increases.

The rationale is depicted in Figure 10, where u_g, w, f, and X denote the wind turbulences, the measurement noise, the vector field of the dynamics, and the state-space vector, respectively.

First, samples $v_i = \{v_{i,1}, \ldots, v_{i,N}\}$ and $v_{k_{j,i}} = \{v_{k_{j,i},1}, \ldots, v_{k_{j,i},N}\}$ are drawn from an independent, identically distributed, random sequence of $v_{i,l}$ and $v_{k_{j,i},l}$. Two-parameter distributions can be used to generate each sample independently (Rubinstein and Kroese, 2004). The Gaussian distributions $N(\bar{v}_i, \sigma_{v_i})$ and $N(\bar{v}_{k_{j,i}}, \sigma_{v_{k_{j,i}}})$ are selected for simplicity purposes.

Second, a batch of N_l simulation runs is carried out for every pair of thresholds $v_l = (v_{i,l}, v_{k_{j,i},l})$, where $l \in \{1, \ldots, N\}$. Each simulation of the batch l is initialized with a pseudo-random seed of the measurement noise and wind turbulence processes, which are detailed in Section 3.6. Results obtained from batch l allow to compute the detection time, the empirical frequency of false alarms, $\hat{p}_F$, and of missed detections, $\hat{p}_{MD}$, which in turn yield $d(t_{\text{detec}}, \hat{p}_F, \hat{p}_{MD}; v_l)$

where $\hat{p}_F$ and $\hat{p}_{MD}$ are substituted for probabilities p_F and p_{MD} in (14), respectively.

Third, the results of the simulations are analyzed to drive the selection, at the next iteration, of new samples v_i and $v_{k_{j,i}}$ that have a tendency to decrease $d(t_{\text{detec}}, p_F, p_{MD}; v_l)$ by modifying $(\bar{v}_i, \sigma_{v_i})$ and $(\bar{v}_{k_{j,i}}, \sigma_{v_{k_{j,i}}})$. To do so, first select the $\lceil \rho N \rceil$ best draws as defined by $\hat{d}_l = d(t_{\text{detec}}, p_F, p_{MD}; v_l)$; that is, $\hat{d}_1 < \hat{d}_2 < \cdots < \hat{d}_{\lceil \rho N \rceil}$. Then, from the $\lceil \rho N \rceil$th-order statistics, update $(\bar{v}_i, \sigma_{v_i})$ and $(\bar{v}_{k_{j,i}}, \sigma_{v_{k_{j,i}}})$ so that the Kullback–Leibler divergence to the optimal density of the threshold is minimized. This task amounts to computing $\bar{v}_i$ and σ_{v_i} as (Rubinstein and Kroese, 2004)

$$\bar{v}_i = \frac{1}{\lceil \rho N \rceil} \sum_{l=1}^{\lceil \rho N \rceil} v_{i,l}$$

$$\sigma_{v_i}^2 = \frac{1}{\lceil \rho N \rceil} \sum_{l=1}^{\lceil \rho N \rceil} (v_{i,l} - \bar{v}_i)^2 \qquad (15)$$

$\bar{v}_i$ and σ_{v_i} are usually smoothed as $y_m = \alpha u_m + (1 - \alpha) y_{m-1}$, where $\alpha \in (0, 1)$. u_m corresponds to $\bar{v}_i$ (resp. σ_{v_i}) computed at iteration m by applying (15), and y_m corresponds to the smoothed update of $\bar{v}_i$ (resp. σ_{v_i}) at m. The same applies to $\bar{v}_{k_{j,i}}$ and $\sigma_{v_{k_{j,i}}}$.

The algorithm is stopped when standard deviations σ_{v_i} and $\sigma_{v_{k_{j,i}}}$ become smaller than a prescribed small constant. The objective here is tantamount to yielding narrow Gaussian distributions centralized on the desired near-optimal thresholds.

The algorithm necessitates initializing $(\bar{v}_i, \sigma_{v_i})$ and $(\bar{v}_{k_{j,i}}, \sigma_{v_{k_{j,i}}})$, which is usually a delicate task. $\bar{v}_i$ and $\bar{v}_{k_{j,i}}$ could be roughly fixed to some low positive values while setting σ_{v_i} and $\sigma_{v_{k_{j,i}}}$ to large positive values so as to visit region of the parametric space where the near-optimal thresholds are likely to be located. Selecting large values for the initial variances may, however, result in slow convergence of the algorithm.

5   SIMULATIONS AND DISCUSSIONS

A string formation of nine small-scale airships, each of which is modeled with a full 6-degree-of-freedom (6-DOF) nonlinear dynamics in closed loop with its autopilots and distributed formation flight controller, is considered to test the integrated health monitoring system. The reader is referred to Léchevin and Rabbath (2007) for details on the model of the systems. Measurement noise and sampling of sensor signals are included in the feedback loops in position global positioning system (GPS) and orientation angles. Each vehicle i is dynamically coupled to a single neighbor $i-1$, as suggested by Figure 11. Models and parameters for the controllers and the two proposed detectors are given in the study by Léchevin and Rabbath (2007, 2009). The formation is required to move around a square area. The time of flight of each edge of the square is 100 s. AFD and NaFD are implemented onboard UAVs so that UAVs 7 and 2 are able to detect a HOF fault in

UAV 6 and a float-type fault in UAV 2, respectively. The adaptation mechanism consists in establishing dynamic couplings with UAVs 1 and 5, once detectors onboard 7 and 2 indicate the presence of a fault. Simulations are first carried out without wind turbulence and aim to demonstrate that all faults are detected despite measurement noise and that moving around the corners of the square does not generate false alarms. Inequality (5) is used to initialize a fine tuning by simulations of threshold $J_{th,i}$ of vehicle i. The following values are used: $J_{th,3} = 0.12$, $J_{th,4} = 0.12$, $J_{th,5} = 0.12$, $J_{th,6} = 0.125$, $J_{th,7} = 0.125$, $J_{th,8} = 0.13$, and $J_{th,9} = 0.13$. Thresholds v_i, $i = 1, \ldots, 9$ are determined by trial and error when the environment is exempt from wind turbulence. This step leads to $v_3 = 0.3$ and $v_{1_{2,3}} = 0.3$, which are used afterwards as initial values to the near-optimal search of thresholds. The search algorithm is presented when turbulence winds are added to the model.

Snapshots of the string formation are shown in Figure 11 at $t = 75$, 150, and 200 s. The faults are triggered at $t = 150$ s, when vehicles 2 and 6 have completed their motion around the corner. The third snapshot shows that the formation has recovered its original shape but without vehicles 2 and 6, which are no longer part of the formation. Detection times for AFD and NaFD are 5 and 10 s, respectively. Figure 12 shows the x-axis trajectories of vehicles that are adjacent to faulty UAVs 2 and 6. Similar trajectories are obtained along the y axis and are not shown for lack of space. UAVs 3 and 7 start to follow the faulty vehicle on $(150\ \mathrm{s},\ 150 + t_d\ \mathrm{s})$ until the faulty behavior is detected at t_d. The two followers then catch up with the rest of the formation, trying to maintain the shape and cohesion of the reduced-size formation.

We now consider the case where mild wind conditions are included in the model and focus on the automated selection of

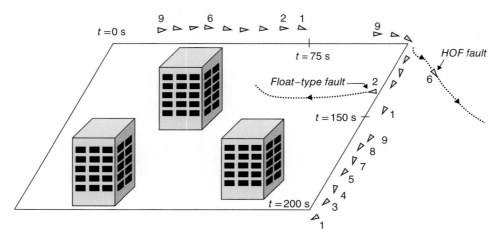

Figure 11. Formation locations at three different time instants.

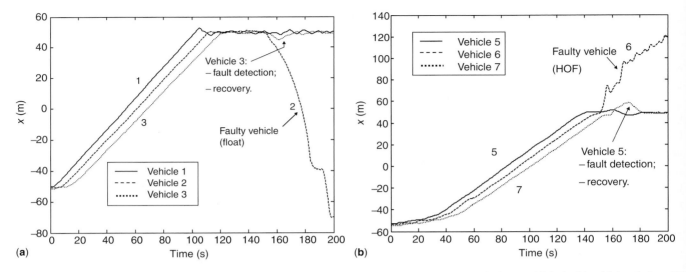

Figure 12. *X*-axis trajectories of triads of vehicles: (a) vehicles 1, 2, and 3 with float-type actuator fault on vehicle 2; (b) vehicles 5, 6, and 7 with hard-over fault on an actuator of vehicle 6.

thresholds v_3 and $v_{1_{2,3}}$. The wind turbulence is modeled by means of a Dryden distribution given by Juang and Cheng (2006)

$$
u_g = u_{gc} + \sigma_u \left(\frac{2a_u}{T_s}\right)^{1/2} \frac{N(0,1)}{(s+a_u)}
$$

$$
u_{gc} = -u_{510}\left[1 + \frac{\ln(h/510)}{\ln 51}\right] \qquad (16)
$$

$$
a_u = \frac{U_0}{600}, \ \sigma_u = 2|u_{gc}|, \ u_{510} = 12 \ \text{ft s}^{-1}
$$

where *s* denotes the Laplace operator. Wind turbulences that fluctuate around 70% of the almost-lighter-than-air vehicle (ALTAV) speed U_0 were modeled.

The iterative search algorithm, whose flowchart is depicted in Figure 10, is applied to finding thresholds v_3 and $v_{1_{2,3}}$ minimizing the cost in (14). Weights α_d, α_F, and α_{MD} are set to 1/3 each. The search is initialized with the values of thresholds used under no wind turbulence; that is, $v_3 = 0.3$ and $v_{1_{2,3}} = 0.3$. The following parameters are used: $N_l = 100$, $N = 50$, $\alpha = 0.7$, $\rho = 0.1$, and $\Delta = 30$ s. Application of the cross-entropy minimization-based search algorithm gives $v_3 = 1.0048$ and $v_{1_{2,3}} = 0.2042$. The average time of detection computed with Monte-Carlo simulations is 13 s. Frequency of missed detections and false alarms are 0 and 2%, respectively, whereas those obtained with $v_3 = 0.3$ and $v_{1_{2,3}} = 0.3$ (no environmental effect) provide a prohibitive level of false alarms and missed detections. The

computation of the thresholds used in NaFD is thus context specific. Operating conditions entailed by other types of environmental effects may necessitate the recomputation of thresholds.

ACKNOWLEDGMENTS

This work was partly funded by grant STPGP 350889-07 from the Natural Sciences and Engineering Research Council of Canada.

REFERENCES

Basseville, M. and Nikiforov, I.V. (1993) *Detection of Abrupt Changes*, Prentice-Hall, Englewoods Cliffs, NJ.

Boskovic, J.D., Bergstrom, S.E. and Mehra, R.K. (2005) Robust integrated control design under failures, damage, and state-dependant disturbances. *J. Guid. Cont. Dyn.*, **28**(5), 902–917.

Daigle, M.J., Koutsoukos, X.D. and Biswas, G. (2007) Distributed diagnosis in formations of mobile robots. *IEEE Trans. Robot.*, **23**(2), 353–369.

Gahinet, P., Nemirovski, A., Laub, A.J. and Chilali, M. (2004) LMI control toolbox – for use with MATLAB. *User's Guide, Version 1.0.9*, The MathWorks, Inc, Natick, MA.

Giulietti, F., Pollini, L. and Innocenti, M. (2000) Autonomous formation flight. *IEEE Cont. Syst. Mag.*, **20**(6), 34–44.

Innocenti, M. and Pollini, L. (2004) Management of communication failures in formation flight. *J. Aerosp. Comput. Inf. Commun.*, **1**, 19–35.

Juang, J.G. and Cheng, K.C. (2006) Application of neural networks to disturbances encountered landing control. *IEEE Intelligent Trans. Syst.*, **7**(4), 582–588.

Léchevin, N., Rabbath, C.A. and Sicard, P. (2006) Trajectory tracking of leader-follower formations characterized by constant line-of-sight angles. *Automatica*, **42**(12), 2131–2141.

Léchevin, N. and Rabbath, C.A. (2007) Robust decentralized fault detection in leader-to-follower formations of uncertain, linearly parameterized systems. *J. Guid. Cont. Dyn.*, **30**(5), 1528–1535.

Léchevin, N. and Rabbath, C.A. (2009) Decentralized detection of a class of nonabrupt faults with application to formations of unmanned airships. *IEEE Trans. Cont. Syst. Technol.*, **17**(3), May 2009, pp. 505–519.

Liang, F., Wang, J.L. and Yang, G.-H. (2002) Reliable Robust Flight Tracking Control: An LMI Approach, *IEEE Trans. Control Syst. Technol.*, **1**(1), pp. 76–89.

Mehra, R.K., Boskovic, J.D. and Li, S.M. (2000) Autonomous formation flying of multiple UCAVs under communication failure. *IEEE Position Location and Navigation Symposium*, San Diego, pp. 371–378. doi: 10.1109/PLANS.2000.838327.

Pollini, L., Giulietti, F. and Innocenti, M. (2000) Robustness to communication failures within formation flight. *American Control Conference*, pp. 2860–2866.

Poor, H.V. (1994) *An Introduction to Signal Detection and Estimation*, Springer, Princeton, NJ.

Rubinstein, R.Y. and Kroese, D.P. (2004) *The Cross-Entropy Method – A unified Approach to Combinatorial Optimization, Monte-Carlo Simulation and Machine, Information Science and Statistics*, Springer, New York.

Scilab 5.1.1 (2009) The Scilab Consortium Scilab Manual, 2009. Online http://www.scilab.org/.

Tuan, H.D., Apkarian, P. and Nguyen, T.Q. (2003) Robust filtering for uncertain nonlinearly parameterized plants. *IEEE Trans. Signal Process.*, **51**(7), 1806–1815.

Valenti, M., Bethke, B., Fiore, G., How, J. and Feron, E. (2006) Indoor multi-vehicle flight testbed for fault detection, isolation, and recovery. *AIAA Guidance, Navigation, and Control Conference and Exhibit*, August 21–24, 2006, Keystone, Colorado. AIAA Paper No. 2006-6200.

Chapter 32

Cooperative Control for Multiple Air Vehicles

Corey J. Schumacher

711 HPW/RH, Wright-Patterson AFB, Ohio, OH, USA

1 INTRODUCTION

The use of unmanned vehicles for various military and civilian missions in the air, sea, space, and on the ground, has received growing attention in the last decade. Although it presents many challenges, there are a variety of advantages to taking the human out of the cockpit. Apart from the obvious advantage of not placing human life in harm's way, the lack of an on-board human operator enables longer endurance. Moreover, working in groups presents the opportunity for new operational paradigms. This chapter will concentrate on the cooperative decision and control problem associated with the operation of a group of unmanned aerial vehicles (UAVs) against multiple ground targets. The concepts that will be presented are general in nature and thus can be extended to other problems involving cooperative unmanned systems or cooperative task execution for manned aircraft.

Unmanned Aircraft Systems. Edited by Ella Atkins, Aníbal Ollero, Antonios Tsourdos, Richard Blockley and Wei Shyy.
© 2016 John Wiley & Sons, Ltd. ISBN: 978-1-118-86645-0.

1.1 Motivation

A group of vehicles working together to achieve a common objective is a cooperative team. The main motivation for team cooperation stems from the possible synergy, as the group performance is expected to exceed the sum of the performance of the individual UAVs. Such cooperation, possible only if the UAVs have a high level of autonomy, can result in improved performance due to global information, team resource management, and enhanced robustness to failures.

Each UAV carries a payload enabling it to sense the environment. By sharing sensor information via a communication network, the entire team can act based on shared, even global situation awareness, instead of the individually available local information. The UAVs can make team decisions regarding flight paths and actions in the uncertain environment, instead of operating independently. Having a cooperative decision algorithm, allows efficient allocation of the overall group resources over multiple tasks. Moreover, sharing information allows the team to compensate for task failures or loss of vehicles. Through cooperation, the team can re-configure its distribution architecture to minimize the performance degradation resulting from vehicle failures.

1.2 Cooperative control structure

Team cooperative decision and control controllers can be characterized in many ways, but one critical question is the information and decision-making structure. Information can be heavily shared, even global, or largely local, in which even case the team is more dependent on predictive model solutions for good performance. The decision architecture can be centralized, which can be thought of as requiring command driven solutions, typically computed at a single location and broadcast to the team, or more decentralized with solutions computed partially by each element of the team.

In a centralized control problem, all of the state information from the distributed vehicles or agents is sent to a centralized agent where it is operated on by a large decision and control program. The resultant individual plans and assignments are disseminated out to the respective vehicles to be executed. This approach can lack significant robustness, is computationally complex, and doesn't scale well. Scaling well means that controller complexity is roughly proportional to team size.

The most decentralized control architecture uses little if any global information. In the limit, there is no communication between the vehicles and information is only inferred about the other vehicles' objectives by measuring their actions through their sensors. In general, this approach leads to an average macro level of performance. For example, the team can maintain some loose cohesion, and a nominal trajectory. The achievable performance is low due to the absence of communication and little global coordinating information. This class of controller performs poorly in cases involving dominant coupling constraints.

1.3   Attributes

Three main attributes associated with cooperative decision and control problems, such as the ones described in this chapter, are complexity, imperfect information, and implementability.

Cooperative decision and control problems are complex. The sheer size of the problem (e.g., number of vehicles, targets, and threats) is one form of complexity. However, in scenarios such as combat ISR, coupling between the completion of the different tasks and coupling between the assignment process and trajectory optimization have the most significant impact on complexity. For example, if the vehicles each have a default task of searching, then performing it cooperatively introduces extensive coupling in their search trajectories. Once a target has been found it may need to be simultaneously tracked by at least two vehicles and attacked by a third, which further imposes coupling between the trajectories of different team members.

Most cooperative teams will operate with limited, imperfect information. The full information state is usually not, in practice, available anywhere in the network. The challenge is to perform the tasks cooperatively to achieve mission objectives under limited information. A specified level of team performance and team coherence requires a minimum level of shared information. This should include the team

objective function, a subset of the events, and a set of functions that represent the ability of the vehicles to perform actions. Ideally, there should be sufficient information to ensure that all the tasks are covered and the tasks are consistent. Also, the vehicles may have state estimation and prediction models to provide information that is not readily available. Information flow imperfections, such as communication delays, may produce different information sets for the different vehicles in the group, which can lead to multiple strategies. The result can be un-coordinated assignments, such as multiple vehicles wrongly assigned to perform the same task on a certain target, while leaving other tasks unassigned.

In the design of cooperative decision and control algorithms, many practical factors must be taken into account to ensure implementability on UAVs in real-life missions. One such factor is computation time. For "real-time" operations, the time cooperative control algorithms must be able to make cooperative control decisions faster than the time scale of mission activity. For instance, given that a mission requires that high value targets be prosecuted immediately upon discovery, then because of travel time, if any of the UAVs are close to the target there is short time to make an assignment decision. The ability to make decisions in the required time frame is a function of the cooperative control algorithms and the capability of the data processors on the UAVs. While finding optimal cooperative decision and control solutions can be computationally intractable, it is possible to implement sub-optimal algorithms that calculate solutions quickly that can then be improved over the available decision time window.

2   COMPLEXITY IN COOPERATIVE TEAMS

In cooperative teams, an interactive decision making process between vehicles takes place, while individual vehicle autonomy is preserved. There is a continuum between centralized and decentralized control. If a fully decentralized team means no communication, then in a cooperative team there is a requirement for the minimum level of globally communicated information that allows the desired level of team performance to be achieved.

A team is here defined as a collection of spatially distributed controlled objects, also known as UAVs, that have objectives in common (Marschak and Radner, 1972; Ho and Chu, 1972). Air vehicles may be a too restrictive term – generically, a team consists of members, or agents, and the team can (and generally will) include humans as operators, task performers (think of target recognition), and/or supervisors. The presence of a common objective forges a team and induces cooperative behavior. If the air vehicles are working together to achieve a common objective, then they are considered a team. At the same time, additional individual objectives of the team members can encourage team members to opt for a weak degree of non-cooperation, mild competition, or outright adversarial action.

When team decision and control problems are discussed, it is important to address the unstructured environment/uncertainty, the organizational structure, the information pattern, and task coupling. Individual operational scenarios can be dominated by one of the above, but will contain elements of all of them. The interaction of these different facets of a team problem cannot be ignored. The vehicles are coupled through the performance/objective function and the task constraints.

In general, the performance of cooperative control can be characterized by task coupling, uncertainty, communications delays, and partial information. The interaction of these dimensions renders cooperative optimal control a complex problem. Currently, no working theory of cooperative systems that takes all these dimensions into account exists. A hierarchical decomposition is normally used to reduce the problem to more digestible bits, but optimality is forfeited in the process. Some degree of coupling is ignored to achieve decomposition. This results in a suboptimal solution but improves solvability and robustness. Many times robustness comes at the expense of optimality, and vice versa. Indeed, the optimal operating point might be sensitive to changes in the problem parameters.

Team control and optimization problems are decomposed in space, time, or along function lines. Forming of sub-teams of UAVs and tasks can also be done by graph theoretic methods (Papadimitriou and Steiglitz, 1982), set partition approaches (Balas and Padberg, 1976), and relative benefit optimization techniques (Rasmussen, Chandler and Schumacher, 2002), as well as by brute force search (Rasmussen *et al.*, 2003). The sub-team optimization problem then reduces to the multiple assignment problem: determine the task sequence and timing, for each team member, that satisfies all the constraints while minimizing an overall team objective function. The individual vehicles then perform their own task planning and send coordinating information, preferably a sufficient statistic, around the network or to a team leader. Algorithms for constrained multiple task assignment include: heuristic search, for example, branch and bound, Tabu search, or genetic algorithms (Balas and Carrera, 1996), generalized assignment (Burkard and Cela, 1998), linear programming (Dantzig, 1963), iterative network flow (Bertsekas, 1992; Goldberg and Tarjan, 1990), and iterative auction (Bertsekas, 1988; Kempka, Kennington, and Zaki, 1991). One of the primary contributors to the complexity of multiple assignment is task coupling in the

face of floating timing constraints–the latter brings in aspects of job shop flow optimization, or scheduling.

2.1 Task coupling

UAV team missions such as suppression of enemy air defenses and wide area search and destroy are dominated by coupled tasks with floating timing constraints. There are a number of issues involved in solving the multiple assignment problem in a cooperative framework. Chief among these is the ability to decouple assignment from path planning for specific tasks. This means that tasks and path plans are generated to determine costs that are then used in the assignment process. The assumption is that these calculations are still valid after the assignment is made. This is even more so for tour sequences. Unless all possible tours are generated, sub-optimality is created when chaining together tasks.

Also, assignment is also decoupled from timing. For example, task tours are assigned first. Then the task order and precedence are enforced. This can be done myopically to set the task time using the earliest task that needs to be done, then the next, and so on, or the task times can be negotiated between the vehicles until a set of task times is arrived at that satisfies all the timing constraints. The assumption here is that these task times will not have a different, closer to optimal assignment. The decomposition assumptions to address coupling may lead to infeasibility, where all the tasks can not be assigned, as well as a significant degree of sub optimality, also known as poor performance. If the task coupling is strong, decentralization more difficult (Rasmussen *et al.*, 2003) – optimality is sacrificed, the algorithm may induce churning, and feasibility is not guaranteed.

2.2 Uncertainty

Some cooperative team problems can be dominated by uncertainty rather than by task coupling. This is true for those missions where the target identification, target localization, threat identification, and threat location are not known in advance. Some of this information may be known, while the rest is estimated using *a priori* given probability distributions. The challenge is to calculate the expected future value of a decision or action taken now. For example, if the UAVs use their resources on targets now, there may be no reserves left for targets that are found later and that have higher value. At the same time, actions taken now might decrease the level of uncertainty. The latter can be gauged using information theoretic concepts. Possible choices are to myopically follow the decision path of least risk, or follow the decision path that maximizes the possible options in the future. Of course, the safest and middle of the road decisions are not generally the best. Furthermore, one critical source of uncertainty is associated with the actions of an adversary in response to an action taken by the UAVs. Possible approaches to account for uncertainty are stochastic dynamic programming and Markov decision processes (Puterman, 2005), Bayesian belief networks (Jensen, 1996), information theory (Gallager, 1968), and, in the case of no information – game theory (Luce and Raiffa, 1989).

2.3   Communication

The basic premise of cooperative control is that the UAVs can communicate whenever and as much as they need to. All networks incur link delays, and if these delays are sufficiently long compared to the time between events, they can completely nullify the benefits of team cooperation (cooperative control). Recall that control system delays in the feedback path are conducive to instability. A critical choice here is whether the team decisions are synchronous or asynchronous. Synchronous implies that there is a common (and up to date) data base accessible to all the vehicles. If an event occurs locally, the event and all associated information is shared across the network and a decision based on the new event cannot occur until this happens. Under this protocol, a single actor can slow down the whole team and compromise time critical tasks. Strategies are needed to maintain team coherence and performance. Synchronous team protocols are frequently used in conjunction with a redundant centralized decision approach where each member of the team solves the same decision problem for the whole team, thus ensuring centralized optimality.

Asynchronous decision protocols however, while more robust to delays, are much more difficult to verify and prove correct operation. They are susceptible to inconsistent information across the network, can lead to decision cycling/churning and infeasibility. The higher the rate of occurrence of events, the more difficult these problems become because the input's frequency exceeds the system's "bandwidth." Some useful protocols include consensus voting (Olfati-Saber and Murray, 2002), parallel computing and load balancing (Bertsekas and Tsitsiklis, 1989), job shop scheduling (Sycara and Liu, 1996), and contract nets (Sandholm and Thomas, 1993). However, with false information and sufficient delays, consensus may never be reached. Indeed, false information strongly negates the benefits of cooperative control. The situation is somewhat analogous to the feedback control situation: feedback action is superior to open loop control, provided the signal to noise ratio in the measurement is high. One then takes advantage of the benefit of feedback. If however the measurements are very noisy, one might be better off ignoring the measurements and instead opt for open loop (feed-forward or model-based) control.

2.4   Partial information

Decentralized control (Kuhn, 1953) as well as cooperative teams are characterized by limited, or partial information. The full information state is not available anywhere in the network. The challenge is to perform the tasks cooperatively and achieve a degree of optimality under limited and distributed information.

A specified level of team performance and team coherence requires a minimum level of shared information, for example, the team objective function, a subset of the relevant events – for example, pop up target information, and part of the state vector, for example, the fuel state of the UAVs. There should be sufficient information to ensure that all the tasks are accounted for and the tasks are consistent. This can be considered the "sufficient statistic." Also, the vehicles may have state estimation and prediction models to provide information that is not available locally. Interestingly, the vehicles may have different objective functions as well as inconsistent and delayed information that can result in conflicts that need arbitration or negotiation. False information, in particular, can induce poor performance in a cooperative team – one may be better off using non-cooperative control.

3   TASK ASSIGNMENT EXAMPLE

The UAV cooperative team problem can be highly complex, even for relatively small teams. Generally, the available theories and approaches can address only one or two aspects of the problem at a time. We are often more interested in a fast, feasible, and robust solution, rather than an optimal one. Since there are many UAV scenarios of moderate size, say 4–8 vehicles, that are of interest, approaches such as MILP, and stochastic dynamic programming may be sufficiently fast at this scale, where a centralized optimal solution is possible. Thus, algorithm scalability may not always be a limiting factor.

If more decentralization is desired, the primary limitation is task coupling. Task coupling can be reduced if a myopic or receding horizon procedure is used where not all tasks are addressed up front. However, this can have a significant impact on mission performance and even feasibility. Also, in the drive for localization, algorithms such as auction and distributed constraint satisfaction can incur extensive message traffic for all but the weakest task coupling. Finally, false information and communication delays can completely negate the benefits of cooperation – similar to losing the benefits of feedback when the sensors are noisy and consequently open loop control is preferable.

This section presents one task assignment method for assigning cooperating UAVs to multiple targets. The scenario of interest involves wide area search munitions (WASMs), where multiple homogeneous UAVs must cooperatively perform multiple tasks on multiple targets. The WASM scenario is a good starting point for the study of cooperative decision and control. This scenario exhibits many of the essential features of multi-agent cooperation that are encountered in UAV teaming problems, without some of the additional complications inherent in combat UAV operations, such as threat avoidance. The assignment problem can be solved using a capacitated transshipment problem formulation to assign, at each stage, at most a single task to each vehicle.

3.1 Wide area search munition scenario

WASMs are small powered UAVs, each with a turbojet engine and sufficient fuel to fly for a moderate duration. They are deployed in groups from larger aircraft flying at higher altitudes. Individually, the munitions are capable of searching for, recognizing, and attacking targets. When employed as a cooperative team instead of individuals, they can execute missions much more efficiently. The vehicles communicate target information whenever this information is updated by any vehicle.

3.2 Required tasks

We begin with a set of N_v simultaneously deployed vehicles, each with a life span of 30 minutes. Let

$$V = \{1, 2, \ldots, N_v\} \qquad (1)$$

be the set of deployed UAVs. Targets that might be found by searching fall into known classes according to the value or "score" associated with destroying them. We index them with n as they are found, so that $n = 1, \ldots, N_t$, where N_t is the number of discovered targets and

$$T = \{1, 2, \ldots, N_t\} \qquad (2)$$

is the set of these targets. Let V_n denote the value of target n. The WASMs are assumed to begin a mission with no

information about specific target numbers or locations. The WASM are assumed to have a laser detection and ranging (LADAR) sensor that can scan a region in front of the vehicle, and automatic target recognition (ATR) software capable of identifying targets with some given probability. The ATR process is modeled using a system that provides a probability or "confidence" that the target has been correctly classified. A potential target cannot be attacked until it has been identified with a high confidence, which may require multiple looks from different viewing angles.

3.3 Task definitions

The WASM are capable of performing four distinct tasks: search, classification, attack, and verification. Which tasks are required at any point in time is dependent on the available target information.

Search involves using the LADAR sensor to find unknown targets. Since cooperative search has been extensively studied (Stone, 2004), it will not be addressed here. For the purposes of our cooperative task assignment work, we assume an a priori specified search pattern, for example the lawnmower search, where the vehicles have pre-assigned search lanes which, if completed, guarantee full coverage of the search area. If a WASM leaves its search path to perform other tasks, when it returns to search, it will return to the point where it left the search pattern, and then continue. Once a potential target, or *speculative*, has been located, a classification task is required. A classification is a second (or third) imaging and ATR of the target, from a different view angle, to raise the confidence level of the target identification. Once a *speculative* has been classified as an actual target, an attack task is needed. To perform an attack, the WASM dives low over the target and explodes, destroying itself and, hopefully, the target as well. The target is then imaged by another WASM to perform verification and confirm that the target has, in fact, been destroyed. If the target is still alive, another WASM may be tasked to attack it. Let

$$M = \{C, A, V\} \qquad (3)$$

be the set of tasks on each target where C, A, and V denote classify, attack, and verify, respectively.

3.3.1 Precedence and task constraints

The three tasks needed to prosecute each target (classify, attack, and verification) must be performed in the specified order. Consequently, the cooperative planning algorithms must ensure that the proper order is achieved, with possible additional delays required between tasks, such as a sufficient

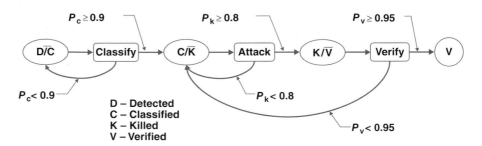

Figure 1. Target state transition diagram.

delay between attack and verification to allow any smoke or debris to clear. Additionally, each task has unique constraints. Thus, it is not possible to simply assign three WASM to a target, and allow each vehicle to perform whatever task is needed at the time it arrives. Classification requires specific look angles, dependent on the target orientation and previous view angle, and has a standoff distance equal to the distance the sensor scans ahead of the WASM. Attack can be performed from any approach angle, but requires the WASM to dive onto the target after locating it with the LADAR sensor. Verification is performed at the sensor standoff distance, similar to classification, but can be performed from any approach angle.

A state transition diagram representation of the general problem is given in Figure 1 where P_c, P_k, and P_v denote the probability of performing a successful classification, attack, and verification; with 0.9, 0.8, and 0.95, respectively, being the thresholds chosen in this example.

3.4 Capacitated transshipment assignment problem formulation

Whenever a new target is found, or an unassigned task is otherwise required (e.g., due to a verification determining that a previously attacked target is not yet destroyed), the assignment algorithm is run to determine which members of the UAV team will perform the required tasks. For the simplest cooperative planning algorithm we will discuss, only a single task is planned for each known target. When that task has been performed, the next task is assigned. Eventually, all required tasks will be performed on a target, unless the target is of sufficiently low priority and the UAVs are instead assigned to continue searching for unknown targets, or the UAVs run out of fuel.

For this version of the assignment problem, we use a linear programming formulation known as a capacitated transshipment assignment problem (CTAP). The CTAP is used as a time-phased network optimization model designed to perform task allocation. The model is run simultaneously on all munitions at

discrete points in time, and assigns each vehicle up to one task each time it is run. The model is solved each time the new information is brought into the system, typically because a new target has been discovered or an already-known target's status has been changed. This linear program can be described by a network flow diagram, as shown in Figure 2. With N_v UAVs and N_t known targets, there are N_v sources, N_t potential non-search tasks, and up to N_v possible search tasks. Each target is responsible for one possible task, and each vehicle is assigned one task, either search, classify, attack, or verification.

This simple assignment algorithm looks only at presently required tasks, with no knowledge of tasks likely to be needed in the future. For example, an attack task is not assigned until after a classify has been successfully performed. Similarly, a verification task is not assigned until after an attack is completed, even though a verification is likely to be desired. The advantage of this myopic approach is that the optimization problem at each stage of the mission can be solved very quickly, allowing real-time implementation. The disadvantage is that known information about future tasks (e.g., the requirement for verification after an attack task) is not being exploited.

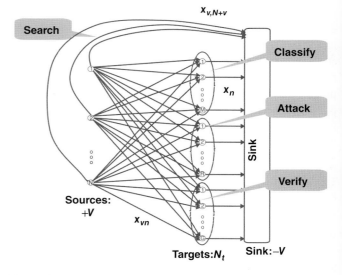

Figure 2. UAV CTAP network flow diagram.

However, the CTAP approach is an effective task assignment strategy for scenarios featuring task precedence constraints because it naturally satisfies task precedence constraints by construction. Such a "myopic" approach is also desirable whenever future tasks are uncertain or dynamic.

The CTAP network optimization model can be expressed as

$$\max J = \sum_{v,n} c_{vn}^k x_{vn} \tag{4}$$

subject to

$$x_{v,N+v} + \sum_{n=1}^{N} x_{vn} = 1, \qquad \forall v = 1, \ldots, N_v \tag{5}$$

$$x_n - \sum_{v=1}^{V} x_{vn} = 0, \qquad \forall n = 1, \ldots, N_t \tag{6}$$

$$\sum_v x_{v,v+n} + \sum_n x_n = N_t \tag{7}$$

where $x \in \{0, 1\}$ is a binary decision variable; the variable c_{vn}^k is the benefit of vehicle v performing task k on target n, where $k = 0$ corresponds to search, $k = 1$ corresponds to classify, $k = 2$ corresponds to attack, and $k = 3$ corresponds to verification. Search can only be paired with $n = 0$, the sink node. Referring back to Figure 2, x_{vn} refer to assignments of vehicles to individual targets, $x_{v,N+v}$ refer to the assignment of a vehicle to the search task, and x_n is used to ensure flow balance constraints are met. Equation 5 guarantees that each vehicle v is assigned only one task. Equation 6 is a flow balance constraint, requiring that for each flow into a task node x_{vn}, there is also one flow out, x_n. Equation 7 requires that N_t tasks be assigned. In combination with the requirement that each x be binary, these constraints guarantee that each vehicle will only be assigned one task, either a search task or a target prosecution task, and also that each target prosecution task will only be assigned to one vehicle. It is possible for a target prosecution task to be left unassigned, if it is of lower value than an available search task.

3.5   Weight calculations

Many approaches could be taken to assigning values or weights, c_{vn}^k, to different tasks. A simplified scheme is presented which does allow trade-offs between tasks based on success probabilities, but which does not use the full probabilistic computation of expected values. The highest value is assigned to killing a target of the highest-valued type,

with other tasks generating less of a benefit. Overall, the chosen weights tend to result in the least possible lost search time for the execution of a particular task. The values of search, classify, attack, and verification tasks are all determined differently.

Suppose that there are P distinct target types, each with a given value V_p and probability of kill P_{k_p} when attacked, with $p = 1, \ldots, P$. Let $\overline{V}$ be the maximum value of $V_p * P_{k_p} \ \forall p$, and $\overline{p}$ be the value of p corresponding to this maximum. Then, $\overline{V}$ is the highest possible value for attacking a target. Note that the value of attacking a target is less than the value of actually destroying it, due to the possibility of failure.

A critical determinant in the resulting vehicle behaviors is the ratio between the value of search, and the value of target prosecution tasks. Since, for the WASM scenario, each vehicle is "used up" when it performs an attack task, the best that a team of N WASM will do is attack N targets of type p. Therefore, the weights should be chosen so that the vehicles will attack any maximum value targets identified, but they will not necessarily attack a lower value target, especially with substantial fuel remaining. Accordingly, the value of search starts equal to the value of attacking the highest value target type, and decreases continually as fuel is used. Let T_f be the remaining available flight time of the munition, and T_0 be the initial endurance of the vehicle when launched. Then the value of search can be set to

$$c_{v0}^0 = \overline{V} * T_f / T_0 \tag{8}$$

At any time after launch, the value of search will always be lower than the value of attacking the highest valued target. Search value decreases linearly with time, so that the decision algorithm will be more willing to conduct attacks and "use up" the available search agents.

Let P_{id_n} be the confidence level with which the ATR process has identified target n. Then, the value of vehicle v attacking target n of type p can be calculated as

$$c_{vn}^2 = P_{id_j} * P_{k_p} * V_p \tag{9}$$

However, this choice for the attack weighting contains a flaw. Assuming heterogeneous vehicles, all vehicles will have the same value for conducting the attack task. A better formulation would give a preference to team members that can perform the task more quickly. Accordingly, we modify Equation 9 as follows. Assume that vehicle v will perform the attack on target n at time t_{vn}. Then let

$$t_n^{\min} = \min_v t_{vn} \tag{10}$$

for $i = 1, \ldots, n$. Then the modified value of vehicle i attacking target j can be expressed as

$$c_{vn}^2 = P_{\mathrm{id}_n} * P_{\mathrm{k}_p} * V_p * \frac{t_n^{\min}}{t_{vn}} \qquad (11)$$

With this modification, the value for the attack with the shortest path is unchanged, while the value of other vehicles performing the task is lowered.

The value of vehicle v attempting to classify target n can be calculated as the assumed probability of classification, multiplied by the expected value of attacking the target, plus the value of the vehicle continuing to search after the classification task has been performed. Including the value of the remaining search time is critical for achieving the proper balance between search and classify task values. Thus, the value of vehicle v attempting to classify target n, believed to be of type p, becomes

$$c_{vn}^1 = P_{\mathrm{cl}} * P_{\mathrm{k}_p} * V_p + \overline{V} * (T_{\mathrm{f}} - T_{\mathrm{cl}})/T_0 \qquad (12)$$

where T_{cl} is the amount of time that will be required for the vehicle to perform the classify task and then return to its search path and P_{cl} is the expected probability of success for the classify task.

The value of a verification task is dependent on the likelihood that the target is still alive after an attack task. The value of vehicle v performing verification on target n of type p can be expressed as

$$c_{vn}^3 = P_v * (1 - P_{\mathrm{k}_p}) * P_{\mathrm{id}_n} * V_p + \overline{V} * (T_{\mathrm{f}} - T_{\mathrm{ver}})/T_0 \qquad (13)$$

where, similar to the classify case, P_{ver} is the assumed probability of successful verification and T_{ver} is the time required to perform the verification task and return to search. A more detailed version of the weightings can be generated if a confusion matrix is used, and multiple target types are possible with different probabilities.

3.6   Simulation results

This section presents simulation results for the CTAP algorithm applied to the WASM problem. In this example, eight vehicles are searching an area containing three targets of different types, and hence of different values. The target information is as follows:

Target	Type	Value	Location (X, Y)
1	1	10	(4000,0)
2	2	8	(3500,−8500)
3	1	10	(18 000,−11 500)

Target (X, Y) location is given in feet. The targets also have an orientation (facing) that has an impact on the ATR process and desired viewing angles, but this will not be discussed as it only affects the the task allocation indirectly, through path length calculations. The search vehicles are initialized in row formation, with fifteen minutes of flight time remaining, out of a possible 30 min. This assumes that the vehicles have been searching for fifteen minutes and then find a cluster of potential targets.

The value of all possible tasks, vehicle, and target assignment combinations are calculated and sent to the capacitated transshipment problem solver. The vehicle and target locations shortly before the first targets are discovered are shown in Figure 3. The black-outlined rectangles represent the sensor footprints of the searching vehicles, and the numbers are the vehicle locations. The lines trailing the vehicles numbers show flight paths. Targets are numbered 1–3.

The next snapshot, shown in Figure 3, is taken at $T = 34$ s. At this point, the searching munitions have discovered two of the targets, nearly simultaneously. When Target 2 was discovered, Vehicle 3 was assigned to classify it. After the classification on Target 2 was complete, Vehicle 3 was also assigned to attack it. Typically, the vehicle that performs the final ATR on a target will be the one assigned to destroy it, but this is not guaranteed. In particular, if one vehicle has significantly less fuel remaining than another, then it is more likely to be assigned to destroy a target. Vehicle 8 is now assigned to image Target 3 and verify that it has been destroyed. Vehicle 4 has been assigned to classify Target 1. In Figure 3, two of the vehicle sensor footprints are shown in grey, to denote that the vehicle is banked and the sensor is thus inactive until the vehicle levels out again. All of the other, unassigned vehicles are continuing to search.

The next snapshot, taken at $T = 84$ s, is Figure 3. At this point, the third target has been detected. Vehicle 4 has classified and attacked Target 1. Vehicle 8 has verified the destruction of Target 2, and is now proceeding to Target 1 to perform the same task. Vehicle 1 is imaging Target 3 to classify it, and will then be assigned to attack it. Finally, Figure 3 shows the complete vehicle trajectories for prosecuting the three targets. Vehicle 2 has verified Target 3, and both Vehicles 2 and 8 have returned to their search patterns.

All discovered targets are fully prosecuted in this example. That will be the case for any discovered target, as long as it is of sufficient value. If the vehicles have a large amount of fuel remaining, resulting in a high value for the search task, then the vehicles will ignore low value targets and search for higher value targets, until their fuel reserve is low. An example of this behavior can be found in Schumacher, Chandler, and Rasmussen (2001).

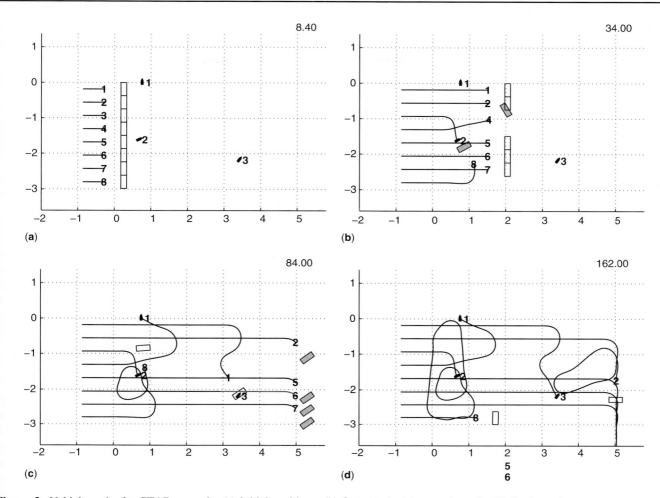

Figure 3. Vehicle paths for CTAP example: (a) initial positions; (b) first attack; (c) second attack; (d) final attack.

3.7 Capabilities and limitations of capacitated transshipment assignment problem formulation

The CTAP formulation presented here for cooperative task assignment is a powerful and flexible tool. Many task assignment problems can be addressed with this method. The resulting optimization problems can be solved very rapidly, making it very amenable to real-time implementations. For the problem sizes that could typically be encountered by eight vehicles, the assignment problem can be solved in a fraction of a second. The CTAP algorithm does, however, suffer from several significant limitations.

The first limitation is that all of the necessary information needs to be gathered in one place, and at one time. The method cannot be implemented in an asynchronous or distributed fashion. Either the decision algorithm has to be run in a centralized manner, where all computations are performed in one location, and assignments are broadcast to the team, or in an implicit coordination architecture, where the necessary information (team and target states, or task values) must be synchronized across all team members, so that they can each run an identical optimization problem.

A second limitation of CTAP is that it has limited flexibility. All of the tasks that enter the assignment problem need to be independent. Only a very limited set of constraints can be included in the formulation. The CTAP is ideal for the classic N on N weapon-target assignment problem, and for the WASM problem when search is considered as a possible task, again resulting in an N on N assignment problem. Some extensions are possible, such as the addition of possible tasks, resulting in an N on M assignment problem, with $M > N$, and only N tasks assigned. Additionally, N can be greater than M, with the problem then inverted and M targets assigned to N vehicles. Or additional tasks can be assigned to each target by the creation of "dummy targets" as done in the previous section. However, the CTAP cannot directly incorporate any timing constraints, and it cannot directly account for coupling between tasks. The most obvious example of this is the task precedence constraint fundamental to the WASM scenario. Classify must be completed before attack, which must be completed before verification. There is no way to include these constraints directly into the CTAP. To incorporate coupling and timing constraints directly into the optimization, more complex methods such as mixed integer linear programming or dynamic programming must be used.

4 CONCLUSIONS

This chapter provides only a limited introduction to cooperative control of unmanned aircraft. Many additional challenges exist which are not addressed here, such as operator interaction, adversary reactions, spoofing, multi-objective collaboration, and so on. Further information on a variety of these issues can be found in the references. Additionally, much of the information for this chapter was taken from (Shima and Rasmussen, 2009) and a much more detailed discussion can be found therein.

REFERENCES

Balas, E. and Carrera, M.C. (1996) A dynamic subgradient-based branch and bound approach for set covering. *Oper. Res.*, **44**, 875–890.

Balas, E. and Padberg, M. (1976) Set partioning: a survey. *Oper. Res.*, **18**, 710–760.

Bertsekas, D.P. (1988) The auction algorithm: a distributed relaxation method for the assignment problem. *Annals Oper. Res.*, **14**, 105–123.

Bertsekas, D.P. (1992) Computational Optimization and Applications. *Comput. Optim. Appl.*, **1**, 7–66.

Bertsekas, D.P. and Tsitsiklis, J.N. (1989) *Parallel and Distributed Computation: Numerical Methods*, Prentice-Hall.

Burkard, R.E. and Cela, R. (1998) Linear Assignment Problem and Extensions. Technical Report 127, Karl-Franzens University of Graz.

Dantzig, G.B. (1963) *Linear Programming and Extensions*, Princeton University Press.

Gallager, R.J. (1968) *Information Theory and Reliable Communication*, Wiley.

Goldberg, A.B. and Tarjan, R.E. (1990) Solving minimum cost flow problems by successive approximation. *Math. Oper. Res.*, **15**, 430–466.

Ho, Y. and Chu, K. (1972) Team decision theory and information structures in optimal control problems - Part I. *IEEE Trans. Autom. Control*, **AC-17**, 15–22.

Jensen, F. (1996) *An Introduction to Bayesian Networks*, Springer.

Kempka, D., Kennington, J.L., and Zaki, H.A. (1991) Performance characteristics of the Jacobi and the Gauss-Seidel versions of the auction algorithm on the Alliant FX/8. *ORSA J. Comput.*, **3**(2), 92–106.

Kuhn, H. (1953) Extensive games and the problem of information. *Ann. Math. Stud.* (**28**), Princeton University Press.

Luce, R.D. and Raiffa, J. (1989) *Games and Decisions: Introduction and Critical Survey*, Dover Publications Inc.

Marschak, J. and Radner, R. (1972) *Economic Theory of Teams*, Yale University Press.

Olfati-Saber, R. and Murray, R.M. (2002) Consensus protocols for networks of dynamic agents. Proceedings of the American Control Conference.

Papadimitriou, C.H. and Steiglitz, K. (1982) *Combinatorial Optimization: Algorithms and Complexity*, Prentice-Hall.

Puterman, M.L. (2005) *Markov Decision Processes: Discrete Stochastic Dynamic Programming*, Wiley Series in Probability and Statistics, Wiley.

Rasmussen, S.J., Chandler, P.R., Mitchell, J.W., Schumacher, C.J., and Sparks, A.G. (2003) Optimal vs. heuristic assignment of cooperative autonomous unmanned air vehicles. *Proceedings of the 2003 Guidance, Navigation, and Control Conference*, AIAA.

Rasmussen, S.J., Chandler, P.R. and Schumacher, C.J. (2002) Investigation of single vs multiple task tour assignments for UAV cooperative control. *Proceedings of the 2002 Guidance, Navigation, and Control Conference*, AIAA.

Sandholm and Thomas. (1993) An implementation of the contract net protocol based on marginal cost calculations. *Proceedings of the Eleventh National Conference on Artificial Intelligence*, AAAI.

Schumacher, C.J., Chandler, P.R. and Rasmussen, S.J. (2001) Task allocation for wide area search munitions via network flow optimization. *Proceedings of Guidance, Navigation, and Control Conference*, AIAA.

Shima, T. and Rasmussen, S. (2009) *UAV Cooperative Decision and Control: Challenges and Practical Approaches*, SIAM.

Stone, L.D. (2004) *Theory of Optimal Search*, 2nd edn, Military Applications Society.

Sycara, K. and Liu, J.S. (1996) Multi-agent coordination in tightly coupled task scheduling. *Proceedings of the International Conference on Multi-Agent Systems*.

Chapter 33
Flight Formation Control

Mario Innocenti[1] and Lorenzo Pollini[2]

[1]*Munitions Directorate, Eglin Air Force Base, Air Force Research Laboratory, FL, USA*
[2]*Department of Information Engineering, University of Pisa, Pisa, Italy*

1 INTRODUCTION

The problem of formation flight has been a topic of research for many years and from different perspectives. The advantages of flying in formation are very clear in manned operations, since sensible drag reduction can be achieved, yielding longer endurance and larger payload capability. The main challenges in the dynamics and control area are in the modeling process, which requires analytical representation of the complex aerodynamic field surrounding the formation, and reliable and robust control systems, for precise position control, and for trajectory tracking. In addition to the control management of a formation, there are other scenarios of interest, which fall in the general category of formation. One in particular is the autonomous aerial refueling problem. This is a very demanding task for the pilots (tanker and aircraft to

be refueled), and for the boom operator (if refueling is not performed by a probe and drogue system). The requirement for tight relative position control in an aerodynamically turbulent environment is very challenging from the flight control system designer standpoint. This becomes more evident when we study formation flight for unmanned air vehicles (UAVs), since pilot override is not possible.

As mentioned above, accurate and reliable controllers for inner loop stabilization and outer loop guidance are necessary when we deal with formation flight of UAVs. UAV missions, in fact, may require a wide variety of formation geometries, tight formations for increased performance, loose formations for safety reasons, changes in formation shapes during the mission, and other cases. In addition, a successful controller must rely on added even redundant sensor suite, in order to accommodate and perform "intelligent" decisions by the vehicles comprising the formation. The literature in this field is very vast, and most of the work relates to UAVs, although formation flight of manned aircraft has been studied as well.

Seminal contributions to the automated flight of a classical leader–wingman configuration and conditions for which proportional integral (PI) control can handle nonlinearities present in the formation dynamics can be found in Dargan, Patcher, and D'Azzo (1992) and Patcher, D'Azzo, and Veth (1996). Aerodynamic characteristics and directional stability properties in air-to-air refueling found in Bloy *et al.* (1993), Bloy and Jouma'a (1995), Giulietti *et al.* (2002), and Wang and Mook (2003a) present analytical and experimental results on aerodynamics modeling. Preliminary aerodynamic considerations as well as the potential use of decentralized linear quadratic regulator (LQR) based control were described in Wolfe, Chichka, and Speyer (1996), with reference to unmanned aerial formations. More recently, at several levels, research efforts are directed and strategic priorities are being given to the use of multiple UAVs for a large spectrum of platforms, from actual-size aircraft (NATO-AGARD AR-31997) to reduced-size UAVs (NATO Advisory Group for Aerospace Research and Development, 1994), and to the implementation of microelectromechanical systems (MEMS) in flight (Brendley and Steeb, 1993).

2   MODELING

A mathematical model of a system for which a controller is designed is always needed. For aerospace applications, mainly for certification purposes, the model should be formalized in terms of physical variables and/or variables directly related to the physics of motion (for this reason, techniques such as neural networks, genetic algorithms, and fuzzy sets are not commonly accepted in the modeling process). As a common practice in the design process, models used for control synthesis are simpler, of lower order, possibly linear, whereas models used for analysis are more complex, nonlinear, and are based on a mix of analytical elements, experimental data, and system identification techniques. The same holds for dynamic modeling of a formation, with the major challenge being the presence of aerodynamic effects (upwash, downwash, and wake interaction) due to the close proximity of several vehicles.

The complexity of the aerodynamic field in the case of formation flight has been studied in the past, however, mostly limited to a two-ship situation. An interesting summary paper, which collects the work of the same authors in previous years, is the one by Patcher, D'Azzo, and Proud (1996). The focus of this work was in developing a control law capable of positioning the wingman aircraft such that maximum drag reduction would be achieved. Aerodynamic interference of the lead aircraft on the wingman was computed using an analogy with electric field produced by an electric current in electromagnetism. The lead aircraft was replaced by a horseshoe vortex model, and the wing aircraft was assumed to have an elliptical lift distribution. Wang and Mook (2003b) studied the same problem by using vortex lattice numerical schemes, also demonstrating definite advantages achievable by the wing's aircraft. Other interesting documentation relative to a two-aircraft formation can be found in Larson and Schkolnik (2004), Hanson *et al.* (2002), and Lavretsky (2002), which describe in some details methodologies and results of NASA's AFF (Autonomous Formation Flight) Program. The literature that deals with modeling of multiple vehicles in a formation, especially unmanned air vehicles, is much more limited, and, to the authors' knowledge; no examples exist of detailed analytical and/or

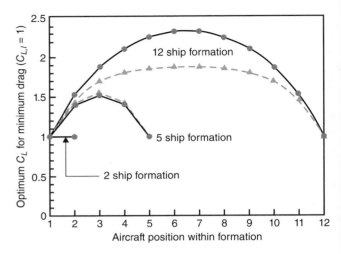

Figure 1. Spatial distribution of formations for maximum lift/drag ratio. (Reproduced with permission from Blake and Multhopp, 1998.)

experimental modeling in this case, except for some early work by Hummel (1996) and Blake and Multhopp (1998), which is summarized in Figure 1.

An interesting effort in developing an organized procedure for the computation of aerodynamic effects in a formation can be found in Giulietti *et al.* (2002). A procedure was designed to model the aerodynamic effects of the downwash and upwash in the formation limited to the case

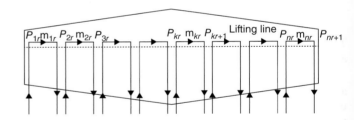

Figure 2. Vortex system segmentation of lifting line for a single lifting body.

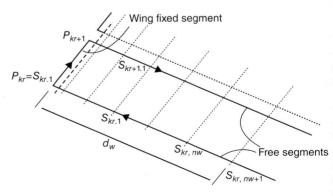

Figure 3. Wake segmentation for each vortex system.

of lifting body surfaces. Figures 2 and 3 show the method used for segmenting each lifting surface (modeled using the lifting line technique) and corresponding wake vortex (see Incompressible Flow over Finite Wings).

Aerodynamic force and moment coefficients can be directly introduced in the equations of motion without further modification. Separation between isolated flight and formation contributions can be useful, especially in control system design problems, in which the wake effects can be seen as disturbances to be rejected by the controller, thus leading to a generic expression shown in Equation 1.

$$\begin{pmatrix} C_F \\ C_M \end{pmatrix} = \begin{pmatrix} C_{Fi} \\ C_{Mi} \end{pmatrix} + \begin{pmatrix} C_{Ff} \\ C_{Mf} \end{pmatrix} \tag{1}$$

If we consider, for example, a standard rigid body six degrees-of-freedom nonlinear model in the aerodynamic reference frame, Equation 1 yields:

$$F_a = \begin{bmatrix} X_a \\ Y_a \\ Z_a \end{bmatrix} = \frac{1}{2}\rho V_a^2 S \begin{bmatrix} -C_D \\ C_Y \\ -C_L \end{bmatrix} \tag{2}$$

$$M_a = \begin{bmatrix} L_a \\ M_a \\ N_a \end{bmatrix} = \frac{1}{2}\rho V_a^2 S\bar{c} \begin{bmatrix} C_l \\ C_m \\ C_n \end{bmatrix} \tag{3}$$

The variation of the critical aerodynamic coefficients for this example is given in Table 1.

As an application example, consider a formation of three aircraft as shown in Figure 4. The relative nominal distances were 3 and 7.5 m along the x and y axes, respectively. Figure 5 shows the wake shape for a number of vortex segments equal to 20, and a number of wake segments equal to 10.

Table 1. Aerodynamic coefficients comparison.

Coefficients (body axis)	Single aircraft	Formation flight		
		Aircraft 1	Aircraft 2	Aircraft 3
Cx	0.025037	0.025422	0.026517	0.026517
Cy	0	0	−0.000082	−0.000082
Cz	−0.692947	−0.695880	−0.703952	−0.703952
Cl	0	0	0.000848	−0.000848
Cm	0.125725	0.125506	0.125106	0.125106
Cn	0	0	0.000124	−0.000124

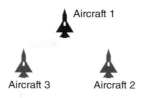

Aircraft 1

Aircraft 3   Aircraft 2

Parameter	Value
Wing area, S,m^2	0.6
Wing span, b,m	3
Root cord, c_r	0.2
Lift curve slope, $C_{L\alpha}$, 1 rad^{-1}	5.878
Gross mass, m, kg	10
Velocity, V_∞, m s^{-2}	19

Figure 4. Test data for aerodynamic forces and moments computation.

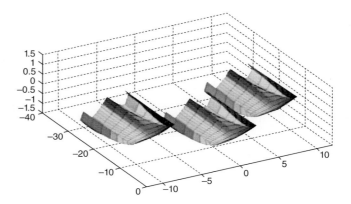

Figure 5. Wake shape graphical computation.

3   CONTROL STRATEGIES

The control problem for a formation of unmanned air vehicles differs not much from the single vehicle problem. The main differences lie in the formation control component, and in the communication capabilities of the formation itself. The formation control component is essentially an outer loop of sorts, which may be incorporated into each vehicle's guidance or navigation loops. Communication capabilities determine how much information can be exchanged by and/or provided to each vehicle, to carry out the formation control implementation.

One early example of bio-inspired modeling can be found in Giulietti, Pollini and Innocenti (2000). For long distance migration flight, birds of several species tend to move in a formation, flying close to each other maintaining a defined geometrical shape. There are two main reasons that birds fly in a formation; the first reason is related to aerodynamic considerations while the other is due to their "social behavior." In terms of aerodynamic effects, birds take advantage of the induced velocity produced by the wingtip vortex phenomenon: the inner wing of each bird in a V formation, for example, gains an increase in lift and then a reduction in induced drag from the upward rising side of the vortex left by the outer wing of the bird ahead. For this reason, it is important that each bird in the group keeps its position;

losing position by one or more birds means to lose a non-negligible portion of aerodynamic efficiency. On the other hand, birds like geese and swans form "family" groups: offsprings remain with parents one or two seasons after the birth and fly together with them during the migration. The elements in such formations know each other and try to remain together independently of aerodynamic advantages while other birds like storks live in nonfamiliar group, but fly together with others just in order to improve the efficiency of the migration (especially during food searching or mating season). To impose these behaviors to a formation of vehicles, the aircraft in a formation are not referring to each other anymore, but they are required to keep a specified distance from an imaginary point called formation geometry center (FGC). The FGC position depends on the relative distances between the aircraft in the formation itself.

$$\dot{V} = g\left(\frac{T-D}{W} - \sin\gamma\right)$$

$$\dot{\gamma} = \frac{g}{V}(n \cdot \cos\phi - \cos\gamma)$$

$$\dot{\chi} = \frac{g \cdot n \cdot \sin\phi}{V \cdot \cos\gamma}$$

$$\dot{x} = V\cos\gamma\cos\chi \tag{4}$$

$$\dot{y} = V\cos\gamma\sin\chi$$

$$\dot{z} = -V\sin\gamma$$

In the presence of disturbances, for instance, if one of the aircraft loses its position, the other senses the change, and departs momentarily from the prescribed trajectory, maneuvering altogether in order to reconstitute the formation geometry.

Once the geometry has been reached again, all aircraft continue to follow the prescribed trajectory. Equation 4 above describes the point mass model, in standard notation, for formation control.

For an N aircraft formation, the FGC dynamics are

$$P_{FGC} = \begin{pmatrix} \dot{x}_{FGC} \\ \dot{y}_{FGC} \\ \dot{z}_{FGC} \end{pmatrix} = \frac{1}{N}\sum_{i=1}^{N}\begin{pmatrix} V_i\cos\gamma_i\cos\chi_i \\ V_i\cos\gamma_i\sin\chi \\ -V_i\sin\gamma_i \end{pmatrix} \tag{5}$$

Integrating Equation 5, the relative position can be found in body-fixed axes and in the kinematic frame:

$$\begin{aligned} d_i^E &= P_{FGC} - P_i \\ d_i^k &= T_{kE}d_i^E \end{aligned} \tag{6}$$

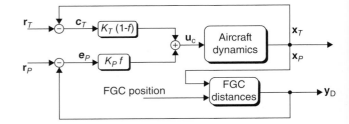

Figure 6. Block diagram for the formation controller.

In order to reproduce a natural behavior of migration birds, a formation controller (FC) was designed referring to a linearized set of equations about a selected flight condition and described by the following state–space model:

$$\dot{x} = \begin{pmatrix} \dot{x}_T \\ \dot{d} \end{pmatrix} = Ax + Bu_c \tag{7}$$

$$y = Cx, \quad u_c = \begin{bmatrix} u_T & u_P \end{bmatrix}^T$$

where x_T is the state vector and d is the vehicles' relative position vector with respect to the FGC. A combined controller can be found by using LQ–servo techniques, yielding the appropriate gain matrices, and a design blending parameter f shown in Equation 7, and in Figure 6.

$$u_c = (1-f)u_T + fu_P = (1-f)K_Te_T + fK_Pe_P \tag{8}$$

The results of a simulation representing a 30° heading change for each aircraft are shown in Figure 7. The plots are relative to several values of the blending parameter, from $f=1$ (position tracking priority) to $f=0$ (trajectory tracking priority).

A more comprehensive study of a formation control development can be found in Giulietti, Pollini and Innocenti (2000). The structure of the controller is depicted in Figure 8. The formation flight controller is implemented by a two-loop system.

The main task of the inner loop controller is to allow tracking of commanded velocity, altitude, and heading $T_C = (V_C, H_C, \psi_C)$ and it actually operates as a preset autopilot for the formation management. Linear quadratic regulation techniques were used for synthesis of the inner loop.

The choice of the methodology was dictated primarily by the relative speed with which suitable gains can be found, and the fact that it is of general use in this context. The model used for the inner loop is a standard derivation of linearization techniques, and separation between longitudinal- and lateral-directional modes. Equation 8 shows the equations

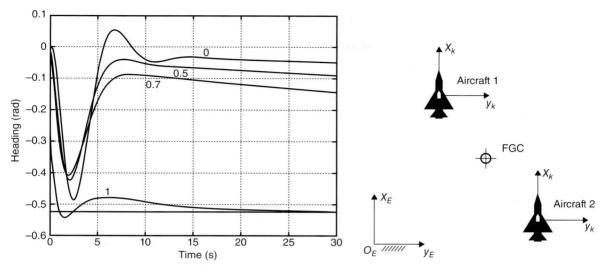

Figure 7. Computer simulation of heading response for a two-ship formation.

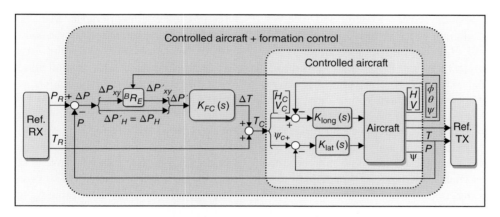

Figure 8. Functional block diagram of a formation controller.

of motion for the longitudinal mode, with the inclusion of a second-order model for the engine thrust. The inner loop gains can be found using a linear quadratic servo optimization given by the minimization of a performance index as in Equation 9.

A similar procedure is applied for the gain computation in the lateral-directional axis.

$$\dot{x}_{\mathrm{long}} = A_{\mathrm{long}}x_{\mathrm{long}} + B_{\mathrm{long}}u_{\mathrm{long}} = A_{\mathrm{long}} \begin{bmatrix} u \\ w \\ q \\ \theta \\ h \\ e_1 \\ e_2 \end{bmatrix} + B_{\mathrm{long}} \begin{bmatrix} \delta_e \\ \delta_{\mathrm{th}} \end{bmatrix} \tag{9}$$

$$\int_{0}^{\infty} \left(y^{\mathrm{T}}Qy + u_{\mathrm{long}}^{\mathrm{T}}Ru_{\mathrm{long}} \right) \mathrm{d}t$$

$$y = [I_5 0] \begin{bmatrix} x_{\mathrm{long}A} \\ x_{\mathrm{long}E} \end{bmatrix} \tag{10}$$

$$u_{\mathrm{long}} = K_{\mathrm{long}}x_{\mathrm{long}}$$

The outer loop formation controller generates a reference path command for the inner loop, in order to follow the desired formation trajectory, and to maintain the aircraft position inside the formation. The main objective of the formation controller, also based on LQR techniques, is to maintain the formation geometry. To compute the distance to

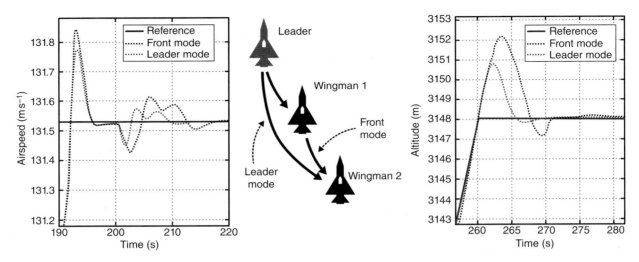

Figure 9. Wingman 2 simulation results (airspeed and altitude commands' responses).

its reference, each aircraft acquires its position $P = (X, Y, H)$ from a GPS-based position sensing system, and receives, through appropriate communication channels, other aircraft's positions $P_R = (X_R, Y_R, H_R)$. The formation controller is also responsible for having each aircraft follow a prescribed path. To validate the overall control system performance, several computer simulations were performed with a three-aircraft diagonal formation shown in Figure 9.

The relative nominal distances between each aircraft were set to be 15 and 10 m along the x and y-axis, respectively, while the altitude was the same for all. With respect to the sequential references chain, two types of simulations were performed with two different strategies.

Leader Mode: Both Wingman 1 and Wingman 2 take the trajectory references from the leader of the formation.
Front Mode: Each aircraft takes its reference from the preceding one. In this case, Wingman 1 is referred to leader and Wingman 2 is referred to Wingman 1. Figure 9 shows the responses of the Wingman 2 to airspeed and altitude commands in *Leader Mode* and *Front Mode*. In the *Front Mode* structure, Wingman 2 presents a poorer transient response, due to error propagation. A formation controller based on *Leader Mode* suggests better transient responses because of the absence of error propagation along the references chain. On the other hand, this type of formation structure may be more critical, because in this case Wingman 2, which is directly connected to the Leader, has no information about its distance to Wingman 1; therefore, it would not be capable of avoiding a collision with Wingman 1.

Inner loop and outer loop controllers are usually overseen by a manager and/or decision maker responsible for high-level functions within the formation. There are many structures possible for this purpose. Management of the aircraft formation can be centralized or decentralized. In the former case, the manager can be one of the aircraft or ground based. The centralized scheme has several disadvantages: if ground based, the amount of communications required between the formation and the manager may be undesirable for some applications; in addition, formation failures must be detected and recovered as rapidly as possible and ground-based management may introduce unacceptable delays. The major advantage of ground-based management is the capability to react to and reach decisions from a possibly higher level of "intelligence" than that achievable by onboard computers.

In a decentralized management scheme, each aircraft has a certain level of decision autonomy in order to achieve formation reconfiguration, navigation, and to meet the mission goals. One of the known issues with distributed management is that the decision-making algorithm must produce deterministic results on all the managed components, that is, on all the aircraft. Conflicting decisions must be avoided for formation safety. There are, however, several advantages, for instance, only interaircraft information is exchanged, except for possible mission updates that can be decided only at the ground control station; the same data channels used by formation-keeping control could be used to exchange management information; low power or alternatively nonradio-based communications, such as optical sensors, can be used because of the small distances among aircraft. This could be very important for military applications.

Information exchange between aircraft can be point to point or broadcast. In the former, not all the aircraft receive dynamic and management data from the others. In the latter, every aircraft receives and sends data to the others. In both the cases, graph theory can be used to describe interaircraft communication.

The aircraft can be thought of as the nodes of a graph. The physical communication channels create the arcs in the graph. These arcs are oriented because, in the most general case, channels are not bidirectional; this is not a limitation, because two nodes can be connected by two opposite direction arcs to model a bidirectional channel. The graph must also be connected because, if two subgraphs exist without any arc connecting them, the aircraft inside the two groups cannot behave like a formation resulting in fact as two separate formations. The communications graph should be redundant from the standpoint of the capability of propagating information. In the event of failures, this redundancy leaves room for reconfiguration. However, having the capability of using a channel does not mean that it must be used at all times.

Optimization of available channels under a cost function constraint can be achieved via graph programming techniques. In an optimization technique, each arc has a weight. The cost function will minimize the total cost of the information paths throughout the formation using the arcs' weights to evaluate the cost of a connection. Weights can be chosen depending on various criteria, for example, the capability of the formation control system to maintain constant interaircraft distances, or formation safety. In general, since the positioning error propagates and increases throughout the formation, the optimization algorithm should then consider optimizing the minimum error propagation path.

The optimization problem solved using graph search techniques requires some assumptions as described in Giulietti, Pollini, and Innocenti (2000), and the optimal communication transfer can be seen as the so-called shortest path problem. In this framework, it is possible to introduce the common concept of virtual leader, as a fundamental element of the graph connectivity tree provided there are no cycles in the graph. Suppose that at least one of the aircraft knows the mission reference trajectory. This reference trajectory can be seen as the effective formation leader, and can be represented in the graph by a node (virtual leader, VL), so that each aircraft that knows the mission trajectory has one communication channel with the VL. If a cycle exists, since the graph must be connected, it must contain all nodes. However, because at least one node has the VL as its only reference, and the VL has no incoming arcs by definition, the VL and all nodes using the mission trajectory as reference cannot be part of a cycle. Thus, the graph cannot have any cycle, and the VL will be the root of the solution tree. A common technique for solving shortest path problems is Dijkstra's algorithm (Hillier and Lieberman, 1993; Innocenti *et al.*, 1999). It has polynomial complexity, it guarantees optimality of the solution, and it is deterministic; that is, starting from the same conditions, all runs lead to the same solution. Each node i has a potential $d(i)$ and a preceding node $p(i)$, and represents a position in the formation, not necessarily an aircraft. The potential is a temporary value used by the algorithm and is initialized to $+\infty$, except for the VL that has zero potential. For all the nodes, the initial value for the preceding node is the VL. Dijkstra's algorithm can also be modified to obtain the optimal redundant channels among those outside the optimal arcs set. Suppose we need m total possible channels for each node, that is, $m-1$ redundant channels. The unmodified algorithm can still be used to find the optimal solution. At the end of the optimization procedure, the nodes' potentials are frozen. Then for all nodes i in the S set, select $m-1$ incoming arcs that have the minimum value of $d'(i) = d(j) + C_{ji}$, where j is a possible redundant preceding node, and order for increasing values of $d'(i)$. This modification computes suboptimal solutions. The arcs chosen with this technique can be considered suboptimal because they are the second-best choices: after removing the incoming arc of a node belonging to the optimal path, the new optimal incoming arc that a new run of the algorithm will produce is the first one of this redundant arcs list.

Figure 10 shows the simulation for the optimal communication path of a sample formation. On the left graph, while the VL is connected with all aircraft, not all possible interaircraft connections are present. On the right graph, the results of the modified Dijkstra's algorithm are given. The optimal path, solid line, and each node potential are shown, together with the redundant channel. When a node switches reference and uses a suboptimal path, its potential increases, and a reconfiguration procedure must be run on all the nodes of the subtree that originates from that node. Since the detected failure and the new node potential after reconfiguration must be propagated to the nodes that belong to the optimal node subtree path only, the same communication channels used for formation-keeping data can also be used to exchange potential updates. Furthermore, since after all nodes have completed the reconfiguration, the new graph is optimal; this procedure can be repeated in case of successive failures, without having to reconsider optimization of the whole graph. This means that any number or combination of successive failures can be managed with this subtree-based technique without compromising whole graph optimality.

A different approach to the decision-making process and applicable to formation flight can be found in Knoll and Beck (2006). The procedure is very different from the one outlined above, in that it concentrates on a somewhat rigid process-based structure, rather than a dynamic one. The aircraft flying in the most forward position is the lead aircraft, as it has no other aircraft interfering with its forward-looking sensors (weather radar, for instance), and can maneuver with comparatively few restrictions. The leader determines the formation to

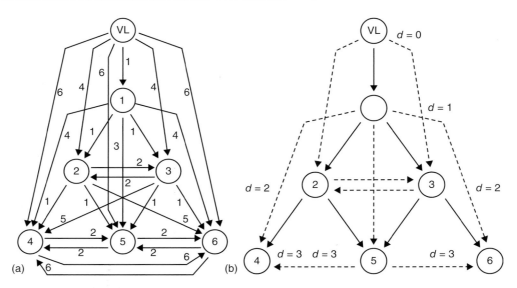

Figure 10. Application of Dijkstra's algorithm to a formation of 6 aircraft: (a) optimal connection; and (b) redundant connection.

be flown. The decision-making process is shown in Figure 11, as specialized to a formation flight problem.

Strategic decisions are made at Level 1 with the following priorities: first of all the aircraft must be flown safely and economically to the destination airport. Be aware of limitations, such as sensitive cargo that require reduced maximum and minimum load factors and airspeed during heavy turbulence. Follow the route, altitude, and airspeed specified by air

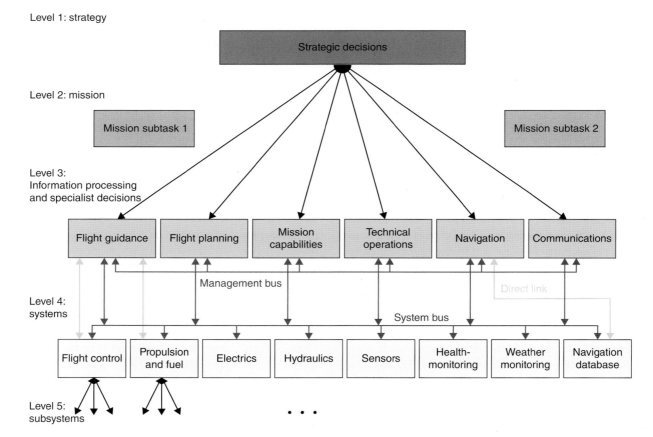

Figure 11. Typical decision-making structure with fixed hierarchical approach.

traffic control and fly over the waypoints for the rendezvous with other aircraft. If possible, use the formation with other aircraft in order to reduce fuel consumption. Whenever Level 1 of the decision structure makes the strategic decision to continue in a formation, Level 2 module performs all decision-making regarding the formation flight itself.

4 RECONFIGURATION

The problem of reconfiguration for a formation of UAVs is essentially twofold. The first aspect deals with changes in formation's shape due to mission requirements, performance issues, obstacle avoidance, and similar aspects. The second aspect refers to changes due to failures of various types occurring to a single vehicle, and/or the formation itself (Innocenti and Giulietti, 2004). This section describes a subset of the latter aspect, while detailed research on the former can be found in the literature relating to path planning and motion control of unmanned vehicles. The main reference is the block diagram in Figure 12. The main issues involved in control and management of an aircraft formation are trajectory tracking and interaircraft relative distance regulation. In the case of decentralized management, each vehicle in the formation needs to exchange data, such as position and trajectory information. Many different configurations could be found for the communication flow within a formation, and not all of these configurations may provide good performance. Thus, the first step is to find an "optimal" one, among different configurations for information exchange. Optimization of available communication channels based on a cost function figure of merit can be achieved using a variety of methods; here, the problem is addressed using graph-programming techniques, as described in Section 3.

In general, since the position error propagates and increases throughout the formation, the optimization algorithm should include the minimization of the minimum error propagation path. Once an optimal solution for the communication flow is found, the next step in the formation management design consists in giving the structure adequate robustness to communication failures. A failure in the communication occurs when one or more aircraft of the formation loses the information exchange capability (send or receive data). After a failure, there may be one or more connections lost and then a new channel configuration must be found. The resulting configuration will not be optimal, because of the loss of one or more channels, but it will be the optimal solution according to the new set of nodes and arcs. The algorithm for the reoptimization of the interaircraft connection is triggered by the fault detection, and it is decentralized for faster reconfiguration time, and information other than that strictly needed for the formation-keeping control system must be exchanged on the data channels. The reconfiguration process must be the same for all aircraft; that is, the local copy of the graph describing the formation communications must be identical in all aircraft at all times. The presence of an outgoing arc in a node in the graph implies the capability of transmitting information, while an incoming arc is related to the capability of receiving information. Two virtual devices model such capabilities: a transmitter (TX) and a receiver (RX). These devices are "virtual" in the sense that failure implies the loss of the device capability, irrespective of what subcomponent has actually failed (antennas, CPU, transmission bus, etc.).

After a generic fault, a fast reconfiguration procedure is run to restore formation-keeping as quickly as possible. When the formation communications are in a safe configuration again, it may be necessary to switch aircraft positions or move an aircraft to an empty node to maximize the formation keeping and safety of all aircraft inside the formation. Since the node-changing decision is decentralized, the algorithm that makes the decision must be deterministic to avoid simultaneous conflicting decisions by more than one aircraft in response to the same post-fault reconfiguration requirements. Even if the available communication channels, those used to exchange flight data for formation control, are sufficient for optimal channel reconfiguration, it is possible that the best aircraft candidate for making the node change decision may not be informed of the fault event. To keep all aircraft informed of active nodes, that is, nodes occupied by an aircraft, an alternative broadcast communication channel is introduced that transmits data at low frequency. Communications on the broadcast channel (BC) are asynchronous, because the formation must react to failure events with the shortest delay possible. After the formation communications have been reoptimized, it may be necessary to move the aircraft inside the formation to fill holes left by a missing aircraft or to exchange two or more aircraft positions to reach desired formation geometry, and to maximize the formation-keeping capability and safety of all aircraft inside the formation. Since formation safety and its precise control capability are measured by the total cost of the communication tree after a generic failure, the new communication cost is greater than

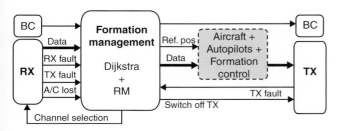

Figure 12. Reference management and formation controller

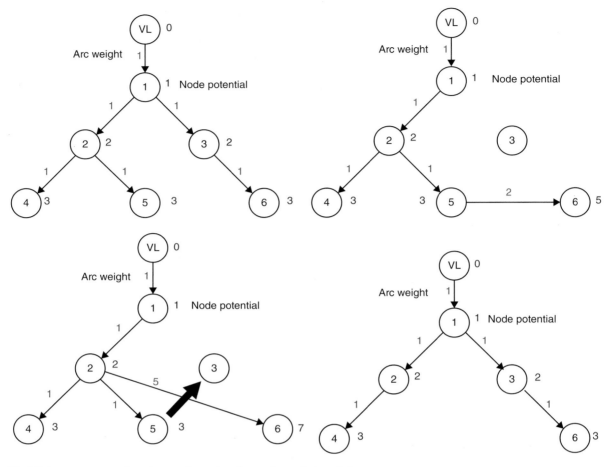

Figure 13. Minimum cost graph and reconfiguration due to loss of aircraft 3.

or equal to the preceding one. By moving and exchanging two or more aircraft inside the formation, some arcs that were assigned a weight of infinity, as unusable arcs because of broken TXs or RXs, could regain their original weight or have assigned a finite weight; thus, it is possible to decrease the total communication tree cost. The introduction of heuristic rules embedded into the reconfiguration process in terms of reconfiguration maps accommodates this situation.

An example of application of the procedure outlined above is shown next. In this example, the execution of the reconfiguration procedure is complete, and it involves Dijkstra's algorithm, broadcast channel information exchange, and the use of reconfiguration maps. The formation has 6 aircraft, with aircraft number 3 experiencing complete TX–RX loss. Figure 13 shows the sequence of events leading first to reestablishing total communication links, and second changing the formation to achieve minimum cost. The software overseeing the entire procedure uses a finite state machine approach, and resides on each vehicle's formation controller.

5   SENSORS

Formation flight, especially when tight or rigid formations are required, relies greatly on the flight controller. However, reliable and additional sensors may be required to achieve controlled performance. Current research directions in the unmanned vehicles' field are devoted to synthetic vision that should empower autonomous vehicles' capabilities of autolocalization, and autonomous operation such as formation flight and aerial refueling and solve the typical problems associated with standard inertial-aided GPS navigation: loss of satellite signal, fast estimation drift in low cost inertial measurement units. One of possible choices in aiding GPS/INS measurements is the use of vision-based sensors, to be used when the aircraft are in close proximity of each other (Pollini, Innocenti, and Mati, 2005). A set of active or passive markers, that is light emitting or reflecting/diffusing light, are attached to the object to be recognized; different choices will be necessary for formation flight or refueling for instance. The structure

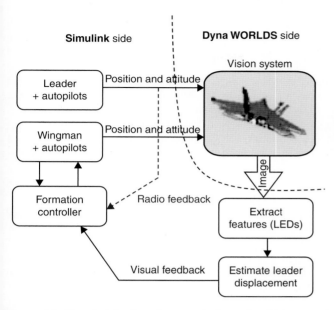

Figure 14. Use of vision-based sensors in formation flight control.

for introducing vision sensing information in the control loop is highlighted in Figure 14.

The application of vision algorithms requires the solution of the so-called pose estimation problem, that is, recovering the relative displacement and attitude between the camera, whose position is known, and a body that as a set of markers in exactly known 3D locations are attached. For applications of vision-based navigation to formation flight, very high accuracy and reliability are required. In Pollini, Innocenti and Mati (2005), a modification of the widely used LHM algorithm was presented in order to take into account situations that a standard LHM cannot resolve, such as the markers 2D acquisition form a nonconvex configuration, and some of the markers are missing.

6 EXAMPLES

Although research work in formation flight has been extensive in the past 10–15 years, actual in-flight tests have been rare. The first example is the Autonomous Formation Flight project run by NASA DFRC, the Boeing Co., and UCLA in the early 2000s. Larson and Schkolnik (2004), Hanson *et al.* (2002), and Lavretsky (2002) present a detailed description of the study and flight test. The AFF project had three planned phases; beginning with autonomous station keeping without drag reduction, then proceeding to the piloted mapping of wingtip vortex locations and effects, and finally culminating with the testing of an autonomous drag reduction

Figure 15. WVU UAV aircraft fleet.

system. The project, however, was cancelled shortly after completion of the second phase.

A second example, achieved with more limited resources but with very interesting results, is the research carried out at West Virginia University relating to the design, manufacturing, and test flight of three unmanned YF-22 models, flying in a formation (Napolitano, 2005; Gu *et al.*, 2007). The ambitious research project started with the design and manufacturing of three replicas of the YF-22 jet fighter aircraft shown in Figure 15. Each model is 3 m long, with a wingspan of about 2 m, a weight of 23 kg, and an electronic payload of about 5 kg. The propulsion system consists of a jet engine manufactured using a single stage centrifugal compressor driven by a single axial flow turbine wheel.

The main objective of this project was to provide a flight demonstration of formation control using UAV research aircraft models. In the selected formation configuration, a radio control (R/C) pilot maintains ground control of the "leader" aircraft, while two autonomous "follower" aircraft are required to maintain a predefined position and orientation with respect to the "leader" aircraft. Each of the "follower" aircraft was essentially an autonomous vehicle once engaged into the formation flight mode. Individual pilots controlled each aircraft from a ground transmitter during takeoff and up to formation engagement, as well as after formation disengagement and landing procedures. Autonomous formation was engaged after all aircraft reached a predetermined altitude at a preselected rendezvous point. In June 2006, a successful test flight was carried out.

7 SUMMARY

This chapter presented some of the issues concerning autonomous formation flight. Although not near complete, typical problems in modeling, control design, integrated sensing, and reconfiguration in the presence of selected failures were described. Two examples of actual flight experiments carried out in the past 3–4 years were also highlighted, underlying the success both at the industrial as well as the academic level.

REFERENCES

Blake, W. and Multhopp, D. (1998) Design, performance and modeling considerations for close formation flight. *Proceedings of the AIAA Atmospheric Flight Mechanics Conference*, August 1998.

Bloy, A.W. and Jouma'a, M. (1995) Lateral and directional stability and control in air-to-air refueling. *ImechE, Part G, J. Aerosp. Eng.*, **209**, 299–305.

Bloy, A.W., West, M.G., Lea, K.A., and Jouma'a, M. (1993) Lateral aerodynamic interference between tanker and receiver in air-to-air refueling. *J. Aircr.*, **30**, 705–710.

Brendley, K. and Steeb, R. (1993) Military applications of micro-electromechanical systems. *RAND Study MR-175-OSD/AF/A*.

Dargan, J.L., Patcher, M., and D'Azzo, J.J. (1992) Automatic formation flight control. *Proceedings of the AIAA Guidance, Navigation and Control Conference*, Hilton Head, SC, August 1992.

Giulietti, F., Pollini, L., and Innocenti, M. (2000) Autonomous formation flight. *IEEE Control Syst. Mag.*, **20**(6).

Giulietti, F., Pollini, L., and Innocenti, M. (2001) Formation flight control: a behavioral approach. *Proceedings of the AIAA Guidance, Navigation, and Control Conference GNC01*, Montreal, Canada, August 2001.

Giulietti, F., Napolitano, M.R., Capetta, A., and Innocenti, M. (2002) Detailed modeling of multiple aircraft within a close formation flight. *Proceedings of the AIAA Atmospheric Flight Mechanics Conference and Exhibit*, Monterrey, CA, August 2002.

Gu, Y., Seanor, B., Campa, G., Napolitano, M.R., Pollini, L., and Fravolini, M.L. (2007) Design and flight-testing of nonlinear formation control laws. *Control Eng. Practice*, **15**, 1077–1092.

Hanson, C.E., Ryan, J., Allen, M.J., and Jacobson, S.R. (2002) An overview of flight test results for a formation flight autopilot. *Proceedings of the AIAA Guidance, Navigation, and Control Conference*, Monterrey, CA, August 2002.

Hillier, F.S. and Lieberman, G.J. (1993) *Introduction to Operation Research*, McGraw Hill, New York.

Hummel, D. (1996) The use of aircraft wakes to achieve power reductions in formation flight. *Proceedings of the AGARD FDP Symposium on the Characterization and Modification of Wakes from Lifting Vehicles in Fluid*, November 1996, pp. 36–1–13.

Innocenti, M., Mancino, G., Garofoli, M., and Napolitano, M.R. (1999) Preliminary analysis of formation flight management. *Proceedings of the AIAA, Guidance, Navigation and Control Conference*, Portland, Oregon, August 1999.

Innocenti, M., Pollini, L., and Giulietti, F. (2004) Management of communication failures in formation flight. *AIAA J. Aerosp. Comput. Inform. Commun.*, **1**, 19–33.

Knoll, A. and Beck, J. (2006) Autonomous decision-making applied onto UAV formation flight. *Proceedings of the AIAA Guidance, Navigation, and Control Conference*, GNC06, Keystone, Co., August 2006.

Larson, G. and Schkolnik, G. (2004) *Autonomous Formation Flight, MIT Course 16.886*, Spring, Boston, MA.

Lavretsky, E. (2002) F/A-18 autonomous formation flight control system design. *Proceedings of the AIAA Guidance, Navigation, and Control Conference*, Monterrey, CA, August 2002.

Napolitano, M.R. Development of Formation Flight Control Algorithms using 3 YF-22 Flying Models, Final Report GRANT AFOSR F49620-01-1-0373, April 2005.

North Atlantic Treaty Organization (1997) Aerospace 2020, *AGARD AR-360*, vols. I and II.

Pachter, M., D'Azzo, J.J., and Proud, A.W. (2001) Tight Formation Flight Control. *AIAA J. Guid., Control Dyn.*, **24**(2), 246–254.

Patcher, M., D'Azzo, J.J., and Veth, M. (1996) Proportional and integral control of nonlinear systems. *Int. J. Control*, **64**(4), 679–692.

Pollini, L., Innocenti, M., and Mati, R. (2005) Vision algorithms for formation flight and aerial refueling with optimal marker labeling. *Proceedings of the AIAA Guidance, Navigation, and Control Conference*, GNC05, August 2005.

NATO Advisory Group for Aerospace Research and Development (1994) Future Use of Unmanned Air Vehicle Systems in the Maritime Environment. *AGARD-AR-307*, vols. 1 and 2.

Wang, Z. and Mook, D.T. (2003a) Numerical aerodynamic analysis of formation flight. *Proceedings of the 41st AIAA Aerospace Sciences Meeting*, Reno Nevada, January 2003.

Wang, Z. and Mook, D.T. (2003b) Numerical aerodynamic analysis of formation flight. *Proceedings of the 41st AIAA Aerospace Sciences Meeting and Exhibit*, Reno, NV, January 2003.

Wolfe, J.D., Chichka, D.F., and Speyer, J.L. (1996) Decentralized controllers for unmanned aerial vehicle formation flight. *Proceedings of the AIAA Guidance, Navigation and Control Conference*, San Diego, CA, August 1996.

PART 9
Airspace Access

Chapter 34

Operational Profiles of Unmanned Aircraft Systems in the Context of the US Regulatory Regime

Eric Mueller

NASA, Moffett Field, CA, USA

ACRONYMS

AC	advisory circular
ADS-B	automatic dependent surveillance-broadcast
AFRL	Air Force Research Laboratory
AGL	above ground level
ALPA	Airline Pilots' Association
ATC	air traffic control
C2	command and control
CBP	Customs and Border Protection
CFR	Code of Federal Regulations
COA	Certificate of Authorization or Waiver
DAA	Detect and Avoid
DARPA	Defense Advanced Research Projects Agency
DoD	Department of Defense
DoE	Department of Energy
FAA	Federal Aviation Administration
FAR	Federal Aviation Regulations
FOIA	Freedom of Information Act
IFR	instrument flight rules
MOA	memorandum of agreement
NAS	National Airspace System (of the United States)
NASA	National Aeronautics and Space Administration
NATCA	National Air Traffic Controllers' Association
NOAA	National Oceanic and Atmospheric Administration
RTCA	Radio Telecommunications Corporation of America
SAC-EC	Special Airworthiness Certificate—Experimental Category
UAS	unmanned aircraft system
US	The United States
USDA	United States Department of Agriculture
UTM	UAS Traffic Management
VFR	visual flight rules

1 INTRODUCTION

Unmanned aircraft systems (UASs) have been used by militaries and hobbyists for decades, but the advent of low-cost and powerful computers, communications technologies and other aerospace systems have supported a recent surge in the number of these types of aircraft that are available to a multitude of users. This increased availability along with their improved reliability and usability has created significant demand for integrating UAS with airspace and air traffic control systems in the United States and internationally. However, a significant set of technical challenges must be resolved before large numbers of these vehicles can be safely operated alongside legacy users of the airspace. The current operational approval process for UAS imposes limitations on the operations they may conduct and limitations that are designed to ensure airspace safety while those technical challenges are addressed. In the meantime, the

Unmanned Aircraft Systems. Edited by Ella Atkins, Aníbal Ollero, Antonios Tsourdos, Richard Blockley and Wei Shyy.
© 2016 John Wiley & Sons, Ltd. ISBN: 978-1-118-86645-0.

process is severely curtailing the number and types of operations being carried out.

This chapter reviews the types of operational profiles that UAS undertake in the current regulatory environment (see Regulatory Policy and Processes: A Moving Landscape) and the profiles they are expected to use when regulations allow routine access to the airspace. In this chapter, an operational profile is characterized by the type of aircraft, the purpose of the operation, and the environment in which the aircraft flies, including its cruise altitude, mission duration, and flight plan. This chapter starts by reviewing the current regulatory framework used in the United States to approve specific UAS operations. Next, descriptions of the operations that have been carried out under the current framework are presented. The path to approve routine operations of UAS that largely comply with existing aviation regulations is described, and examples of the operations that a diversity of public and private interests are likely to undertake when UAS become certified are presented. Finally, descriptions of alternate regulatory frameworks are given and the types of operations they could entail are outlined. The scope of this chapter is confined to public and civil applications of UAS, so other than a brief description of the US military's use of UAS in domestic airspace, the operational profiles of defense-related applications are not discussed.

2 OPERATIONS UNDER EXISTING REGULATORY FRAMEWORK

A detailed description of how UAS can obtain approval for operations in the US National Airspace System (NAS) is available through FAA Notice JO 7210.891 (FAA, 2015). The next section summarizes this process, and the following section explains how the public, civil, and hobbyist communities have operated UAS under these rules.

2.1 Regulatory approval framework

The regulatory approval process that UAS must follow in order to receive permission to fly in the NAS greatly affects the types of operations that are approved (see Regulatory Policy and Processes: A Moving Landscape). Certain operations, for example those at low altitude over populated areas, are generally prohibited and so few examples of them exist today. However, changes in technology, policy, or procedures could enable those in the future, so it is important to distinguish between operations not conducted today because there is no scientific, economic, or other benefit to them versus those that do not fall within the current regulatory

structure. Without an understanding of the process, it will be difficult to separate these two cases.

Generically, a UAS must receive both airworthiness approval and operational approval to fly in the NAS. The former requirement pertains to the integrity and safety of all components of the UAS, but it does not grant approval to actually operate the aircraft in any manner. The latter approval is necessary for each type of operation intended for the UAS. In other words, the airworthiness process ensures the vehicle itself is safe, and the operational approval process ensures that the things to be done with that aircraft are safe in the context of the airspace and for people and property on the ground. Two sets of approval processes are currently in place for UAS depending on whether they are being operated by public agencies (federal, state, or local government) for noncommercial purposes or private ones. The distinction between the two types of operations is not always clear, so the FAA has issued guidance to assist in making the differentiation (Advisory Circular (AC) 00–1.1A, 2014).

The process of gaining regulatory approval for public entities is simpler than for civil ones because the FAA does not certify the airworthiness[1] of public aircraft. The public entity must only obtain operational approval from the FAA to fly UAS in the NAS, for which they require a public aircraft certificate of authorization or waiver ("public Certificate of Authorization or Waiver (COA)[2]") or a memorandum of agreement (MOA). This public COA applies to aircraft used only for the US government or owned by the government and operated for crew training, equipment development, or demonstration (AC 00–1.1A, 2014). If the government intends to operate a civil aircraft, operates a public aircraft for commercial purposes, or has no government employees in the crew, then it must follow the process for a civil aircraft COA. Although no airworthiness approval is required from the FAA, the government entity conducting the public aircraft operation must comply with all regulations applicable to UAS operating in the NAS and is responsible for ensuring the aircraft is airworthy. Large organizations such as the US Department of Defense (DoD) and NASA have thorough procedures in place to evaluate and ensure airworthiness, while smaller organizations such as a local police department can ensure airworthiness by purchasing an off-the-shelf

---

[1] "Airworthiness" is the measure of an aircraft's suitability for safe flight (http://en.wikipedia.org/wiki/Airworthiness).
[2] The FAA defines a COA as "an authorization issued by the Air Traffic Organization to a public operator for a specific (unmanned aircraft) activity. After a complete application is submitted, FAA conducts a comprehensive operational and technical review. If necessary, provisions or limitations may be imposed as part of the approval to ensure the UA can operate safely with other airspace users."

system for which the appropriate process has been followed and documented.

The FAA-issued public COA is "an authorization issued by the Air Traffic Organization to a public operator for a specific unmanned aircraft activity."[3] The COA application describes aspects of the UAS and its intended operation relevant to determining whether other users of the airspace or the public will face an unreasonable threat to their safety: the control and communications link, lost-link procedures, operating maps, launch and recovery procedures, and so on. The approval then specifies the operations that may be carried out, for example below a particular altitude, in the vicinity of a particular airport, with a specified aircraft, for a given purpose, and for no more than 2 years. It may also require additional safety restrictions be placed on the operations to ensure the UAS meets requirements that apply to all aircraft operating in the NAS: use of visual observers to ensure separation from other airspace users, minimum pilot and observer qualifications, allowable weather conditions, etc. This process is designed to accommodate aircraft that do not meet the usual requirements for flying in the NAS by specifying alternative mitigations.

All UAS operations that do not fall under the requirements of a public aircraft COA must obtain both a civil aircraft COA and comply with the FAA's airworthiness requirements. Until late 2014, there were only two ways to receive airworthiness approval: through the issuance of a special airworthiness certificate in the experimental category[4] (SAC-EC) or by obtaining a UAS type and airworthiness certificate in the restricted category. The former method is frequently used to obtain approval for amateur-built aircraft or kit aircraft for which no type certificate is available. To receive SAC-EC, the applicant must describe the design and construction of the vehicle, software development processes, quality assurance procedures, and how and where they intend to fly.[5] If the FAA inspector is satisfied that the system can be safely operated, then that particular aircraft is approved; however, even identically constructed aircraft will require separate approval. Aircraft that receive an experimental airworthiness certificate may be used for research and development, crew training, and market surveys only, they may not be used to carry passengers for hire or put to other commercial purposes. The restricted category approval process is

designed to allow civil- or military-derived aircraft to be operated in ways for which they were not originally certified and to permit exceptions to airworthiness requirements the FAA finds inappropriate for the special purpose of the aircraft (FAA, 2008). This approval method has been used rarely for UAS, because, for the most part, UAS have not been converted from previously certified aircraft.

In September 2014, the FAA began to exercise its authority under Section 333 of the FAA Modernization and Reform Act of 2012 to approve the airworthiness of UAS on a case-by-case basis (the approval is called a "Section 333 exemption"). These approvals do not carry the limitations on commercial operations that the experimental approval does, thus have allowed a narrow range of applications to be approved. UAS with airworthiness approvals may either pursue an individual COA, largely along the lines described in the previous paragraph, or operate under a "blanket COA" issued by the FAA in March 2015.[7] The operational limitations of this latter approach are summarized in Table 1. Although most UAS operations are expected to occur well away from airports, on airport operations are permitted if a letter of agreement is signed between the UAS operator and the appropriate airport authority (FAA, 2015). The blanket COA, along with summary approval of Section 333 exemption applications for operations substantially similar to previous applications, has enabled a rapid rise in the overall number of approved civil UAS operations to 2451 as of November 25, 2015.[6] This figure is the total since the FAA approved the first exemption on September 25, 2014.

Table 1. Blanket COA operational limitations.[a]

Parameter	Limits
Altitude	200 ft AGL
Distance from observer	Visual (unaided) line of sight (usually <0.5 nmi)
Maximum velocity	100 mph, 87 kts
Maximum takeoff weight	55 lb
Weather conditions	Visual meteorological conditions
Hours of operation	Daylight
Distance from airports	2–5 nmi from airports, depending on presence of an operational tower or published instrument flight procedures
Other limitations	Outside restricted areas, away from densely populated areas (e.g. cities), outside national parks

[a]*Source:* www.faa.gov/news/updates/?newsId=82245

---

[3] http://www.faa.gov/about/office_org/headquarters_offices/ato/service_units/systemops/aaim/organizations/uas/coa/
[4] From faa.gov: "A special airworthiness certificate in the experimental category is issued to operate an aircraft that does not have a type certificate or does not conform to its type certificate and is in a condition for safe operation."
[5] FAA Order 8130.34C.

[6] https://www.faa.gov/uas/legislative_programs/section_333/

Table 2. Categories of public entity applicants for COAs.

Applicant category	No. of Applications	% Applications	Example applicants
Public Universities	32	43.2	University of Colorado
Local Government – Law Enforcement	17	23.0	Miami-Dade Police Department
Local Government – Operations	3	4.1	Hays County, Kansas Emergency Services Office
State Government – Operations	3	4.1	California Department of Forestry and Fire Protection
Federal Government – Law Enforcement	3	4.1	Customs and Border Patrol, FBI
Federal Gov. – Research	6	8.1	NASA, NOAA
Federal Government – Operations	4	5.4	US Department of Agriculture
DoD – Research	2	2.7	Air Force Research Laboratory
DoD – Operations	4	5.4	US Air Force

2.2   Current UAS operations

No comprehensive source of information about current operations of UAS is available; however, the publicly released applications for COAs and Section 333 exemptions do indicate the types of operations that are intended for these new aircraft. It should be noted that simply because an application was submitted does not mean that it was approved or that the intended operation is taking place, but it does mean that the operation was of sufficient interest for the applicant to spend a significant amount of time and money seeking its approval.

2.2.1   Public UAS operations under a public aircraft COA

The largest and most recent source of information on public entity COA applications since 2006, when UAS COA were first made available, comes from a 2012 Freedom of Information Act (FOIA) request.[7] A summary of the categories of the applicants, the number of applications in each category, and example applicants reported in the FAA's response is shown in Table 2.

The largest number of applications was submitted by public universities. It is noteworthy that "public entities," which most people associate with government organizations such as the DoD and NASA, do include public, but not private, universities. The FAA released a clarification on May 4, 2016 that they consider UAS operations for educational purposes to be equivalent to recreational uses. An endnote with a reference to this update would be worthwhile. The public university applications were usually for operations to study technological or procedural requirements for the integration of UAS with the airspace system, but also included research on UAS technologies themselves (flight

control system, command and control link, etc.) or research about how to use UAS for other applications (e.g., aerial surveying). The limitations of the operations the FAA would approve under the current COA regime meant that the approved operations in this category were substantially similar: altitude limits of no more than 1000 ft above ground level (AGL) and frequently limited to 400 ft, at least 5 nmi from airports, in visual meteorological conditions during the day and within the pilot's unaided visual line of sight. With the exception of the maximum operating altitude, the proposed operations closely conform to the operational requirements under the blanket COA. The UAS operations proposed by public universities are a good example of the regulatory and safety processes taking precedence in determining the operational profile over the requirements of the mission itself.

The second largest number of applications was in the category of local law enforcement, usually either a city police department or a county sheriff's office, though in one case also including a university public safety office. A wide variety of applications were proposed in this category: developing or verifying operational procedures, determining training plans and requirements, support of "tactical" situations including hazardous material spills on highways or railroads, research on the law enforcement applications of UAS in collaboration with local universities, video relay for search and rescue, and as a demonstration in preparation for larger investments in UAS. At least several applications specifically indicated UAS would not be used for routine patrols. The operational parameters of most of these missions are very similar to those of the university applications and in line with the blanket COA, but they are likely to take place in different geographic areas. While the university applications and the law enforcement research or training applications usually take place over rural, sparsely populated locations, the tactical response and crash site investigations are likely to occur over urban areas and transportation networks. Applicants will have a harder time ensuring the safety of these

---

[7] https://www.faa.gov/uas/public_operations/foia_responses/

operations because of obstacles to navigation and the higher likelihood that a mishap could cause damage or injury to property and people on the ground.

Six local and state government entities have applied for COAs, most of which are for applications related to public safety and disaster response. All of the applications indicated operations would be within line of sight and remain in class G airspace below a maximum of 1000 ft, though in most cases an even lower ceiling of 400 ft was specified. Typical missions included fire support, disaster mitigation, and search and rescue. The interesting exceptions to these missions were from the Ohio and Washington State Departments of Transportation: they were seeking to use UAS to conduct aerial photography in support of construction project lifecycle management, including planning, design and quality control, and to evaluate the cost effectiveness of UAS for avalanche control, respectively. Overall, the lack of resources available to local and state governments to support the lengthy COA application process means that the missions they have proposed are very similar to those enabled by the blanket COA.

The five remaining categories of applicants, all representing branches of the US Federal Government, comprise only about a quarter of the public entities, but they represent a much wider range of operational profiles. The Federal Government is able to conduct these additional operations for two reasons: Federal agencies have the resources to go through the FAA's lengthy COA approval process; and agencies like NASA and the DoD are able to self-certify their own aircraft as airworthy and qualify their own pilots rather than ask the FAA to do so. Unfortunately, all COAs related to UAS operations of Federal law enforcement entities, including the Department of Homeland Security, Federal Bureau of Investigation, and Customs and Border Protection (CBP), are nearly entirely redacted so it is difficult to confirm what aircraft they are using or which missions they carry out. These three entities appear to rely on commercially available UAS rather than developing their own vehicles. It has been widely reported in the media that CBP conducts regular Predator-B flights over the United States–Mexico border, but little public information about those flights is available. The US Department of State and Department of Agriculture (USDA) have applied for COAs for low-altitude, line-of-sight operations. The former agency planned to train employees in the use of UAS for international convoy protection and site surveillance, though most of the COA application is redacted. The USDA Forest Service planned to use a small UAS, an RQ-11 Raven, equipped with a thermal camera to gather intelligence on wildfires threatening populated areas. These operations largely fall in line with the blanket COA restrictions with the exception of the maximum altitude, but all would remain within Class G airspace.

Several Federal agencies applied for COAs to conduct research on UAS or their potential applications: NASA, the National Oceanic and Atmospheric Administration (NOAA), the National Institute of Standards and Technology (NIST), and three Department of Energy (DOE) National Laboratories. NASA has the largest variety of UAS operations among these agencies, including many that operate in controlled airspaces (Classes D, E, and A): a series of very high-altitude flights with a pair of Global Hawk aircraft for severe storm monitoring (Braun *et al.*, 2013); operation of a medium-sized Sierra UAS to investigate the marginal ice zone in Northern Alaska (Bradley *et al.*, 2015); and high-altitude real-time wildfire imaging in connection with fire-fighting agencies (Ambrosia *et al.*, 2011). Good overviews of NASA's use of UAS for remote sensing and earth science applications may be found in Watts, Ambrosia, and Hinkley (2012) and Albertson *et al.* (2015), respectively. NOAA planned to use small UAS to survey pack ice in the Bering Sea, and to locate derelict fishing gear at sea and assist ships in removing the dangerous materials. The three DOE laboratories indicated all operations would be with small UAS (<55 lbs) and occur only over federally owned facility lands up to an altitude of 1200 ft. Although many of the mission descriptions were redacted, particularly for the Idaho National Lab, several applications are described: detecting fugitive methane emissions, atmospheric sampling, sensor research and development, and support of security programs.

The US DoD and associated research institutions, the Defense Advanced Research Projects Agency (DARPA), and Air Force Research Laboratory (AFRL) possess by far the largest fleet of UAS among the COA applicants and conduct the greatest number of operations. However, the types of operations they conduct outside their designated test ranges, and for which public COA information is available, fall into only a few categories. Most large UAS operations in domestic US airspace are transits between air bases and dedicated DoD areas for operational applications (homeland defense, homeland security, and defense support of civil authorities), training missions (pilot or unit readiness), and support missions (UAS development and testing, acceptance testing, and postmaintenance check flights) (DoD, 2011). These flights may be in airspace Classes G, E, and A. For example, AFRL conducts direct transit between Grey Butte Field, CA, a site for development and testing of military UAS, and the airspace surrounding Edwards Air Force Base. They have requirements not to climb above 2500 ft or leave a 2.5 nmi radius around the airfield without two-way communications with air traffic control and must employ either airborne "chase" observers or ground observers for safe separation and collision avoidance. The US Navy has applied for a COA to ferry Global Hawk aircraft to warning areas in oceanic airspace, all at altitudes above 50 000 ft. Finally, a number of small UAS applications, such as sensor development, autonomous controls (Berrios *et al.*, 2014), and obstacle field navigation (Hubbard *et al.*, 2007), have been approved for low-altitude and line-of-sight operations. These limited examples of DoD operations are largely constrained by the current operational approval requirements and are not representative of the broad spectrum of missions that they would prefer to carry out (DoD, 2011).

2.2.2   Civil UAS operations under Section 333 exemptions

Most civil UAS operations, along with some public operations, are conducted under a Section 333 airworthiness exemption approval and the blanket COA. While a comparatively small number of UAS have received special airworthiness certificates in the experimental category, most publicly available information is related to the 333 airworthiness exemptions. The operations proposed by civil operators that are discussed in this section were drawn from the FAA's list of 2451 exemptions granted through November 25, 2015.[8] The operations are usually required to conform to the blanket COA restrictions, with regular exceptions to allow flights at altitudes up to 400 ft. The most important distinguishing factor among these exemption requests, then, relates to the proposed UAS missions. An analysis of the key words specified in the "Operation/Mission" section of the FAA's website resulted in the top thirteen missions shown in Table 3. Exemption requests could specify more than one mission so the total does not add up to 2451, and the granting of summary approvals based on similarity to previously approved requests likely increased the number of petitioners who cited aerial photography and aerial videography. This list is valuable, however, for its identification of the types of activities for which commercial operators would employ UAS under the blanket COA constraints (see Table 1).

2.2.3   Hobbyist operations

The model aircraft hobbyist community has operated safely in the United States for more than three decades under the succinct guidelines specified in the FAA's AC 91-57 (FAA, 1981). These voluntary guidelines, which are similar to those requirements established under the blanket COA, were revised on September 2, 2015 primarily to clarify that only hobbyists, not commercial operators, were authorized to operate in the specified way. A major factor contributing to the need to clarify the rules is the explosion in a number of

Table 3. Top UAS applications cited on Section 333 authorizations.

Proposed application	Number of authorizations citing this mission/operation
1. Aerial photography	1215
2. Aerial videography	999
3. Aerial surveying	587
4. Inspections	342
5. Cinematography	276
6. Search and rescue	256
7. Real estate photography	203
8. Aerial data collection	202
9. Training	126
10. Agriculture	104
11. Construction	91
12. Research	77
13. Special events	53

small UAS that are easy to fly "out of the box." Previously, the complexity of building and safely operating model aircraft required enthusiasts to work within the hobby community, where they would also learn about the restrictions called for in AC 91-57 and the duties of a responsible aviator. The clarifications and additional restrictions contained in the updated document, AC 91-57A, should allow the continued operation of model aircraft in a responsible way and differentiate those operations from more tightly regulated civil operations.

Hobbyist flights, whether of "model aircraft" or "drones," (no distinction has been made) complying with FAA guidelines will generally occur at least 5 nmi from an airport, 2 nmi from a heliport, under 400 ft AGL, and will be made by aircraft weighing less than 55 lbs. They will be conducted away from "prohibited areas," "special flight rule areas," sensitive areas (including stadiums, power plans, dams, and national parks) and will comply with temporary flight restrictions (TFRs). They will also not fly farther than the remote pilot can observe them with unaided sight (usually regarded as less than about ½ nmi), and the pilot may not use vision-enhancing devices such as binoculars, night vision goggles, or "first-person view" devices under the definition of "unaided sight."[9] The FAA's notice of proposed rulemaking for small UAS[10] includes an airspeed restriction of 87 kts, though AC 91-57 has no such restriction.

Compliance of hobbyists with these restrictions is generally hard to evaluate because enforcement is not applied evenly across the community. However, a large amount of data have been uploaded by individual model aircraft and

---

[8] https://www.faa.gov/uas/legislative_programs/section_333/333_authorizations/

[9] https://www.faa.gov/uas/media/model_aircraft_spec_rule.pdf
[10] https://www.faa.gov/uas/nprm/

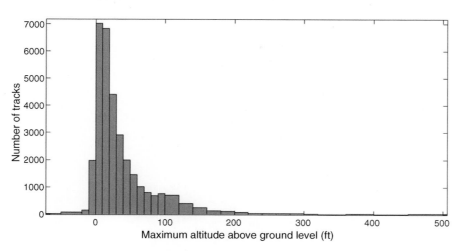

Figure 1. Maximum altitude of UAS flights (US flights only).

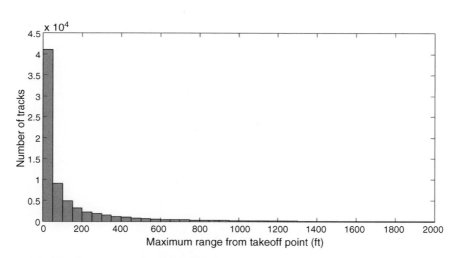

Figure 2. Maximum range of UAS from takeoff point (US and international flights).

drone enthusiasts to an online community called Drone-share.[11] Detailed flight information including position, altitude, airspeed, and other aerodynamic parameters from 75 000 individual flights (tracks) around the world are available for download and analysis (Vela, 2016). These flights are not statistically representative of all drone flights because sharing is voluntary, but they do provide some insight into how hobbyists use their aircraft. Charts of the distribution of altitudes, maximum ranges from takeoff, and total flight distance derived from the publicly available Droneshare.com data are shown in Figures 1–3, respectively.

The hobbyist data is interesting from several perspectives. First, as indicated in Figure 1, the overwhelming majority of

the 32 000 flights in the United States (the other 43 000 were international for which no terrain altitudes were available) were conducted at a maximum altitude under 100 ft. Only 0.67% of these flights ever reached an altitude greater than 400 ft AGL. The maximum range from the point of takeoff for all aircraft, those inside and outside the United States, is plotted in Figure 2. That chart shows that during most flights the aircraft never got far from the operator. About 2.39% of flights reached a maximum range of more than 0.5 nmi, and 1.65% ventured more than 1 nmi; the difficulty of seeing a small UAS at these ranges implies that an alternate method of control, for example first-person video, was employed during these flights. The percentages given here are likely below the actual number of flights to reach such a maximum range because flying beyond the point of being able to detect an

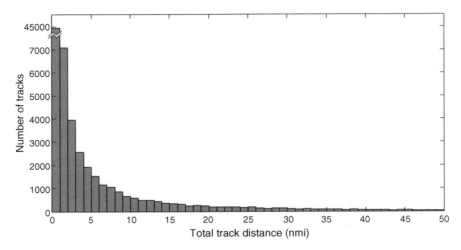

Figure 3. Total track distance flown by UAS (US and international flights).

aircraft with unaided vision is illegal is the United States and some other countries, so is likely underreported. Finally, the total track distance of each flight is shown in Figure 3. This chart indicates that the distance covered by flights is considerably farther than one might expect from the maximum range metric. Although only 1.65% of flights ventured more than 1 nmi from takeoff, 8.59% of flights covered more than 15 nmi. This indicates that the trajectories followed by the aircraft are dominated by short passes back-and-forth over the operator, consistent with the requirements of the blanket COA and hobbyist rules specified in AC 91-57A.

3 OPERATIONS UNDER FUTURE REGULATORY FRAMEWORKS

Estimates of the demand for UAS and their economic impact over the next 10 years vary widely according to the organizations making the predictions (Jenkins and Vasigh, 2013; Teal Group, 2014), at least in part because the regulatory structure under which UAS will operate has not been determined (FAA, 2014). Restrictions on UAS akin to certification standards of manned aircraft would keep their numbers relatively low, while redesign or segregation of airspace with more homogeneous users could result in much more widespread use of UAS than is seen in 2015. A discussion of the future uses of UAS is therefore incomplete without a discussion of the potential technological and regulatory frameworks that will govern their operations.

Several methods have been proposed to support more seamless access of UAS to the airspace and larger numbers of

these aircraft flying in close proximity to each other (Lacher *et al.*, 2010): direct or alternative means of compliance with existing federal aviation regulations; segregation of UAS operations from legacy users through airspace redesign; and a hybrid approach that would manage UAS through a parallel air traffic system without prohibiting legacy users from accessing the same airspace.

3.1 Future regulatory frameworks

3.1.1 Airspace integration

The most conservative approach to integrating UAS with the NAS is to require them to meet all existing federal aviation regulations (FARs) either through direct compliance or through alternative means if direct compliance is challenging. For example, UAS airframes could be subject to the same airworthiness requirements as those to which manned aircraft are subject, but the requirement that onboard pilots "see and avoid" neighboring aircraft could be replaced by an electronic means of accomplishing the same function. This approach has multiple benefits: UAS will operate in the airspace in a way that is largely indistinguishable from existing users so their disruption to the NAS will be minimal; major existing airspace stakeholders such as the FAA, airline pilots association (ALPA), aircraft owners and pilots association (AOPA), and the national air traffic controllers association (NATCA) favor this approach; and a set of requirements that would enable at least a subset of UAS operations—those favoring large, expensive aircraft—is expected to be available by late 2016. As an example of a long-term integration concept implied by this approach,

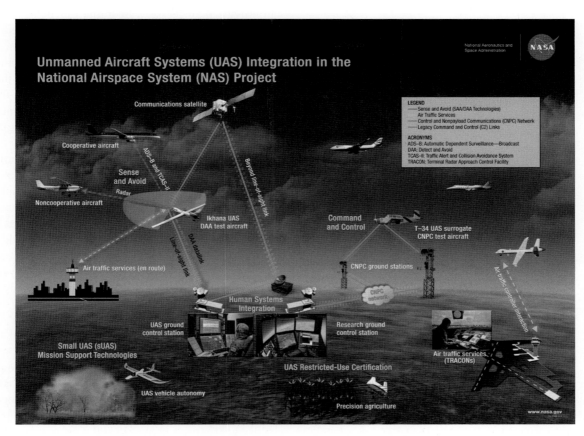

Figure 4. Concept for full integration of UAS with the NAS. (Reproduced with permission from NASA Image. www.nasa.gov/centers/armstrong/news/FactSheets/FS-075-DFRC.html.)

Figure 4 shows the technologies and interactions NASA believes will be necessary to integrate UAS with the NAS.

The drawbacks to seamless airspace integration, however, are significant: the requirements for alternative means of compliance with all regulations are difficult to determine (the see-and-avoid requirements alone will have taken dozens of engineers more than 5 years to complete by late 2016); complying with all existing FARs will likely be possible only by very large UAS capable of equipping with heavy, expensive, and power-intensive sensors and processors; most promising UAS applications will be infeasible from a technological and economic perspective; and some of the most useful places for UAS to operate (e.g. under 1000 ft over populated areas, see 14 CFR 91.119(b)) are expressly forbidden in the FARs. Sensor technologies, communications systems, autonomous decision-making capabilities, operator training requirements, and verification and validation processes necessary to enable seamless integration for mid-sized UAS could be more than a decade away.

The largest portion of US domestic UAS–NAS integration research is currently devoted to this approach, with much of

that effort supporting the RTCA[12] Special Committee 228 work on detect-and-avoid (DAA) and command-and-control (C2) communications requirements. The first phase of that group's effort, targeted to be complete by December of 2016, will provide the FAA with recommended performance requirements for those two critical systems to allow UAS to transition through Class D, E, and G airspace on their way to Class A airspace. The FAA plans to use these requirements to inform the certification of manufacturers' DAA- and C2-related equipment, which would provide a partial path for UAS manufacturers to build aircraft certified to operate without the restrictions described in Section IIA some years after the 2016 requirements deadline. Operators of large UAS, such as the US DoD and NASA, will then be permitted to transition their largest unmanned aircraft to and from high altitudes, but the vast majority of operators and operations will not be permitted.

---

[12] Radio Telecommunications Association of America, an aviation standards organization.

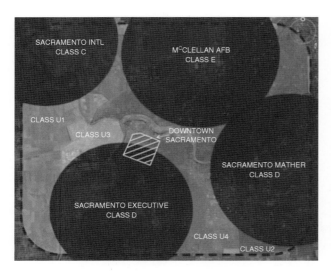

Figure 5. New UAS airspace classes within existing airspace classification system. (Reproduced with permission from NASA Image. ntrs.nasa.gov/archive/nasa/casi.ntrs.nasa.gov/20150006814.pdf.)

3.1.2 Airspace adaptation

The disadvantages of full and seamless airspace integration may be avoided by building a parallel air traffic system that would provide airspace services to participating UAS, handle the transition of aircraft to and from the legacy airspace system and continue to allow nonparticipating aircraft to enter the airspace controlled by the new system. A detailed proposal for how such a system would work, called the UAS Traffic Management (UTM) system, is detailed in Kopardekar (2014). Many of the challenges of integrating UAS with the airspace, from separation assurance, contingency management, and surveillance to traffic flow management and privacy concerns would be addressed by shifting responsibility from onboard systems to a centralized command and control system. While development and certification of that centralized system would be a difficult feat, if successful it could allow many more UAS to operate in a wider variety of ways than would a decentralized system that required significant equipage onboard every aircraft.

An example of the way the UTM system would differentiate the requirements for operating in airspace environments with different risk-based classifications is shown in Figure 5. The boundaries of the areas would be determined by the jurisdiction for providing air traffic services and the services the UTM system itself would provide: in Figure 5 the classes U1–U4 are bounded by existing terminal (airport) area traffic control authorities. The outer limits of the UTM airspace are determined by connectivity to the UTM system and where it

can provide the necessary air traffic functions. The classes themselves would be defined by four risk-based criteria: population density, density of man-made structures, likelihood of encountering manned aircraft, and the number of planned UTM operations. It should be noted that these classes relate only to the services provided by the UTM system in existing Class G airspace; they are not intended to redefine the way airspace is classified in the existing air traffic system.

While the concept of a separate authority for provision of air traffic services is unfamiliar in the United States, other countries contract out these functions to commercial entities, so the precedent does exist for the government's creation of such a system in principle, if not in practice. Perhaps the most significant benefit of such a concept is that it would not rely on as many unproven technologies, such as lightweight and low-power noncooperative intruder detection systems, as concepts that require the UAS itself to equip fully for safe operation in the existing airspace.

3.1.3 Airspace segregation

The concept of temporal and geographic airspace segregation is employed in the NAS today to increase the safety and efficiency of the system. The DoD controls firing ranges and prohibits any aircraft from entering those areas during exercises. The airspace above 18 000 ft is reserved for those aircraft filing an instrument flight rules (IFR) flight plan, carrying a transponder, and receiving separation services from ATC, among other requirements. Aircraft conducting loitering or repeating pattern operations such as aerial refueling can request an "altitude reservation" for a volume of airspace that air traffic controllers subsequently prevent other aircraft from penetrating. In a similar way, the low-altitude airspace that is currently mostly, but not completely, off limits to manned traffic[13] could be set aside for UAS operations. Existing users of that airspace, such as helicopters, hang gliders, and powered parachutes, would either be forbidden from operating in the designated airspace or would have to comply with additional requirements to enter it (e.g., carry an automatic dependent surveillance-broadcast (ADS-B) transmitter). Segregation of airspace could simplify requirements for UAS to operate in those set-aside areas because the mix of users would be more homogeneous, potentially allowing certain operations (e.g. precision agriculture, power line, and pipeline inspection) to occur much sooner than they could under an airspace integration or modification concept.

_____
[13] 14 CFR 91.119.

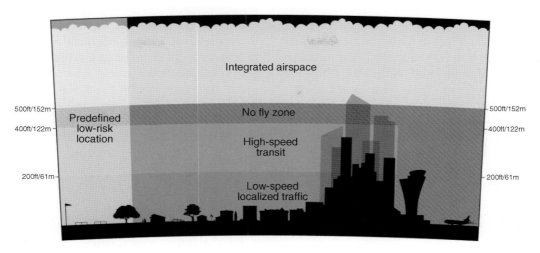

Figure 6. Airspace segregation proposal from Amazon.com. (Reproduced with permission from Amazon, 2016. © Amazon. images-na.ssl-images-amazon.com/images/G/01/112715/download/Amazon_Revising_the_Airspace_Model_for_the_Safe_Integration_of_sUAS.pdf.)

Segregating UAS operations from legacy airspace users at low altitudes has received significant media attention.[14] A well-publicized proposal[15] from Amazon.com Inc. would segregate the airspace below 500 ft for UAS operations only: altitudes between ground level and 200 ft would be reserved for "low-speed" operations like delivering packages; altitudes from 200 to 400 ft would be reserved for "high-speed" transit; and altitudes between 400 and 500 ft would be a buffer zone to ensure safe separation from existing airspace users. See Figure 6 for a diagram of the Amazon-proposed airspace classification. Airports with legacy manned aircraft would be off limits to UAS, except when authorized through specific agreements. This airspace redesign would support certain operations, and specifically it would allow Amazon to meet its stated objective to deliver packages to businesses and residences in rural, suburban, and urban areas, but it would not enable many other UAS operations. In addition, it would be inconsistent with the FAA's proposed small-UAS rule, which permits low-speed operations up to 400 ft.[16]

The concept of airspace segregation to enable wider UAS airspace integration has not received significant research attention for several reasons. One of the most important is that the FAA has indicated that it prefers to move from the current paradigm in which the air traffic system "accommodates" UAS to one in which UAS are "integrated" (FAA Roadmap, 2013). A second reason to avoid segregation is that existing examples of segregated airspace were suitable for only a narrow class of operations. Enabling a segregated airspace of this magnitude would be more akin to designing an entirely new type of airspace, which would entail a set of operational, performance, and equipage requirements potentially more complex than those required to integrate with existing airspace types. Finally, segregated airspace designs would likely not be suitable for most proposed UAS operations. The problem of how to integrate UAS into the nonsegregated airspace would remain. Segregated airspace proposals will likely continue to be proposed and covered in the media because they are relatively straightforward to comprehend, but their disadvantages and lack of support from the air transportation system regulator means they are unlikely to be a major component of the solution to UAS–NAS integration.

3.2 Future UAS operations

UAS have been proposed for use in a wide variety of areas, many of which are already underway as described in previous sections, but the regulatory framework under which they will operate will be a major factor in determining whether the technological, economic, and public policy hurdles will be low enough that UAS will be preferred over existing alternatives. A permissive regulatory environment may allow early adoption of UAS for a particular application, but the public perception backlash that could accompany an accident

---

[14] See, for example, http://aviationweek.com/technology/amazon-google-want-changes-low-altitude-airspace-uas
[15] https://images-na.ssl-images-amazon.com/images/G/01/112715/download/Amazon_Determining_Safe_Access_with_a_Best-Equipped_Best-Served_Model_for_sUAS.pdf
[16] https://www.faa.gov/uas/nprm/

might constrain long-term operations. A stricter regulatory environment could stifle innovation and have a lasting detrimental impact on the industry. It is not clear which approach will best support the growth of the UAS operations, so this section will instead focus on the operations that end users of the technology desire. These users do not actively seek out UAS to fulfill their operational need, instead they have a particular goal in mind and existing methods for achieving that goal, but if UAS can support progress toward the goal they are willing to consider its use. Several studies have examined the potential for a variety of UAS operations and predicted the demand for those applications in the coming years (FAA Roadmap, 2013; Teal Group, 2014; Volpe, 2013).

Future UAS operations may be identified by examining existing operations that occur outside of a regulatory environment. Those operations could be allowed once the appropriate regulations have been put in place. For example, a research group studying arctic sea ice with an autonomous underwater vehicle (AUV) needed to keep track of the ice-margin zone so that the AUV could safely return to the surface, however tracking that margin in real time is risky and time consuming (Lehmenhecker and Wulff, 2013). Instead of using a fixed transmitter, the team deployed a UAS to land on ice rafts at the margins and relay its location. With appropriate regulations, UAS could be used as flexible location tracking systems in higher aircraft density environments, not just in the remote arctic.

Another example in which remote area demonstration could be adapted to domestic applications is monitoring of the environment. An environmental research group developed a small UAS to monitor forest cover, species distributions, and carbon stocks in Indonesia, finding its combination of spatial resolution and geographic coverage to be more cost effective than the satellite or ground-based alternatives (Kuh and Wich, 2012). The prototype UAS was designed to be operable by a conservation researcher with limited engineering expertise in a developing country, but to be inexpensive enough ($2000) and have a long enough endurance and range (25 min and 15 km) to be useful for wide-area surveying. Applications like these are likely to spread to many other areas in which the ecosystem is threatened, whether by poaching, human activities, or climate change, including eventually to places in which human populations and other airspace users currently make such flights impractical.

The most in-depth analysis of the applications end users would pursue if UAS were economically competitive and allowed in controlled airspace was published by Wieland *et al.* (2014). That detailed report, which is summarized in Ayyalasomayajula *et al.* (2015), used interviews with subject matter experts in 19 different civilian and commercial applications along with socioeconomic modeling to develop tens of thousands of future UAS flight plans. The number and location of flight plans for each application are a function of season and the future year in which the application would occur; the future year is a proxy for the degree to which technological and regulatory progress would enable the application rather than a direct estimate of the year in which a given application would actually be feasible. The following sections describe a subset of the operations identified as high-priority uses for UAS. It should be noted that no comparable analysis of UAS applications in uncontrolled, low-altitude airspace has been done.

3.2.1   Wildfire monitoring

One of the most promising UAS applications is the strategic and tactical monitoring of potential and ongoing wildfires. NASA flew a series of monitoring missions between 2006 and 2010 to collect and relay real time thermal imagery to fire-fighting personnel on the ground (Ambrosia *et al.*, 2011). This demonstration of the benefit of UAS for wildfire monitoring provides a justification for their use in this area, but the lack of a regulatory system to allow such uses and the difficulty of obtaining approval for such an operation under a COA means that such flights will not be routine for at least several years. However, consultations with representatives from the US Geological Survey and US Forest Service identified the monitoring of areas with significant historical rates of wildfires as an application that would benefit from use of a UAS. The aircraft would loiter above the high-burn probability regions and provide early alerts to fire fighters when a wildfire begins. Early detection of these fires could significantly decrease the cost and risk of fighting them.

The strategic fire-monitoring mission would be carried out by large UAS flying at high altitudes in order to maximize the area that can be scanned during each aircraft pass (Ayyalasomayajula *et al.*, 2015). The sensors used to detect the nascent fires are sensitive enough for positive detection from an altitude of 30 000 ft and could cover the high-probability burn areas every 1–2 h over large regions of the United States with between 75 and 325 aircraft. The historical burn probabilities upon which the flight plans are based are shown in Figure 7, while the set of flight plans that would provide routine coverage of every area that has experienced a wildfire with greater than 1% probability each year is shown in Figure 8. The cost of this surveillance using current large UAS operating expenses would be between $14.7 and $60.7 million per year depending on the number of aircraft, a figure that would depend on the evolution of UAS technologies and procedures and could be feasible given that the cost of

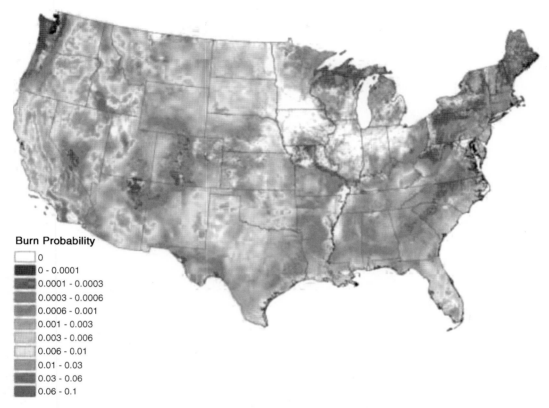

Burn Probability

- 0
- 0 - 0.0001
- 0.0001 - 0.0003
- 0.0003 - 0.0006
- 0.0006 - 0.001
- 0.001 - 0.003
- 0.003 - 0.006
- 0.006 - 0.01
- 0.01 - 0.03
- 0.03 - 0.06
- 0.06 - 0.1

Figure 7. Wildfire burn probability map. (Reproduced from Missoula Fire Science Laboratory, NASA, Wieland, 2014 with permission from NASA.)

Figure 8. UAS strategic wildfire monitoring flight plans. (Reproduced from NASA, Wieland, 2014 with permission from NASA.)

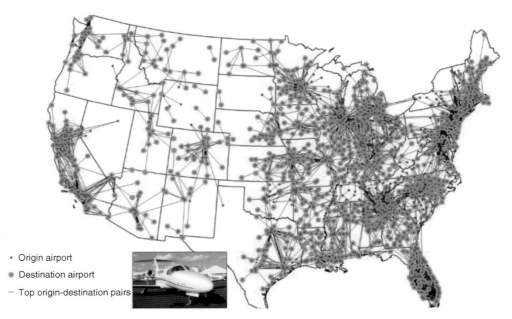

Figure 9. On-demand air taxi origin–destination airport pairs. (Reproduced from NASA, Wieland, 2014 with permission from NASA.)

fighting a single large wildfire can reach $1 million per day. The total property losses attributed to wildfires in the decade ending in 2014 was approximately $11 billion, and the 10 largest wildfires in US history created losses between $214 million and $2.6 billion, so the preventive capacity of UAS in a strategic wildfire-monitoring mission could be economically justified.

3.2.2   On-demand air taxi

Demand for some UAS applications can depend primarily on socioeconomic factors such as population distribution and demand for transportation services rather than domain-specific ones. A good example of this type of application is the use of UAS for on-demand air taxi services, which compete with ground transportation modes and scheduled commercial air travel and are subject to public perceptions about the safety of small aircraft and autonomous flight operations. An analysis of the demand for transportation services and alternative modes of travel was conducted by Ayyalasomayajula *et al.* (2015), along with the lifecycle costs of building and operating remotely piloted or autonomous air taxi aircraft. The number of flights per day was calculated as a function of the ability to fly in all weather conditions and the degree of public acceptance of this type of operation. The potential origin–destination airports for such services is shown in Figure 9, though in reality only a small subset of these routes would be serviced by an on-demand air taxi on a particular

day. A UAS with similar performance to a Cessna Mustang (twin-jet engine aircraft with four-passenger capacity) could be a competitive mode of transportation, traveling at 340 kts and 30 000 ft and reaching perhaps 3500 daily flights if the public fully accepted autonomous air travel.

4   CONCLUSIONS

This chapter described the types of operations carried out by UAS in domestic US airspace in recent years and reviewed planned or desired UAS operations in the future. The selection of UAS applications is highly constrained by the current and future regulatory environment under which these aircraft will operate, so detailed descriptions of the existing environment and several potential future environments are described. Without this context, an observer of the state of UAS today might conclude that commercial operators only want to fly small drones at low altitudes and in the immediate vicinity of where they took off, and that the only entities interested in operating larger UAS in a larger set of operational profiles are organizations within the US Federal Government.

The regulatory regime for UAS and the technologies supporting improvements in their capabilities are evolving at a rate unprecedented in the civilian aviation world. Within just a few years, there will likely be many new applications approved for civil UAS operators and the rules under which they operate are likely to have been relaxed significantly from the current conservative approach. These developments should fundamentally alter the nature of the national airspace system and its uses, in many cases bringing significant benefits that have yet to be identified even by UAS proponents.

REFERENCES

AC 00–1.1A (2014) Public Aircraft Operations, Federal Aviation Administration, US Department of Transportation Advisory Circular, Washington DC, Feb. 14, 2014.

Albertson, R., Schoenung, S., Fladeland, M., Cutler, F., and Tagg, B. (2015) Enabling earth science measurements with NASA UAS capabilities. *The International Archives of the Photogrammetry, Remote Sensing and Spatial Information Sciences*, vol. XL-7/W3.

Ambrosia, V.G., Wegener, S., Zajkowski, T., Sullivan, D.V., Suechel, S., Enomoto, F., Lobitz, B., Johan, S., Brass, J., and Hinkley, E. (2011) The Ikhana Unmanned Airborne System (UAS) Western States Fire Imaging Missions: from Concept to Reality (2006–2010). *Geocarto Int.*, **26**(2), 85–101.

Ayyalasomayajula, S., Sharma, R., Wieland, F., Trani, A., Hinze, N., and Spencer, S. (2015) UAS Demand Generation using Subject Matter Expert Interviews and Socio-Economic Analysis. *AIAA Aviation Conference*, Dallas, TX.

Berrios, M.G., Tischler, M.B., Cicolani, L.S., and Powell, J.D. (2014) Stability, control and simulation of a dual lift system using autonomous R-MAX helicopters. *American Helicopter Society 70th* Annual Forum, *Montreal, Québec, Canada.*

Bradley, A.C., Palo, S., LoDolce, G., Weibel, D., and Lawrence, D. (2015) Air-Deployed Microbuoy Measurement of Temperatures in the Marginal Ice Zone Upper Ocean during the MIZOPEX Campaign. *J. Atmos. Oceanic Technol.*, **32**(5), 1058–1070 doi: 10.1175/JTECH-D-14-00209.1.

Braun, S.A., Kakar, R., Zipser, E., Heymsfield, G., Albers, C., Brown, S., Durden, S.L., Guimond, S., Halverson, J., Heymsfield, A., Ismail, S., Lambrigtsen, B., Miller, T., Tanelli, S., Thomas, J., and Zawisla, J. (2013) NASA's genesis and rapid intensification processes (GRIP) field experiment. *J. Am. Meteor. Soc.*, **94**(3), 345–363.

FAA (2014) Aerospace Forecast: Fiscal Years 2014–2034, Federal Aviation Administration, Washington DC.

Federal Aviation Administration (2015) Unmanned Aircraft Operations in the National Airspace System, US Department of Transportation Notice N JO 7210.891, Nov. 25, 2015.

Hubbard, D., Morse, B., Theodore, C., Tischler, M., and McLain, T. (2007) Performance Evaluation of Vision-Based Navigation and Landing on a Rotorcraft Unmanned Aerial Vehicle, *IEEE Workshop on Applications of Computer Vision*, Austin, TX.

Integration of Civil Unmanned Aircraft Systems (UAS) in the National Airspace System (NAS) Roadmap, 1st edn, Federal Aviation Administration, Washington DC, 2013.

Jenkins, D. and Vasigh, B. (2013) The Economic Impact of Unmanned Aircraft Systems Integration in the United States, *Association of Unmanned Vehicle Systems International Economic Report,* Washington DC.

John, A. Volpe National Transportation System Center (2013) Unmanned Aircraft System (UAS) Service Demand 2015–2035, US Department of Transportation Technical Report DOT-VNTSC-DoD-13-01, Sept. 2013.

Kopardekar, P. (2014) Unmanned Aerial System (UAS) Traffic Management (UTM): Enabling Low-Altitude Airspace and UAS Operations, NASA/TM 2014–218299, Moffett Field.

Kuh, L.P. and Wich, S.A. (2012) Dawn of drone ecology: low-cost autonomous aerial vehicles for conservation. *Trop. Conserv. Sci.*, **5**(5), 121–132.

Lacher, A., Zeitlin, A., Maroney, D., Markin, K., Ludwig, D., and Boyd, J. (2010) Airspace Integration Alternatives for Unmanned Aircraft, *Association of Unmanned Vehicle Systems International's Unmanned Systems Asia-Pacific Conference*, Singapore.

Lehmenhecker, S. and Wulff, T. (2013) Flying Drone for AUV Under-Ice Missions. *Sea Technology*, Compass Publications, **55**(2), 61–64.

Model Aircraft Operating Standards, Federal Aviation Administration, US Department of Transportation Advisory Circular 91-57, Washington DC, June 9, 1981.

"Restricted Category Type Certification," Federal Aviation Administration, US Department of Transportation Notice 8110.56A, September 30, 2008.

Teal Group (2014) *World Unmanned Aerial Vehicle Systems: 2014 Market Profile and Forecast*, Teal Group Corporation, Fairfax, VA.

Unmanned Aircraft System Airspace Integration Plan Version 2.0, US Department of Defense, UAS Task Force Airspace Integration Integrated Product Team, Washington DC, 2011.

Vela, A. (2016) A comparative analysis of the behaviors and risks of small UAVs operating within the NAS, *AIAA Aviation and Aeronautics Forum and Exposition*, Washington DC.

Watts, A.C., Ambrosia, V.G., and Hinkley, E.A. (2012) Unmanned aircraft systems in remote sensing and scientific research: classification and considerations of use. *Remote Sens.*, **4**(6), 1671–1692.

Wieland, F., Ayyalasomayajula, S., Trani, A., Hinze, N., and Spencer, T. (2014) UAS Demand Generation and Airspace Performance Impact Prediction, NASA SBIR Phase 2 Final Report, Contract Number NNX13CA07C.

Chapter 35

High Altitude: Among and Above Commercial Transport

Matthew R. Rabe, Brandon R. Abel, and R. John Hansman

International Center for Air Transportation, Massachusetts Institute of Technology, Cambridge, MA, USA

1 INTRODUCTION

Unmanned aircraft systems (UAS) have seen increased use in military operations, which has required limited operations in the US national airspace system (NAS). This experience provides insight into requirements for more widespread UAS operations among or above the typical US commercial transport altitudes. This altitude regime is within or above the Class A airspace structure. As shown in Figure 1, Class A airspace begins at 18 000 ft mean sea level (MSL) and continues to flight level (FL) 600 (60 000 ft at standard sea-level pressure of 1013.25 hPA (hectoPascal)), although the commercial traffic usually operates on Jet Routes at altitudes at or below FL 450. For the UAS to enter the controlled Class A

airspace, it must first transit some portion of Class E or G uncontrolled airspace. Transiting this uncontrolled airspace is a significant challenge for medium-to-high altitude UAS.

The three most commonly operated UAS among and above commercial transports are the Northrup Grumman RQ-4 Global Hawk, the General Atomics MQ-9 Reaper, and the General Atomics MQ-1 Predator (Figures 2 and 3). This chapter will summarize the requirements associated with such operations and discuss some of the challenges that need to be addressed as UAS continue to integrate in the US NAS. These challenges include sense-and-avoid requirements, lost-link operations, divert/contingency points and flight termination points, mission differences, and weather limitations.

2 HIGH-ALTITUDE UAS MISSION REQUIREMENTS OVERVIEW

Since the pilot or operator of an unmanned aircraft is dislocated from the cockpit, the UAS must depend on a datalink to send commands to the aircraft. The lack of an onboard human perspective to see and avoid other aircraft forces the UAS operator to rely on alternative methods for traffic deconfliction. These are two central challenges to UAS operations. Since the UAS is controlled by an operator from the ground control station (GCS) rather than from a traditional cockpit, a datalink is required to communicate the operator's commands to the aircraft and retrieve the aircraft status for display in the GCS. During operations close to or on the airport, this datalink is typically established via a line-of-sight (LOS) pathway that communicates directly between the aircraft and the GCS. This

Unmanned Aircraft Systems. Edited by Ella Atkins, Aníbal Ollero, Antonios Tsourdos, Richard Blockley and Wei Shyy.

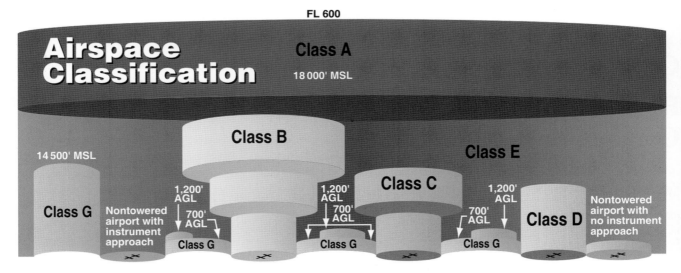

Figure 1. Classes of airspace (FAA, 2016).

is depicted with a dotted bidirectional arrow in Figure 4. Operations dependent on LOS datalink require the necessary facilities at or near the airport location; however, this is usually the preferred datalink for takeoff and landing because of the reduced latency and increased bandwidth compared with beyond-line-of-sight (BLOS) communications that involve satellite relay. *En route* operations usually rely on a BLOS datalink, such as that depicted with a dashed bidirectional arrow in Figure 4. This traditional reliance on LOS datalinks for takeoff and landing and the associated infrastructure requirement can limit potential alternative or emergency airports during *en route* operations.

Maintaining one of these datalinks is crucial to aircraft control; however, regardless of datalink reliability, planning is required to ensure the aircraft continues to fly in a "lost-link" scenario. An aircraft is considered "lost-link" if the datalink is interrupted long enough that the aircraft must revert to an alternative flight plan. This alternative flight plan is known as the lost-link mission. This will be discussed further in the Lost Link section of this chapter, but this is an important concept regarding UAS operations.

Because current UAS operations have developed locally, different procedures have developed for each aircraft system and location. However, each of these has addressed the "see-and-avoid" limitations of UAS. Since a UAS does not have a

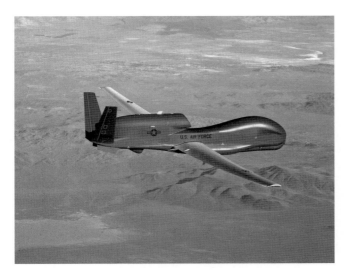

Figure 2. USAF RQ-4 Global Hawk (USAF, 2014).

Figure 3. USAF MQ-9 Reaper (USAF, 2015).

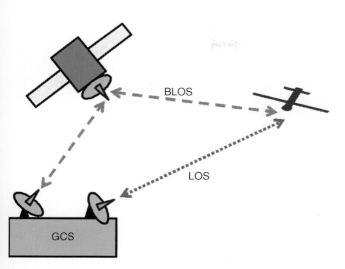

Figure 4. UAS datalink architecture.

pilot on board, it cannot comply with the Code of Federal Regulations (CFR) 14 CFR 91.113 requirement of "vigilance shall be maintained by each person operating an aircraft so as to see and avoid other aircraft."

This see-and-avoid requirement can be satisfied through several available alternatives:

1. Airspace control measures
   a. Restricted area
   b. Temporary flight restriction (TFR)
2. Human observers to spot traffic and issue deconfliction instructions
   a. Airborne observers in a chase ship
   b. Ground observers along the route of flight
3. Ground-based sense-and-avoid (GBSAA) system
4. Airborne-based sense-and-avoid (ABSAA) system

At the time of this chapter most high-altitude UAS operations rely on airborne observers on a manned chase aircraft until the UAS is established in controlled or restricted airspace. There are also several operations within restricted or TFR areas from takeoff until controlled airspace, and even some which rely on GBSAA systems. The reliance on chase aircraft or TFR areas could represent a significant inefficiency as UAS operations increase in frequency. There are significant efforts to develop adequate ABSAA systems to reduce UAS reliance on outside assistance, or on special airspace, for traffic deconfliction.

2.1 Mission overview

This section will outline the challenges associated with each phase of flight for a typical high-altitude UAS mission.

2.1.1 Ground operations

Ground operations also require an alternative to the typical onboard human deconfliction. Here, the human pilot is typically responsible for clearing the taxi path of an aircraft to avoid obstacles, personnel or other aircraft. For a UAS, this service could be provided by a ground vehicle following the aircraft, video sensors mounted on the aircraft, or additional procedural responsibility to the air-traffic control tower personnel. Current operations do not put this responsibility solely on the airport control tower. While the tower is responsible for clearing the taxi path before issuing taxi clearance, the ultimate responsibility still remains with the aircraft pilot. For this reason, the ground observer, via a chase car or truck, and the aircraft video sensors are the two primary means to clear a taxi path. Most aircraft with an available video sensor rely on this to clear their own path and do not require a chase vehicle; however, aircraft that do not have a gimbaled camera typically require a chase vehicle to provide visual clearance of the taxi path. Coordination with a chase vehicle, usually on a second radio, can cause additional workload during taxi operations.

2.1.2 Takeoff, climb out, descent, and landing

With the potential requirement to transition through uncontrolled airspace, the takeoff, climb out, descent, and landing portions of the mission can present additional challenges that depend on which technique is used to replace the typical human pilot see-and-avoid capability.

Airspace control measures If these operations use airspace control measures, such as a TFR or Restricted Area to takeoff and climb to Class A airspace, the pilot can rely solely on Air Traffic Control (ATC) to navigate through the lower altitude airspace. This technique can restrict scheduling of takeoff and landing since this airspace often has limited availability. This could even lead to evening or overnight UAS operations to satisfy scheduling constraints.

Human observers Use of human observers requires a chase ship because ground observers are unable to monitor the flight path of an aircraft all the way up to 18 000 feet, the bottom of Class A airspace. Coordination between the UAS and the chase aircraft is an additional complication. Such coordination usually involves a second radio for the UAS to communicate with ATC and the chase pilot. The chase ship will usually perform an "airborne pickup" that requires the UAS to receive takeoff clearance from the control tower and then wait on the runway until the chase ship, which is circling over the field, signals the appropriate time for takeoff. This procedure allows the chase ship to rejoin just after the UAS takeoff. Although this extended time on the runway is usually only 1–3 min, it can have a significant effect on busy airports.

For the remainder of the UAS climb out to Class A airspace or a restricted area, the pilot must always plan maneuvers in conjunction with the chase ship. While the chase ship is acting as the UAS pilot's see-and-avoid capability, most traffic deconfliction will still be planned and initiated by the UAS pilot. The GCS often has a display available with a complete ATC radar picture. This gives the UAS pilot a complete picture of all cooperating traffic (aircraft with active transponders) through the entire route of flight. This level of information is rarely available in a manned aircraft, including the chase ship.

Sense-and-avoid   As planned, an ABSAA system would provide both a situational awareness display of upcoming traffic and recommendations for traffic avoidance maneuvers with impending conflicts. This information would be available directly to the pilot and not require any outside assistance. In this way, the pilot could transition through the uncontrolled airspace in a manner similar to a manned aircraft, but with one additional critical system dependency, the ABSAA system.

No current systems include an ABSAA capability; however, there are GBSAA systems being introduced for UAS operations in uncontrolled airspace. These systems can provide the situational awareness and traffic avoidance functions of the planned ABSAA systems, but they may require additional outside coordination because they rely on ground-based facilities not necessarily collocated with the GCS. Additionally as the ABSAA system would travel with the aircraft and give a similar capability throughout a flight, a GBSAA system has geometric limitations based on its location on the ground and its relation to the planned route of flight. This restricts the available flight paths allowable for the UAS during the climb to Class A airspace.

Any of these options would present the same challenges during the descent and landing portions. However, on the descent phase, delays are more costly than the takeoff phase of the flight because the waiting aircraft would require holding airspace and the necessary fuel to wait.

2.1.3   En route

Once established in or above the Class A airspace structure where all aircraft are controlled by ATC (14 CFR 91.215, 2015), the see-and-avoid requirement is often relieved, so UAS operations here are quite similar to manned aircraft operations.

As with some manned aircraft, current UAS are limited in their use of Class A airspace because they typically lack standard equipment for a reduced vertical separation minima (RVSM) capability. This capability is required above FL 290 (=29000 ft altitude at standard sea-level pressure of 1013.25 hPa) that limits the UAS when requesting to operate at higher flight altitudes. Unfortunately, these higher flight altitudes are well suited for some UAS operations, so crews must wait for a gap where ATC can allow the UAS to fly above the FL 450 jet routes.

All of the UAS that operate in this region have onboard radios for communication directly with the different ATC sector controllers. During lost link or lost communication (radio) emergencies, the pilot would also have access to a telephone if landline communication was required. This represents an increase in capability over most manned aircraft operations.

These current UAS typically do not have any traditional navigational aids (NavAids). They are reliant on GPS and inertial navigation system (INS) for guidance. For this reason, they are not designed to execute the same flight routes as manned aircraft. Although it is possible to configure UAS to follow simulated routing to NavAids, and pilots could manually fly such routes, that ability is not typically designed into the UAS control architecture. This can increase the workload on ATC controllers and pilots.

Additionally, lost-link planning, divert considerations, mission differences, and weather restrictions, which all represent *en route* challenges, are discussed in the following sections.

3   LOST-LINK OPERATIONS

Perhaps the most significant difference between manned and unmanned aircraft operations at or above commercial transport is the need for lost-link flight planning. As mentioned in the overview section, the possibility that the pilot may lose control of the aircraft during a loss of datalink necessitates a procedure for autonomous aircraft operations during such conditions. The FAA does maintain oversight of potential lost-link flight plans in the NAS; however, because procedures have developed locally and individual UAS manufacturers have designed their systems with different capabilities, there is no standardized lost-link procedure for all UAS. Considerations influencing these procedures include the constant update requirement of the lost-link flight plan, real-time deviations from the original flight plan, protection of infrastructure and personnel on the ground, and lost communication procedures from manned aircraft operations.

From takeoff to landing, the pilot must continually monitor and update the lost-link flight plan. Since the pilot has no warning of an impending lost-link situation, he or she must strive to update the lost-link flight plan as often as he or she updates the real-time flight plan of the aircraft. To assist with this burden, some manufacturers have designed this responsibility into the onboard autopilot. These systems work well when there are no changes to the planned route of flight. As shown in the example in Figure 5, with an aircraft flying along the three waypoints indicated by black triangles, an autonomous system could update the next lost-link waypoint from 1 to 2 when waypoint 1 is passed. With manual control of this function, the pilot must remember to maneuver the aircraft to fly to the next waypoint and then immediately send a command to change the next lost-link waypoint as well. While the automation can reduce the manual workload on the pilot, some manual adjustments may still be required. If changes are made to the planned route or altitude, the pilot will need to initiate a corresponding change to the lost-link mission. In a sense, the UAS operator is managing the flight path of two vehicles. First, the real-time flight of the UAS, and second, the backup lost-link flight plan for the UAS to follow in the event of a lost datalink. Either could be critical at any given moment.

The top example in Figure 5 illustrates the ease with which an automated system could update the lost-link flight plan (dotted line) given the aircraft crosses the first waypoint; however, the bottom portion begins to illustrate how automation may have difficulty updating a lost-link mission if it deviates from the preplanned flight plan. In this example, the aircraft is flown around a weather obstruction, but this could also represent a deviation for traffic or other unanticipated conflicts. As the aircraft maneuvers off course, the lost-link flight plan must be redesigned and sent to the aircraft so that the aircraft does not fly to the original waypoint 2 during a lost-link mission. Current air operations assume that pilots have this sort of flexibility to reroute around unexpected obstacles or obstructions, but not all UAS are designed for dynamic inflight changes to the lost-link flight plan. Future

regulations may require a capability for real-time manual updates to any lost-link mission even if the system has an autonomous update capability.

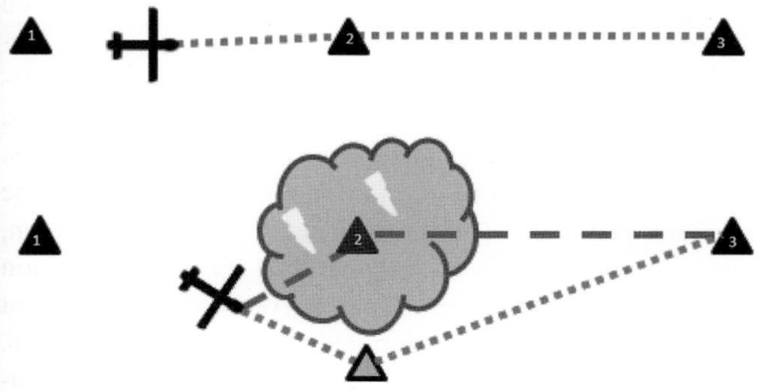

While UAS follow similar inflight procedures to manned aircraft in controlled airspace operations, the lost-link mission plan still represents a significant difference. The UAS must plan for potential divert/contingency points (DCP) or flight termination points (FTP) that are alternative landing locations and ditch locations, respectively that the aircraft could fly to without human oversight. This lack of oversight has led to additional restrictions governing the lost-link flight plan. These can include restrictions about populated areas, airports, certain facilities, and even avoidance of major roads by crossing perpendicular to traffic. These concerns may have contributed to the current lack of standardized instructions for lost-link flight plans and may limit reconciliation of these procedures with manned aircraft lost communication procedures.

Manned aircraft operations have standardized lost communication procedures established so that the pilot knows where to fly and the controller knows what to expect in the event of aircraft radio failure (14 CFR 91.185, 2015). These procedures establish a routing and altitude, and also a transponder code (7600) for the aircraft, which a UAS pilot could use as the basis for a lost-link flight plan. While the UAS pilot often still has the ability to reach ATC via a ground radio or telephone, he or she has no ability to adjust the flight path of the aircraft, because there is no datalink to control the aircraft during lost link. For this reason, these situations are similar to manned aircraft lost communication scenarios where ATC has no ability to change the routing of these aircraft. Given the locally developed procedures and varying capabilities of UAS, lost communication routes are not the standard lost-link procedures.

Standardizing lost link procedures could decrease pilot or ATC confusion during lost-link situations and allow UAS to integrate more smoothly into commercial aircraft operations.

4   DIVERT/CONTINGENCY AND FLIGHT TERMINATION POINTS

Two major considerations with lost-link operations are the DCPs and FTPs mentioned in the previous section. A DCP is an alternative landing/recovery site to be used in the event of an abnormal condition that requires a precautionary landing (FAAN 8900.227, 2013), similar to an emergency divert landing in a manned aircraft. DCPs are available for UAS that have the ability to land fully automated, that is accomplish a normal landing during a lost-link condition. For mission operations conducted in Class A airspace or Class E airspace

Figure 5. Lost-link flight plan changes.

Figure 6. Divert/contingency point *en route.*

above FL 600, DCPs must be designated no further than 1 h away from the aircraft's current position *en route* (FAAN 8900.227, 2013). For instance, Figure 6 shows a possible DCP requirement during an *en route* mission where the aircraft is further than 1 h away from both the departure (DEP) and arrival (ARR) airfields. However, current guidelines for DCP approval include not only communication with the ATC facility managing the airfield through descent and landing but also a plan for ground operations, securing and parking the aircraft on the ground, control station launch capability, communication equipment, and adequate power sources for retrieval operations (FAAN 8900.227, 2013). These requirements likely limit the ability for a UAS to divert to locations without current UAS operations of the same type already on location. If these DCP requirements cannot be met, FTPs must be designated.

Flight termination is the intentional and deliberate process of performing a controlled flight into terrain (FAAN 8900.227, 2013). FTPs must be designated for unmanned aircraft without automated landing capability during a lost-link scenario or for any unmanned aircraft that cannot reach its departure airfield, arrival airfield, or DCPs upon fuel exhaustion or a complete loss of engine scenario. In fact, FAAN 8900.227 (2013) requires FTPs to be located within power-off glide distance of the aircraft during all phases of flight. For instance, Figure 7 shows a possible FTP where the aircraft, upon engine failure, could not reach the departure, arrival, or DCP during engine-out glide. Therefore, a FTP is required. As UAS performance continues to improve and UAS transit cross-country, each flight plan will require numerous FTPs. In addition, FTPs are to be located no closer than 5 nmi from any airport, heliport, airfield, NAVAID, airway, populated area, major roadway, oil rig, power plant, or any other infrastructure (FAAN 8900.227). The designation of FTPs during cross-country flights may become a barrier to UAS integration in the NAS.

Currently, the FAA requires site surveys every 6 months and desires landowner prior permission for all FTPs (FAAN

Figure 7. Flight termination points *en route.*

8900.227, 2013). In addition, users assume full risk and all liability associated with the selection and use of any designated FTP (FAAN 8900.227, 2013). For UAS capable of automated landings during lost-link operations, or UAS not capable of automated landings during a complete engine loss situation, FTPs are required when DCP requirements cannot be met. Here, UAS will be terminating aircraft in remote locations that could otherwise land themselves on suitable runways. As the cost of unmanned aircraft increases, and for the unmanned aircraft that are already expensive, attractive alternatives to remote-location FTPs are current airports and airfields. However, in order to use airports that do not meet the requirements of DCPs, UAS users must ensure a similar or greater level of safety for people and structures at and around the airport facility. This would include site surveys for approved approaches to the runway, communications procedures that ensure separation between the UAS and other traffic during an emergency, logistics required to ensure continued airfield operations after the divert, and contingencies baring system failures beyond lost link, among other items.

Not only is the consideration of the cost of the unmanned aircraft asset a concern for users (already over $200 million for some aircraft in this altitude class), but also the risk and liability of termination points in remote and unpopulated areas is still large (Gertler, 2012). Although currently most FTPs lay within restricted areas or government land, future FTPs will venture beyond these categories. While getting surveyed points and seemingly unpopulated areas are tractable, ensuring this area is clear when required may not be. Proponents of using current runways for these points argue that these areas are already surveyed, clear from the public, and can be controlled real time with the proper logistical system in place. Regardless, ensuring UAS have appropriate, safe, yet efficient DCPs and FTPs is a major concern for future high-altitude UAS users as full NAS integration becomes a reality.

5   INTEGRATING UAS MISSION DIFFERENCES INTO THE NAS

The largest component of the NAS, the Air Route Traffic Control Center (ARTCC) provides ATC service to aircraft operating primarily on IFR flight plans during the *en route* phase of flight (FAA-H-82014). This *en route* structure consists of the airway and route system and special use airspace (SUA). While the primary purpose of the ATC system is to prevent a collision between aircraft and is to provide a safe, orderly, and expeditious flow of traffic

(FAAO JO 7110.65 V CHG 3, 2015), controllers typically manage traffic flows on the airway and route system. Here, aircraft are generally transporting people and cargo from one destination to another, with an emphasis on maximizing efficiency with respect to cost and time. Conversely, controllers do not typically control aircraft inside SUA, where military and other government operations occur. These missions may involve loitering for extended periods of time or executing large horizontal, vertical, or speed changes to their flight path. SUA was thus designed to segregate disparate and incompatible types of aviation.

The FAA first authorized UAS in the NAS in 1990 (Dorr and Duquette, 2014), and much of the early UAS operations were flown exclusively in SUA (usually restricted airspace). However, unmanned aircraft systems have brought increased capabilities to users that want to operate along the standard airway and route system. In fact, UAS currently perform a wide variety of missions in the United States operating from COAs allowing them to conduct missions outside of SUA and among the airway and route system of the NAS. These include scientific studies, agricultural monitoring, border surveillance, search and rescue missions, disaster response, and military training (Paczan *et al.*, 2012; Joint Planning and Development Office (JPDO), 2013). Yet many of the missions UAS perform are atypical compared to standard operating procedures along the airway and route system. This inconsistency is due to differences in lateral maneuvering requirements, endurance capabilities, airspeed, and flight plan interfaces.

Many current UAS missions including border surveillance, scientific studies, and agricultural monitoring require different types of lateral navigation than what usually transpires at higher Flight Levels. These missions require lateral navigation to ensure airborne sensors track specific parameters on the ground – whether along nonlinear borders, cutting swaths across geographic regions, or other specific patterns – but not along airways and route structures like most current users of the NAS. Additionally, search and rescue, disaster response, and military training flights may require extensive holding in a location, but holding in certain airspace is impractical. For example, air traffic control sectors with large blocks of SUA and many flows of traffic to multiple airports make it difficult for holding airspace to exist there (FAA Air Traffic Organization, 2009). UAS lateral maneuvering requirements also pose significant challenges for the NAS due to controller complexity and flight plan interfaces. For instance, controllers use both automation tools and mental projection to evaluate future separation of aircraft. Mental projection becomes more challenging when aircraft are not flying in a straight flight path, increasing the complexity of the

controllers' task. Also, as discussed later in this section, automation tools are not as effective due to input limitations of the system.

The missions performed by these UAS may also last for hours, days, or weeks. The USAF RQ-4 Global Hawk has an advertised endurance of "more than 34 h" and currently operates outside of SUA in the NAS (US Air Force, 2014). The Airbus Zephyr, a UAS sensor platform, flew more than 336 h (14 days) in a single flight in 2010 (Airbus, 2015). One major advantage gained from lack of crew limitations on the aircraft is the ability to stay aloft for extreme amounts of time. Future UAS users could take advantage of this capability to better perform a variety of missions. Yet increased time on station may put a strain on ATC capacity, manning requirements, and logistics.

One performance trade-off of the ability to stay aloft for days at a time is airspeed. UAS operate with widely varying performance characteristics that do not necessarily align with manned aircraft performance (FAA, 2013). Speed variance can have a large impact on the integration of aircraft along airways and route structure. For example, a Department of Transportation (DOT) audit report (2014) stated that controllers at one ATC facility handling UAS always maneuver manned aircraft away from UAS because of mission and performance characteristics, such as differing airspeeds and rates of climb. These unique lateral maneuvering requirements, long-endurance capabilities, and airspeed variances pose independent problems for current ATC operations. Controllers mitigate these problems by using decision support tools and automation to better project conflict situations on their displays. Each aircraft's flight plan provides an input to these decision support tools. The three unique problems of UAS mentioned above combine to create another problem with the controllers' interface.

ATC controllers rely on automated decision support tools, such as the *En Route* Automation Modernization (ERAM) system, to help project future conflicts between aircraft. However, current ERAM implementation cannot handle the large flight plans of long-endurance UAS missions or their navigational complexity (Hampton, 2014). This system cannot handle the number of waypoints required for long-endurance missions and does not incorporate an adequate interface for orbits or other nondirect flight paths. Research conducted by the MITRE Corporation investigated issues associated with FAA, International Civil Aviation Organization (ICAO), Department of Defense (DOD), and the North Atlantic Treaty Organization (NATO) flight plan interfaces for UAS (Paczan *et al.*, 2012). Issues include limitations inserting delays based on number of orbits or fuel rather than time, the maximum number of route points, reusing the same route point, maximum time length of the flight plan, loiter configurations, and contingency route planning (Paczan *et al.*, 2012). To better aid controllers in future UAS–NAS integrated operations, decision support tools such as ERAM must fully handle the complexities and differences of UAS and have standardized input mechanisms for various lateral navigation routes. When these procedures and automation tools are standardized and UAS friendly, better flight-path projection and warning systems will be available for UAS integration (Figure 8).

Long-endurance UAS missions pose unique challenges to the pilot–controller relationship. Such extended flights in the same geographic region or sector may not require radio frequency changes or frequent ATC communications. Long-duration manned aircraft flights would require regular radio frequency changes and communication with ground controllers. In addition, long missions will require crew breaks and crew changes by both the UAS operator and ATC controller. The voice one hears may be different from one radio call to the next. Also, both the UAS operator and ATC controller may have higher workload from other tasks, whether it is sensor tasks on the UAS or sector airway management for the controller. Infrequent radio calls, controller and operator crew changes, and other mission tasks could exacerbate complacency issues when ATC controllers and UAS operators need to contact each other during long-endurance missions. Procedural and/or design changes may be required to help mitigate the increased likelihood of

missed radio calls and confusion during these long-endurance missions.

While many opportunities exist for UAS to conduct missions previously unimaginable by manned aircraft, thoughtful integration of these two categories of aircraft will be crucial to continued success. Inconsistencies currently exist between standard operating procedures along airways and route systems and UAS requirements due to differences in lateral maneuvering, endurance, airspeed, and ATC flight plan interfaces. Flight plan interfaces must be compatible with UAS requirements in order to achieve the increased safety and decreased workload required in future NAS environments. Beyond these maneuvering differences, infrequent radio calls, controller and operator crew changes, and other mission tasks could exacerbate complacency issues inside the ATC-to-UAS communication structure.

6   UAS WEATHER LIMITATIONS

Unmanned aircraft possess performance capabilities previously unattainable. Although weather conditions above commercial transport typically are favorable, poor weather conditions are a large and continuing limitation for UAS during climbs, descents, and among commercial transport. Risks include airframe icing, high winds, turbulence, instrument and navigation limitations, and certification/regulatory restrictions. These obstacles continue to limit full UAS–NAS integration.

Mission goals and development factors have led to unmanned aircraft weather susceptibility, including rapid designing and prototyping, lightweight structures, small airframes, and high-aspect ratio wings. In fact, a GAO report (2005) highlighted that unmanned aircraft are more likely to be grounded in inclement weather than manned aircraft due in part to their lighter weight. Lighter weight structures could preclude the engineering capacity to install features such as de-icing capability, higher loading tolerances, or additional navigational equipment. Specifically, airframe icing has caused a number of accidents, including two accidents involving the Hunter unmanned aircraft (US Army) and three crashes of the Predator unmanned aircraft (US Air Force) as of 2002 (GAO, 2005). In addition, the US Army restricts UAS operations when winds exceed 50 knots at altitude (FMI 3–04.155, 2006). But as of 2005, the DOD had not determined whether all-weather capability was worth the trade-off of potential performance reductions (GAO, 2005); as of 2013, the largest issue with UAS remains an inability to operate in bad weather (DOD, 2013).

Turbulence presents a unique challenge for UAS designers and operators. While operating manned aircraft, pilots

Figure 8. ERAM R-Side Sector (FAA, 2015).

rely on physical sensory cues, such as vestibular and proprioceptive stimuli, to perceive unusual or potential dangerous flight conditions. These cues are not present for UAS operators. Although some unmanned aircraft provide visual warnings to operators during periods of turbulence, this perception of a turbulence caution may be more latent and less transferable from normal pilot perception of turbulence. The risk due to icing, high winds, or turbulence is increased and possibly unavoidable with unmanned aircraft due to the possibility of lost link. Some UAS may have the ability to loiter above poor weather in an attempt to land during more favorable forecasted weather conditions. However, lost-link procedures may include a descent to the destination airfield or a DCP, resulting in unplanned weather penetration.

Weather limitations also decrease UAS autonomy and usability. The Global Hawk and the Scan Eagle possess significant automated flight capabilities but their degree of autonomy is reduced based on their need for continuous operator intervention in poor weather conditions (Gertler, 2012). Certification standards for UAS have also decreased usability. For example, most UAS have no ground-based navigational capabilities such as an instrument landing system (ILS), VHF omnidirectional range (VOR), or distance measuring equipment (DME), but rely primarily on satellite-based GPS. Current UAS cannot execute published instrument approach procedures (FAA, 2013). Although initial plans for UAS integration at airports include segregation from mainstream air traffic, UAS launch windows, special airports, or off-airport locations, airport integration will eventually be required (FAA, 2013). Rules and regulations may need to be adjusted to allow for satellite-only navigation vehicles as the FAA moves away from ground-based navigational aids.

UAS are also generally restricted from operating in clouds in the NAS. FAAN JO 7210.873 (2014) states "all operations must be conducted under visual meteorological conditions (VMC)" for UAS in the terminal areas of Class C, D, E, and G and the *En Route*/TRACON areas of Class E and G. Currently, some UAS are allowed to operate without see-and-avoid capabilities in Class A airspace; however, this could be restricted to a "clear of clouds" requirement even in Class A airspace. Full integration of UAS flight through clouds should be based on airframe and instrumentation limitations, not restrictions based on a lack of see-and-avoid capability. In clouds, see-and-avoid offers no increase to safety. Here, similarly equipped and certified UAS are just as likely to avoid collisions as manned aircraft due to ATC-provided separation and Airborne Collision Avoidance System (ACAS) systems.

7 CONCLUSION

Several UAS currently operate among and above commercial aircraft in the NAS; however, current system and procedural inconsistencies have made integration challenging. As the number of UAS increases in the NAS, these challenges will also increase unless system requirements and procedures are standardized nationwide. Current challenges associated with high-altitude operations among and above commercial aircraft include individual coordination requirements for each airframe and organization, significant differences in lost-link flight planning between aircraft systems, financial and safety concerns surrounding mandated flight termination points, specific complexities involved with airspace integration, and weather capability restrictions. Full integration will require standard approaches to lost-link planning and execution, lost communication planning and execution, flight termination points, divert/contingency points, and ATC controller automation and display interfaces.

REFERENCES

Airbus: Defence and Space (2015) Zephyr Unmanned Aircraft System UAS. Available at http://militaryaircraft-airbusds.com/Aircraft/UAV/Zephyr.aspx.

Department of Defense (DOD) (2013) Unmanned Systems Integrated Roadmap: FY2013-2038.

Department of Transportation (2014) Office of Inspector General Audit Report: FAA Faces Significant Barriers to Safely Integrate Unmanned Aircraft Systems into the National Airspace System. Federal Aviation Administration Report Number: AV-2014-061.

Dorr, L. and Duquette, A. (2014) *Fact Sheet – Unmanned Aircraft Systems (UAS)*. Federal Aviation Administration, Retrieved on https://www.faa.gov/news/fact_sheets/news_story.cfm?newsId=14153.

FAA-H-8083-16 (2014) Instrument Procedures Handbook. U.S. Department of Transportation: Federal Aviation Administration, Flight Standards Service.

FAAN 8900.227 (2013) Unmanned Aircraft System (UAS) Operational Approval. U.S. Department of Transportation, Federal Aviation Administration, National Policy.

FAAN JO 7210.873 (2014) Unmanned Aircraft Operations in the National Airspace System (NAS). U.S. Department of Transportation, Federal Aviation Administration, Air Traffic Organization Policy.

Federal Aviation Administration (2013) Integration of Civil Unmanned Aircraft Systems (UAS) in the National Airspace System (NAS) Roadmap, 1st edn.

Federal Aviation Administration (2015) Federal Aviation Administration Order (FAAO) JO 7110.65V Change (CHG) 3: Air Traffic Control.

Federal Aviation Administration (2015) En Route Automation Modernization (ERAM). Retrieved on https://www.faa.gov/air_traffic/technology/eram/.

Federal Aviation Administration (2016) Pilot Handbook. Chapter 14: Airspace. Retrieved on https://www.faa.gov/regulations_policies/handbooks_manuals/aviation/ pilot_handbook/media/PHAK%20-%20Chapter%2014.pdf.

Federal Aviation Administration (FAA) Air Traffic Organization (2009) Traffic Flow Management in the National Airspace System.

Field Manual Interim (FMI) 3–04.155 (2006) Army Unmanned Aircraft System Operations. Headquarters, Department of the Army.

GAO (2005) Unmanned Aerial Vehicles: Improved Strategic and Acquisition Planning Can Help Address Emerging Challenges. GAO-05-395T.

Gertler, J. (2012) U.S. Unmanned Aerial Systems. Congressional Research Service (CRS). Retrieved on https://www.fas.org/sgp/crs/natsec/R42136.pdf.

Hampton, M.E. (2014) Office of Inspector General Audit Report: FAA Faces Significant Barriers to Safely Integrate Unmanned Aircraft Systems Into the National Airspace System. Federal Aviation Administration Report Number: AV-2014-061.

Joint Planning Development Office (JPDO) (2013) Unmanned Aircraft Systems (UAS) Comprehensive Plan: A Report on the Nation's UAS Path Forward.

Paczan, N.M., Cooper, J., and Zakrzewski, E. (2012) *Integrating Unmanned Aircraft Into NextGen Automation Systems*. The MITRE Corporation.

United States Air Force (USAF) (2014) RQ-4 Global Hawk. Retrieved on http://www.af.mil/AboutUs/FactSheets/Display/tabid/224/Article/104516/rq-4-global-hawk.aspx.

United States Air Force (USAF) (2015) MQ-9 Reaper. Retrieved on http://www.af.mil/AboutUs/FactSheets/Display/tabid/224/Article/104470/mq-9-reaper.aspx.

Chapter 36

Low-Altitude Rural to Urban Unmanned Aircraft System Operations

Ella M. Atkins[1] and Pedro F.A. Di Donato[1,2]

[1]*Department of Aerospace Engineering, University of Michigan, Ann Arbor, MI, USA*
[2]*National Civil Aviation Agency–Brazil (ANAC), São José dos Campos, Brazil*

1 INTRODUCTION

Small unmanned aircraft systems (sUASs) are rapidly becoming affordable and reliable platforms for carrying sensors and small packages into the sky. Due to their limited range and endurance, sUASs are expected to fly local area missions at low altitudes. Per tradition, model aircraft enthusiasts worldwide continue to enjoy sUAS flight at low altitudes above ground level (AGL) within the line-of-sight (LOS). However, many of the most compelling applications in airborne inspection, surveillance, and small package delivery will require sUAS to safely fly beyond line-of-sight (BLOS) of any ground-based operator. BLOS flight will in turn require a new level of autonomy to support safe flight, since any operator's situational awareness will be less than or equal to the information collected and shared by the sUAS and ground-based sensor systems managed by an air traffic management (ATM) or UAS traffic management (UTM) (Kopardekar, P.H.) system.

Regulators are struggling to keep pace with technology; ongoing sUAS policy and associated issues are discussed in detail in Chapters eae1137 and eae1144 and thus are not a focus this chapter. This chapter instead focuses on defining low-altitude sUAS use cases and describing how these applications translate to technology and system-wide information management (SWIM) requirements and constraints.

Information and operating requirements for low-altitude sUAS missions are critically dependent on the environment in which the sUAS flies. At high altitudes, the specific characteristics of the overflown property are not important, yet the sUAS operating at low altitude, for example, 0–500 ft ($\approx$150 m) above ground level (AGL),[1] will necessarily impact that property by supporting desired missions but also in terms of potential negative impacts. While increased risk of personal and material damage due to sUAS crashes

---

[1] The limit of 500 ft is based in recent Federal Aviation Administration (FAA) Notice of Proposed Rulemaking No. 15-01, FAA-2015-0150. Current FAA policy in Advisory Circular 91-57 establishes a limit of 400 ft for recreational use.

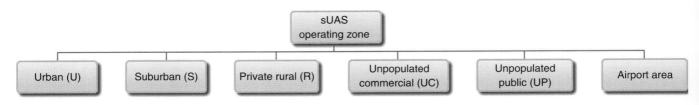

Figure 1. Low-altitude sUAS operating zones.

and emergencies is one of the most discussed of those negative impacts (Brown *et al.*, 2015), environmental impacts (Efroymson *et al.*, 2001), annoyance, and loss of personal and property rights should also be evaluated and considered. This chapter is therefore structured around the low-altitude operating environments or *zones* in which envisioned sUAS missions will be conducted. Figure 1 provides a high-level classification of potential low-altitude sUAS flight zones.

The zones depicted above classify sUAS operational profiles by overflown region population density as well as property ownership. sUAS flights over urban and suburban areas inherently pose more risk to the overflown populous, vehicles, and structures than do flights over rural and other unpopulated areas. This implies that urban (U) and suburban (S) sUAS flights must be conducted with focus on protecting the overflown population as well as managing noise, annoyance, and other concerns to the local community. As will be discussed below, an urban or suburban sUAS mission must be assured to operate at a high level of integrity to ensure safety for an overflown population. Note that flight near airports, acknowledged in the figure, is not the focus of this chapter, as associated policy is still in flux and may vary by airspace class, runway layout, traffic types, patterns and density, and regulator and local airport manager preferences. Section 8 briefly discusses sUAS operations near airports.

sUAS flight over less densely populated areas may pose little risk to people and property of value even when the sUAS itself may have some risk of being ditched or crashed so long as the flight and any associated risk are determined to be acceptable (and worthwhile). To balance risk with benefit of flight, this chapter further distinguishes low-population properties based on traditional "zonings" that are carried today in land maps, for example in the United States. First, private rural property (R), which has traditionally been an accepted, safe environment for model aircraft operations, offers an infrequently used low-altitude airspace environment where the likelihood of annoying or bringing harm to entities on the ground is typically quite low. "Unpopulated" commercial properties (UC), such as large-acreage warehouses, shipyards, commercial construction sites, and

manufacturing facilities, also offer a single organization or shared facility team benefit from low-altitude sUAS flights that will pose little risk to the populous.[2] Similarly, "unpopulated" public properties (UP), such as state and national forests and parks, (rural) public waterways and lakes, and military exercise/training ranges, offer the public entity managing the property the ability to extend their reach into the air for benefit, while also protecting sensitive areas from low-altitude flights that might be disruptive or damaging to the facility.

This chapter begins with an overview of low-altitude sUAS. sUAS can be divided into two basic classes: *high-integrity* operations that pose acceptable levels of risk to an overflown population and *low-integrity* operations that enable low-cost and experimental flight opportunities of potential benefit to the property owner despite the chance the sUAS itself may be damaged or lost. Low-altitude sUAS mission types specific to each property zone shown in Figure 1 will be presented along with use cases illustrating how high-and/or low-integrity sUAS can effectively complete each mission type.

While some sUAS missions can be strictly conducted at low altitude over a single property or at least within a single operating zone, other missions will require that the sUAS transit between properties, between zones, and in some cases climb to and later descent from higher-altitude airspace shared with manned aircraft. This chapter therefore discusses how airspace transitions can be safely handled to accommodate the suite of envisioned sUAS missions, followed by a brief conclusion.

2   sUAS TECHNOLOGIES: CAPABILITIES AND CLASSIFICATIONS

The variety of low-altitude sUAS use cases and operating environments motivate the development and operation of two distinct sUAS classes: *high integrity* and *low integrity*.

---

[2] Properties labeled as unpopulated could be occupied by business day and/or shifts of workers, for example, but such workers would be aware of company-supported sUAS activities.

Figure 3. Moderate-cost quadrotor with redundant motor assemblies (http://store.dji.com/product/spreading-wings-s1000-plus).

Figure 2. Low-cost low-integrity quadrotor (http://www.horizonhobby.com/chroma-camera-drone-with-st10-and-3-axis-gimbal-for-gopro-hero-blh8660).

Low-integrity sUAS already exist and are being sold in large quantities (millions) primarily to hobbyists who just want to explore flight or take "cool videos." Low-integrity sUAS has also enabled new small businesses to pursue a variety of airborne imaging applications. The low-integrity sUAS is composed of low-cost subsystems that function capably so long as no components fail and the operator enters appropriate commands, but the low-integrity sUAS is likely to crash when any component fails or erroneous user/pilot commands are input. A typical quadrotor or quadcopter as shown in Figure 2 is classified as low-integrity because failure of any safety-critical component such as a motor, motor (speed) controller, battery, or stability sensor will result in loss of control (LOC) followed by a crash. Loss of the communication link in low-cost quadrotors will typically also result in loss of control. In more sophisticated low-integrity quadcopters, lost link is managed by onboard software that either gracefully lands where link is lost or returns to a base waypoint based on Global Positioning System (GPS), inertial measurement unit (IMU), and pressure (altitude) sensor data. While LOC is the least desirable lost-link behavior, a preset lost-link behavior may be inadequate for densely populated areas as the slow descent to landing might occur on a crowded city street or sidewalk, and a return to a base waypoint might involve attempted flight through a building, tree, or power line. The low-integrity sUAS is therefore primarily appropriate to operate within line-of-sight of an operator, away from people and structures of value.

While commercial transport and large UAS have the triply redundant safety-certified systems characteristic of high-integrity aircraft, research and development activities are still underway to develop and mass-produce the high-integrity sUAS. Simply applying manned aircraft certification standards is not adequate for a variety of reasons (Hayhurst et al., 2006; Clothier et al., 2011). For the sUAS to remain compact and cost-effective, redundancy will necessarily be limited, and each operation cannot require oversight by multiple human operators. In fact, even one full-time human operator/pilot per vehicle may be cost-prohibitive for applications such as small package delivery, as will be discussed below. Early high-integrity sUAS will likely be a direct extension of the low-integrity sUAS, for example, multicopters with more redundancy and resilience to anomalies and failures. The octocopter configuration shown in Figure 3 offers redundancy and sufficient lifting capacity to carry redundant processors and sensors, but to date such lifting capacity has been primarily used to carry larger payloads as well as handle the specific thruster failure case. Novel high-integrity sUAS designs using a combination of lifting surfaces and thrusters to maintain flight are also under development to enable new sUAS missions with greater range and endurance; those intended for BLOS flight in populated areas will need meet high-integrity performance standards.

sUAS including the octocopter now offer lifting (thruster) redundancy but also need to demonstrate sufficient reliability and robustness in sensing, software, link, and operational models before they can be approved as high-integrity platforms. Level of integrity required ultimately is related to risk, so the definition of "high" is directly related to any risks that must be mitigated. Several methods have been proposed to compute the risk of UAS operations in populated areas ranging from general assessment (Melnyk et al., 2014; Lazatin, 2014) to mission-specific (Ford, and McEntee, 2010; Lum et al., 2011). Computed risks can be used in combination with a target level of safety (TLS) to define required sUAS reliability. Such reliability must be then verified by identification of failure modes and evaluation of its probabilities (Olson, and Atkins, 2013). Traditional hazard and safety risk assessment methods will need to evolve to overcome current limitations (Cox, 2008) particularly as the complexity of airspace increases with the introduction of UAS (Oztekin, and Luxhøj, 2008). Risk to overflown regions can also be mitigated by considering overflown area risk during flight planning (Rudnick-Cohen, Herrmann, and Azarm, 2015) or by sUAS autonomy that can update flight plans to minimize risk during anomalous or emergency situations (Olson, Ten Harmsel, and Atkins, 2014). Structural health monitoring (SHM) can also contribute to overall high integrity (Brown *et al.*, 2015). A more general application of safety management system (SMS) principles focusing on developing a regulatory baseline for sUAS is presented in (Oztekin and Wever, 2015).

Figure 4 shows a general architecture for high-integrity sUAS operation. The guidance, navigation, and control sensors and software are analogous to those embedded in low-integrity sUAS to stabilize the platform and follow a waypoint sequence. A suite of onboard and cloud-based database, processing, and sensing resources collectively maintain acceptable risk levels (e.g., achieve high-integrity levels) by fusing the most effective algorithms and data to autonomously complete preflight and real-time decision-making activities. Vision, lidar, radar, and other information-rich sensors characterize the local environment along with more traditional GPS, air data, and IMU sensors used by an autopilot. Cooperative (multivehicle) sensing can provide larger area information. UTM in cooperation with existing ATM must offer an unprecedented level of system-wide cooperative information collection and exchange through the cloud. Such data will enable each sUAS to accurately account for atmospheric, traffic, and even ground-based events of interest (e.g., population clusters) in-flight planning and in-flight replanning. Cloud-based data storage and distribution can occur in near-real-time given the high-bandwidth link availabilities expected at least in urban and many suburban areas.

Information-rich sensor data processing for situational awareness, onboard database storage, redundant link management, and airborne access to cloud and ATC / UTM datalinks will distinguish the high-integrity sUAS from the low-integrity sUAS that offers little link redundancy and requires an observer/pilot team to handle all anomaly and exception conditions. Once the suite of Figure 4 capabilities is in place, urban and suburban sUAS operations likely exposes an individual to much lower risk levels than does the drive to work.

3   URBAN (U) ZONE OPERATIONS

Urban centers are densely populated regions occupied day and night. A combination of high-rise and smaller structures cover the landscape and is connected with surface streets and freeways to facilitate personal transit. Urban populations expect to share resources and rely on public services to build and maintain the significant infrastructure required to comfortably and safely support residents and businesses.

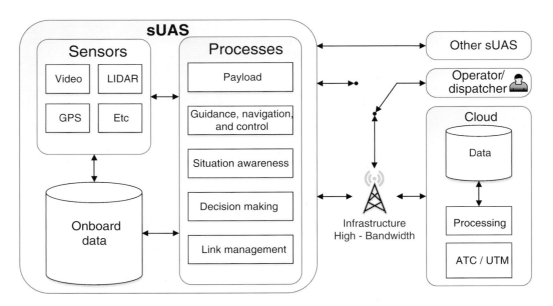

Figure 4. sUAS architecture.

Most urban areas are proximal to major airports to provide convenient air transportation. sUAS must not interfere with manned aircraft operating out of these airports requiring that low-altitude sUAS traffic avoid airport and heliport no-fly zones, for example, low-altitude approach and departure paths. sUAS traffic must also remain sufficiently low to maintain separation from manned aircraft transiting over the region. sUAS flight in urban centers might occur just above all buildings, above most buildings, or flights might require traversal through the urban canyon environment. A high-integrity sUAS will be required because urban areas present the highest risk both of sUAS collision with another aircraft and sUAS impact into a person or property of value should an unexpected landing or ditching occur. Urban canyon flight will require novel sensor fusion as GPS may not be available but the environment will be visually informative and numerous network access points may be within range (Rufa, and Atkins, 2015).

Ultimately technologies such as high-integrity sUAS systems, UTM or ATM services, and assured geofencing or containment (Hayhurst *et al.*, 2015) can allow sUAS to operate in urban areas such that they do not climb too high or enter into sensitive airport traffic areas and so they minimize risk to people and properties with a combination of map and sensor information each time a normal recovery or unexpected landing/ditching occurs. With assurance that restricted areas will not be entered, a *high-integrity* sUAS can then safely conduct a variety of missions or applications with benefit to the city and its residents.

Figure 5 shows different sUAS operation types expected to proliferate once technology and policy converge to support such operations. Public agencies might use sUAS to inspect infrastructure, conduct surveillance in support of law enforcement, and providing an "eye in the sky" to provide critical early detection and monitoring capability for events of interest, for example, building fires and car accidents, followed by rapid supply delivery via emergency sUAS despite potential surface street gridlock.

A host of civil operations are also envisioned. Contractors will provide inspection capabilities for public and civil infrastructure, while news gatherers demand the ability to rapidly collect video and images from events of public interest. Urban residents, unlike suburban and rural area residents, do not have an expectation of privacy so long as they are in public spaces, yet residents expect that sUAS will not point cameras into apartment windows, follow a pedestrian down the sidewalk, or engage in other disorderly behaviors that annoy or pose risk. Hobby, education, and training are listed in the set of envisioned urban sUAS applications because urban citizens are already enjoying sUAS in record numbers, yet it remains unclear whether cities will allow even micro-scale UAS to be flown wherever a hobbyist chooses. Instead, urban communities may benefit from creating "drone parks" that offer sUAS enthusiasts a great place to fly for hobby, (student) education, or flight training. The presence of such parks may also reduce unwanted sUAS flights in more sensitive areas since the city will offer residents a safe place to fly.

4   SUBURBAN (S) ZONE OPERATIONS

Suburban areas are characterized by single-family dwellings with modest yards and multifamily dwellings including or connected to outdoor recreation facilities and natural areas for community residents to enjoy. A network of streets link community businesses ranging from gas stations and strip malls to large package stores with housing clusters. Landowners in suburbia will typically not have appreciable acreage, so communities tend to have local ordinances to promote a high quality of life (and safety) for residents.

sUAS offers substantial benefit to suburban residents, yet sUAS proliferation is also scary for many due to concerns over loss of privacy, safety risk, and annoyance. Figure 6 shows public and civil sUAS applications. Use cases are divided into public and civil to recognize that the benefits and

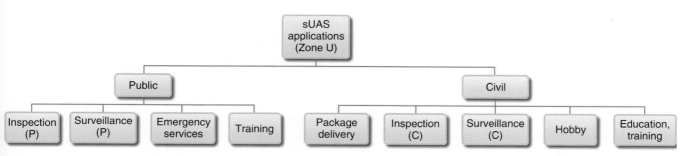

Figure 5. Urban (U) zone sUAS applications.

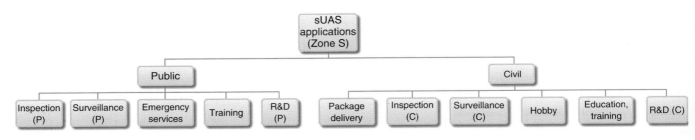

Figure 6. Suburban (S) zone sUAS applications.

risks offered by each are distinct as are the processes by which these types of operations are planned, approved, conducted, and funded.

sUAS applications operated by and of benefit to the *public* range from inspection and surveillance to emergency services, training, and potentially research and development (R&D). A public sUAS operator might routinely inspect critical infrastructure such as an aging bridge or public above-ground utility line. Such inspection flights would typically pass low over multiple suburban properties would be recognized as beneficial to the community. Surveillance is distinguished from inspection by its relation to identifying incidents such as crimes or automobile accidents and potentially following persons of interest. sUAS may improve the effectiveness of law enforcement, firefighting, and other critical public agencies at modest cost. Also envisioned are sUAS emergency responders that can rise above traffic and rapidly access regions where an airborne camera and/or supply delivery (e.g., medication) can help save lives. Finally, training and limited R&D might be possible in suburban areas provided such activities are not conducted too close to bystanders. While the suburban environment poses more risk than open rural areas, sUAS operators and planners will require *in situ* testing and training to assure effectiveness during actual time and safety-critical missions.

Suburban residents will also benefit from a variety of *civil* use cases. Operations can be divided into three basic types: payload or package delivery, aerial sensing, and experimental (model) flight for hobby, education, or R&D. Airborne sensors can be used to inspect residents' properties, for example, a roof or gutter system, and communities need to resolve where and how hobbyists and students who want to learn about sUAS must fly. Community-based organizations (CBOs) may develop specific guidelines at the local, region/state, or national level, but certainly common sense must lay the foundation for rules. Experimental sUAS platforms and/or new sUAS operators must be offered open flight areas to encourage responsible behaviors. sUAS operators must not buzz their neighbors or distract drivers even if they fly high-integrity platforms with sufficient skill to avoid a crash.

Inspection and surveillance by civil operators must be done in a manner that protects suburban resident privacy; after all one reason people move from urban areas to suburbia is to maintain a more quiet and private living environment than might be available in the city closer to work. Civil low-altitude inspection and surveillance missions must therefore be conducted with permission of the overflown property owner (public or private), and guidelines, ranging from "Know before you Fly" (http://knowbeforeyoufly.org/) to regulatory directives (http://www.faa.gov/uas/regulations_policies/), must ultimately balance demand for new operations with community preferences. Ultimately, some communities may ask sUAS to fly above a certain altitude unless over streets or with specific property owner permission. Other communities may agree on acceptable flight areas and altitudes and share these with the UTM or ATM entities that inform all civil and public sUAS entities what permissions they can acquire for local low-altitude operations or transits.

The package delivery or "courier" sUAS must fly over multiple suburban properties to carry its package to the intended recipient. While "package delivery" has received substantial attention in popular press, it may be the most difficult mission type to safely and efficiently realize. The courier sUAS first must transit the airspace between package pickup and drop-off sites, typically returning to the launch site without human handling to repair, charge, or launch. The courier sUAS might complete "pizza or sandwich delivery" missions strictly in the local suburban (S) airspace. However, more typically, the sUAS would transit between a package distribution center to the suburban recipient, requiring transit between different airspace zones and potentially higher into the airspace shared with manned aircraft. Ultimately, coordination between sUAS operator and recipient or ground-based infrastructure will be required to support the courier sUAS mission. With the infrastructure solution, controlled sUAS package drop-off stations might be constructed near neighborhood mailbox banks, for example. With the coordination solution, the sUAS might deliver the package to a site marked by the user's cell phone given appropriate guarantees (and additional sensor confirmation) that the marked delivery location is safe for final delivery.

In summary, only a *high-integrity* sUAS would typically be approved to transit freely over the streets, businesses, and [approved] private properties in a Suburban (S) environment. The high-integrity sUAS would need to consistently interact with UTM or equivalent to plan and receive approval for each flight, and would need to maintain communication with other sUAS to ensure separation is maintained. Civil and public operators must be careful to never store or disseminate data with potential to invade privacy, unless specific approval, for example, a search warrant, authorizes data collection. The authors of this chapter cannot predict how different communities will organize their efforts to enable but control sUAS in a manner that is generally agreed upon to provide the most benefit to the local residents. However, it is clear that such rational and thoughtful actions are needed to ensure community-based organizations can promote safety and acceptance for the sUAS operations both residents and businesses are beginning to demand.

5   PRIVATE RURAL (R) ZONE OPERATIONS

Privately held rural properties have a variety of uses, including agriculture, timber harvesting (logging), energy extraction and delivery (oil and gas), and leisure (hiking, hunting, boating, fishing, etc). Rural residential properties may have small acreage but are typically not positioned near neighboring homes. Agricultural and residential tracts are improved by property owners. Timberland and natural (leisure) properties are typically unimproved apart from maintaining access for people, their vehicles, and related commercial equipment.

The small UAS offers a variety of benefits to the rural property owner or their representative designated by lease or rental agreement. Figure 7 summarizes the suite of sUAS applications envisioned to offer appreciable benefit to the rural property owner or their representative. Perhaps the most publicized application is sUAS and even larger UAS support for agriculture (Zhang and Kovacs, 2012; Shahbazi, Théau, and Ménard, 2014). Frequent crop inspection will allow water and pesticides to be more precisely applied and only when required. A camera-equipped sUAS can fly over a field faster and with lower carbon footprint than a farmer can drive through all parts of a field. Even a low-integrity sUAS can capture video of fence lines and track livestock to improve operation efficiency. Larger rotorcraft UAS are now used to apply pesticides for precision farming in some areas of Asia, though regulatory restrictions have limited their proliferation worldwide.

Rural property holders and companies engaged in timber and energy extraction also can benefit from the use of low-altitude sUAS. Managed timberland is often difficult to access on the ground, especially in mountainous regions. Such areas can be more effectively managed by deploying low-altitude sUAS to provide imagery for identifying mature trees to harvest as well as diseased trees to treat. sUAS can survey access roads, distribution pipelines, and utility poles and lines more rapidly and cost-effectively than is possible from the ground.

Low-altitude sUAS operations near mountainous and forested terrain present a number of challenges. sUAS must follow the terrain through fusion of elevation map and onboard sensor data. While a multicopter might accurately follow steep slopes and navigate canyons, a fixed-wing sUAS with longer range requires maneuvering space. Winds are also important to consider, particularly given the limited ability of an sUAS with lifting surfaces to accurately follow a flight plan given strong updrafts, downdrafts, and local turbulence.

While sUAS support for leisure does not offer substantial societal benefit, a major benefit of rural property ownership is enjoyment of scenery and outdoor leisure activities. Numerous recreational sUAS uses have been proposed, ranging from the "skier following drone" to the "ice fisherman beer

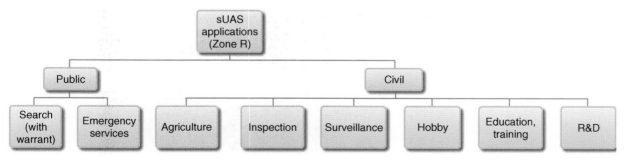

Figure 7. Private rural (R) zone sUAS applications.

delivery drone." Creative uses such as these provide motivation for people of all ages to learn about sUAS and to challenge their imaginations.

The model aircraft or hobbyist community associated with the Academy of Model Aeronautics (AMA) in the United States (http://www.modelaircraft.org/) and similar organizations worldwide have typically relied on rural property holders to provide safe open fields for member flight activities, demonstrating long term that such activities are safe and do not interfere with manned aircraft traffic. Education and research and development activities can similarly be supported on rural properties with little risk to the populous or to overflying manned aircraft so long as the sUAS operators remain at sufficiently low altitudes, typically within 400 ft of the ground.

Public sUAS operators will not often need to fly at low altitudes over private rural properties. However, there are two specific scenarios where such flights are needed. First, public entities engaged in emergency service activities such as search and rescue and firefighting may require low-altitude access. Such operations would typically be welcomed by rural property holders; coordination with low-integrity sUAS would simply be to request that the property holder's sUAS land and remain on the ground until the emergency service was completed. Law enforcement also may employ low-altitude sUAS over private property, but at least in the United States such search is likely to be considered an invasion of privacy without a court-issued search warrant. Note that public infrastructure inspection is not shown in Figure 7 because such inspections would be conducted by sUAS operating on publicly held properties such as rural roadways, bridges, and shared waterways.

As is evident from the above discussion, private rural property offers important use cases for existing (low-integrity) sUAS without the need for the additional support infrastructure and more costly sUAS equipage to enable high-integrity low-altitude sUAS operations over populated areas. This distinction motivates the need to organize low-altitude sUAS operations by property zones rather than forcing a unified set of technology and operational requirements on all low-altitude sUAS.

6   UNPOPULATED COMMERCIAL (UC) ZONE OPERATIONS

Large-acreage properties with no residents may also be used for a variety of commercial purposes and include warehouses, factories, shipyards and distribution centers, mining operations, construction sites, and more. Figure 8 describes civil and public sUAS mission types for zone UC. So long as workers are separated from sUAS flights, sUAS inspection and surveillance activities can be safely conducted at low altitudes in large-acreage UC zones with low-integrity platforms. Commercially operated mines, farms, and construction sites in particular can benefit from routine sUAS aerial inspection flights; these flights need not leave the UC zone from which they originated given the nature of their missions yet rural (R) and UC flights will be most beneficial when geofencing/containment systems provide sufficient assurance that range departure will not occur to reduce or eliminate the need for continuous operator oversight. Large-acreage commercially owned properties also offer an ideal environment for the owning entity or their representative (lessee) to conduct sUAS training/education and R&D flights. While training activities might involve hobbyists and students, the owning or leasing entity operating at low altitude over UC zones will also be able to safely conduct sUAS flights for training or validating new sUAS technologies or concepts of operation. Civil UC operations that remain in the property holder's UC zone are therefore analogous to rural (R) flight operations described above. Similarly, the need to support infrequent public agency search and emergency service flights must be acknowledged and supported by simply communicating company contact information.

As will be discussed further in Section 9.1, high-integrity sUAS may be the dominant occupant of distribution center

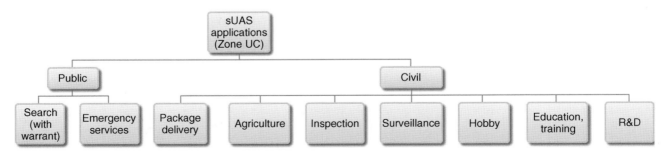

Figure 8. Unpopulated commercial (UC) zone sUAS applications.

(UC) property zones, since the sUAS must be capable of consistently operating with the highest level of integrity required over the full course of its mission. A large distribution center might internally desire to mix low-and high-integrity sUAS operations in its UC zone, but since this center would be responsible for the safety of its employees and facility, it would also be responsible for segregating or otherwise deconflicting all UC zone sUAS operations.

7 UNPOPULATED PUBLIC (UP) ZONE OPERATIONS

Large-acreage public or government-owned properties offer areas for government entities to conduct activities of national interest as well as providing a combination of improved and natural environments visitors can enjoy. National and state parks, forests, and waterways are all UP zones, as are publicly owned and controlled bodies of water. Public-owned roads and bridges in rural areas are also classified as UP zones. In some countries, the majority of land is publicly owned. For any UP zone, the agency managing the property must negotiate with the airspace regulatory agency to ensure compatible land and airspace use policies that make sense. The US Department of Interior (DOI) manages national parks among other entities; the DOI has negotiated a Memorandum of Understanding (MOU) with the FAA (http://www.doi.gov/sites/doi.opengov.ibmcloud.com/files/uploads/DOI_FAA_MOA_Class_G_09112015.pdf) to gain authority both to conduct and restrict low-altitude sUAS flights over park properties when such properties are in uncontrolled (Class G) airspace. This MOU process provides some guidance for how public agencies might negotiate authority and use of the low-altitude airspace between the ground (including water) and the high-altitude airspace manned aircraft are authorized to transit.

Figure 9 summarizes sUAS applications for UP property zones. Because both the airspace and underlying property are government-owned, civil operations are typically allowed only with authority from the airspace and/or property controlling agency. Numerous public use cases are envisioned, on the other hand. Public agencies remain cognizant of the need for search and rescue capabilities for stray hikers whether these people are authorized or not. Emergency services ranging from firefighting to search and rescue support are critical to support. Public sUAS entities may be interested in a variety of land uses, ranging from improved land and natural resource management (Shahbazi, Théau, and Ménard, 2014) to infrastructure inspection and surveillance. For UP property zones, surveillance missions may not necessarily be directed at identifying and monitoring people, but instead may monitor wildlife or look for wildfires. sUAS may be regularly deployed for inspection missions such as monitoring aqueduct water-level monitoring, inspecting infrastructure (e.g., dam/levee) integrity, and rapidly identifying anomalous events such as rock or landslides. Though not discussed in detail here, large military training installations would typically be classified under the UP zone. Special support for airspace segregation (restriction) agency-specific governance of defense agency flight operations has been long recognized as essential to their success.

UP zones can offer a variety of education, training, and R&D activities associated with sUAS. The sUAS themselves can be thoroughly tested in a controlled low-altitude (UP) environment where they pose little hazard to people or property. New sensors and algorithms can be validated, and new concepts of operation can be validated. Scientists and engineers can work with public authorities to monitor the (UP) property atmosphere, water, biological entities, and soil/land characteristics. Such sUAS missions can provide safe and essential support for advancing our knowledge of the sUAS mission while offering unprecedented access for airborne data collection. Of course the engineering, operational validation, and science sUAS missions discussed here could also be performed in rural (R) and unpopulated commercial (UC) zones. R&D and science missions are emphasized because large-scale technology development and

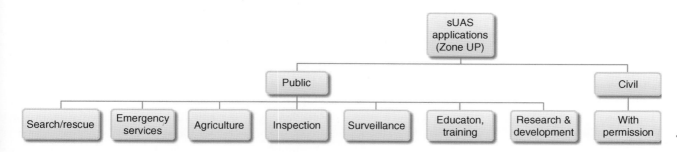

Figure 9. Unpopulated public property (UP) zone sUAS applications

fundamental science and engineering research is often supported by government agencies.

8   AIRPORT AREA OPERATIONS

As discussed in Section 1, integration of sUAS in airport terminal areas may introduce risk of collision with manned aircraft. This higher collision risk has resulted in outright bans or severe restrictions on number and type of sUAS operations particularly near busy airports. High-integrity sUAS offer the most potential for sharing the airspace. Current policies on restricting flights without specific permission in the vicinity of airports will continue to be applicable in the near-term. Datalink UTM/ATM plus geofencing/assured containment technologies (Stevens, Coloe, and Atkins, 2015; Hayhurst *et al.*, 2015) could together provide sufficient guarantees that each sUAS is aware of specific flight authorizations and restrictions. Restricted areas are currently circular (cylindrical in three dimensions), defined by a fixed radius from a specific airport. In the future, a less constraining rectangular no-fly region may be aligned with active runway approach and departure paths to only restrict flight as necessary.

There are a large number of infrequently used airports particularly in the United States. Those airports may offer a large area for sUAS operations, and sUAS may offer an important economic opportunity for small airports struggling to establish a viable business model. Small airports electing to support sUAS could be used similarly to UC zones with airport manager permission considering that manned aircraft traffic is restricted during such operations. Operational rules that allow combined use of manned and unmanned aircraft are desirable to encourage facilities to support both operation types. Without such flexibility, a number of small public airports might close to become "drone ports" with UC or (R) zone designations if the drone port eventually becomes more economically attractive to the airport owner and/or manager.

9   AIRSPACE ZONE TRANSIT: CROSSING BOUNDARIES

sUAS operations will not be necessarily restricted to one zone. Two types of transitions are possible. The first is simple low-altitude transition between different low-altitude zones, for example, a sUAS flight launched from a farm in an R zone delivering produce to a house in zone (S). The second consists of sUAS transitions from low-altitude airspace to higher altitude airspace layers then returning back to a low-altitude zone. sUAS operation in higher layers may be necessary due to desired mission specifications (e.g., high-altitude photo required) or due to low-altitude no-fly zones. Since low-altitude airspace has close interaction with the overflown population, it is foreseeable that some low-altitude flights may be restricted over certain areas for certain sUAS. High-altitude flight will be independent of landowner restrictions as per current manned aircraft operations, providing a viable means of passage for the sUAS operator.

9.1   Transit between low-altitude zones

Transition between low-altitude zones will require the sUAS to be authorized to operating in both zones. If any part of the flight requires *high-integrity* sUAS, the mission must be executed with high-integrity equipage. Each zone's local regulations must be followed when the sUAS is in that zone, and transit between neighboring zones must be coordinated.

In sparsely populated flight zones (R), (UC), or (UP), sUAS transit approval must involve property owners or appropriate public agency authorization, for example, a search warrant. Ultimately, transit between zones R and UC can be accomplished as if they were combined into one larger zone assuming the operator has approval to operate similarly in both zones. Different operators may require additional coordination for sUAS to cross the boundaries of such different ownership and population density zones. Such coordination could range from simply communicating with a responsible park ranger to entering each flight request into a cloud-based information system linked with UTM.

More complex coordination such as described above is envisioned for transit to and from the more densely populated zones S and U. The high volume of diverse sUAS operations ultimately expected in these areas may also require coordination with ground-based authorities, particularly when the sUAS is engaged in a surveillance, emergency, or delivery activity requiring direct interaction with people on the ground.

9.2   Transit between low-altitude zones and higher layers

While sUAS will primarily occupy low-altitude airspace, transition to higher altitude airspace shared with manned aircraft may also be beneficial. The focus of this chapter is low-altitude operations, so this discussion is restricted to two models: an sUAS coordinating a climb above a low-altitude region for transit, and temporary "extension" of a lower layer airspace zone to a higher altitude when necessary.

Suppose a sUAS needs to climb over a low-level altitude layer to transit. Excessive traffic or no-fly directives may be reasons to justify this climb. Transition complexity in terms of clearance, coordination with other traffic, and sUAS system and operation requirements are dependent on class and traffic density in the high-altitude airspace to be traversed. During high-altitude transit, a trusted detect-and-avoid capability would be required. The current USA National Airspace System (NAS) implements a layered approach to assuring the target TLS for collision avoidance (Lacher, Maroney, and Zeitlin, 2007). See-and-avoid (SAA) is one layer, but coordination using ground-based infrastructure such as UTM and connection with ATM can also provide mitigation of collision risk. Definition of a "well clear" concept for sUAS is one of several concepts adapted from manned aircraft to allow sUAS NAS integration (Weibel, Edwards, and Fernandes, 2011). sUAS can be deployed from a large range of locations, resulting in traffic entering controlled airspace from uncontrolled airspace. This poses a challenge to the current ATM (Hayhurst et al., 2006) system. Rural (R) and unpopulated commercial (UC) zones, for example, may need very simple or no coordination. More densely populated zones, however, will require coordination through ground-based infrastructures such as UTM.

sUAS operations also may be contained within one lateral property area or zone but may require temporary excursion to altitudes higher than 500 ft, for example, to capture a wide area photo. Such sUAS operations can be conducted as is currently done for acrobatic or parachute operations: temporary airspace segregation. Airspace reservation requests can be handled through UTM and in turn ATM, yet such requests may not be approved unless the sUAS is operating in an area with low-airborne traffic density.

10   CONCLUSIONS

Small unmanned aircraft systems (sUAS) will primarily operate in low-altitude airspace to support a variety of airborne sensing and small package delivery missions. Manned aircraft seldom uses low-altitude airspace except near airports and heliports, so the primary safety risks are to people and property on the ground. Low-altitude sUAS flights increase the potential to interact with overflown areas inspiring exciting new applications while also presenting negative impacts to the overflown area, particularly in populous areas, due to safety, privacy, and annoyance concerns.

Because of the proximity of low-altitude sUAS to the underlying property, this chapter first characterized overflown property types or zones. Six overflight "zones" were defined: urban, suburban, private rural, unpopulated commercial, and unpopulated public and airport areas. sUAS capabilities were then outlined and classified with respect to level of operational integrity. High-integrity UAS are required for overflight of populated areas and near high-traffic airports to manage risk, while low-integrity low-cost sUAS can be safely deployed in rural and other unpopulated regions away from airports. The following conclusions are drawn from this chapter:

- Distinction of public and civil applications is motivated by current regulatory processes as well as underlying mission motivators. Although mission types overlap, operational paradigms vary greatly. For private rural properties where low sUAS flights primarily benefit property holders, a variety of economic models motivate sUAS flight ranging from agriculture to leisure. Publicly held properties typically require (civilian) visitors to obtain specific permission to fly. Public and civil sUAS will share urban and suburban airspace, and public operations such as emergency services will receive highest priority access as is the case on the roads today.
- "Low-integrity" sUAS exist today and are flown routinely. However, sUAS have not yet achieved the performance levels required to qualify as "high-integrity" platforms. "High-integrity" sUAS is not needed for most operations in sparsely populated regions, but rural and unpopulated zones offer substantial opportunities to thoroughly develop and validate high-integrity platforms and operational paradigms in preparation for the exciting sUAS applications envisioned for urban and suburban communities.
- sUAS flights will not necessarily be limited to one zone only. Transit between different zones will require the sUAS to comply with the most stringent requirements of all overflown zones. sUAS may also climb to higher altitude layers shared with manned aircraft. Entry into higher altitude airspace will require additional equipage and operational oversight. Such operation may be beneficial, for example, if low-altitude overflight of certain areas is not possible, or during long transit periods over uneven terrain.

Substantial work remains to further develop and validate high-integrity UAS along with efficient and safe concepts of operation for the spectrum of envisioned missions. UAS traffic management and system-wide information management capabilities also must mature to support sUAS missions. Careful consideration of risks and performance requirements based on overflight zone as well as mission type will ensure region and mission-appropriate constraints are imposed. Such considerations will ultimately enable the proliferation of safe sUAS operations over rural and unpopulated areas in the near-term. Rural flight testing can also prove high-integrity UAS can be trusted to safely and reliably operator. In time, sUAS can migrate into urban and suburban areas in a manner that respects privacy and property rights while offering value to each community.

NOMENCLATURE

AGL	above ground level
AMA	Academy of Model Aircraft
ATM	air traffic management
BLOS	beyond line-of-sight
CBO	community-based organization
DOI	US Department of Interior
FAA	Federal Aviation Administration
GPS	Global Positioning System
INS	inertial measurement unit
LOC	loss of control
LOS	line of sight
MOU	Memorandum of Understanding
NAS	national airspace system
(R)	private rural zone
R&D	research and development
(S)	suburban zone
SAA	see-and-avoid
SHM	structural health management
SMS	safety management system
sUAS	small unmanned aircraft system
SWIM	System-Wide Information Management
TLS	target-level of safety
(U)	urban zone
(UC)	unpopulated commercial zone
(UP)	unpopulated public zone
UAS	unmanned aircraft system
UTM	UAS traffic management

ACKNOWLEDGMENT

The authors would like to thank CAPES Brazil for supporting the second author's graduate studies at the University of Michigan.

REFERENCES

Brown, J.M., Coffey, III, J.A., Harvey, D., and Thayer, J.M. (2015) Characterization and prognosis of multirotor failures, in *Structural Health Monitoring and Damage Detection*, vol. **7**, Springer, pp. 157–173.

Clothier, R.A., Palmer, J.L., Walker, R.A., and Fulton, N.L. (2011) Definition of an airworthiness certification framework for civil unmanned aircraft systems. *Saf. sci.*, **49**(6), pp. 871–885.

Cox, A.L.T. (2008) What's wrong with risk matrices? *Risk Anal.*, **28**(2), pp. 497–512.

Efroymson, R.A., Suter, I., Glenn, W., Rose, W.H., and Nemeth, S. (2001) Ecological risk assessment framework for low-altitude aircraft overflights: I. Planning the analysis and estimating exposure, *Risk Anal.*, **21**(2), pp. 251–262.

Ford, A. and McEntee, K. (2010) Assessment of the risk to ground population due to an unmanned aircraft in-flight failure. *Proceedings of the 10th AIAA Aviation Technology, Integration, and Operations (ATIO) Conference, AIAA, Fort Worth, TX*, AIAA 2010–9056.

Hayhurst, K.J., Maddalon, J.M., Miner, P.S., DeWalt, M.P., and McCormick, G.F. (2006) Unmanned aircraft hazards and their implications for regulation. *25th Digital Avionics Systems Conference*, IEEE, 2006, pp. 1–12.

Hayhurst, K.J., Maddalon, J.M., Neogi, N.A., and Verstynen, H.A. (2015) A case study for assured containment. *2015 International Conference on Unmanned Aircraft Systems (ICUAS)*, IEEE 260–269.

Kopardekar, P.H. (2016) Unmanned Aerial System (UAS) Traffic Management (UTM): Enabling Low-Altitude Airspace and UAS Operations, Tech. rep., NASA Ames Research Center, NASA/TM-2014-218299. Available at http://ntrs.nasa.gov/search.jsp?R=20140013436 (February 24, 2016).

Lacher, A.R., Maroney, D.R., and Zeitlin, A.D. (2007) Unmanned aircraft collision avoidance -technology assessment and evaluation methods. *The 7th Air Traffic Managment Research & Development Seminar*, Barcelona, Spain, 2007.

Lazatin, J. (2014) A method for risk estimation analysis for unmanned aerial system operation over populated areas. *14th AIAA Aviation Technology, Integration, and Operations Conference*, AIAA 2014–2284.

Lum, C.W., Gauksheim, K., Kosel, T., and McGeer, T. (2011) Assessing and estimating risk of operating unmanned aerial systems in populated areas. *11th AIAA Aviation Technology, Integration, and Operations (ATIO) Conference*, AIAA 2011–6918.

Melnyk, R., Schrage, D., Volovoi, V., and Jimenez, H. (2014) A third-party casualty risk model for unmanned aircraft system operations. *Reliab. Eng. Sys. Saf.*, **124**, pp. 105–116.

Olson, I.J. and Atkins, E.M. (2013) Qualitative failure analysis for a small quadrotor unmanned aircraft system. *Proceedings of the AIAA Guidance, Navigation, and Control (GNC) Conference*, AIAA 2013–4761.

Olson, I.J., Ten Harmsel, A.J., and Atkins, E.M. (2014) Safe landing planning for an energy-constrained multicopter, *2014 International Conference on Unmanned Aircraft Systems (ICUAS)*, IEEE 1225–1235.

Oztekin, A. and Luxhøj, J.T. (2008) Hazard, safety risk, and uncertainty modeling of the integration of unmanned aircraft systems into the national airspace. *26th Congress of International Council of the Aeronautical Sciences, Anchorage, Alaska*, vol. 1419.

Oztekin, A. and Wever, R. (2015) Development of a regulatory safety baseline for UAS sense and avoid, *Handbook of Unmanned Aerial Vehicles*, Springer, pp. 1817–1839.

Rudnick-Cohen, E., Herrmann, J.W., and Azarm, S. (2015) Risk-based path planning optimization methods for UAVs over inhabited areas. *ASME 2015 International Design Engineering Technical Conferences and Computers and Information in Engineering Conference*, American Society of Mechanical Engineers, doi: 10.1115/DETC 2015-47407.

Rufa, J.R. and Atkins, E.M. (2015) Unmanned aircraft system navigation in the urban environment: a systems analysis. *J. Aerosp. Inf. Syst.*, **12**(12), pp. 710–727.

Shahbazi, M., Théau, J., and Ménard, P. (2014) Recent applications of unmanned aerial imagery in natural resource management. *GISci. Remote Sens.*, **51**(4), pp. 339–365.

Stevens, M., Coloe, B., and Atkins, E. (2015) Platform-independent geofencing for low altitude UAS operations. *Proceedings of the 15th Aviation Technology Integration, and Operations Conference*, Dallas, TX, USA, pp. 22–26.

Weibel, R.E., Edwards, M.W., and Fernandes, C.S. (2011) Establishing a risk-based separation standard for unmanned aircraft self separation, *AIAA Aviation Technology, Integration, and Operations (ATIO) Conference*, AIAA 2011–6921.

Zhang, C. and Kovacs, J.M. (2012) The application of small unmanned aerial systems for precision agriculture: a review. *Precis. Agric.*, **13**(6), pp. 693–712.

Chapter 37

UAS in the Terminal Area: Challenges and Opportunities

Alessandro Gardi,[1] Subramanian Ramasamy,[1] Trevor Kistan,[1,2] and Roberto Sabatini[1]

[1]*RMIT University, Melbourne, Australia*
[2]*THALES Australia, Melbourne, Australia*

1 INTRODUCTION

Unmanned Aircraft Systems (UAS) are at the forefront of current Research, Development, Test, and Evaluation (RDT&E) efforts in both the civil and the military aviation communities. In addition to their innate suitability for operating in conditions that are hazardous and hostile to human crews such as dull, dangerous and dirty environments, UAS increasingly show advantages in terms of economic, environmental and social factors for a growing number of applications. Nevertheless, the increasing adoption of UAS poses technological and regulatory challenges, which have hindered a quickly growing number of UAS operators, national regulators and international aviation organizations. Since the

Unmanned Aircraft Systems. Edited by Ella Atkins, Aníbal Ollero,
Antonios Tsourdos, Richard Blockley and Wei Shyy.
© 2016 John Wiley & Sons, Ltd. ISBN: 978-1-118-86645-0.

early UAS adoption stages, military operators have overcome most issues successfully by opportune implementations of airspace segregation. In particular, combined operations of military UAS and manned aerial and terrestrial vehicles have been introduced to minimize the reciprocal threats and introduce adequate safety margins. A similar approach has been pursued in the civil jurisdiction whenever access to commercial and recreational UAS operators has been granted to limited portions of the airspace (where conventional manned aircraft operations are typically denied), such as in proximity of the ground. In these cases, the risk of Mid-Air Collisions (MAC) between manned aircraft and UAS is effectively removed, thanks to the reciprocal segregation in place. Following this approach, a progressive integration of UAS by means of a gradual exploitation of the conventional civil airspace structure and regulations has been considered. In particular, the safety of manned aircraft would be substantially unaffected if UAS were to be granted access to regions where conventional aircraft operations are very sparse. From this perspective, an integration of UAS starting from less restrictive classes of airspace would be favored. Civil airspace is organized in up to seven classes from "A" to "G" in terms of decreasing level of Air Traffic Services (ATS) offered (ICAO, 2001). Correspondingly, decreasing requirements in terms of Communication, Navigation, Surveillance (CNS) and Avionics (CNS+A) equipment, and in terms of compliance to Air Traffic Management (ATM) clearances are defined. The less demanding requirements, especially in classes "F" and "G," are certainly valued by UAS operators, as their reliability is still far from the levels offered by manned aircraft (Clough, 2005). Nevertheless, separation in these airspace regions is the responsibility of the pilots and based on visual detection and deliberate execution of avoidance manoeuvres according to the rules of flight (ICAO, 1990). These aspects involve substantially higher technological challenges for UAS developers, as separation relies uniquely upon onboard sense-and-avoid (SAA) systems and decision-making processes, hence higher levels of onboard autonomy are required to mitigate the risks arising from the possible failures to the Command and Control (C2) loops involving the ground pilot.

2   TERMINAL-AREA OPERATIONS

By definition, *control areas* are regions of airspace delineated so as to encompass sufficient airspace to contain the flight paths of instrument flight rules (IFR) flights or portions thereof for which it is desired to provide the applicable parts of the ATS (ICAO, 2001). Control areas also include ATS routes (airways) and Terminal Control Areas (TCA), frequently known as Terminal Manoeuvring Areas (TMA). Control zones complement control areas in their purpose and are specifically conceived to contain arrival and departure IFR flight paths in close proximity of the ground. For minor and isolated airports with limited IFR traffic, an appropriate combination of control zones and ATS routes is usually sufficient to contain IFR traffic safely. This is due to the fact that tactical deconfliction by ATM operators is almost never required and occasional peaks of traffic can be successfully managed by exploiting multiple approach and departure procedures and en-route holdings. Larger airports or combinations of multiple minor airports in close proximity to each other require more frequent tactical ATM interventions in terms of path stretching and separation measures. Whenever additional allowance for these tactical measures is necessary, TCA/TMA are established. In order to maximize safety of IFR traffic, TCA/TMA are most commonly granted the highest levels of ATS, corresponding to airspace classes "A", "B," or, less frequently, "C." Due to their distinctive nature, TCA/TMA have become the paradigm of dense air traffic scenarios, where arriving, departing, and overflying aircraft intersect leading to frequent conflict situations. The highest levels of ATS assigned to TCA/TMA implicates that most of the responsibilities for conflict detection, separation maintenance, sequencing and spacing lies on the ATM side. This is effectively the opposite of recreational airspace assigned classes "G" and "F", leading to important consequences for UAS developers.

In the United States, the National Aeronautics and Space Administration (NASA) is leading research and development efforts toward the introduction of systems specifically designed to manage UAS – the so-called UAS Traffic Management (UTM) system (Kopardekar, 2015; NASA, 2015). As part of the NASA UTM program, commercial demands and proposals by major industrial operators are being reviewed in order to set the requirements and identify the optimal operational configurations (Amazon Inc., 2015a,b; Google Inc., 2015; Kopardekar, 2015). NASA's UTM system is only slated for full implementation in controlled, nonsegregated airspace in 2030 (Stevenson, 2015). The need for interim solutions can be gauged from a recent study that identified 327 potentially hazardous encounters between UAS and manned aircraft in the United States during the period from December 2013 to September 2015, 90 of which involved commercial jets (Gettinger, 2015). Prior to the introduction of specifically designed UTM systems, relatively simple modifications to current generation ATM systems would allow them to monitor UAS operations and manage potential conflicts between UAS and with conventionally piloted aircraft in nonsegregated controlled airspace. Lockheed Martin has already demonstrated the concept using the automated conflict probe of its En Route Automation Modernization (ERAM) ATM system software (Carey, 2015).

3   KEY CHALLENGES AND OPPORTUNITIES

Due to their intrinsic nature, TCA/TMA are typically characterized by significantly higher traffic densities and levels of ATS compared to other airspace regions. These specificities dictate requirements that are very different from the ones associated with operations in other classes of airspace, where the responsibility for detecting conflicts and maintaining separation from other traffic is on the aircraft side. In particular, as deconfliction duties within the TCA/TMA are within the ATM jurisdiction, aircraft are expected to execute ATM directives after very limited review. In particular, under high levels of ATS, the analysis of the ATM clearance by pilots is restricted to:

1. feasibility in terms of safe operational envelope;
2. suitability in terms of the short/medium-term objectives of the mission.

As an example, if an arising conflict prompts an ATM operator to vector a commercial transport aircraft direct to a specific waypoint and overfly it at a specific altitude and time (4D path constraint), the flight crew will assess with the assistance of onboard Decision Support Systems (DSS) whether (1) the manoeuvring required to attain the 4D constraint is within the operational envelope of the aircraft and (2) if the altitude and airspeed constraints are compatible with the current phase of flight (i.e. if a descent and no significant airspeed increases are involved in the arrival phase or, vice-versa, if a climb and no significant airspeed reductions are involved in a departure phase). Pilots will subsequently execute the ATM directives and monitor aircraft systems for any anomalies arising. In such an operational scenario, the algorithmic complexity necessary to grant UAS the required autonomy is very limited, as UAS are expected to execute ATM directives after very limited analysis. Conversely, in recreational airspace where deconfliction is responsibility of aircraft, substantially higher levels of expert processing are required at all times to grant the necessary autonomy (Clough, 2005).

Autonomy can be simply defined as the capability of making decisions without assistance by human operators. Autonomy in UAS is essential to both increase the utility of the UAS and the safety of its operation as it allows the unmanned aircraft to avoid hazardous situations if failures occur in the C2 loop (Clough, 2002). Levels of autonomy were initially defined by Sheridan (Sheridan, 2003). Sheridan's 10-level scale of autonomy is based on the decision maker (human or system) and on how the decisions are executed. Different implementations of Sheridan's scale were considered. For instance, in the ATM context UAS can be configured for autonomous, overridable, aided, and deliberate operations as listed in Table 1. These levels are based on information acquisition, information analysis, decision and action selection, and action implementation functions.

The expert processing required to grant UAS the appropriate levels of autonomy can be attained either by knowledge-based algorithms or by more advanced forms of machine intelligence. In both cases, the algorithm size and complexity can grow considerably based on the number of factors and variables to be taken into account and on the inclusion of *what–if* scenario analysis processes. With respect to machine intelligence, the eventual integration of machine learning processes may also introduce a behavior that is a priori nondeterministic, which poses further concerns in view of certification. In terms of trusted autonomy for unrestricted UAS operations in all classes of airspace, some of the most pressing challenges that UAS developers are facing are summarized by Clough (Clough, 2002):

- Significant complex flight critical software needs to be developed, tested, certified and executed onboard without disruptive impacts on UAS development costs.
- Certification of systems with adaptive behavior must be possible.
- The reliance of flight critical processes on data links susceptible to security and integrity issues.

In particular, researchers estimated that software size necessary to successfully integrate the autonomy required for unrestricted operations is well in excess of one million lines of code, which is an order of magnitude greater than manned aircraft equivalents (Clough, 2002). Effective and certified failure management is also required to attain the desired level of safety for unrestricted operations.

Notwithstanding, it shall be noted that the autonomous decision making in UAS is not limited to safety-critical purposes, and usually accounts for other features, such as performance optimization. Therefore, the costs of certification could be managed by reducing the number of subsystems/lines of code subject to certification requirements. Licensing already-certified software modules could also help manage these costs. However, unlike software safety, in terms of system safety requirements, new systems deploying *certified algorithms/software* still need to be certified themselves in very specific contexts, as the software may be running on different hardware, be configured differently, be employed in different operational contexts etc.

Table 1. Classification of autonomy levels.

Automation mode	Operator control	Sheridan scale	Outline of the human–systems interaction
Autonomous	Executive	9–10	The system performs all aspects of decision-making and informs the operator after execution, if required, per preplanned criteria or on operator's request.
Overridable	Supervisory	6–8	The system generates decision alternatives and a preferred option for execution and informs the operator in time for override intervention.
Aided	Consent-based	3–5	The system generates decision alternatives and recommends one to be carried out – but only after operator's approval. The operator may select alternative options.
Deliberate	Manual	1–2	The system executes commands initiated by the operator (the system may provide and/or recommend decision alternatives to the operator).

(Adapted from (Sullivan *et al.*, 2004)).

In summary, the high level of ATS in TMA can be seen as an opportunity as it relieves UAS developers from the requirement of introducing very high levels of autonomy, with all the associated computational complexity. Notwithstanding, the limited jurisdiction of aircraft in airspace classes characterized by high levels of ATS does not remove the requirements for UAS to equip a certified SAA system, as visual detection and avoidance is also prescribed as a last resort in these classes of airspace. Another opportunity offered by the integration of UAS in TMA and other airspace regions with high levels of ATS is associated with potential reductions in the longitudinal separation standards, which would reduce the demand insisting on route and airspace capacities. The minimum longitudinal separation distances are in fact correlated to the wake turbulence category of the preceding aircraft and on the ratio between this and the wake turbulence category of the following aircraft. Unmanned aircraft most commonly feature lower take-off weights (TOW) than conventional manned aircraft, as they are usually required to outperform conventional manned aircraft in terms of lower size, weight, and power (SWaP) in order to offer economic savings that promote their adoption. Analogous benefits to the route/airspace capacities are being pursued within the framework of Wake Vortex Re-Categorization (RECAT) initiatives in Europe and globally. Since longitudinal separation dictates the spacing between traffic leaving a *feeder fix*, any improvement in that respect would result in traffic demand reductions within TMA. One possible challenge for operation within TMA is posed by groups or swarms of UAS, which would likely saturate ATM capacity if considered as separate aircraft. In this perspective, regulations are already accommodating formation flying of manned aircraft (ICAO, 1990), and formations of UAS can be analogously considered as a single aircraft from the ATM perspective.

Comprehensively, in terms of granting the required levels of operational safety in airspace with high levels of ATS, the emphasis is clearly on CNS+A equipment that can allow the UAS to reliably comply with ATM directives and be constantly monitored in their execution, rather than integrating high levels of autonomy with the associated computational complexity. In the long term, unmanned aircraft will not require onboard speech recognition as ATM instructions will consist exclusively of digitally codified 4DT directives. In the medium term, unmanned aircraft may however implement simple speech synthesis in order to respond to ATM instructions, as the maturity attained by this technology is adequate. For instance, successful experimental flight trials have already been reported with speech recognition on the ATM side converting spoken controller instructions into data link-mediated directives (Bouwmeester *et al.* 2015).

4 CNS+A CONCEPT OF OPERATIONS

In line with the evolutions initially envisaged by the Future Air Navigation System (FANS) special committee of the ICAO back in the 1980s (Bradbury, 1991), the innovative operational concepts being developed by the research community prescribe the evolution of ATM into a highly automated, integrated, and collaborative system, allowing a more flexible and efficient management of the airspace resources through higher levels of information sharing, automation and more accurate navigation systems. The key CNS+A concepts and capability advances include (Sabatini *et al.*, 2015b)

- four-dimensional (4D) trajectory-based operations (TBO);
- enhanced line-of-sight (LoS) and beyond line-of-sight (BLoS) aeronautical communications involving a substantial exploitation of data links;
- Ground, Avionics, and Satellite-Based Augmentation Systems (GBAS/ABAS/SBAS) enhancing Global Navigation Satellite System (GNSS) performance in all flight phases;
- enhanced ground-based and satellite-based surveillance, including Automatic Dependent Surveillance (ADS) and self-separation;
- System Wide Information Management (SWIM);
- Collaborative Decision-Making (CDM) to allow all stakeholders to participate in the enhancement of system performance by utilizing more accurate information from airborne systems;
- Dynamic Airspace Management (DAM) for an optimized exploitation of airspace resources.
- Improved Human–Machine Interface and Interaction (HMI[2]), interoperability, and higher levels of automation.
- Role shifting of ground-based ATM command and control-oriented units to a highly automated collaborative decision-making system in an interoperable environment, based on user preferred 4D trajectories.

In order to operationally integrate these enhanced concepts and capabilities, new airborne and ground-based CNS+A DSS are required, supporting human operators in processing the higher amounts of information, performing optimal decisions, and keeping a continuously updated understanding of the situation. A CNS+A enabled concept of operation was introduced in Sabatini *et al.* (2015). In the scope of this research activity, the key enabling CNS+A DSS include a Next-Generation Flight Management System (NG-FMS) onboard manned and unmanned aircraft and a next-generation ATM DSS on the ground. The CNS+A concept of operation integrating conventional aircraft and UAS in nonsegregated airspace is illustrated in Figure 1, with a special focus on the TMA, the control area (CTA), and the Aerodrome Traffic Zone (ATZ) (Gardi, Ramasamy, and Sabatini, 2015; Gardi *et al.*, 2015; Ramasamy *et al.*, 2015). To meet the stringent operational requirements set for strategic and tactical online ATM transactions, the automatic 4DT planning, negotiation, and validation functionalities shall support rerouting and rescheduling in real time. The 4DT intent data include 4D waypoints (latitude, longitude, altitude, and time), leg and turn information, as well as performance criteria and constraints. Real-time air–ground transactions ensure the validated 4DT are updated frequently and forcibly when any change in operational conditions emerges.

The UAS NG-FMS generates 4DT intents consisting of a number of flyable optimal trajectories in order of priority that are subsequently transmitted to the ground-based 4DT Planning, Negotiation, and Validation (4-PNV) system via Next-Generation Aeronautical Data Links (NG-ADL). The provision of multiple trajectory options decreases the transaction duration and reduces the reliance on remotely calculated optimal trajectories from the 4-PNV system. Once the globally optimal conflict-free 4DTs have been identified, the

4-PNV system instructs the traffic to fly the validated 4DT and the NG-FMS send a confirmation to the ground. When feasible 4DTs cannot be identified among the NG-FMS intents, the 4-PNV system acknowledges the NG-FMS and, based on the shared aircraft state and the collaboratively agreed optimality criteria, calculates a new family of optimal 4DTs and uplinks them to the respective aircraft. The onboard NG-FMS then identifies the preferred 4DT and sends a confirmation to the 4-PNV system. A subcase of the latter is when a new negotiation loop is initiated by the 4-PNV system (e.g., due to conflicting air traffic and/or direct inputs from the ANSP/AOC).

5   CNS+A TECHNOLOGIES FOR UAS

The introduction of a new regulatory framework and CNS+A DSS for TBO involves several important research activities including the development of 4DT algorithms for conflict-detection, planning, negotiation, and validation that permit unrestricted access of UAS to civil airspace. The first recommendation paper in the public domain that addressed the operational and certification issues for civil UAS was JAA CNS/ATM pp026 issued by the JAA CNS/ATM Steering

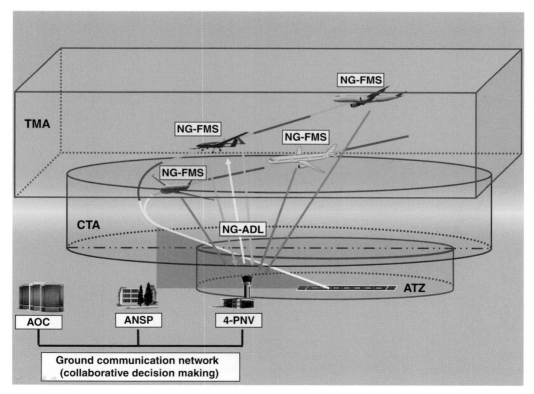

Figure 1. CNS+A concept of operations in the terminal area.

Group. The successive detailed report provided technical and strategic recommendations to the UAS research community (JAA/EUROCONTROL, 2005). In line with the Steering Groups, EUROCAE Working Group (WG-73) was established to address (Amato, n.d.)

- UAS operations and SAA functions,
- command, control, communication, spectrum, and security issues, and
- airworthiness and continued airworthiness.

The architectures, interfaces, communication protocols, data elements, and message formats for operation of UAS are defined in the NATO Standardisation Agreement (STA-NAG) 4586 (NATO, 2005). In terms of general applicability, the Joint Architecture for Unmanned Systems (JAUS) provides a better perspective than STAGNAG 4586. The JAUS standard Domain Model (DM) and Reference Architecture (RA) provide mechanisms for UAS interoperability, including integration into the airspace, architecture framework, message formats, and a set of standard messages (FAA, 2013). UAS support is one of the key performance improvement areas identified by the ICAO for the Aviation Systems Block Upgrade (ASBU) (ICAO, 2013). In the 4D-TBO context, the following are included:

- Initial integration of UAS into nonsegregated airspace: implementation of basic procedures and functions including SAA for operating UAS.
- UAS integration in traffic: implementation of defined procedures addressing lost link as well as enhanced SAA functions.
- UAS transport management: implementation of UAS operations on the airport surface and in nonsegregated airspace similar to manned aircraft.

The integration of UAS into the nonsegregated airspace requires recommendations from standardisation committees (RTCA SC-203, ASTM F 38, EUROCAE WG 73, and others), with the aim of developing Minimum Aviation System Performance Standards (MASPS). From the CNS+A perspective, the key elements are given in the following:

- BLoS communication systems.
- High-integrity airborne and ground-based UAS navigation systems and integrated failsafe avionics architectures.
- The adoption of fused cooperative/noncooperative surveillance systems incorporating collision avoidance and collaborative conflict resolution capabilities in a network centric operational scenario.
- Integration of SAA capabilities in the NG-FMS.

- The interactions between Guidance, Navigation and Control (GNC) and Track, Decision and Avoidance (TDA) loops.

Other issues that are addressed for a safe integration of the UAS in the airspace include the definition of automation functions and standards for Human-in-the-Loop (HITL) interactions, operational contingency procedures and certification processes. Additionally, the air safety nets that can be used as a last resort necessity have to be clearly defined for all UAS types to address the emergency scenarios rising in a CNS+A context. Cooperative and noncooperative SAA performance-based requirements are currently developed and have to be certified to support the UAS operational improvements (ICAO, 2013). The initial integration of UAS requires capabilities including ground-based SAA systems and the adoption of a combination of policies, procedures, and technologies intended to facilitate safe airspace access. The SAA technology will be integrated in the flight management system of the UAS to meet collision and hazard avoidance responsibility and to provide situational awareness. UAS integration in traffic requires the development of an airborne SAA system, which must be able to fulfill the requirements for MAC avoidance in nonsegregated airspace for both cooperative and noncooperative targets (Ramasamy and Sabatini, 2015). In particular, this technology will cope with air and ground obstacles of various characteristics (natural and man-made) including long and thin structures (e.g. electrical cables, poles) and aerial obstacles such as other UAS and manned aircraft (Ramasamy *et al.*, 2015). Significant impacts are also expected in the areas of SAA airworthiness and design standard evolutions. The CNS+A systems and sub-systems required for UAS are illustrated in Figure 2.

6 UAS FLIGHT MANAGEMENT SYSTEMS

The avionics equipment primarily responsible for providing automated guidance and navigation services onboard manned aircraft is the Flight Management System (FMS). In particular, the guidance, navigation, and control (GNC) functionalities integrated in the FMS are

- lateral and vertical navigation (state determination, position estimation, navigation radio tuning, data fusion, and polar navigation),
- trajectory computation, estimation, and optimization,
- lateral and vertical guidance (information for autopilot and flight director),

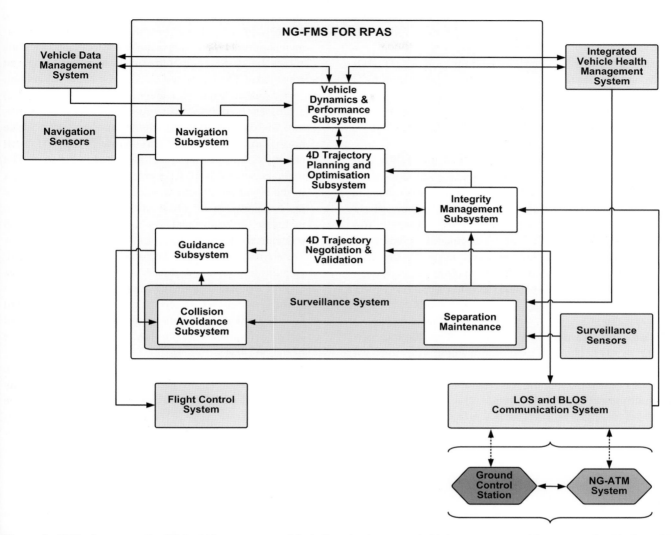

Figure 2. CNS+A systems for UAS. Airborne systems (shaded) and subsystems (white) are represented by rectangular blocks, while ground-based systems are represented by hexagonal blocks.

- performance predictions,
- determination of input commands for the autothrottle (A/THR) system, and
- continuous monitoring and correction of the flight path.

Onboard manned aircraft, the FMS also integrates suitable pilot and copilot interfaces. Current FMS provides area navigation (RNAV) services including required navigation performance (RNP) levels in all the flight profiles ensuring accuracy, availability, continuity and integrity by constantly providing performance monitoring of the position vector (Cramer *et al.*, 2010). There are a number of integrated navigation and guidance system architectures, which are currently implemented in the FMS architecture for meeting the required performance level (Sabatini *et al.*, 2015;

Cappello *et al.*, 2015a,b; Cappello, Ramasamy, and Sabatini, 2015). The key components of a typical FMS architecture for manned aircraft are the Flight Management Computer (FMC), Multi Control Display Unit (MCDU), Flight Control Unit (FCU), Electronic Flight Instrument System (EFIS), Multi-Function Display (MFD) and Navigation Display (ND). Flight crews have the ability to input information including flight plan, Zero Fuel Weight (ZFW), zero fuel center of gravity, cost index, cruise flight level, predicted wind data, and Required Time of Arrival (RTA). Based on the crew entered data, the position, Estimated Time of Arrival (ETA), constraints/limits on speed, and altitude are determined for every waypoint in the flight plan.

Next-Generation Flight Management Systems (NG-FMS) are FMS that fulfill the operational, safety and environmental requirements to deploy the 4D-TBO capabilities. The key functionalities of NG-FMS for TBO are given in the following:

- Multi-Objective 4D Trajectory Optimization (MOTO-4D) for both flight planning and real-time ATM operations.
- 4D Trajectory monitoring.
- 4D Trajectory negotiation/validation with 4-PNV systems.
- Real-time rerouting and information updating.

NG-FMS are currently being developed for 4DT-Intent-Based Operations (IBO) in combination with ground-based 4-PNV systems and NG-ADL. The NG-FMS architecture is primarily based on the core functionalities, namely, flight planning (FPLN), localization and state determination, trajectory optimization (TRAJ), performance predictions (PRED) and guidance. Additionally, the FMS provides autothrottle input commands and automation-assisted transactions with novel ground-based DSS. The primary NG-FMS modules are the following:

1. *Trajectory planning/optimization* – This module performs 4DT planning and optimization functions for pretactical, tactical, and emergency situations. The trajectory planning module is based on multi-objective and multi-model 4DT optimization algorithms. The 4DT

optimizer includes suitable models, databases, and a number of predefined operational/economic/environmental optimization criteria in the form of cost functions and enables the users to introduce an arbitrary number of constraints. The integrated databases include navigation, performance, magnetic deviation, and environmental databases. Predefined cost functions include minimization of fuel consumption, flight time, operative costs, noise impacts, emissions, and contrails. The 4DT planner and optimizer blocks are illustrated in Figure 3. The implementation of 4DT optimization algorithms as part of the identified system architecture modules allows for the development of TBO aspects. The NG-FMS also includes cooperative and non-cooperative SAA function software modules.

2. *Trajectory monitoring* – Performs state estimation, calculating the deviations between the active 4DT intents and the estimated/predicted aircraft states.

3. *Path correction* – Corrects the path deviation in terms of lateral, vertical, and time profiles and the generated steering commands are provided to the guidance module of the NG-FMS.

4. *Trajectory negotiation and validation* – It carries out the process of negotiation that can be initiated by the pilot via the NG-FMS, by making use of the information

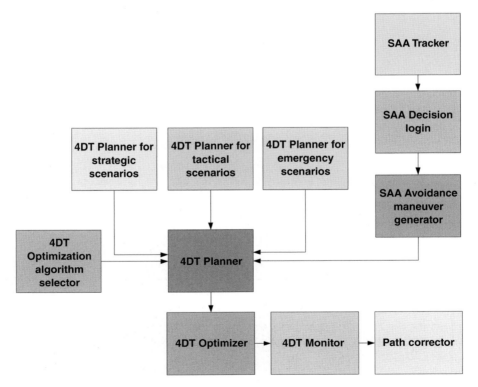

Figure 3. NG-FMS 4DT planner and optimizer.

available on board, or by the 4-PNV system supervised by the air traffic control operator (ATCO).

5. *FMS performance manager* – Monitors the active 4DT intents for potential performance violations, to address integrity requirements. The integrity analysis module is based on RNP, RCP, and RSP managers.

6. *FMS integrity manager* – This module is used to generate integrity caution (predictive) and warning (reactive) flags based on inputs from different sensors/ systems and predefined decision logics. A loss of data leads to reinitializing of the trajectory planning and subsequently the 4DT generation and optimization process. For instance, the main causes of GNSS signal outage and degradation in flight, namely antenna obscuration, multipath, fading due to adverse geometry, and Doppler shift are identified and modeled to implement integrity thresholds and guidance algorithms in the Avionics-Based Integrity Augmentation (ABIA) system (Sabatini *et al.*, 2013a, 2013b, 2014, 2015).

Each aircraft equipped with NG-FMS generates 4DT intents, defined according to the Flight Management Computer (FMC) ARINC 702A-3 characteristic as a string of 4D points that define the predicted trajectory of the aircraft along with the point type and turn radius associated with the flight path transition (ARINC, 2006). Intent data are updated in situations of a change in the nominal flight path, addition of new sequencing points, and weather data. The trajectory computation and optimization component of the NG-FMS is reconfigurable with that of the ground-based counterpart to enable negotiation and validation updates in real-time. Additionally, the intents are recomputed according to flight plan revisions, weather updates, guidance mode modification, cost index modification, and corrections for position uncertainties in real-time based on the 4D trajectory optimization algorithms described earlier. The efficiency and effectiveness of 4DT planning, negotiation, and validation functionalities of the NG-FMS are directly driven by the nature of information sharing. Subject to various in-flight changes, trajectory calculations are refreshed to maintain consistency and downlinked to the 4-PNV system via the NG-ADL.

The negotiation and validation of 4DT intents by the NG-FMS/4-PNV systems is dependent on the following:

- Onboard validation based on synchronization, sufficient fuel, compliance with dynamics (time performances, turn performances, speed, and altitude), obstacle separation, locally sensed weather, and compliance with the integrated vehicle health management (IVHM) system regarding aircraft health status and other aspects.

- Ground-based validation based on air traffic separation (lateral, vertical, and longitudinal), sector occupancy, airspace restrictions (special use areas), and time based restrictions (night time noise abetment procedures).

With the increasing levels of onboard automation, integrity monitoring and augmentation systems have occupied a fundamental role. Errors affecting the CNS+A systems (e.g., pseudo-range GNSS observables) are taken into account in evaluating the CNS performance. As illustrated in Figure 4, the performance manager module provides inputs to the integrity flag generator (IFG) based on the errors affecting the CNS systems. The IFG uses a set of predefined caution and warning integrity flags (CIF/WIF) threshold parameters to trigger the generation of both caution and warning flags associated with CNS performance degradations. In case a warning flag is generated, a recapture command is used to trigger the 4DT regeneration and optimization process.

Figure 5 is a schematic block diagram of the NG-FMS performance management modules. The performance management tasks are defined for all CNS+A parameters. These modules receive data from the 4DT planner/optimizer module. The RNP, RSP, and RCP integrity performance management modules provide data to the integrity management blocks that in turn generate CIF/WIF. The CNS+A

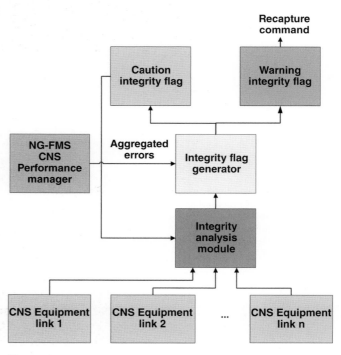

Figure 4. NG-FMS integrity flag generation.

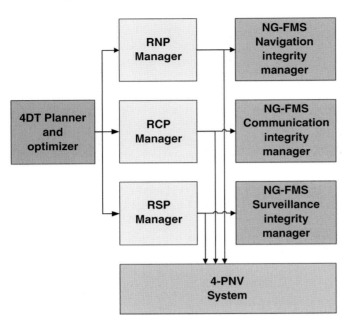

Figure 5. NG-FMS performance management.

performance management blocks are interfaced with the 4-PNV system.

Figure 6 is a schematic block diagram of the CNS+A systems for UAS. The optimization of 4DT trajectories is performed onboard by the NG-FMS. A number of UAS equipped with NG-FMS are controlled by the ground command, control and intelligence system aided by LOS, and

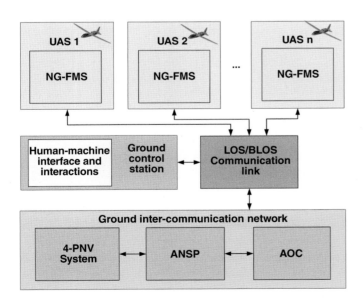

Figure 6. UAS CNS+A systems.

BLOS communication links. Human–machine interface and interactions at the ground control station (GCS) are equipped with navigation, tactical, health management, and engine management displays. The ground intercommunication system consists of a ground-to-ground communication network between the 4-PNV system, Air Navigation Service Provider (ANSP), and the Airline Operation Centre (AOC).

The TDA loop consists of the following functions:

- *Track:* A group of sensors collect the required data from the environment. Tracking is accomplished by the continuous acquisition of obstacle/intruder data.
- *Decide:* As the intruder aircraft/obstacle is tracked, suitable decision logics are employed for identifying the possibility of collisions.
- *Avoid:* Once a possibility of collision is detected, then the onboard computers determine an action to avoid the collision by regenerating the 4DT and optimizing it against the set constraints and performance parameters.

The lateral path is constructed in terms of straight and turning segments and is generally based on the required course change and the predicted ground speed of the aircraft during turns. A turn is constructed based on the maximum ground speed of the UAS during a course change, so that the turn radius is given by:

$$T_R = \frac{GS^2}{g \cdot \tan(\phi)} \qquad (1)$$

where GS is the maximum predicted UAS ground speed during the turn, ϕ is the selected bank angle, and g is the gravity acceleration module. The turn arc length is given by:

$$T_{AL} = \Delta TRK \cdot T_R \qquad (2)$$

where ΔTRK is the difference between the final and the initial ground tracks in radians.

The NG-FMS computes turn radius and speed based on the selected altitude by taking into account the predicted wind at that altitude. The bank angle is selected based on UAS dynamics and airspace configurations. In order to construct the vertical profile, a number of approximated energy balance equations are typically adopted leading to nominal climb/descent, fixed gradients climb/descent, intermediate speed changes, and level flight configurations. The vertical profile is obtained from the energy method given by

$$\frac{dh}{dt} = \frac{(T-D) \cdot V_T}{GW \cdot \left(1 + \frac{V_T}{g} \cdot \frac{dV_T}{dh}\right)} \qquad (3)$$

where V_T is the true air speed, GW is the gross weight. The data involved in the energy balance equations come from the airframe/engine dependent thrust, fuel flow, drag, and air speed schedule models stored in the performance database (PerfDB). The integration steps are constrained by altitude, speed, and time restrictions imposed as part of the mission profile as well as UAS performance limitations such as speed and buffet limits, maximum altitude, and thrust limits. The NG-FMS receives the controlled time of arrival target defined by the 4-PNV system, which becomes the required time of arrival (RTA) to be used by the NG-FMS in determining the optimal trajectory states (final time). The estimated time of arrival (ETA) may be computed at multiple fixes along the flight path. In general, a specific performance objective can be defined for each route segment. This performance objective is a multiobjective generalization of the cost index (CI) that allowed an optimal selection of calibrated air speed (CAS)/Mach number based on time and fuel costs only. The cost index allows the mission operators to weigh time, fuel, emissions, noise, and other costs. The time cost J_{time} is given by

$$J_{time} = K_t t_f \tag{4}$$

For UAS adopting internal combustion engines, fuel consumption optimization is achieved by implementing a fuel flow (FF) model. In terms of emissions, although engine design and other factors may influence total amounts, pollutant emissions are considered as a function of fuel flow multiplied by specific emission factors, ϱ. Hence, the mathematical description of the emission performance index, $J_{emission}$, defined with respect to emissions, e and time, t is given by

$$J_{emission} = \int_{t_0}^{t_f} \frac{de}{dt} = (m(t_f) - m(t_0))\varrho \tag{5}$$

In designing a lateral track control strategy for the UAS, a viable control strategy is based on the following relation (Marius, 2001):

$$\frac{\dot{X}_{track}}{k \times X_{track}} = \frac{\dot{Y}_{track}}{Y_{track}} \tag{6}$$

where X_{track} and Y_{track} are the current track position of the UAS with respect to a TCP. The along-track and cross-track velocities are obtained from the airspeed and wind speed velocity vectors. Proportional-integral-derivative (PID) and fuzzy logic based controllers are generally implemented for eliminating the error. In order to study the effects of uncertainties on the generated 4DT, a detailed error analysis is

performed. The errors might be due to database accuracy degradations, system modeling errors, atmospheric disturbances, and subsystem errors. The random errors, which are unpredictable, are quantified to estimate the overall error associated with the position of the aircraft. The system states are modified with the addition of the stochastic term, e(t) and can be represented by

$$\dot{x}(t) = f[x(t), u(t), e(t), t] \tag{7}$$

In order to perform a sensitivity analysis, the sensitivity of a trajectory attribute, J including time, fuel, emissions, etc. is considered with respect to a model parameter, Δ and is given by

$$\Delta = \Delta^n + d\Delta \tag{8}$$

The open-loop sensitivity of J is given by

$$\frac{\partial J}{\partial \Delta_{OL}} = \frac{J^{OL}(\Delta^n + d\Delta) - J^{OL}(\Delta^n)}{d\Delta} \tag{9}$$

The closed-loop sensitivity of J is given by

$$\frac{\partial J}{\partial \Delta_{CL}} = \frac{J^{CL}(\Delta^n + d\Delta) - J^{CL}(\Delta^n)}{d\Delta} \tag{10}$$

The values of J^{OL} and J^{CL} are optimized based on the 4DT optimization algorithms employed in the NG-FMS. Each performance index provides a quantitative measure of the attainment of a specific objective and different objectives are typically conflicting, and thus the optimization in terms of two or more objectives typically leads to a number of possible compromise choices, which are still optimal. Therefore, a trade-off is generally introduced in the context of multi-objective trajectory optimization. In the aviation domain, single and biobjective optimization techniques have been exploited for decades but they accounted only for flight time-related costs and fuel-related costs. These techniques have also been implemented in a number of current generation FMS in terms of the Cost Index, which is a scalar value to balance the relative weighting of fuel and time costs. In the NG-FMS, the weightings are varied dynamically among the different flight phases of the flight. The uncertainties associated with the position, velocity, and attitudes of the UAS depend on the uncertainty of navigation data propagated through the ADM. Since the ADM equations are nonlinear functions of the navigation variables, a suitable linearization shall be introduced and this can be conveniently performed by a second-order Taylor series expansion, wherein S signifies sine and C signifies cosine of an angle:

$$\sigma_{\dot{V}} = \sqrt{(-g \cdot C\gamma)^2 \, \sigma_\gamma^2 + \left(\frac{T_{\text{norm}}}{m}\right)^2 \sigma_\tau^2} \tag{11}$$

$$\sigma_{\dot{h}} = \sqrt{(S\gamma)^2 \, \sigma_V^2 + \sigma_{V_{w_h}}^2 + (V \cdot C\gamma)^2 \, \sigma_\gamma^2} \tag{12}$$

$$\sigma_{\dot{\chi}} = \sqrt{\left(\frac{-N \cdot g \cdot S\phi}{C\gamma \cdot V^2}\right)^2 \sigma_V^2 + \left(\frac{N \cdot g \cdot C\phi}{V \cdot C\gamma}\right)^2 \sigma_\phi^2 + \left(\frac{N \cdot g \cdot S\phi \cdot S\gamma}{\gamma \cdot C\gamma^2}\right)^2 \sigma_\gamma^2} \tag{13}$$

$$\sigma_{\dot{\gamma}} = \sqrt{\left[\frac{g(-N \cdot C\phi + C\gamma)}{V^2}\right]^2 \sigma_V^2 + \left(-\frac{N \cdot g \cdot S\phi}{V}\right)^2 \sigma_\phi^2 + \left(-\frac{g \cdot S\gamma}{V^2}\right)^2 \sigma_\gamma^2} \tag{14}$$

$$\sigma_{\dot{\varphi}} = \left(\begin{aligned} &\left[\frac{1}{(R_M + h)}\right]^2 \sigma_{V_{w\varphi}}^2 + \left(\frac{-V \cdot S\gamma \cdot S\chi + V_{W_\varphi}}{(R_M + h)^2}\right)^2 \sigma_h^2 + \left(\frac{-V \cdot S\gamma \cdot S\chi}{(R_M + h)}\right)^2 \sigma_\gamma^2 \\ &+ \left[\frac{C\gamma \cdot S\phi}{(R_M + h)}\right]^2 \sigma_V^2 + \left(\frac{V \cdot S\gamma \cdot C\chi}{(R_M + h)}\right)^2 \sigma_\chi^2 \end{aligned} \right)^{1/2} \tag{15}$$

$$\sigma_{\dot{\lambda}} = \left(\begin{aligned} \sigma_{\dot{\lambda}} = &\left[\frac{C\gamma \cdot C\chi}{C\varphi(R_T + h)}\right]^2 \sigma_V^2 + \left[\frac{-V \cdot C\gamma \cdot S\chi}{C\varphi(R_T + h)}\right]^2 \sigma_\chi^2 + \left[\frac{1}{C\varphi(R_T + h)}\right]^2 \sigma_{V_{W_\lambda}}^2 \\ &+ \left[\frac{-V \cdot S\gamma \cdot C\chi}{C\varphi(R_T + h)}\right]^2 \sigma_\gamma^2 + \left[\frac{-C\varphi(V \cdot S\gamma \cdot C\chi + V_{W_\lambda})}{C\varphi(R_T + h)^2}\right]^2 \sigma_h^2 + \left[\frac{S\varphi \cdot C\varphi(R_T + h)(V \cdot C\gamma \cdot C\chi + V_{W_\lambda})}{C\varphi(R_T + h)^2}\right]^2 \sigma_\varphi^2 \end{aligned} \right)^{1/2} \tag{16}$$

where N is the load factor, ϕ is the bank angle, m is the aircraft (variable) mass, φ is the geodetic latitude, λ is the geodetic longitude, h is the altitude, V is the true air speed, γ is the flight path angle, χ is the heading, R_M is the meridional radius of curvature, R_T is the transverse radius of curvature, V_W is the wind velocity, T_{norm} is the axial thrust, and g is the nominal acceleration due to gravity of the Earth. The resulting uncertainties in aircraft position and kinematics are conveniently described by the associated uncertainty volumes. For cooperative and noncooperative obstacle avoidance and safe-separation maintenance, the overall uncertainty volumes are obtained by combining the navigation error with the tracking error and then translating them to unified uncertainty volumes (Ramasamy et al., 2015). In this unified SAA approach, both navigation error of the host UAS platform and tracking error of other traffics are combined as necessary in order to obtain the volume of airspace that needs to be considered for separation maintenance and collision avoidance. The fundamental principle is to express the SAA sensor/system error in range and bearing uncertainty descriptors. The correlation between the SAA sensors/ systems employed is analyzed and the individual uncertainties are characterized as either uncorrelated, covariant, or contravariant. In the current CNS/ATM context, the general case is that of multiple traffics performing either cooperative or noncooperative surveillance. In this scenario, potential conflicts are defined as close encounters in the 4D space–time domain. Close encounters are typically evaluated as part of an intermediate step for pruning the full set of potential conflicts. Such 4D close encounters are assumed to occur when the relative distance – that is the norm of the 3D relative position vector – between the nominal positions of a pair of traffics at a certain time instant is below a specified threshold. For all identified close encounters, the uncertainty volume associated with the host and other platforms are determined. Due to bandwidth limitations affecting air-to-air communication systems, a compact and versatile parameterization of the uncertainty volume is highly functional to extrapolate its actual shape and size at close encounter points with minimal data link and computational burden. Therefore, the combined navigation and tracking uncertainty volumes can be conveniently described using spherical harmonics and the associated parameters are communicated to the air and ground nodes of the network (Ramasamy *et al.*, 2015).

In particular, for each and every intruder, the SAA function of the UAS NG-FMS processes the trajectory information. Criticality analysis is carried out to prioritize (i.e., to determine if a collision risk threshold is exceeded for all tracked intruders) and to determine the action commands. In order to estimate the overall avoidance volume, navigation error of the host platform and tracking error of obstacles/intruders are extracted. The variation in the state vector (host and intruder) is expressed as

$$\delta\left(X_i(t)\right) = \left[\frac{\delta X}{\delta p}\right]_t \cdot \sigma_{p_j} \tag{17}$$

where p is the position of the UAS and t is the time of measurement. Let R, α and ε be the range, azimuth, and elevation, respectively, obtained from the SAA non-cooperative sensors or cooperative systems. Let R_0, α_0 and ε_0 be the nominal range, azimuth, and elevation values, respectively. Let σ_R, σ_α, and σ_ε as standard deviations of the error in range, azimuth, and elevation, respectively. The error ellipsoids are given as

$$\frac{(R - R_0)^2}{\sigma_R^2} + \frac{(\alpha - \alpha_0)^2}{\sigma_\alpha^2} + \frac{(\varepsilon - \varepsilon_0)^2}{\sigma_\varepsilon^2} = 1 \tag{18}$$

In a static noncooperative cases, the errors in range, azimuth, and elevation are given by

$$\delta R = R_0 + \sigma_R \cdot \sin \psi \tag{19}$$

$$\delta \alpha = \alpha_0 + \sigma_\alpha \cdot \cos \phi \cdot \cos \psi \tag{20}$$

$$\delta \varepsilon = \varepsilon_0 + \sigma_\varepsilon \cdot \sin \phi \cdot \cos \psi \tag{21}$$

where R_0, α_0, ε_0 are the nominal range, azimuth, and elevation measurements, respectively; and $\{\phi, \psi\}$ are parameterization factors. The transformation of $\{R, \alpha, \varepsilon\}$ to $\{x, y, z\}$ is given by

$$x = R \cdot \cos \alpha \cdot \cos \varepsilon \tag{22}$$

$$y = R \cdot \sin \alpha \cdot \cos \varepsilon \tag{23}$$

$$z = R \cdot \sin \varepsilon \tag{24}$$

In the case of dynamic obstacles, the uncertainty volume is obtained based on a confidence region given by

$$\delta v_0 = v_0 + \sigma_{v_0} \cdot \sin \psi \tag{25}$$

$$\delta \nu_0 = \nu_0 + \sigma_{\nu_0} \cdot \cos \phi \cdot \cos \psi \tag{26}$$

$$\delta \upsilon_0 = \upsilon_0 + \sigma_\upsilon \cdot \sin \phi \cdot \cos \psi \tag{27}$$

where v_0, ν_0, υ_0 are the nominal velocity measurements. When an error in elevation and azimuth is present, a resultant cone is obtained at the estimated range. The kinematic relationships are

$$v_x = v \cdot \cos \nu \cdot \cos \upsilon \tag{28}$$

$$v_y = v \cdot \sin \nu \cdot \cos \upsilon \tag{29}$$

$$v_z = v \cdot \sin \upsilon \tag{30}$$

These equations are governed according to the following laws of motion

$$x = x_0 + v_x \cdot t \tag{31}$$

$$y = y_0 + y_y \cdot t \tag{32}$$

$$z = z_0 + v_z \cdot t \tag{33}$$

The errors in $\{x, y, z\}$, used to calculate the actual shape of the avoidance volumes are given by

$$\sigma_x^2 = (\sigma_{x0}^2 + \sigma_{v_x}^2 \cdot t + 2 \, \sigma_{x0 \, v_x} \cdot t) \tag{34}$$

$$\sigma_y^2 = (\sigma_{y0}^2 + \sigma_{v_y}^2 \cdot t + 2 \, \sigma_{y0 \, v_y} \cdot t) \tag{35}$$

$$\sigma_z^2 = (\sigma_{z0}^2 + \sigma_{v_z}^2 \cdot t + 2 \, \sigma_{z0 \, v_z} \cdot t) \tag{36}$$

As an example, a deconfliction scenario in the context of an arrival point merge in the TMA is illustrated in Figure 7. In this scenario, the predicted trajectory of an unmanned aircraft (traffic A) is conflicting with another unmanned aircraft (traffic B), a commercial airliner (traffic C), and a general aviation aircraft (traffic D) at three distinct points. Traffic A is equipped with both noncooperative sensors/cooperative systems for SAA and traffic B has onboard noncooperative sensors only. Traffic C is equipped with cooperative traffic collision detection and avoidance systems (e.g., based on ADS-B), whereas the general aviation aircraft (traffic D) features a noncooperative collision avoidance system (e.g., machine-vision/radar-based). Data link technologies are employed for real-time surveillance data exchange with the ground ATM from properly equipped platforms (ADS-B, TCAS, etc.). Based on the traffic data exchanged or sensed, three distinct avoidance volumes are generated for traffic A. In this example, the volumes associated with traffics B and C are merged into a combined volume (avoidance volume 1) when the algorithm detects they are overlapping. A similar process is repeated for all four traffics in the TMA and, in this case, the TMA control service defines the traffic sequencing based on first-come first-served and/or

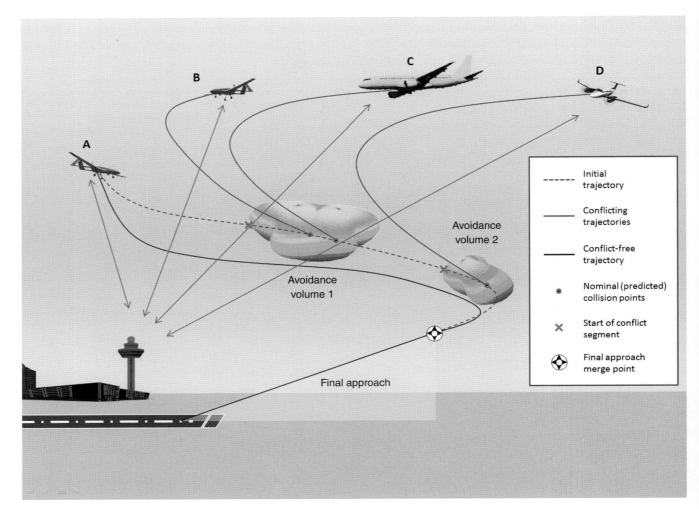

Figure 7. ATM-assisted deconfliction for an arrival point merge in the TMA.

best-equipped best-served criteria. The resulting path constraints are uplinked to the 4DT planner and optimizer modules of the NG-FMS in order to generate a set of feasible conflict-free trajectory solutions. These intents are then sent to the 4-PNV system for final validation.

In the example described above, the reliance on ATM services represents a bottle neck and, potentially, a single point of failure in the overall SAA network strategy (e.g., intentional/unintentional data link signal degradations or ground equipment malfunctions). Therefore, the example scenario is modified to consider a fully autonomous deconfliction approach. In this approach, the uncertainties in navigation and tracking associated to each manned/unmanned platform (as seen by all other conflicting traffics) are combined to generate avoidance volumes surrounding each traffic in the TMA. The uncertainty volumes are computed at discrete time intervals as a function of traffic relative dynamics. Figure 8 shows a conceptual representation of the variable avoidance volume associated with traffic A obtained by combining (in a statistical sense) the navigation and tracking errors related to cooperative and noncooperative SAA observations by all other traffics at sequential time epochs. In this case, the NG-FMS of all manned/unmanned aircraft exchanges real-time avoidance information updates and computes the optimal 4DT to be executed by each platform. The final intents are communicated to the ground ATM for surface movement coordination purposes but no direct deconfliction, sequencing, or validation function is accomplished by the 4-PNV system.

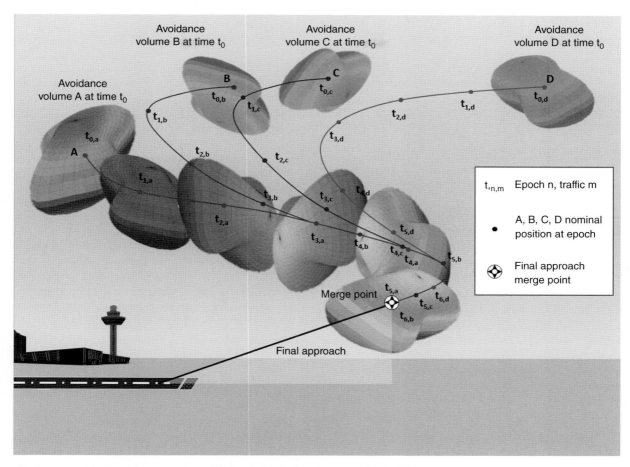

Figure 8. Concept of fully autonomous deconfliction for arrival point merge in the TMA.

7 CONCLUSIONS

In this chapter, we discussed the opportunities and challenges associated with the operation of unmanned aircraft systems (UAS) in terminal areas. The specificities of this category of airspace are analyzed in detail to extract the key factors affecting UAS in terms of their equipment and operation. Challenges, in particular, are associated with the frequently intersecting flight paths of departures arrivals and overflights, resulting in an increasing need for tactical interventions by ATM operators. Conversely, some advantages are envisaged in relation to the very high level of jurisdiction on the ATM side, which diminishes the requirements for expert processing capabilities onboard the UAS. Additionally, capacity benefits associated with lower longitudinal separations are mentioned. Based on the analysis, requirements for unrestricted operation of UAS in the TMA are translated to communication, navigation, surveillance, ATM (CNS/ATM) and avionics (CNS+A) equipment upgrades. This chapter therefore describes in detail the current state-of-the-art in CNS+A technologies and the research activities in progress to integrate the required capabilities onboard unmanned aircraft. Some theoretical models underlying key functionalities of the NG-FMS are presented, particularly including 4DT prediction and optimization algorithms currently being developed (Gardi *et al.*, 2015; Ramasamy *et al.*, 2015). Despite the dense air traffic, TMA can present opportunities for UAS integration as the expert processing required is considerably lower and this has the potential to simplify the algorithm complexity to integrate sense-and-avoid (SAA) capabilities, leading to significant economic and safety benefits.

REFERENCES

Amato, G. EUROCAE WG-73 on Unmanned Aircraft Systems.

Amazon Inc. (2015a) Determining Safe Access with a Best-Equipped, Best-Served Model for Small Unmanned Aircraft Systems (White Paper), Amazon Inc.

Amazon Inc. (2015b) Revising the Airspace Model for the Safe Integration of Small Unmanned Aircraft Systems (White Paper), Amazon Inc.

ARINC (2006) Advanced Flight Management Computer System - ARINC Characteristic 702A-3, Airlines Electronic Engineering Committee, Aeronautical Radio Incorporated (ARINC).

Bouwmeester, L., Clothier, R., Sabatini, R. and Williams, G. (2015) Autonomous communication between air traffic control and remotely piloted aircraft, in *16th Australian International Aerospace Congress (AIAC16)*, Melbourne, Australia. IEA, pp. 48–57.

Bradbury, J.N. (1991) ICAO and future air navigation systems, in *Automation and Systems Issues in Air Traffic Control*, Springer, pp. 79–99.

Cappello, F., Ramasamy, S., and Sabatini, R. (2015) Low-cost RPAS navigation and guidance system using square root unscented Kalman filter, *SAE Technical Paper 2015-01-2459*, doi: 10.4271/2015-01-2459.

Cappello, F., Ramasamy, S., Sabatini, R., and Liu, J. (2015a) Low-cost sensors based multi-sensor data fusion techniques for RPAS navigation and guidance, in *Proceedings of the 2015 International Conference on Unmanned Aircraft Systems (ICUAS '15)*, Denver, CO. doi: 10.1109/ICUAS.2015.7152354.

Cappello, F., Sabatini, R., Ramasamy, S., and Marino, M. (2015b) Particle filter based multi-sensor data fusion techniques for RPAS navigation and guidance, in *Proceedings of the 2015 International Workshop on Metrology for Aerospace (MetroAeroSpace 2015)*, Benevento, Italy. doi: 10.1109/MetroAeroSpace.2015.7180689.

Carey, B. (2015) Lockheed Martin: simple ATC mods would allow drone flights, *Aviation International News – AINonline,* December 8, 2015.

Clough, B.T. (2002) Chapter 3: Unmanned aerial vehicles: autonomous control challenges, a researcher's perspective, in *Cooperative Control and Optimization* (eds P. Murphey and P.M. Pardalos), Kluwer Academic Publishers, The Netherlands, pp. 35–53.

Clough, B.T. (2005) Unmanned aerial vehicles: autonomous control challenges, a researcher's perspective. *J. Aerosp. Comput. Infr. Commun.*, **2**, 327–347.

Cramer, M. R., Herndon, A., Steinbach, D., and Mayer, R. H. (2010) Modern aircraft flight management systems. *Encycl. Aerosp. Eng.*

FAA (2013) Integration of Civil Unmanned Aircraft Systems (UAS) in the National Airspace System (NAS) Roadmap, Federal Aviation Administration (FAA).

Gardi, A., Ramasamy, S., and Sabatini, R. (2015) 4-Dimensional trajectory generation algorithms for RPAS mission management systems, in *Proceedings of the 2015 International Conference on Unmanned Aircraft Systems (ICUAS '15)*, Denver, CO. doi: 10.1109/ICUAS.2015.7152314.

Gardi, A., Sabatini, R., Kistan, T., Lim, Y., and Ramasamy, S. (2015) 4-Dimensional trajectory functionalities for air traffic management systems, in *Proceedings of the Integrated Communication, Navigation and Surveillance Conference (ICNS 2015)*, Herndon, VA. doi: 10.1109/ICNSURV.2015.7121246.

Gettinger, D. and Michel, A.H. (2015) *Drone Sightings and Close Encounters: An Analysis,* Center for the Study of the Drone, Bard College, Annandale-on-Hudson, NY.

Google Inc. (2015) Google UAS Airspace System Overview (White Paper), Google Inc.

ICAO (1990) Annex 2 to the Convention on International Civil Aviation – Rules of the Air, The International Civil Aviation Organization (ICAO), Montreal, QC, Canada.

ICAO (2001) Annex 11 to the Convention on International Civil Aviation – Air Traffic Services, International Civil Aviation Organization (ICAO), Montreal, QC, Canada.

ICAO (2013) Working Document for the Aviation System Block Upgrades – The Framework for Global Harmonization, The International Civil Aviation Organization (ICAO).

JAA/EUROCONTROL (2005) UAV TASK-FORCE Final Report – A Concept for European Regulations for Civil Unmanned Aerial Vehicles (UAVs), JAA/EUROCONTROL.

Kopardekar, P.H. (2015) Safely enabling UAS operations in low-altitude airspace, Presented at the NASA UAS Traffic Management (UTM), Moffett Field, CA.

Marius, N. (2001) Lateral track control law for AEROSONDE UAV, in *Proceedings of the 39th Aerospace Sciences Meeting and Exhibit*, American Institute of Aeronautics and Astronautics.

NASA (2015) *UTM: Air Traffic Management for Low-Altitude Drones*, National Aeronautics and Space Administration (NASA) NASA Facts NF-2015-10-596-HQ, Washington DC, USA.

NATO (2005) Standardisation Agreement – Standard Interfaces of UAV Control System (UCS) for NAT UAV Interoperability, STANAG-4586, NATO Standardization Agency.

Ramasamy, S. and Sabatini, R. (2015) A unified approach to cooperative and non-cooperative sense-and-avoid, in *Proceedings of the 2015 International Conference on Unmanned Aircraft Systems (ICUAS '15)*, Denver, CO. doi: 10.1109/ICUAS.2015.7152360.

Ramasamy, S., Gardi, A., Liu, J., and Sabatini, R. (2015) A laser obstacle detection and avoidance system for manned and unmanned aircraft applications, in *Proceedings of the 2015 International Conference on Unmanned Aircraft Systems (ICUAS '15)*, Denver, CO. doi: 10.1109/ICUAS.2015.7152332.

Ramasamy, S., Sabatini, R., and Gardi, A. (2015a) Novel flight management systems for improved safety and sustainability, in the CNS+A Context, in *Proceedings of the Integrated Communication, Navigation and Surveillance Conference (ICNS 2015)*, Herndon, VA. doi: 10.1109/ICNSURV.2015.7121225.

Ramasamy, S., Sabatini, R., and Gardi, A. (2015b) A novel approach to cooperative and non-cooperative RPAS detect-and-avoid, SAE Technical Paper 2015-01-2470, doi: 10.4271/2015-01-2470.

Sabatini, R., Cappello, F., Ramasamy, S., Gardi, A., and Clothier, R. (2015a) An innovative navigation and guidance system for small unmanned aircraft using low-cost sensors. *Aircr. Eng. Aerosp. Technol.*, **87**, 540–545.

Sabatini, R., Gardi, A., Ramasamy, S., Kistan, T., and Marino, M. (2015b) Modern avionics and ATM systems for green operations, in *Encyclopedia of Aerospace Engineering* (eds R. Blockley, and W. Shyy). doi: 10.1002/9780470686652.eae1064.

Sabatini, R., Moore, T., Hill, C., and Ramasamy, S. (2015c) Assessing avionics-based GNSS integrity augmentation performance in UAS mission- and safety-critical tasks, in *Proceedings of the 2015 International Conference on Unmanned Aircraft Systems (ICUAS '15)*, Denver, CO.

Sabatini, R., Moore, T., and Hill, C. (2013a) A new avionics-based GNSS integrity augmentation system: Part 1 – Fundamentals. *J. Navig.*, **66**, 363–384.

Sabatini, R., Moore, T., and Hill, C. (2013b) A new avionics-based GNSS integrity augmentation system: Part 2 – Integrity flags. *J. Navig.*, **66**, 501–522.

Sabatini, R., Moore, T., and Hill, C. (2014) Avionics-based GNSS augmentation for unmanned aerial systems sense-and-avoid, in *Proceedings of the 26th International Technical Meeting of the Satellite Division of the Institute of Navigation (ION GNSS+ 2014)*, Tampa, FL.

Sheridan, T.B. (2003) *Telerobotics, Automation, and Human Supervisory Control*, MIT Press, Cambridge, MA.

Stevenson, B. (2015) *NASA Plans Next Phase of UAV ATM Integration*, Flight Global (accessed October 23).

Sullivan, D., Totah, J.J., Wegener, S.S., Enomoto, F.Y., Frost, C. R., Kaneshige, J., and Frank, J.E. (2004) Intelligent mission management for uninhabitated aerial vehicles, in *Proceedings of the SPIE 5661: Remote Sensing Applications of the Global Positioning System*. doi: 10.1117/12.582446.

Chapter 38

Unmanned Aircraft Systems Operations in US Airspace

Brian M. Argrow

Department of Aerospace Engineering Sciences, Research and Engineering Center for Unmanned Vehicles, University of Colorado Boulder, Boulder, CO, USA

1 INTRODUCTION

The global demand for unmanned aircraft systems (UAS) continues to grow as militaries, law enforcement groups, weather and climate researchers, emergency responders, cinematographers, and others recognize the utility of UAS for a wide range of applications. Airspace regulatory agencies in countries around the world continue to struggle with the pressure from these groups, and from UAS manufacturers

to begin the widespread integration of UAS into the airspace systems of countries around the world. While military operations have seen tremendous growth in the deployment of UAS in regional conflicts, where airspace safety is not a primary concern, there has also been an increasing number of civilian applications of UAS operating under strict constraints of regulatory agencies such as the US Federal Aviation Administration (FAA).

This chapter focuses on UAS classification and regulatory requirements for flight tests and operations in the National Airspace System (NAS) of the US. UAS classifications, based on measures of the airframe size and performance, and levels of autonomy, are discussed. This is followed by an introduction to the requirements for UAS operations, including a discussion of the certification process for "public" UAS. An exploration of military and civilian operations is presented, including a discussion of the regulatory requirements that enabled the Tempest UAS to be deployed for supercell thunderstorm research near populated areas in the central US. Flight test procedures, including suggestions for facilities appropriate for current UAS development and research, as well as procedures to ensure flight safety are also investigated. The chapter concludes with a presentation of recent issues and challenges related to the public perception of UAS operations.

2 UAS CLASSIFICATION

A UAS is composed of an airborne component, the unmanned aircraft (UA) that operates without an onboard

Unmanned Aircraft Systems. Edited by Ella Atkins, Aníbal Ollero, Antonios Tsourdos, Richard Blockley and Wei Shyy.
© 2016 John Wiley & Sons, Ltd. ISBN: 978-1-118-86645-0.

human pilot, and with a communication, command, and control (C3) system that enables the UA to be remotely operated or managed.[1] A growing number of countries have developed and deployed UAS; however, the United States remains the leader in UAS development and applications, and Israel is also recognized as a leader in the development and deployment of UAS, primarily for military applications. Comprehensive global surveys of current UAS are published annually by professional organizations such as the American Institute of Aeronautics and Astronautics (AIAA) (Wilson, 2011). The US Office of the Secretary of Defense periodically publishes "roadmaps" that provide a detailed, publicly accessible overview of UAS employed by the US military and its allies. The most recent roadmap (Office of the Secretary of Defense, 2009) is now an overview of integrated systems that include unmanned ground vehicles and unmanned maritime systems.

2.1   Levels of autonomy

The degree to which a human pilot or operator has direct control of the UA in flight is determined by the level of autonomy of the UAS. The minimum level of autonomy is for the UA to be flown by radio control (RC) by a pilot with direct visual sight of the UA, as demonstrated by RC hobbyist, where the pilot moves the UA control surfaces and throttle in proportion to the manual movement of joysticks on a hand-held radio console. In this case, the remote human pilot is "in the control loop" and relies on visual cues while maneuvering the UA to maintain the appropriate heading, speed, altitude, and attitude. UAS with the highest levels of autonomy are commanded by a human operator to perform specific tasks or missions. The operator then monitors the UAS as the tasks or mission are carried out automatically, or autonomously if the UAS is capable of independent decision making and planning without intervention by the operator. In this case, the human operator is "on the control loop" to monitor the UAS and to re-task the UAS or make mission decisions, and to ensure safety.

Increased levels of autonomy are achieved with increased system automation. This begins with aircraft stability augmentation where sensors (typically rate gyros and accelerometers) onboard the UA are used to adapt the manual control inputs from the RC pilot according to the predetermined dynamic flight characteristics of the UA. An onboard autopilot enables increasing levels of UAS operational autonomy by taking additional sensor inputs such as airspeed to estimate the UA dynamic state, an instantaneous estimate of the position, speed, and attitude of the UA and the rates at which these quantities are changing. The autopilot then automatically issues control surface and throttle commands based on an attempt to maintain a desired vehicle state, or seek some prescribed new state. This is the same function as the autopilot in a manned aircraft. As the level of autonomy of the UAS increases, the automatic control functions of the autopilot increase. This is usually accompanied with an increase in the type and number of sensors, and the integration of a "flight computer" to process the increased demands for state determination, maneuvering, and to manage sensors for the autopilot and other onboard subsystems, and possibly payloads. For these intermediate levels of autonomy, the UAS operator is removed from the inner control loop associated with low-level tasks, such as commanding control surfaces and throttle, and is generally more focused on higher level tasks such as path planning and navigation, typically accomplished by sending GPS waypoints to the flight computer, or accomplished in advance with preloaded GPS waypoints. For the highest levels of autonomy, the mission objectives might be loaded into the flight computer, with all functions associated with carrying out the mission performed automatically by onboard or remote systems. In this case, the operator might only issue commands to the UAS to change the mission objectives, or to intervene for emergencies.

2.2   Airframe size and performance

The classification of UAS by some combination of physical characteristics, performance, and flight management, remains a topic of debate. UAS are often classified as *large*, *medium*, *small*, or *micro*, based on the physical size of the UA. There is no consensus, however, for an official division of UAS into those categories. With a physical size on the order of a small wide-body airliner, the Global Hawk shown in Figure 1 is unquestionably a large UAS. The MQ-1 Predator shown in Figure 2 is generally classified at the lower-end of the large class or the upper-end of the medium class, with a physical size on the order of a small general aviation aircraft. Usually categorized on the lower-end of the medium UAS with a maximum takeoff weight of about 250 lb (113 kg) the US Army's Shadow 200 is shown in Figure 3. The Scan Eagle in Figure 4 is in the small size category, consistent with a large hobby RC model, and the Raven shown in Figure 5 is small enough to be launched by hand. The Wasp shown in Figure 6 is generally categorized as a micro aerial vehicle (MAV), even though it is a bit larger than the original MAV definition where no single dimension exceeds 6 inches (Masoud, 1999).

In 2009, a FAA Aviation Rulemaking Committee (ARC) recommended that small UAS (sUAS) be classified as in Table 1 in groups according to the physical characteristics of

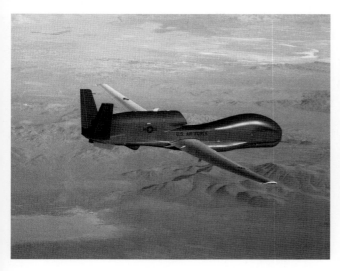

Figure 1. An RQ-4 Global Hawk UAS soars through the sky to record intelligence, surveillance, and reconnaissance data. Accessed 5 February 2012. Photo reproduced from U.S. Air Force. Available at http://www.af.mil/shared/media/photodb/photos/090304-F-3192B-401.jpg.

the UAS that include weight, frangibility (ability to breakup on impact), maximum speed, and type of flight control system (Small Unmanned Aircraft System Rule Making Committee, 2009). The classification by weight and speed accounts for the kinetic energy of the airframe, and the flight management system generally refers to whether the UA is flown with an autopilot that provides some level of automatic control of the UA in flight. The weight classification for the largest sUAS category was chosen consistent with that for the maximum weight of the airframe according to the guidelines

Figure 2. An MQ-1 Predator UAS (right) and F-16 Fighting Falcon return from an Operation Iraqi Freedom combat mission. Accessed 5 February 2012. Photo reproduced from U.S. Air Force/1st Lt. Shannon Collins. Available at http://www.af.mil/information/factsheets/factsheet.asp?fsID=122.

Figure 3. The Shadow UAS has flown 505 000 combat flight hours, making it a significant contributor to the Army's unmanned aircraft fleet surpassing 1 million combat flight hours. Accessed 5 February 2012. Photo reproduced from U.S. Army. Available at http://www.army.mil/media/164647.

established in the FAA Advisory Circular for unregulated RC hobby aircraft (Department of Transportation Federal Aviation Administration, 1981). The ARC provided no classification recommendations for lighter-than-air sUAS, and at the time of this writing the ARC recommendations have yet to be approved.

Advances in the miniaturization of sensors, microprocessors, and high-performance computers, in terms of both

Figure 4. A Scan Eagle UAS is launched from a catapult. Accessed 5 February 2012. Photo reproduced from U.S. Air Force. Available at http://www.af.mil/information/factsheets/factsheet.asp?fsID=10468.

Figure 5. A US Army soldier preparing to hand-launch a Raven UAS. Accessed 5 February 2012. Photo reproduced from Michael Allison. Available at http://www.army.mil/media/108015.

physical size and power consumption, have enabled sUAS to be operated with levels of automation and sophistication comparable to that of large UAS. The physical size of a UA is now determined almost entirely by performance requirements, cost, and the specific mission (e.g., speed, range, and payload capacity) and not by the onboard control and processing capabilities. Given these capabilities, sUAS are not only used for missions where their size is determined by the mission requirements, they now frequently serve as surrogates at test sites to explore missions and operations concepts in place of larger, more expensive UAS with proportionally larger logistical requirements.

Figure 6. The underside of the Wasp micro-air vehicle shows the breakaway fins and rudder, control servos, skid plate, camera payload, skid bar, and carbon-fiber leading-edge wings. Accessed 5 February 2012. Photo reproduced from Gunnery Sgt. Frank Patterson. Available at http://www.marines.mil/Pages/PhotoDetails .aspx?ItemUrl=http://www.marines.mil/unit/hqmc/PublishingImages/ Wasp%204.JPG#.T6Gr99X0k8B.

Table 1. FAA Aviation Rulemaking Committee's proposed classification of sUAS

Physical characteristic	Group I	Group II	Group III	Group IV
Weight	2 kg (4.4 lb)	2 kg (4.4 lb)	9 kg (20 lb)	25 kg (55 lb)
Frangibility	Required	Not required	Not required	Not required
Max speed	<30 kt (mph)	<60 kt (mph)	<87 kt (mph)	<87 kt (mph)
Manual flight control	√	√	√	√
Automatic flight management	⊗	√	√	√

√ Allowed, ⊗ Not permitted

Small Unmanned Aircraft System Rule Making Committee (2009).

3 REGULATORY COMPLIANCE

3.1 UAS certification

In the United States, as part of the Department of Transportation, the FAA develops and administers airworthiness standards, which must be satisfied for an aircraft to be certified for flight. The FAA also develops and maintains standards for pilot certification. Standards for the certification of aircraft, pilot certification, and regulations for the operation of aircraft in the NAS are codified in the Federal Aviation Regulations (FARs), Title 14 of the Code of Federal Regulations (CFR). As of this writing, there are no FARs specific to the certification or operation of UAS, nor for their remote pilots. Therefore, UAS must be operated according to the FARs developed primarily for manned aircraft. The *see-and-avoid rule* in the General Operating Rules of FAR Part 91 has presented the greatest challenge for the operation of UAS in the NAS. Although there has been progress in the development of *sense-and-avoid* technologies that often rely on non-vision-based sensing systems and automatic decision-making systems, no sense-and-avoid system is yet certified to satisfy the see-and-avoid rule. In lieu of a certified see-and-avoid system, a UAS can only fly in the airspace regulated by the FAA with special provisions that compensate for the inability to directly satisfy the see-and-avoid requirement from the UA. This might include an observer in a chase plane, or a ground observer who can provide the visual function to an "equivalent level of safety" of a manned aircraft. Currently, these observers must have a class-2 medical certification, which is a greater requirement than the class-3 medical requirement for conventional general aviation pilots in a manned aircraft.

The FAA does not regulate airspace in active Restricted, Prohibited, or Warning areas (Davis, 2008). Operations in these areas are conducted based on the rules of the test-range manager. Similar to manned aircraft, UAS are broadly categorized as (U.S. Code 40102 Title 49, 2012) (i) *civil UAS* that are operated for commercial purposes and (ii) *public UAS* that are operated by the US or state governments, and their associated agencies. Civil UAS require a special airworthiness certificate in the experimental category for operation in the NAS, and public UAS operations require a certificate of authorization or waiver (COA). The COA enables a proponent to "self certify" a UAS with the submission of an airworthiness statement prepared according to an accepted standard as described in the reference Davis (2008). Therefore, the certification/authorization process, and access to the NAS, is currently simpler for public UAS. As of this writing, there are few civil UAS operating with a special airworthiness certificate; there are many public UAS operations currently conducted by the Department of Defense (DoD), Department of Homeland Security (DHS), the National Aeronautics and Space Administration (NASA), the National Atmospheric and Oceanic Administration (NOAA), and several public universities. Because they are under the oversight of state governments, public (state) universities are allowed to operate public UAS with a COA. Private universities, however cannot, unless they are contracted to work under a COA obtained for a public UAS.

While most member states in the United Nations manage their airspace according to the accepted standards of the International Civil Aviation Organization (ICAO), there are no specific ICAO standards for UAS. As of this writing, several ICAO countries have similar policies to the United States, but generally pursue their own civil aviation interests with processes developed for allowing UAS in their specific airspaces.

3.2 The COA application process

Elston *et al.* (2011a, 2011b) describe the COA application process for the Tempest UAS. As previously stated, in the application process, the airworthiness statement is used to establish that the UAS has been constructed according to an acceptable standard (the Air Force Military Handbook 516B was used for the Tempest UAS) and that it can be safely operated with acceptable risk. Figure 7 shows the annotated aviation chart submitted as part of the WSA-2008-COA application that shows the 20 mi × 20 mi (32 km × 32 km) boundary of a COA for the tempest UAS located in the Pawnee National Grasslands in northeast Colorado. Figure 8

is an annotated aviation chart that illustrates the process for constructing COA boundaries. *Victor airways* (broad lines designated by the letter "V" followed by a number) are corridors in Class E airspace for aircraft flying under instrument flight rules (IFR) that extend from 1200 ft AGL to 18 000 ft AGL. The victor airway labeled V160 is indicated by an arrow in Figure 8. The victor airways are observed to establish the situational awareness of the UAS flight crew for potential air traffic. To minimize the risk of collisions with people or property on the ground, the COA boundaries for this application were chosen a minimum of 5 mi from population centers and 1 mi from major highways and interstates. Other requirements for the airworthiness statement include a summary of the airframe performance capabilities (rate of climb, turning rate, etc.), an overview of the communications and control systems including the autopilot systems, and a detailed listing of emergency procedures and contingencies for situations where communications or control of the UA is lost. This document, along with additional information about pilot medical and operational certification, and observer requirements, are entered directly into the online COA application interface (COA, 2012). The submitted application is then reviewed by FAA safety and air-traffic offices, followed by coordination with any aviation facilities (e.g., airports) that might be affected by UAS operations. The nominal turnaround time from submission (assuming no corrections or modifications are required) is 60 working days. The nominal duration of the COA is 1 year, and it can be renewed for an additional year.

4 UAS PLATFORM DESIGNS

UAS airframes are usually manufactured with the materials and construction techniques common to conventional aircraft in their size and performance class. For example, the Global Hawk (Figure 1) is constructed using aluminum and carbon-composite materials with a structural design similar to recently fielded military and commercial aircraft, while many sUAS airframes are constructed with off-the-shelf materials and components originally developed for RC hobby aircraft. Because the physical appearance of some sUAS is indistinguishable from a hobby aircraft, for example, the Raven (Figure 5) and the Wasp (Figure 6), there continues to be a lingering problem with illegal operation of UAS without a COA or an experimental certificate; the FAA guidelines for hobby aircraft (Department of Transportation Federal Aviation Administration, 1981) are often mistakenly cited as giving permission to fly sUAS. This has created persistent tension between legitimate UAS operators, RC hobbyist, and the FAA.

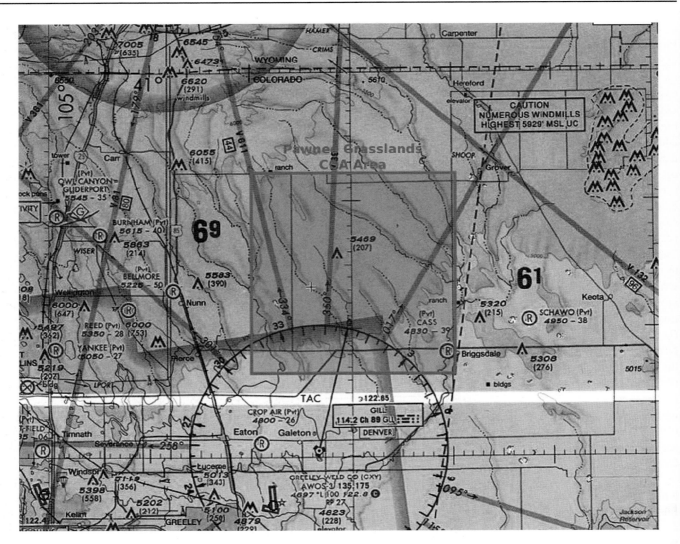

Figure 7. Annotated aviation chart submitted for 2008-WSA-82-COA. The shaded box in the center of the chart designates the COA airspace (Elston *et al.* 2011).

The physical size and performance of the UAS must be considered in the certification process, for both special experimental certification and the COA, and are primary factors in the design of flight experiments. Figure 9 compares the performance of "fixed-wing" UA from representative UAS. Each plot shows the clustering of the UA according to propulsion type. All but the smallest UA are powered by internal combustion (IC) engines (piston, rotary) or gas-turbine engines similar to, or the same as, those used in manned aircraft and RC hobby aircraft. Many of the smallest UA employ electric propulsion systems powered by batteries,[2] and fuel cell powered sUAS are becoming more common (Flinn, 2006). A comparison using the traditional aircraft performance metrics shows that the largest and fastest UA are propelled by turbofans, although UA propelled by IC engines flying at slower speeds can provide endurance comparable to turbofans. Even the most efficient batteries have far less energy storage capacity and propulsive-power per unit weight compared to hydrocarbon-fueled IC and gas-turbine propulsion systems. This is reflected in Figure 9a–c by the clustering of the small UA propelled by electric motors in the lower left corner. Electric propulsion becomes more attractive for the smallest UA in part because of the increased inefficiency and relatively poor performance and reliability of small IC and turbine engines.

An increasing number of UAS operators, particularly university and industrial research groups, employ small rotorcraft (helicopters) that include "multi-rotor" aircraft that use more than one rotor to provide lift and directional control. The ability to hover is the primary feature that distinguishes rotorcraft from fixed-wing aircraft. This can be particularly important, for instance, in applications that require a sensor to

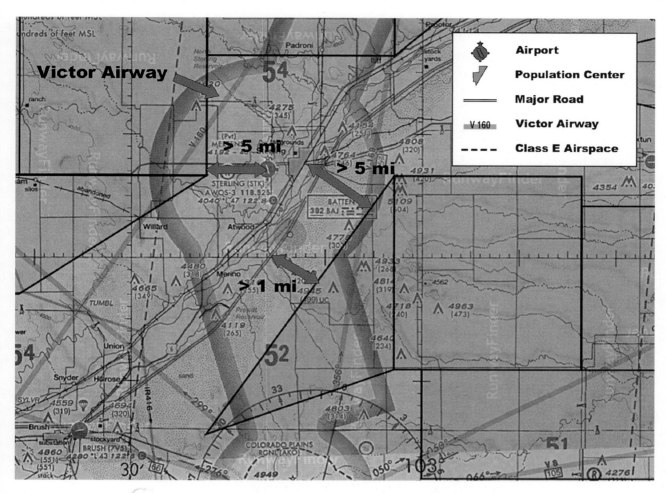

Figure 8. Annotated aviation chart illustrating the procedure for establishing COA airspace boundaries (Elston *et al.* 2011).

"stare" at a target from some fixed location, or be able to follow or track a target or phenomenon, and keep some fixed distance when the target is moving or is stationary.

In recent years, the quadrotor (four rotors) has become one of the most used rotorcraft configurations. They can generally be flown in relatively small areas and indoors, usually with quiet battery-powered motors. They have become the sUAS of choice for many researchers because when flown indoors they are not subject to FAA oversight, and they are particularly useful as a test bed for multi-aircraft communications and cooperative UAS teams.

5 MILITARY APPLICATIONS

Most large UAS currently in operation were developed for military applications; each of the examples shown in Figures 1–6 have military origins. The ability of large UAS to fly high, far, and fast, and to carry large and powerful sensor payloads have made them an indispensable part of the current US warfighting capability, and a growing capability of other nations. With the ability to cruise above 60 000 ft, and with an endurance of over 35 h, the Northrup Grumman Global Hawk, shown in Figure 1, is in a performance class of its own. In 2001, a RQ-4A Global Hawk made the first trans-Pacific flight by a UAS, when it was launched from Edwards Air Force Base in California, to land at the Royal Air Force Base in Edinburgh, South Australia. The RQ-4B Global Hawk is being developed with increased payload and endurance compared to the RQ-4A to enhance its intelligence, surveillance, and reconnaissance (ISR) mission capabilities.

In addition to the traditional ISR mission, some medium-to-large UAS have been armed to carry out lethal missions. The MQ-1A Predator, manufactured by General Atomics Corp., was originally deployed as an ISR platform during the second Iraq War, it was armed with the Hellfire missile and became the first armed UAS to carry out precise, lethal missions that combined its persistent surveillance capability

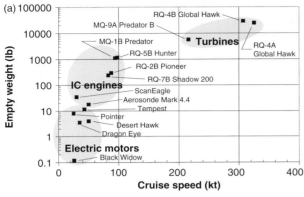

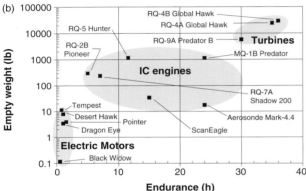

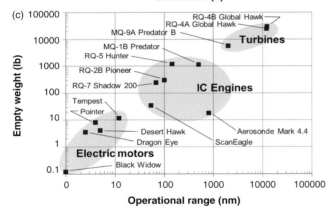

Figure 9. Comparison of UAS performance for representative UAS grouped by propulsion type, (a) cruise speed (kt), (b) endurance (h), and (c) range (nautical miles).

with the precise target engagement capability of the Hellfire missile. Figure 2 shows an armed Predator MQ-1B with a manned F-16 Fighting Falcon in the background.

The Joint-Unmanned Combat Aircraft System (J-UCAS) program, terminated in 2006, employed UAS technologies developed from the Boeing X-45A unmanned combat air vehicle (UCAV) technology demonstrator that first flew in 2002. Technologies from these programs continue to be developed as part of the X-47 N-UCAS program managed

by the US Navy. These UAS are being developed as large, heavily armed, high-performance systems that will conduct the most dangerous missions currently performed by manned aircraft, particularly the suppression of enemy aircraft defenses (SEAD) mission. Several of the medium and small UAS discussed earlier have been deployed, some in large numbers, in Iraq and Afghanistan to conduct medium-range ISR missions.

6 CIVILIAN APPLICATIONS

UAS are being applied in an increasing number of civilian applications to cover a range of missions that include Arctic sea-ice studies, border patrol, investigations of severe weather phenomena, law enforcement, wildfire detection and monitoring, and pollutant detection and tracking. For many of these applications, a UAS is employed for the same reasons that UAS have proven so useful in military applications. They can operate in conditions and environments too dangerous or risky for manned aircraft and they provide persistence for targeted observations that is often superior to that of manned aircraft.

In October 2007, the Symposium for Civilian Applications of Unmanned Aircraft Systems (CAUAS) was convened in Boulder, Colorado. This meeting brought together major stakeholders for civilian applications of UAS in the NAS (Argrow, *et al.*, 2009). Symposium attendees concluded that the top three civilian applications areas for the decade 2008–2017, in order of priority are (i) disaster response, (ii) homeland security, and (iii) climate change. As discussed earlier, FAA policies make it more difficult to certify UAS for civil operations in the NAS than for public applications. Consequently, it was concluded that the federal agencies (e.g., NASA, NOAA, DHS) should lead the effort to integrate civilian UAS operations into the NAS. Although there is great demand for civilian applications of UAS, and manufacturers are anxious to develop systems for a potential market, current UAS cannot satisfy the FAA regulatory requirements developed for manned aircraft, as discussed earlier. Therefore, actual deployments to conduct civilian missions continue to be few in number.

Although most of the UAS plotted in Figure 9 were developed and deployed for military applications. Several military systems have been "repurposed" for civilian missions, or a variant of the original military design has been developed for civilian applications. As an example, the characteristics of the Global Hawk UAS that make it an unparalleled long-range, high-altitude ISR platform for the US Air Force also make it an ideal platform for a number of civilian remote-sensing applications. Figure 10 shows a repurposed Global

Figure 10. A NASA Global Hawk undergoes systems testing while parked on the ramp at NASA's Dryden Flight Research Center. Accessed 5 February 2012. Photo reproduced from NASA. Available at http://www.nasa.gov/centers/dryden/multimedia/imagegallery/Global_Hawk/index.html.

Hawk with the NASA logo being prepared for a civilian, high-altitude remote sensing mission and Figure 11 shows the flight plan for a NASA Global Hawk that overflew tropical depression Frank during the 2012 hurricane season.

Derived from the General Atomics Predator B, the NASA Ikhana is a medium-altitude, long-endurance UAS. Ikhana is currently an experimental test-bed system that NASA has deployed for a number of missions, including to image wildfires in the western United States. Figure 12 shows Ikhana with an instrument pod clearly visible under the left wing and Figure 13 shows infrared imagery captured by Ikhana during the Gap Fire in Santa Barbara County, California on 8 July 2008 (NASA, 2008). Different shading in this rendering corresponds to burned and unburned areas. The Department of Homeland Security has also deployed fleets of Predator B UAS to patrol large portions of the US borders with Mexico and Canada.

Although the US Office of Naval Research provided initial support starting in 1995, by the time it became operational in the late 1990s, the Aerosonde UAS was being developed primarily for civilian meteorological missions for

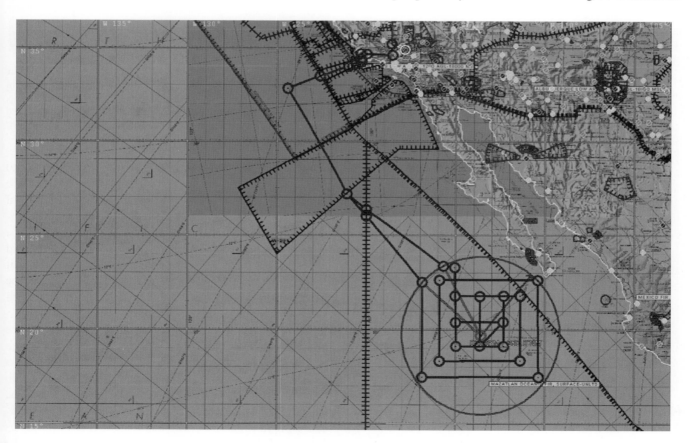

Figure 11. Planned flight path of NASA's Global Hawk high-altitude environmental science aircraft during its 28 August 2010 flight over tropical depression Frank off the southwestern tip of Baja California. Accessed 5 February 2012. Photo reproduced from NASA. Available at http://www.nasa.gov/centers/dryden/multimedia/imagegallery/Global_Hawk/index.html.

Figure 12. Ground crew prepares NASA's Ikhana remotely piloted research aircraft for a flight on 2 July 2008. The infrared imaging sensor pod is visible under the left wing. Accessed 5 February 2012. Photo reproduced from NASA/Tony Landis. Available at http://www.nasa.gov/centers/dryden/multimedia/imagegallery/Ikhana/index.html.

the study of severe weather (Holland *et al.*, 1998). In 1998, the Aerosonde *Laima* flew for 26 h 45 min, and more than 2030 mi (3270 km) to become the first unmanned aircraft to cross the North Atlantic Ocean. At about 30-lb (14-kg) take-off weight, early versions of the Aerosonde are in the sUAS

Figure 13. The Autonomous Modular Scanner carried aboard NASA's Ikhana unmanned aircraft captured this image of the Gap Fire in Santa Barbara County, California, 8 July 2008. Accessed 5 February 2012. Photo reproduced from NASA/Google. Available at http://www.nasa.gov/topics/earth/features/ikhana_images.html.

Figure 14. The Aerosonde unmanned aircraft is released from its transport vehicle on the runway at the NASA Wallops flight Facility, Wallops Island, Virginia, to fly into and take measurements of Tropical Storm Ophelia. Accessed 5/2/2012. Photo reproduced from NASA. Available at http://www.nasa.gov/vision/earth/lookingatearth/aerosonde.html.

Group-IV category of Table 1. Weighing up to 55 lb (25 kg) at takeoff, the latest version of the Aerosonde, the Mark 4.7, pushes the maximum Group IV weight limit, with much of this growth in size to accommodate military applications. In September 2005, NASA and NOAA teamed-up to fly the Aerosonde (Figure 14) into Hurricane Ophelia, to conduct the first observations of the near-surface, high-wind hurricane environment, flying as low as 500 ft (152 m) above the waves (NASA, NOAA and Aerosonde Team Up on Hurricane Observation Milestone, 2005).

A recent example of a low-cost sUAS used for severe weather research is the 12-lb (5.4-kg) Tempest UAS developed in the University of Colorado's Research and Engineering Center for Unmanned Vehicles (RECUV) shown in Figure 15. The 59 COAs in Colorado, Kansas, Nebraska, and Wyoming, obtained for the spring 2010 VORTEX2[3] field deployment are indicated by the light-colored polygons shown in Figure 16. The smaller unlabeled boxes are the locations for test flights conducted during the field deployment, and boxes designated by dates are the locations for

Figure 15. The tempest unmanned aircraft returns to base after a supercell storm intercept mission in northwest Kansas. Reproduced by permission of University of Colorado/Jack Elston.

actual supercell-intercept flights. On 10 June 2010 near Last Chance, Colorado, the Tempest UAS, flown by a team from the University of Colorado Boulder and the University of Nebraska Lincoln, made the first UAS intercept of a tornado-producing supercell thunderstorm. During the design of the airframe, supercomputer simulations were used in which a simulated tempest UA was flown to establish the UAS performance and design requirements for the supercell-intercept mission. The 59 COAs that allowed Tempest UAS operations over about 24 000 sq mi (62 000 sq km) of the western Great Plains was considered a major accomplishment. A typical flight required that a Notice to Airmen (NOTAM) be filed 2 h before launch and a phone call to the local Area Route Traffic Control Center (ARTCC) 30 min before launch. This large area and short-time requirement allowed the fast-moving storms to be targeted and tracked before each launch of the Tempest UA.

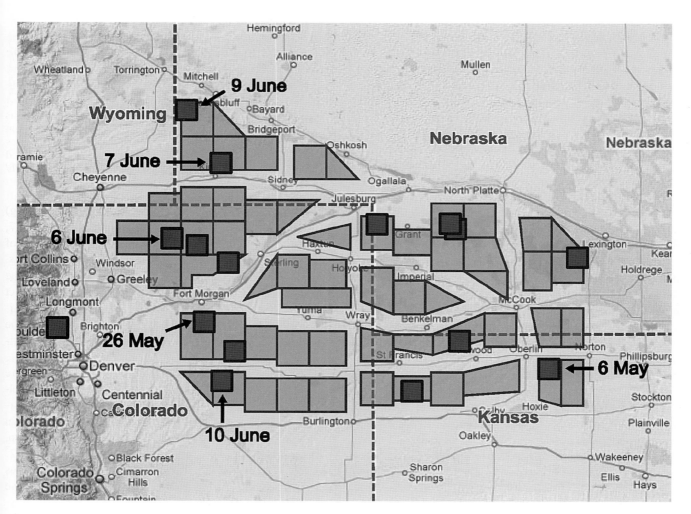

Figure 16. Fifty-nine COAs obtained for the VORTEX2 deployment of the tempest UAS, 1 May–11 June 2010.

The see-and-avoid rule required that the RECUV team apply C3 and tracking technology derived from earlier military applications. For the storm-intercept mission, the UA was programmed to track a WiFi node (a conventional 2.4-GHz link to a laptop computer) located in a ground "tracker vehicle" (a sport utility vehicle). The Tempest UA was programmed to orbit the WiFi node on the tracker vehicle to keep the UA within the observation distance prescribed in the COAs. Inside the tracker vehicle, a mobile operator could issue simple UAS commands to assist an observer who scanned the skies to satisfy the see-and-avoid requirement. A mobile ground station that was stationary during the flights maintained a 900-MHz radio link directly to the UA so that primary C3 could be maintained in case the less reliable WiFi link was dropped. More details of the Tempest UAS missions, and an earlier storm research mission conducted in the Pawnee National Grassland with the RECUV NexStar UAS, are provided in Elston et al. (2011, 2011) and Houston et al. (2012).

7   UAS FLIGHT TEST PROCEDURES

Flight test plans for UAS depend on several factors that include the UA size and performance, the C3 infrastructure, payload, and the regulatory environment. While specific details differ based on these factors, the overall procedures for UAS testing are general and scalable from large, high-performance systems, to battery-powered sUAS with miniature payloads. This discussion will focus on flight tests in the US NAS, so one must start with designing experiments within the constraints of the Federal regulatory environment. For instance, if operations are confined to a prescribed test site, the COA will specify the rules for UAS operations, with *provisions* that spell out additional operational constraints not specifically articulated in the FARs. For example, there might be specific language for the see and avoid requirement that might require either a manned chase aircraft, or ground observers positioned to assist the pilot in command of the UAS to meet the see and avoid rule. The provisions will require the proponent to issue a NOTAM and might also include a requirement to notify local airports, TRACON (Terminal Radar Approach Control), and ARTCC facilities some time in advance of the UA launch.

The aircraft size and performance must be taken into account during the COA application process to ensure that there is enough areal coverage inside the boundaries for the UA to maneuver with minimal risk of crossing the boundaries. The size and performance also directly determine any flight-termination plan. Even in the worst situation, a sUAS constructed of foam and weighing a few ounces poses little risk if its flight is terminated and it impacts a person or property. However, the termination of the flight of a Global Hawk or Predator B obviously could pose great risk to life and property. These risks must be accounted for in the airworthiness statement and in the provisions and emergency procedures articulated in a COA.

If communications are planned over licensed frequencies, there must be some coordination with the Federal Communications Commission (FCC) for civilian radio frequencies. If government frequencies are to be used, coordination with the National Telecommunications and Information Administration (NTIA) might also be required. The application process for FCC and NTIA permissions are carried out independently of the COA application process.

Preflight checklists should include verification of the weights and balances of the UA, particularly the location of the center of gravity, and a sequential check of all UAS subsystems. A sample of the preflight checklist is generally required in a COA application. The preflight check should include calibration of sensor transducers and connections, and the GPS unit that deliver data to the autopilot system. If the UAS will carry a payload that payload should be powered as part of a full-system check. It is good practice to ensure that the payload power system can be isolated from the critical propulsion and C3 subsystems to ensure that the critical flight systems will remain operational after a payload failure. It is also critical that the payload be fully powered during the preflight check to verify that radio frequency (RF) interference is not present. Interference can be generated by RF emissions from computer components such as a motherboard, so can be present even if there are no dedicated radio transmitters in the payload or flight computers. If the UA will fly with an onboard autopilot, hardware-in-the-loop simulations are advised to discover and correct potential problems that might not be otherwise detected.

A survey of current UAS research literature, for example, the references Houston et al. (2012), Elston et al. (2011a) Stachura (2011), and Ostler et al. (2009), shows that most current UAS research is focused on networked C3 infrastructures, and mobile sensing and data collection. Therefore, it is important that facilities are in place to record the flow of data and information flow through the C3 infrastructure, as well as autopilot flight data. An example of a flight test range design, illustrated in the Figure 17 schematic, is at the Table Mountain Test Site owned by the US Department of Commerce and operated by the NTIA's Institute for Telecommunications Science (ITS). RECUV developed and operates the flight test site through a cooperative research and development agreement (CRADA) with ITS. The site is located on the Table Mountain mesa 8-mi (10 km) north of the University of Colorado Boulder campus. The mesa top is approximately oval, running about 2.5 mi (4.0 km) north to south and

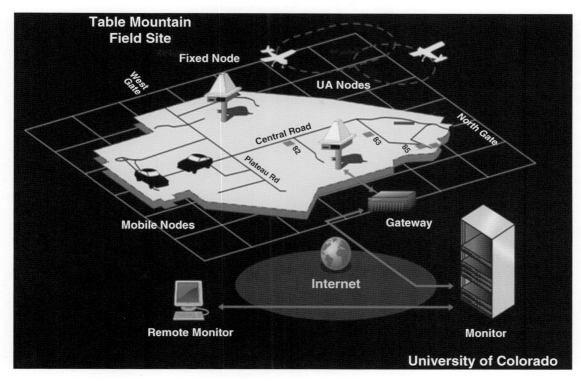

Figure 17. Schematic of the University of Colorado Table Mountain Test Site.

1.5 mi (2.4 km) east to west. The schematic highlights the C3 and data infrastructure that enables data collection to be routed through the Internet, where it is archived on a server on the Boulder campus. This Internet routing also allows real-time, remote access to flight-test activities from any participant with an Internet connection.

8 PUBLIC PERCEPTION OF UAS

Recent military activity in Iraq, and ongoing activities in Afghanistan and Pakistan have brought attention to US military operations employing UAS – images of the Global Hawk and armed Predator UAS have become synonymous with the global reach of the US military. At the 2007 CAUAS Symposium (Argrow and Weatherhead, 2009), it was reported that the US Armed Forces community has a very high opinion of UAS, and that there is a building expectation that the skills acquired by military personnel to operate UAS during deployments will be valuable for those returning from war. Because of the slow pace of regulatory reform to integrate UAS into the NAS, the market for UAS to perform civilian applications remains almost nonexistent.

An additional discussion at the Symposium was the public perception of robotic aircraft. In addition to flight-safety concerns, the potential for invasion of privacy is a major concern.

With their now well-known capabilities for surveillance, many citizens are concerned about how this capability might be applied.

Although automation, such as autopilots and collision avoidance systems have unarguably increased the safety of manned aircraft operations, there is some undefined threshold above which the public, particularly aircraft pilots, become uncomfortable with increased automation (Murphy and Argrow, 2009). This sense of the potential loss of the decision-making capability of a human pilot is not only a concern of pilots and the general public, but also it is reflected in the current regulatory policies of the FAA. Issues related to human–machine interaction are at the forefront of research to develop trusted autonomous systems, and will be a major focus of UAS-related research for the foreseeable future.

9 CONCLUSION

The demand for UAS is already well established for military applications. These systems are continuing to increase the capability to conduct war remotely, which has raised a number of ethical issues with the potential to constrain future military development (Singer, 2009). The demand for civilian applications is also great, and increasing. However, regulatory constraints will continue to severely limit the number, type, and frequency of civilian missions. Even though a significant number of UAS have proven their ability to be safely operated in the NAS, a few spectacular failures remind us of the motivation for airspace regulatory agencies to constrain UAS operations (Peck, 2003). One of the problems of UAS compliance with current regulatory policy is the notion that UAS must be operated with a level of safety at least equivalent to manned systems. Many of these requirements are based on expected human capabilities, such as vision requirements, where the standards are difficult to define, and are therefore difficult to replicate for UAS. Such requirements also often ignore that some systems can exceed the capabilities of any human in detecting and tracking aircraft. Some consider the detection and tracking problem as the "easy" part of a flight safety solution, with the difficult part in the requirement to develop reliable, FAA-certifiable, decision-making systems that can operate with a high level of autonomy to maneuver a UAS to maintain flight safety.

An additional technological challenge to the integration of UAS into the NAS is the frequency and bandwidth demand for communications. In many instances, a radio link between the UAS control station and the UA must be guaranteed for safe operations. Multiplying this requirement for a large number of UAS operating simultaneously easily creates a scenario where radio interference could disrupt missions and operational safety. This is currently an area of research that is being funded, in part, by the FAA, and solutions to meet the radio communications problem must be developed before there is widespread integration of UAS into the NAS.

Regardless of the pace of airspace integration, UAS will eventually become common occupants of the sky. An often quoted example of the first widespread use of UAS in a commercial application is that of large cargo transport aircraft that fly long distances with no passengers other than the onboard pilots, whose function is to essentially monitor the automatic controls, except for takeoff and landing. Automatic takeoff and landing has been a capability of some manned aircraft for many years, so the use of onboard pilots to control the aircraft during takeoff and landing is often viewed as an artificial requirement, thus suggesting that the entire mission is suitable for a UAS. One of the most ambitious questions asked at the 2007 CAUAS Symposium was "When can I expect to have a UAS deliver a pizza to my door?"

ACKNOWLEDGMENTS

The author recognizes the contributions of Professors Timothy Brown and Eric Frew, and current and former research assistants Thomas Aune, Anthony Carfang, Cory Dixon, Jason Durrie, Jack Elston, Joshua Fromm, Jason Roadman, and Maciej Stachura.

NOTES

1. Note that the C3 sequence begins with communication, a necessary requirement for UAS command and control. This is distinguished from the command, control, and communication sequence used in military parlance.
2. As of this writing, there are no certified electric propulsion systems for manned aircraft.
3. The second Verification of the Origin of Rotation in Tornadoes Experiment (VORTEX2) was funded by the National Science Foundation (NSF) and the National Oceanic and Atmospheric Association (NOAA) for field deployments in the spring of 2009 and 2010.

REFERENCES

Argrow, B, Weatherhead, E. and Frew, E. (2009) Real-Time Participant Feedback from the Symposium for Civilian Applications of Unmanned Aircraft Systems. *J. Intell. Robot. Syst.,* **54**, No. 1–3, 87–103.

Certificate of Authorization or Waiver (COA), Federal Aviation Administration, http://www.faa.gov/about/office_org/headquarters_offices/ato/service_units/systemops/aaim/organizations/uas/coa/ (accessed February 5, 2012).

Department of Transportation, Federal Aviation Administration, Model Aircraft Operating Standards, Advisory Circular AC 91–57, Jun (1981).

Davis, K.D. UAS Interim Operational Approval Guidance 08-01, Unmanned Aircraft Operations in the U.S. National Airspace System, FAA Aviation Safety Unmanned Aircraft Program, Mar (2008). Available at http://www.faa.gov/about/initiatives/uas/reg/media/uas_guidance08-01.pdf (accessed February 5, 2012).

Elston, J., Roadman, J., Stachura, M., Argrow, B., Houston, A. and Frew, E. (2011) The tempest unmanned aircraft system for in situ observations of tornadic supercells: design and VORTEX2 flight results. *J. Field Robot.,* **28**(4), 461–483 .

Elston, J., Stachura, M., Argrow, B., Frew, E. and Dixon, C. Guidelines and Best Practices for FAA Certificate of Authorization Applications for Small Unmanned Aircraft, AIAA Infotech@Aerospace Conference, Mar (2011).

Flinn, E. (2006) Hydrogen fuel cell powers UAV. *Aerosp. Am.,* **44**(11), 28–29.

Holland, G., McGeer, T., Tyrrell, G. and McGuffie, K. (1998) Development of the Autonomous Aerosonde, Final Report, Bureau of Meteorology Research Centre, Melbourne, Australia, Award Number: N-00014-94-1-0493.

Houston, A.L., Argrow, B., Elston, J., Lahowetz, J., Frew, E.W. and Kennedy, P.C. (2012) The Collaborative Colorado-Nebraska Unmanned Aircraft System Experiment. *Bull. Am. Meteorol. Soc.,* **93**(1), 39–54.

Masoud, R.-R. and Hicks, G.R. (1999) Multidisciplinary design and prototype development of a micro air vehicle. *J. Aircr.,* **36**(1), 227–234.

Murphy, R. and Argrow, B. (2009) UAS in the National Airspace System: Research Directions. *Unmanned Syst.*, **27**(6), 23–28.

NASA, NOAA and Aerosonde Team Up on Hurricane Observation Milestone (2005) http://www.nasa.gov/vision/earth/lookingatearth/aerosonde.html (accessed February 5, 2012).

NASA, News Topics (2008), http://www.nasa.gov/topics/earth/features/fire_and_smoke.html (accessed February 5, 2012).

Office of the Secretary of Defense, Unmanned Systems Roadmap (2009–2034), Apr. 2009.

Ostler, J., Bowman, J., Snyder, D. and McLain, T. (2009) Performance Flight Testing of Small Electric Powered Unmanned Aerial Vehicles, *Int. J. Micro Air Veh.*, **1**(3), 155–171.

Peck, M. (2003) Global Hawk Crashes: Who's to Blame? *National Defense* http://www.nationaldefensemagazine.org/archive/2003/May/Pages/Global_Hawk3871.aspx (accessed February 5, 2012).

Singer, P. (2009) *Wired for War: The Robotics Revolution and Conflict in the 21st Century*, Penguin Press.

Small Unmanned Aircraft System Rule Making Committee, Comprehensive Set of Recommendations for sUAS Regulatory Development, Apr. (2009).

Stachura, M. and Frew, E. W. (2011) Cooperative target localization with a communication aware unmanned aircraft system. *AIAA J. Guid., Control Dyn.*, **34**(5), 1352–1362.

U.S. Code 40102 Title 49. Available online at http://www.gpo.gov/fdsys/pkg/USCODE-2011-title49/pdf/USCODE-2011-title49-subtitleVII.pdf (accessed February 5, 2012).

Wilson, J.R. (2011) UAV Roundup 2011. *Aerospace America*, American Institute of Aeronautics and Astronautics, Reston, VA, **49**(3), 22–31.

Chapter 39
Aircraft Communications and Networking

James M. Rankin[1] and David W. Matolak[2]

[1]*Avionics Engineering Center, School of Electrical Engineering and Computer Science, Russ College of Engineering and Technology, Ohio University, Athens, OH, USA*
[2]*Department of Electrical Engineering, University of South Carolina, Columbia, SC, USA*

1 INTRODUCTION

1.1 Background and scope

From the earliest days of powered human flight, communication to and from the aircraft has been essential. This communication is necessary, primarily for safety of flight, but is also necessary, or at least very useful, for a number of other functions. Example functions include relaying of reconnaissance information for search and rescue operations, and allowing ground-based operators control of remotely piloted vehicles. For obvious practical reasons, any communication to/from aircraft beyond very short distances must be wireless.

Our focus in this chapter is on radio communication to/from aircraft.

As is widely known, terrestrial wireless communications have made great advances in the past few decades, and this is continuing. The same can be said of aeronautical communications, although progress in that field is dramatically slower, primarily for reasons of stringent safety requirements and economics. As appropriate, we connect with terrestrial communication systems when discussion of that is useful for understanding. In this chapter we refer to communication to/from aircraft as aeronautical communications. We are also concerned in this chapter only with aeronautical communications that actually take place directly to/from an aircraft, and not with any wired "relay" or "ground links" that may be used in the transfer of messages from one aircraft to another, or in transfer of messages from an aircraft to a distant ground-based recipient unreachable by a direct aircraft-to-ground-recipient communication link. The discussion in this chapter pertains to the lowest three layers of the communication protocol stack: the physical, data link, and network layers. Finally, we restrict our discussion to commercial aeronautical communications, and do not discuss the specialized (yet fairly rich) field of military aeronautical communications. Unmanned Aircraft Systems (UAS) which cover both commercial and military applications are discussed in the Technical and Policy Issues section.

Unmanned Aircraft Systems. Edited by Ella Atkins, Aníbal Ollero,
Antonios Tsourdos, Richard Blockley and Wei Shyy.
© 2016 John Wiley & Sons, Ltd. ISBN: 978-1-118-86645-0.

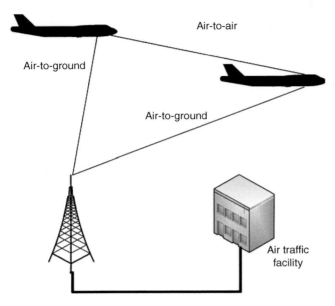

Figure 1. Air-to-ground and air-to-air communication links.

1.2 Aircraft communication systems

In this chapter we consider two primary types of communication systems: air-to-ground (AG) and air-to-air (AA). For the AG case, in most instances discussion applies equally to communication in the opposite direction, that is, ground-to-air (GA). Note that in many current and future aeronautical systems, a given message may traverse both these types of links, for example, a message may be transmitted from one aircraft to another (AA), then from the recipient aircraft to a ground station (AG). Except when analyzing networking performance, these links can be addressed separately. For reasons of simplicity, most often communication systems for AA and AG links employ common techniques, so that from the perspective of the physical and data link (and sometimes network) layers, the AA and AG links can be described simultaneously, with differences between the two identified as needed. Figure 1 displays the air-to-ground and air-to-air communication links.

1.3 Propagation basics for aircraft communications

For the vast majority of aeronautical communications, an optical line of sight (LOS) is present between aeronautical transmitter and ground-based or air-based receiver. Some frequency bands (e.g., high-frequency, 3–30 MHz) do not require this, but for most current and future systems with operation at very high frequency (VHF) bands and higher, a LOS is preferred if not essential.

In the lower atmosphere, specifically the troposphere, propagation of radio energy is via so-called "space waves" (Parsons, 2000), which include an LOS component and reflected components. These reflections are commonly from the ground, or from large structures on the ground, for example, buildings. In this case, a good model for the wireless medium is that of "free space." In this environment, the signal energy spreads with distance, and this spreading causes reduction of signal strength – attenuation – which is often called path loss. This loss is a power loss, and is typically expressed in decibels (dB) as

$$L_{fs}(\text{dB}) = 20 \log_{10}(4\pi d/\lambda) \qquad (1)$$

where d is the distance between transmitter and receiver, and λ is the signal wavelength, $\lambda = c/f$, with c the speed of light and f the signal frequency. As an example, for a link distance of 166 km ($\sim$103.6 miles, or 90 nautical miles) and a carrier frequency of 120 MHz, $L_{fs} = 118$ dB. This means that the signal power at the receiver is reduced by a factor of $10^{11.8}$ from that at the transmitter. These path losses can ultimately be the limiting factor in achievable link distances for a given received signal quality. This is because most communication system performance is a strong function of signal-to-noise ratio (SNR); the larger the SNR, the better the receiver's ability to reproduce the transmitted message accurately.

Another factor in system quality of service (QoS) for aeronautical communications that results from wireless propagation is channel dispersion. In short, this phenomenon most often results when the signal traverses multiple paths in going from transmitter to receiver and this is in general called multipath propagation. When the multiple replicas of the transmitted signal that reach the receiver are spread (dispersed) in time, this can cause severe distortion, which generally must be compensated by some means. Even when a strong LOS signal component is present, if the multipath components are strong enough and delayed enough with respect to the LOS component, system performance can be poor even if SNR is very large. For most aeronautical communications, the multipath replicas emanate from ground reflections, but depending upon location, large buildings can also produce them. As signal bandwidth increases, corresponding to larger data rates, channel dispersion can be significant. For very narrowband signals, distortion caused by dispersion is usually negligible except in cases where the path length difference between the LOS and reflected signal is large, for example, a strong reflection from a distant mountain.

1.4 Chapter contents

In this chapter we provide a brief introduction to the rich and fascinating topic of aircraft communications and networking. Section 2 provides an overview of the fundamentals of aircraft communications, including analog voice communications, a taxonomy of communication types, and aeronautical communications spectrum. Section 3 covers current voice and data link aeronautical communication systems. Section 4 describes future systems, including the use of data links in Next Generation communication, navigation, and surveillance services. Technology and policy issues related to bandwidth limitations, encryption and security, and required communications performance are covered in Section 5.

2 FUNDAMENTALS OF AIRCRAFT COMMUNICATIONS

2.1 Analog voice communications

As with terrestrial telephony, transmission and reception of human voice was important in past aeronautical communications. This continues today. The prevalent transmission technique for these systems was double-sideband amplitude modulation (DSB-AM). Virtues of DSB-AM include simplicity and its ability to operate with signals of multiple bandwidths simply by adjustment of receiver filter bandwidth (the modulated signal bandwidth is approximately 8 kHz). The original DSB-AM systems employed a 100 kHz channel bandwidth (Stacey, 2008). In order to increase the number of available channels in the limited aeronautical VHF spectrum, reductions down to 50 kHz channels were then made, and since the 1970s, a 25 kHz bandwidth has been used in the United States. Most recently, a channel bandwidth of 8.33 kHz has been employed in Europe. The analog voice systems are still in use, worldwide, as of 2009, and will continue to be used for the foreseeable future, even as digitally modulated VHF systems are deployed in a (currently small) set of frequency channels.

2.2 A "taxonomy" of aeronautical communication types

Aircraft communications can be separated into several categories related to air traffic control (ATC), airline operations (AOC), and Flight Informational Services (FIS). ATC communications are two-way communications used to provide positive control to aircraft to ensure safe and orderly operation in the National Airspace System (NAS). Example uses of ATC communications are to provide the aircraft with altitude assignments, transponder codes (for active radar identification), enroute routing, transfer of control to another ATC facility, clearance to take off or land, and taxi instructions on the airport surface. Facilities involved in ATC communications can include ground control, control tower, approach and departure control, and enroute Air Route Traffic Control Centers (ARTCC). These ATC communications use AM in the VHF aeronautical band, from 118–137 MHz (Spence, 2007).

AOC messages are between an airliner and its operations center. AOC messages relate to aircraft and flight crew information that allow the airline to operate more efficiently. Typical AOC communications concern flight status, maintenance issues, flight planning, fuel loading, and gate assignments.

Flight Information Services (FIS) communications provide advisory information that assists the flight crew to operate safely in the NAS. FIS data are relayed to the aircraft via a broadcast or two-way VHF communication channel. FIS data can include local and remote weather, airport status, communication and navigation system status, and notices to airmen (NOTAMs). Aircraft receive FIS data through two-way communication with Automated Flight Service Stations (AFSS) or from automated airport broadcasts (Spence, 2007).

2.3 Aeronautical spectrum allocations

As is the case with most other radio systems, spectral allocation is important to reliable, "interference-free" operation. Current ATC aviation communication channels are in the VHF frequency spectrum from 118.0 to 137.00 MHz. Each channel is allocated a 25 kHz (or 8.33 kHz) spectrum "slice." There are also aeronautical bands in the high frequency (HF) range from approximately 2.85–28.35 MHz. These bands are used for low-data-rate communications when aircraft are over the ocean and poles; HF propagation employs ionospheric "bounce" for trans-global signaling. In addition, spectrum in the ultra-high frequency (UHF) band, from 960 to 1215 MHz has recently been used for aeronautical communication. This band has traditionally been used for navigation (and is shared with military use), but several current systems, and a number of proposed systems, may employ this band. Similarly, spectrum formerly allocated for airport landing systems in the 5 GHz band has been authorized for communications on airport surface areas. Finally, several aeronautical satellite bands (1544–1545 MHz, 1645.5–1646.5 MHz, and 14–14.5 GHz) are also used. As with the HF bands, these satellite services are mostly used for oceanic and polar communications. Figure 2 illustrates the major aeronautical communications bands.

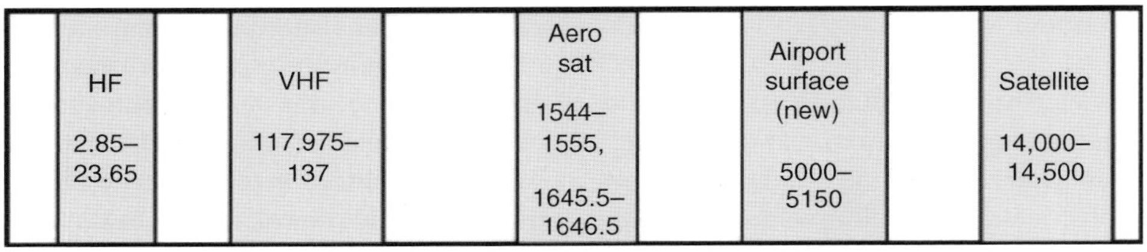

HF 2.85– 23.65		VHF 117.975– 137		Aero sat 1544– 1555, 1645.5– 1646.5		Airport surface (new) 5000– 5150		Satellite 14,000– 14,500

Figure 2. Major aeronautical communication frequency bands (in MHz).

3   CURRENT COMMUNICATION SYSTEMS

3.1   Voice communications

Air Traffic Control (ATC) communications continue to use analog voice communications (DSB-AM). Each ATC facility or function such as an enroute center or approach control has multiple frequency channels assigned for their use. As aircraft transit their airspace control volume, the aircraft are assigned a frequency channel. This may be thought of as frequency division by spatial (airspace) volume. The number of aircraft using a channel may be few or many, and the channel is shared via a "manual carrier sense multiple access" (CSMA) protocol, in which aircraft and ground controller listen before speaking. This is sometimes called "listen before push-to-talk." In situations where there are many aircraft on a channel, it may be difficult for a pilot to find an opening in which to talk. This channel congestion is one of the major problems with analog voice communications. Another problem with analog voice communications is the available spectrum. In Europe, the 25 kHz channel has been separated into three 8.33 kHz channels in order to nearly triple the number of communication channels.

Airlines use voice communications as one method for the flight crew to communicate with their AOC, or dispatch, as well as airline operations at the destination airport. Each airline maintains its own unique frequencies and may also utilize the nationwide VHF network provided by service providers such as ARINC. The AOC frequency channels are in the same 118.0–137.000 MHz band as ATC communications, but are assigned to specific (unique) frequencies.

Non-control, or advisory, FIS information is communicated to the aircraft through continuous VHF broadcasts or through two-way communications. Examples of broadcast FIS information systems are the Automated Terminal Information System (ATIS) or Automated Weather Observing System (AWOS) broadcasts, whereas two-way communication with an Automated Flight Service Station (AFSS) or Enroute Flight Advisory Service (EFAS) can provide pilot

requested FIS information. Local weather conditions, runway and taxiway status, the instrument approach procedure in use, and NOTAMs can be broadcast to aircraft within 60 nautical miles (nm) via the Automated Terminal Information System (ATIS). The AWOS, which is used at non-towered airports, broadcasts real-time weather information to aircraft within 25 nm. The AFSS can provide updates on enroute and destination weather, open and close flight plans, and provide facility information based on the flight crew's request. EFAS or "Flight Watch" provides aircraft flying at least 5000 feet above ground level the capability of requesting weather information (Spence, 2007). Both Flight Watch and AFSS accept pilot reports (PIREPs) of weather that the aircraft is encountering.

3.2   Digital datalink communications

As digital datalinks appeared in aviation, initial systems consisted of a digital radio that used a stand-alone digital Control Display Unit (CDU) on an analog flight deck. With the advent of the Boeing 757/767 in the 1980s as the first "digital" airliner, the aircraft flight deck incorporated digital technology to handle flight controls, displays, and radios. This transition from analog to digital provided the opportunity to use digital datalinks for aircraft communications and surveillance. It also allowed the various systems to be integrated so that the digital datalink messages can be transferred to other systems such as the Flight Management System (FMS) (see Modern Aircraft Flight Management Systems).

3.2.1   AOC datalink

The Aircraft Communications Addressing and Reporting System (ACARS) was developed in the 1970s to allow airliners to link information to their AOC. Typical information sent over ACARS were aircraft out of the gate/off the ground/on the ground/in the gate (OOOI) time reports, updated estimated time of arrival, winds aloft measurements, aircraft maintenance and fuel status (ARINC2010). Initially, the ACARS system used a character-oriented protocol over a 2400 bit per second VHF communications channel. The communications service provider, ARINC or SITA, was required to maintain a network of VHF ground stations across the continents to communicate with aircraft. Later enhancements allowed ACARS over the High Frequency (HF) and Satellite Communications (SATCOM) channels, which allowed transoceanic and over the pole flights to stay in AOC contact. In an effort to relieve congestion on the slow ACARS channels, the ACARS system is being transitioned to a VHF Digital Link (VDL) "mode 2" (VDLM2) datalink that has an increased data rate of 31500 bits per second (Roy, 1998). Modulation is differential 8-ary phase shift keying (D-8PSK), and two levels of forward error correction are used (Helfrick, 2009). This scheme also uses a bit-oriented protocol instead of a character-oriented protocol. The bit-oriented protocol allows the airline industry to define new messages and applications that take advantage of the new technology. For airlines that want to support legacy (character-oriented) messages and applications, the communication service providers offer ACARS over AVLC (Aviation VHF Link Control), or AOA. AOA allows the character-oriented ACARS messages to be transmitted across the VDLM2 datalink.

3.2.2 FIS datalinks

Aircraft preparing to depart most airports must receive a departure clearance from Air Traffic Control. The departure clearance details the flight path to the destination, the initial altitude, departure control frequency, and the transponder code. It is possible that the routing may be "Cleared as Filed" which means that the route in the filed flight plan (before takeoff) was accepted. However, when the airport area is congested or when weather is causing deviations, the flight path or route may use many airways and intersections which make the route very complicated to understand. In a non-datalink (i.e., voice-only) environment, the flight crew requests and receives the clearance over the voice communications channel, which can lead to errors if communication link quality is marginal, or if the speech is simply unclear due to speech rate, speaker accent, or listener experience level. With the Pre-Departure Clearance (PDC) datalink system, the flight crew sends a datalink clearance request through their Airline Operations Center via ACARS. Air Traffic Control then uplinks the departure clearance to the aircraft using the ACARS. The departure clearance is then entered into the FMS, either automatically from the ACARS system or manually by the flight crew.

The flight crew receives local airport information such as weather conditions, the instrument approach in use, taxiway

and runway closure information, and facility frequency assignments through the Automatic Terminal Information Service. Initially, ATIS was a continuous broadcast of information over a VHF communications frequency. Each time that the information is changed, the International Civil Aeronautics Organization (ICAO) phonetic alphabet letter assigned to the information is incremented. The first time that communication is established with an air traffic controller at the airport, the flight crew provides the ICAO letter assigned to the information, for example, "with Information Tango," to ensure that the crew has the latest information. Many airports are now using Digital ATIS (D-ATIS). The D-ATIS information is datalinked to the aircraft where the flight crew is able to read the information on a Control Display Unit (CDU). An important advantage of D-ATIS over ATIS is that the flight crew can request D-ATIS information for their destination as well as for the departure airport; the ATIS system can only broadcast the local information.

The Federal Aviation Administration's (FAA) Tower Data Link Service (TDLS) consolidates weather, facility, and clearance information provided in PDC and D-ATIS messages. The messages are relayed to the aircraft using a third party communications service provider. The communications service provider and associated datalink used for PDC and D-ATIS messages can depend on the aircraft type. Airliners typically use their ACARS datalink through their AOC. General aviation aircraft use a Flight Information Services Datalink (FISDL). The FISDL is operated by an FAA contracted service provider and uses two 25 kHz communication frequencies in the 136.450 and 136.475 MHz spectrum.

3.2.3 Surveillance datalinks

ATC surveillance of airborne traffic has historically relied on enroute and terminal radar systems. The radar system is comprised of a primary radar which detects reflections from the aircraft's surface ("skin paint") and secondary surveillance radar (SSR) which interrogates a transponder on-board the aircraft. The SSR uses a 1030 MHz interrogation signal and receives a 1090 MHz reply signal. Early transponders replied with the assigned ATC "squawk code" (Mode A) or with code plus altitude (Mode-C). With the development of Mode-S SSR, which allowed "selective" or addressable transmissions to a specific aircraft, the SSR system incorporated a basic datalink capability. The Mode-S has 56- and 112-bit messages that facilitate data exchange between ATC and the aircraft. A use of the Mode-S datalink by ATC is to uplink position information about other aircraft traffic in the area. This is called Traffic Information Services (TIS). The TIS data is displayed on a Cockpit

Display of Traffic Information (CDTI) display to facilitate the flight crew's "see and avoid" responsibility. The Mode-S downlink capability allows aircraft to transmit 24-bit ICAO address, flight identification, latitude, longitude, altitude, heading, speed, and air/ground status.

4 FUTURE AERONAUTICAL COMMUNICATION SYSTEMS

Digital data link communications are the backbone for aviation's communication, navigation, and surveillance (CNS) systems in the future. Modernization is critical to the continued success and safety of commercial flight, and aeronautical communications plays a key role in ensuring this. New aeronautical communications capabilities are needed in both the near- and long-terms. One reason for this need is the high demand for use of the civil aviation spectrum in the VHF band from 118–137 MHz, but this is not the only motivation. New communications requirements are expected for a number of other reasons, including (i) the desire for new services for multiple applications, (ii) the need for airlines to increase efficiency, and (iii) the need to manage the growing size and complexity of the civil air transportation system.

The US Congress and the Executive Office formed the Joint Planning and Development Office (JPDO) in 2003 for the Next Generation Air Transportation System (NGATS), now referred to as "NextGen." The JPDO has members from multiple federal agencies, including the Departments of Transportation (DOT), Homeland Security (DHS), Defense (DOD), and Commerce (DOC), plus the National Aeronautics and Space Agency (NASA), FAA, and White House Office of Science and Technology Policy. Industry and private sector members also participate in the JPDO. The JPDO's initial charter identified eight key capabilities for which research and development were cited as critical to modernization. These capabilities are: network-enabled information access; performance-based services; weather assimilated into decision making; layered, adaptive security; broad area precision navigation; aircraft trajectory-based operations; equivalent visual operations; and super density operations. The first capability – network-enabled information access – pertains to aeronautical communications in all phases of flight.

4.1 NextGen communications

In the realm of air traffic control (ATC) communications, analog voice communications will be complemented by the Data Communications data link system (DATACOMM). The DATACOMM system will be used to handle non-time critical data between the aircraft and ground. The analog voice channels will continue to be used to convey time critical information. The DATACOMM system uses a VDLM2 datalink channel to convey messages between ATC and the aircraft. Controller-Pilot Data Link Communications (CPDLC) and Tower Data Link Services are two services that will be provided by DATACOMM.

Many pilot-controller communications involve routine messages such as altimeter setting, change of altitude, or change of frequency channel instructions. The CPDLC system was developed to handle these messages automatically through a digital datalink instead of using voice, thus freeing pilots for more important tasks. A set of preformatted message codes were developed to handle the routine messages. The VDLM2 data link allows messages to be "addressed" to a specific aircraft. When a controller sends a CPDLC message to an aircraft, the message is received and displayed, typically, on the aircraft's Flight Management System (FMS). The flight crew has the option to accept or reject the message. If the flight crew would like to send a message requesting an action, the FMS is used to send a request downlink to the controller.

In NextGen, the Tower Data Link Services (TDLS) will not operate through a third party service provider, but will instead be direct from the FAA tower to the aircraft. The Pre-Departure Clearance (PDC) service will be replaced by Departure Clearance (DCL). The flight crew datalinks its clearance request directly to the tower. The clearance will be delivered directly to the aircraft and once the flight crew accepts the clearance, it will be automatically programmed into the Flight Management System. The TDLS will also include a D-TAXI service which datalinks airport surface taxi routes directly to the aircraft. The taxi route will be a sequence of taxiways and intersections that the aircraft should follow from the landing runway to the gate or from the gate to the departure runway.

The NextGen FIS services will still include "weather to the cockpit," which provides graphical and textual weather information to the flight crew. The Digital Operational Terminal Information Service (D-OTIS) will combine the current ATIS and NOTAM systems together to provide on-demand information for any airport.

The aeronautical band from 960–1215 MHz is used primarily for navigation (distance measuring equipment, DME), but with the recent need for enhanced communications, this "L-band spectrum" is being planned for communications use. A current navigation system (Universal Access Transceiver, UAT, described subsequently) also employs this band, as do secondary surveillance radar transponders. The latest proposals for new systems in this band are called L-band Digital Aeronautical Communication System (L-DACS), and the leading contender system, L-DACS1, employs a multicarrier orthogonal frequency division multiple-access (OFDMA) approach. This scheme, which employs multiple narrow bandwidth subcarriers, can potentially avoid interference with DME and other L-band transmissions. The L-band links are envisioned to carry data (not voice), with rates from approximately 200 kbps to 1.3 Mbps.

4.2 NextGen navigation

Even navigation will utilize datalink services in NextGen. Until the late 1990s, aircraft navigation was conducted using radio navigation signals from ground-based navigation aids (navaids). The navigation information was contained in the radio signal. Even with the advent of GPS and the Wide Area Augmentation System (WAAS) Differential Global Positioning System (DGPS), aircraft continue to navigate using the ground- or satellite-based radio signals. In the NextGen system, the Instrument Landing System (ILS) precision approach systems will be replaced with Local-Area Augmentation System (LAAS) DGPS which uses a datalink. The VHF Data Broadcast (VDB) datalink conveys differential GPS corrections and approach path information to the aircraft. The VDB uses the range 108.00–118.00 in the navigation frequency spectrum.

4.3 NextGen surveillance

Automatic Dependent Surveillance – Broadcast (ADS-B) is the NextGen surveillance system that relies on each aircraft broadcasting its GPS position. ADS-B is automatic in that pilot action is not required and it is dependent because it relies on the aircraft to determine its position and communicate it. The ADS-B broadcast allows other aircraft in the area and ATC to know the aircraft's current position (latitude, longitude, and altitude) as well as its heading and velocity. ADS-B requires a digital datalink to communicate for the air-to-air and air-to-ground communications. In the FAA NextGen architecture, ADS-B uses both the 1030/1090 datalink defined for SSR transponders and the 978 MHz Universal Access Transceiver (UAT). UAT employs a 1 MHz channel, and uses binary frequency shift keying. Transmissions are synchronized by 1-s frames, and aircraft listen to uplink broadcasts from a ground station, and transmit (in CSMA fashion) on available "message start opportunities."

The 1030/1090 datalink is modified to add "Extended Squitter" messages that contain the ADS-B data. The term

"squitter" refers to the capability of the transponder to periodically transmit its information without receiving an interrogation. Airliners and higher end business aviation aircraft are typically equipped with 1030/1090 Mode-S transponders, which minimizes the equipage needs for ADS-B. Lower-end general aviation aircraft that do not have Mode-S transponders are more likely to equip with the UAT system. Because aircraft are not on the same datalink, the FAA rebroadcasts any ADS-B messages that it receives on the other datalink. For example, an ADS-B report on UAT is rebroadcast on the 1030/1090 channel and an ADS-B message on the 1030/1090 channel is rebroadcast on the UAT channel.

Since not all aircraft may be equipped with ADS-B, ATC uplinks traffic information on non-cooperative targets. The traffic uplink through the ADS-B system is a derivative of the TIS service provided by the radar system and is called TIS-Broadcast or TIS-B. TIS-B collects traffic information from the ADS-B and 1030/1090 radar system to generate the traffic uplink.

The ADS-B system provides a tactical view of aircraft position while the ADS-C (or Contract) service provides a strategic view. Through ADS-C, the aircraft links to ATC its current position and the future waypoints that comprise its route. At the beginning of the flight, the aircraft negotiates a "contract" with ATC that specifies when and where the aircraft will make position reports. The position reports can be automatic, such as passing a waypoint on the route, or periodic, such as every 15 minutes. The initial use of ADS-C is for transoceanic routes where ATC does not have radar coverage. The transoceanic use of ADS-C combined with the use of CPDLC for datalinking ATC instructions permit aircraft to reduce their horizontal separation from 50 nautical miles to 30 nautical miles. Both the ADS-C and CPDLC systems will utilize satellite communication channels due to the inability of VHF channels to operate over the ocean. In the long-term, ADS-C will be used over continental areas that currently have radar coverage.

4.4 NextGen airport surface operations

As identified in the Joint Planning and Development Office (JPDO) "Integrated Plan," enhancing airport operations is essential for increasing efficiency, capacity, and safety of the airspace system (JPDO,). Wireless networking at airports is clearly a critical contributor to enhancing airport operations, as numerous applications could employ such an airport network. This includes use of the network by pilots, air traffic controllers, airport security, airport operations and maintenance personnel, and many other entities. The likely candidate technology for airport surface networking is the IEEE 802.16e Standard (dubbed WiMax by industry) (Andrews, Ghosh, and Muhamed, 2007). The airport system will be a "tailored" version of 802.16. As with L-DACS, 802.16e uses OFDMA, and has multiple possible values of channel bandwidth and data rate (up to approximately 40 Mbps).

5   TECHNOLOGY AND POLICY ISSUES

5.1   Bandwidth limitations

Ultimately, the capacity of any radio link is upper-bounded by Shannon's well-known capacity formula (Shannon, 1948). This formula for an additive white Gaussian noise channel provides an upper limit to capacity in terms of signal-to-noise ratio and bandwidth. With many types of services vying for rights to spectrum, dedicated spectrum for aeronautical communications will always be limited.

The channel bandwidth of 25 kHz is minute in comparison to channel bandwidths for commercial systems (200 kHz–5 MHz for cellular, 10–20 MHz for LANs), and the 8.33 kHz allocation is likely the smallest of any allocation made in decades for any communications use. Yet non-technical constraints will keep these allocations in place for the foreseeable future. Due to technology limitations in terms of filtering and out-of-band emissions, adjacent channel interference means that the "full" band (VHF) cannot be filled with active channels in any one region. Hence the total capacity is less than one can compute using the given value of bandwidth. Gains in system capacity can thus be attained through spatial confinement (directional antennas) and spatial re-use, higher-order modulation and advanced error control coding, and advanced signal processing (e.g., active interference cancellation), all within the given bandwidth limits.

One additional area for bandwidth expansion is the optical area, where free-space optics (FSO) can provide enormous data rates, under conditions of line of sight (and no intervening clouds). The DOD has demonstrated a hybrid aeronautical radio/FSO system in its ORCLE program (Kenyon, 2004). This may prove usable for some aeronautical communication needs in the future.

5.2   Encryption and security

Existing aeronautical datalink technology does not use any encryption or any other security measures. This leaves the datalinks open, which allows third parties to listen to transmissions (Adams, 2006). The current system is also susceptible to spoofing, which is an unauthorized party pretending to be air traffic control or airline operations and sending false messages.

In the next generation of datalink services, security measures will be added to ensure that the message is authentic and that the message content can be trusted. It will also prevent unintended parties from eavesdropping on the communication. ARINC, RTCA, EUROCAE, and AEEC standards organizations are currently working in collaboration to define the future security technology and protocols.

5.3   Required communications performance

As the global air transportation system transitions to performance-based operations, air traffic management services and access to airspace will be provided to aircraft that are adequately equipped. An aircraft will be designated by its Required Total System Performance (RTSP) capability. RTSP is comprised of Required Communications Performance (RCP), Required Navigation Performance (RNP), and Required Surveillance Performance (RSP).

The RCP will be determined by the aircraft's communications capabilities. These capabilities include the ability to communicate via voice and data links through various spectra and media such as VHF, HF, and satellite. Communication performance factors such as latency, availability, and integrity affect the RCP value. Since the benefits related to RCP depend on the communications performance to be operational, the aircraft's RCP value changes during flight if a failure causes the loss of communication performance.

In 2007, the FAA and the European air traffic control organization (EUROCONTROL) published the second version of the Communications Operating Concept and Requirements (COCR) for the Future Radio System (EUROCONTROL/FAA, 2007). This document establishes communication requirements in support of global air traffic management, and also identifies sets of the most appropriate technologies to meet these requirements.

The COCR presents a timeline for achieving objectives, divided into two phases, with phase 1 extending to near the year 2030, and phase 2 to 2035 and beyond. The phases overlap beginning around 2020. In addition to defining operational concepts, environments, and key representative "scenarios," the COCR also presents specific performance requirements for communication systems in multiple domains of flight. These requirements include voice and data, for broadcast and point to point modes, for a large number of air traffic service types. The scope of the COCR and the Future Radio System is shown in Figure 3. Key requirements include message latency, reliability, and integrity.

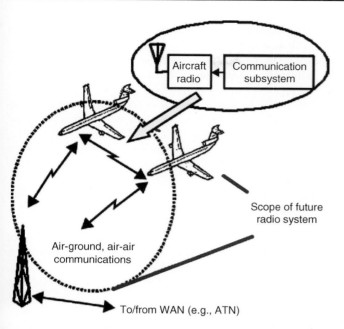

Scope of future
radio system

Air-ground, air-air
communications

To/from WAN (e.g., ATN)

Figure 3. Illustration of scope of Future Radio System. (Based on EUROCONTROL/FAA, 2007.)

5.4   Unmanned aircraft systems (UAS)

Unmanned aircraft systems (UAS) are increasingly being used for civilian, homeland security, and defense applications. Each UAS is operated by either a ground-based user or an airborne user in another aircraft. A command and control communications channel is required to operate the UAS. Depending on the application, the UAS may have data telemetry channels to download video and sensor data to remote operators.

As UAS are approved to operate in the National Airspace System, they will be required to equip with at least a subset of the communication equipment mentioned in this chapter. Communication equipment will be needed that allows the operator to communicate with the various enroute and terminal ATC facilities. The UAS will need to carry transponder and/or ADS-B equipment to interact with ATC surveillance equipment. The UAS will be required to have its own surveillance equipment sufficient for "see-and-avoid" operation.

6   SUMMARY

Early aircraft communications used analog radios to transmit voice communications for air traffic control and airline

operations. As technology advanced, digital radios began replacing the analog communications radios. The digital radios allowed both voice and data to be transmitted between the aircraft and ground systems. The data channels, or datalinks, gave rise to a host of applications such as weather, maintenance, traffic, and airport operational information that could be transmitted directly to the aircraft. Among the benefits provided by the datalinks, the flight crew could request information when needed and the information could be routed directly to the flight management system or displays without having the pilot manually listen and then enter the data.

In the Next Generation Air Transportation System, digital datalinks are the enabling technology that allows new communication, navigation, and surveillance systems to work. Datalinks also enable uplink of Flight Information Services (enroute weather, airport conditions, taxi instructions, and predeparture clearance).

NOMENCLATURE

AA	air-to-air
ACARS	aircraft communications addressing and reporting system
ADS-B	automatic dependent surveillance – broadcast
ADS-C	automatic dependent surveillance – contract
AFSS	automated flight service stations
AG	air-to-ground
AOA	ACARS over AVLC (Aviation VHF link control)
AOC	airline operations or airline operational control
ARTCC	air route traffic control centers
ATC	air traffic control
ATIS	automated terminal information system
AWOS	automated weather observing system
CDTI	cockpit display of traffic information
CDU	control display unit
CNS	communication, navigation, and surveillance
CPDLC	controller-pilot data link communications
CSMA	Carrier-sense multiple access
D8PSK	differential 8-ary phase shift keying modulation
DATACOMM	data communications (data link system)
DCL	departure clearance
DME	distance measuring equipment
DSB-AM	double-sideband amplitude modulation
D-OTIS	digital operational terminal information service
D-TAXI	digital airport taxi route datalink service
EFAS	enroute flight advisory service
FIS	flight informational services
FISDL	flight informational services data link
FMS	flight management system
GA	ground-to-air
GHz	gigahertz

HF	high frequency
ICAO	International civil aeronautics organization
JPDO	joint planning and development office
kHz	kilohertz
LAAS DGPS	local area augmentation system differential GPS
LDACS	L-band digital aeronautical communication system
LOS	line of sight
MHz	megahertz
NAS	National airspace system
NGATS	next generation air transportation system
NOTAM	notice to airmen
OFDMA	orthogonal frequency division multiple-access
PDC	predeparture clearance
PIREP	pilot report
QoS	quality of service
RCP	required communications performance
RNP	required navigation performance
RSP	required surveillance performance
RTSP	required total system performance
SATCOM	satellite communications
SNR	signal – noise ratio
SSR	secondary surveillance radar
TDLS	tower data link service
TIS	traffic information service
UAS	unmanned aircraft systems
UAT	universal access transceiver
UHF	ultra high frequency
VDB	VHF data broadcast
VDLM2	VHF data link – mode 2
VHF	very high frequency
WAAS DGPS	wide area augmentation system differential GPS

REFERENCES

Adams, C. (2006) Securing ACARS: Data Link in the Post-9/11 Environment. *Avionics Magazine*, June 2006. http://www.aviationtoday.com/av/categories/military/Securing-ACARS-Data-Link-in-the-Post-911-Environment955.html (accessed March 8, 2010).

Andrews, J.G., Ghosh, A. and Muhamed, R. (2007) *Fundamentals of WiMAX: Understanding Broadband Wireless Networking*, Prentice Hall, Upper Saddle River.

ARINC (2010) *Airline Operations Communications (AOC) Standardization Subcommittee*. http://www.aviation-ia.com/aeec/projects/aoc/index.html.

EUROCONTROL/FAA. Communications Operating Concept and Requirements for the Future Radio System (COCR) v.2.0. May 2007.

Federal Aviation Administration (FAA) (2004) Use of Cockpit Displays of Digital Weather and Operational Information, Advisory Circular No: 00-63.

Federal Aviation Administration (FAA). National Airspace System (NAS) Enterprise Architecture. http://nasea.faa.gov (accessed December 10, 2009).

Helfrick, A. (2009) *Principles of Avionics*, 5th Edn, Avionics Communications, Leesburg, VA.

JPDO (2009) *Next Generation Air Transportation System: Integrated Plan*.

Kenyon, H.S. (2004) Light and radio harmonized on one channel. *Signal Magazine*, published online, November 2004. http://www.afcea.org/signal/articles/templates/SIGNAL Article Template.asp?articleid=505&zoneid=111 (accessed March 8, 2010).

Parsons, J.D. (2000) *The Mobile Radio Propagation Channel*, 2nd edn, John Wiley & Sons, Inc., New York.

Roy, A. (1998) ACARS to VDL Transition Plan, AEEC Data Link User Forum, January 28–29.

Shannon, C.E. (1948) A Mathematical Theory of Communication. *Bell Syst. Tech. J.*, **27**, 379–423.

Spence, C.F. (2007) *AIM/FAR 2007: Aeronautical Information Manual/Federal Aviation Regulations*, McGraw-Hill, New York.

Stacey, D. (2008) *Aeronautical Radio Communication Systems and Networks*. John Wiley & Sons, Inc., New York.

Chapter 40

Sense-and-Avoid System Based on Radar and Cooperative Sensors

Stéphane Kemkemian and Myriam Nouvel

Thales Airborne Systems, Elancourt, France

1 INTRODUCTION

UAS (unmanned aircraft systems) can carry various payloads mainly intended for surveillance purposes, and sometimes for combat tasks. Up to now, these systems are mainly operated in military theaters or crisis theaters (i.e., in restricted areas): they do not have to follow all the Rules of the Air met by all the aircraft when flying in the general air traffic. The navigation systems, which they are provided with, are intended to ensure the operational success of the mission in particular by reducing the risk of fratricide losses. They are not safety systems in the strict meaning.

Unmanned Aircraft Systems. Edited by Ella Atkins, Aníbal Ollero, Antonios Tsourdos, Richard Blockley and Wei Shyy.
© 2016 John Wiley & Sons, Ltd. ISBN: 978-1-118-86645-0.

However, in the near future, the UAS will be deployed more and more for new emerging applications, mainly intended for safety and security tasks in peacetime (fire surveillance, border or costal surveillance, pollution surveillance, etc.). These new applications need the UAS to be flown in the general airspace without interfering with other aircraft, or endangering people or property on the ground. In other words, they need to be fitted with a sense-and-avoid system ensuring a high level of reliability.

In this chapter, we will not address most of the current civilian applications (or even "consumer" applications) such as movie filming, engineering structure inspection, agriculture, and leisure. In these particular cases, the aircraft in question do not need autonomous navigation systems over long distance because they legally ought to be operated under visual control from ground and out of the regulated air traffic areas.

2 SENSE-AND-AVOID SYSTEM REQUIREMENTS

The safe introduction of UAS into the airspace requires a specific approach taking into consideration the regulatory framework, the correct application of air operation procedures, the evolution of the worldwide sky in the future, and the capability of available or soon to be available technology. One key issue remains in the requirements, definition, and development of a sense-and-avoid system, necessary to replace the human presence on board the vehicle, in order to detect the presence of any other traffic or obstacles and to avoid any potential collisions, in both controlled or uncontrolled airspaces.

A sense-and-avoid system has two separate functions (Figure 1):

- The *sense* function whose objective is to assess the situation around the UAS and to deliver intruders' trajectory prediction.
- The *avoid* function whose objectives are first to provide information for separation (for the ground control station to negotiate with air traffic controller the future trajectory of the UAS) and second to provide the automatic flight management system with the maneuver to be followed.

Both functions interact to provide relevant information to the UAS pilot by the GCS (ground control station). Whenever possible, the UAS control is carried out by the GCS:

- The separation assistance function helps the UAS pilot to maintain separation relative to other traffic when separation is not provided by ATC (air traffic control).
- The collision avoidance (CA) function provides an automatic last instant maneuver with the objective of avoiding an imminent collision.

The avoid function is aided by a host aircraft model in order to account for maneuver capability.

2.1   Sense function

2.1.1   Cooperative and noncooperative sensors

It might seem easy to use only cooperative sensors such as ADS-B (automatic dependent surveillance–broadcast). The main problem with this approach is its lack of reliability, especially in VFR (visual flight rules) airspace: indeed, all the aircraft encountered in this space are not always fitted with such cooperative systems.

Only systems with at least one noncooperative (autonomous) sensor can provide a sufficient level of functional reliability and a consistent level of safety, which is mandatory to deploy UAS in the general air traffic. Among the noncooperative sensors, the radar is the most suitable sensor for several reasons:

- It directly provides 4D raw measurements (angles, distance, and closing rate).
- It provides nearly all weather situation awareness at long range.

In favorable cases for optics, additional electro-optical (EO) sensor(s) can improve the angular measurements using data fusion methods.

However, when information from cooperating sensors such as ADS-B is available, it is most often of better quality than that provided by noncooperative sensors.

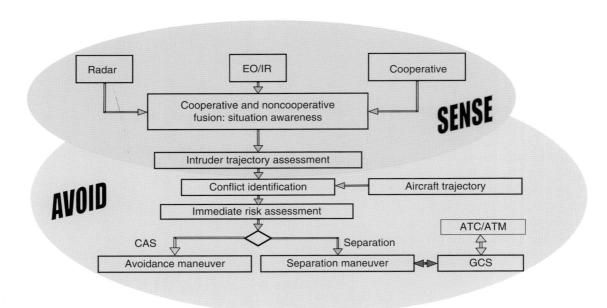

Figure 1. Functional architecture.

2.1.2 Sensor fusion

The purpose of the data fusion component is precisely to merge the measurements coming from all sensors for elaborating the best assessment of surrounding traffic. The result of the data fusion may be used by the sensor management unit to send feedback to the sensors and enhance their performances.

2.1.3 Intruder trajectory estimation

It addresses the question of how a collision or a noncollision trajectory can be predicted, keeping in mind the need to obtain the best balance between detection of real collision risk and false alarm, which leads to a useless avoidance maneuver.

2.2 Avoid function

We will say a few words about this function, which contains two tasks.

2.2.1 Traffic separation

Under "normal" conditions, the actual Rules of the Air for traffic separation are

- a vertical spacing greater than 500 ft (150 m);
- a horizontal spacing greater than 0.5 nautical mile (925 m).

Only one of the two conditions must be fulfilled. The vertical spacing is, by far, the most constraining one in terms of angular localization accuracy to decide whether to carry out an avoidance maneuver.

2.2.2 Emergency avoidance

Under "emergency" conditions, the system shall avoid collisions with other aircraft. A safety cylindrical area with a radius and a height both equal to 500 ft (150 m) is defined around each aircraft. A collision or a quasi-collision (air miss) occurs if an aircraft enters in the safety area of another one. The maneuver related to the collision avoidance is an emergency one carried out at "short range." In this case, long-range detection is not required. If the traffic insertion/separation function has properly worked out and if all surrounding aircraft have followed the Rules of the Air, no emergency maneuver should occur.

3 SENSOR REQUIREMENTS

The main requirements for the sense-and-avoid function are as follows:

- *Widefield of view* (FOV) equivalent to the FOV of a pilot in a cockpit: at least $\pm110°$ in azimuth and $\pm15°$ in elevation, which corresponds to a solid angle of $\Omega \approx 2$ sr.
- *Fast revisit rate* (about 1 s): The reactivity of the system has to be similar to that of a "human pilot" to face a sudden threat.
- *About 20 s of advance warning* (time-to-go: "TTG"): This value allows a fight path correction taking into account the UAS maneuverability. The value of 20 s typically applies for fixed-wing tactical UAS, which are neither very powerful nor very maneuverable. For such aircraft, the fastest separation maneuver is usually in the horizontal plane. Indeed, the avoidance maneuver consisting in climbing may be very long (several tens of seconds) and it is not always possible to go down because of the terrain. The typical avoidance maneuver is indicated in Figure 2. It is formed of a couple of successive inverse turns with a maximum roll of 45° (constant turn rate). In case of a closing rate of about 400 m s^{-1}, the maneuver requires being executed 8 km before the crossing (TTG = 20 s). This value yields an order of magnitude of the required tracking range.
- *Sufficient tracking accuracy* is required for making the right decision to maneuver or not. Even if the separation takes place mostly in the horizontal plane, the decision to maneuver depends mainly on good measurement of the altitude of the intruder. An accuracy of 150 m in altitude requires an elevation accuracy of 1° at 8 km.
- The weight, consumption, cooling, and aerodynamic disturbances of the system must be limited too.

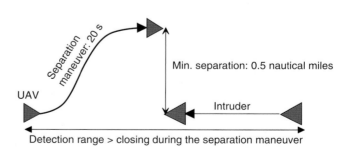

Figure 2. Separation maneuver in the horizontal plane.

4   RADAR TRADE-OFFS

The radar design should take into account the previous requirements and make trade-off between conflicting requirements.

4.1   Angle accuracy

The achievable angle accuracy depends mainly on the antenna baselines and the equivalent signal-to-noise ratio (SNR). The standard deviations of the azimuth and elevation measurements depend on the beam widths $\theta_{AZ,EL}$ (after beamforming, cf. next section):

$$\sigma_{AZ,EL} \approx \frac{\theta_{AZ,EL}}{2\sqrt{2\,SNR}} \quad \text{with} \quad \theta_{AZ} \approx \frac{\lambda}{L} \quad \text{and} \quad \theta_{EL} \approx \frac{\lambda}{H} \quad (1)$$

The baselines L and H are, respectively, the length and the height of the antenna system (assuming a rectangular shape). In fact, the equivalent SNR_{EQ} depends on the detection conditions according to Equation 2:

$$SNR_{EQ} = \log P_{FA} / \log P_D \quad (2)$$

P_D is the probability of detection (>0.5 in the usable range) and P_{FA} is the probability of false alarm (generally chosen $<10^{-6}$). This yields $SNR_{EQ} > 20$ ($+13\,dB$). As the SNR, that is, the maximum range, is determined at fairer for cost and consumption issues, we can only play on the ratios λ/L and λ/H. In practice, the minimum reachable standard deviations are of the order of a tenth of beam width. Thus, an accuracy of $1°$ at 2σ requires beam widths narrower than about $5°$.

The value of L and H being limited for a size reason, the higher the frequency, the more accurate the angle measurements.

4.2   Scanning technologies

There are mainly three candidate technologies:

- Mechanical scanning (M-SCAN)
- Electronic scanning (E-SCAN)
- Digital beamforming (DBF)

4.2.1   Scanning issues

The first two methods (M-SCAN and E-SCAN) utilize sequential scanning of the angular domain Ω. The number of beams (Equation 3), the dwell time on target (Equation 4),

and the scanning rate (Equation 5) are

$$N_{BEAM} > \frac{\Omega}{\omega} \quad \text{with} \quad \omega = \frac{\pi}{4}\theta_{AZ}\theta_{EL} \quad (3)$$

$$T_{DWELL} < T\frac{\omega}{\Omega} \quad (4)$$

$$V_{SCAN} > \frac{\Omega}{\max(\theta_{AZ}, \theta_{EL})\,T} \quad (5)$$

The beam is necessarily narrow (due to measurement accuracy requirements). For instance, if $\theta_{AZ} = \theta_{EL} = 100$ mrad, $\Omega = 2$ sr, and $T = 1$ s, we get

- $N_{BEAM} > 200$;
- $T_{DWELL} < 5$ ms;
- $V_{SCAN} > 20$ rad s^{-1} ($\approx 1200°$ s^{-1}).

These values lead to the following conclusions:

- The M-SCAN is hardly conceivable. It is not possible to use a sectored scanning since it requires rapid inversion of the scanning direction ($\pm 1200°\,s^{-1}$) and would probably be of poor reliability.
- In addition, the conventional E-SCAN implies very short dwell time on target due to the scanning rate ($<5\,ms$), which makes problematic the implementation of an efficient Doppler processing.

The solution retained in this study is to use a hybrid technology "E-SCAN + digital beamforming."

4.2.2   What is digital beamforming?

The following sections explain the main differences between E-SCAN and DBF.

4.2.3   E-SCAN

The array antenna is made of a large number of receiving channels, each of them comprising successively (Figure 3)

- a radiating element;
- a low-noise amplifier (LNA);
- a phase shifter.

The summation of all the channels is performed at the RF level. After amplification and frequency conversion in the baseband, the signal is digitized. We thus obtain a narrow beam with high gain; however, as the phase shifters can take a single state at a given moment, only a single direction can be observed at a given time.

The same thing occurs at transmission (not represented at Figure 3): the RF signal is divided into multiple channels, and

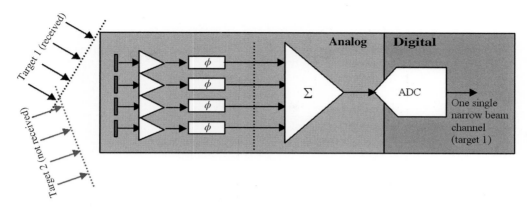

Figure 3. E-SCAN antenna (reception part).

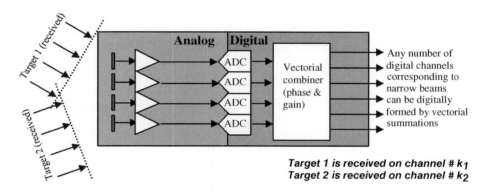

Figure 4. Digital beamforming at reception.

each channel is phased, amplified via high-power amplifiers (HPAs), and finally radiated. The radiated fields add in the far field yielding a narrow but high-gain transmitted beam.

4.2.4   Dbf

The front end at reception is the same as in the E-SCAN case. Each receiving channel comprises successively (Figure 4)

- a radiating element;
- a low-noise amplifier;
- an individual receiving channel (analog);
- an individual analog-to-digital converter (ADC).

The summation of the channels is now carried out at the digital level; thus, as many narrow beams at reception as we want can be formed using digital computation only. It is thus possible to observe several directions at the same time (not sequentially), which relaxes the constraints on the dwell time.

In contrast, we have to use a wide beam at transmission that encompasses all directions we want to observe in a single look at reception (Figure 5). A wider beam necessarily has a lower gain, so to maintain the same power budget, we have to compensate the loss of antenna gain by increasing the signal processing gain. This is achieved by increasing the signal integration time, which is theoretically possible since we no longer need to share the observation time between several directions.

In the case of radar for a sense-and-avoid system, this scanning technique is the only way to get a good angular accuracy while enabling an efficient Doppler processing.

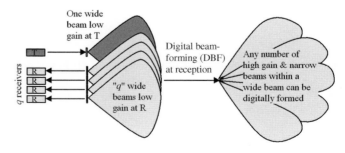

Figure 5. Digital beamforming radar.

Table 1. Atmospheric attenuations for a target at 5 nautical miles (dB)

Band	Dry air	Rain 1 mm h^{-1}	Rain 4 mm h^{-1}	Rain 16 mm h^{-1}	Fog 1 g m^{-3}
X (10 GHz)	0.04	0.23	1.2	6	1.85
Ku (16 GHz)	0.1	0.75	3.5	15	3.7
Ka (35 GHz)	2.2	4.5	18	No vis.	5.5

4.3 Power budget

4.3.1 Power/aperture trade-off

The detection range R_{MAX} at given P_D depends on the average transmitted power P_{AVG}, the surface area of the receiving antenna A_R, the angular coverage Ω, the revisit rate T, the radar cross section (RCS) of the target σ_T, and the atmospheric losses L_{ATM} (Melvin and Scheer, 2012):

$$R_{MAX}^4 < R_{MAX\_OPT}^4 \propto \frac{P_{AVG} A_R}{\Omega T L_{ATM}} \sigma_T \qquad (6)$$

The symbol "$\propto$" denotes "proportional to." This equation yields an upper bound of the detection range regardless of the scanning method or the detection method used (Kemkemian et al., 2009). It is worth noting that Equation 6 does not explicitly take into account the wavelength of the signal. The angular coverage Ω, the revisit rate T, and the RCS of the target σ_T depend on the operational context (cf. Equation 3), so Equation 6 reduces to Equation 7:

$$R_{MAX}^4 < R_{MAX\_OPT}^4 \propto \frac{P_{AVG} A_R}{L_{ATM}} \qquad (7)$$

P_{AVG} increases the cost of transmitter, the consumption, and the cooling requirements. On the other hand, the antenna surface A_R is necessarily limited to issues of integration on aircraft. Jaska (2003) provides a discussion on this trade-off.

4.3.2 Atmospheric losses

The wave attenuations may be significant in the millimeter bands (Ku and Ka bands) under adverse weather conditions, while they moderately affect the X-band (Table 1). Other radar bands are not considered here:

- Low-frequency bands are not a good trade-off in terms of angular accuracy versus size (Section 4.1).
- High-frequency bands suffer from prohibitive attenuations and technological costs considering the required range.

Therefore, the weather conditions to consider for the radar design must be consistent with the flight envelope of the UAS. Strong rain rates above 10 mm h^{-1} are generally associated with stormy or windy conditions and only the largest UAS can be flown under these conditions.

4.4 What is the best operating frequency?

At first glance, the best choice for accuracy is to increase the operating frequency until the atmospheric attenuation becomes too large. However, the cost of the technologies required for E-SCAN or DBF dramatically increases with frequency. At the time of writing, the Ka band is not affordable for DBF and the technologies required for the Ku band are barely mature for large "power × aperture" systems.

The best choice is the Ku band if one focuses mainly on performance and looks for a compact system and the X-band if one looks for a good balance between cost and performance.

5 EXAMPLE OF RADAR SYSTEM

The proposed radar system is implementing both E-SCAN and DBF (© Thales, patented). This architecture is based on the general coherent "MIMO" principles (Le Chevalier, 2008; Li and Stoica, 2009). This radar system operates in the X-band and it is extensively described by Kemkemian et al. (2010).

5.1 Architecture

It is highly desirable to minimize the cost and the complexity of the transmission hardware. Thus, the chosen waveform is continuous (minimal peak power) and linearly modulated in frequency (FMCW) for allowing distance measurements. This involves using two separate antennas for transmit and receive functions (Figure 6) because it is not possible to transmit and receive simultaneously on the same antenna with the required power. The transmission part and the reception part do not need to be colocated on board the UAS.

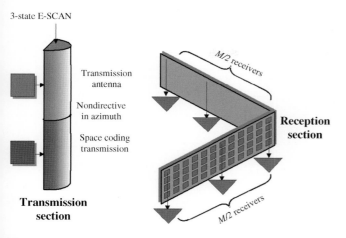

Figure 6. Architecture of the proposed radar system.

5.1.1   Receiving antenna

In order to cover the ±110° azimuth FOV, two panels are needed: the first one steered toward the right side of the UAS and the second one steered toward the left side of the UAS. Figure 7 shows an example of a receiving panel. The four elements of each column are summed and feed a receiving channel (one receiver per column) ended by digitization. This yields a beam width of about 30° in elevation.

The two panels form an angle $\alpha \approx 80°$ between them, which allows covering the FOV while having a relatively constant gain around the axis. Indeed, the gain in a direction depends on the projected area in the direction of the assembly of two panels.

The receiving antenna operates in the DBF mode (only in the azimuth plane). It can therefore measure the direction of a target in azimuth; however, this antenna alone is unable to determine the direction of arrival (DOA) in elevation. The elevation measurement of a target is done using the principle of the colored transmission, in other words, using the principle of coherent MIMO.

5.1.2   Transmitting antenna

Thanks to colored transmission, the radiated signal depends on the direction in elevation from which it is observed (Figure 8, left). At reception, signal processing allows us to retrieve the elevation where the target was illuminated. We call it "monopulse at transmission."

To obtain sufficient angular accuracy in elevation, the beam width is about 10°. This width is lesser than the area to be covered (about 30°). Thanks to E-SCAN, the beam is steered toward successively three directions in elevation to cover the whole FOV. In contrast, the radiating elements of the transmitting antenna have a wide coverage in azimuth in order to illuminate the whole azimuth domain (±110°).

Antenna Architecture. The antenna is made of a linear vertical array of 12 sources divided into two subarrays for MIMO capability. The additional E-SCAN is performed by switching delay lines etched on printed board before the HPAs (Figure 8, right). Each HPA radiates about 2 W. The total radiated power is therefore 24 W and provides sufficient range detection on small targets (>8 km on target RCS $\sigma_T = 1$ m^2).

Elevation Measurements. Two orthogonal waveforms are generated by two waveform generators (WFG1 and WFG2). These two waveforms are then radiated, each of them by one-half antenna. ("Orthogonal" means that the intercorrelation between the two waveforms is null.) Thanks to this property, the received signals coming from the two half-antennas can be separated. The elevation angle θ is retrieved from the phase difference $\Delta\varphi$ between signals (Equation 8). The phase centers of each half-antenna are spaced out by the vertical length $a \approx H/2$.

$$\Delta\varphi = 2\pi \frac{a}{\lambda} \sin\theta \qquad (8)$$

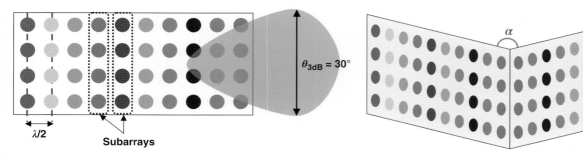

Figure 7. Receiving panels.

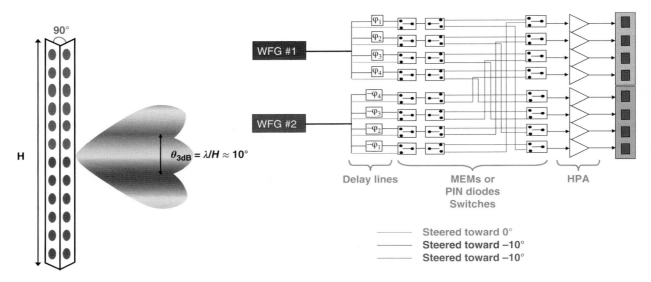

Figure 8. Transmitting MIMO + E-SCAN array.

Note: In Figure 9, the receiving array is drawn at the center of the transmitting antenna. However, the receiving array can be placed at any other convenient place on the UAS.

5.1.3 Waveforms and processing

The range resolution is $\delta r = c/2B$ (B is the FM excursion, cf. Figure 10, left). Each half-antenna radiates a specific FMCW signal (Figure 10, left). The two FMCWs are slightly offset so that

$$\Delta F = k/T_R \qquad (9)$$

The couple of FMCWs are orthogonal provided that k is integer. As large targets may occupy several range gates, the two transmitted signals will be fully separable if $\Delta F > L_{MAX}/\delta r T$, where L_{MAX} is the maximum radial extension of a target. In the X-band, even more so in the Ku band, it is not possible to get a waveform unambiguous in both distance and velocity. We choose a waveform

unambiguous in velocity (ambiguous in distance). Indeed, such a waveform maximizes the detection range on fast targets (maximum TTG) since they are observed in a clutter-free region.

The proper functioning of Doppler processing using such a range-ambiguous FMCW assumes the following:

- The phase rotation of the transmitted wave during one ramp must be a multiple of 2π.
- A sampling rate exactly equal to ΔF, so the two spectral lines corresponding to δF and $\Delta F - \delta F$ are folded one over the other.

The range compression is then carried out for each T_R interval, and then the Doppler processing is performed along a burst of N_{FFT} ramps. The range ambiguities are then removed by using classical methods.

5.2 Example embodiment

5.2.1 Receiving part

Each receiving panel may contain one, two, or more building blocks. The greater the number of building blocks, the better the antenna gain but the more complex and costly the design. A building block (Figure 11) is made of six subarrays (six vertical columns), each column feeding a single LNA. The so-produced signal is then downconverted (mixers) to intermediate frequency. Signals from each subarray are then sampled (ADC) and feed a FPGA for

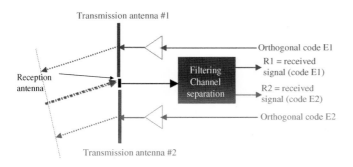

Figure 9. "Monopulse" at transmission.

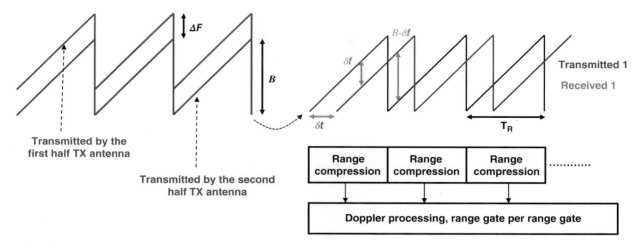

Figure 10. Waveform and processing.

preprocessing. Interface between FPGA and processor is an optical fiber link.

Description of a building block. The first printed board contains

- the radiating elements on its front side;
- the RF sections (LNA and mixers) on its back side;
- OL signal distribution and power connector in the internal layer.

The second printed board contains

- the IF amplifiers on its front side;
- the ADC and a FPGA (pulse compression and Doppler processing) on its back side.

5.2.2   *Transmitting part*

The studied design includes RF parts and radiating elements on a single printed board, which leads to a very compact architecture (see example in Figure 12):

- the radiating elements on the front side;
- HPAs and power connector on the back side;
- signal distribution circuits and fixed delay lines (vertical E-SCAN) in the internal layers.

6   COOPERATIVE SENSORS

Among the possible cooperative sensors, ADS-B is particularly suitable for implementing a sense-and-avoid function.

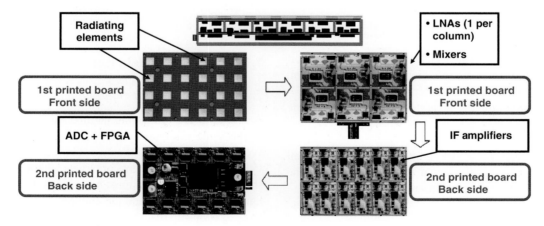

Figure 11. Example of the receiving panel architecture.

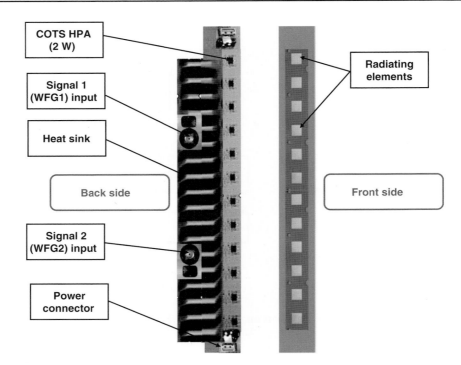

Figure 12. Example of the transmitting antenna architecture.

ADS-B stands for automatic dependent surveillance–broadcast (http://www.skybrary.aero/index.php/ADS-B):

- *Automatic*: there is no interrogation needed to start the data or squitter coming from the transponder.
- *Dependent*: it relies on on-board navigation and broadcast equipment to provide information to other ADS-B users.
- *Surveillance*: automatic surveillance and traffic coordination.

Recent Mode S transponders can provide the "extended squitter" message incorporating ADS-B information. An ADS-B-equipped aircraft determines its own state vector and other information and broadcasts them at a rate of 2 Hz. The most useful data for the sense-and-avoid function are

- the position data, which are usually derived from GPS or, if not available, from an aircraft's inertial reference system;
- the true altitude data that are derived from GPS;
- the barometric altitude that is derived from barometric sensors.

If the surrounding aircraft are all fitted with ADS-B as well as the integrity of the GPS signal can be guaranteed, the ADS-B provides information that is

- *available at long range* (provided that aircraft are in line of sight),

- *all weather*, and
- *much more accurate* than any other noncooperative sensor.

One of the main problems in using ADS-B is linked to knowledge of the integrity of received data.

7   DATA FUSION AND INTRUDER TRAJECTORY ESTIMATION

The functioning of the data fusion is explained in the case where the UAS has both a radar and an ADS-B equipment as sense-and-avoid sensors.

7.1   Principles of data fusion

7.1.1   Measurement referential

Any data fusion processing requires all data to be expressed in the same referential. ADS-B directly provides its state vector in geodesic coordinates. Radar and EO sensors provide relative measurements of targets with respect to the own ship. The geodesic data must be converted into relative ones knowing the position and the attitude angles (roll, pitch, and yaw) of own ship that should be provided by the inertial/GPS reference system.

7.1.2   Architecture and processing

The fusion process has to combine heterogeneous sensor data (Bar-Shalom, 1989). There are two steps in the fusion process, which are successively executed:

- first, the data association;
- second, the data fusion.

To avoid excessive raw data streams between the sensors and the fusion core, the proposed design of the fusion of heterogeneous data is a hybrid hierarchical process.

At Sensor Level (Before Fusion).

- Each sensor is responsible for its own tracking. There are two main reasons for this choice:
  - Each sensor can operate in stand-alone mode, which facilitates the system integration.
  - The output data rate is minimal since only the measurements aggregated and associated with consistent tracks are transmitted to the fusion component.
- Each sensor provides the fusion with
  - the state vector corresponding to each sensor track (output of tracking);
  - the raw measurements corresponding to the last detection(s) that has been associated with each sensor track. In the case of radar, these raw measurements are the range, the closing range rate (Doppler), the azimuth, and the elevation of the target.

At system level (after fusion).   The result of the data fusion is to provide the "avoid" function (see Figure 1) with relevant "system" trajectory estimators resulting from the fusion of data from different sensors.

The output of the data fusion may generate high-level commands to the sensors, for instance, including the feedback of fusion estimates toward sensors in order to improve their processing (e.g., detection threshold, refresh rate, and range-Doppler resolution).

Data association (between different sensors and "system" tracks).   This process attempts to associate the data provided by different sensors with the existing "system" tracks (trajectory estimators). It is based on statistical distance between common axes of correlation, for instance, by using the "Mahalanobis" distance (Equation 10):

$$D^2 = \left(\hat{X}_1 - \hat{X}_A\right)^{\mathrm{T}} \cdot \left(P_1 + P_A\right) \cdot \left(\hat{X}_1 - \hat{X}_A\right) \qquad (10)$$

where $\hat{X}_1$ and $\hat{X}_A$ are, respectively, the state vectors issued from the sensor track #1 and the "system" track #A, and P_1

and P_A are, respectively, the covariance matrices associated with sensor track #1 and "system" track #A.

A threshold on D^2 allows rejecting or accepting the association. As the raw data coming from the various sensors are asynchronous and may have very different refresh rates, the association test is carried out between the "sensor" tracks and the "system" tracks (so not between raw measurements). There are two cases:

- No likely association can be established: a new "system" track is initialized.
- At least one sensor provides a likely association with an open track. This track is then refreshed with the latter raw measurement issued from the sensor.

Data fusion.   Once a "sensor" track is associated with a "system" track, the latter is actualized with the raw measurement data attached to the "sensor" track (and not directly with the state vector corresponding to the "sensor" track). This method avoids problems on trajectory estimation that may arise because the "system" (common) state vector is not expressed in the same referential as those used by the various sensors (Blackman and Popoli, 1999; Bar-Shalom, 1989).

7.2   Multiple-sensor tracking algorithm

7.2.1   Target kinematic model

Most of the dynamic behavior that UAS is supposed to encounter is from civil aviation, but we must also take into account some maneuvering targets such as combat aircraft or UCAV.

- For "civilian" aircraft, two kinds of behavior are considered: constant velocity (CV) model and coordinated turn (CT) model. These models are the most frequent models in the air traffic control. For the CT model, left and right turns and two amplitudes for the turns – $2° \, \mathrm{s}^{-1}$ (typical turn) and $6° \, \mathrm{s}^{-1}$ (assumed as maximum value for civilian aircraft) – are considered.
- For a maneuvering target (especially to deal with the possible presence of military targets), an extra CT model (left and right turns) at a rate of $9° \, \mathrm{s}^{-1}$ has been added.

7.2.2   Tracking algorithm

As mentioned earlier, the civilian aircraft either fly mostly at a constant velocity or take a turn with constant speed.

- Since the objective of the tracking is to detect potential collision between the UAS and any surrounding aircraft, the tracking algorithm is expected to estimate the "target" velocity vector as accurately as possible. Indeed, a colliding target is characterized by a null or a close to zero relative angular velocity.

- For that reason, it is preferable to select a multiple-model approach with adapted dynamic model, instead of a single-model algorithm tuned to cover a large variety of dynamics.

Interacting multiple model. Interacting multiple model (IMM) is the state of the art of tracking algorithms when multiple dynamic behaviors have to be considered. The most important properties of the IMM are

- competing dynamic models and evaluation of their probability through likelihood computation and Bayes' theorem;
- interaction between models (also called mixing), in which prior information promotes or inhibits transition from one model to another, through a Markov chain.

The description of the IMM algorithm can be found in Blom and Bar-Shalom (1988).

Variable structure IMM. Although IMM is designed to handle multiple kinematic behaviors, its performance gets worse as long as the number of competing models increases. In fact, information gathered by the sensors diffuses across the multiple models, instead of being limited to major models. That is why the variable-structure IMM (VS-IMM), proposed in the literature (Li, 2000), is well suited for this application. The basic idea is to design supervision, over the IMM, which is able to select the most appropriate models and parameters according to the current situation.

7.3 Simulations of data fusion

7.3.1 Scenario for simulations

To illustrate multiple-sensor tracking facing intruders that perform maneuvers, we choose the following scenario (cf. Figure 13):

- *Intruder 1*: low maneuvering UAV, at $60 \, \mathrm{m \, s^{-1}}$, going straight at constant speed, and then performing a left turn at $3° \, \mathrm{s^{-1}}$.
- *Intruder 2*: maneuvering UCAV at $260 \, \mathrm{m \, s^{-1}}$, going straight at constant speed, and then performing a left turn at $7° \, \mathrm{s^{-1}}$.
- *Own ship UAV* with multiple-sensor tracking algorithm, going straight at $60 \, \mathrm{m \, s^{-1}}$, toward intruders.

7.3.2 Main parameters of algorithm

IMM with seven models: CV, left and right CT at $2° \, \mathrm{s^{-1}}$, $6° \, \mathrm{s^{-1}}$, and $9° \, \mathrm{s^{-1}}$.

Configuration 1. Radar sensor only with following error statistics (rms):

- $50 \, \mathrm{m}$ in range, $1°$ in azimuth, elevation, and $2.5 \, \mathrm{m \, s^{-1}}$ in closing velocity, refresh rate of $2 \, \mathrm{s}$.
- Track initialization on the first four radar detections.

Configuration 2. Radar and EO sensors with radar error statistics as provided in configuration 1 and for the EO sensor:

- $0.2°$ (rms) in azimuth and elevation.
- No radial measurement.
- Refresh rate of $10 \, \mathrm{Hz}$.

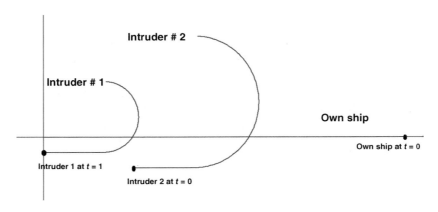

Figure 13. Geometry of the scenario.

The case of a data fusion between a cooperative sensor (such as ADS-B using GPS localization) and a non-cooperative sensor is not considered because it leads to trivial results:

- *If all present platforms are equipped with a cooperative localization mean*, the purpose of the data fusion is only to check the integrity of the data issued from this cooperative mean. If the integrity is verified, the data coming from the cooperative mean can be used alone because their accuracy is far better than that of the noncooperative sensor.
- *If a platform is not equipped with the cooperative mean*, or if the cooperative information is not available aboard the UAS for any reason, the accuracy depends only on the noncooperative means.

7.3.3   Simulation results

Configuration 1 (radar only). Figure 14 shows that the IMM tracking is able to detect and follow both intruders' maneuvers, even if their turn rate does not match any one of the IMM models. However, the target heading has some fluctuations, especially during the first part of the trajectory. This is mainly due to the time required by the Kalman filter to converge, which is itself caused by the relatively low refresh rate provided by the radar sensor (2 s).

Configuration 2 (radar and EO sensor). It can be seen (Figure 15) how accurate is the joint radar–EO tracking, during both nonmaneuvering and maneuvering phases. (This result is obtained thanks to complementarities of both sensors: range, closing rate, and angles provided by the radar combined with very accurate angular data at high refresh rate

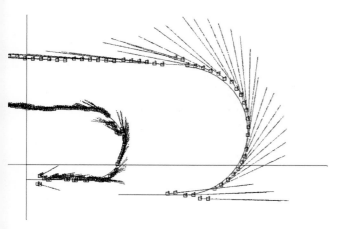

Figure 14. First configuration: radar-only tracking.

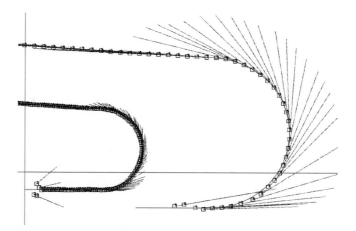

Figure 15. Second configuration: radar and EO tracking.

provided by the EO sensor.) Both position and velocity vectors are very accurate, allowing a much more reliable trajectory prediction.

8   CONCLUSION

Two topics, relative to the implementation of a sense-and-avoid function aboard UAS, have been discussed in this chapter:

The first part of the chapter was dedicated to the description of "MIMO"-based radar and the relative technologies for the "sense-and-avoid" function. The described solution operates in the X-band because this band leads to the best trade-offs between performance and cost. The design provides a good angular accuracy and sufficient detection range on fastest closing targets. It is a static solution and can be integrated within the UAS airframe without extra volumes.

The second part has explained why a multiple-sensor approach, based on the sensing of different physical parameters (electromagnetic, optic, information broadcast, etc.), is interesting, since they have specific and complementary contributions that ensure reliability for safety, which is required for the introduction of UAV in controlled and uncontrolled airspaces. The principle of the multiple-sensor fusion, using the multiple-model approach, is explained and illustrated through simulation.

REFERENCES

Bar-Shalom, Y. (1989) *Multitarget–Multisensor Tracking: Advanced Applications*, Artech House, Boston, MA.

Blackman, S. and Popoli, R. (1999) *Design and Analysis of Modern Tracking Systems*, Artech House, Boston, MA.

Blom, H.A.P. and Bar-Shalom, Y. (1988) The interacting multiple model algorithm for systems with Markovian switching coefficients. *IEEE Trans. Autom. Control*, **33**(8), 780–783.

Jaska, E. (2003) Optimal power–aperture balance. *International Radar Conference*.

Kemkemian, S., Nouvel-Fiani, S., Cornic, P., Le Bihan, P., and Garrec, P. (2009) Radar systems for "sense and avoid" on UAV. *International Radar Conference*, October 2009, Bordeaux, France.

Kemkemian, S., Nouvel-Fiani, M., Cornic, P., and Garrec, P. (2010) A MIMO radar for sense and avoid function: a fully static solution for UAV. *Proceedings of the 11th International Radar Symposium (IRS-2010)*, June 16–18, 2010, Vilnius, Lithuania.

Le Chevalier, F. (2008) Space–time transmission and coding for airborne radars. *Radar Sci. Technol.*, **6**(6), 411–421.

Li, J. and Stoica, P. (2009) *MIMO Radar Signal Processing*, John Wiley & Sons, Inc., Hoboken, NJ.

Li, X.R. (2000) Chapter 10: Engineer's guide to variable-structure multiple-model estimation for tracking, in *Multitarget/Multisensor Tracking: Applications and Advances* (eds. Y. Bar-Shalom and D.W. Blair), Vol. **III**, Artech House, Boston, MA, pp. 499–567.

Melvin, W.L. and Scheer, J.A. (eds.) (2012) *Principles of Modern Radar*, Vol. **1**, The Institution of Engineering and Technology.

Chapter 41

Standards and Interoperability: A Systems Engineering Perspective

Andy Yu, Linas Mockus, and Dan DeLaurentis

School of Aeronautics and Astronautics, Purdue University, West Lafayette, IN, USA

Lack of [UAS] interoperability has had a real-life impact on U.S. operations . . . there have been cases where a Service's [UAS], if it could have gotten data to another Service, another component, it may have provided better situational awareness on a specific threat in a specific area that might have resulted in different measures being taken.

– Dyke Weatherington (2004)

Unmanned Aircraft Systems. Edited by Ella Atkins, Aníbal Ollero, Antonios Tsourdos, Richard Blockley and Wei Shyy.
© 2016 John Wiley & Sons, Ltd. ISBN: 978-1-118-86645-0.

1 INTRODUCTION

During the twentieth century, assembly lines, popularized by Ford, were used to speed up production and reduce costs by using a division of labor. Instead of a single person, multiple people contributed interchangeable parts toward the manufacture of a product through an organization of workers, tools, and parts. This reduced the need for specialization in the overall product and allowed more focus on product parts. Today, a similar methodology is used to manufacture products from simple toys to complex systems. This concept, however, would not have been possible if interchangeable parts were not held to a standard and a level of interoperability. This chapter will examine interoperability and standards related to unmanned aerial systems (UAS). While most examples will reference US military applications, it is only one of many UAS use cases. The intended audience is readers who seek a background into the history and current state-of-the-art in standards and interoperability concepts with a focus on UAS. This chapter is by no means a comprehensive document characterizing the detailed workings of either topic. The chapter proceeds by first reviewing the history behind the concepts of each topic; focus then shifts to the importance, challenges, and opportunities for improvement from a systems perspective.

2 STANDARDIZATION

While the adoption of standards largely depend on socio-technical and socioeconomic aspects, most technical standards have *evolved* from a set of ad-hoc guidelines to a set of tightly held and regulated procedures, practices, and specifications regarding testing, manufacturing, and operation (Grösser, 2013). While some standards are voluntarily adopted by communities of interest (such as the QWERTY keyboard), others are required by law (aircraft operations).

The most widely adopted standard to measure physical properties are the International System (SI) units. Although some standards are straightforward, others are not. The process of implementing standards for building and operating complex systems, such as aircraft in the air transportation system, can be very challenging. Therefore, several regulatory bodies have been established to focus solely on identifying and setting these standards so that organizations may refer to these administrations for assistance. Administrations in the United States with a focus on UAS include the American Society for Testing and Materials (ASTM), Federal Aviation Administration (FAA), Radio Technical Commission for Aeronautics (RTCA), and the US Defense Standard (MIL-SPEC). International administrations include the International Organization for Standardization (ISO), International Civil Aviation Organization (ICAO), and the Association for Unmanned Vehicle Systems International (AUVSI). The primary activities for these organizations include, but are not limited to, development, promotion, revision, interpretation, and production of technical standards that are intended to address the needs of adopters.

In the United States, the RTCA and FAA develop comprehensive and industry-approved recommendations for the US government on issues ranging from technical performance to operational concept standards for air transportation. One such standard is the FAA's Federal Aviation Regulations (FARs). The FARs govern all aviation activities in the United States ranging from activities such as aircraft design and maintenance, airline flights, pilot training, model aircraft operation, and more (United States Code of Federal Regulation Title 14, 2015).

In the international arena, ICAO has been selected to serve as the cornerstone for global interoperability and to coordinate the development of civil UAS standards and recommended practices. Their goal would be to harmonize the development of a strategic guidance document that would support emerging work in the regulatory evolution (ICAO Cir 328 - Unmanned Aircraft Systems (UAS), 2011). The RTCA has also released several standards to regulate aircraft systems at an international level and is currently establishing document SC-203 that details the standards, certification, and procedure for sense-and-avoid systems. Document SC-203 will help ensure the safe, efficient, and compatible operation of UAS with other vehicles within the National Air Space (NAS) (RTCA, 2015).

In recent years, UAS have emerged as innovative solutions for numerous commercial and public service missions. Although recreational and military UAS existed for many years, the push for civil and commercial use (and expanded military use inside the United States) has brought them under scrutiny due to new risks and uncertainties associated with their integration into the NAS. ASTM, NASA, and the FAA have been approached by the Department of Defense (DoD) to develop these regulations in order to help guide that path (ASTM Standards to Support New FAA Regulations for Small Unmanned Aircraft Systems, 2010; ASTM, 2003).

The most prominent interoperability standard for UAS currently employed is NATO Standardization Agreement (STANAG) 4586. While there are many documents that promote interoperability standards, such as the Joint Architecture for Unmanned Systems (JAUS) and MIL-STD-1760, none of them are as comprehensive, well established, or adopted regarding UAS (Pedersen, 2007). The primary goal of STANAG 4586 is to improve operation and management of multiple existing and future UAS in a complex environment such as those between various command, control, communications, computers, and intelligence (C4I) ground and air segments. STANAG 4586, a living document now in its third edition, includes the 4586 standard and four volumes of supporting documentation. These documents define the architectures, interfaces, communication protocols, data elements, and message format requirements of all UAS operating in a NATO Combined/Joint Service Environment (North Atlantic Treaty Organization, 2004). Although a majority of STANAG 4586 is dedicated toward listing requirements and standards, key interoperability concepts and terms are identified. As the reader progresses through this section, more of these concepts and terms are described.

3 INTEROPERABILITY

There are numerous definitions of *interoperability* in academic, professional, and government documents. While a seemingly straightforward concept, interoperability definitions vary extensively in meaning and context. A survey report from 2007 has identified over 34 distinct definitions of interoperability (Ford *et al.*, 2007). Most definitions originate from defense and computing, while others come from business, art, and healthcare.

Despite the variation, these definitions share certain commonalities that enable grouping into four distinct domains: technical, semantic/data, organizational, and legal/environmental. Each domain has its own unique, and equally important, contribution to interoperability and contextual approach. The European Commission has outlined each domain in the

Table 1. Four interoperability domains.

Interoperability	Description	Consists of
Technical	Planning of technical issues involved in linking computer systems and services	The data communication between systems
Semantic (Data)	Precise meaning of exchanged information that is preserved and understood by all parties	Syntax and semantics
Organizational	Coordinated processes in which different organizations achieve a previously agreed and mutually beneficial goal	Business processes, cultural, and economic
Legal (Environmental)	Aligned legislation so that exchanged data is accorded proper legal weight	Policy, legality, legislation, and so on

Source: Reproduced with permission from European Commission.

European Interoperability Framework (EIF) that explain in further detail the context of each one (European Interoperability Framework for European Public Services, 2010). The EIF describe *technical interoperability* as the planning of technical issues involved in linking computer systems and services. *Semantic interoperability* details the precise meaning of exchanged information, which is preserved and understood by all parties. *Organizational interoperability* covers coordinated processes in which different organizations achieve a previously agreed and mutually beneficial goal. Finally, *legal interoperability* covers legislation, so that political exchanges are given the proper legal weight. A reproduction of these four categories is shown in Table 1.

The broadest, most widely accepted and used term (which we will adopt) is provided by the 1999 Joint Chiefs of Staff Publication 1–02 (US Joint Chiefs of Staff, Joint Doctrine; Department of Defense,): "The ability of systems, units, or forces to provide services to and accept services from other systems, units, or forces, and to use the services so exchanged to enable them to operate effectively together." This definition will be used to explore interoperability in multiple dimensions, its relative importance, operational and tactical perspectives, complexities, and modeling techniques for UAS applications.

Technical interoperability can also be associated with various levels. MITRE Corporation identifies five maturity levels of systems interoperability in their Levels of Information Systems Interoperability (LISI) model. The five levels of LISI with increasing maturity are isolated (nonconnected), connected (electronic connection), functional (minimal common functions), domain (shared data), and enterprise (interactive manipulation) (Levels of information systems interoperability (LISI), (1998)). While higher levels imply higher capabilities and flexibility, no level is better than the other. Each level has its own attributes that should be used strategically depending on the system's needs (Table 2).

Table 2. Interoperability levels from LISI.

Level type definition	Requirements	Examples
Level 4: Enterprise Interactive manipulation Shared data and applications	Advanced collaboration: Cross-domain information and applications sharing	Event triggered database update
Level 3: Domain Shared data "Separate" applications	Sophisticated collaboration: Shared databases	Common operational picture
Level 2: Functional Minimal common functions Separate data and applications	Basic collaboration: Heterogeneous product exchange	Annotated imagery or maps with overlays
Level 1: Connected Electronic connection Separate data and applications	Homogenous product exchange	Voice, tactical data links, text files, messages, email
Level 0: Isolated Nonconnected	Manual gateway	Flash drives, hard copy exchange, and so on

Source: Reproduced with permission from MITRE.

The LISI model defines each level as follows:

- *Level 0 – Isolated* Interoperability in a manual environment: This level of interoperability concerns isolated, or stand-alone, systems. At this level, there are no connections to the system of focus and all data extraction and analyses are done manually by the user, information transfer takes place through the use of removable media such as thumb drives and diskettes.
- *Level 1 – Connected* Interoperability in a peer-to-peer environment: At this level, systems are capable of providing simple electronic exchanges. Although homogenous data types are shared, there is little capacity to fuse information.
- *Level 2 – Functional* Interoperability in a Distributed Environment: Functional interoperability concerns the ability of locally connected systems to transfer data sets. Data is typically heterogeneous and may contain information from various similar formats merged together (e.g., an image with annotations). Although logical and physical data models are shared at this level, each program generally defines its own physical data model.
- *Level 3 – Domain based* Interoperability in an Integrated Environment: At this level, systems are connected via wide area networks that share information between independent applications using domain-based models. Systems are also capable of implementing business rules and processes that facilitate direct database-to-database interactions. This enables the system to support group collaboration on fused information.
- *Level 4 – Enterprise based* Interoperability in a Universal Environment: Systems are capable of utilizing a distributed global information space across several domains. Shared data has a common interpretation regardless of format. Multiple users must be capable of simultaneously fusing and interacting with complex data. Applications are also shared and can be distributed to support information fusion. This is the highest level of technical interoperability, where data is electronically delivered to the user regardless of the access method he uses, handheld devices, or workstations, and the location where he uses his device.

At the heart of higher interoperability is the meaningful exchange of information between systems. At such levels, operational and managerial controls of these systems are often independent of one another and are geographically distributed, as implied at the Enterprise level. This type of communication and control is similar to those in a system of systems (SoS) (Maier, 1998). The US Global Information Grid is one good example of an enterprise information system that aims to achieve this technical vision. The outcome of SoS operational interoperability can be broadly categorized as coexistence, cooperation, or collaboration.

Coexistence is a prerequisite for both cooperation and collaboration (Wah *et al.*, 2008). For example, during coordinated strikes, different groups of shooters may coexist with each other. By coordinating their maneuvers, the shooters are able to attack different sets of targets within short-time intervals.

Cooperation can be understood as the process of each system doing different task(s), sequentially or concurrently, to contribute to a shared outcome. One example is cooperative sensing, where several tactical UAS provide surveillance in different parts of an area of operation. Another example is cooperative engagement, where unmanned combat aerial vehicles (UCAVs) decide and execute among themselves the most efficient plan of attack. A single UCAV, if given sufficient time, would be able to attack all the targets over a number of sorties. This approach, however, may not be as efficient and effective, especially if the targets were capable of concealment or self-defense, as multiple UCAVS cooperating in the same environment (Wah *et al.*, 2008).

Collaboration refers to the process where the system elements work together to achieve a desired outcome, with each system element being incapable of achieving it alone, even when given sufficient time. One example is the integrated air defense architecture, where various sensor systems provide real-time surveillance and cues to the shooter systems. Another example can be found in nature, where ants swarm a larger insect or animal and attack until the prey dies. In both examples, each system element is unable to achieve the outcome alone.

4 IMPORTANCE OF INTEROPERABILITY AND STANDARDIZATION

Identifying proper interoperability and standards methods has implications in both the private and public sectors. Computer technology can apply similar interoperability frameworks to leverage future and legacy components. Businesses can employ a similar technique to improve their competitiveness or provide better services. Healthcare could integrate cross-manufacturer medical devices (Baldwin, 2010). Governments and militaries could also use interoperability to quickly adopt and integrate the variety of systems (manned and unmanned as well as ground, air, and naval systems) and tools already at their disposal (Hura *et al.*, 2000).

UAS is one category of systems with growing interest in defense establishments. As UAS continue to become one of the predominant and diverse set of systems across nearly every echelon and service of most major militaries, the need to coordinate, share, and integrate into a more substantial force is vital (Gimenes *et al.*, 2014), as reflected in the Joint Chief's definition quoted above.

As such, the operation of both manned and unmanned air vehicles requires a high level of interoperability and clear standards to ease training, logistics, integration, and CON-OPs (Weatherington and Deputy, 2005). An example of the potential of interoperability for manned and unmanned teaming (MUM-T) aerial systems can be seen in the United States Army Apache helicopter (Richard, 2014). As of 2014, the US Army has been able to obtain a higher level of interoperability between their Apache helicopters, Gray Eagle UAS, and RQ-7 UAS. In the past, only surveillance videos were shared between the helicopter pilots and UAS via a ground control system. Efforts in interoperability, however, have enabled the capability for manned aerial systems to handle the payloads, sensors, and flight of the UAS directly. Apache pilots are now capable of seeing over the horizon, remove battlefield unknowns, and reduce potential risks (Figure 1).

While the DoD has identified many military benefits of UAS interoperability, AUVSI has recognized the economic benefits across multiple private sectors (Jenkins and Vasigh, 2013). Precision agriculture and public safety, accounting for 90% of the private market share, are expected to make the biggest impact. One example of the benefits of interoperability for precision agricultural applications would be data management between aerial and ground systems. A relatively small and lightweight UAS could survey large acres of land quickly, identify locations of interest, and then relay those locations to slower, but more capable ground systems for crop treatment. Japan has been using UAS for agricultural purposes (mainly spraying of pesticides) for nearly two decades. During the last decade, Japanese agricultural UAS have expanded into geological, infrastructure, and even environmental monitoring (Sato, 2003).

At higher interoperability levels, consolidation of individual databases into a collective could provide the information to analyze and model weather, pest, and other crop management factors that span the region (National Research Council. Committee on Assessing Crop Yield: Site-Specific Farming, 1997). These efforts could provide the capabilities to predict, preemptively treat, and consolidate treatment efforts for crops over a wide area. Another application could be used in urban planning/development or even the tracking of environmental and geological changes over time. The key issue, however, is developing a way to not only consolidate but also harmonize standards developed in the United States,

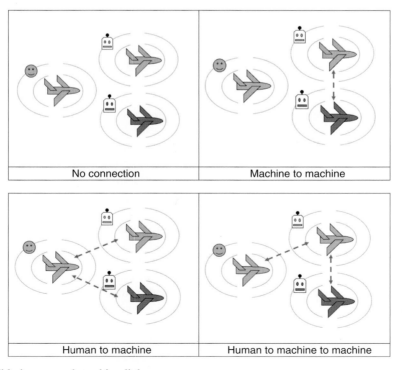

Figure 1. Example of possible human and machine links.

Canada, Europe, Australia, and Asia (Dalamagkidis *et al.*, 2012).

5 SYSTEMS ENGINEERING, ENABLED BY MODELING AND SIMULATION, FOR UAS INTEROPERABILITY AND STANDARDIZATION

This section will focus on the use of systems engineering, specifically SoSE, to analyze the communications and data exchanges of interoperability. Later sections will focus more broadly on mechanical and electrical exchanges. As mentioned before, the heart of higher interoperability is the meaningful exchange of information between systems. The increasing complexity of SoS necessitates a more meaningful communication between constituent systems in order to achieve operational goals. To this extent, interoperability needs to be taken into consideration to guide SoS design decisions.

The System of Systems Interoperability (SOSI) model, developed by Edwin Morris *et al.*, identifies the following types of interoperability in his report (Morris *et al.*, 2004):

- *Programmatic:* Interoperability between different program offices, that is activities that manage the acquisition of a system.
- *Constructive:* Interoperability between the organizations that are responsible for the construction (and maintenance) of a system, that is activities that develop or evolve a system (e.g., use of standards and off-the-shelf products, architecture).
- *Operational:* Interoperability between the systems, that is activities within the executing system and between the executing system and its environment, including the interoperation with other systems.

The report states that multiple models on SoS interoperability exist (i.e., LISI, OIM, NC3TA, LCIM, LCI, and SOSI). While all of them are successful at partial representation of some aspect of interoperability, a set of compatible models that collectively tackles every dimensions of interoperability is still desired. Other reports also acknowledge that processes to address SoS interoperability issues, support to coordinate planning among systems prior to integration, and as the components evolve, are still lacking (Sledge, 2010).

To address these gaps, Elizabeth Jones-Wyatt provides a brief overview of interoperability metrics and proposes a flexible, intuitive interoperability measure of system pairs within a potential architecture performing a set of resource exchanges (Jones-Wyatt *et al.*, 2013). The architecture draws from reliability theory to incorporate system requirements and to link the interoperability of a SoS with operational metrics of performance. Wyatt also provides a methodology for measuring system pair interoperability, SoS interoperability, and enables measurement of network metrics using interoperability inputs. This allows decision makers to evaluate and compare interoperability architectures for system pairs and collaborate on single and multiple resource exchange levels.

Modeling and Simulation (M&S) is a key tool for systems engineering of UAS systems and their interoperability. The capabilities and applications for UAS continue to grow as their cost decreases. It is now practical to use several UAS in large, distributed networks for a variety of missions. As this trend continues, M&S has grown increasingly important for designers, researchers, and policy makers to have tools that allow them to simulate, control, and prototype distributed networks of UAS in real time (Jang *et al.*, 2005).

5.1 Modeling and simulation

The DoD Net Centric Strategy has outlined four data interoperability foundations (visibility, accessibility, understandability, and trust) to support data interoperability and goals (Stenbit, 2003).

- The first foundation, *visibility*, makes data visible through the creation and registration of metadata. Registration of metadata such as Extensive Markup Language (XML) components, database segments, and data dictionary elements enable users to quickly sift through and identify properties critical to their goals. This opens opportunities for future applications of data collected or generated by a system (Interoperability, 1999).
- The second foundation, *accessibility*, focuses on the interfaces between systems. Identification of key interfaces between systems enables users to quickly access and make use of the data provided. This allows interested parties to focus their energy on the interfaces of interest and at the appropriate level of emphasis and analysis. This decentralized and distributed approach helps to reduce the burden of complex systems and the commitment of effort and resources on a single investor.
- The third foundation, *understandability*, focuses on the use of format-related metadata. Data, while accessible and visible can still sometimes be misinterpreted or misrepresented. Identification of the size, dimension, type, physical requirements, and context of the data is

critical for understandability. This helps to identify which tools and capabilities are required to support the data structure and process it in a meaningful way.

- Finally, the fourth foundation, instill *trust* and compliance with net-centric interface standards. As systems become more complex and distributed, a certain level of trust is required to ensure that data assets will contain the proper visibility, metadata, and accessibility required for interoperability. Compliance with the standards set by the Net Centric Strategy helps to ensure that a certain pedigree of data is achieved for those systems. In doing so, various differences, complications, and redundancies between systems of interest can be reduced and enable faster deployment or acquisition.

High-level architecture (HLA) has been a widely used framework in the defense community to simulate interoperability and reusability of systems. It consists of three components: a framework, an object model template (OMT), and a federate interface specification (IEEE Specification (1516.1-2000), 2001; IEEE (1516.2-2000), 2001; IEEE Specification (1516.1-2000), 2001). OMT defines HLA as comprised of both federation object models (FOM) and simulation object models (SOM). SOM specifics that certain information types could provide to or receive from other HLA federations, while FOM specifies the information exchanged at runtime to achieve federation objectives.

Researchers have used HLA, or variations of it, for simulating UAS communication, coordination, and mission planning (Jones *et al.*, 2003). However, development of newer technology and a need for a more dynamic net-centric data for the Global Information Grid (GIG) has brought

many deficiencies, such as those in composability, of HLA to light (Davis, 2003; Tolk, 2010). Andreas Tolk and James Muguira have developed a model to close some of these gaps in HLA (Tolk and Muguira, 2003). Tolk's model, the Levels of Conceptual Interoperability Model (LCIM), helps to address the need for implementation, integration, or federation between conceptual and technical designs. A figure of the seven levels with increasing capability is shown below (Figure 2).

Tolk argues that attempts to create interoperable systems without proper coordination at the conceptual level fail to address the cause-effect chains of "meaningful" interoperability. Therefore, LCIM acts as an outline to validate whether the coordination and composability of conceptual models is possible (Tolk and Muguira, 2003). A reproduction of Charles Turnitsa's and Andreas Tolk's description is shown in Table 3.

Despite the use of better modeling and simulation methods, most still lack a common language among various systems and models (Gustavson and Chase, 2007). While its absence may not pose any real threat for simulations confined to one system, it does make it difficult for multiple distributed systems to reach an agreement. One example is the use of the acronym "AA." It could mean a battery type, attack assessment, or antiaircraft.

An emerging standard called Battle Management Language (BML) developed by SISO hopes to fill in the gap of information ambiguity regarding architectures, components, and data representations. BML is defined as "the unambiguous language used to command and control forces and equipment conducting military operations and to provide for situational awareness and a shared, common operational

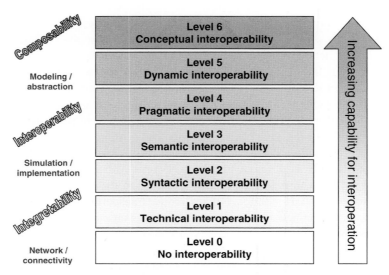

Figure 2. Levels of conceptual Interoperability. (Reproduced with permission from Tolk and Muguira (2016).)

Table 3. Description of LCIM levels.

Levels	Description of interoperability at this level
L6 (Conceptual)	Interoperating systems at this level are completely aware of each other's information, processes, contexts, and modeling assumptions
L5 (Dynamic)	Interoperating systems are able to reorient information production and consumption based on understood changes to meaning, due to changing context as time increases
L4 (Pragmatic)	Interoperating systems will be aware of the context (system states and processes) and meaning of information being exchanged
L3 (Semantic)	Interoperating systems are exchanging a set of terms that they can semantically parse
L2 (Syntactic)	Have an agreed protocol to exchange the right forms of data in the right order, but the meaning of data elements are not established
L1 (Technical)	Have technical connection(s) and can exchange data between systems
L0 (None)	No Interoperability

Source: Reproduced with permission from Tolk and Muguira (2003).

picture (Carey *et al.*, 2001)." Although developed for the DoD, BML can be generalized for multidiscipline groups or organizations to report findings or communicate ideas.

BML was built upon four key principles. The first and second principle outline that BML must not be ambiguous nor constrain the full expression of intent. The third principle outlines that BML, when possible, must use the existing Command, Control, Communications, Computers, Intelligence, Surveillance, and Reconnaissance (C4ISR) data representations. Finally, the fourth principle outlines that BML must allow all elements to communicate information pertaining to themselves, their mission, and their environment in order to create situational awareness and shared, common operational picture. Although these principles can be initially difficult to unite, integration difficulties or conflicts in core concepts further down a project can be even harder to resolve. While this standard is still relatively new in the Command and Control (C2) systems, BML has shown promise in the facilitation of coalition interoperability and C2 of UAS (Heffner and Hassaine, 2010).

One of the major proponents of the net-centric strategy has been the Network Centric Operations Industry Consortium (NCOIC). The NCOIC has developed several tools, recommendations, and resources used in several government agencies and organizations to lower engineering costs, to speed program implementation, to increase capability, and to reduce risk. One resource of interest is the NCOIC Interoperability Framework (NIF™). The NIF is different from many of the frameworks proposed by acting as a type of umbrella meta-framework (Osvalds *et al.*, 2008). NIF acknowledges that a single monolithic framework may not achieve systems interoperability or enable various systems capabilities. The NIF helps to identify the benefits and shortcoming of various standards, frameworks, and requirements from different system domains (aerial, ground, etc.). A

diagram comparing two different example standards is shown in Figure 3.

The top of the framework (also known as the overarching framework) is used to identify services that support capabilities for key missions, while lower levels provide key infrastructure capabilities. This process guides the user through various levels of abstraction to select the ideal framework, components, and standards necessary for a net-centric interoperable solution.

6   KEY MODULAR DESIGN TRADE AREAS

Building interoperability into UAS operations implies the design of open, modular systems. This section will explore key trades in the design of UAS architectures that consist of common interfaces that enable flexible component additions, upgrades, or swapping. While the following examples still look at UAS, several references will be made toward the use of UAS with next-generation general-purpose munition (i.e., Guided Bomb Unit and Air-to-surface Guided Missile). Key trades for such a UAS include the number of modules, support equipment, payload integration and standards, vendor competition, and transportation and assembly. Each of these is discussed in more detail below.

Number of modules: From a logistics and operations perspective, the number of modules needs to be minimized to reduce transportation and assembly costs. However, in order to achieve mission flexibility, many interchangeable modules are desirable.

Support equipment: From a maintenance perspective, the amount of supporting equipment needs to be minimized to reduce the maintenance and personal cost. However, as noted

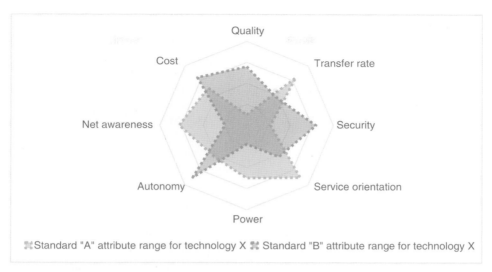

Figure 3. Mission Guided Standard. (Reproduced with permission from Network Centric Operations Industry Consortium, NCOIC Interoperability Framework (NIF™) and Patterns Overview (2008).)

above, to increase flexibility, more modules are needed, which in turn requires more supporting equipment and therefore higher maintenance. The durability and integrity of parts will be needed to offset these challenges in order to reduce maintenance costs.

Payload integration and standards: In single-purpose missions, the closed architecture is the cheapest alternative. However, in order to accommodate multiple purposes, such as anti-tank, the open architecture and the ability to swap interchangeable modules are required. Further, from a load-out perspective, while smaller diameter and shorter length munition is better – operational costs are less, a larger diameter may be required by some types of warheads and a longer weapon length may be desirable for better propulsion performance. Similar trades exist for payload integration in UAS design, including issues surrounding standardization among sensor capability and power requirements. One civilian example is the use of a single automotive platform (chassis, axles, steering, suspension, and engines) for different models, sizes, types, and price segments (Global Automotive Modular Platform Sharing Market Analysis 2013–2023, 2014). Automotive companies such as Toyota and Lexus have already implemented a system that allows them to build upon a highly interoperable platform for lower costs. The single-platform approach has allowed automotive manufacturers to customize and develop new models of vehicles based on similar underpinnings (Figure 4).

Competition at subsystem and integration level: Traditionally, only a single contractor wins a contract for developing what are typically single-purpose munitions. However,

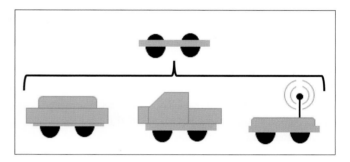

Figure 4. Use of a single platform for multiple vehicles.

this paradigm does not allow the incorporation of the strengths of multiple contractors to develop a general-purpose munition where each entity focuses only on the module (s) that correspond to its strengths. This has opened a market where a small but growing number of commercial vendors are offering a multivendor UAS development environment. One example is Airware's Aerial Information Platform that connects airframes, actuators, payloads, and application-specific software (Airware).

Transportation and assembly: Smaller and/or lighter munition generally incur lower transportation costs. However, from a holistic perspective the mission goals are better achieved by having the flexibility to assemble the required weapon configurations whenever they are needed. From the loader perspective, a ready-to-go weapon is easier to load, while an Air Operations Center may prefer the wider array of choices offered by open architecture.

The open architecture concept enables modular and flexible design but relies on technology being automatically synced up when added to the core of the weapon (analogous to building a PC where one can replace/install new video or memory cards without modifying the rest of the system).

7　MECHANICAL AND ELECTRICAL EXCHANGES

At the center of interoperability and effective open systems are standards. These standards provide the necessary foundations to define how systems communicate or interface with one another. As such, standards form an integral component in designing interoperability for mechanical, electrical, and information systems.

All components should be easily interchangeable as long as one uses components that support the same standard interface as the original component. The concept is to exchange equipment with standardized interfaces to provide rapid conversions from one role to that of another. This concept provides the ability to change the role configuration (surveillance, combat, civilian, etc.) rapidly to adapt to new situations. Examples of rapidly adaptable technologies can be observed in civilian sectors using modular function deployment (Clemen, 2006) to develop products that shorten the design cycle time and production costs by using preexisting elements as a common base for different products.

An emerging commercial application of the modular approach is the aforementioned Airware integration system that can be installed in commercial UAS to connect data from various components, such as sensors, cameras, actuators, and communication devices. The software claims to allow manufacturers to select which components to add and allows users to plan and monitor missions in real time. The key component is the autopilot device responsible for flying the vehicle in a safe and reliable manner acting as a hub for the other components. The system is customized by using software to select third-party drone vehicles and components – such as sensors, cameras, actuators, and communication devices.

As more UAS manufacturers enter the arena, the likelihood of multiple vendor systems collaboratively performing an operation is increasing. Daniel Gonzales and Sarah Harting have identified several interoperability shortcomings stemming from mixing vendor-specific systems (Gonzales, 2014). Many communication channels, control stations, and encryption technologies exist for UAS; however, these systems often consist of a single-vendor proprietary solution. These vendor-specific systems are often required to use similar proprietary interfaces that often share little to no compatibility with other vendor systems. While these architectures have been optimized for their applications, the lack of interoperability not only in resources but also in communication has limited the potential to leverage available capabilities and surplus resources with external users. To overcome such issues, Gonzales and Harting have proposed the UAS Modular technical reference model (TRM). TRM provides a foundation to identify the standards, specifications, and technologies that support secure delivery and construction of components and their interfaces. TRM is also used to identify the layers of a component-based architecture and the required technology to support it (Gonzales, 2014). The modular TRM identifies the essential modular components that support multiple levels of autonomy. These components, which are consistent with the UAS Pattern-Oriented Software Architectures (POSA) framework, improve unmanned system interoperability and can enable simpler autonomous capabilities integration into unmanned systems after they are developed.

8　CHALLENGES IN UAS INTEROPERABILITY AND STANDARDIZATION

As summarized in our review, interoperability comes in various shapes and sizes, and can occur at several levels for both civilian and military applications. Benefits are wide ranging and can enable system effectiveness in ways that were previously unfeasible. Despite attempts to standardize and implement interoperability, challenges remain and this is certainly the case in areas related to UAS development and use.

Better promotion and adoption of interoperability standards is still needed. Despite the availability of capable systems, a high degree of synchronization and information sharing are still absent for unmanned vehicle (air, ground, maritime) cooperation (Tolk, 2013). Many communication channels, control stations, and encryption technologies exist for UAS. However, these systems often consist of a single-vendor proprietary solution. This challenge goes beyond software, having implications in hardware components such as payload, power plant, avionics, and manufacturing. Additional effort could be focused on leveraging advanced manufacturing techniques such as additive manufacturing or using multipurpose parts. A diagram of a system with a high level of integrability made with additive manufacturing designed at Purdue University is shown in Figure 5.

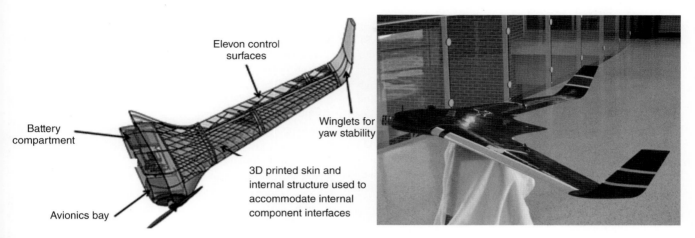

Figure 5. 3D Printed Plane developed at Purdue University.

However, designing toward higher levels of interoperability also has implications in identifying the conceptual model (Tolk, 2013). As new systems are developed and implemented for interoperability, verification and validation of conceptual models is critical to guarantee correct interoperability.

Research should continue to develop metrics that measure and identify tradeoffs related to interoperability. Information is needed to identify how the four types of interoperability (technical, semantic, organizational, and legal) are measured and what tradeoffs are made as the level of interoperability increases. Questions regarding the ideal interoperability level, security vulnerabilities, data transfer, and schema conflict resolution remain.

REFERENCES

Airware (Producer). airware.com. Retrieved from http://www.airware.com

ASTM (2003) Sharing the sky: unmanned air vehicle industry seeks standards. *ASTM Standardization News*, pp. 28–29.

ASTM Standards to Support New FAA Regulations for Small Unmanned Aircraft Systems (2010) *ASTM Standardization News*, **38**(1), 6–7.

Baldwin, C.Y. (2010) *When Open Architecture Beats Closed: The Entrepreneurial Use of Architectural Knowledge*, vol. 10–063, Harvard Business School.

Carey, S., Kleiner, M., Hieb, M., and Brown, R. (2001) Standardizing battle management language: a vital move towards the army transformation. *IEEE Fall Simulation Interoperability Workshop*.

Clemen, G. (2006) Application of the modular function deployment tool on a pressure regulator.

Dalamagkidis, K., Valavanis, K.P., and Piegl, L.A. (2012) *On Integrating Unmanned Aircraft Systems into the National Airspace System: Issues, Challenges, Operational Restrictions, Certification, and Recommendations*, 2nd ed., vol. 54, Springer.

Davis, P.K. (2003) *Improving the Composability of Department of Defense Models and Simulations*, Rand, Santa Monica, CA.

Department of Defense (1998) Levels of information systems interoperability (LISI).

Department of Defense (2002) Dictionary of Military and Associated Terms.

European Interoperability Framework for European Public Services (2010) European Commission.

Ford, T.C., Colombi, J.M., Graham, S.R., and Jacques, D.R. (2007) *Survey on Interoperability Measurement*, PN.

Gimenes, R., Vismari, L., Avelino, V., Camargo, J., Almeida, J., and Cugnasca, P. (2014) Guidelines for the integration of autonomous UAS into the global ATM. *J. Intell. Robot. Syst.*, **74**(1), 465–478.

Global Automotive Modular Platform Sharing Market Analysis 2013–2023 (2014) New York.

Gonzales, D. (2014) *Designing Unmanned Systems with Greater Autonomy: Using a Federated, Partially Open Systems Architecture Approach*, RAND Corporation, Santa Monica, CA.

Grösser, S.N. (2013) *Co-evolution of Standards in Innovation Systems: The Dynamics of Voluntary and Legal Building Codes*, Physica, Heidelberg.

Gustavson, P. and Chase, T. (2007) Building composable bridges between the conceptual space and the implementation space, IEEE, pp. 804–814.

Heffner, K. and Hassaine, F. (2010) Using BML for command & control of autonomous unmanned air systems, IEEE/SISO Fall 2010 Simulation Interoperability Workshop, Orlando, FL, pp. 305–317.

Hura, M., McLeod, G., Larson, E., Schneider, J., and Gonzales, D. (2000) A Broad Definition of Interoperability, in

Interoperability: A Continuing Challenge in Coalition Air Operations, Rand Corp, Santa Monica, CA, pp. 7–15.

ICAO Circular 328 (2011) Unmanned Aircraft Systems (UAS). Retrieved from http://www.icao.int/Meetings/UAS/Documents/Circular%20328_en.pdf

IEEE Standard for Modeling and Simulation (M&S) (2001) High Level Architecture (HLA): Federate Interface Specification (1516.1-2000), IEEE.

IEEE Standard for Modeling and Simulation (M&S) (2001) High Level Architecture (HLA): Object Model Template (OMT) Specification (1516.2-2000), IEEE.

IEEE Standard for Modeling and Simulation (M&S) (2001). High Level Architecture (HLA): Federate Interface Specification (1516.1-2000), IEEE.

Interoperability. (1999) *Realizing the Potential of C4I: Fundamental Challenges*, National Research Council, pp. 64–278.

Jang, M.-W., Reddy, S., Tosic, P., Chen, L., and Agha, G. (2005) *An Actor-Based Simulation for studying UAV Coordination.* Paper presented at the 15th European Simulation Symposium.

Jenkins, D. and Vasigh, B. (2013) *The Economic Impact of Unmanned Aircraft Systems Integration in the United States*, Association for Unmanned Vehicle Systems International (AUVSI).

Jones, E.D., Roberts, R.S., and Hsia, T.C.S. (2003) STOMP: a software architecture for the design and simulation of UAV-based sensor networks, *Proceedings of IEEE International Conference on Robotics and Automation*, vol. **3**, IEEE, pp. 3321–3326.

Jones-Wyatt, E., Domercant, J.C., and Mavris, D.N. (2013) A reliability-based measurement of interoperability for systems of systems, IEEE, pp. 408–413.

Maier, M.W. (1998) Architecting principles for systems-of-systems. *Syst. Eng.*, **1**(4), 267–284. doi: 10.1002/(SICI)1520-6858 (1998)1:4<267::AID-SYS3>3.0.CO;2-D.

Morris, E., Levine, L., Meyers, C., Place, P., and Plakosh, D. (2004) System of Systems Interoperability (SOSI): Final Report. Carnegie-Mellon University, Pittsburgh, PA.

National Research Council and Committee on Assessing Crop Yield: Site-Specific Farming (1997) *Precision Agriculture in the 21st Century: Geospatial and Information Technologies in Crop Management*, National Academy Press, Washington, DC.

North Atlantic Treaty Organization (2004) STANAG 4586 – Standard Interfaces of UAV Control System for NATO UAV Interoperability. NATO Standardization Agency.

Osvalds, G., Bowler, M., Jones, A., Noble, J., Yanosy, J., Schylberg, L., and Lebas, F.-X. (2008) NCOIC Interoperability Framework, NCOIC.

Pedersen, J. (2007) Interoperability Standards Analysis (ISA).

Richard, T. (2014) Army touts interoperability of Apache helos, unmanned aircraft.

RTCA (2015) Sunsetted Committees.

Sato, A. (2003) Civil UAV applications in Japan and related safety & certification. Retrieved from http://uvs-international.org/phocadownload/03_5ac_Relevant_Information/Applications_Civil-UAV-Applications-in-Japan.pdf

Sledge, C.A. (2010) Reports from the Field on System of Systems Interoperability Challenges and Promising Approaches. Carnegie-Mellon University, Pittsburgh, PA.

Stenbit, J.P. (2003) DoD Net-Centric Data Strategy, Department of Defense.

Tolk, A. (2010) Avoiding another green elephant: a proposal for the next generation HLA based on the model driven architecture.

Tolk, A. (2013) *Interoperability, composability, and their implications for distributed simulation: towards mathematical foundations of simulation interoperability.* Paper presented at the *Proceedings of the 2013 IEEE/ACM 17th International Symposium on Distributed Simulation and Real Time Applications.*

Tolk, A. and Muguira, J.A. (2003) The levels of conceptual interoperability model. *Proceedings of the 2003 Fall Simulation Interoperability Workshop,* p. 7.

US Code of Federal Regulation. (2015). Title 14, Electronic Code of Federal Regulations. Retrieved from http://www.ecfr.gov/cgi-bin/text-idx?tpl=/ecfrbrowse/Title14/14tab_02.tpl

Wah, S.K., Jin, F.J., and Boon, D.C. (2008) *Realising System of Systems Interoperability,* DSTA Horizons.

Weatherington, D. and Deputy, U. (2005) Unmanned Aircraft Systems Roadmap, 2005–2030. UAV Planning Task Force.

PART 10
Integration Issues: Safety, Security, Privacy

Chapter 42

Unmanned Aircraft Systems (UAS) – Regulatory Policy and Processes: A Moving Landscape – A US Perspective

Daniel P. Salvano

Aviation Consultant, Safety, Certification and CNS Systems, Haymarket, VA, USA

1 INTRODUCTION

Unmanned aerial vehicles (UAVs) have had a long recorded history that reaches back before powered flight. One of the first documented uses was by the Austrians, in August 1849, when they launched some 200 pilotless balloons mounted with bombs against the city of Venice, Italy, in an effort to retake control of the city (Monash University, n.d.). Nearly a century later, from the late 1944 until early 1945, the Japanese launched thousands of fire balloons, called Fusen Bakudan, across the Pacific Ocean, using the jet stream, toward North America (the United States and Canada) (National Geographic, 2013). Despite the high hopes of their designers, the balloons were relatively ineffective as weapons.

It was not until the aeronautics revolution, fueled by World War I (WWI), that powered UAV were developed. The first military UAV in the United States was the rail-launched "Kettering Aerial Torpedo" developed during WWI (see Figure 1).

While UAVs have been flying in the United States for more than 80 years, their use in civil aviation is relatively new. The current market for civil UAS is relatively small, but will be one of the fastest growing aviation segments over the next 10 years (Federal Aviation Administration, 2015). A Teal Group 2014 market study estimates that UAV spending will nearly double over the next decade from current worldwide UAV expenditures of $6.4 billion annually to $11.5 billion, totaling almost $91 billion in the next 10 years (Teal Group, 2014). The most active and commercially viable segment of the UAS market is small UAS (sUAS) that weigh 55 lb or less. For the purposes of this chapter, the terms UAV, UAS, and drones are interchangeable. They are defined in Appendix A.

As new technologies are introduced, there are misunderstandings or myths surrounding their implementation. This chapter will separate the facts from myth. It will also describe today's regulatory and policy environment and the

Unmanned Aircraft Systems. Edited by Ella Atkins, Aníbal Ollero, Antonios Tsourdos, Richard Blockley and Wei Shyy.
© 2016 John Wiley & Sons, Ltd. ISBN: 978-1-118-86645-0.

Figure 1. Kettering Aerial Torpedo. (Photo courtesy of the Museum of the US Air Force.)

development of new internationally harmonized technical standards to support the changing UAS requirements.

2   UAS LEGAL, REGULATORY, AND POLICY DOCUMENTS AND THE CAUSE FOR CONFUSION

Part of the mistrust and angst between the government regulators and the UAS community, manufacturers, and operators is confusion over fact versus myth with respect to terminology, process, and governance of UAS.

Note: Additional information concerning UAS fact versus myth can be found at www.faa.gov/news.updates/?newsid=76381.

2.1   Facts versus myth

Myth: UAS are not aircraft and therefore not regulated by the FAA.

Fact: UAS are aircraft and subject to FAA regulation. The US law describes an unmanned aircraft as an aircraft that is operated without the possibility of direct human intervention from within or on the system (Public Law 112–95, 2012). The Code of Federal Regulations defines an aircraft as a device that is used or intended to be used for flight in the air (Code of Federal Regulations, n.d.).

Myth: Model aircraft are not "real" aircraft and are not regulated by the FAA.

Fact: US law defines model aircraft as an unmanned aircraft that is capable of sustained flight in the atmosphere, flown within visual line of sight (VLOS) of the person operating the aircraft, and flown only for hobby or recreational purposes

(Public Law 112-95, n.d.). Under this definition, model aircraft are "aircraft" that are regulated by FAA.

FAA Advisory Circular (AC), AC 91-57A, has been recently updated that provides guidance to operators of unmanned aircraft for hobby or recreational purposes that meet the statutory requirements of a "model aircraft" (Advisory Circular 91–57A, 2015). In order to operate as a model aircraft, you must meet the following conditions:

1. The aircraft is flown strictly for hobby or recreational use.
2. The aircraft operates in accordance with a community-based set of safety guidelines and within the programming of a nationwide community-based organization (CBO).
3. The aircraft is limited to not more than 55 lb, unless otherwise certified through a design, construction, inspection, flight test, and operational safety program administered by a CBO.
4. The aircraft operates in a manner that does not interfere with, and gives way to, any manned aircraft.
5. When flown within 5 miles of an airport, the operator of the model aircraft provides the airport operator or the airport air traffic control tower (when an air traffic facility is located at the airport) with prior notice of the operation. Model aircraft operators flying from a permanent location within 5 miles of an airport should establish a mutually agreed upon operating procedure with the airport operator and the airport air traffic control tower (when an air traffic facility is located at the airport).

Note: While the advisory circular does not specifically speak about operations near heliports or helipads, the FAA has recently developed a "smartphone application" to provide situational awareness to model aircraft owners as to where they can or cannot fly. The application includes all charted airports, heliports, and helipads. Information on this application can be found at www.faa.gov/uas/b4ufly.Most model aircraft weigh less than 55 lb. But there are some that are larger. Figure 2 is such an aircraft. This 66% scale model of an Icelandic homebuilt called an "Aerokot" weighs 67 lb with a wingspan of 13 ft.

Myth: The FAA has no authority to regulate the airspace over private property.

Fact: The Paris Convention of 1919 sought to solve the issue of whether a state has the right to claim sovereignty over the airspace overlying its territory or is the airspace open to all users and no

Figure 2. Aerokot. (Photo taken by Dean and Pete Coxon and published in *Model Airplane News*. Reproduced with permission from Peter Coxon, 2016. © Peter Coxon.)

state can claim sovereignty (The Paris Convention, 1919). One of the principles agreed to was that each nation has absolute sovereignty over the airspace overlying its territories and waters. The Protocol of June 15, 1929 amended the Paris Convention by adding a subparagraph to Article 15 stating: "No aircraft of a contracting State capable of being flown without a pilot shall, except by special authorization, fly without a pilot over the territory of another contracting State" (ICAO Document 10019, 2015). These principles have been incorporated into the Chicago Convention of 1944 that founded the International Civil Aviation Organization (ICAO) (The Convention on International Civil Aviation, 1944).

Note: The term Contracting State is used when reference is made to the signatory countries to the Chicago Convention versus the term Member State that is used in connection with ICAO membership.

In the United States, the struggle over private property owners rights and need for the unrestrictive flow of air commerce was decided by the US Supreme Court (The United States, 1946). In that case the court rejected the British and US common law concept of Cujus solum ejus est usque ad coelum, that is, he that owns the land owned the sky out to the heavens. The court held that a landowner owns as much space above the ground as he can occupy or use in connection with the land (Id. At 264, 1936). The use of UAVs in civil airspace has reopened certain unanswered questions: Where is the dividing line between the landowners space used in connection with the land and public domain airspace. Can "navigable airspace" (Title 14, n.d.), defined as airspace at and above the minimum safe flight altitudes, including airspace needed for safe takeoff and landing, intersect airspace belonging to the landowner?

Myth: A drone is different from a UAS and a remotely piloted aircraft (RPA).

Fact: ICAO has differentiated between an RPA and an autonomous aircraft. ICAO Document 10019 (ICAO Document, 2015) defines the superset "Unmanned Aircraft" as containing three subsets: remotely piloted aircraft, autonomous aircraft, and model aircraft (see Figure 3).

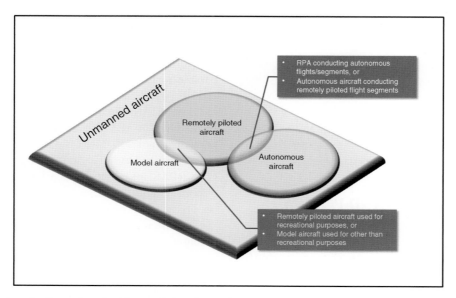

Figure 3. Unmanned aircraft. (Reproduced with permission from ICAO, 2016. © International Civil Aviation Organization.)

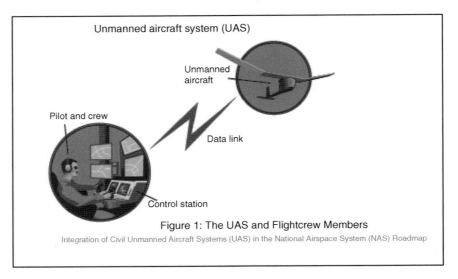

Figure 1: The UAS and Flightcrew Members
Integration of Civil Unmanned Aircraft Systems (UAS) in the National Airspace System (NAS) Roadmap

Figure 4. UAS. (Reproduced from www.faasafety.gov.)

In the United States, an unmanned aircraft vehicle is defined as an aircraft that is operated without the possibility of direct human intervention from within or on the aircraft (Public Law 112–95, 2012). Drone is the informal name for a UAV or an RPA that can be flown autonomously or with a remote pilot. The term has become a common descriptor since the casual observer cannot tell which method is used to operate the aircraft from an external visual examination.

Finally, the term drone, UAV, or RPA only refers to that part of the system that actually flies. A UAS or remotely piloted aircraft system (RPAS) is a system that includes the following as a minimum:

- the remotely piloted aircraft,
- the remote pilot station (RPS) that contains the equipment used to pilot the aircraft, and
- Command and control (C2) link that connects the RPA and the RPS for the purpose of managing the flight (Figure 4).

For the purpose of ease of understanding terminology, I will use the US terminology listed below, unless referencing a particular term is required for understanding.

3 UAS REGULATIONS AND POLICY TODAY

Today, UAS are flying in the US National Airspace (NAS) under very controlled conditions with operations ranging from essentially ground level to 50 000 ft with vehicle sizes ranging from being able to fit in the palm of your hand to the size of a Boeing 737. All these civil UAS operations are authorized, by the FAA, on a case-by-case basis, in specific airspace, using "Certificates of Waiver or Authorization" (COA).

The FAA goal is that "Ultimately, UAS must be integrated into the NAS without reducing existing capacity, decreasing safety, negatively impacting current operators, or increasing the risk to airspace users or persons and property on the ground any more than the integration of comparable new and novel technologies." (Integration of Civil Unmanned Aircraft Systems, 2013).

The ICAO goal with respect to UAS ". . . is to provide an international regulatory framework through Standards and Recommended Practices (SARPs), with supporting Procedures for Air Navigation Services (PANS) and guidance material, to underpin routine operation of UAS throughout the world in a safe, harmonized and seamless manner comparable to that of manned operations." (ICAO Document 10019, 2015). In 2015, ICAO released the RPAS Manual, replacing UAS Circular 328 issued in 2011. The manual looks only at RPAS and not at the entire superset of unmanned aircraft. These two goals are identical and can be summarized by saying it is their goal that the introduction UAS/RPAS into any airspace (en route or terminal) or at airports should not increase safety risks to manned aircraft, people on the ground or in the air, or property.

The difference between remotely piloted vehicles and an autonomous aircraft becomes critical when we consider the technical, policy, regulatory, and societal issues between airspace integration of these two types of aircraft. To fully integrate into the NAS, UAS need to be able to bridge the gap from existing systems requiring accommodations to future systems that are able to obtain a standard airworthiness certificates (FAA, Integration of Civil Unmanned Aircraft Systems, 2013). The FAA, in its role as both an aviation safety regulator and an air traffic service provider, has addressed the integration of UAS as an evolutionary process. This process has three perspectives – *accommodation, integration*, and *evolution* – that examine the complex relationship of activities necessary to integrate UAS into the NAS. Each perspective is discussed below.

Perspective 1: Accommodation – Take current UAS and apply special mitigations and procedures to safely facilitate limited access to the NAS. UAS operations in the NAS are considered on a case-by-case basis. The FAA has streamlined the issue of Certificates of Waiver or Authorization (COAs) using its new Section 333 exemption authority provided by the US Congress (FAA Modernization and Reform Act, 2012). This process is being used to allow UAS to perform commercial operations in low-risk, controlled environments such as power line and pipeline inspection, precision agriculture, mining and railroad surveys, and aerial photography. As of August 2015, the FAA has granted more than 1000 COAs under the provisions of Section 333 of the law (FAA News Release, 2015). Both the Section 333 and original COA processes exist and either can be used at the applicant's discretion. The COA process is comprehensively covered in Operational Profiles of Unmanned Aircraft Systems in the Context of the United States Regulatory Regime. Accommodation will predominate in the near term, and while it will decline significantly as integration begins and expands in the mid-term, it will continue to be a viable means for NAS access with appropriate restrictions and constraints to mitigate any performance shortfalls.

Perspective 2: Integration – Establishing threshold performance requirements for UAS that would increase access to the NAS is a primary objective of integration. During the mid- to far-term, the FAA will establish new or revised regulations, policies, procedures, guidance material, training, and understanding of systems and operations to support routine NAS operations. FAA key policy documents are listed below:

- In February 2007, the FAA issued a policy stating that no person may operate an unmanned aircraft system (UAS) in the NAS without specific authority (Federal Register, 2007).
- In February 2015, the FAA issued a Notice of Proposed Rulemaking (Federal Register, FAA, n.d.). The FAA proposal offers safety rules for small UAS (below 55 lb) conducting nonrecreational operations. The rule would limit flights to daylight and visual line of sight operations. It also addresses height restrictions, operator certification, optional use of a visual observer (VO),

aircraft registration and marking, and operational limits. The proposed rule also includes extensive discussion of the possibility of an additional, more flexible framework for "micro" UAS that are less than 4.4 lb. The comment period closed in April 2015 and at a Congressional Committee hearing the FAA deputy administrator testified the final rule is expected to be issued by June 2016 (FAA Deputy Administrator, 2015). He also stated that the timing of any large UAS rulemaking would be driven by commercial demand and technology requirements.

- On June 18, 2014, the FAA issued a Notice of Interpretation with request for comments (Federal Register Notice of Interpretation and Request for Comments, n.d.).

  This action provides interested persons with the opportunity to comment on the FAA's interpretation of the special rule for model aircraft established by Congress in the FAA Modernization and Reform Act of 2012. In this interpretation, the FAA clarifies that model aircraft must satisfy the criteria in the Act to qualify as model aircraft and to be exempt from future FAA rulemaking action; and consistent with the Act, if a model aircraft operator endangers the safety of the National Airspace System, the FAA has the authority to take enforcement action against those operators for any safety violations.

- FAA Order 8130.34C, Airworthiness Certification of Unmanned Aircraft Systems and Optionally Piloted Aircraft, was published on August 2, 2013. This order establishes procedures for issuing special airworthiness certificates in the experimental category or special flight permits to UAS, optionally piloted aircraft (OPA), and aircraft intended to be flown as either a UAS or an OPA under the designation "UAS/OPA."

- September 2, 2015, the FAA issued Advisory Circular AC 91-57A, Model Aircraft Operating Standards. This AC provides updated guidance for persons operating unmanned aircraft meeting the requirements defining model aircraft, for hobby or recreational use, on their safe operation in the NAS.

- August 4, 2015, the FAA issued National Policy 8900.313 (FAA Order 8900, n.d.). The policy provides guidance to FAA inspectors on the process of contacting and educating individuals who are the subject of an inquiry relating to unauthorized operation of UAS in the NAS.

- (*Note:* All of the documents listed above, plus other pertinent documents can be found at www.faa.gov/uas/regulations_policies/)

At the time this chapter was being written, pilot reports of UAS sightings in 2015 were double the rate of 2014, averaging approximately 100 per month. Pilots have reported seeing drones at altitudes up to 10 000 ft, or as close as half-a-mile from the approach end of a runway. There were reports of the presence of multiple UAS in the vicinity of wild fires in the western part of the United States that prompted fire-fighters to ground their aircraft on several occasions. These UAS operations are unsafe and illegal (FAA – Pilot Reports of Close Calls with Drones Soar, 2015). As a result, the DOT/FAA announced to reconsider its past practice of exercising discretion with respect to requiring UAS to be registered, consistent with statutory requirements of 49 U.S.C. 44101–44103, and had determined that registration of all UAS is necessary to enforce personal accountability while operating an aircraft in the NAS. A task force was created to develop recommendations for a registration process for UAS. The task force was composed of diverse representatives from the UAS and manned aviation industries, the federal government, and other stakeholders. The group advised the Department of Transportation (DOT)/FAA concerning which aircraft should be exempt from registration due to low safety risk, including toys and certain other small UAS. The task force also explored options for a streamlined system that would make registration less burdensome for commercial UAS operators. They submitted their report on November 20, 2015 so that an emergency rulemaking could be issued by the Christmas. Effective from December 21, 2015, anyone who owns a small aircraft that weighs between 0.55 and 55 lb must register with the FAA (Registration and Marking Requirements for Small Unmanned Aircraft Interim Final Rule, n.d.). Nearly 300 000 owners have registered their small unmanned aircraft in the first 30 days after the FAA registration system went live (FAA Press Release, 2016).

Perspective 3: Evolution – All required policy, regulations, procedures, guidance material, technologies, and training are in place and routinely updated to support UAS operations in the NAS operational environment as it evolves over time. It should be noted that the technical, policy, regulatory, economic, and societal challenges between an unmanned aircraft that is remotely piloted, an RPAS, and an autonomous aircraft, although they are today considered UAS, are significant. Also, the types of UAS missions and vehicle size impact the safety/risk assessments required, which will impact the implementation. These challenges will affect their full integration into the National Airspace.

4   TYPE CERTIFICATION (TC) AND AIRWORTHINESS APPROVALS

Historically, for manned aviation, the aircraft is the primary entity in which all aircraft components are integrated.

Table 1. ICAO manual remotely piloted aircraft systems.

Aeronautical products that may have a type certificate	
Manned aircraft	Remotely piloted aircraft
Aircraft (on which other products are mounted)	RPA
Engine	Engine
Propeller	Propeller
—	RPS

Therefore, the airworthiness approach focuses on the aircraft. Due to the distributed nature of unmanned aviation, the aircraft (UAV/RPA) requires ingenuity in trying to align their requirements with the manned TC process as far as practicable. In both the ICAO and US TC processes, the UAV/RPA is designated as the entity to receive the type design approval. It then becomes the TC holder's responsibility for the integration of all the components that support the safe operation of the UAV/RPA, for example, UAS/RPAS and any other needed equipment for operation. Table 1 provides a representation of the aeronautical products defined for manned aviation versus unmanned aviation.

In the United States, two manufacturers Boeing Insitu (ScanEagle) and AeroVironment (PUMA) have received TCs for their products (Figure 5). A copy of the type

Figure 5. ScanEagle. (Reproduced from US Navy.)

certificate data sheet (TCDS) for the Insitu ScanEagle is shown in Appendix B. You will note the UAV was certificated under the provisions of 14 CFR Part 21.15, Restricted Category Aircraft. Also, note the flight restrictions, on the TCDS, that are provided in the FAA-approved flight manual (Salvano, 2010).

5   DEVELOPMENT OF TECHNICAL STANDARDS

Two of the most difficult technical issues in the way of fully integrated UAS/RPAS operations in civil airspace are developing and approving the minimum performance specifications for detect and avoid (DAA) and the command and control (C2) link. In the United States, the general operating and flight rules (US Code of Federal Regulations, n.d.) require that vigilance shall be maintained by each person operating an aircraft so as to *see and avoid* other aircraft. The physical separation between the aircraft and the pilot of an unmanned aircraft require a robust C2 link, connecting the two. This is one of the most safety-critical systems in the UAS/RPAS.

In the United States, the RTCA (formerly known as the Radio Technical Commission for Aeronautics (www.rtca.org) is an FAA advisory committee that is chartered to develop consensus-based standards. The European Organization for Civil Aviation Equipment (EUROCAE) is its European counterpart. Working together they published the following:

- A DAA Minimum Operational Performance Standard (MOPS) that provides initial standards for a collection of airborne sensors whose outputs are combined using software on the unmanned aircraft to provide the unmanned aircraft system pilot in command (PIC) awareness of proximate traffic as well as suggestive guidance on how to avoid any loss of traffic separation.
- The C2 Data Link MOPS contains standards for the UAS command and control non-payload communication link system. C2 refers to the information exchanges needed to support the pilot in command to safely maneuver the UA on the ground and in the air. The document is focused on C-Band and L-Band terrestrial data links (RTCA, 2015).

Note: Final release of the documents is expected in 2016 after extensive verification and validation testing.

ICAO also has requirements for DAA and C2 Link in its manual or remotely piloted aircraft systems.

6   EUROPEAN REGULATIONS AND POLICY

The European Aviation Safety Agency (EASA) was created in 2008 and tasked by the European Commission to develop a set of European rules for "drones." Current aviation regulations (Regulation (EC) No. 216/2008) require that drones (also called "unmanned aircraft") above 150 kg be regulated in a similar way to manned aircraft. Those below that weight are to be regulated by each Member State of EASA as they see appropriate. EASA issued a Notice of Proposed Amendments to the Rules (A-NPA) number 2015-10. The A-NPA proposes the creation of common European safety rules for operating drones regardless of their weight. It proposes a proportional and operation-centered approach that focuses more on "how" and under "what conditions" the drone is used rather than only on the design characteristics of the drone as done in the type certification of manned aircraft. The full A-NPA can be found at www.easa.europa.eu/drones. The comment period closed September 25, 2015.

7   SUMMARY

Unmanned aircraft systems are a "disruptive innovation" similar to the introduction of the Model T Ford or the personal computer. At a recent US Senate hearing on UAS, the FAA Administrator stated that we are at a "Wright Brothers moment in Aviation" (US Senate Appropriations Subcommittee on Transportation, 2015).

Given the economic and political pressure surrounding small UAS (less than 55 lb in the United States and less than 150 kg in Europe), regulation and certification standards should be completed in 2016.

Larger UAS, especially those large enough to carry ADS-B, terrain awareness systems, and so on would follow the path forged by small UAS paced by user demand and technology.

Autonomous aircraft operations may take up to 10 years after large UAS integration into the NAS. The National Aeronautics and Space Administration (NASA) are currently performing research under the Safe Autonomous Systems Operations (SASO) project (www.aeronautics.nasa.gov/programs-aosp.htm).

Critical to the growth of UAS in the world market will be international harmonization of regulations and airworthiness standards. These efforts are underway between working groups at the FAA, EASA, and ICAO.

There are certain non-safety of flight issues that may delay the full integration of UAS into the National Airspace.

They are the fundamental right to personal privacy and possible encroachment of privacy by UAS, provision of controls to ensure that UAS are not used as weapons or for other illegal purposes, and finally a third party liability of an insurance system to account for UAS accidents. (This may be decided through the courts.)

APPENDIX A – DEFINITIONS

Aircraft   A device used or intended to be used for flight in the air, including unmanned aircraft (UA).

Airworthiness   A condition in which the UAS (including the aircraft, airframe, engine, propeller, accessories, appliances, and control station (CS)) conforms to its type certificate (TC), if applicable, and is in condition for safe operation.

Airworthiness Certification   A repeatable process that results in a documented decision that an aircraft system has been judged to be airworthy. It is intended to verify that the aircraft system can be safely maintained and safely operated by pilots within its described and documented operational envelope.

Airworthiness Statement   Document required from public UAS applicants during a Certificates of Waiver or Authorization (COA) application process that confirms aircraft airworthiness.

Certificates of Waiver or Authorization (COA)   An FAA grant approval for a specific operation. COAs may be used as an authorization, issued by the Air Traffic Organization (ATO) to a public operator for a specific UA activity. COAs for civil and commercial operations are only for aircraft that have received an airworthiness certificate from Aircraft Certification Service (AIR). Provisions or limitations may be imposed as part of the approval process to ensure the UA can operate safely with other airspace users.

Chase Aircraft   A manned aircraft flying in proximity to a UA that carries a qualified observer and/or UA pilot for the purpose of seeing and avoiding other aircraft and obstacles.

Civil Aircraft   Aircraft other than public aircraft.

Congested Area   A congested area is determined on a case-by-case basis. The determination must take into consideration all circumstances, not only the size of an area and the number of homes or structures (e.g., whether the buildings are occupied or people are otherwise present, such as on roads).

Cooperative Aircraft   Aircraft that have an electronic means of identification (i.e., a transponder or Automatic Dependent Surveillance – Broadcast (ADS-B) transceiver) operating onboard.

Crew Member (UAS)   In addition to the crew members identified in Title 14 of the Code of Federal Regulations (14 CFR) Part 1, a UAS flight crew member includes pilots, sensor/payload operators, and visual observers, but may also include other persons as appropriate or required to ensure safe operation of the aircraft.

Crew Resource Management (CRM)   The effective use of all available resources, including human, hardware, and information resources.

Daisy-Chaining   The use of multiple, successive observers to extend the flight of a UA beyond the direct visual line of sight of any other pilot in command or VO.

Disruptive Innovation   An innovation that creates a new market by applying a different set of values, which ultimately (and unexpectedly) overtakes an existing market (e.g., the personal computer). This term of art is attributed to Clayton Christensen.

Due Regard   A phase of flight wherein an aircraft commander of a state-operated aircraft assumes responsibility to separate his or her aircraft from all other aircraft.

Experimental Certificate   A type of Special Airworthiness Certificate issued for the purposes of research and development (R&D), crew training, exhibition, and market survey as defined in 14 CFR Part 21, § 21.191(a), (c), and (f).

Note: According to 14 CFR Part 91, § 91.319(a)(2), experimental aircraft may not be used for carrying persons or property for compensation or hire.

R&D Aircraft   Aircraft testing new design concepts, equipment, installations, operating techniques, or uses for aircraft. Any UAS, including an optionally piloted aircraft, is eligible for an experimental certificate under this purpose. The proponent may conduct operations only as a matter of research or to determine whether an idea warrants further development.

Crew Training   The process of bringing a person or persons to an established standard of proficiency. Crew training is limited to the number of flight crew required by the operator to conduct UAS aircraft operations.

Market Survey   Aircraft may be used for the purposes of conducting market surveys, sales demonstrations, and customer crew training of the manufacturer's customers, as provided in § 21.195.

External Pilot   A UAS pilot who flies from outside a control station with direct visual contact with the aircraft.

FAA-Recognized Equivalent   An FAA recognition that a public agency may exercise its own internal processes regarding airworthiness and pilot, aircrew, and maintenance personnel certification and training; furthermore, the agency has determined that its UAS is capable of safe operation in the NAS when conducting public aircraft operations under Title 49 of the United States Code (49 U.S.C.) §§ 40102(a)(41) and 40125.

Flight Termination   The intentional and deliberate process of performing controlled flight into terrain (CFIT). Flight termination must be executed in the event that all other contingencies have been exhausted, and further flight of the aircraft cannot be safely achieved, or other potential hazards exist that require immediate discontinuation of flight.

Flyaway   An interruption or loss of the control link, or when the pilot is unable to effect control of the aircraft and, as a result, the UA is not operating in a predicable or planned manner.

Internal Pilot   A UAS pilot who flies from inside a control station without direct visual contact with the aircraft.

Lost Link   The loss of command-and-control link contact with the remotely piloted aircraft such that the remote pilot can no longer manage the aircraft's flight.

Missile   A nonrecoverable, powered, guided munition that travels through the air or space. Ballistic missiles follow a ballistic trajectory. Cruise missiles generate lift. Guided missiles are launched from a ship or aircraft and serve as a self-contained precision bombs.

Noncooperative Aircraft   Aircraft that do not have an electronic means of identification (e.g., a transponder) aboard or that have inoperative equipment because of malfunction or deliberate action.

Observer   A trained person who assists a UAS pilot in the duties associated with collision avoidance and navigational awareness through electronic or visual means. Collision avoidance includes, but is not limited to, avoidance of other traffic, clouds, obstructions, terrain, and navigational awareness. A visual observer is a trained person who assists the UAS pilot by visual means in the duties associated with collision avoidance. A VO includes the OPA pilot when the OPA is being operated as a UAS.

Off-Airport   Any location used to launch or recover aircraft that is not considered an airport (e.g., an open field).

Optionally Piloted Aircraft   An aircraft that is integrated with UAS technology and still retains the capability of being flown by an onboard pilot using conventional control methods (see OPA Safety Pilot, below).

OPA Safety Pilot   The PIC that is responsible for ensuring the safe operation of an optionally piloted aircraft, whether under remote control or onboard control, for the purposes of overriding the automated control system in the case of malfunction or any other hazardous situation.

Pilot Duty Period   The period beginning when a flight crew member is required to report for duty with the intention of conducting a flight and ending when the aircraft is parked after the last flight. It includes the period of time before a flight or between flights that a pilot is working without an intervening rest period.

Pilot in Command   The person who has final authority and responsibility for the operation and safety of flight, has been designated as PIC before or during the flight, and holds the appropriate category, class, and type rating, if applicable, for the conduct of the flight. The responsibility and authority of the PIC as described by 14 CFR § 91.3 apply to the UA PIC. The PIC position may rotate duties as necessary with equally qualified pilots. The individual designated as PIC may change during flight.

Note: The PIC can only be the PIC for one aircraft at a time. For an OPA, the PIC must meet UAS guidance requirements for training, pilot licensing, and medical requirements when operating an OPA as a UAS.

Public Aircraft   An aircraft operated by a governmental entity (including Federal, State, or local governments, and the US Department of Defense (DOD) and its military branches) for certain purposes as described in 49 U.S.C. §§ 40102(a)(41) and 40125. Public aircraft status is determined on an operation-by-operation basis. Refer to Part 1, § 1.1 for a complete definition of a public aircraft.

Public Operator   An operator that is classified as government and/or otherwise qualifies for public aircraft operation under 49 U.S.C. §§ 40102(a)(41) and 40125. Not all flights by a public aircraft operator qualify as a public aircraft operation under the statute. Public aircraft operation status is not automatic for flights conducted by a government entity or a contractor to a government entity.

Remotely Piloted Aircraft   An unmanned aircraft that is piloted from a remote pilot station.

Remotely Piloted Aircraft System   A remotely piloted aircraft, its associated remote pilot station(s), the required command and control links, and any other components as specified in the type design.

Safety Risk Management (SRM)   A formalized, proactive approach to system safety. SRM is a methodology that ensures hazards are identified; risks are analyzed, assessed, and prioritized; and results are documented for decision-makers to transfer, eliminate, accept, or mitigate risk.

Scheduled Maintenance (Routine)   The performance of maintenance tasks at prescribed intervals.

Segregation   Setting apart from other air traffic operations in the NAS. Segregation is not synonymous with required air traffic separation standards. Therefore, segregation does not prescribe or mandate criteria such as vertical, lateral, or longitudinal distances.

Supplemental Pilot   Pilots assigned UAS flight duties to augment the PIC. It is common for operators to have both an internal and an external UAS pilot. The supplemental pilot can assume either of these positions. The supplemental pilot may also assume duties of the PIC if the specified qualifications are met.

Unmanned Aircraft   A device used or intended to be used for flight in the air that has no onboard pilot. This device excludes missiles, weapons, or exploding warheads, but includes all classes of airplanes, helicopters, airships, and powered-lift aircraft without an onboard pilot. UAs do not include traditional balloons (refer to 14 CFR Part 101), rockets, and unpowered gliders.

Unmanned Aircraft System   A UA and its associated elements related to safe operations, which may include control stations (ground-, ship-, or air-based), control links, support equipment, payloads, flight termination systems (FTS), and launch/recovery equipment.

Unscheduled Maintenance (Nonroutine)   The performance of maintenance tasks when mechanical irregularities occur.

Visual Line of Sight   Unaided (corrective lenses and/or sunglasses exempted) visual contact between a PIC or a VO and a UA sufficient to maintain safe operational control of the aircraft, know its location, and be able to scan the airspace in which it is operating to see and avoid other air traffic or objects aloft or on the ground.

APPENDIX B – TYPE CERTIFICATE DATA SHEET (TCDS)

DEPARTMENT OF TRANSPORTATION
FEDERAL AVIATION ADMINISTRATION

Q00017LA
Revision 3
Insitu Inc.
ScanEagle X200
June 26, 2014

TYPE CERTIFICATE DATA SHEET No. Q00017LA

This data sheet, which is part of Type Certificate No. Q00017LA, prescribes conditions and limitations under which the product for which the type certificate was issued meets the airworthiness requirements of the 14 Code of Federal Aviation Regulations (14 CFR).

Type Certificate Holder
Insitu Inc.
118 East Columbia River Way
Bingen, WA 98605
USA

I. Model ScanEagle X200 (Restricted Category UAS) Approved July 19, 2013 (See NOTES Section)

UAS
This is an Unmanned Aircraft System (UAS) that is comprised of the air vehicle and the transportable ground control station.

Unmanned Aircraft Dimensions
Wingspan 10.2 Ft (3.11 m)
Length 4.5 Ft (1.37 m)

Engine
(1) Northwest UAV, Block D Hush
FAA Engine Type Certificate: None
Engine type: Normally-aspirated, carbureted, two-stroke, direct drive, air cooled, single cylinder engine.

Optional engine:
    Northwest UAV, HFE-EFI – see NOTE 15

Fuel
High Octane C-10
    Fuel for optional engine – see NOTE 15

Oil
STIHL HP Ultra 2-cycle oil
    Oil for optional engine – see NOTE 15

Fuel-Oil Mix Capacity
2.5 gallons

Engine Limits

Max Takeoff Power	1.75 HP at 8500 RPM
Max Continuous Power	1.75 HP at 8500 RPM
Max Cylinder Head Temperature (CHT)	180 °C
Min CHT for flight	50 °C

Engine limits for optional engine – see NOTE 15

Page No.	1	2	3	4	5	6
Rev. No.	3	3	3	3	3	3

Q00017LA   2

Propeller and Propeller Limits.	(1) APC, Propeller Model LP3-15013W FAA Propeller Type Certificate: None Propeller Type: 3-blade, chopped fiberglass and resin, 15 x 13 fixed pitch pusher Diameter (Nominal): 15 inches (38 cm) Pre-flight Static rpm requirement: Engine must successfully achieve the I-MUSE checklist item: Engine Performance Checked Propeller for the optional engine: APC, Propeller Model LP16014P – see NOTE 15
Electric Generator	20 Volts, nominal 6.0 Amps, maximum 95 Watts, maximum
Backup Battery	19.2 Volts, nominal 1100 mA Hrs.

Airspeed Limits

V_{NE}	98 KTAS (181 km/hr)
V_{NO}	85 KTAS (157 km/hr)
V_A (Maneuvering Speed)	85 KTAS (157 km/hr)
Landing Speed (Closing Speed) – minimum	25 KT (46 km/hr)
Landing Speed (Closing Speed) – maximum	52 KT (96 km/hr)

Center of Gravity (C.G.) Range	Center of gravity from datum Minimum: -1.97 in (-50 mm) at any weight Center of gravity from datum Maximum: -2.76 in (-70 mm) at any weight Reference Datum Location: 8.07 in (205 mm) forward from the aft edge of the fuselage module
Empty Weight C.G. Range	None
Datum	Located on centerline of airplane at wing trailing edge intersection: positive in the x direction to the nose; positive in the y direction to the right wing; and positive in the z through belly of the aircraft. (Insitu ScanEagle Datum Reference Drawing, 15Jul2013)
Mean Aerodynamic Chord (MAC)	9.5 in (241.3 mm) long with leading edge: x = -0.55 in (14.0 mm) from datum y = 27.8 in (706.7 mm) from datum.
Leveling Means	When level—in the aircraft stand—and not moving, accelerometers should read: X-axis: 0.0 G+/-0.03 G Z-axis: -1.0 G+/-0.05 G

Maximum Weights

Ramp	44 lb (19.96 kg)
Takeoff	44 lb (19.96 kg)
Landing Weight	44 lb (19.96 kg)

Empty Weight	29.5 lb (13.4 kg) Empty weight using optional engine and propeller – see NOTE 15

Data UP-Link Frequencies	1.3 GHz, Commands for: aircraft control, sensor control, video control
Data DOWN-Link Frequencies	1.3 GHz, Report status on: aircraft, sensors, and video
Video Down-Link Frequency	2.4 GHz

NOTE: FCC license is required to utilize the above frequencies.

Computer Software	I-MUSE Software Version 5.6.13 Computer software for optional engine – see NOTE 15
Minimum Crew	(1) UAS pilot at the Ground Control Station (2) Personnel for launch and recovery
Number of Seats	(0) Not Applicable

Fuel Capacity

Fuel System: 12.3 lb of fuel (5.6 kg)
Unusable Fuel: 0.2 lb of fuel (0.1 kg)
    See NOTE 15 for fuel capacity when using optional engine

Fuel/Oil mixture ratio: 50:1 – by volume
    See NOTE 15 for Fuel/Oil mixture ratio when using the optional engine

NOTE: Fuel capacity includes the oil mixed with the fuel

Oil Capacity	Not Applicable
Max. Operating Altitude	2000 ft. AGL (610 M)

Control Surface Movements
Deflections are +/- 2 degree

Outboard Elevon	Up 30°	Down 30°
Inboard Elevon	Up 30°	Down 30°
Rudder	Left 25°	Right 25°

Flight Endurance   18.5 Hrs.

Flight Limitations

1. Day Visual Flight Rules (VFR) in visual meteorological conditions (VMC)
2. Flight through visible moisture: PROHIBITED
3. Flight operations in icing conditions at assigned operational altitudes: PROHIBITED
4. Flight Pitch Attitude: $^+/_- 45°$
5. Flight Bank Angle: $^+/_- 44°$
6. Ambient Outside Air Temperature (OAT)
   a. Maximum OAT: 120°F / 49°C
   b. Minimum OAT at Altitude: -4°F / -20°C
7. Wind. See Note 5
8. Flight Operations. See Note 4
9. For this operation only one ScanEagle can be airborne at any given time
10. Over water operation: PERMITTED
11. Over land operation: PROHIBITED, except PERMITTED for:
      a) Ingress/egress routes for access to over water operations to and from
         coastal launch and recovery sites;
      b) Airspace defined in Arlington Certificate of Waiver and Authorization
         (COWA) 2014-AHQ-101

12. An authorization for the specific location of operation issued by the Administrator is required and must be available at the control station. AFM number FAA-01-AFM, dated July 16, 2013 or later FAA approved revision, and certificate of airworthiness (C of A) must be available at the control station (reference FAA Memorandum, "Certification of Unmanned Aircraft", from AAL-7 to ANM-100L, dated June 19, 2013). Additionally, any certificates of authorizations or waivers must be available at the control station.
13. Only for operation in the designated Arctic Area as defined by the FAA Modernization and Reform Act of 2012, Section 332(d)(1).
14. Operation with inoperative instruments and equipment: PROHIBITED

Serial Numbers Approved:	Air Vehicle: 11-1313, 11-1453, 11-1458, 11-1459 Ground Control Station (GCS): TGCS274, GCS143
Certification Basis	Restricted Category Only 14 CFR part 21.25(a) (2) for the special purpose of aerial survey, 14 CFR part 36, amendment 29, Appendix G
Production Basis	None
Ground Control Station:	P/N 900-201124-001 P/N 900-201125-002
UAS Support Equipment:	Launcher: Insitu P/N 090-000200R00. See Note 6 and Note 10. Skyhook: Insitu P/N 900-200402-005. See Note 6 and Note 10.
NOTES:	
NOTE 1	Current weight and balance data, loading information, and a list of equipment included in the empty weight must be provided for each UAS at the time of original certification.
NOTE 2	Placards Required: None
NOTE 3	This UAS must be maintained in accordance with Unmanned Aerial Systems Maintenance Handbook, Version 2.0, dated September 2007, Document Number 026-010019, or later FAA accepted revision.
NOTE 4	UAS shall be operated under 14 CFR part 91, operating requirements, as mitigated. Operations shall be conducted in accordance with a waiver of flight regulations applicable to the operation, including but not limited to 14 CFR § 91.113, issued by the Administrator and specific to the intended operation, including geographical limitations.

NOTE 5

Wind Limitations:
Ship launch wind over deck conditions:
(a) Wind over deck conditions shall be determined by shipboard wind
    measurement and indication system.
(b) Max gusts for launch and recovery: 5 KT (5.75 mph, 9.26 kph)
(c) Launches (including gusts):
    1. 10 KT from +/- 30° relative to the launcher centerline.
    2. 20 KT from +/- 20° relative to the launcher centerline.
    3. Launches with tailwinds: PROHIBITED.
(d) Recoveries (including gusts):
    1. Port recoveries:
       a. 20 KT from 320° to 350° relative to the ship centerline.
       b. 30 KT from 320° to 330° relative to the ship centerline.
    2. Starboard recoveries:
       a. 30 KT from 10° to 40° relative to ship centerline.
    3. Recoveries with tailwinds: PROHIBITED.
(e) Wind limitations during flight:
    1. Max winds (sustained plus gusts): 40 KT
    2. Max gust component (gusts are considered any wind variations above the
       measured sustained value): 10 KT

NOTE 6

Personnel Keep Out Zones. Typical exclusion zones apply for Launch and
Recovery (SkyHook) as described in AFM and UAS Operations Manual.

NOTE 7

This Type Certificate Data Sheet (TCDS) is the principal document for
ScanEagle Operation. For any operational discrepancies among the TCDS,
AFM, Insitu ScanEagle Ops. HDBK, etc., this TCDS takes precedence.

NOTE 8

Restricted category aircraft may not be operated in a foreign country without the
express approval of that country.

NOTE 9

This aircraft has not been shown to meet the requirements of the applicable
comprehensive and detailed airworthiness code as provided by Annex 8 of the
Convention on International Civil Aviation. This aircraft meets 14 CFR
§ 21.25(a)(2).

NOTE 10

For this restricted category type certificate, the part numbers of the Launcher
and Skyhook must be those listed under UAS Support Equipment of this Type
Certificate Data Sheet.

NOTE 11

Operations shall be conducted by properly certificated airmen who have
completed training, checking, currency, and recency of experience requirements
as approved by the Administrator.

NOTE 12

The Flight Standards Board (FSB) report is available on request. Contact the Long
Beach AEG (LGB-AEG-NM17).

NOTE 13

No aircraft may be manufactured under this approval.

NOTE 14

Compliance to §21.25 (a) (2) was shown by following the deviation Memo dated July 5,
2013, in lieu of Order 8110.56 Revision A.

Q00017LA 6

| NOTE 15 | The following items and limits apply when using the optional engine. All items and limits must be used when the optional engine is installed. |

Optional Engine:

Northwest UAV, HFE-EFI (Heavy Fuel Engine–Electronically Fuel Injected)
FAA Engine Type Certificate: None
Engine type: Normally-aspirated, electronically fuel injected, two-stroke, direct drive, air cooled, single cylinder engine.

Fuel: JP-5 (MIL-DTL-5624U), or JP-8 (MIL-DTL-83133H)
Oil: Bel Ray H1R 2-Cycle Oil
Fuel/Oil mixture ratio: 32:1 – by volume

Engine Limits:

Max Takeoff Power	3.35 HP at 7000 RPM
Max Continuous Power	3.35 HP at 7000 RPM
Max Cylinder Head Temperature (CHT)	170 °C
Min CHT for flight	110 °C

Propeller and Propeller Limits:
APC, Propeller Model LP16014P
FAA Propeller Type Certificate: None
Propeller Type: 2-blade, chopped fiberglass and resin, 16 x 14 fixed pitch pusher
Diameter (Nominal): 16 inches (40.64 cm)
Pre-flight Static rpm requirement: Engine must successfully achieve the I-MUSE checklist item: Engine Performance Checked

Computer Software: I-MUSE Software Version 5.9.05 or 5.10.08

Empty Weight: 31.6 lb (14.3 kg)

Fuel Capacity: Fuel system: 10.2 lb of fuel (4.6 kg)

--END--

REFERENCES

Advisory Circular 91-57A – Model Aircraft Operating Standards, September 2, 2015.

Code of Federal Regulations, Title 14, Chapter 1, Subchapter A, Part 1 – Definitions and Abbreviations, Section 1.1 General Definitions.

Statement by FAA Deputy Administrator Michael G. Whitaker before the House Transportation and Infrastructure Committee, Aviation Subcommittee on October 7, 2015.

FAA (2013) *Integration of Civil Unmanned Aircraft Systems (UAS) in the National Airspace System (NAS) Roadmap*, 1st edn.

FAA Modernization and Reform Act (2012) Public Law 112-95, Title III, Subtitle B, Unmanned Aircraft Systems.

FAA News Release dated August 4, 2015. Available at http://www.faa.gov/news/updates.

FAA Order 8900, vol. 16, Unmanned Aircraft Systems.

FAA – Pilot Reports of Close Calls with Drones Soar in 2015. (www.faa.gov/news/updates/?newsld=83445).

FAA Press Release – January 22, 2016.

Federal Aviation Administration (2015) Aviation Forecast Fiscal Years 2015–2035.

Federal Register (2007) FAA, Notice of Policy, Unmanned Aircraft Operations in the National Airspace, February.

Federal Register, FAA, Notice of Proposed Rulemaking, Operation and Certification of Small Unmanned Aircraft.

Federal Register Notice of Interpretation and Request for Comments, Interpretation of the Special Rule for Model Aircraft.

ICAO Document 10019 (2015) Manual on Remotely Piloted Aircraft Systems, 1st edn.

Id. At 264 (citing *Hinman v. Pacific Air Transport*, 84 F.2d 755 (9th Circuit Court 1936).

FAA (2013) Integration of Civil Unmanned Aircraft Systems (UAS) in the National Airspace System (NAS) Roadmap, 1st edn.

Monash University, Australia. Available at www.ctie.monash.edu.au

National Geographic (2013) Japan's secret weapon: Balloon Bombs, May 27, Johnna Rizzo.

Public Law 112-95, Title III, Subtitle B-Unmanned Aircraft Systems, Section 331, Definitions (FAA Modernization and Reform Act of 2012).

Public Law 112-95, Title III, Subtitle B – Unmanned Aircraft Systems (FAA Modernization and Reform Act of 2012).

Public Law 112-95, Title III, Subtitle B – Unmanned Aircraft Systems, Section 336, Special Rules for Model Aircraft.

Registration and Marking Requirements for Small Unmanned Aircraft Interim Final Rule. Available at https://www.federalregister.gov/articles/2015/12/16/2015-31750/registration-and-marking-requirements-for-small-unmanned-aircraft.

RTCA media release dated October 2, 2015.

For additional information on the type certificate process, see Salvano, D.P. (2010) Aircraft safety certification, in *Encyclopedia of Aerospace Engineering* (eds. R. Blockley and W. Shyy), John Wiley & Sons, Ltd., Cinchester, UK, pp. 5363–5376.

Teal Group, (2014) World Unmanned Aerial Systems, Market Profile and Forecast.

The Convention on International Civil Aviation, December 7, 1944, Chapter 1, Article 1.

The Paris Convention of June 1919, formally called "The Convention Relating to the Regulation of Aerial Navigation."

The United States v. Causby, 328 U.S. 256 (1946).

Title 14, United States Code, Chapter 1, Subchapter A, Part 1, Section 1.1.

US Code of Federal Regulations, Title 14 Aeronautics and Space, Chapter 1, Subchapter F – Part 91.113.

US Senate Appropriations Subcommittee on Transportation, Housing and Urban Development on UAS and Integration into our National Airspace, October 28, 2015.

Chapter 43
Requirements: Levels of Safety

Charles H. Patchett

School of Engineering, University of Liverpool, Liverpool, UK

1 INTRODUCTION

UAS first came to prominence in military operations in the Balkans in 1994. Since then their use and their importance to military capability has escalated almost exponentially. This rise has not gone unnoticed in the civil aviation world and there is an increasing demand for the integration of UAS into national airspace.

However, there remain significant barriers for this to occur. First is the problem of UAS certification and a "Catch 22" situation exists where the system designers require firm requirements from the regulators and the regulators want a

Unmanned Aircraft Systems. Edited by Ella Atkins, Aníbal Ollero, Antonios Tsourdos, Richard Blockley and Wei Shyy.
© 2016 John Wiley & Sons, Ltd. ISBN: 978-1-118-86645-0.

system to examine to determine its certifiability. Second, operational regulations have evolved over the last 100 years; those for UAS are still to be determined and finalized. Finally, there is the issue of safety. The overwhelming majority of UAS operations are flown in segregated airspace. Before UAS can be released into nonsegregated airspace, questions over the safety of other air users and the general population have to be satisfactorily answered. This chapter discusses those issues.

2 DEFINITIONS AND ASSUMPTIONS

The acronyms for Unmanned Air Systems have changed regularly over the years. Remotely Piloted Air Systems (RPAS) is the latest and is meant to underline the fact that the pilot is in control and legally responsible for the conduct of the flight. The term UAS on the other hand reflects that the system comprises an aircraft and a remote control station and that the remote pilot may not always be in contact with the aircraft that may be under the (temporary) control of an onboard autonomous system. Whatever is preferred, the term UAS is used throughout this chapter.

The desire for safety increases as the UAS get larger: some small UAS are actually disposable. This chapter addresses UAS that are the subject of certification, namely those weighing more than 150 kg.

Most UAS to date have been operated by the military in restricted environments. In addition, they have frequently been put into production based on experimental prototypes. Cost, weight, performance, and time into service, not reliability or safety, were, and perhaps still are, seen as the most important aspects. This chapter really only considers civil operations in unrestricted airspace where safety is paramount.

3   THE CONTEXT FOR UNMANNED SYSTEMS

3.1   The meaning of safety

Generally "safety" means "an absence of danger." Absolute safety, or absolute absence of danger, is generally an unachievable or at least a very expensive goal. Therefore, the concept of acceptable safety has been adopted in risk-bearing industries, including aviation. The term "acceptable risk" describes an event with a probability of occurrence and consequences acceptable to society, that is, the society is willing to take or be subjected to the risk that the event might bring.

This concept of safety being a combination of acceptable consequences and probability of occurrence is at the heart of aviation certification. A system is declared to be (acceptably) safe if the consequence of a fault, failure, or event, expressed as a category such as "catastrophic," is inversely proportional to its probability of occurrence. As we shall see later, a "catastrophic" event is only acceptable with an extremely low probability.

Safety, however, is not just a matter of producing a certified design. How the aircraft is flown or operated is just as important and there are many regulations covering this aspect. The phrase *Safe to Fly – Flown Safely* captures these twin aspects.

3.2   UAS modes of control

At first appearance, the certification problem seems simple. Treat the UAS as a normal aircraft and certificate it according to current, manned aircraft regulations. Unfortunately, this approach is hindered by several basic facts, the major one is that UAS are fundamentally different and the current regulations are sometimes inappropriate. For example, current operational rules require air users to *See and Avoid* other aircraft. Another is to remain "clear of cloud" or "within sight of the ground." These rules and regulations began in the early days of flight and set standards for construction, operation, and licensing in order to improve safety. These have since evolved to a huge amount of regulations that have to be conformed to, and which are, daily, by thousands of aircraft. These regulations, with an evolution of over a hundred odd

years, have a single common factor. They all presume there is a pilot on board and in control.

Of all the differences between manned and unmanned flight, how the vehicle is controlled represents the biggest aspect. UAS can be flown in a variety of control modes all of which have in common the fact that the pilots, commander, or operator(s) are remote from the vehicle, either within sight or beyond line of sight (LOS) of the aircraft.

There are many concepts of control modding. Some of these are as follows:

- *Manually piloted.* The aircraft is flown manually from the ground control station at all times in a manner somewhat similar to a radio-controlled model. Such operation requires a high level of skill. In addition to the skilled operator, a communication link of high integrity and negligible latency is required. This can often only be achieved with line-of-sight communication links. The pilot is in direct (and constant) control of the vehicle.
- *Supervised automation.* The aircraft is flown under full autopilot control as specified by a human supervisor. Such operation allows use of beyond-line-of-sight (BLOS) data links since real-time feedback control is not now necessary.
- *Autonomous operation.* The aircraft is flown according to an agenda of objectives that can be specified by the pilot, the on-board autonomous control system, or jointly by both in cooperation. The pilot is now a manager of the autonomous system, which is controlling the vehicle. It should be mentioned that, at present, the International Civil Aviation Authority (ICAO) does not envisage such operation for the foreseeable future.

These control modes have elements that are common to each other but different to manned aircraft. The immediacy and reliability of the conventional pilot–vehicle link in manned aircraft is absent in UAS. This inherent remoteness of the pilot or operator of a UAS introduces safety-related problems, some subtle, others not so, as will be discussed.

3.3   Equivalence and transparency

The concepts of Equivalence and Transparency are frequently used in the UAS community. Equivalence is taken to mean Equivalent in adherence to operational regulations as other air users (not to be confused with human performance), while Transparency means Transparent in operation to an external viewer (such as ATC) in comparison with other air users.

4   CERTIFICATION – AEROSPACE STANDARDS FOR SOFTWARE AND HARDWARE

There are a variety of standards and guidelines for software and hardware in the aerospace industry and the more important of these are briefly mentioned below.

4.1   CS 1301 and 1309

What is deemed to be safe, or more accurately, tolerably unsafe, for flight operations is defined by the appropriate regulatory authority. In the United States, for civil aviation, this is the Federal Aviation Administration (FAA) and they publish regulations, known as the Federal Aviation Regulations (FARs) that cover the whole arena of flight operations. In the United Kingdom, the equivalent body is the Civil Aviation Authority (CAA). However, they are part of the Joint Aviation Authorities (JAA) of European states and they publish, broadly equivalent, Joint Aviation Requirements (JARs). In this document, FAR and JAR are regarded as the same and the term JAR is used consequently.

General airworthiness regulations for civil aircraft are covered in Sections 23 (Light/Commuter) and 25 (Transport) of the JARs. Briefly, each aircraft is covered by Section 23 or 25 according to a variety of criteria, for example, size, weight, kinetic energy, use and so on. As this chapter is aimed at medium-to-large UAS, we will assume that medium UAS are covered by an equivalent document, yet to be produced, to JAR 23 and large ones, JAR25.

There are two particularly relevant paragraphs in the JARs: 1301 and 1309.

Para 1301, Function and Installation states that installed equipment must "Be of a kind and design appropriate to its intended function . . . and function properly when installed." In other words, it must be "fit for purpose." This paragraph is designed to prevent systems from other vehicles being used, without proper modification and subsequent recertification, in aircraft.

Para 1309 Equipment, Systems, and Installations, states simply, *this requires justification that all probable failures, or combinations of failures, will not result in unacceptable consequences.* This requires the identification of failure probabilities, including multiple failures, by detailed analysis of essential systems and evaluation of the consequences of those failures. In particular, it requires that the frequency of occurrence (probability) of system failures must be inversely proportional to the severity of the effects. The definitions of these are set out in the ACJs to JAR23/25.1309 Equipment Systems and Installations (Interpretative Material and Acceptable Means of Compliance) (FAA, 2014a) and are repeated in the Table 1.

An abridged version of the failure condition classification is given in Table 2 noting that the equivalent FAA table includes the category "No Safety Effect."

It should be emphasized that there is no current UAS.1309 (although there are several drafts in circulation) and it is assumed that if there were, it would follow the above in principle at least. As such there are a few problems with the above when UAS are under consideration.

1309 is specifically concerned with maintaining safety by keeping the aircraft airborne or landing safely. It is also clearly aimed at passenger aircraft (according to the above definitions). A UAS is unlikely at this present time to carry passengers, nor, by definition, does it carry a crew. In addition, there is a distinct possibility that a UAS can be designed to crash safely (some are already), and they can certainly be designed to land automatically and safely.

4.2   Systems development and design

Compliance with 1309 must be proved for the system to be certified and although the principle behind 1309 is very neat in theory, there are practical problems such as

- The full consequences may be dependent on the situation the system finds itself in. These situations may be close to infinite and the consequent safety case analysis is

Table 1. Acceptable probability of failure, their effects, and their definition.

Classification of failure conditions	Minor	Major	Hazardous	Catastrophic
Allowable qualitative probability	Probable	Remote	Extremely remote	Extremely improbable
Allowable quantitative probability per flight hour on the order of	$<10^{-3}$	$<10^{-5}$	$<10^{-7}$	$<10^{-9}$
Required software development assurance level	D	C	B	A

Table 2. Definition of failure condition severities.

Minor	No significant reduction in aircraft safety. It may involve crew actions that are well within their capabilities: • Specifically, a slight reduction in safety margins or functional capabilities • A slight increase in crew workload (e.g., flight plan changes) • Some physical discomfort to passengers or cabin crew
Major	Reduce the capability of the aircraft or the ability of the crew to cope with adverse operating conditions. Specifically: • A significant reduction in safety margins or functional capabilities • A significant increase in crew workload • Discomfort to the flight crew • Physical distress to passengers or cabin crew, possibly including injuries
Hazardous	Reduce the capability of the aircraft or the ability of the crew to cope with adverse operating conditions. Specifically: • A large reduction in safety margins or functional capabilities • Physical distress or excessive workload such that the flight crew cannot be relied up on to perform their tasks accurately or completely • Serious or fatal injury to a relatively small number of the occupants other than the flight crew
Catastrophic	Failure conditions that could result in multiple fatalities and/or hull loss.

invariably inexhaustive. The usual work around is to limit the analysis to "worst case" scenarios.

- If the system is being applied in a manner that is unanticipated by the designer, the safety analysis is likely to breakdown completely. This brings into focus the role, if applicable, of the operator and his effect on system safety.

- For safety critical systems, the probability of occurrence may have to be very low – for catastrophic failures typically less than 10^{-9} per flight hour. How does one prove a level of assurance in such a case?

- Probability of occurrence may be difficult to quantify, particularly over the lifetime of the system. Faults may be random or systematic and, for hardware, may vary over time. Random faults can be assessed mathematically (to a degree) but software, it could be argued, only has systematic failures and these failures are always the result of design faults.

In order to overcome the above and provide a practical route to certification, a variety of important documents have been produced to provide guidance when contemplating the systems design and proof of compliance with 1309.

These documents present recommended practices that should not be construed to be formal requirements as the regulatory bodies recognize that alternative methods to the processes described or referenced in them may be available to an organization desiring to obtain certification. However, and this is not stated by any of these documents, if the guidelines, which are well understood by regulators, are not followed, the burden of proof that these alternative approaches are sufficient to achieve certification is on the organization

developing the systems and not the regulator. The recommendation here is that it will almost certainly be simpler to follow the guidelines rather than try to circumnavigate them.

A schematic showing how these documents fit together in the system development life cycle is shown in Figure 1, taken from Aerospace Recommended Practice (ARP) 4754A.

Four of these documents are now outlined.

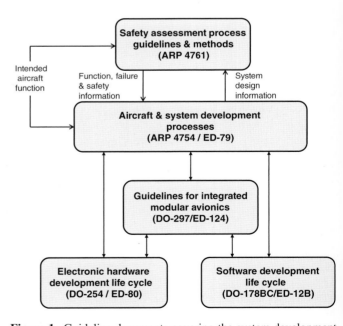

Figure 1. Guideline documents covering the system development phase.

4.3 Arp 4761 Guidelines and Methods for Conducting the Safety Assessment Process on Civil Airborne Systems and Equipment

ARP 4761 (SAE International, 1996) describes guidelines and methods of performing the safety assessment for certification of civil aircraft. It is primarily associated with showing compliance with FAR/JAR 25.1309. It presents guidelines for conducting an industry accepted safety assessment consisting of Functional Hazard Assessment (FHA), Preliminary System Safety Assessment (PSSA), and System Safety Assessment (SSA). It also presents information on the safety analysis methods needed to conduct the safety assessment. These methods include the Fault Tree Analysis (FTA), Dependence Diagram (DD), Markov Analysis (MA), Failure Modes and Effect Analysis (FMEA), Failure Modes and Effects Summary (FMES), and Common Cause Analysis (CCA), including each of the sub–analyses of Zonal Safety Analysis, Common Mode Analysis, and Particular Risk Analysis.

4.4 ARP 4754A guidelines for development of civil aircraft and systems

ARP 4754A discusses the development of aircraft systems taking into account the overall aircraft-operating environment and functions. This includes validation of requirements and verification of the design implementation for certification and product assurance. Its main aim is to provide a means of assuring that likely sources of error or failure within complex or highly integrated systems are identified and mitigated. To do so, it provides practices for showing compliance with the regulations. The guidelines given are directed toward systems that support aircraft-level functions and have failure modes with the potential to affect the safety of the aircraft. ARP 4754A notes that typically, aircraft systems involve significant interactions with other systems in a larger integrated environment and that significant elements of these systems are developed by separate individuals, groups, or organizations. These systems require added design discipline and development structure to ensure that safety and operational requirements can be fully realized and substantiated. It recommends that a top-down iterative approach from aircraft level downwards is key to initiating the processes outlined.

While ARP 4754A gives guidance for system development, it recognizes that, generally, functionality is divided between software- and hardware-based functions. Frequently (almost invariably) these functions are "complex," that is, the purpose of the function cannot be established by direct observation. Thus, how these functions are derived and

delivered within the system becomes of the utmost importance. Two documents cover these processes: RTCA DO-178B/C and DO-254A.

4.5 RTCADO-178B/C software considerations in airborne systems and equipment certification

This document provides guidance for determining, in a consistent manner and with an acceptable level of confidence that the software aspects of airborne systems and equipment comply with airworthiness requirements (RTCA, 1996). It is important to note that DO-178B does not seek to impose how the software is developed but rather states objectives that must be achieved to reach the satisfactory rigor of the software development life cycle and the traceable transition from one level to the next. This life cycle covers the following:

- Planning
- Requirements – high level and low level
- Design
- Code
- Integration
- Test

DO-178C extends DO178B by clarifying several issues. In particular, it provides the following supplements to DO-178B:

- DO-330 Software Tool Qualification Considerations
- DO-331 and DO 278 concerning Model-Based Development and Verification
- DO-332 Object-Oriented Technology and Related Techniques Supplement to DO-178C and DO-278A
- DO-333 Formal Methods Supplement to DO-178C and DO-278A

4.6 RTCADO-254 design assurance guidance for airborne electronic hardware

This document is intended to help aircraft manufacturers and the suppliers of aircraft electronic systems ensure that electronic airborne equipment safely performs its intended function. The document identifies design life cycle processes for hardware that includes line replaceable units, circuit board assemblies, application-specific integrated circuits (ASICs), programmable logic devices, and so on. It also characterizes the objective of the design life cycle processes and offers a means of complying with certification requirements.

4.7 CAP-722 unmanned aircraft system operations in UK airspace – guidance

Although not discussed here as it is not of a global nature, it is of relevance to mention a publication by the UK CAA entitled CAP 722 – Unmanned Aircraft System Operations in UK Airspace – Guidance (CAA, 2012).

5 SYSTEM DESIGN CONSIDERATIONS

While the above documents provide detailed guidelines for the successful certification of an aircraft and its systems, and the serious student of aircraft safety should become completely familiar with them, there are a few fundamental issues worthy of more detail. These are now discussed.

5.1 Safety assessment

The general approach to the achievement of (acceptable) system safety is to identify and eliminate faults or at least reduce their impact or likelihood. It should be performed throughout the development life cycle and should consider not only the identification of all relevant failure conditions but also those related to interdependencies that result from systems integration. One of the means to do so is the Preliminary System Safety Assessment (PSSA). This assessment takes place once the preliminary design has been achieved and is intended to generate safety requirements based on the analysis of the system architecture and functional requirements. One form of assessment is the (functional) Fault Failure Analysis (FFA). This considers each function in terms of the effect on safety for three test conditions:

- If the system performs the function when it is *not* required to do so
- If the system does *not* perform the function when it *is* required to do so
- If the system performs the function incorrectly

Such an assessment, even though simple in concept, can be a valuable aid toward proving the system safety requirements and that the architecture can meet the considerations of DO-178B.

In the FFA, human interaction, such as output verification or instrument reading cross monitoring, is sometimes used as a means of failure mitigation. However, the following should be always borne in mind if this is used:

- Humans are not formally part of the certification process, which stresses determinism as a formal requirement.

- They may not perform as the designer intended and they may not even be aware of the expectation required of them.
- While human skills and training are adept at avoiding accidents, human error is responsible for 80% of aircraft accidents. It is indeed possible that they could make the situation worse rather than better.
- As referred to earlier, in a UAS they are remote. This means they may not be able to
  - Perceive or recognize the problem. They may even believe something completely different due to limited cues and thus make the situation worse rather than better.
  - Provide the necessary response in a timely manner, a situation that could be made worse by the latency of satellite command links.
  - Allocate sufficient priority to the failure either through lack of appreciation of the situation or competing workload.

5.2 Development/design assurance levels

In Table 1, the Design/Development Assurance Levels (DAL) associated with particular failure outcomes were stated. These define the degree of design rigor that must be applied to mitigate the failure event. They are determined as a result of outputs from the PSSA process and consideration of the system architecture. At the highest level, Level A, the utmost rigor is required to develop the system, historically referred to as a Safety Critical System. Such systems are very expensive to implement and test for compliance with 1309. AP 4754 proposes means to reduce the DAL requirement using the system architecture to mitigate or control the degree to which an item contributes to a specific failure condition. It is essential to this process that the items under consideration are contained wholly within the system boundary. These architecture choices include (SAE):

- *Redundancy.* The provision of multiple and separate implementations of a function. This is the equivalent to a hardware "AND" gate, whereby both sides of the gate must fail to produce the fault condition. Redundancy is the primary consideration for the fail-safe protection from Catastrophic events.
- *Dissimilarity.* The implementation of the same function using substantially different means. It is a basic technique against what is known as Common Mode Failures. This may be achieved by using two separate and independent design teams and/or software/hardware components.
- *Partitioning.* Separating or isolating functions either in hardware or software.

5.3   Validation and verification

A major aspect for the elimination of software faults is the notion of software correctness or reliability, which are taken here to mean an absence of faults. There are two major aspects to this. First, does the software deliver what it was required to do and in accordance with its original specification; a process known as Validation. Second, has the software been produced correctly, a process known as Verification. These are sometimes identified as the following:

- *Validation* – Have we built the right thing
- *Verification* – Have we built it right

In order to properly validate the software, it is first necessary to properly specify the software requirements. In the early days of military aviation software (admittedly nonsafety critical), it was common to express the requirements in plain English, mainly by pilots who expressed themselves in operational terms. This led to a number of software problems. Firstly, the required behavior was incomplete and designers had to invent the missing links. Second, such expressions are open to ambiguity – most software designers are not pilots. Final, such use precludes complex behaviors. To summarize, we need a software specification that is complete, unambiguous, and capable of handling complexity. If the specification writing was assisted by automated tools for checking for these properties, then the number of software faults could be considerably reduced. In fact, if the specification was itself was mathematically expressed, the resulting software should be able to be proved to be correct using mathematical tools. This is the basis for what is now called formal methods but the concept is not new, Alan Turing wrote of such things in 1947.

Another problem with software requirements is that they tend to change as the system life cycle progresses. This is a real problem for the "waterfall" design process and has led to the development of the spiral process where one starts with a rough requirement set and an early prototype. This is progressively refined until the system requirement set is considered complete. At this point, any software produced to date should be scrapped and the now formal specification be properly engineered. Such a process helps requirements stability but cannot counter the lack of it completely.

It is for all the above reasons that DO-178B, described above, was written to aid the software design process.

6   OPERATIONAL ISSUES

Given that a UAS.1301/1309 will eventually exist, and assuming that much of it will be based on the current JARs 23/25 for medium-to-large aircraft, there seems to be little doubt that a future UAS can be produced which is "Safe to Fly." However, there still remains the "Flown Safely" requirement for full certification and acceptable safety levels in operation. There are a number of barriers to these aspects and these are briefly discussed.

6.1   Rules of the air

On December 7, 1944, the majority of the world's nations became signatories to the "Chicago Convention," the aim of which was to assure the safe, orderly and economic development of air transport. ICAO sets out, in the terms of the convention, the rules, regulations, and requirements to which each signatory must adhere (ICAO, 1944–2006). In the United States, the FAA details these rules in the Code of Federal Regulations (CFR), Title 14, Part 91 "Aeronautics and Space." In the United Kingdom, they are stated in CAP 393 – "Air Navigation: The Order and the Regulations."

Within these publications are the Rules of the Air that are set out in Rules of the Air is ICAO Chicago Convention, Annex 2. These cover general operations and specifically the Instrument Flight Rules (IFR) and Visual Flight Rules (VFR). These Rules apply to all civil air users and UAS operators will also have to conform where appropriate. It is possible that special regulations may be required for UAS operations but the regulating authorities are loathe to have such regulations, as all air users will have to comply with them. An alternative, and more likely route, is for UAS operators to demonstrate an "Acceptable Means of Compliance" (AMC) with the regulations. For UASs, a particular flight rule worthy of note is the requirement for the pilot of an aircraft to "See and Avoid" other aircraft. The AMC for this is the fitment, mandatory for Beyond Visual Line of Sight operations, of a certified Detect and Avoid System. The requirements for such a system, including the acceptable level of performance, have been the subject of much recent research and debate.

6.2   Airspace considerations

A full description of airspace is beyond the scope of this chapter. Suffice it to say, generally there is segregated or nonsegregated airspace. For safety reasons, UASs have usually operated in the former, and from which other air users are excluded. However, for commercial reasons, ways of safely operating in nonsegregated airspace need to be developed. Nonsegregated airspace is of two broad types (albeit with several subdivisions). Controlled airspace, where the control of the aircraft is vested in the ground-based Air

Traffic system, and noncontrolled airspace where the on-board pilot is responsible for safe navigation. The vast majority of traffic within noncontrolled airspace operate according to the VFR. These rules generally assume that a pilot is on-board and is in visual contact with other traffic.

The route toward airspace integration for UAS is likely to be operations under IFR within controlled airspace initially. Operations under VFR within uncontrolled airspace are very much in the future and little work has been done to progress this.

Despite the above, there are many changes to the management of Air Traffic services in the pipeline and in general, they appear to ease the problems of airspace integration for UAS and conventional aircraft despite increasing traffic. Items worthy of further study include the following:

- *Cockpit Pilot Data Link Control (CPDLC)*: This provides a data link alternative to VHF communications. It also offers far higher data integrity because of encryption, virtually guarantees reception at the targeted receiver and prevents garbled or misunderstood voice messages. It is, however, not yet in widespread use and for thousands of aircraft, VHF speech will be the primary communications medium for many years to come.
- *Single European Sky (SES) and SESAR*: This is designed to organize ATM in Europe around traffic flows rather than national boundaries to increase network capacity. The SES ATM Research (SESAR) program will deliver the technologies required for this, including 4D Trajectory planning that is aimed at the synchronization of airborne held and ground-held trajectories for the same flight in order to allow a more accurate and efficient planning and sequencing of flights (EUROCONTROL, 2008).
- *NextGen (FAA, 2014b)* – The United States future ATM program broadly equivalent to SES.

6.3   Autonomy and automation

In the Chicago Convention (ICAO, 1944–2006), ICAO states that the pilot in command of a remotely piloted aircraft has precisely the same responsibilities as an on-board pilot-in-command, that is, the safe operation of the aircraft.

So the big issues are, how do we

- maintain safe flight in the event of malfunctions and failures, particularly partial or full communications failure;
- maintain sufficient Situational Awareness for the remote pilot to perform his tasks.

Modern aircraft, even smaller ones, feature a wide range of automated functions, generally invoked upon pilot command, even if somewhat indirectly. Unfortunately, automatic cockpit systems tend not to have intuition, intelligence, discernment, and decision capacity. In fact the majority of such systems are limited, by design, in their knowledge to the control at hand and not the wider picture. This need thus falls on the pilot, who for UASs is remote and may be sometimes out of the loop, especially in times of loss of communications, high workload, stress, or problem fixation.

Given that ICAO demand that the aircraft is operated safely at all time, even if communications are lost, an alternative is to give cockpit systems some the above characteristics – so, called autonomous systems, that is, those that are capable of making, independently or in conjunction with the remote pilot, decisions after consideration of its situation.

As stated earlier, ICAO has ruled out autonomous operations currently (ICAO, 2011); however, partial autonomy, in the form of preassigned authority to make decisions or act, may be acceptable providing that a clear improvement on overall safety is the result. However, there are also barriers to autonomous operations, particularly in the design of the human machine interface, as the remote pilot needs to have sufficient trust and confidence to invoke any autonomous modding.

6.4   Health management and emergency handling

One impact of removing the pilot to the ground is the effect on correct and rapid fault diagnosis and emergency handling. The pilot can be viewed as a multisensor and information processing system. The most important of those sensors, the eyes, we make up for in the design of a UAS by the inclusion of a "Detect and Avoid" subsystem that replaces the "See and Avoid" responsibility we would normally assign to the pilot of a vehicle. On-board pilots can be highly proactive in investigating faults – off-board pilots may not even be aware that a fault condition exists. In addition, humans are naturally predisposed to associate data from different sensors, occasionally incorrectly, thus reinforcing perceptions to promote a single cause from the combined effects – vibration and noise mean more together than separately, as does heat and smoke, fire and light, aircraft movement and vestibular responses, thunder and lightning, and so on. Consider the Airbus A320-214 incident (NTSB, 2010) when it ditched in the Hudson River after sustaining multiple bird strikes. The crew noticed the birds just before impact, felt the effect of the impact, heard both engines spooling down and correctly diagnosed a double engine failure within 7 s. There must be considerable doubt that a remote pilot would even be aware of the bird strike, would certainly not have felt the impact and may have even taken considerable time to notice the engine winding down in the absence of any audio clues. In automatic flight, the first he would probably be aware of is the aircraft losing height for what would appear to be unknown reasons. There have been several instances of exactly this general sequence of events in Predator accidents reports (USAF, 2014).

So, in remotely locating the pilot,

- the vehicle is being denied important sensor information that the pilot normally provides (for free) together with his sophisticated sensor fusion/processing and proactive diagnostics;
- the pilot is denied important vehicle information such as peripheral vision, tactile feedback, sound, light, feel, and so on.

These factors indicate a much higher priority for sophisticated health management systems for a UAS to retain an appropriate level of safety.

6.5 Communications

Communications is a huge differentiator between manned aircraft and UA operations. Situating the pilot remotely inherently sets the requirement for an external communication link between the controller and the vehicle – the Command and Control link. Such a link is, of course, entirely absent in manned aircraft where the pilot maintains the connection between the aircraft's current status (the displays) and his demands (the controls). More importantly, in manned aircraft, this link is virtually instantaneous, and highly reliable. The fact that this is not the case in UAS is the cause of many an accident. In addition, in manned aircraft, the pilot's awareness of aircraft state and status from his displays is augmented by that from his own senses. Not so for UAS. To operate anywhere away from the local area, Satellite Communications (SatComms) is required. Unfortunately, current SatComms has a few undesirable characteristics:

- *Bandwidth limitation*. UAS can consume a large amount of bandwidth, especially when several are in the air at once, and this may mean the medium may become unavailable at short notice.
- Latency is an issue. It is commonplace for transmissions to be received seconds late or even suffer short-term outage due blanked antennas.

- Satellite constellations are vulnerable to physical phenomena such as sunspots and other radiations.

In summary, controllers of UAS (Air Traffic and pilots) require a communications architecture that is required to be safe, reliable, and capable of integrating within the present and future communications architectures and with existing air users. It should also operate at an equivalent level of performance to manned aircraft in terms of connectivity, availability, and bandwidth. Such an architecture is likely to be complex and difficult to design in order to satisfy these requirements. Consequently, it is imperative that the UAS should be able to operate independently of its controller, at least for short intervals.

The necessary communications data link also provides an opportunity for unauthorized control. It is essential that means to prevent this are provided both physically and electronically. Equally, should authorized control be lost, provision for regaining it must be made.

6.6 Ground control station

Consideration must also be given to the safety aspects of the ground control station (GCS). Given that it is an integral part of the system, it needs to be similarly certified to the aircraft element. Operational procedures and regulations also need to be in place to prove the necessary level of Equivalence to human operation. However, these latter factors do not appear to be too problematic.

7 ACCIDENT ANALYSIS

UAS have, historically, a poor accident record. They first came to the fore in the Bosnian war of 1992–1995 and the losses during that time could not often be confidently ascribed to being accidents or as the result of enemy action. In addition, these early, and therefore immature, systems were very much prototypes rushed into service. Finally, the weather conditions in which these prototypes had been developed (sunny California) were very much different to the far more hostile climate of the Balkans. All these factors therefore make the analysis of UAS accidents prior to 1999 extremely difficult if not impossible.

Having said that, UAS are now much more mature and all aircraft tend to become safer over time due to incremental improvements to the design, better and more practiced operating procedures and improved operator skills and knowledge. A comparison of the accident rates (Class A Mishaps per 10 000 flight hours) of three major UAS types

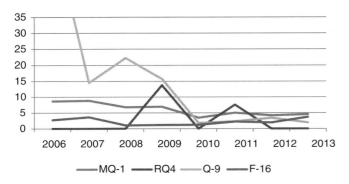

Figure 2. A comparison of UAS accident rates with that of the F16.

operated by the USAF, *Predator* MQ1, *Global Hawk* RQ4, and *Reaper* Q9 (UASF), is shown in Figure 2 together with that of a manned aircraft, the F16.

The maturity of the F16 is indicated by its flat rate over the last 7 years and the line for the MQ1 shows how its rate is being improved each year. Overall, the accident rate for these types is generally about 4 per 100 k flight hours. This figure can be compared with that of noncommercial fixed wing General Aviation (GA) in the United States, which has remained constant at about 6.4/100 k flight hours (AOPA, 2013) over several years. Medium/large airliners have accident rates two orders lower. Thus, it can be said that UAS are about twice as safe as GA but are still considerably worse than large commercial aircraft.

8   CONCLUDING REMARKS

The integration of UAS into nonsegregated airspace remains elusive despite many research programs being in place. An understanding of what needs to be achieve to certify the aircraft and its systems, including the GCS is progressing well. However, there is still much to be done to prove the operational aspects can provide the requisite levels of safety.

ACKNOWLEDGMENTS

The decision to include much of the certification material presented here originates from conversations with Philippa Moore, Avionics and Electrical Systems Surveyor, Safety and Airspace Regulation Group of the UK CAA and her valuable input is hereby acknowledged.

REFERENCES

AOPA (2013) The Nall Report 2013 is available at http://www.aopa.org/Pilot-Resources/Safety-and-Technique/Accident-Analysis/Joseph-T-Nall-Report.

CAA (2012) CAP722. Available at https://www.caa.co.uk/cap722.

EUROCONTROL (2008) Initial 4D-Trajectory Data Link (4DTRAD) Concept of Operations. Available at http://www.faa.gov/about/office_org/headquarters_offices/ato/service_units/techops/atc_comms_services/sc214/current_docs/version_h/media/4DTRAD_Ops_Concept_v1_0_final.pdf.

FAA (2014a) Further information about compliance is given in AC 25.1309-1A – System Design and Analysis published by the FAA. Available at: http://www.faa.gov/regulations_policies/advisory_circulars/index.cfm/go/document.information/documentID/22680.

FAA (2014b) NextGen Implementation Plan, Federal Airspace Administration. Available at http://www.faa.gov/nextgen/library/media/NextGen_Implementation_Plan_2014.pdf.

ICAO (1944–2006) Convention on International Civil Aviation - Doc 7300 otherwise known as the Chicago Convention Available at http://www.icao.int/publications/pages/doc7300.aspx (also see Article 8, "Pilotless Aircraft").

ICAO (2011) Circular 393, Unmanned Aircraft Systems, ICAO, 2011. Available at: http://www.icao.int/Meetings/UAS/Documents/Circular%20328_en.pdf.

NTSB (2010) Aircraft Accident Report NTSB/AAR-10/03, PB2010-910403: Loss of Thrust in Both Engines After Encountering a Flock of Birds and Subsequent Ditching on the Hudson River, US Airways Flight 1549, Airbus A320-214, N106US, Weehawken, New Jersey January 15, 2009.

RTCA (1996) DO-178B/C. Available from the RTCA online store http://www.rtca.org/.

SAE International (1996) ARP 4761. Available at SAE International at http://standards.sae.org/arp4761/.

USAF (2014) accident reports and summaries for all USAF air accidents between 2000–2011. USAF Main Website at http://www.afsec.af.mil/organizations/aviation/aircraftstatistics/index.asp.

Chapter 44

Insurance as a Mission Enabler

Joseph J. Vacek

Department of Aviation, University of North Dakota, Grand Forks, ND, USA

1 INTRODUCTION

Insurance is a significant consideration at all levels in aviation generally, and the same holds true for those involved at every level in the unmanned aerial systems (UASs) industry within aviation. As much as we would like to think our technical expertise, operational experience, or other professional credentials will prevent mistakes, wisdom and experience lead us to conclude that insuring against both ordinary and unforeseen risks is necessary. This chapter will discuss how insurance works in the UAS industry, starting with a general overview of insurance principles to ensure every reader possesses the basic information required to understand the methods of managing risk. Specifics relevant to managing risk related to UAS will be explained in detail, with special consideration given to problems insuring UAS due to their unique capabilities.

Most readers are undoubtedly familiar with insurance in the automobile realm. In the United States and in most developed nations with civil legal codes, automobile owners and drivers are required to obtain some kind of vehicle insurance to legally operate their vehicle. That requirement provides a level of protection for both unforeseen and ordinary risks related to driving, for both the owner/driver and other people in the society. The most basic kind of vehicle insurance is *liability insurance*, which primarily provides financial protection against bodily injury or damage resulting from traffic collisions.

Additional insurance may be obtained that provides financial protection from theft or damage to the vehicle from things other than traffic collisions, such as hail damage, for example. Insurance is not only available to automobile owners and drivers, but manufacturers and dealers may also obtain insurance for accidents, employee mistakes, unforeseen losses, or ordinary risks in their line of business. Such commonly offered insurance types include *products liability* insurance, *commercial* insurance, or *business* insurance. Similarly, additional types of insurance are available in the aviation industry and to UAS designers, manufacturers, component manufacturers, commercial operators, pilots, and others involved in the industry. An example of such an additional type of insurance for UAS would be *hull loss* insurance, which insures for the loss of the aircraft structure and associated components.

In order for insurance to be available for a specific activity, there must be at least two parties who are willing to contract with each other for the purpose of indemnity. Legally, *indemnity* is an obligation by one person or business to compensate another person or business for a particular loss. For an indemnity contract to be valid, several legal requirements must be met. Those requirements will be discussed in more detail later.

Problematically for the UAS industry, UAS is still relatively new in terms of civil and commercial operations.

Unmanned Aircraft Systems. Edited by Ella Atkins, Aníbal Ollero,
Antonios Tsourdos, Richard Blockley and Wei Shyy.
© 2016 John Wiley & Sons, Ltd. ISBN: 978-1-118-86645-0.

Therefore, the risk of unexpected and ordinary loss is not yet well known, and an insurance principle from probability theory (Cardano, 1663) called the law of large numbers (Poisson, 1837) currently limits the extent of the UAS insurance market. While at least four major aviation insurers offer insurance coverage for some UAS operations at the time of this chapter, a review of several available example UAS insurance policies strongly resembles established and tested insurance contracts from other, manned aviation activities, and is therefore narrower in scope than might otherwise be the case. More options will undoubtedly become available as UAS operations become more common. The law of large numbers applied to insurance works to reduce risk by aggregating enough similar risk units (cars, houses, airplanes, etc.) that the individual losses become collectively predictable. An easy example to illustrate the operation of the law of large numbers is a coin toss. Tossing a coin only a few times could result in a particular side turning up nearly every time. But tossing the coin thousands of times will result in each side turning up about 50% of the time, which is the underlying probability of the event. The law of large numbers is a bedrock principle of insurance, and means that mature industries such as ocean shipping – where the insurance industry was born – or automobile operation enjoy lower insurance premiums due to much larger numbers of them compared to new industries like UAS.

2   HOW DOES INSURANCE WORK?

2.1   Risk

The term *risk* has different definitions depending on the context in which it is used, but for insurance the use of the term *risk* focuses on uncertainty and loss – in fact, the very definition within the insurance industry of risk is uncertainty about financial loss (Stemple, 2014). In aviation, the terminology is more precisely defined: *risk* is defined as "the future impact of a hazard that is not controlled or eliminated" or "future uncertainty created by the hazard" (FAA, 2009). A hazard is defined as "a present condition, event, object, or circumstance that could lead to or contribute to an unplanned or undesired event such as an accident" (FAA, 2009). In the aviation insurance context, the parameters of risks stemming from common hazards are well known and multiple methods exist to mitigate those risks. Some of those methods will be examined in more detail later. First, however, it is necessary to classify risk into two categories: static risk and dynamic risk (Willett, 1951).

<u>Static risk</u> is a risk from normal, natural hazards and human failure. Static risks are fairly regular occurrences and

therefore predictable. For example, every aircraft landing involves risk, and pilots train extensively to minimize those risks on every flight. *Dynamic risk* is coupled with external changes not associated with hazards, such as financial market fluctuations or technological innovations. As such dynamic risks are difficult to prepare for or predict. Because of that, most insurance situations in aviation and UAS focus on predictable static risks and ignore dynamic risks.

Second, the category of static risk is further subdivided into *fundamental* and *particular* risk (Vaughn and Vaughn, 2003). The main distinction here is that fundamental risks are risks that affect large groups all at once. An example of a fundamental risk would be a major storm that destroys all the hangars at an airport. Particular risks, on the other hand, are risks that affect individuals. An example of a particular risk would be a hangar fire. Together, these risk classifications and subdivisions help determine which risks are insurable and which are not. Generally speaking, risks that are both static and particular are fairly straightforward to insure against, whereas risks that are dynamic and fundamental are not, and insurance premiums for those are consequently very expensive or simply unavailable. There are further subdivisions of these categories and divisions, but they are outside the scope of this chapter.

2.2   Indemnity

Indemnity simply means compensation for loss (Blacks, 2009). While owners of damaged or destroyed property certainly have the legal right to seek repayment for their loss if another person is responsible for causing the loss, the potential for loss resulting from a natural disaster where there exists no responsible person, the cost of litigation, and the uncertainty of recovery against a responsible party lead most individuals and business to contract for indemnity with an insurance provider. Within the definition of indemnity comes a major principle of indemnity–compensation. *Compensation* means reimbursement and nothing more. Thus, a person may not insure her 10-year-old car worth $8000 for the value of a new car worth $35 000. Second, a major requirement from the Statute of Frauds (1677) for any indemnity contract to be valid is that it must be in writing. That compensation can be paid to one of the contracting parties or a third, unrelated party, depending on the contract. The contract itself has a few other limitations on it as well.

2.2.1   The requirement of an insurable interest

For an indemnity contract to be valid, there must be an insurable interest. An insurable interest prevents "gambling" by disinterested parties. The requirement developed from English marine insurance, to prevent the placing of bets as to whether a certain ship would successfully cross the ocean (Jerry and Richmond, 2012). From a social policy perspective insurance transactions serve to protect assets, whereas gambling transactions do not. Further, motivation to destroy the asset to win the bet cuts against the very purpose of insurance protection of that asset. Courts and legislatures rapidly developed prohibitions against such "gambling" contracts and the requirement of an insurable interest resulted. Simply understood, having an insurable interest means having a stake in the outcome. If a person or business entity faces personal or financial risk or has a legal interest in the outcome, there is likely an insurable interest. In the context of aviation and UAS insurance, this applies when insuring a physical asset against damage or loss and liability resulting from operating it. Commonly called "hull" insurance, the insurable interest in an aircraft and its systems must exist at the time of loss. It need not before then, however. Liability insurance automatically meets the requirement of an insurable interest because a person or business entity always has a legal interest in protecting itself against liability. Therefore, an insurance policy for hull and liability insurance for UAS may be purchased prospectively, that is, before the person or business owns the UAS, and still be valid. That may be an enabling condition if, for example, a UAS business is bidding on a service contract to inspect infrastructure and must show evidence of insurance before being awarded the contract. The business could secure insurance for the operation before purchasing additional UAS assets, which is likely a far less expensive business proposition.

2.2.2 *What is the meaning of "insured"?*

It seems almost too obvious to ask the question, and the threshold definition is indeed rather obvious: the person or entity whose loss requires the insurer to pay (Jerry and Richmond, 2012). However, beneath the surface of that definition complications can quickly threaten to drown those unaware. In the context of UAS insurance, where property and liability are at issue, the insurance contract will identify on the declarations page the named insured. That can be an individual, a business entity, or a class of persons or entities. It is important to consider how multiple listed persons or entities might share the insurance proceeds should a loss occur. A simple listing presumes all in the list are to share the proceeds equally, whereas a phrase such as "as their interests may appear" means that each insured shares the proceeds proportionally to the losses incurred. That simple definition of "insured" can have large financial effects.

A second issue in defining the meaning of "insured" is the problem of an insured allowing a noninsured to use its assets. Good insurance contracts should contain an *"omnibus clause"* which covers noninsureds using an insured's asset with permission. An example of this is from the automobile insurance industry where such omnibus coverage is mandated by state law as a matter of public policy. Omnibus clauses are less common in aviation insurance and when they exist they are typically much narrower in scope (e.g., a noninsured pilot must hold minimum qualifications in order to be covered). A problem arises when a noninsured person then gives permission to another noninsured person. If such secondary permission is given (like a noninsured pilot allowing a third noninsured sensor operator to operate an insured's UAS asset), most likely the original permission extends to the secondary permission if the secondary use was for the same purpose as the first (*Clayton v. S. Gen Ins. Co.*, 2010; *Conklin v. Acceptance Indem. Ins. Co.*, 2010).

2.3   The Principle of Subrogation

Subrogation is the legal right of the insurance company to pursue the person or entity responsible for causing damage (the tortfeasor) to the insured. In subrogation, the insurance company first pays the insured under the terms of the insurance contract. Upon satisfaction of that contract, the insurance company obtains the legal rights of the insured and may elect to sue the tortfeasor. While the right of subrogation is an efficient tool for insurance companies to minimize their own risk exposure to expensive claims, the larger legal policy goal of ensuring fairness means subrogation functions to prevent the insured from unjustly enriching themselves by recovering twice–once from the insurance company under the insurance contract, and again later by suing the tortfeasor. Subrogation prevents this by transferring the insured's legal rights to damages to the insurance company. Subrogation actions can also be brought in limited circumstances against the insured, which is generally an unpleasant surprise. The law limits these circumstances to when an insured intentionally violates the terms of the insurance contract, a claim is made for liability damages by a third party, and the insurer pays those damages.

An example of that occurring in the UAS context may be an insured commercial UAS entity that violates a Federal Aviation Regulation in the context of doing business and consequently injuring someone. The insurance contract will exclude from coverage activities that violate the law or applicable regulations, and if the injured party presents a damage claim to the insurance company that is paid, the insurance company has the right to subrogate against its own insured. A similar result occurs if the insurance company simply refuses to cover the damage claim due to the insured violating the terms of the insurance contract. Either way, operating outside the scope of the insurance contract allows the insurance company to avoid paying claims made from such operation and is equivalent to not having insurance.

3   MANAGING RISK IN UAS

3.1   Overview of Risk Management in Aviation

Risk management in aviation is well understood, and fortunately most of the principles translate fairly well to UAS applications. *Risk management* is a systematic approach to identify exposure to loss risk and determine the most effective ways to mitigate that exposure. Application of the principles of risk management occurs in two overlapping contexts – first, when the insurance contract is negotiated, and second in normal, everyday operations.

During negotiations, the insurance contract will likely assess five major types of aviation risk: (1) property loss, (2) loss of use of that property, (3) liability loss, (4) criminal loss, and (5) key person loss. Assessing and controlling risk in these areas is a key to safe, successful UAS missions.

The first major type of aviation-specific risk is property loss, which includes both real property and chattels. *Real property* means land and *chattels* means all property that is not land, which includes both *tangible* (physically existing) and *intangible* (not physically existing) or *intellectual property*. Real property and tangible property are fairly straightforward in loss calculation and valuation. Intangible or intellectual property requires more thought – especially in UAS, since the rapid development of cutting edge technology results in rapidly changing valuations for intangible things such as system design and software architecture or hardware in the pilot/sensor operator interface, for example. Accurate valuation of such intangible property is necessary to ensure that the insurance premiums are adequate for new developments as well as not excessive for dated technologies.

The second major type of aviation-specific risk, loss of use of that property, is an *indirect loss* from the primary loss of the property itself. A common indirect loss of use of property in aviation is revenue losses resulting from damaged (and unusable) aircraft. Some UAS assets are designed with a higher risk of loss in mind, as a mitigation method for "dull, dangerous, or dirty" missions where the risk of loss of a human life is too high. Loss of the UAS asset may be acceptable in that context, but the obvious result is that the UAS asset is gone, which would lead to a loss of use of that property claim. If the nature of the mission is such that asset loss is probable, it becomes a fundamental risk by definition,

and the insurance premium cost may approach the value of the UAS asset itself. Insuring for consequential losses in such a case would require careful consideration and carefully drafted insurance contract provisions in order to prevent costly litigation since this kind of risk is dynamic. In some cases, insurance for loss of use of the property will likely be unavailable.

The third major type of aviation-specific risk, *liability loss*, is usually a significant part of the premium value and insurer's risk calculus because passenger-carrying aircraft can cause significant damage to persons or property on the ground due to the combination of high speeds, mass, and volatility of aircraft fuels involved in accidents. In the UAS context, especially for small, low-mass assets, liability loss is less. A review conducted by the author of current UAS insurance contract applications does not differentiate between larger aircraft capable of carrying passengers or cargo and small UAS with significantly less mass. Development of contractual distinctions in this area is likely to occur soon as insurers accrue enough data to differentiate risk categories between different sizes of UAS assets.

The fourth major type of aviation-specific risk, *criminal loss*, can occur both from outsider criminal activity and insider criminal activity. While theft of large passenger-carrying aircraft is relatively infrequent because of the requirements that all aircraft be registered with a national database plus the rather obvious consideration that they are large and difficult to conceal, the opposite is the case for smaller UAS assets. Their relatively high value coupled with their small size and modularity means that the risk of loss from outsider criminal activity is higher. Loss from insider criminal activity is a possibility as well, but not likely significantly different from the aircraft industry as a whole.

The fifth major type of aviation-specific risk, *key person loss*, is pertinent mostly to smaller business entities where the owner, majority owner, or other key person is inseparable from the success of the business. The loss of that key person can mean the end of the business. This risk area is not significantly different in the UAS context than from other aviation business entities.

Considered together, these five major risk areas combine to provide insurers a relatively coherent picture of the maximum static and dynamic risk exposures and consequent probable and possible losses. Each area can be quantified individually and assessed qualitatively. Then, using a risk matrix, the sum of all risks for the venture can likewise be assessed in the risk matrix. A generic, basic risk matrix is provided by FAA and encouraged for adaptation by aviation industry users (FAA, 2009) with severity of risk as the *x*-axis and likelihood of event as the *y*-axis (see Figure 1; FAA, 2013).

The greater the severity and more likely the event the more the total risk exposure and the consequent need to mitigate. A

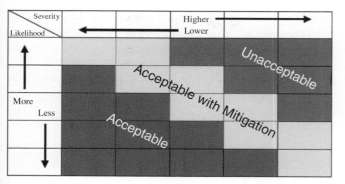

Figure 1. Sample risk assessment matrix.

generic matrix such as FAA's is adaptable to many different situations within aviation, and specific descriptive terms for risk likelihood or severity may be assigned by the user to fit the appropriate situation. Specific values for potential losses may also be assigned. The general idea of the matrix remains constant; however, where the further to the upper left quadrant of the matrix a particular risk is assessed, the greater the need to minimize it. Generally, insurance availability, coverage, and premiums increase in proportion with the matrix values, where a high likelihood of a probable, catastrophic loss is uninsurable and should be avoided at the root, versus covered by insurance. While a matrix is useful for assessing the risk in planning and operations, another model is needed for risk reduction after the assessment.

3.2   How to Reduce Risk

A generally accepted model in aviation for risk reduction includes an iterative four-step process. First, identify the risk. Best practices in risk identification include the entire system versus focusing solely on the shop floor or the flight line, for example. In aviation, system-wide risk is reduced through a *safety management systems* (SMSs). Second, determine the maximum possible loss. The five types of risks explained above factor into this determination. Third, manage the risk. This step is usually both a hands-on control approach and a passive financial approach through an insurance contract. The last step is the evaluation of the results, which, as best practices indicate is not actually the last step but rather a pause in the process before revisiting the model from step one. Each step is further described below.

3.2.1   Identify the risk

The possible sources of risk in UAS are as broad or broader as in manned aviation. For illustration of this model for this chapter, assume that the model is assessing a small UAS operation where a small business entity has been hired to conduct aerial imagery of a highway bridge. A time-tested risk identification for aircraft operations is the PAVE model (FAA, 2009), which stands for Pilot, Aircraft, enVironment, and External Pressures. It applies to the example UAS operation in that the UAS has a pilot/operator (and likely an additional sensor operator), the UAS asset is an aircraft, the environment includes the bridge, the highway, the traffic on it, the weather, surrounding terrain, any overlaying airspace and aircraft within it, obstructions to visibility or line of sight, etc. External pressures would include the contract terms for performing the bridge inspection, such as a deadline, or particular requirements limiting the UAS operator such as altitude restrictions or other regulatory restrictions. Within the PAVE model, each area can have one or more checklists guiding operators to a conclusion that the risk is acceptable or not.

There are a plethora of these checklists widely available in aviation, and those unfamiliar with them are usually overwhelmed with the multitude of checklists. A common question asked is whether requiring a checklist for each and every step in the operation is really necessary. Accident and incident data show that procedural discipline and following checklists are effective at risk reduction (AOPA, 2014). Here, within the P (pilot) element of PAVE, a commonly accepted checklist, known as the IMSAFE checklist, would be an acceptable tool for assessing a UAS operator's fitness to fly: IMSAFE stands for Illness – any symptoms?; Medication – any prescription or over-the-counter drugs?; Stress – how are stress levels from job, finances, family, or health?; Alcohol – any drinks within 24 hours?; Fatigue – adequate rest?; Emotion – upset? Any one or a combination of elements of IMSAFE can indicate unacceptable risk. If the IMSAFE checklist indicates acceptable levels of risk, the UAS operator would move on to the next element of PAVE.

Tribal Knowledge versus Systematic Approach. The PAVE and IMSAFE examples above are a systematized approach to risk reduction. Best practices require that those models and checklists be formalized, standardized, and used. First, having a safety program that exists only in form is not a safety program, meaning that an organization may have formalized safety procedures but those procedures are ignored in day-to-day operations. Second, reliance upon tribal knowledge where risk areas are informally passed along from the more experienced to the less experienced is ineffective for effective risk reduction. A safety management system is the most effective method to systematize risk reduction. "SMS is the formal, top-down business-like approach to managing safety risk, which includes a systemic approach to managing safety, including the necessary organizational structures, accountabilities, policies, and procedures" (FAA Order 8000.367, 2012). The promotion of a culture of safety is one of the main goals of a safety management system, and assessments of SMS programs find the benefit of accident and incident reduction is generally worth the cost (Adjekum, 2014). Additionally, an SMS program will likely result in reduced insurance premiums.

3.2.2 Determine maximum possible loss

The major areas from which losses typically occur were outlined above. Insurers must know with reasonable levels of certainty the maximum values of those potential losses.

Property valuation is generally a straightforward calculation. For real property, valuation can simply be taken from a tax assessment or a commercial depreciation schedule. For physical things, similar depreciation schedules exist (e.g., machinery and tooling would have a known purchase price, and the value is reduced as it ages). UAS equipment – including the airframe, associated sensors, and ground control equipment – would likely be classified as depreciable assets. Intellectual property resulting from UAS activities is more difficult to value, but should be considered.

Loss of use is also a relatively straightforward value, as long as there are data already existing or there is a fair comparison. To use the same UAS infrastructure inspection model as above, the existence of a previous contract that had been performed would be strong evidence for valuing a loss of use claim. The less certain the business venture or cost of services the weaker the valuation for loss of use would be. If the loss of use valuation approaches speculation, insurance will not be available for that area.

Liability valuation is fairly straightforward. While liability has inherent in it an element of unpredictability, entities offering aircraft insurance already have enough loss data and the ability to predict liability-related losses from aircraft accidents and mishaps that insurance contracts for UAS liability will simply borrow almost verbatim from those insurance contracts. Therefore, any aviation insurance contract for similar size and scope operations can be used as a starting point to evaluate the maximum potential UAS-specific liability.

Crime or fraud losses are not easy to predict, but such losses can be fairly easily estimated. For example, if an employee (insider) quits and leaves the premises with a small UAS, several sensors, and a laptop computer, the value of those physical assets can be known with certainty. More difficult to value would be theft of proprietary data or processes, which are kinds of intellectual property. However, insurers do write policies covering losses for theft or fraud, so several scenarios involving insider and outsider theft or fraud can be prepared and the valuations discussed during insurance contract negotiations.

Personnel losses are probably the most difficult to value. When any key person is lost there is a cost associated with it. A general rule is that the more important the person is to an income or profits, or the more unique the skill set that person possesses, the more valuable that person is. Key person losses can most easily be insured by the purchase of a separate life insurance policy for the value of that key person's contribution.

3.2.3 Managing the risk

Once all the risk variables are accounted for and their maximum potential losses calculated, there remains but one decision to be made: whether to manage the risk by controlling it or by insuring for it. The title of this chapter likely leads the reader to conclude that insurance is the preferred method, but controlling the risk internally is by far more effective, both in terms of physical risk reduction to life and property and in terms of reducing the financial burden of the insurance premium. Many aviation safety and aviation insurance texts address specifics of risk management – see, for example, *Risk Management Handbook* (FAA, 2009) and *Introduction to Aviation Insurance and Risk Management* (Wells and Chadbourne, 2007). The most effective risk management programs are formalized in writing, directed by an independent manager or corporate officer not responsible for meeting financial or operational quotas or goals, and accessible and accountable to all parties.

3.2.4 Evaluate the results

To recap, the steps in thinking like an insurer to best enable a UAS operation are (1) identify the risk, (2) determine the maximum possible loss, (3) manage the risk, and (4) evaluate the results. This final step is not really the final step, but rather the time when all elements are reviewed, evaluated, or assessed in the iterative process. The time for evaluation can be formally calendared or tied to a fiscal year, or simply performed each time the insurance premium is due in accordance with the insurance contract and in consultation with the insurance carrier's agent or representative.

4 SPECIAL CONSIDERATIONS IN UAS

Many of the insurance companies currently offering UAS insurance treat UAS risk essentially the same way as manned aircraft insurance risk (J VanMeter, Allianz Company, personal communication, 12 October 2015). However, several special considerations apply, which may be addressed in an endorsement or rider to the contract. Those special considerations include federal regulatory compliance, additional state law requirements, strict liability issues, uninsurable lawsuits, and various kinds of business insurance.

4.1 Statutory and Regulatory Requirements

Under current law and regulation in the United States, UAS operations are differentiated into three categories – public aircraft, civil aircraft, and model aircraft. Insurance coverage for modelers is available through membership in the Academy of Model Aeronautics (AMA). Liability insurance coverage limits of up to $2.5 million in general coverage, $25 000 for accident or medical coverage, $10 000 for accidental death coverage, and $1000 for loss coverage are available to members. That coverage, however, is only for amateur, noncommercial hobbyist operations, and those relatively low coverage limits reflect the relative risk posed by amateur, noncommercial, hobbyist operations. Professional services for commercial purposes entail a different risk analysis, as outlined in Section 3.2. By way of example, a hobbyist operating a small drone in a public park may run the risk of injuring one or a few bystanders, which is contemplated by the AMA coverage described above. But a commercial operator inspecting bridge infrastructure will need to manage far greater risk in operations — a UAS loss of control accident could cause a multicar collision on the bridge, for example. Commercial operations by their very nature require a different risk calculus. A threshold requirement for insurance coverage for any commercial operation is compliance with the statutory and regulatory requirements for those operations. Section 333 of the FAA Modernization and Reform Act of 2012 allows UAS operations that do not pose undue hazard to the public, and FAA has defined those operations generally as those that are conducted below 200 feet, away from airports and approach paths, in daytime line of site, and the UAS asset must weigh less than 55 lbs. The FAA has streamlined the process for obtaining permission to operate a UAS under Section 333 Exemption, and many insurance companies have adopted, as their own internal requirement, an FAA Section 333 Exemption as a condition for insurability (J VanMeter, Allianz Company, personal communication, 12 October 2015).

For larger UAS or public aircraft operated by governmental agencies the 333 Exemption does not apply, and those operators must obtain an individualized Certificate of Authorization from FAA to operate their UAS in compliance with the laws and regulations. The state of the US regulatory structure is still uncertain in terms of a final rule for UAS operations, as is the case in many other nations in the world. By way of example, some nations have similar "special permission required" regulations, such as

Australia, Canada, and Belgium. Some nations, such as Japan, France, and Poland, as examples, have finalized regulations and allow certain UAS operations without special permission. Finally, some nations, such as Uzbekistan, have completely banned the operation of UAS within their borders. The requirement to comply with national laws is usually required in the insurance contract, and intentional noncompliance will generally void insurance coverage. Accidental or negligent noncompliance may be covered, however, as long as the operator has done due diligence and is acting reasonably.

4.2 State Laws May Impose Additional Requirements

National laws and regulations are only the first step in compliance for insurance coverage. The state or local jurisdiction may require additional elements. While the general rule is that inconsistent state or local law is preempted by federal law or regulation, this preemption only applies to airborne operations. On the ground, state and local law and regulations have full force and effect. For example, some states have statutory requirements that the owner of an aircraft shall insure such aircraft (Minnesota Statute 360.59(10)). Other requirements may be requirements to register a business entity with the Secretary of State, or obtain a local business or merchant's license.

Whatever the case may be, if there are applicable state laws or local regulations a UAS operator must follow them to ensure insurance coverage, the same as for national laws or regulations. Again, intentional noncompliance generally voids insurance coverage, while accidental or negligent noncompliance may be covered as long as the operator has done due diligence and is acting reasonably.

4.3 Strict Liability

Multiple states define liability for aircraft accidents causing damage to persons or property to be absolute or strict. *Absolute liability* is imposed by state legislation using language such as "The owner of every aircraft which is operated over the lands or waters of the state is absolutely liable for injury or damage to persons or property on the land or water beneath . . ." (see, e.g., Minnesota Statute 360.12, 2015). Statutes such as the Minnesota absolute liability example above are valid because they regulate aircraft owner behavior on the ground versus aircraft procedures in flight. Such laws apply to UAS as well since UAS are defined as "aircraft" (*Huerta v. Pirker*, 2014).

Absolute or strict liability applies in a second UAS context as well. Manufacturers of UAS are subject to strict liability under the law of products liability. Generally, a manufacturer who sells goods to the public is strictly liable for harm those products cause. This means that traditional defenses such as contributory negligence are unavailable. Strict liability in products liability results from a balancing of public safety and fairness with the economic burden of such strict liability. That applies to UAS. Consider the bridge inspection example: A person who is injured from a malfunction of the UAS equipment used in infrastructure inspection would not be required to engage in traditional tort or negligence litigation against the manufacturer of the malfunctioning UAS. The injured person could recover for her injuries much more efficiently using a products liability lawsuit instead. What that means for UAS manufacturers is that a similar process as above must be used to minimize risk. In manufacturing, attention to risk identification is important because the law has imposed three subtypes of products liability categories on producers of goods: design flaw, manufacturing defects, and failure to warn. While the specifics of each of those areas is beyond the scope of this chapter, they all operate to impose strict liability on the manufacturer if a product is deficient in any of the areas and specific mitigation methods exist for each.

4.4   Ultrahazardous Activities

A final special consideration for UAS is whether the operation may be defined legally as an *ultrahazardous activity*. An ultrahazardous activity, also known as an "inherently dangerous activity," is an activity that is so inherently dangerous that a person engaged in such an activity can be held strictly liable for injuries to third parties even if every reasonable precaution has been taken. Essentially, it is a balance between social utility and economic benefit to the operator. Activities that include the use of explosive, volatile, unstable, or radioactive substances have been found to be ultrahazardous, as the testing of experimental aircraft. For UAS operations, injuries from the use of an experimental device or an unstable battery, especially over or too near uninvolved bystanders, could result in liability based on the ultrahazardous activity doctrine.

4.5   Uninsurable Activities

Several of the above-discussed types of liability are, by their very nature, uninsurable. Not only will insurance be unavailable specifically to cover those activities, but "umbrella" or excess insurance, group insurance, or other kinds of broad coverage will exclude those activities in the contract or endorsements to the contract. Additionally, the law precludes certain activities from being insurable. If a business activity results in "ill-gotten gains" as a result of misrepresentation, illegality, fraud, or outright theft, the later loss of those profits may be found to be uninsurable after the fact. For UAS operations, this particular issue may arise if profits are made from the sale of imagery obtained in violation of state privacy laws or in violation of federal UAS regulations. Even though insurance may have been purchased to cover the operation and the operation did not violate the insurance contract, later litigation can result in liability imposed by law outside insurance coverage because of the earlier violation.

4.6   Business Insurance

At this point, the reader may have observed that so many exceptions to coverage and insurance protection exist in the law that the value of purchasing any insurance coverage at all is questioned. The vast majority of insurance claims are eventually settled for a fair result, and only the unusual claims or the rare claim that violates the insurance contract or the law will proceed to litigation. But that kind of litigation attracts the most attention, unfortunately. Following the best practices to risk reduction outlined above is the best way to ensure that the mission is insurable (and to obtain insurance at the best price), but one can never prevent a lawsuit. The final piece to the puzzle, therefore, is to purchase some kind of professional liability insurance, sometimes called errors and omissions insurance. Typically, such insurance covers the cost of defending a negligence claim from a business client, and would cover such lawsuits as a software failure, inaccurate advice, or consequential damages from a product or operational failure.

5   CONCLUSION

While the basic principles of insurance as a mission enabler in UAS are similar to common insurance principles, the unique nature of UAS operations and manufacture requires additional consideration and due diligence to limit risk exposure and ensure insurability. By following established risk reduction models and formalizing their implementation and review, UAS operators, manufacturers, and contractors can minimize their risk exposure and consequently their insurance premiums. Additional considerations must be taken into account, depending on the particular situation, and those considerations include regulations and laws specific to aviation and UAS and whether strict liability applies to the activity. Finally, including some kind of business liability insurance to cover the business enterprise is the final piece in enabling the UAS mission.

REFERENCES

Adjekum, K. (2014) Safety management systems in aviation operations in the United States: is the return on investment worth the cost? *Prime J. Business Admin. Manage. (BAM)*, **4**(4) 1442–1450.

Aircraft Owners and Pilots Association (AOPA) (2014) Air Safety Report, 24th ed.

Garner, B.A. (ed.) (2009) *Blacks Law Dictionary* 9th ed.

Cardano, Gerolamo. (1663) *Liber de ludo aleae* ("Book on Games of Chance").

Clayton v. S. Gen. Ins. Co., 702 S.E.2d 446 (Ga Ct. App. 2010).

Conklin v. Acceptance Indem. Ins. Co., 702 S.E.2d 727 (Ga. Ct. App. 2010).

Federal Aviation Administration (FAA) (2009) *Risk Management Handbook*, Federal Aviation Administration Flight Standards Service.

Federal Aviation Administration (FAA) (2012) *Order 8000.367 Safety Management System Requirements*, AVS Services.

Federal Aviation Administration (FAA) (2013) *Flight Standards Information Management System*, fsims.faa.gov.

Jerry, R.H. and Richmond, D.R. (2012) *Understanding Insurance Law*, 5th ed., Matthew Bender & Co., Massachusetts.

Poisson, S.D. (1837) *Probabilité des jugements en matière criminelle et en matière civile, précédées des règles générales du calcul des probabilitiés*, Bachelier, Paris, France.

Statute of Frauds (29 Car 2 c 3) (1677) An Act for Prevention of Frauds and Perjuryes.

Stemple, J.W. (2014) *Stemple on Insurance Contracts*, 3rd ed., Wolters Kluwer, Maryland.

Vaughn, E.J. and Vaughn, T.M. (2003) *Fundamentals of Risk and Insurance*, 9th ed., John Wiley & Sons Inc., New York.

Wells, A.T. and Chadbourne, B.D. (2007) *Introduction to Aviation Insurance and Risk Management*, 3rd ed., Kreiger Publishing Co., Florida.

Willett, A.H. (1951) *The Economic Theory of Risk and Insurance*, Richard Irwin, Inc., Illinois.

Chapter 45

Fail-Safe Systems from a UAS Guidance Perspective

Christoph Torens and Florian-Michael Adolf

German Aerospace Center (DLR), Department of Unmanned Aircraft, Institute of Flight Systems, Braunschweig, Germany

1 INTRODUCTION

Unmanned aerial systems (UAS) are highly automated systems that will in the future populate segregated and also unsegregated airspace. The aerospace domain is safety critical and the utilization of such highly automated systems in safety-critical areas imposes high quality and safety standards onto these systems. As a result, manned aircraft are designed to be fail-safe to reduce risk of harm for the crew, passengers, or people on the ground. Of course, the same should apply for unmanned aircraft. The risk of harm for people should be minimized and a UAS should be designed so that if it fails, it fails safely. Conceptually, heterogeneous aircraft systems, for example, from the size of an insect up to the size of Global Hawk, all fall into this category. This work discusses the question of how unmanned aircraft can be made fail-safe when software complexity is growing and the human pilot is removed from the onboard system. For unmanned aircraft, there are several challenges to create a fail-safe system. Questions are as follows: What does fail-safe mean in context of an unmanned system? How safe is safe enough for an unmanned aircraft? How can software-intense systems be made fail-safe?

The DLR is researching these questions with its ARTIS platform (Autonomous Research Testbed for Intelligent Systems), shown in Figure 1. Recent activities focus on verification and certification aspects, especially for the software modules that implement high-level behavior, for example flying in unknown terrain, and thus replaces the onboard human pilot (Torens and Adolf, 2013).

The remainder of this chapter is structured as follows: Section 2 describes the ARTIS research platform as background. In Section 3, related work is presented on fail-safe systems, especially software systems. The basics of fail-safe systems are presented in Section 4, discussing special features of UAS and software intense systems. Section 5 discusses the importance of comprehensive and correct requirements specification for the design and operation of fail-safe systems and shows how formal requirements were elicited for ARTIS. Next, Section 6 shows how software design influences system safety and proposes using formal methods for design validation. Verification and validation procedures are presented in Section 7. Section 8 discusses the important topic of runtime monitoring for unmanned aircraft and the influence of runtime monitoring for fail-safe systems. Finally, Section 9 concludes this chapter.

Unmanned Aircraft Systems. Edited by Ella Atkins, Aníbal Ollero, Antonios Tsourdos, Richard Blockley and Wei Shyy.

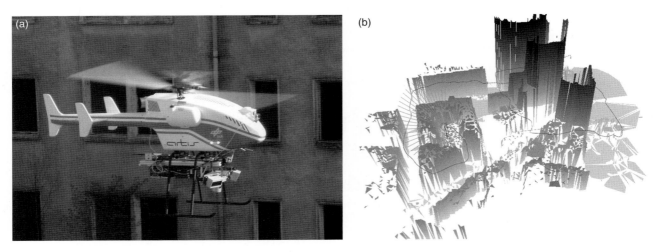

Figure 1. A DLR unmanned rotorcraft (a) and a simulation result of online planning in urban terrain for our rotorcraft with onboard perception (b): path flown (black curve), virtual distance sensor (gray lines), obstacles mapped (grayscale). (Copyright © 2015 by DLR Institute of Flight Systems.)

2   ARTIS SYSTEM DESCRIPTION

ARTIS represents a family of unmanned rotorcraft testbeds developed by the German Aerospace Center (DLR) for unmanned aircraft research purposes. Since 2006, ARTIS has been equipped with a Mission Planning and Execution (MiPlEx) software framework that comprises real-time mission plan execution, 3D world modeling, and algorithms for combinatorial motion planning and task scheduling. The framework is based on a decoupled approach for path planning, trajectory generation, trajectory following, and inner

loop flight control (Lorenz and Adolf, 2011). The rotorcraft's guidance algorithm has been evaluated in flight tests with respect to closed-loop motion planning in obstacle-rich environments. (Adolf and Thielecke, 2007) describe a control architecture behind the guidance layer. It achieves hybrid control by combining the main ideas from a behavior-based paradigm (Brooks, 1990; Flanagan et al., 1995) and a three-tier architecture (Bonasso et al., 1996). The behavior-based paradigm reduces system modeling complexity for composite maneuvers (e.g., land/takeoff) as a behavior module that interfaces with the flight controller (Figure 2). The three-tier

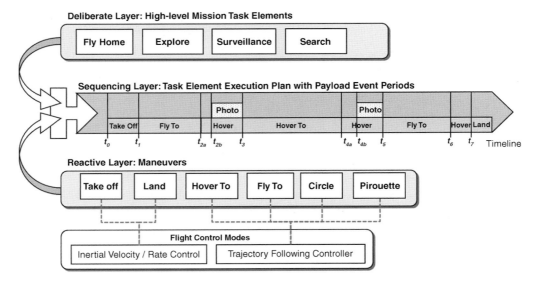

Figure 2. Mapping mission shown in context with the control architecture: High-level behaviors use task-specific planners (deliberate layer), behaviors are compiled into plans (sequencing layer), and movement primitives (reactive layer) interface with the flight controller. (Copyright © 2015 by DLR Institute of Flight Systems.)

architecture has the advantage of different abstraction layers that can be interfaced directly such that each layer represents a level of system autonomy.

UAS operators have an intrinsic need for reduced mission planning complexity when specifying collision-free paths and mission tasks (e.g., search covering a given area). With the onboard control system, it is possible to design and execute mission plans. This planning system automates the translation of user-specified sets of waypoints into a sequence of parameterized behavior commands. The path planner is able to find collision-free paths in an obstacle-constrained three-dimensional space. The task planner determines a near-optimal order for a given set of tasks. Moreover, a task planner can solve specialized problems, for example specifying the actual waypoints for an object search pattern within an area of interest.

As a result, the mission planner must build three-dimensional, safe, and unobstructed paths quickly while optimizing task assignments and task orderings for multiple UAS, as shown in Figure 3. In the remaining work, this system will be used to demonstrate fail-safe systems development. The novel aspect of unmanned aircraft as an application is the missing onboard pilot, an aspect the public is most concerned about. Increased effort in verification and validation to achieve fail-safe guarantees are therefore required. We concentrate this chapter on this high-level software, the guidance component.

3 RELATED WORK

Even though safety requires a holistic approach, introducing software into safety-critical systems requires building safe software. A comprehensive analysis researching safety of software was done by Leveson and Harvey (1983). It defines software safety and describes a methodology to analyze software safety by using software fault tree analysis, an analogous technique to hardware fault tree analysis, with the goal of showing that the logic contained in the software will not produce safety failures. However, Leveson and Harvey (1983) state that safety analysis techniques must be combined with runtime-safety techniques, to catch errors or environmental conditions that are not recognized or anticipated during *a priori* safety analysis.

Furthermore, Leveson (1986) specifically discusses the problems of introducing complex software into safety-critical systems. It is argued that many safety techniques developed for electromechanical systems do not apply when software is introduced. One reason is that these techniques often address random failures that are not a part of software. Traditional testing approaches cannot cope with this complexity, since testing cannot cover the complete state space. Furthermore, testing can only find errors according to the given specification. But wrong, misinterpreted, or inconsistent specifications can still lead to failures.

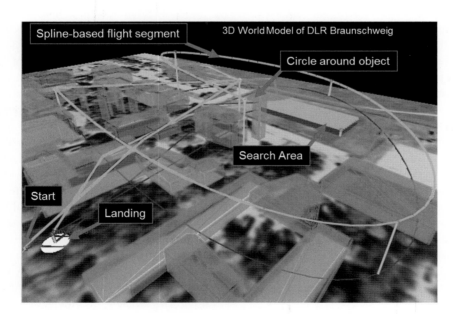

Figure 3. Automated mission with nonlinear flight segments, a search area task, and a horizontal inspection circle. (Copyright © 2015 by DLR Institute of Flight Systems.)

Leveson finally gives a general view about computer systems and approaches for achieving safety (Leveson, 1995). This book discusses risks, hazard analysis, safety requirements, and elements of a safe software design as well as verification aspects and also human factors. From a design point of view, there are two approaches to creating fail-safe systems: applying standards, codes of practice, and checklists into development and verification processes that reflect expert knowledge and lessons learned from previous accidents. The second is to use hazard analysis during design. For achieving the latter, there are four types of design techniques to achieve a fail-safe system: hazard elimination, hazard reduction, hazard control, and damage minimization. This technique should be used in the listed precedence order for a specific hazard.

Another overview of dependable computing systems is given by Geffroy and Motet (2002). A fail-safe system is described as a system that may fail, but when it does, either the failure consequences must not be "dangerous" or the probability of a "dangerous" failure must be smaller than a given acceptance level. As such, it is important for the design of a fail-safe system to define failure condition categories and set corresponding acceptable risks. Techniques to create fail-safe systems are based on intrinsic safety and safety designed by use of structural redundancy. Intrinsic safety utilizes a subset of technological development solutions that are known to be safe.

The work from Jhumka et al. (Jhumka et al., 2004; Jhumka and Suri, 2005) gives the theoretical background on generating fail-safe systems with a given safety specification. Any program specification can be considered formally as the intersection of a so-called liveness specification and a safety specification. They show that it is possible to generate a fail-safe program from a fault-intolerant program by using formal detectors for the safety specification. The generated, so-called, perfect multitolerant program is, in the absence of faults, equivalent to the fault-intolerant program, but detectors remove transitions from the program's state space that are inconsistent with the given safety specification.

An overview of state-of-the-art and future research directions for safety-critical software engineering is given by Lutz (2000). Identified future research topics include integration of informal and formal methods, safe reuse of software, verification and runtime monitoring, education, and collaboration with related fields. Similarly, Knight (2002) discussed that development of safety-critical systems requires significant effort as well as advances in the areas of requirements specification, architecture, verification, and process management.

In summary, creating fail-safe systems requires knowledge from different areas, for example requirements specification, architecture, and verification. These are also the areas that are specifically discussed in this chapter (Sections 5, 6 and 7, respectively). Software failures are systematic; as such verification of software is crucial. However, software testing alone cannot cope with verification of complex software systems, since testing cannot cover the complete state space and not all environmental conditions can be anticipated. Therefore, the application of formal methods is discussed as a promising approach throughout the literature, especially for checking consistency with high-level requirements, as discussed in Section 5. Finally, runtime monitoring (Section 8) can be used to catch failures that arise despite careful software testing.

4 FAIL-SAFE SYSTEMS

Informally, a safety-critical system should not endanger people and thus should be fail-safe. Before discussing further details of fail-safe systems, it is necessary to provide a proper definition of this term.

A failure is an event that is inconsistent with the system specification.

A safety-critical failure is an event that is inconsistent with the system safety specification. A safety-critical failure can lead to a hazard and a potentially catastrophic event.

A fail-safe system is a system that is designed so that the occurrence of a safety-critical failure is mitigated in a safe way. Remaining system failures are either non-safety critical or the hazard probability is below an acceptable risk level.

It is worth mentioning that the definition of fail-safe systems does not claim that such systems do not fail. Instead, a system is expected to fail, but any system failure must occur in a safe way. Furthermore, fail-soft or fault-tolerant systems, which in case of a fault will still be able to perform most tasks, must not be mistaken for fail-safe systems. Fault-tolerant systems try to minimize all failures and not only safety-related failures, which also considers usability. A system can in fact be fail-safe but not fault-tolerant.

Fail-safe systems have a long and successful history. The automotive, railway, and aerospace domains are classical examples of mass-produced fail-safe systems. Due to its safety-critical nature, commercial aircraft have also been designed as fail-safe systems. The aerospace domain uses regulations and standards, for example DO-178C (Radio Technical Commission for Aeronautics (RTCA), 2011), ARP4754a (The Engineering Society for Advancing Mobility in Land, Sea, Air, and Space, 2010), ARP4761 (The Engineering Society for Advancing Mobility in Land, Sea, Air, and Space, 1996), and accreditation by federal authorities to ensure these principles. An aircraft is designed to be redundant; for instance, commercial aircraft have multiple engines. Furthermore, multiple controls and hydraulics ensure that the aircraft does not fail, even if one of these redundant systems fails. By applying such design strategies, it is possible for aircraft to achieve an overall failure rate of 10^{-9} per flight hour for safety-critical failures per flight hour.

From an abstract point of view, there are two types of errors that can cause a system to fail, systematic or design failures and nonsystematic failures, that is, the so-called infant mortality, random, and wear-out failures. Furthermore, there are two types of origins: a failure can originate from software and/or hardware. Systematic failures result from problems during system design. Each phase of the system development process possibly introduces an error that would lead to a later failure of the system. A systematic hardware error could be a wrongly designed coupling. A nonsystematic failure could result from an error during production, or from inherent properties of the item. A wear-out failure would be a correctly designed coupling that breaks toward the end of its life cycle due to mechanical stress.

A systematic software failure would be a wrongly designed software function that produces a wrong output each time a specific set of data is used as input for that function. A wear-out software failure, on the other hand, is not possible as in hardware. On the contrary, software actually matures with operating service hours, since each service hour can be viewed as a passed test case.

Essentially, there are three questions to answer to create a fail-safe software system:

- How can a (software) system fail?
- What strategies can be used to eliminate or mitigate these (software) failures?
- How can these safety strategies be ensured practically?

To answer the first question, a safety assessment is necessary to analyze specific origins of failures. For example, a failure hazard analysis (FHA), fault tree analysis (FTA), and failure modes and effect analysis (FMEA) can be used to determine possible system failures and their criticality (The Engineering Society for Advancing Mobility in Land, Sea, Air, and Space, 1996). This question also requires an analysis of possible hazards and their criticality, which in turn requires asking the question, how safe is safe enough? To answer this, it is important to create a specific set of safety requirements (Rierson, 2013). However, there are some general concepts or properties that can be discussed with respect to unmanned systems. We address this problem in Section 4.1. The problem of how a system failure can be mitigated, specifically for software intensive systems, for example design fix or runtime mitigation, is addressed in

Section 4.2. The remaining sections of this chapter answer the third question and address how fail-safe design techniques can be used as mitigation for hazards to ensure a target level of safety.

4.1 Fail-safe UAS

A UAS is an aircraft with spatially distributed guidance by a pilot who acts on the physical world remotely. This requires onboard guidance to be fail-safe. We will specifically look at mission management and will highlight points that differentiate manned from unmanned aircraft. From a UAS perspective, the design of a fail-safe system is a challenge for several reasons:

1. The meaning of fail-safe for unmanned aircraft is not evident, since there is no pilot or passengers on board.
2. Strongly heterogeneous types and classes of unmanned aircraft exist.
3. There is no onboard pilot to complete specific tasks.
4. There is no onboard pilot to supervise specific tasks.
5. Software-intensive high-level functions are hard to verify.

We learned from Leveson (1995) and Geffroy and Motet (2002) that when speaking about fail-safe it is necessary to define an acceptable risk level for specific hazard categories. For a manned aircraft, for example CS25, this is strictly defined, and since people are onboard, a catastrophic failure has to be extremely improbable, which relates to a failure rate of 10^{-9}. Some have proposed use of the same acceptable risk level for unmanned aircraft, since it might operate in the same airspace as manned aircraft. But, on the other hand, for an unmanned aircraft the risk of fatal casualties is lower. Only a midair collisions with a manned aircraft or a crash in which two people are on the ground would result in fatalities.

Furthermore, consider the very different types of UAS. There exist efforts to create and use unmanned aircraft literally from the size of an insect, weighing only a few grams, up to the size of very large aircraft like Global Hawk, weighing thousands of kilograms. Between these extremes of unmanned aircraft, there is an almost continuous variation of unmanned aircraft sizes and also several different types of configurations. It would be very strict to assume the same risk categories for the multitude of unmanned aircraft, or to demand the same amount of rigor for a fail-safe system across all UAS. As a result, the acceptable risk should be assessed differently from that of manned aircraft. However, although efforts to create regulation for unmanned aircraft are increasing (European Aviation Safety Agency, 2015a; European Aviation Safety Agency, 2015b), there is no established definition for these acceptable risks yet.

UAS clearly differ from manned aircraft with respect to the pilot's distance to the aircraft, his situational awareness, and his ability to bring resilience to unforeseen events into the guidance and control loop. The onboard pilot is replaced by a command and control data link that sends specified data from and to a pilot sitting at a ground control station. Because of these reduced capabilities, software has to take over functions that the pilot would usually perform onboard, resulting in software systems of increasing complexity. For the sake of brevity, we will call this software component the guidance component of the unmanned aircraft. Writing and verification of software in safety-critical areas is already a problem for today's aircraft. Additionally, it is possible to build unmanned aircraft that are a lot smaller, that is, no pilot can occupy the vehicle. But small systems, dependent on the actual size of the system, can be very limited in their capability to carry computing and energy resources. As a result, it is especially challenging to add redundancy to such systems. Another factor is that, due to the distance to the aircraft, the pilot has reduced situational awareness. Therefore, the pilot may notice problematic behavior of the aircraft only after it is too late to recover.

4.2   Fail-safe software

Hardware can be designed to be intrinsically fail-safe, for example trains use braking systems where pressure is constantly applied to open the brake and loss of pressure results in automatic activation of the brake. Therefore, a full stop, the safe state, for a train can always be ensured. In software (Leveson, 1986), different techniques are required and a safe-state often cannot be entered from within the software. For example, software that uses dynamic memory allocation can behave unpredictably in case of error.

Section 3 stated that software errors are strictly systematic. Special care is to be taken during software development and verification to reduce the probability that software errors remain in the final product. It is necessary to eliminate as many errors as possible, because any hazard that results from a software failure in fact was not detected during this activity. Therefore, the standard for developing software in the aerospace domain (Radio Technical Commission for Aeronautics (RTCA), 2011) is based on strict conformity to such development and verification processes and objectives. But on the other hand, software does not wear out, so no errors will be introduced during the system lifetime.

Furthermore, in contrast to hardware, it makes no sense to replace a faulty software system. As long as the error is not fixed in the software design, the same erroneous software will reside in that system. The same problem arises, when trying to use redundancy to increase safety. To actually achieve redundancy, the same software cannot be used twice, instead software diversity must be achieved. The software needs to be completely designed, implemented, and verified a second or even a third time in independent development branches, essentially doubling and tripling the costs. Furthermore, each of these implementations is dependent on the quality of software verification activities to ensure that no errors exist. However, there are several approaches for using redundancy of independently developed software components. To realize fail-safe guarantees, an independent subsystem must be able to detect a failure in an active system. In principle, there either has to be some vote of n out of m systems, which is very costly and not trivial, or some kind of checks have to be performed on the output of the system via a runtime monitor. It is therefore possible to deactivate the failed system and switch to an independent system that re-establishes a safe system behavior, possibly with reduced capabilities. A more detailed depiction of software safety tactics is presented in Wu and Kelly (2004).

The remainder of this chapter shows the best practice approach used for ARTIS to achieve high software quality. Our approach uses three layers of safety: checking requirements and design using formal methods, a multitude of verification techniques with different properties for the software implementation, and runtime monitoring to find errors at runtime and enable fallback strategies. As mentioned, software standards like DO-178C focus on development quality and conformity to development processes. Although this is an important aspect of creating fail-safe systems, this is not discussed further in this work. For details on conformance of ARTIS development with DO-178C, refer to Torens and Adolf (2013).

5   FAIL-SAFE UAS REQUIREMENTS

The basis of fail-safe system design is a set of functional, nonfunctional, and safety requirements. These requirements are also used for verification and validation and runtime monitoring system design. Whereas the safety requirements are specific to fail-safe design, formally modeled safety properties could specify the behavior to ensure consistency and formally guarantee the intended behavior is possible, for example can always reach a safe state. Model checking is a technique to ensure the consistency and safety properties of such specifications.

Table 1. Requirements in a tabular semiformal format.

ID	When	System	Obligation	Activity
3.4	When the safety pilot intervenes	the mission-manager	must	go into pause state immediately
⋮	⋮	⋮	⋮	⋮
6.1	When the UAS is landing and gets a stop command and the UAS is already on ground	the mission-manager	must	go to status onground immediately
6.2	When the UAS is landing and gets a stop command and the height is above the safe minimum height	the mission-manager	must	go into StandbyAir mode immediately
6.3	When the UAS is landing and gets a stop command and the height is below the safe minimum height	the mission-manager	must	cancel landing immediately and rise to safe minimal height

Source: Copyright © 2015 by DLR Institute of Flight Systems.

```
1   LTLSPEC G(Status=Landing&Command=Stop&Position=Ground -> X Status=StandbyGround); --R6.1
2   LTLSPEC G(Status=Landing&Command=Stop&Position=Air& MinHeight -> X Status=StandbyAir); --R6.2
3   LTLSPEC G(Status=Landing&Command=Stop&Position=Air&!MinHeight -> X Status=AbortLandTakeoff); --R6.3
```

Figure 4. Functional requirements in LTL. Each line is concluded with a comment that states the requirements reference, for example R6.1. (Copyright © 2015 by DLR Institute of Flight Systems.)

```
1   -- safety: something bad will never happen
2   -- safety-pilot requirements
3   LTLSPEC G (SafetyPilot) -> X ( ManagerState=pause_state); --R3.4
```

Figure 5. Safety property in LTL, that is, the safety pilot shall always have the option to control the UAS, or in other words, he shall not ever lose the ability to control the UAS. (Copyright © 2015 by DLR Institute of Flight Systems.)

5.1   Requirements formalization

A semiformal method was used to write down requirements for ARTIS. The tabular template was sufficiently easy to support directly writing down requirements, while it was formal enough to facilitate the use of keywords and a fixed structure. The template also was helpful in formulating the requirements and ensuring requirements were considered with all relevant aspects. The used template is shown in Table 1 and is an adaptation of a *requirements template with conditions* (Pohl, 2010). The disadvantages of using natural language for requirements can thus be minimized. Fields like "Activity" are used in this example mainly to fill natural language parts and the other fields, for example, "System" and "Obligation" are positioned to make the resulting row more readable.

After an initial learning phase, it was relatively easy to transform the tabular requirements to a formal specification in most cases. For the formal part, LTL (linear temporal logic) (Figure 4) and CTL (computation tree logic) (Figure 6)

were used. It was easier to express conditions in LTL if possible. CTL seemed a bit harder to get used to. The used model checking tool NuSMV is able to handle both languages and the expressiveness of these languages is different, so both languages were used.

The examples show a clear correspondence of keywords on the left side of the property (left of the implication "→") and the "When" column of the tabular requirement specification, for example landing, stop, ground, MinHeight. The same observation can be made about correspondence of keywords on the right side of the property and the "Activity" column of the tabular requirement, for example, StandbyGround, Slowdown, CancelStart. Furthermore, in most cases the requirement usually is globally true; therefore, the structure is `LTLSPEC G(lefthand -≫ righthand)`. Not all, but a big part of the requirements could be handled with this left hand, right hand scheme. The requirements shown in Table 1 are transformed with this process into a formal specification (see Figure 4). With the act of formalizing these requirements, the first objectives of DO-178C

```
 1  -- liveness: something good eventually happens
 2  -- all states always reachable
 3  SPEC AG EF(ManagerState=landing); --R6.0 -- CTL
 4  SPEC AG EF(ManagerState=standby_air); --R3.5 -- CTL
 5  SPEC AG EF(ManagerState=standby_ground); --R3.6
 6  SPEC AG EF(ManagerState=takeoff); --R5.0 -- CTL
 7  SPEC AG EF(ManagerState=abort_takeoff); --R6.4, R5.2 -- CTL
 8  SPEC AG EF(ManagerState=pause_state);  --R3.4 -- CTL
 9  SPEC AG EF(ManagerState=slowdown);  --R1.1 -- CTL
10  SPEC AG EF(ManagerState=man_trans_command); --R1.4, R1.5, R7.2 R7.0 -- CTL
11  SPEC AG EF(ManagerState=execute_mission); --R2.0 -- CTL
12
13  -- all states reachable infinitely often
14  LTLSPEC G F(ManagerState=landing); -- LTL
15  LTLSPEC G F(ManagerState=standby_air); -- LTL
16  LTLSPEC G F(ManagerState=standby_ground); -- LTL
17  LTLSPEC G F(ManagerState=takeoff); -- LTL
18  LTLSPEC G F(ManagerState=abort_takeoff); -- LTL
19  LTLSPEC G F(ManagerState=pause_state); -- LTL
20  LTLSPEC G F(ManagerState=slowdown); -- LTL
21  LTLSPEC G F(ManagerState=man_trans_command); -- LTL
22  LTLSPEC G F(ManagerState=execute_mission); -- LTL
23  SPEC AG AF(ManagerState=execute_mission); -- CTL
```

Figure 6. Some live-ness properties in LTL and CTL, that is, all states shall always be reachable and occur infinitely often in an endless run. (Copyright © 2015 by DLR Institute of Flight Systems.)

(Radio Technical Commission for Aeronautics (RTCA), 2011), respectively the formal supplement DO-333 (Radio Technical Commission for Aeronautics (RTCA), 2011), are already fulfilled, thus supporting safety (Torens and Adolf, 2015).

5.2   Safety, liveness, and fairness

In addition to translating the requirements into a formal model, some specific requirements are also to be modeled to ensure general model safety and liveness properties. In this context, a safety property is formally defined. A safety property essentially states that "something bad never happens" (Figure 5), whereas liveness properties state that "something good eventually happens" (Figure 6), more details for the definitions of these properties can be found in the respective literature (Clarke, Grumberg, and Peled, 2000; Baier and Katoen, 2008). These liveness properties are necessary to prove that the model is not trivial. For example, a UAS could never takeoff, then trivially all safety-critical requirements would automatically hold.

A fairness constraint further allows only those paths in the execution of the model that allow the fairness constraint to hold infinitely often. In this example the fairness properties ensure that in an infinite execution each command will be triggered (in a suitable state) infinitely often.

6   FAIL-SAFE SYSTEMS DESIGN

System architecture and design can affect the safety of a system in two independent ways. First, it can directly affect safety by facilitating or limiting safety as an integral part of the architecture. Furthermore, architecture and design have to be compliant to requirements, as required by software standards DO-178C (Radio Technical Commission for Aeronautics (RTCA), 2011). In this section we will discuss the former aspect briefly on the design considerations and revisit a similar approach in more detail later. The second aspect will be detailed by showing a prototype approach to verifying software design.

In Ref. Cherepinsky and Pinto (2012) a safety-critical architecture is shown, suitable for autonomous behavior of unmanned aircraft. High-level functionality executed with reduced criticality ensures support for sophisticated functionality and behavior. On the other hand, a second software module with higher integrity and priority can act as a backup with lower level, basic functionality to always ensure a safe state.

6.1   Design verification

We utilize formal methods to check software design and facilitate model-based design approaches. For the creation of

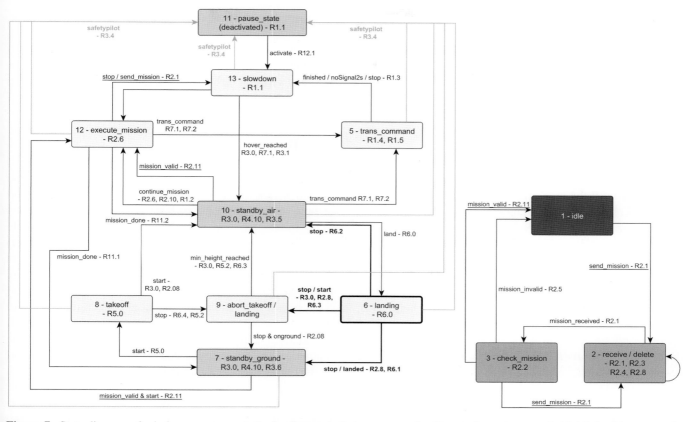

Figure 7. State diagram of mission management; the landing state that represents the discussed requirements is highlighted by arrows in bold, underlined transitions are used to communicate and synchronize the two separate FSMs. (Copyright © 2015 by DLR Institute of Flight Systems.)

the formal model, at first a state diagram (Figure 7) was used to visualize thoughts and further clarify requirements as an intermediate step. Figure 7a shows the operational mode of the mission manager, including "safe" states standby_ground, standby_air, and a pause_state. The remaining states represent different kinds of movement, for example, landing. Figure 7b shows a separate task that can receive and validate missions transferred from the ground control station.

This model was then transformed into a formal NuSMV (Cimatti *et al.*, 2002) specification. NuSMV uses a textual input language and allows definition of a behavioral model, that is, a finite state machine (FSM), and properties to be validated on the model in both CTL and LTL. A FSM can be specified via an `assign` declaration by giving an initial value `init(var)` and state transitions `next(var)`, which are defined by boolean expressions. The different FSMs are then "executed" concurrently by NuSMV. A model-checking algorithm tries to find counterexamples for the properties that could occur during parallel execution of the FSMs. Eclipse was used as an editor, which featured syntax highlighting

using an available NuSMV plug-in[1] and a derivative plug-in distribution[2]. Execution of model-checking could be done directly from within Eclipse using the external tools option.

The two FSMs of Figure 7 can be translated to NuSMV syntax straightforwardly. The transitions are guarded by a logical expression, for example check actual state, and can then be executed by assigning the next value. To retain general comprehensibility, the produced expressions mimic a simple syntax: state + event = next-state. But in some cases, additional state variables further alter the behavior, as in the following example:

```
ManagerState=execute mission &
Command=mission done & Position=air
        :standby air; --R3HL
```

---

[1] http://code.google.com/a/eclipselabs.org/p/nusmv-tools/
[2] http://code.google.com/a/eclipselabs.org/p/nuseen/

This example shows that it is generally possible to translate a previously developed FSM into a formal model.

While building the model, the properties can be used to test it. Therefore, it is most important to have clearly specified requirements. Also, a complete traceability greatly supports a correct incremental build of the model.

6.2 Model-checking results

Conceptually, the model and its properties are developed independently. If the properties and the model are valid and the model also flawlessly represents the software, an error found by the model checker would always reveal a flaw in the software. In practice however, these parts usually debug each other mutually, as stated in the last section. The cause of an error may be one or a combination of a modeling error, a property formalization error, incorrect or missing environmental constraints, or errors in the requirements themselves. This is one of the key benefits of formal methods. It is therefore possible to find errors in the requirements while developing the formal model, which in turn can reduce development costs.

Temporal logic statements are evaluated as true for the specified model. A detailed discussion about conformance to DO-178C (Radio Technical Commission for Aeronautics (RTCA), 2011) and DO-333 (Radio Technical Commission for Aeronautics (RTCA), 2011) is discussed in (Torens and Adolf, 2015). However, it was invaluable to do the model checking to be able to validate the requirements themselves. A missing link still exists today between reality (code) and model checking. This gap is only filled by human expert review. Future work will enable assured conformance of the formal model and the source code or the object code.

7 FAIL-SAFE SYSTEMS VERIFICATION

Comprehensive system verification is a complex task; therefore, there are several layers of testing for the ARTIS system. Details for development and verification processes can be found in Ref. Torens and Adolf (2013). In this work, the MiPlEx testing efforts are discussed along with categories of testing. Here, static tests, dynamic tests, Software-in-the-Loop and Hardware-in-the-Loop simulations, flight tests, and model-based tests will be elaborated as the different testing layers in an integrated test strategy. The presented verification approach utilizes these methodologies in a way that the tests start from inner, low-level aspects (source code level) and go to outer, high-level aspects of the system (integration and system level). Different test layers have

different test characteristics and different costs. A good test design should combine these techniques to achieve a maximum coverage of all test characteristics. The key is to find errors at the earliest possible moment. This considers the fact, that software errors tend to cause exponentially growing costs, the latter they are found during software development.

7.1 Test methodologies

The first identified test dimension is the size of the specific system under test (SUT) that is used by the verification technique, for example what is being tested. Possible values range from single lines of code, software functions, a whole software module, interaction of different modules, and software systems up to the complete embedded system. Test effort (TeE) defines the costs of a specific test method. The test or scenario complexity (ScC) is low if a mathematical function is tested standalone, increases with combination and interaction of functions, and is highest if a complex scenario is tested or simulated. The coverage (Cov) assesses the theoretical state space that can be covered by the method or the code coverage if that is not applicable. The Feedback time (Fet) describes how much time it takes to get the result from the test back to the developer. A short feedback time is good, as it allows faster development cycles. Finally, the level of automation (Aut) describes if and to what degree a test can be automatized. These test dimensions are evaluated for the ARTIS approach and finally visualized for comparison in Figure 8.

A system engineering process starts with acquiring requirements. If the models can be formalized to a substantial level, then formal approaches can be utilized to analyze that specification. Extensive sets of high-quality test cases can also be generated from a formal model.

There are several different techniques that can be identified by static analysis. For ARTIS, a large number of static assertions were added inside the implementation. This ensures a feedback to the programmer directly during compile time. In particular, BOOST (Boost Unit Test Suite, 2012) (C++ source libraries) concept checks are used, which define and test type requirements during compile time. Furthermore, static code checkers are used to assess the code quality. The tools CCCC (CCCC - C and C++ Code Counter, 2013), Cppcheck (Marjamäki, 2013), and CppLint (Google, 2013) are used for the ARTIS project. Also, Doxygen (van Heesch, 2013) warnings are used to ensure code quality by identifying missing comments.

Dynamic analysis refers to running the executable, or parts of it to perform tests. This means unit tests, but also coverage and memory leak tests. In particular, a lot of effort

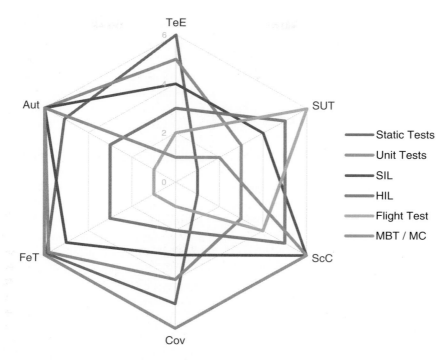

Figure 8. Schematic of *test dimensions*: Example of different testing techniques applied to the ARTIS rotorcraft. (Copyright © 2015 by DLR Institute of Flight Systems.)

was put into creating an extensive set of unit tests. Units are being tested on various integration levels. As a last step, high-level functionality is tested, for example mission manager, mission planner, behavior sequences, the corresponding sequence controller, the flight mechanics, flight controller, and roadmap features.

Simulations can be categorized into Software-in-the-Loop (SIL) and Hardware-in-the-Loop (HIL) simulations. SIL simulations are used to test the software integration level, that is, the main parts of the software system as a whole, including the interfaces to other software and hardware components. For this purpose, a simplified software vehicle simulator has been written. The simulation can act in a variable simulation time, where calls are executed synchronously. This enables the MiPlEx framework to perform fast functional tests. Asynchronous execution of tests is possible to test real-time behavior. For the next test step, HIL tests are run on the actual flight hardware. This test setup additionally embeds the target platform, actuators, and sensor fusion into the tests. On this level, system integration is tested up to the actual hardware. The execution is done asynchronously to enable real-time behavior. HIL tests can be used to assess computing time and real-time performance. More details on the simulation framework, which is able to integrate different kinds of vehicles in various scenarios, can be found in Ref. Dauer and Lorenz (2013). Finally, flight tests in a controlled

environment validate the system. The flight test represents a system test in the test taxonomy.

7.2   Test dimensions

In this section we discuss test characteristics to provide a comparative view on the different testing methodologies that have different strengths and drawbacks. Static tests, unit tests, scenario tests, SIL/HIL, flight tests, and model-based tests and verification techniques are compared qualitatively by using an easy-to-use visual approach. Within a good test strategy, the different aspects of test methodologies should be considered. The testing methods are all integrated to identify strengths and weaknesses of the different approaches.

Figure 8 shows a visualization of the different ARTIS test dimensions as a star diagram. The axes are size of the specific system under test (SUT), test effort (TeE), scenario complexity (ScC), state space coverage (Cov), feedback time (Fet), and the level of automation (Aut). The different test methodologies are for Figure 8a unit tests, SIL/HIL, and flight tests. Figure 8b features static tests, model-based tests, and model-checking methods (MBT/MC) and again SIL for a better comparison of the dimension values. Note that a high test effort results in a low score value for test effort in the diagram. Low scores are in the inner region, high scores are in the outer region of the star diagram. This means, the larger the area covered by a test methodology, the better. Each testing methodology is given a unique color. The diagram has been split into two sections, to better view the values. This diagram shows that testing greater artifacts for the SUT as well as increasing the scenario complexity is difficult and always increases test effort (and thus lowers the test effort score). Nonetheless, the SIL/HIL and flight tests cannot be replaced by other test methods, because they offer an increasing degree of fidelity (at increasing costs) that is not matched otherwise. The HIL simulations are necessary test steps before going into flight test, but they require manual integration overhead.

Notice that for ARTIS development, the SIL simulations are extremely useful, because they cover relatively large artifacts of the SUT at an excellent feedback time and can handle a wide variety of scenarios. Furthermore, these high-level tests are automated and the mentioned benchmarks can be utilized to assess the simulations, which gives invaluable feedback for development. The used static tests are extremely cheap, compared to other methodologies. Combined with the possible level of automation, static tests should be used, even if the achieved test complexity is low. Furthermore, studies show that the benefit of model-based techniques (Utting, 2005; Utting and Legeard, 2006) and model-checking (Clarke, Grumberg, and Peled, 2000) also comes from the necessity to create a formal model and thus find errors in specifications or requirements. Additionally, such techniques promise a systematic coverage of the state space and thus seem to be an interesting extension to the usual unit and scenario tests.

In the provided visualization, ranges for specific values cannot be visualized for different characteristics of a flight test. Also, the specific results of this analysis may differ from project to project. For example, unit tests can achieve a different degree of scenario complexity with different efforts. However, in comparison, the conclusion usually remains the same. System or flight level test, integration tests, and unit tests are all necessary test categories. The test dimensions show the different properties to understand the use of model-based approaches. In general, test automation has shown to be of utter importance in the ARTIS testing strategy. Automatic tests and tests with earlier feedback information are often utilized in addition or, whenever possible, as a substitution for manual tests. In the overall view, the additional effort is usually worth the earlier feedback and thus lowers effort for development. Furthermore, all the identified test dimensions are covered by the presented strategy. Still, several improvements are possible for the current test strategy.

8   FAIL-SAFE SYSTEMS RUNTIME MANAGEMENT

As mentioned earlier, some systems can be created that are intrinsically fail-safe. This means that a safe state will be entered as soon as a system failure has been detected. In the case of software, the system has to be monitored to check whether the system has failed or not. In earlier sections, we mentioned the importance of runtime monitoring for fail-safe systems. However, the monitoring approach is a work in progress for ARTIS, although some concepts have been shown (Dauer, Goormann, and Torens, 2014).

Generally, if the monitor assesses a system fault, then a mitigation action must be triggered. The monitor could either initiate an action by itself, reset, and reactivate the faulty system or deactivate it and activate a backup system. The backup system could have reduced capabilities but would just be able to maintain or ensure a safe state. As a result, the monitoring of systems and subsystems is the key approach to design fail-safe software systems. A very important property of a monitoring system is that of independence from the system under observation, so no problem that addresses the system under observation may also obscure the monitoring of this system. Otherwise, the monitoring would not add structural redundancy, and thus not make the system safer. The work from Refs Schumann et al. (2015); Reinbacher, Rozier, and Schumann (2014) introduces three additional principles that should be achieved by a runtime monitor: Unobtrusiveness, which means no influence or alteration of the actual runtime behavior of the system, should be imposed by the monitor, so that the system does not need to be recertified with the addition or alteration of the monitoring, responsiveness, meaning the monitor must be able to continuously supervise the system and deviations must be noticed within a defined interval, and finally realizability, which ensures the possibility to use the monitoring as a kind of plug and play system that can be easily adapted, changed, and introduced into a new system.

One important benefit of runtime monitoring is that it is possible to capture and formalize observed properties into temporal logic. This gives engineers a powerful tool for verification and debugging. Furthermore, it is possible to make assumptions of system design explicit, therefore such invalid assumptions can be found easily. Not every invalid assumption will cause a failure, because it can be masked (dormant) until specific additional conditions are fulfilled. Without monitoring, such inconsistencies would not be noticed. Since these formal requirements can be synthesized into monitors, it is possible to directly verify the implementation against these properties. Our goals are similar to the above-mentioned work, but our approach is different that we explicitly want to go on a high level for supervision of these properties. There are several categories of monitoring properties we consider:

1. Safety properties
2. Functional requirements
3. Performance (nonfunctional) requirements, including anomalies in performance
4. Situational awareness properties, to check for anomalies in environmental conditions

Safety properties can be inferred from the system safety assessment. Functional and nonfunctional properties can be reused from the requirements analysis phase. If the requirements have been formalized for consistency checks with the model-checking tool, as it was proposed in Section 5, the formalized requirements can be directly reused. This reduces errors from implementing complicated monitoring properties and ensures conformance to the initial requirements. Situational awareness can further be categorized into three levels. The first level is perception, a comparison of input data with threshold values. For the second level, input data are combined with additional context. Finally, in level 3 projection, the data are projected into the future and warnings about upcoming hazards can be issued. This can be seen as a countermeasure to spatial separation of the pilot from the unmanned aircraft.

As a result, the MiPlEx software component takes over active tasks that an onboard pilot should normally perform. To complement this, the runtime monitor takes over the supervisory tasks of the onboard pilot. As such, the runtime monitor not only supervises mere functionality but also acts as an intelligent component that ensures high-level decisions and actions do not cause a catastrophic situation and are consistent with known environmental conditions.

9 CONCLUSION AND OUTLOOK

This chapter discussed the various challenges in the domain of unmanned aircraft to design and create fail-safe systems. Because of unfinished regulations, the near endless multitude of aircraft sizes, and the missing pilot onboard, the risk to harm people must be assessed different for unmanned aircraft. To determine if a system is safe, acceptable risks must be established. Without the above data, reasoning about safety of unmanned aircraft remains problematic. This situation will persist until corresponding regulations are finished.

Furthermore, existing approaches to make software systems fail-safe have been addressed. Two approaches complement each other: safety accomplished at design time through safe software design and thorough verification activities on one hand and detection of failures at runtime to enable appropriate corresponding mitigation strategies on the other. Software failures are systematic failures. Therefore, verification activities play a crucial role to make a system safe. The use of formal methods is important to check system requirements and design. But these techniques are not yet common, even for designing safety-critical systems. Even then, formal methods are only one aspect of a holistic verification strategy, that combines complementary techniques. However, while comprehensive verification should always be the final goal, not all software failures can be found by verification activities, and not all environmental constraints can be anticipated. Therefore, runtime monitoring is a key aspect to make software-intense systems fail-safe.

The ARTIS UAS research platform enables the development and test of autonomous software functions. We have used this platform to demonstrate the described techniques within our best practice approach for designing fail-safe systems. Our approach uses three layers of safety; The first layer is checking requirements and the design using formal modeling of requirements and model-checking techniques. To eliminate software errors, a holistic testing and evaluation approach for our mission planning and execution (MiPlEx) framework was presented as the second layer. The different testing dimensions were described with their different properties, such as cost and test coverage. Furthermore, the way they interact and complement each other was analyzed by decoupling the different test dimensions and by comparing results of this classification. With this argumentation, there is high confidence for the MiPlEx software to be correct, because there are a large number of tests run on different layers before progressing to the final flight testing. For remaining errors and environmental conditions that could not be anticipated, a runtime monitoring approach is used as a third layer. The monitor supervises high-level behavior, safety, performance, and situational awareness. Monitoring adds structural redundancy to the system without actually having to implement the function twice, because backup functionality can have reduced capabilities. However, monitoring must not be seen as simplified implementation, instead it adds a formal perspective to the system that is focused on safety. Therefore, monitoring can be a relatively inexpensive way to achieve fail-safe system design with the added benefit of a formal tool for debugging. The benefit is even higher, when formal properties can be reused from earlier phases of checking requirements consistency. Future work will focus on refining the approach and implementation of our monitoring framework. As a result, we presented a theoretically founded best practice approach for holistic fail-safe high-level software of a UAS.

REFERENCES

Adolf, F. and Thielecke, F. (2007) A sequence control system for onboard mission management of an unmanned helicopter, in *AIAA Infotech @ Aerospace*, AIAA SciTech, No. AIAA 2007-2769, American Institute of Aeronautics and Astronautics.

Baier, C. and Katoen, J.-P. (2008) *Principles of Model Checking*, vol. 26202649, MIT Press, Cambridge.

Bonasso, R.P., Kortenkamp, D., Miller, D.P., and Slack, M. (1996) Experiences with an architecture for intelligent, reactive agents, in *Intelligent Agents II Agent Theories, Architectures, and Languages: IJCAI'95 Workshop (ATAL) Montréal, Canada, August 19-20, 1995 Proceedings*, Springer, Berlin, pp. 187–202.

Boost Unit Test Suite (2012) www.boost.org/doc/libs/1 50 0/libs/test/ (last accessed June 24, 2013).

Brooks, R.S. (1990) A robust layered control system for a mobile robot. *IEEE J. Robotics Automat.*, **2**(1), 204–213.

CCCC - C and C++ Code Counter. Web Resource. http://cccc.sourceforge.net/ (last accessed June 24, 2013).

Cherepinsky, I. and Pinto, A. (2012) Stringent safety design and verification methods for VTOL unmanned aerial vehicles, in *American Helicopter Society 68th Annual Forum*.

Cimatti, A., Clarke, E., Giunchiglia, E., Giunchiglia, F., Pistore, M., Roveri, M., Sebastiani, R., and Tacchella, A. (2002) Nusmv 2: an opensource tool for symbolic model checking, in *Computer Aided Verification*, Springer, pp. 359–364.

Clarke, E., Grumberg, O., and Peled, D. (2000) *Model Checking*, MIT Press.

Dauer, J.C., Goormann, L., and Torens, C. (2014) Steps towards scalable and modularized flight software for unmanned aircraft systems. *Int. J. Adv. Robot. Syst.* doi: 10.5772/58363.

Dauer, J.C. and Lorenz, S. (2013) Modular simulation framework for unmanned aircraft systems, in *AIAA Modeling and Simulation Technologies Conference*, Boston, MA, August 19–22, 2013.

European Aviation Safety Agency (2015a) Advance Notice of Proposed Amendment 2015-10: *Introduction of a Regulatory Framework for the Operation of Drones*.

European Aviation Safety Agency (2015b) *Concept of Operations for Drones: A Risk Based Approach to Regulation of Unmanned Aircraft*.

Flanagan, C., Toal, D., Jones, C. and Strunz, B. (1995) Subsumption architecture for the control of robots, in *Proceedings Polymodel-16*, Sunderland, UK, pp. 150–158.

Geffroy, J.-C. and Motet, G. (2002) Fail-safe systems, in *Design of Dependable Computing Systems*, Springer, The Netherlands, 451–468.

Google (2013) *Google-Styleguide: style guides for Google-originated open-source projects*. Web Resource. https://code.google.com/p/google-styleguide/ (last accessed June 24, 2013).

Jhumka, A., Hiller, M. and Suri, N. (2004) An approach for designing and assessing detectors for dependable component-based systems, in *8th IEEE International Symposium on High-Assurance Systems Engineering (HASE '04)*, March 25-26, 2004, Tampa, FL, pp. 69–78.

Jhumka, A. and Suri, N. (2005) Designing efficient fail-safe multito- lerant systems. *Formal Techniques for Networked and Distri-buted Systems (FORTE '05)* (ed. F., Wang), vol. 3731 of Lecture Notes in Computer Science, Springer, Berlin, pp. 428–442.

Knight, J.C. (2002) Safety-critical systems: challenges and directions, in *Proceedings of the 24th International Conference on Software Engineering*, IEEE, pp. 547–550.

Leveson, N.G. (1986) Software safety: why, what, and how. *ACM Comput. Surveys*, **18**, 125–163.

Leveson, N.G. (1995) *Safeware: System Safety and Computers*, ACM.

Leveson, N. and Harvey, P. (1983) Analyzing software safety. *IEEE Trans. Softw. Eng.*, **SE-9**, 569–579.

Lorenz, S. and Adolf, F.M. (2011) A decoupled approach for trajectory generation for an unmanned rotorcraft, in *Advances in Aerospace Guidance, Navigation and Control* (eds. F., Holzapfel and S., Theil), Springer, Berlin, pp. 3–14.

Lutz, R.R. (2000) Software engineering for safety: a roadmap, in *Proceedings of the Conference on the Future of Software Engineering (ICSE '00)*, New York, NY, ACM, pp. 213–226.

Marjamäki, D. (2013) *Cppcheck: A Tool for Static C/C++ Code Analysis*. http://cppcheck.sourceforge.net/ (last accessed June 24, 2013).

Pohl, K. (2010) *Requirements Engineering: Fundamentals, Principles, and Techniques*, Springer.

Radio Technical Commission for Aeronautics (RTCA) (2011) DO-178C/ED-12C Software Considerations in Airborne Systems and Equipment Certification, RTCA, Washington, DC.

Radio Technical Commission for Aeronautics (RTCA) (2011) DO-333/ED-216 Formal Methods Supplement to DO-178C and DO-278A, Washington, DC.

Reinbacher, T. Rozier, K.Y. and Schumann, J. (2014) Temporal-logic based runtime observer pairs for system health management of real-time systems, in *Proceedings of the 20th International Conference on Tools and Algorithms for the Construction and Analysis of Systems (TACAS)*, vol. 8413 of Lecture Notes in Computer Science (LNCS), Springer, pp. 357–372.

Rierson, L. (2013) *Developing Safety-Critical Software: A Practical Guide for Aviation Software and DO-178c Compliance*, CRC Press.

Schumann, J., Rozier, K.Y., Reinbacher, T., Mengshoel, O.J., Mbaya, T. and Ippolito, C. (2015) Towards realtime, on-board, hardware-supported sensor and software health management for unmanned aerial systems. *Int. J. Prognostics Health Manage*, **6**(1), 27.

The Engineering Society for Advancing Mobility in Land, Sea, Air, and Space (1996) *Guidelines and Methods for Conducting the Safety Assessment Process on Civil Airborne Systems and Equipment ARP4761*, SAE International.

The Engineering Society for Advancing Mobility in Land, Sea, Air, and Space (2010) *Guidelines for Development of Civil Aircraft and Systems 4754A*, SAE International.

Torens, C. and Adolf, F. (2013) Software verification considerations for the ARTIS unmanned rotorcraft, in *51st AIAA*

Aerospace Sciences Meeting Including the New Horizons Forum and Aerospace Exposition, American Institute of Aeronautics and Astronautics, January.

Torens, C. and Adolf, F. (2015) Using formal requirements and model-checking for verification and validation of an unmanned rotorcraft, in *AIAA Infotech @ Aerospace*, AIAA SciTech, No. AIAA 2015-1645, American Institute of Aeronautics and Astronautics, January.

Utting, M. (2005) Position Paper: model-based testing. *Verified Software: Theories, Tools, Experiments*, vol. **2**, ETH, Zürich, IFIP WG.

Utting, M. and Legeard, B. (2006) *Practical Model-Based Testing: A Tools Approach*, Morgan Kaufmann Publishers Inc., San Francisco, CA.

van Heesch, D. (2013) Doxygen: Generate Documentation from Source Code. Web Resource. http://www.stack.nl/dimitri/doxygen/index.html (last accessed June 24, 2013).

Wu, W. and Kelly, T. (2004) Safety tactics for software architecture design, in *Proceedings of the 28th Annual International Computer Software and Applications Conference 2004 (COMPSAC'04)*, vol. 1, pp. 368–375.

Chapter 46
UAS Reliability and Risk Analysis

Christopher W. Lum and Dai A. Tsukada

William E. Boeing Department of Aeronautics & Astronautics, University of Washington, Seattle, WA, USA

1 INTRODUCTION

The past several decades have seen significant advances in UAS technology (see Unmanned Aerial Vehicles (UAVs)). In the last 10 years, there has been a corresponding increase in their use by military organizations around the world. Recently, the utilization of these technologies has begun to grow beyond the military domain with an increased interest in civilian and commercial applications. Recent market analysis shows evidence for exponential growth and utilization of UAS in the future (US Department of Transportation, 2013). This increased usage will result in complex interactions between UAS and general aviation and commercial flights. UAS missions must be able to achieve an acceptable level of safety and reliability when accessing the National Airspace System (NAS). Reliable and realistic methods of evaluating risk must be developed in order to allow further development and use of UAS while ensuring public safety. After examining several risk factors, this chapter will present a simplified model to assess and predict the risk associated with a given UAS operation.

Several efforts have been made in the past to analyze the risk of a UAS operation. One of the first efforts involved modeling midair collisions of manned aircraft using random collision theory and comparing results to historical data (Anno, 1982). Similar work was performed by McGeer with extensions involving regulatory policy and economics of these systems (McGeer, 1994). More recently, the focus has shifted toward integrating UAS into the NAS. Weibel and Hansman performed risk analysis of UAS operation in the NAS by combining the severity of the hazard and its likelihood of occurrence (Weibel and Hansman, 2005). A risk-based approach to analyze the safety of UAS operations was examined at North Carolina State University in the development of the System-Level Airworthiness Tool (SLAT) (Burke, 2010). Groups such as Clothier *et al.* have developed models such as the barrier-bow-tie model to identify and manage risk (Clothier *et al.*, 2015a, 2015b, 2015c; Williams *et al.*, 2014). A simplified risk assessment framework and tool was developed to enable UAS manufacturers and operators to quantitatively evaluate risk of a mission in terms of human safety (Lum and Waggoner, 2011; Lum *et al.*, 2011). In the previous works, the authors focus on the expected number of fatalities per flight hour as the primary safety metric.

2 MOTIVATION FOR RISK ANALYSIS

It is generally perceived that there are a number of obstacles to the full integration of UAS into the NAS. The Federal Aviation

Unmanned Aircraft Systems. Edited by Ella Atkins, Aníbal Ollero, Antonios Tsourdos, Richard Blockley and Wei Shyy.
© 2016 John Wiley & Sons, Ltd. ISBN: 978-1-118-86645-0.

Administration (FAA) has identified, in the UAS–NAS integration roadmap, that the most pressing technological challenges are "sense-and-avoid" (SAA) capability and command-and-control (C2) link reliability (Federal Aviation Administartion, 2013). Since the operator of a UAS is not able to provide the "see-and-avoid" ability of which an onboard pilot is capable, the development of reliable SAA technology is essential for UAS to gain full airspace access (Anon, 2011).

Although the most UA possess low-level autonomy, a reliable communication link between the UA and the ground control station is often necessary for high-level control such as navigation, tasking, and air-traffic control. In addition to improving the C2 link reliability, regulations and protocols must be established to ensure safe and predictable behavior in the case of a lost link due to situations such as equipment failure or malicious jamming.

Thoroughly addressing these issues so that UAS may be routinely and safely incorporated throughout the NAS will take years. In the meantime, standards and tools need to be developed that will "enable the widest range of activity that can be safely conducted within the shortest rulemaking timeframe" (ASTM F38 Committee).

The risk assessment tool presented here aims to provide UAS operators and airspace regulators with a simplified and trustworthy method of evaluating the safety of proposed UAS operations. The model in this framework is first and foremost concerned with estimating the potential risk to human safety both aboard other aircraft and on the ground, and does not take into account the potentially significant economic risk associated with a mission.

3 RISK FACTORS

There are numerous ways in which a UAS may fail and many incidents are the result of multiple factors. These causes may be grouped into several categories such as operator error, improper maintenance, equipment failure, weather, and bird strike. Understanding each risk factor and its ramification is necessary to conduct an accurate risk assessment. Having a thorough understanding of risk factors also helps to improve the reliability of the UAS, as it allows operators and regulators to address each factor individually, and understand how failure rates might be lowered over time. In the following section, several risk factors specific to certain operational time and flight phase are discussed.

3.1 System failure

System failure is a broad term that may encompass several factors. A hardware or mechanical failure (including engine failure, loss of link, or damage to control surface) could lead to unintended or abnormal system behavior. Hazards could also arise from software failures such as a flight computer failure or a ground control station failure. Given the wide taxonomy of UAS, enumerating and evaluating each individual failure is not practical and therefore we adopt the use of the general system failure designation. Referring to the historical data is one approach to estimate system failure rate of the UAS. For example, the Air Force Class A Aerospace Mishap records, maintained by the Judge Advocate General's office, are a useful resource for tracking the distribution of mishap causes over time for a particular aircraft system (Accident Investigation Board, n.d.). Another way of estimating failure rate is to note the failure rate of each subcomponent using reliability analyses such as failure modes, effects, and criticality analysis (FMECA) (US Department of Defense, 1980). These analyses are effective especially when the system is relatively new and no historical data is available.

3.2 Human error

Human error is another major risk factor for UAS operation. This includes inadequate operator response, mission planning error, or improper maintenance of the UAS. A recent study of Predator mishaps conducted by the Air Force Research Laboratory revealed that after system failures that happen in the first several years of operation are addressed and mitigated, the dominant risk factor becomes various human errors (Herz, 2008). These risks can be mitigated by refocusing on the training of the new and current operators.

3.3 Bird strikes

Although the model considered here does not distinguish between mishaps in different phases of flight, it is noteworthy that bird strike is one of the greatest risk for aircrafts during taxi, takeoff, and landing phase, with 80% occurring below 305 m (1000 ft) AGL and 96% occurring below 1542 m (5000 ft) AGL (Dolbeer et al., 2009). For general aviation and commercial flights, these altitudes are only encountered during landing and takeoff. Combining this with the fact that operating areas during these phases of flight are typically near airports or otherwise controlled areas, it is reasonable to assume that bird strikes only pose a nontrivial threat to people onboard aircraft as opposed to those on the ground. This assumption may not be valid for UAS operations given that the majority of these operations are currently below 400 AGL. While the current model neglects bird strikes, this factor should be considered as a potentially nontrivial risk associated with UAS operations in the future.

4   RISK MODEL

The risk to human safety in this model stems from two potential causes: midair collisions and ground collisions. For instance, if a UAS collides with a transient aircraft (e.g., commercial flights, regional jets, and general aviation), it may injure or kill the people onboard the transient aircraft. Both of these vehicles will then create debris that has potential to affect bystanders on the ground. The model aims to quantify the risk to human life by estimating fatalities per flight hour due to these factors.

4.1   Midair collisions

A midair collision is further separated into two categories. The first category models collisions of a UAS with other transient aircraft (denoted as transient collisions) and the second is collision of UAS with other UAS within their same fleet (denoted in-fleet collisions). For both cases, unmitigated collisions rates are modeled using a Maxwell molecule formulation (McQuarrie and Simon, 1997). This theory was similarly applied to air traffic in prior literature (Anno, 1982; McGeer, 1994; Vagners *et al.*, 1999).

4.1.1   Transient collisions

The collision frequency between a single UA and transient air traffic is a product of the transient aircraft density, the combined frontal areas, and the velocity of both the UA and the transient aircraft. We define ρ_O to be the density of transient aircraft per km^3, ϕ_O and ϕ_{ua} to be the frontal area in km^2 of the transient aircraft and the UA, and V_O and V_{ua} as the velocity in $km\ h^{-1}$ of the transient aircraft and the UA. In

order to average the risk of a midair collision over all orientations, the frontal areas of the UA and the transient aircraft are recast as circles of radii $R_{ua} = \sqrt{\phi_{ua}/\pi}$ (km) and $R_O = \sqrt{\phi_O/\pi}$ (km). A collision occurs if the centers of the aircraft are within a distance $R_{ua} + R_O$. The instantaneous collision area is therefore,

$$\phi_{col} = \pi(R_{ua} + R_O)^2 = \phi_{ua} + \phi_O + 2\sqrt{\phi_{ua}\phi_o} \quad (1)$$

We start with a simplification that the transient aircraft are stationary ($V_O = 0$), in which case the volume of collision airspace that the UA sweeps out in a time ΔT is simply $V_{col} = \phi_{col} V_{ua} \Delta T$. The number of collisions is a product of the collision volume and the transient aircraft density. Dividing by the time ΔT gives the expected collision rate for a single UA with stationary transient aircraft.

$$\widehat{F}_{transient} = \rho_o \phi_{col} V_{ua} \quad (2)$$

To correct for the fact that the transient aircraft are not stationary ($V_O \neq 0$), V_{ua} is replaced with a relative velocity. In order to develop conservative model of collisions, we assume that all transient aircraft are flying directly at the UA that gives us the maximum (and conservative) relative velocity of $V_{rel} = V_{ua} + V_O$. Assuming that UA collisions are independent of each other, the total collision rate for the fleet of UA is simply obtained by multiplying Equation 2 by the number of UA in the fleet and replacing V_{ua} with V_{rel}

$$\widehat{F}_{transient} = N_{ua}\rho_o \phi_{col} V_{rel} \quad (3)$$

Collision avoidance capabilities gained from the airspace structure, procedural separation, or SAA technologies are incorporated in the collision model using the parameter ε, the probability that a given aircraft will avoid and imminent collision with another aircraft. This framework is explained in Figure 1. With this framework, the expected collision rate of the UA fleet and transient aircraft with collision avoidance

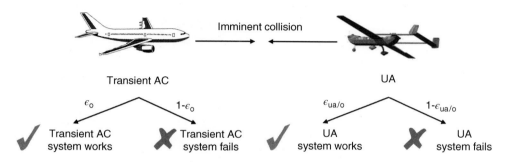

Figure 1. Collision avoidance framework. Both transient aircraft and UA collision avoidance must fail in order for a midair collision to occur.

taken into account is given by

$$F_{\text{transient}} = N_{\text{ua}}\rho_{\text{o}}\phi_{\text{col}}V_{\text{rel}}(1 - \varepsilon_{\text{ua/o}})(1 - \varepsilon_{\text{o}}) \qquad (4)$$

4.1.2   In-fleet collisions

A similar analysis can be performed to estimate collision rates of the second category, collision between UA within their own fleet. This involves using the previous equations with substitutions of $V_{\text{O}} = V_{\text{ua}}$, $\phi_{\text{O}} = \phi_{\text{ua}}$, and $\rho_{\text{O}} = \rho_{\text{ua}}$. Previously, the density of transient aircraft was assumed to be uniformly spread over the operating volume. With in-fleet UA operations, there may be missions where the fleet of UAS are in close proximity to each other or otherwise spaced such that a uniform density over the entire operating volume is not a reasonable assumption. In order to take these factors into account, once the mission has been selected, the appropriate volumes are calculated using $\eta_{\text{ops}} = A_{\text{ops}}(z_{\text{max}} - z_{\text{min}})$ and $\eta_{\text{fleet}} = A_{\text{ops,fleet}}(z_{\text{max,fleet}} - z_{\text{min,fleet}})$. The expected in-fleet collision rate is therefore

$$F_{\text{fleet}} = N_{\text{ua}}\rho_{\text{ua}}(4\phi_{\text{ua}})(V_{\text{rel}})(1 - \varepsilon_{\text{ua/ua}})^2 \qquad (5)$$

The total number of midair collisions, α (both between UA and transient aircraft and between UA and other UA), during a mission is simply the sum of $F_{\text{transient}}$ and F_{fleet} multiplied by the mission duration, M_{L}.

$$\alpha = M_{\text{L}}(F_{\text{transient}} + F_{\text{fleet}}) \qquad (6)$$

4.2   Ground collisions

Midair collisions are only a portion of the analysis. After midair collision or general system failure occurs, a risk to

pedestrians or bystanders on the ground still exists as the UA will fall to the ground and either strike a person or a building (the two scenarios considered in this model). The risk of ground collisions from crashes due to systems failures is found assuming that upon failure, the UA glides to the ground at maximum L/D (worst-case scenario) with glide angle γ. The associated geometry is shown in Figure 2.

The risk of ground collisions from midair collisions assumes upon midair collision, the UA will approach the surface in vertical free fall. The expected number of building and pedestrian strikes is composed of two calculations that take each case (glide and free fall) into account. For example, if the UA has a system failure and glides to the ground at the best glide angle, the collision areas in km^2 that the UA may strike are given by

$$A_{\text{L}_{\text{H}_{\text{p}}}} = (w_{\text{ua}} + 2R_{\text{p}})\left(L_{\text{ua}} + \frac{H_{\text{p}}}{\tan\gamma} + 2R_{\text{p}}\right) \qquad (7)$$

$$A_{\text{L}_{\text{H}_{\text{b}}}} = (w_{\text{ua}} + w_{\text{b}})\left(L_{\text{ua}} + \frac{H_{\text{b}}}{\tan\gamma} + w_{\text{b}}\right) \qquad (8)$$

In a similar fashion, if the UA sustains a midair collision, it is assumed that it will fall vertically to the ground. In this case, the collision areas in km^2 become

$$A_{\text{L}_{\text{V}_{\text{p}}}} = \pi\left(\frac{\max(w_{\text{ua}}, L_{\text{ua}})}{2} + R_{\text{p}}\right)^2 \qquad (9)$$

$$A_{\text{L}_{\text{V}_{\text{b}}}} = \pi\left(\frac{\max(w_{\text{ua}}, L_{\text{ua}})}{2} + \frac{w_{\text{p}}}{2}\right)^2 \qquad (10)$$

In Equations 7–10, w_{b} is the average building width in km (defined as $w_{\text{b}} = \sqrt{A_{\text{b}}}$, where A_{b} is the average building size

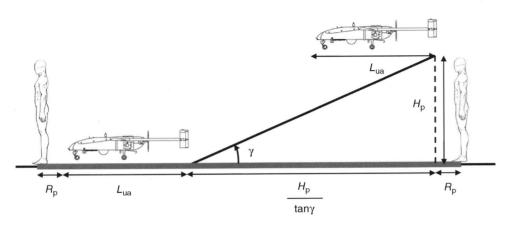

Figure 2. Geometry showing affected distance covered by UA during a horizontal, gliding crash. The total affected area is this distance multiplied by the wingspan of UA plus $2R_{\text{p}}$.

in km^2), H_b is the average building height in km, R_p is the radius of a person, H_p is the height of a person, γ is the UA glide angle without power, w_{ua} is the UA wingspan in km, and L_{ua} is the UA length in km.

The numbers of aircraft crashes are a function of both $F_{transient}$ and F_{fleet}. Since a single midair collision affects two aircraft, the rate of aircraft crashes (and subsequent ground strikes) per hour becomes

$$C_{midair} = 2(F_{transient} + F_{fleet}) \qquad (11)$$

The number of pedestrian and building strikes per hour is a combination of system failures and midair collisions.

$$F_{ped} = F_{ped,p} + F_{ped,midair} = N_{ua}\lambda\sigma_p A_{L_{H_p}} + C_{midair}\sigma_p A_{L_{V_p}} \quad (12)$$

$$F_{bldg} = F_{bldg,p} + F_{bldg,midair} = N_{ua}\lambda\sigma_b A_{L_{H_b}} + C_{midair}\sigma_b A_{L_{V_b}} \quad (13)$$

In these expressions, λ is the UAS midair failure rate per hour from all sources for a single UA. This can be estimated by examining risk factors associated with UAS operation as described in previous section, or cited by manufacturers as the mean time between failures. σ_p and σ_b are the building and pedestrian densities (respectively) per km^2.

A successful risk assessment must communicate the results in a way that provides the user with a tangible sense for the risk involved. The most important result is the number of fatalities expected. Using previously obtained parameters, the expected number of fatalities per hour becomes

$$F_{fat} = F_{fat,p} + F_{fat,midair} \qquad (14)$$

where

$$F_{fat,p} = F_{ped,p}D_{ped} + F_{bldg,p}D_{bldg}$$

and

$$F_{fat,midair} = F_{ped,midair}D_{ped} + F_{bldg,midair}P_O$$

In Equation 14, D_{ped} is the fatality rate for a pedestrian strike. It is defined as the average number of fatalities incurred when a UA strikes a pedestrian and is therefore in range of [0,1]. D_{bldg} is the fatality rate for a building strike (in range of [0,all people in building]). This allows versatility in modeling hard structures where people are more protected versus softer structures such as residential homes. P_O is the average number of passengers on a transient aircraft. The model assumes that a collision between a UA and a transient aircraft causes the death of all passengers aboard the transient aircraft.

The risk model presented here was designed to be a conservative and easily accessible method to estimate the risk to human life incurred from a given UAS operation. The

capabilities of this model can be further extended by adding extra functionalities. For example, one can utilize this risk model to focus on assessing risk of a location specific mission by incorporating the probability distribution function (PDF) for impact near the operating area and the local bystander distribution obtained from census data or satellite imagery (Lum *et al.*, 2011). Although this requires an accurate understanding of the system through high-fidelity simulation or experimental data, the outcome will give a site-specific risk assessment.

5 EXAMPLE CALCULATIONS

This section presents example calculations of risk assessment for two scenarios. The first scenario is a mission that shows a potentially viable operation for UAS. The second scenario illustrates an operation that does not appear to be a good fit for utilizing UAS.

5.1 Scenario 1: Environmental monitoring

Environmental monitoring has been a popular civilian application for UAS in recent years. UAS have been widely used to gather environmental data, assess damage from natural disasters, monitor wildfires, and perform aerial surveys (Lum *et al.*, 2005, 2015; Lum and Vagners, 2009). As an example case, we will consider a team of small UAS taking part in environmental mapping for precision agriculture using multispectral camera over cropland. The risk assessment will be for mapping application, but the process is essentially the same for wildfire detection, search and rescue, or other low-altitude operations.

5.1.1 UAS properties, operating area, and transient aircraft

For this scenario, we assume that the operator uses a Skywalker 1900 airframe with customized flight controller to operate as a UAS. By referencing pictures and diagrams of the Skywalker 1900, seen in Figure 3, the frontal area was estimated. The frontal area approximation is depicted in Figure 4.

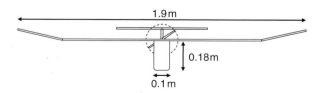

Figure 3. Skywalker 1900 frontal area geometries.

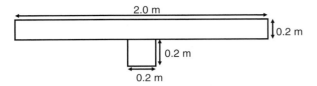

Figure 4. Skywalker 1900 frontal area approximation.

The Skywalker 1900 is dynamically stable and therefore without power, it will continue gliding at an estimated glide angle of 5°. We assume a mission length of 20 min with two aircraft operating simultaneously. In order to simulate the worst-case scenario for system failure, we consider a case where flight computer fails and commands full throttle with a wings-level condition with a fully charged battery. We also assume that after the loss of battery, the aircraft continues gliding. Given that the nominal operational altitude of UAS is 122 m AGL, these give a maximum impact distance (assuming no wind) of 23.6 km. The impact distance geometry is seen in Figure 5. Based on the flight history, the system failure rate was conservatively estimated to be 0.1 per flight hour. Since it is a small aircraft, we assume no collision avoidance capability is available ($\varepsilon_{ua/o} = \varepsilon_{ua/o} = 0$). Collision avoidance capability from transient aircraft is also assumed to be small ($\varepsilon_o = 0.05$) as the Skywalker 1900 is

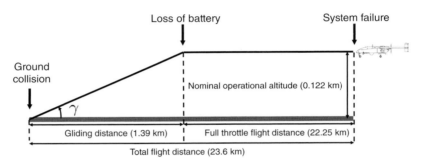

Figure 5. Maximum impact distance geometry for the worst-case scenario.

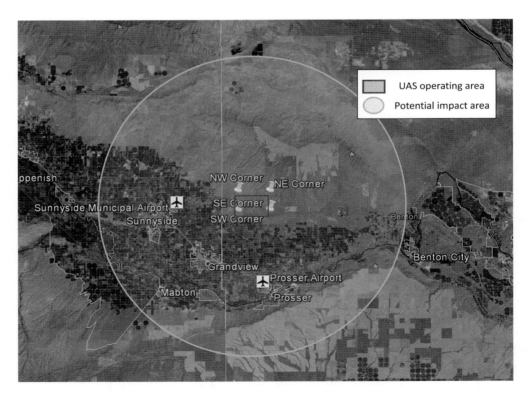

Figure 6. UAS operating area and potential impact area with respect to nearest airports (Sunnyside Municipal Airport and Prosser Airport).

not equipped with any type of transponder and is generally difficult to see.

The example operating area for this mission is near the city of Sunnyside, Washington in the United States. Figure 6 illustrates the operating area as well as potential impact area calculated for the worst-case scenario due to the general system failure.

To determine the population density near the UAS operating area, we will use the data available from the US Census Bureau. The potential impact area has $1749.7\,\text{km}^2$. Population and housing unit density calculated in the area is therefore 19.89 people km^{-2} and 5.93 housing-units km^{-2}, respectively. Estimating that the average person spends not more than 10% of their time outdoors, 10% of the population will be considered pedestrians, and the remaining 90% will be divided among the housing units. This estimation gives a pedestrian density of 1.99 people km^{-2}, and the average housing unit has 3.02 people in every building. The forward-facing areas of the UA (area likely to impact an obstacle) are mostly foam with no sharp surfaces, therefore we assume fatality when UA collides with a pedestrian and building to be 5×10^{-2} and 1×10^{-3} (aircraft penetrates 10% of the time and in those cases fatality is of people inside is 1×10^{-2}), respectively.

The operating area is within the proximity of both Sunnyside Municipal Airport (K1S5) and Prosser Airport (KS40), so there are potential interactions between the UAS and the transient aircraft near the airport. From the airport database, the transient aircraft category was assumed to be general aviation only. Although the operating altitude of the Skywalker 1900 is relatively low compared to the typical airport-approach altitude profile, we still consider a portion (50%) of air traffic to be within the operation altitude of Skywalker 1900. This is further justified by the fact that some of the aircraft in this area are used for agricultural use such as crop dusting and therefore operate at low altitudes. Based on the air traffic information from two airports, air traffic density is estimated to be 3.54×10^{-4} aircraft km^{-3}.

The input parameters necessary for this risk assessment are summarized in Table 1.

5.1.2   Risk assessment results

Using these parameters, the model predicts values shown in Table 2. The first two columns are the values as described in the risk model section. Recall that these are defined as a per hour rate of occurrence. Because the mission persists for M_L hours, the number of occurrences during the mission can be obtained by multiplying by M_L. The resulting per mission rate can then be inverted to obtain the number of

Table 1. Risk assessment inputs for the environmental monitoring scenario.

Parameter	Value
UAS	
V_{ua}	$20.6\,\text{m s}^{-1}$
ϕ_{ua}	$0.44\,\text{m}^2$
w_{ua}	$1.9\,\text{m}$
L_{ua}	$1.2\,\text{m}$
γ	$5°$
λ	0.1
$\varepsilon_{\text{ua/o}}$	0
$\varepsilon_{\text{ua/ua}}$	0
$z_{\text{max,fleet}}$	$122\,\text{m}$
$z_{\text{min,fleet}}$	$0\,\text{m}$
M_L	$0.3\,\text{h}$
Operating area	
N_{ua}	2
A_{ops}	$1749.7\,\text{km}^2$
σ_b	5.93 housing-units km^{-2}
A_b	$200\,\text{m}^2$
H_b	$5\,\text{m}$
D_{bldg}	1×10^{-3} fatalities strike^{-1}
σ_p	1.99 people km^{-2}
H_p	$1.75\,\text{m}$
R_p	$0.25\,\text{m}$
D_{ped}	5×10^{-2} fatalities strike^{-1}
Transient aircraft	
ρ_o	3.54×10^{-4} aircraft km^{-3}
V_o	$222\,\text{m s}^{-1}$
ϕ_o	$80\,\text{m}^2$
P_o	45 people aircraft^{-1}
ε_o	0.05

missions between occurrences. This value is shown in the third column.

The interesting result is the order of magnitude difference in fatalities due to midair collisions (1 every 65 641 years) and general system failures (1 every 1.28×10^6 years). Although the operation expects 1 fatality every 62,433 years, the cause of this is most likely due to a midair (either transient or in-fleet) collision rather than a general system failure. Tracing the cause further back, in the already unlikely situation of a midair collision causing a fatality, this midair collision is mostly a transient aircraft collision instead of an in-fleet collision ($F_{\text{transient}}$ is more than three times higher than F_{fleet}). This stems from the fact that operating area is near airports and potential interaction between transient aircraft and UA is more likely to happen.

These results suggest that more lives can be saved by spending more time and effort into collision avoidance technologies than making UAS more robust and less susceptible to general system failures. Installing a transponder would be an effective solution to mitigate relatively high

Table 2. Risk assessment outputs for the environmental monitoring scenario.

Parameter	Value	Equivalent representation
$F_{transient}$	7.62×10^{-6} collisions h^{-1}	1 transient collision every 1.97×10^5 missions
F_{fleet}	2.45×10^{-6} collisions h^{-1}	1 in-fleet collision every 6.13×10^5 missions
$F_{ped,p}$	2.07×10^{-5} strikes h^{-1}	1 pedestrian strike due to general failure every 72 358 missions
$F_{ped,midair}$	1.81×10^{-10} strikes h^{-1}	1 pedestrian strike due to midair collision every 8.28×10^9 missions
$F_{bldg,p}$	0.0014 strikes h^{-1}	1 building strike due to general failure every 1088 missions
$F_{bldg,midair}$	2.41×10^{-8} strikes h^{-1}	1 building strike due to midair collision every 6.22×10^7 missions
F_{ped}	2.07×10^{-5} strikes h^{-1}	1 pedestrian strike every 72 357 missions
F_{bldg}	0.0014 strikes h^{-1}	1 building strike every 1088 missions
$F_{fat,p}$	1.17×10^{-6} fatalities h^{-1}	1 fatality due to general failure every 1.28×10^6 missions
$F_{fat,midair}$	2.29×10^{-5} fatalities h^{-1}	1 fatality due to midair collisions every 65 641 missions
F_{fat}	2.40×10^{-5} fatalities h^{-1}	1 fatality every 62 433 missions

transient aircraft collision rate as it aids other aircraft in sensing the UA and avoiding collisions.

5.2   Scenario 2: Urban patrol

The previous example represented the type of operation for which UAS is a reasonably safe solution. The following example will demonstrate why many other suggested uses for UAS are expected to be less viable in near term from the safety perspective. While they may have technical merit and excellent potential benefits, risk analysis reveals significant safety issues that will prevent regulatory approval and public acceptance. One such application is the use of UAS to patrol urban environments to provide traffic monitoring, law enforcement surveillance, antiterrorist intelligence, and other services. The idea is that UAS can be used to provide persistent surveillance or monitor over areas of interest such as harbors, airports, and highways. This application has been proposed and explored by some governmental agencies (McCormack, 2008).

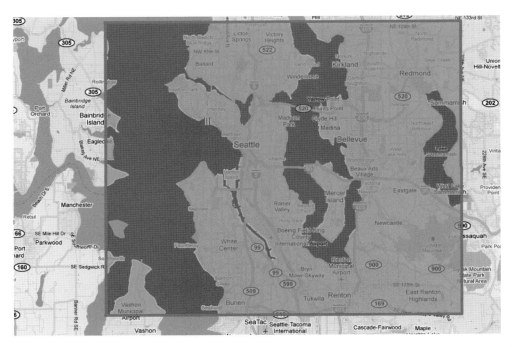

Figure 7. Total operating area for urban patrol UAS team. Risks are separately computed on land-based environments and marine environments.

5.2.1 UAS properties, operating area, and transient aircraft

Several case studies and trials have selected MLB Company's BAT 3 UAV for traffic monitoring projects, so its specifications will be used here in this scenario. For the operating area, the city of Seattle, Washington in the United States will be used as an example city. The total operating area will be broken in two areas: land based environments and marine environments. These areas are shown in Figure 7. To cover this area, it is assumed that four UAS are dedicated to patrolling the total area 24 h a day, 7 days a week, for 1 year. Therefore, the definition of a single mission for this scenario is four vehicles operating continuously for 1 year. The team will be operated in an altitude range of 152 m (500 ft) to 914 m (3000 ft). The population information of the operating area will be based on data available from US Census Bureau. Finally, the air traffic densities are found from Flight Explorer Personal Edition (real-time air traffic tracking database).

The specifications as well other relevant input parameters necessary for this risk assessment are tabulated in Table 3.

5.2.2 Risk assessment results

The results of this risk assessment are summarized in Table 4.

The most striking result of this analysis is the high level of risk this operation incurs. Due to the high density of people and buildings, both pedestrians and building strikes are virtually guaranteed over the course of 1-year mission duration. In total, there are over two fatalities expected each mission. From the analysis, it was revealed that the causes of these fatalities are mostly due to the unreliability of the UAS rather than midair collisions ($F_{fat,p} \approx 9F_{fat,midair}$).

It should be noted that the above scenario assumed a set of nonredundant vehicles that are operated over sensitive areas with no planned emergency procedures. Perhaps

Table 3. UAS specification for urban patrol risk assessment.

Parameter	Value
V_{ua}	26 m s^{-1}
ϕ_{ua}	0.11 m^2
w_{ua}	1.83 m
L_{ua}	1.43 m
γ	3.18°
λ	0.001
$\varepsilon_{ua/o}$	0
$\varepsilon_{ua/ua}$	0.9
$z_{max,fleet}$	914 m (3000 ft)
$z_{min,fleet}$	152 m (500 ft)
$A_{ops,fleet}$	970.9 km^2

unsurprisingly, this results in a predicted high level of fatalities. The results of this analysis perhaps motivate a more structured and planned approach to integration of UAS into operations over populated areas such as this. Carefully selecting flight paths and operating areas along with additional safety checks to increase UAS reliability could mitigate these risks significantly. The conclusion to draw from this analysis is not that unmanned systems are infeasible for this type of mission but rather that UAS with higher reliability and more carefully planned operating procedures should be utilized to bring risk levels down to acceptable levels. This illustrates how this risk assessment tool and framework can be used to identify ways to increase safety for a given UAS mission.

6   CONCLUSION

In this chapter, motivations for risk analysis, risk factors associated with its operation, and finally methodologies to

Table 4. Risk assessment outputs for the urban patrol scenario.

Parameter	Value	Equivalent representation
$F_{transient}$	2.54×10^{-7} collisions h^{-1}	1 transient collision every 448 missions
F_{fleet}	4.01×10^{-7} collisions h^{-1}	1 in-fleet collision every 284 missions
$F_{ped,p}$	8.28×10^{-5} strikes h^{-1}	1 pedestrian strike due to general failure every 1.3 missions
$F_{ped,midair}$	5.88×10^{-10} strikes h^{-1}	1 pedestrian strike due to midair collision every 193 926 missions
$F_{bldg,p}$	0.01 strikes h^{-1}	1 building strike due to general failure every 0.01 missions
$F_{bldg,midair}$	9.62×10^{-8} strikes hr^{-1}	1 building strike due to midair collision every 1185 missions
F_{ped}	8.28×10^{-5} strikes h^{-1}	1 pedestrian strike every 1.4 missions
F_{bldg}	0.01 strikes h^{-1}	1 building strike every 0.01 missions
$F_{fat,p}$	0.0002 fatalities h^{-1}	1 fatality due to general failure every 0.43 missions
$F_{fat,midair}$	2.97×10^{-5} fatalities h^{-1}	1 fatality due to midair collisions every 3.8 missions
F_{fat}	0.0003 fatalities h^{-1}	1 fatality every 0.4 missions

quantitatively assess the potential risk to the public were presented. Although it is difficult to predict the exact form of regulations in the future, it is safe to say that the primary goal of those regulations will always be to ensure the safety of the public.

Development of effective SAA and C2 technologies are critical to ensure the safe interaction between manned aviation and UAS. Collision avoidance or separation systems such as TCAS and ADS-B may also be required for certain types of UAS operation. As seen in example scenarios, incorporating these elements could serve to ameliorate concerns about manned and unmanned aircraft coexisting in shared airspace as they can lower the risks for potential midair collisions and subsequent fallout during the operation.

Finally, from a certification and engineering standpoint, to ensure safe integration of UAS in the NAS, a risk analysis of the critical hazards such as midair collisions and ground impacts must be considered. The risk assessment framework presented in this chapter is designed to be a conservative and easily accessible method to estimate the risk to human life incurred from a given UAS operation. Although, the model presented here has several limitations (such as assuming the use of single type of UAS for the entire operation) the model can be given even greater flexibility by adding functionalities to accommodate missions that are more complex. A higher fidelity risk analysis should include the fact that aircraft are typically following predefined and carefully selected flight paths to mitigate some of the risk. In addition, for flights beyond 5 nmi from an airport, it is unlikely that midair collisions will be a factor due to low traffic density. The current risk model conservatively assumes a uniform density distribution and spatial flight spacing. In any situation, it is important for an operator or regulator to develop and use an appropriate model depending on the objectives and requirements for the risk analysis.

NOTATION AND NOMENCLATURE

A_b	average building area
ADS–B	Automatic Dependent Surveillance – Broadcast
$A_{L_{H_b}}, A_{L_{H_p}}$	collision area for buildings and pedestrians in a horizontal crash (due to system failure)
$A_{L_{V_b}}, A_{L_{V_p}}$	collision area for buildings and pedestrians in a vertical crash (due to midair collision)
A_{ops}	operating area
C_{midair}	rate of aircraft crashes due to midair (transient and in-fleet) collisions
C2	command and control
D_{bldg}	expected number of fatalities when a UA collides with a building
D_{ped}	expected number of fatalities when a UA collides with a pedestrian

FAA	Federal Aviation Administration
F_{fat}	fatalities per flight hour
$F_{fat,p}$	fatalities due to system failures
$F_{fat,midair}$	fatalities due to midair collisions
F_{ped}	collision rate of UA fleet with pedestrian per hour
$F_{bldg,midair}$	collision rate with buildings due to midair collision
$F_{bldg,p}$	collision rate with buildings due to system failure
$F_{ped,midair}$	collision rate with pedestrians due to midair collision
$F_{ped,p}$	collision rate with pedestrians due to system failure
F_{bldg}	collision rate of UA fleet with buildings per hour
F_{fleet}	collision rate of UA fleet of other UA within fleet per flight hour
$\tilde{F}_{transient}$	collision rate of a single UA w/o avoidance & stationary transient aircraft
$\tilde{F}_{transient}$	collision rate of UA w/o avoidance & moving transient aircraft
$F_{transient}$	collision rate of UA with transient aircraft per hour
H_b	average building height
H_p	average pedestrian height
L_{ua}	length of UA
M_L	mission length
NAS	National Airspace System
N_{ua}	number of UA in fleet
P_O	average number of passengers on a transient aircraft
R_p	radius of a pedestrian
SAA	sense and avoid
TCAS	Traffic Alert and Collision Avoidance System
UAS	unmanned aircraft system
w_b	average width of buildings
w_{ua}	wingspan of UA
z_{max}, z_{min}	maximum and minimum altitude of operating area
α	number of midair collisions predicted by the risk model
γ	glide angle of UA
ρ_O, ρ_{ua}	density of transient aircraft and UA, respectively
ϕ_{col}	Instantaneous collision area
ϕ_O, ϕ_{ua}	frontal area of a transient aircraft and UA
ε_O	ability of transient aircraft to avoid collisions with UA
$\varepsilon_{ua/O}$	ability of UA to avoid collisions with transient aircraft
$\varepsilon_{ua/ua}$	ability of UA to avoid collisions with other UA in fleet
η_{ops}	volume of entire operating space of mission
η_{fleet}	volume of only operating space where UA fleet exists
λ	UAS midair system failure rate
σ_b, σ_p	buildings and pedestrian densities

REFERENCES

Accident Investigation Board (n.d.) *United States Air Force Class A Aerospace Mishaps*, SL-1 Accident Investigation Board.

Anno, J. (1982) Estimate of human control over mid-air collisions. *J. Aircr.*, **19**, 86–88.

Anonymous (2011) *Code of Federal Regulations, Title 14 Aeronautics and Space, Part 91 General Operating and Flight Rules*, US Government, Washington DC.

Burke, D. (2010) *System Level Airworthiness Tool: A Comprehensive Approach to Small Unmanned Aircraft System Airworthiness*, North Carolina State University.

Clothier, R.A., Williams, B.P., Coyne, J., Wade, M., and Washington, A. (2015) *Challenges to the Development of an Airworthiness Regulatory Framework for Unmanned Aircraft Systems*, Australian Aerospace Congress, Melbourne.

Clothier, R.A., Williams, B.P., and Fulton, N. L. (2015) Structuring the safety case for unmanned aircraft system operations in non-segregated airspace. *Saf. Sci.*, **79**, 213–228.

Clothier, R., Williams, B., and Washington, A. (2015) *Development of a Template Safety Case for Unmanned Aircraft Operations Over Populous Areas*, SAE International,

Dolbeer, R., Wright, S., Weller, J., and Begier, M. (2009) *Wildlife Strikes to Civil Aircraft in the United States 1990–2008*, Animal and Plant Health Inspection Service and Federal Aviation Administration.

Federal Aviation Administartion (2013) *FAA's Roadmap for Integration of Civil Unmanned Aircraft Systems (UAS) in the National Airspace System (NAS)*, Federal Aviation Administartion, Washington, DC.

Herz, R. (2008) *Assessing the Influence of Human Factors and Experience on Predator Mishaps*, Northcentral University.

Lum, C.W., Carpenter, B., Rodriguez, A., and Dunbabin, M. (2015) Automatic wildfire detection and simulation using optical information from unmanned aerial systems, *Proceedings of the 2015 SAE Aerotec Conference*. Seattle.

Lum, C.W., Gauksheim, K.R., Vagners, J., and McGeer, T. (2011) Assessing and estimating risk of operating unmanned aerial systems in populated areas, *Proceedings of the AIAA Aviation Technology, Integration, and Operations Conference*.

Lum, C.W., Rysdyk, R.T., and Pongpunwattana, A. (2005) Autonomous airborne geomagnetic surveying and target identification, *Proceedings of the AIAA Infotech@Aerospace Conference*, Arlington.

Lum, C.W. and Vagners, J. (2009) A modular algorithm for exhaustive map searching using occupancy based maps, *Seattle, Proceedings of the AIAA Infotech@Aerospace Conference*, Seattle.

Lum, C.W. and Waggoner, B. (2011) A risk based paradigm and model for unmanned aerial systems in the national airspace, *Proceedings of the AIAA Infotech@Aerospace Conference*, St. Louis.

McCormack, E.D. (2008) *The Use of Small Unmanned Aircraft by the Washington State Department of Transportation*, Washington State Transportation Center (TRAC), Seattle.

McGeer, T. (1994) *Aerosonde Hazard Estimation*, Aerovel Corporation.

McQuarrie, J. and Simon, D. (1997) *Physical Chemistry: A Molecular Approach*, University Science Books.

US Department of Defense (1980) *MIL-STD-1629A – Procedures for Performing a Failure Mode, Effects and Criticality Analysis*, US Department of Defense, Washington, DC.

US Department of Transportation (2013) *Unmanned Aircraft System (UAS) Service Demand 2015–2035: Literature Review & Projections of Future Usage*, US Department of Transportation.

Vagners, J., McGeer, T., and Newcome, L. (1999) *Quantitative Risk Management as a Regulatory Approach to Civil UAVs*, International Workshop on UAV Certification, Paris.

Weibel, R. E. and Hansman, R. J. (2005) *Safety Consideration for Operation of Unmanned Aerial Vehicles in the National Airspace System*, MIT International Center for Air Transportation, Cambridge.

Williams, B.P., Clothier, R.A., Fulton, N., Lin, X., Johnson, S., and Cox, K. (2014) Building the safety case for UAS operations in support of natural disaster response, in *Proceedings of the AIAA Aviation Technology, Integration, and Operations Conference*, Atlanta.

Chapter 47

Sense and Avoid: Systems and Methods

Martina Orefice, Vittorio Di Vito, and Giulia Torrano

Air Transport Sustainability Department, Head, CIRA Italian Aerospace Research Center, Capua, Italy

1 INTRODUCTION

In the last decade, the unmanned aerial vehicles (UAVs) development has grown exponentially and a lot of applications proved UAV reason for existence. Most of these applications are military, but more and more civil and commercial opportunities are opening for UAVs. In fact, the size of UAVs is extremely variable and this allows performing some tasks impracticable for manned aircraft, such as detecting, monitoring, and measuring the evolution of natural disasters, like forest fires or landslips.

Especially in the United States, numerous efforts have been made by government and industries to integrate UAVs in the NAS (National Airspace System) and extensive research has been carried out, especially in the framework of navigation and control techniques regarding UAVs. Nevertheless, some lacks still remain in terms of safety: Indeed, the major obstacle for the integration of UAVs in

the NAS is the sense-and-avoid (SAA) capability and the consequent possibility of midair collisions avoidance. Therefore, although the efficiency of those systems has been proved under different and varied conditions, their safety, reliability, and compliance with aviation regulations still remain to be proven.

Certainly, the fundamental difference between an UAV and a manned aircraft is the physical onboard absence of the pilot, who in the UAV case maintains the interaction with the aircraft from the ground. Essentially, an UAV is remotely piloted, although it is capable of numerous automated operations. It implicates that the pilot has no direct situational awareness; therefore, one of the most significant challenges is the replacement of the "see-and-avoid" capability with the "sense-and-avoid" one.

2 SAA SYSTEMS REGULATORY ASPECTS

2.1 Regulatory state of the art

As emphasized above, since the pilot remotely controls the UAV, it is necessary to replace the "see-and-avoid" capacity with the "sense-and-avoid" capacity. The US regulatory survey of sense and avoid, carried out by Marshall *et al.* (2007), gives an indication about the challenges related to the integration of UAVs into the NAS.

The cornerstone of the current visual flight rules (VFR), that is, the concept of "see and be seen," had its first appearance in a Federal regulation, in the Air Commerce Act of 1926. Only in 1955 the Civil Aeronautics Board

Unmanned Aircraft Systems. Edited by Ella Atkins, Aníbal Ollero, Antonios Tsourdos, Richard Blockley and Wei Shyy.
© 2016 John Wiley & Sons, Ltd. ISBN: 978-1-118-86645-0.

(CAB), the predecessor of Federal Aviation Administration (FAA), inserted the sentence "see and be seen" in a document (CAB, 1954) that stated "the philosophy behind the Visual Flight Rules is that aircraft being flown in accordance with these rules are operated in 'see and be seen' weather conditions permitting the pilots to observe and avoid other traffic." Starting from this, in 1968, the FAA published an amendment confirming the pilot's responsibility and now the Amendment 91 Code of Federal Regulations (CFR) states "When weather conditions permit, regardless of whether an operation is conducted under instrument flight rules or visual flight rules, vigilance shall be maintained by each person operating an aircraft so as to see and avoid other aircraft" (FAA, 2014a). In addition, pilots are responsible to not "operate an aircraft so close to another aircraft as to create a collision hazard" (FAA, 2014b).

Over the years, many efforts have been made in order to develop sense and avoid requirements and, in 2004, the Radio Technical Commission for Aeronautics (RTCA) Special Committee 203 (SC-203) was formed. It had the task of producing the Minimum Aviation System Performance Standards (MASPS) for several systems, including the MASPS for SAA. The quantitative performance standards for a SAA system (MASPS) were declared to have published on December 2013, but the document indeed has not been issued.

Between December 2008 and March 2009, the FAA organized several workshops in order to define the capabilities that a SAA system should have to be compliant with the current rules governing the "see-and-avoid." The workshop published a document in October 2009 (FAA, 2009), where the sense-and-avoid concept was defined as "the capability of [an unmanned aircraft] to remain well clear from and avoid collisions with other airborne traffic." Moreover, the workshop defines that a SAA system would be characterized by two components:

- A *self-separation* component that ensures a safe separation based on a variable time-based threshold. In this way, the aircraft remain "well-clear" of each other.
- A *collision avoidance* component that operates when the safe separation is lost and an extreme maneuver is needed to prevent a collision, that is, penetrating the collision volume. In fact, for the collision avoidance maneuver, a distance-based threshold is considered.

The SC-203 terminated the activities on June 2013 and, contextually, the SC-228 was created. The committee has the task of producing the MOPS (Minimum Operational Performance Standards) for unmanned aircraft.

Currently, the FAA Regulation states that an UAV "must provide equivalent levels of safety, comparable to see-and-avoid requirements for manned aircraft" (FAA, 2004) and the UAV that wants to operate in US NAS must obtain Certificates of Authorization. An equivalent level of safety to the see capabilities of manned aircraft implies that the SAA system must be able to detect "other aircraft within a range of $\pm15°$ elevation and $\pm110°$ azimuth and respond in sufficient time so that a collision is avoided by a minimum of 500 ft. The 500 ft margin of safety derives from what is commonly defined as a near midair collision" (Pellebergs, 2010).

In the European framework, relevant effort is devoted to support the definition of suitable standards allowing the integration of UAS into the civil airspace. EDA (European Defense Agency) funded the ongoing project MIDCAS (Mid-Air Collision Avoidance System), started in 2009 and expected to be completed by 2015. The specific aim of MIDCAS (Pellebergs, 2010) is to identify adequate technology, contribute to standardization, and demonstrate a SAA system for UAS able to fulfill the requirements for traffic separation and midair collision avoidance in nonsegregated airspace. The MIDCAS SAA system is currently in the final test campaign, using the UAV in real-world environment, and the project findings are shared with European regulatory bodies to provide the technical background for them to establish SAA standards. Therefore, the outcomes of the MIDCAS project will be used as baseline input for the process of standardization of UAV integration into nonsegregated airspace.

2.2 Certification challenges of SAA systems

In order to define a set of SAA standards, for the certification and operational approval of UAVs, the FAA Workshop of 2009 identified five subfunctions that a SAA system shall perform:

- Detection of conflicting traffic
- Determination of the right of way
- Analysis of the flight paths
- Maneuvering
- Communication

For each subfunction, a set of requirements are identified, related to the functionalities of detect, sense, and avoid. The above-quoted set of requirements, categorized by subfunction, can be found in the document issued by the FAA "Sense and Avoid Technology for Unmanned Aircraft Systems" (Hottman, Hansen, and Berry, 2009).

A wide range of possible solutions are available and a trade-off between the subfunctions is needed in order to take into account, for example, the traffic characteristics of the interested airspace class, the aircraft performance, and the size, weight, cost, and performance of the sensors. Moreover, the architecture of the SAA system also needs to be taken into account. An issue to be considered is the pilot control latency, the communication link, and how the UAV pilot remotely controls the UAV. The pilot, on the ground, would receive surveillance data from the UAV, evaluate the situation, and communicate the avoidance maneuver onboard after the decision on when and how avoid the threat has been taken. But another solution would be to use an automatic SAA algorithm onboard the UAV with no communication with the pilot.

Another trade-off is related to the interaction between the sensors and the avoidance maneuvers. Indeed, the accuracies and performance of the sensing system are strictly related to the used sensors; therefore, the collision avoidance maneuver could comply with the minimum required distance or consider an extra size in order to compensate the possible measurement errors (Zeitlin, 2010).

A review about possible sensors solution and SAA architecture is reported in the following (Orefice, 2015).

3 SAA AVAILABLE TECHNOLOGIES

The statement "Currently, there is no recognized technology solution that could make these aircraft capable of meeting regulatory requirements for see-and-avoid and command and control" by Nick Sabatini (associate FAA administrator for aviation safety) was made before the House Committee on Transportation and Infrastructure, Subcommittee on Aviation on Unmanned Aircraft Activities in 2006 (Sabatini, 2006). The situation however has not changed much ever since, due to the complexity of sense-and-avoid technologies. For this reason, a great effort has been made, during the last years, by industry and agencies in order to identify a technological solution that could satisfy an equivalent level of safety of manned aircraft.

The technologies that have been used, during the years, can be divided into two macroareas:

- *Cooperative technologies* that typically require a transponder onboard the aircraft; they require other aircraft to equip the same devices when sharing the same airspace.
- *Noncooperative technologies* that identify all the aircraft not equipped with a transponder or, for example, gliders, hot air balloons, and so on; they do not require other

aircraft to equip the same devices when sharing the same airspace.

It is worth noting that in a multisensor approach, a data fusion system is required to integrate the best features of the dissimilar sensors while ensuring high reliability and limiting the computational burden so as to enable real-time software implementation.

3.1 Cooperative technologies

3.1.1 TCAS

Traffic alert and collision avoidance system (TCAS) (RTCA, Inc., 2008) is the principal collision avoidance system and it uses transponder in order to transmit information. Therefore, it generates alerts for the pilot for potential collision threats related to transponder-equipped aircraft. In addition to traffic advisories (TA), the TCAS II can provide resolution advisories (RA) supporting the pilot in the conflict resolution. The suggested collision avoidance maneuver is generated in a cooperative manner with the other aircraft. TCAS is mandated on all aircraft with 10 seats or more. Nevertheless, this system was never intended to replace see-and-avoid and, moreover, a safe horizontal maneuver is not guaranteed due to the low accuracy of bearing measurements.

3.1.2 ADS-B

Automatic dependent surveillance–broadcast (ADS-B) is a relatively new technology and it was developed in order to support aircraft operations and overcome the ground-based radar surveillance. Actually, it allows both ground station and pilots to detect other ADS-B-equipped aircraft with more precision than ever.

ADS-B consists of two different services: ADS-B OUT and ADS-B IN. In a typical application, aircraft equipped with ADS-B OUT technology compute their own precise position through satellite-based GPS (Global Positioning System). This information, along with others such as altitude, velocity, and identification, and additional info are transmitted in broadcast via a digital frequency data link. That information can be received by other aircraft equipped with ADS-B IN technology or by ground station, improving the awareness of pilots about the surrounding traffic conditions and reducing the risk of misleading controller orders due to stress condition. The main expected outcome of ADS-B technology is the improvement of the separation assurance function, and in the future the ADS-B will enable pilots to perform self-separation assurance maneuvers (SESAR, 2008). The introduction of ADS-B will provide specific benefits to support the integration of UAV into civil airspace. Moreover, general aviation aircraft will be provided with a system that will ensure a remarkable increase in the overall situational awareness and a reduction in the number of collision threats.

3.1.3 Application of cooperative technologies to UAVs

Cooperative technologies are widely used on manned aircraft due to their proved reliability. Moreover, these systems have been already certified and approved for use. Nevertheless, there are some disadvantages that must be taken into account when the cooperative sensors are intended to use on UAVs. First of all, cooperative technologies work only when all the aircraft in the shared airspace possess and utilize them. They provide no SAA capability against ground obstacles, that is, terrain and mountains, and they were developed assuming that a pilot would be in the loop evaluating warnings and taking the appropriate maneuvers. Moreover, some of these systems, such as TCAS, might be cost-prohibitive for some users. For these reasons, a recertification might be needed for use in UAVs, in order to maintain the equivalent level of safety of manned aircraft.

3.2 Noncooperative technologies

3.2.1 Active microwave sensors

Active microwave sensors represent a suitable option to provide the required situational awareness in the case of medium/large UAV platforms that have to attain a reliable full autonomy from ground. In fact, airborne radars provide direct and typically accurate range estimates (also range rate if Doppler processing is used). Moreover, they can guarantee large detection range, low levels of missed or false detections (ground echoes have to be properly filtered), and cannot be much affected by weather conditions, so that all-time all-weather operation can be guaranteed.

It is worth noting that, in the choice of wavelength, maximizing detection range, minimizing sensor dimensions to enable installation onboard a lightweight aircraft, and improving as much as possible angular resolution are contradicting requirements.

In fact, radars operating at low frequencies are relatively unaffected by atmosphere, but are large in size and unable to provide required spatial resolution, due to main lobe width, which is directly proportional to operating wavelength:

$$\sigma_{dB} = K \frac{\lambda}{l}$$

The parameter K is a coefficient whose value depends on the considered aperture and feeding and l is the antenna length in the considered direction. In conventional architectures, the main lobe width coincides with the achievable angular resolution.

A higher frequency radar, instead, is smaller in size and provides better resolution for given aperture size, but it is more susceptible to atmospheric and weather effects, in particular to rain, as it results, if we consider atmospheric attenuation produced by fog and rain.

Frequencies ranging from C-band (about 6 GHz) to Ka and W band (35 and 94 GHz, respectively) have been used and/or proposed in sense and avoid applications.

Besides angular resolution, other important performance parameters are the detection range, the range and Doppler resolution, the achievable field of regard, and scan rate.

The detection range for a given target can be calculated in probabilistic terms on the basis of achievable signal-to-noise ratio (SNR) and the number of impulses integrated to perform target detection (Skolnik, 1990).

Given a field of regard, the achievable scan rate depends on the radar pulse repetition frequency (PRF) (which in its turn influences average power consumption and maximum unambiguous range), number of integrated pulses for each resolution cell, and main lobe width.

Compared with mechanically scanned systems, electronically scanned arrays have the significant advantage of beam agility, that is, the beam can be pointed adaptively without the constraints of mechanical inertia. Thus, track update rate for a given target can be increased without significant effects on the revisit rate in the rest of the sensor field of regard.

However, electronic scanning allows beam pointing within angular limits that are smaller than typical sense and avoid requirements. In general, stand-alone radar architectures are typically characterized by coarse angular resolutions (about 1°) and low update rates (about 1 Hz), since finer resolution essentially implies larger antenna dimensions. In general, radars are demanding in terms of cost, size, weight, and required electric power, so they do not represent an affordable sensing solution for small unmanned platforms, considering current technological levels. However, increasing efforts are being made toward miniaturization and adaptation to small UAVs. In particular, currently in use is the synthetic aperture radar (SAR) that synthesizes an antenna of wide aperture through the aircraft motion.

3.2.2 Laser: LIDAR, laser identification detection, and ranging

Laser systems work similar to conventional radar: Laser scans are taken at regular interval and processed by echo analysis software. The obstacles and intruders can be used as input to automated collision avoidance systems (SELEX ES, 2015). Due to their high configurability, laser systems can be used in several atmospheric conditions, reducing the false alarms. Moreover, they are capable of detecting small obstacles up to 5 mm of diameter and large obstacles such as buildings and bridge.

3.2.3 Electro-optical systems

Electro-optical (EO) sensors are largely used in the framework of collision avoidance systems for small UAV, thanks to the low cost, power consumption, and weight. In particular, they are often used as stand-alone systems or in integrated architectures, comprising radar or other systems to produce an estimate of vehicle states through a multisensor fusion.

In terms of wavelength, visible band sensors are usually exploited to detect the sunlight scattering from other aircraft, while during nighttime, when no sun scattering is available, the best solution is to use a thermal infrared sensor that can detect the energy emitted by the same object.

Important parameters relevant to EO detection and tracking performance are related to the available field of view and angular resolution.

In general, stand-alone EO systems require heavy computational resources in order to fulfill real-time full image detection of obstacles, and their output can suffer from a high false alarm rate since background removal processing is less accurate as the image size increases. Moreover, EO detection range is very much affected by weather and illumination conditions and it can be poor.

3.2.4 Acoustic sensors

Acoustic sensors can be used to detect and track aircraft basing on the signal emitted from a propeller-driven aircraft that comprises a strong narrowband tone imposed onto a broadband random component.

3.3 Cooperative and noncooperative technologies summary

The advantages and drawbacks of existing technologies are indicated in Table 1, which is an extract of the accurate analysis conducted by Yu and Zhang (2015). Moreover, the detection ranges of the introduced sensors are illustrated in Figure 1 (which is extracted from Yu and Zhang (2015) analysis).

The main features of the most used sensors in SAA can be summarized as follows:

- Optical sensors (pixel/visual):
  - low cost, size, and weight;
  - suffer atmospheric disturbances;
  - a visual radar is highly comparable to a human's ability to observe (equal level of safety); and
  - to achieve the required FOV (field of view), sensors have to be arrayed in various positions on the aircraft, taking up valuable external area.
- Infrared sensors:
  - higher cost than EO;
  - low size and weight;
  - able to conduct nighttime operations;

Table 1. The characteristics of sensor technologies (Yu and Zhang, 2015). (Reprinted from Progress in Aerospace Sciences, Yu, X., Zhang, Y, Sense and avoid technologies with applications to unmanned aircraft systems: Review and prospects © (2015), with permission from Elsevier.)

	Information provided	VMC	IMC	SWAP	Cost	Others
TCAS	Range altitude	√	√	×	×	Well proven, widely used
ADS-B	Position altitude velocity	√	√	×	×	Well proven
SAF	Range bearing	√	√	×	×	Typically poor accuracy
LIDAR	Range	√	√	×	×	Easy configuration, narrow FOV
EO system	Azimuth	√	×	√	√	Data link required
	Elevation					Lack or direct range
Acoustic system	Azimuth	√	×	√	√	Data link required
	Elevation					Lack of direct range
IR system	Azimuth	√	×	√	√	Data link required
	Elevation					Lack of direct range

Note: SWAP: size, weight, and power; √: favorable/applicable; ×: arid, not favorable/applicable.

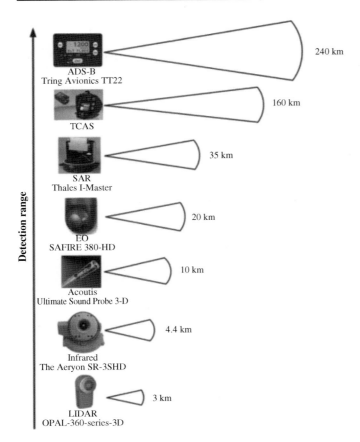

Figure 1. Detection range of typical sensors (Yu and Zhang, 2015). (Reprinted from Progress in Aerospace Sciences, Yu, X., Zhang, Y, Sense and avoid technologies with applications to unmanned aircraft systems: Review and prospects © (2015), with permission from Elsevier.)

— operate under harsh weather conditions;
— to achieve the required FOV, sensors have to be arrayed in various positions on the aircraft, taking up valuable external area;
— unable to pick up objects lacking some type of heat signature (cables or gliders); and
— development and integration would be very costly.
• Microwave radar (MMW radar):
— very mature technology;
— detect intruder aircraft at great distances; and
— high size and weight.
• Laser (LIDAR):
— the size of the cone is very small and it makes possible to target a specific obstacle;
— the revisit rate is poor and it takes multitude of laser sensors to achieve the same rate as a microwave radar;
— extremely underdeveloped;

— high cost; and
— inadequate in adverse weather (light can be absorbed and reflected by the conditions).

4   SAA ARCHITECTURES AND METHODS

Regarding the "sense" function, for a sense-and-avoid system on UAVs, MIT Lincoln Laboratory developed the Airborne Sense and Avoid (ABSAA) radar panel that is a unique lightweight sensor performing quick and repeatable scanning of the search region. The radar solution meets the all-weather and day/night requirements (Lester, Cook, and Noth, 2014). This prototyping effort was focused on the General Atomics Predator B that nominally could carry two or three separate radar arrays to cover a total of 220° in azimuth and 30° in elevation (Owen, Duffy, and Edwards, 2014). Other radar approaches for the sense function are presented by Itcia *et al.* (2013) and Moses, Rutherford, and Valavanis (2011). In the first paper, prototype radar for mini-UAV is presented. This radar is able to differentiate other miniature rotorcraft by their Doppler signature. Moreover, a performance analysis related to the signature matching algorithms is presented. The second paper introduces a radar technology and shows the tests that have been performed in order to evaluate the performance of a digital beam forming concept associated with flood light illumination: It allows combining wide angle coverage, high velocity resolution, and high refresh rate.

Several different approaches have been considered in literature for vision-based flying object detection, ranging from optical flow to morphological filtering (Lai, Mejias, and Ford, 2011). An emerging technology based on the active electronically scanned array (AESA) couples the radar-based technology and EO systems: The EO system scans and records images, while the radar shifts through its various modes (Barnhart *et al.*, 2012).

A visual approach is proposed by Zarandy *et al.* (2012): Their prototype uses five pieces of 1.2 megapixel miniature cameras, an FPGA board with a Spartan 6 (XC6SLX45T), and a 128 Gbyte solid-state disk drive for recording raw video data. The paper focuses on image processing algorithms and it proves that the designed system is able to identify 10 m-sized aircraft from at least 2000 m under regular daylight image conditions.

A trade-off analysis of EO sensors, used to provide a sense-and-avoid capability for Global Hawk, is reported by Griffith, Kochenderfer, and Kuchar (2008). It is assumed that Global Hawk has three cameras, whose coverage does not overlap, that provide a FOV of ±100° by ±15°. The analysis suggested that the EO system is suitable for detecting larger aircraft, but may not be ideal for detecting smaller aircraft with enough lead time for Global Hawk to avoid them.

Detection and tracking strategies based on acoustic array are proposed by Case, Zelnio, and Rigling (2008). These systems use array of microphones onboard an aircraft and a combination of narrow- and broadband processing techniques to characterize the temporal variation of the received tone of an approaching aircraft and estimate its propeller blade rate, together with its speed and the time and distance to the closest point of approach.

Scientific Applications and Research Associates, Inc. (SARA) proposed an acoustic sensor for use on small UAVs. The Passive Acoustic Noncooperative Collision Alert System (PANCAS) is characterized by a series of microphones mounted in order to compute bearing information for sound at each frequency. A proprietary algorithm is considered in order to minimize false alarms, due to fixed and random errors (atmospheric effects, wind effects, and signal processing errors), and to determine the threshold to apply a collision avoidance maneuver (Milkie, 2007).

The sense-and-avoid system proposed by Ramasamy, Sabatini, and Gardi (2014) considers cooperative and noncooperative sensors and includes

- visual camera,
- thermal camera,
- LIDAR,
- MMW radar,
- acoustic sensors,
- transponder,
- ADS-B, and
- TCAS/ACAS.

With reference to noncooperative sensors, a high-level tracking detection is performed by using a Kalman filter starting from the continuous cameras detection and range information provided by LIDAR. The Track-to-Track (T^3) algorithm is used for sensor fusion. This method combines the estimates instead of the observation from different sensors. The ADS-B system is used to obtain the state of the intruders and an Interacting Multiple Model (IMM) algorithm is considered for data fusion. The risk of collision is, then, evaluated considering the probability of a near midair event for the predicted trajectory over the time horizon by employing Monte Carlo approximations. Finally, the volume that must be avoided by the host UAV can be obtained computing and combining the navigation and tracking error ellipsoids.

Another multisensor data integration for an autonomous sense-and-avoid system is reported by Chen et al. (2011) and

Graham et al. (2011). The suite of SAA sensors used is composed by Ads-B, TCAS, Radar/LIDAR, and EO/IR. The proposed system is called "Multisensor Integrated Conflict Avoidance" (MuSICA) and the data integration is performed by an extended Kalman filter (EKF) and a measurements-to-track association. The collision avoidance algorithm is called "Jointly Optimal Conflict Avoidance" (JOCA) and it computes an optimal avoidance maneuver considering hierarchical constraints in order to make the maneuver as human-like as possible.

A SAA algorithm based on the Laser Obstacle Avoidance Marconi (LOAM) system is proposed by Sabatini, Gardi, and Richardson (2014). LOAM system, developed and tested by SELEX-ES and the Italian Air Force Research and Flight Test Centre (SELEX ES, 2015), is a low-weight/volume navigation aid system for rotary wing/UA platform specially designed for detecting potentially dangerous obstacles placed in or nearby the flight trajectory and providing the crew with warnings and information of the detected obstacles. A laser beam periodically scans the area around the flight trajectory in the FOV and using a dedicated signal processing algorithm, optimized for low-level obstacle detection, the system provides obstacle shapes. Measurement uncertainties are taken into account, adding a Gaussian error to every data and computing a statistic of the position error for obstacles near and far from the aircraft. If a collision risk is established by the impact warning processing, a collision avoidance maneuver is computed having the smaller possible correction and which is compatible with a safe flight plan. Here too, an ellipsoidal avoidance volume is associated with the obstacle considering the two-sigma standard deviation of the total obstacle detection and tracking errors.

A SAA algorithm based on the surveillance data provided by electro-optical sensors and an airborne radar is proposed by Fasano et al. (2008). The conflict detection criterion is based on the calculation of the closest point of approach and the resolution maneuver is computed considering the minimum variation from the original path. In particular, the collision volume is assumed to be spherical and the resolution maneuver is computed considering the tangent to that sphere. The sphere radius is related to the NMAC parameter.

A sense-and-avoid system, which uses an ADS-B Transceiver, is reported by Martel et al. (2009). The collision detection algorithm considers the GPS position obtained by the ADS-B device and a threat is declared if an aircraft or a fixed obstacle is predicted to enter a collision or near-collision course with the ownship aircraft within a certain time frame. The collision avoidance algorithm considers a behavior-based approach derived from a guidance method developed for unmanned maritime vehicles. In particular, it represents a multiobjective optimization problem using a set of behaviors that may include "reach target," "avoid small threats," "avoid large threats," and "follow right-of-way rules." These behaviors may or may not produce objective functions for which a priority weighting is assigned. Objective functions are defined considering a set of explicit constraint constructs representing the dynamic characteristics of the UAV: horizontal velocity, vertical velocity, and direction. The behaviors are based on the closest point of approach. Hence, the sense-and-avoid algorithm uses interval programming (IvP) methods to balance objective functions for each behavior (Mullins *et al.*, 2013). Concerning the avoidance volume computation, one possible assumption considers the turning and climb–descendent performance, creating a cylindrical volume.

A range-based method used to create dynamic alerting thresholds is reported by Huiyao, Zhihao, and Yingxum (2013). The relative dynamics of the incoming aircraft and the "sense" feature is assumed to be evaluated through the ADS-B system. The alerting thresholds are defined based on the geometric relationship of the encounter and the UAV's maneuvering ability and the four kinds of alarms include "dangerously close," "perform maneuver," "vertical maneuver," and "super maneuver only."

A collision avoidance approach, based on conflict probing, is presented by Tadema and Theunissen (2009). Conflict probing consists of predicting the future separation between ownship and hazards for a set of ownship velocity vectors, up to a predefined prediction horizon. The probing data indicate which velocity vectors will lead to a future conflict and the related time to conflict. Conflict probing can provide a common framework for the computation of coordinated conflict avoidance maneuvers that include integration of multiple types of hazards and constraints such as vehicle performance and right-of-way rules.

Another approach relying on the use of ADS-B data for trajectory prediction and conflict detection is proposed by Baek and Bang (2012). The methodology addresses the problem from a probabilistic point of view aimed at assessing the conflict probability based on the approximation of the conflict zone by a set of blocks.

A further approach is the one proposed by the MIDCAS project, where a SAA system based on the data coming from a set of sensors comprising EO, IR, radar, ADS-B, and transponder has been designed and tested. The collision volume can be defined in two ways, starting from the NMAC parameter: the first one defines the collision volume as a spheroid, with vertical half-axis of 350 ft and horizontal half-axis of 500 ft; the second defines the collision volume as a cylindrical volume centered on the UAV with a horizontal radius of 500 ft and a vertical height of 200 ft. The system shows threats information display and it computes an automated collision avoidance maneuver (Sellem-Delmar and Farjon, 2011).

Finally, in Europe, the general aviation and UAVs are currently experiencing the introduction of other and cheaper cooperative means for conflict detection (as alternative to TCAS), such as FLARM (FLight AlaRM) (AAVV, 2015). Each FLARM device evaluates its position and altitude with highly precision GPS receiver. Based on other information, such as speed, acceleration, heading, and track, a flight plan can be calculated and sent over a radio channel to all nearby aircraft equipped with FLARM too. Therefore, a motion prediction algorithm calculates a collision risk for each received aircraft based on an integrated risk model. The FLARM device gives alerts to the pilot who can take resolute actions. The newer FLARM incorporates a very accurate ADS-B and transponder (SSR i.e. Secondary Surveillance Radar) Mode-C/S receiver in order to include all the transponder-equipped aircraft in the collision prediction algorithm.

5   CONCLUSIONS

In this chapter a comprehensive analysis of the state of the art about sense-and-avoid systems for unmanned aerial vehicles application has been provided. The analysis emphasized the constantly growing importance of these systems in order to support the integration of the UAVs in the civil airspace and also to support the safe use of these platforms for the various applications for which they are currently considered. Main issues related to sense-and-avoid systems have been emphasized, with particular reference to the need of ensuring the compliance of these systems with the multiple and conflicting constraints that affect the application of these systems on UAVs, whose size can be very variable depending on the application for which the vehicles have been considered. The pros and cons of the most relevant sense-and-avoid technologies have been indicated and, finally, some commercially available solutions, mainly supporting cooperative sense and avoid, have been described.

REFERENCES

AAVV (2015) www.flarm.com/technology/.

Baek, K. and Bang, H. (2012) ADS-B based trajectory prediction and conflict detection for air traffic management. *Int. J. Aeronaut. Space Sci.* **13**(3), 377–385.

Barnhart, R.K., Hottman, S.B., Marshall, D.M., and Shappee, E. (2012) *Introduction to Unmanned Aircraft Systems*, Taylor & Francis Group.

CAB, (1954) Part 60-Scheduled Air Carrier Rules, 19 *Fed. Reg.* 6871.

Case, E.E., Zelnio, A.M., and Rigling, B.D. (2008) Low-cost acoustic array for small UAV detection and tracking, in *Proceedings of the IEEE National Aerospace and Electronics Conference,* Dayton, OH.

Chen, R.H., Gevorkian, A., Fung, A., and Chen, W.-Z. (2011) Multi-sensor data integration for autonomous sense and avoid. *Infotech @ Aerospace 2011 Conference,* St. Louis, MO, March 29–31, 2011.

FAA (2004) Order 7610.4K, Special Military Operation.

FAA Sense and Avoid (SAA) for Unmanned Aircraft Systems (UAS), October 2009.

FAA (2014a) CFR: Title 14–Aeronautics and Space, Part 91 General Operating and Flight Rules, Section 113, Right-of-Way Rules: Except Water Operations.

FAA (2014b) CFR: Title 14–Aeronautics and Space, Part 91 General Operating and Flight Rules, Section 111, Operating Near Other Aircraft.

Fasano, G., Accardo D., Moccia, A., Carbone, C., Ciniglio, U., Corraro, F., and Luongo, S. (2008) Multi-sensor-based fully autonomous non-cooperative collision avoidance system for unmanned air vehicles. *AIAA J. Aerosp. Comput. Inform. Commun.,* **5**, 338–360.

Graham, S., Chen, W.-Z., De Luca, J., Kay, J., Deschenes, M., Weingarten, N., Raska, V., and Lee, X. (2011) Multiple intruder autonomous avoidance flight test. *Infotech @ Aerospace 2011 Conference,* St. Louis, MO, March 29–31, 2011.

Griffith, J.D., Kochenderfer, M.J., and Kuchar, J.K. (2008) Electro-optical system analysis for sense and avoid. *AIAA Guidance, Navigation and Control Conference and Exhibit,* Honolulu, Hawaii, August 18–21, 2008

Hottman, S.B., Hansen, K.R., and Berry M., (2009) Literature review on detect, sense, and avoid technology for unmanned aircraft systems. Federal Aviation Administration DOT/FAA/AR-08/41, September, 2009.

Huiyao, W., Zhihao, C., and Yingxun, W. (2013) UAVs autonomous collision avoidance in non-segregated airspace using dynamic alerting thresholds. *AIAA Infotech@Aerospace Conference,* Boston, MA, August 19–22, 2013.

Itcia, E., Wasselin, J.P., Mazuel, S., Otten, M., and Huizing, A. (2013) FMCW radar for the sense function of sense & avoid systems onboard UAVs. *Proceedings of SPIE,* Vol. 8899, Emerging Technologies in Security and Defence; and Quantum Security II; and Unmanned Sensor Systems X, 889914, October 23, 2013.

Lai, J., Mejias, L., and Fordn J.J. (2011) Airborne vision-based collision-detection system. *J. Field Robot.,* **28**(2), 137–157.

Lester, T., Cook, S., and Noth, K. (2014) *USAF Airborne Sense and Avoid (ABSAA) Airworthiness and Operational Approval Approach,* Version 1.0, MITRE Corporation, January 31, 2014.

Marshall, D.M., Trapnell, B.M., Mendez, J.E., Berseth, B.L., Schultz, R.R., and Semke, W.H. (2007) Regulatory and technology survey of sense-and-avoid for UAS. *AIAA*

Infotech@Aerospace Conference and Exhibit, Rohnert Park, CA, May 7–10, 2007.

Martel, F., Schultz, R.R., Semke, W.H., Wang, Z., and Czarnomski, M., (2009) Unmanned aircraft systems sense and avoid avionics utilizing ADS-B transceiver. *AIAA Infotech@Aerospace Conference,* Seattle, WA, April 6–9, 2009.

Milkie, T., (2007) Passive Acoustic Non-Cooperative Collision Alert System (PANCAS) for UAV Sense and Avoid, unpublished white paper by SARA, Inc.

Moses, A., Rutherford, M.J., and Valavanis, K.P., (2011) Radar-based detection and identification for miniature air vehicles, in *Proceedings of the 2011 IEEE International Conference on Control Applications,* Denver, CO.

Mullins, M., Holman, M., Foerster, K., Kaabouch, N., and Semke, W. (2013) Dynamic separation thresholds for a small airborne sense and avoid system. *AIAA Infotech@Aerospace Conference,* Boston, MA, August 19–22, 2013.

Orefice, M. (2015) ADS-B based sense and avoid applications for general aviation/unmanned aircraft. PhD thesis. University of Naples Federico II.

Owen, M.P., Duffy, S.M., and Edwards, M.W.M. (2014) Unmanned aircraft sense and avoid radar: surrogate flight testing performance evaluation, in *Proceedings of the IEEE Radar Conference,* Cincinnati, OH.

Pelleberg, J. (2010) The MIDCAS Project, ICAS 2010. *27th International Congress of the Aeronautical Sciences,* Nice, France, September 19–24, 2010.

Ramasamy, S., Sabatini, R., and Gardi, A. (2014) Avionics sensor fusion for small size unmanned aircraft sense-and-avoid. *IEEE Metrology for Aerospace (MetroAerospace),* Benevento, Italy, pp. 271–276.

RTCA, Inc. (2008) *Minimum Operational Performance Standards for Traffic Alert and Collision Avoidance System II (TCAS II),* Version 7.1, DO-185B, June, 2008.

Sabatini, N. *Testimony–Statement of Nicholas A. Sabatini.* Statement of Nicholas A. Sabatini, Associate Administrator for Aviation Safety, March 29, 2006.

Sabatini, R., Gardi A., and Richardson, M.A., (2014) LIDAR obstacle warning and avoidance system for unmanned aircraft. *Int. J. Mech. Aerosp. Ind. Mechatron. Eng.,* **8**(4), 718–729.

SELEX ES (2015) *LOAM: laser obstacle avoidance system.* Available at http://www.selexelsag.com/internet/?open0=5338&open1=5350&section=COMM&showentry=17089.

Sellem-Delmar, S. and Farjon, J. *D2.2.2-1 MID-Air Collision Avoidance System (MIDCAS) Concept of Operations (CONOPS),* EDA, February 15, 2011.

SESAR Consortium, Work Programme for 2008–2013, Doc. SESAR Consortium No. DLM-0710-002-01-00, Bruxelles, Belgium, 2008.

Skolnik, M.I. (1990) *Radar Handbook,* McGraw-Hill Professional, New York, NY.

Tadema, J. and Theunissen, E. (2009) An integrated conflict avoidance concept for aviation. *IEEE/AIAA 28th Digital Avionics Systems Conference,* Orlando, FL, October 23–29, 2009.

Yu, X. and Zhang, Y. (2015) Sense and avoid technologies with applications to unmanned aircraft systems: review and prospects. *Prog. Aerosp. Sci.*, **74** 152–166.

Zarandy, A., Zsedrovits, T., Nagy, Z., Kiss, A., and Roska T. (2012) Visual sense-and-avoid system for UAVs, in *Proceedings of the 13th International Workshop on Cellular Nanoscale Networks and Their Applications*, Turin, pp. 1–5, 2012.

Zeitlin, A. (2010) *Progress on requirements and standards for sense & avoid*, MITRE Corporation.

Chapter 48

System and Cyber Security: Requirements, Modeling, and Management

Inseok Hwang and Cheolhyeon Kwon

School of Aeronautics and Astronautics, Purdue University, West Lafayette, IN, USA

1 INTRODUCTION

As the technological capabilities of automated systems and the interconnectivity of network nodes increase, the use of unmanned aircraft systems (UASs) for traditionally dull and dangerous manned missions becomes more feasible in both military and civilian applications. However, these benefits come at the cost of increased susceptibility to cyber attacks, as UASs greatly rely on their onboard autopilot and inter-communications to function. Indeed, all automated aviation systems, including UASs, have vulnerabilities in their cyber assets (e.g., malicious code insertion, packet spoofing, and network protocol exploits), which can result in potential cyber security threats. The growth of future UAS usage relies on understanding cyber threats to the UAS network, identifying possible security breaches, and managing security risks.

1.1 Recent Cyber Attack Incidents on UAS

During the past decade, several incidents have made clear that UASs are vulnerable to cyber attacks and that appropriate care needs to be taken for their security. These incidents have occurred in both the military and civilian fields. In December 2009, a US military UAS's video feed was intercepted by Iraqi militants (Arthur, 2009). Although this incident did not threaten the mission or the safety of any soldier, the most alarming part of the incident was that the Iraqi militants had used off-the-shelf software to capture the video feed. Specifically, the SkyGrabber software (2013) was used, which is an off-the-shelf product mainly used to intercept satellite feeds of music, TV, and videos. Figure 1 shows how SkyGrabber was used to tap the video feeds from the UAS.

At that time, a large proportion of US military UAS was not properly encrypted and thus was easily exploitable. The US military had not addressed the issue because they were researching for a new UAS platform that would incorporate encryption. Although the Pentagon said that the proper precautions had been taken since the incident, this event diminished trust that people had in UAS and made it evident that UAS can also be targets in cyberwarfare.

Another notable incident occurred in 2011, when a virus infected a computer on an Air Force Base (Technical Report, 2011). In September 2011, a keylogger virus was discovered on the computers in Creech Air Force Base that was functioning as

Unmanned Aircraft Systems. Edited by Ella Atkins, Aníbal Ollero,
Antonios Tsourdos, Richard Blockley and Wei Shyy.
© 2016 John Wiley & Sons, Ltd. ISBN: 978-1-118-86645-0.

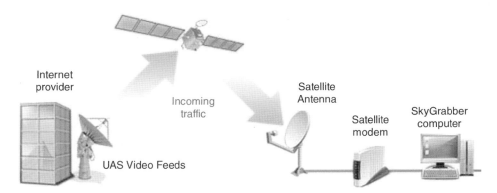

Figure 1. UAS video feed capture via SkyGrabber.

a ground control station for the Predator and Reaper UAS Communication Network. Later, it was ascertained that the virus was not very intrusive and did not feed information to any outside sources, although it could have potentially been a significant threat to national security. The virus entered the network by means of external hard drives or unauthorized use of computers. Although these computers operated isolated from the Internet, they were not completely invulnerable. The keylogger virus tracked every stroke made by the military personnel on the computers. These computers were specifically used to manage UAS over Iraq and Afghanistan. The military was fortunate that this incident did not jeopardize any missions or the lives of military personnel.

Civilian UAS are more vulnerable to cyber attacks than military UAS owing to insignificant monitoring and security measures. One of the major civilian UAS vulnerabilities is Global Positioning System (GPS) spoofing that deceives a GPS receiver by broadcasting counterfeit GPS signals, structured to resemble a set of normal GPS signals, or by rebroadcasting genuine signals captured elsewhere or at a different time. As demonstrated by the University of Texas at Austin's Radio Navigation Lab (Kerns *et al.*, 2014), a spoofing attack can be achieved with easily procured equipment. When spoofing a UAS, the attacker takes over and controls the incoming GPS signals that the UAS receives. This can be done covertly or overtly, and in both cases the user loses total control of the UAS and may be unable to regain it. GPS spoofing equipment is expensive but readily accessible, making spoofing a real threat to civilian UAS. The prime reason a spoofing attack is easy is that the civilian GPS is not encrypted and the documentation for it is publicly available. Encrypting the civilian GPS would require a severe modification to the infrastructure, and such a large investment could take years to implement.

Interestingly, UAS can be used to compromise civilian electronic devices, reversing the attack path considered above. A distributed tracking and profiling software has been developed that can be put on a UAS and can access any connected network (Snoopy, 2014). Then, the UAS can successfully obtain all the data transmitted over the network, such as personal information and the person's whereabouts, from, say, the cell phone network. Studies show that by using botnets (a number of Internet-connected computers communicating with other similar machines in an effort to complete repetitive tasks and objectives) and employing a controlling botmaster (who operates the botnets for remote process execution), it is possible to break into Wi-Fi networks using a UAS and the identified vulnerabilities, as the security techniques used in Wi-Fi networks are known to be unsecure and unreliable (Reed, Geis, and Dietrich, 2011). With greater funding and skill, the damage of cyber attacks on UAS is expected to become a significant concern, necessitating in-depth research into cyber threats and vulnerabilities.

1.2   Research Challenges to UAS Security

Traditionally, most cyber security studies on UAS have analyzed risks from the computer security perspective, focusing on issues such as the trustworthiness of data flow without rigorously considering the system's physical processes such as its real-time dynamic behaviors (Javaid *et al.*, 2012). A typical UAS network is *not* similar to a generic computer network. Although significant research has investigated security modeling for various kinds of wireless networks including sensor networks and mobile ad hoc networks, these models cannot be directly applied to a typical UAS network. Modeling a UAS communication network is tougher than and different from modeling other networks because of increased complexities and a significant disparity in certain properties. Differing channels of communication, range of communication (short/long), different power requirements for different devices, different types of data flows (commands/video/audio/image), and different integrity and confidentiality requirements are some of the characteristics that make UAS security requirements different from security requirements of other existing networks.

Although computer security components are key elements in the hardware/software layer, these methods alone are not sufficient to diagnose the healthiness of a UAS's physical behavior. For example, if the attacker gains and accesses control of the application layer at the source node of the UAS, then link-layer encryption does not protect the UAS (Flying Operations of Remotely Piloted Aircraft Unaffected by Malware, 2011). There are also multiple sensor attacks that can corrupt the state of a UAS without breaking encryption (Warner and Johnston, 2003). Such scenarios do not require the UAS computing system to be entirely compromised. It is also important to note that conventional computer security techniques are unable to ensure the system's safety when physical tampering occurs. This is mainly due to the unpredictable nature of cyber attacks and an inability to distinguish between attack features and common disturbances and faults. Thus, UAS security issues require a systems point of view to analyze cyber attacks within the context of a unified physical and logical process model of the UAS. This analysis ensures that the UAS is not only secure on the cyber level but is also sufficiently robust to counter and account for damage by a cyber attack. The difficulty lies in conjoining and coordinating the cyber–physical resources of a UAS since UAS differs from the majority of other dynamical systems in that (1) a UAS makes more intelligent use of its surroundings through a comparatively complex network of sensors and actuators and (2) a UAS has a substantially more complex dynamic response and complex interactions with other systems and with the environment.

2 UAS CYBER THREAT MODEL

Cyber threat analysis is the first priority for ensuring the security of a UAS, as it can lead to the discovery of vulnerability in the system. This section first describes a common architecture and communication network for a UAS and then discusses the identification of cyber threats posed to a UAS from both the computer security and the systems perspectives.

2.1 UAS Architecture

In a single unmanned aircraft (UA), common components found in the onboard autopilots are shown in Figure 2.

Although the specific equipment shown here is meant for a small, low-cost UA, the components share the same

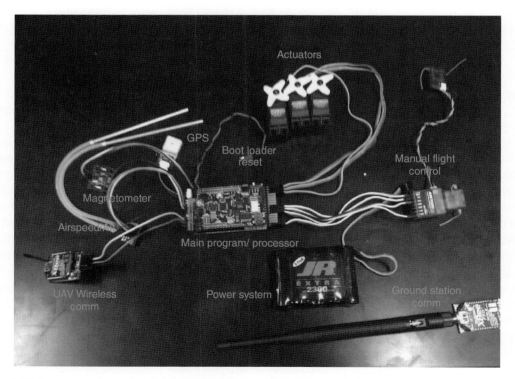

Figure 2. Onboard UAS components used at Purdue Hybrid Systems Lab (Kim *et al.*, 2012).

functionality as is found in larger and more expensive UAs. For example, more expensive UAs might have more powerful processing units, satellite communication links, and sensor redundancy, but suffer from the same security problems. Figure 3 shows the UA components in a data flow diagram.

In this diagram, the components of the onboard autopilot can be divided into three major parts that support guidance, navigation, and control functions:

- *Guidance:* It determines the path based on the UA's state, waypoints, mission objective, avoidance maneuvers, target tracking, and so on, and involves trajectory generation (center block in Figure 3).
- *Navigation:* It determines the UA's state using sensory data (left block in the lower panel in Figure 3).
- *Control:* It keeps the UA in the safe and stable state in the presence of disturbances, and steers the UA to the reference state, based on the information from the guidance and navigation blocks (right block in the lower panel in Figure 3).

In Figure 3, the left block in the upper panel represents communication in and out of the UA, the middle block in the upper panel represents the automatic dependent surveillance-broadcast (ADS-B) communication technologies, and the right block in the upper panel represents the sensors that can aid the guidance of the UA. Note that other physical components of UA such as airframe, payloads, and so on are not included as they are not explicitly related to autopilot.

2.2   UAS Communication Network

In a UAS network that consists of multiple UAs and ground control stations (GCS), the different components of the UAS rely on wireless communication links with each other. The GCS can be of two types: regional (typically stationary) and mobile. The regional GCS is located at a local or central command and control (C2) center that can cover a large area of operation and manage multiple UAs simultaneously, while the mobile GCSs such as portable digital assistants (PDAs), smartphones, or rugged laptops each typically manage a single UA operating within a small area. Although the communication channels between each of these nodes seem similar, there are significant differences between the channels with respect to security (Rudniskas, Goraj, and Stankunas, 2009). The link between a satellite and a UA is line-of-sight (LOS) radio communication, while UA-UA, PDA-UA, and regional GCS-UA can be LOS radio communication or general packet radio service/enhanced data rates

for global evolution-based communication using existing communication infrastructure. Threats to each of these communication links and components are different and have different security requirements. Components such as satellites and regional GCS might have certain threats, but the existing security measures make the communication network less prone to an attack.

UAS communication transfers different types of UAS aviation data: (1) flight operational and planning data (e.g., UA trajectory, navigation performance data, and ADS-B-Out messages on air–ground communications); (2) airspace data (e.g., ADS-B-In, terrain maps, weather radar, and traffic information shared on air–ground and air–air communications); (3) position, navigation, and timing data (e.g., GPS and ADS-B-Out); (4) aeronautical information services and meteorological data (e.g., real-time updates on meteorological conditions and airport conditions, and emergencies and restrictions that limit airspace use); (5) controller–pilot messages (e.g., two-way communications that replace voice communications with a data link of automated messages and receipts); and (6) security-relevant data (e.g., digital certificates, keys, credentials, and passwords that protect UA information). Communications are also coupled with software, protocols, algorithms, and memory used in hardware and infrastructure in UAS handling data. All of these assets have properties and requirements that constrain their use and management in aviation, and thus may serve as weaknesses or strengths for security.

2.3   UAS Computer Security Cyber Threats

Cyber threats can result from possible attacks that target UAS cyber components, and include the consequences of casual or intentional attacks, such as malware, maliciously corrupted data, unauthorized access to systems, and denial of service (DoS) of networks and computers, employing access and the use, disclosure, disruption, modification, or destruction of data and/or data interfaces (Bishop, 2004). All these threats corrupt or misuse data and disrupt cyber components, impacting operational capabilities and performance of the UAS. More severely, they may cause intentional or accidental violations of UAS safeguards such as (1) logical, physical, and/or organizational inhibitors in UAS cyber–physical operations and maintenance; and (2) errors in safety-critical UAS cyber component design, development, and testing.

Within the traditional computer security domain, possible attack paths pertaining to such threats can be categorized by three main computer security properties: confidentiality, integrity, and availability, as shown in Figure 4 (Bishop, 2004).

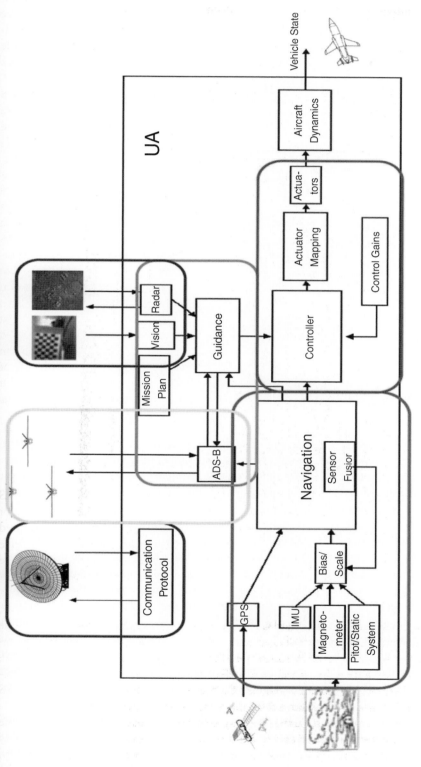

Figure 3. UAS autopilot data flow diagram (Kim *et al.*, 2012).

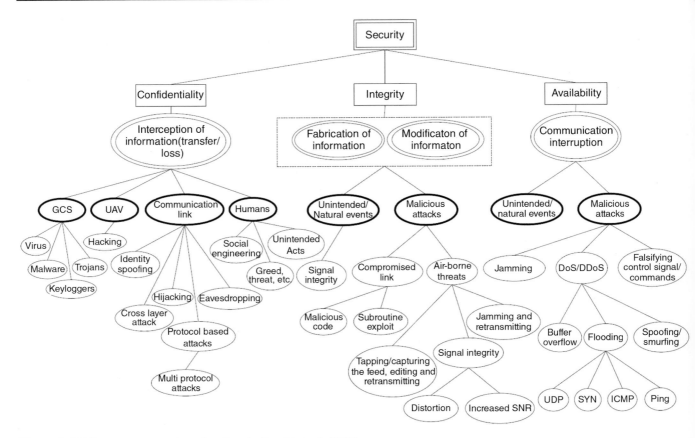

Figure 4. UAS computer security cyber threats (Javaid *et al.*, 2012).

- *Confidentiality Attacks:* This primarily deals with unauthorized access to information, and the most common way of compromising the security of this property is the interception of information. The four major components of the UAS that are vulnerable to this class of attack are the UA, the GCS (all types), communication links, and UA operators. Threats to the GCS are mostly software based, such as viruses, malware, Trojans, and keyloggers. It should be understood that software-based threats can affect the UA as well, but there are in general fewer ways to install them onto the UA. A GCS security breach or a UA security breach itself may lead to other threats to the entire UAS, but the need to address these threats can be readily fulfilled if addressed at the GCS level.

    The primary source of compromising the security of communication links between various UAS components is through network attacks such as eavesdropping, identity spoofing, cross-layer attacks, and multiprotocol attacks. Clearly, all these attacks might not be applicable to each of the links available in the UAS, but the aim of including all of these attacks in one group is to identify all possible threats to the communication link involved instead of identifying the threat to each link type separately. When putting proper mitigation measures in place, one must identify the attacks that actually affect these links and deploy mitigation measures accordingly. As for the human operator, the increasing trend of social and business networking has caused a rise of a new kind of threats, such as social engineering, "fake" online competitions, blackmailing, or a behavioral exploit.

- *Integrity Attacks:* The modification of existing information and the fabrication of new information can compromise the integrity of the UAS. Modification aims at altering the data during transit or while in storage. Natural events such as lightning, magnetic pole shifts, and sun flares can cause some integrity loss and add unwanted noise to the signal. However, these natural events are rare, and most communication protocols and equipment can take care of issues caused by these.

    In airborne threats, there are three broad subcategories: jamming, compromising signal integrity, and tapping or capturing the feed/signal. Jamming aims at disrupting communication through interference before reception. Many defense strategies have been proposed to handle jamming attempts in wireless networks. For compromising signal integrity, distortion or increasing the signal-to-noise ratio is the most common approach. The third method, tapping or capturing the feed, is the most difficult type of attack to launch as it requires substantial intelligence in terms of transmission signal frequency, range, and so on. Interestingly, a separate branch of communication engineering called Signal Intelligence involves the study of these attacks. Furthermore, fabrication or modification of information can be done by using malicious code or existing subroutines of the system. Subroutine exploits involve attacking the system by searching for vulnerabilities in the code, and once the adversary has enough information, the vulnerability is exploited through a planned or brute force attack (Poisel, 2003).

- *Availability Attacks:* Primary cyber attacks that might affect availability are jamming, falsifying signals, and DoS attacks (Poisel, 2003). As discussed, transmitting false commands or control signals requires substantial signal intelligence. This can be a major threat to UAS availability as false signals can actually make the UAS land or attack elsewhere.

DoS or DDoS (distributed DoS) attacks are mainly based on network congestion or overflow in the network card of the system so that the system appears to be unavailable. During such an attack, the system or network is actually busy serving other "fake" requests. Three ways of launching such an attack are flooding, spoofing/smurfing, and buffer overflow. Flooding completely occupies the network with one or more kinds of network packets by sending multiple packets to the host system. Usually, synchronization (SYN), user datagram protocol (UDP), Internet control message protocol (ICMP), or Ping packets are used in such an attack. Another type of attack belonging to this class is buffer overflow, which aims at overflowing the buffer memory of network cards on the devices used in the system. Smurfing involves flooding the system using spoofed broadcast network packets so that it appears to the target system that all packets are coming from different addresses (Pelechrinis, Iliofotou, and Krishnamurthy, 2011).

2.4 Specific Attack Scenarios from the Systems Perspective

Note that the above cyber attack classifications are based on computer security, which does not account for the attack's impact on the dynamic behaviors of the UAS. Therefore, it is suitable to recategorize the cyber threats into the following two attack scenarios from a systems point of view, as shown in Figure 5. In particular, the systems approach does not look

deeper into the methods of cyber attacks; it rather focuses on studying the resulting damage to the UAS due to the malfunctioning components under the assumption that there are possibilities of corrupting the data flow within the UAS through cyber attacks such as buffer overflows. This framework models the broader impact of cyber threats, enabling a comprehensive UAS vulnerability and risk analysis.

- *Control System Security:* Attacks that attempt to prevent the hardware/CPU from behaving as programmed. Some examples of this type of attacks include a buffer overflow exploit through some input device, a forced system reset to load malicious code, or a hardware change of the system.
- *Application Logic Security:* Attacks that manipulate the sensor or the environment data, thereby providing false data to the control system. In this case, the control system behaves as programmed without any fault, but some or all of the inputs to the system are corrupted. Some examples of this type of attacks include sensory data manipulation, system component state data manipulation, navigational data manipulation, and command and control data manipulation. Figure 5 shows the most likely type of vulnerabilities for each component of the UAS.

3 UAS VULNERABILITY ANALYSIS AND RISK ASSESSMENT/MITIGATION

To heighten the security of a UAS, it is essential to identify and mitigate vulnerabilities before any attack actually occurs. This can be achieved by appropriate UAS vulnerability analysis and risk assessment, followed by mitigation. Risk assessment schemes are generally employed for most types of software and hardware components. However, such assessment schemes could be ineffective for UASs, requiring a more sophisticated design due to the complex cyber–physical nature of the UAS. Moreover, the integration of risk assessment results needs a promising high-level approach to make both cyber and physical components aware of their decision impacts on one another, and make cyber and physical components capable of sensing and responding to misbehaviors of the other (Sampigethaya and Poovendran, 2013).

3.1 Proactive Risk Assessment

Technically, the vulnerability of a UAS is an aspect of the system that reflects the probability of malfunction due to specific cyber threats. Depending on the severity of the malfunction, ranging from the complete loss of control/

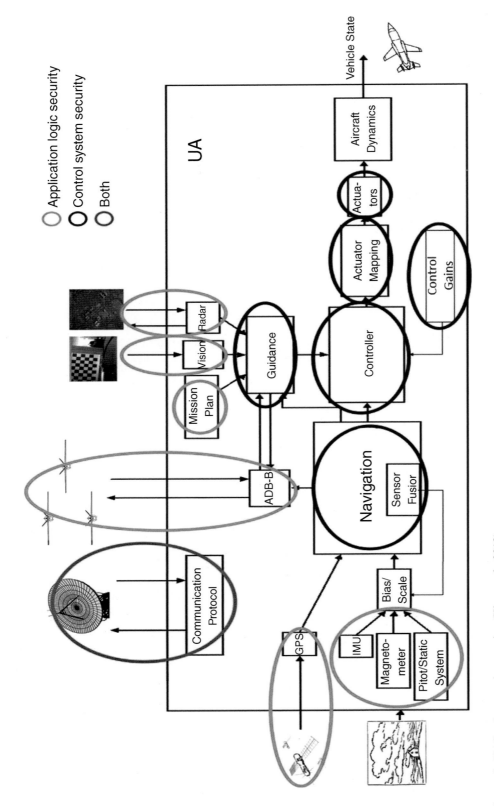

Figure 5. UAS cyber attack scenarios (Kim *et al.*, 2012).

destruction of the UAS to mere data or trajectory tracking errors, the risk is an index of cyber threats to the UAS security. Thus, in terms of system security, a risk is a combination of the severity of the impact of the attack on the UAS security, weighted by its probability (or likelihood) of occurrence. Thus, the risk assessment quantifies the possible severity and likelihood of attacks. This value is crucial for any security-critical computer system (Bishop, 2004). The risk of a security violation of UAS can be assessed based on individual UAS component models. Then, the overall risk assessment of a UAS can be posed as the summation of its components' risk assessment.

Figure 6 represents a prototype UAS risk assessment scheme, which is a component-wise, probability-based evaluation of the confidentiality, integrity, and availability of the UAS (Hartmann and Steup, 2013).

The scheme assesses the risk based on the type of security needed. For example, a result with a high-risk value in the confidentiality block corresponds to a severe risk regarding the loss of confidentiality. This scheme provides information on the susceptibility to cyber attacks, the risk analysis of the integrity, confidentiality, or availability of the individual components, respectively, and in totality of the UAS. Since multidimensional risk assessment considers the different

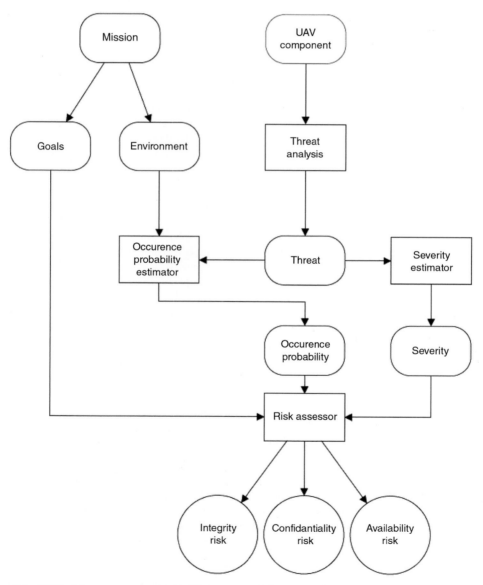

Figure 6. Overview of the UAS risk assessment scheme (Hartmann and Steup, 2013).

requirements of UAS, different aspects of security play varying roles and must be weighed accordingly. Therefore, the risk assessment of UAS is always mission bound, that is, strongly dependent on operational goals.

- *Environment:* The environment influences the UAS's sensors, its communication links, and its avionics. It is important to distinguish between political and physical factors of the environment as these influence security aspects differently. In the simplest case, two types of landscape (lowland and mountainous) or two political states (friend or enemy) can be considered. The physical factors are in general not capable of influencing UAS confidentiality or integrity. However, other factors such as weather conditions, altitudes, and so on may influence integrity. A UA moving in enemy territory may lose its availability due to a heightened threat of destruction, takeover, signal disturbances, and so on. Additionally, the UA is exposed to the threat of confidentiality or integrity loss due to the risk of takeover, theft, or manipulation. Thus, the change in political factors influences the properties of system security.
- *Communication Links:* As explained earlier, there are several types of communication links used in UAS. It is important to note that, although all communication links impose security threats to all aspects of security, the degree of susceptibility greatly varies due to their different specifications.
- *Sensors:* Sensors may be classified according to the type of reference used. References can be external or internal. An external reference is, for example, a GPS satellite. Inertial navigation systems, on the other hand, rely only on internal references of physical parameters such as acceleration and/or angular rates. To determine the risks of the individual sensors in the UAS, the characteristics of the sensor, the importance of the observed aspect, and the mechanisms to detect spoofed or false sensor values must be considered. Sensors with external references are in general more susceptible to jamming and spoofing than sensors with internal references.

Aspects of the environment that are crucial to the correct execution of the mission must be observed correctly and reliably. If such an aspect is observed solely by a single sensor with an external reference, a risk for the integrity and the availability of the system may emerge. A UAS relying on GPS-based navigation is prone to attacks on the GPS sensors, which may be jammed or spoofed. In this case, due to the reliance on the external reference and the lack of control and coping mechanisms, the correct autonomous behavior of the UA cannot be guaranteed.

The redundancy mechanisms used to compensate sensor values may additionally contribute to improve the system's security. If several, different types of sensors are used to observe one aspect of the environment, the acquired values are considered more reliable. It is less likely that multiple sensors that acquire the parameter through different methods can be jammed or spoofed collectively. Therefore, it may be concluded that single sensor observations impose an additional threat to the system's integrity. If only one sensor observes a crucial value, such as flight attitude, this poses a threat to the availability of attitude information system-wide.

- *Data Storage:* The risk assessment of data storage mechanisms considers three main aspects: volatility, encryption, and signature. The usage of volatile storage imposes a risk to the availability of the stored data. If appropriate coping strategies are lacking, this may also lead to an inconsistent storage state and hence result in a loss of the integrity of the stored data. The use of encryption mechanisms preserves the confidentiality of the stored data. A lack of encryption generally raises the risk of confidentiality loss. Encryption mechanisms do not prevent the stored data from being overwritten, which implies a risk for data integrity. To secure the integrity, mechanisms such as signatures or forgery detection must be integrated. These mechanisms have no influence on the confidentiality or availability of the data (Hartmann and Steup, 2013).
- *Fault-Tolerant System:* Fault-tolerant systems are difficult to assess concerning their "usefulness" in terms of security aspects. Although it is intuitive that a "good fault-tolerant system" should improve the UAS's overall security, the connection between the fault and UAS security is not clear. This is a common restriction of the computer-security-oriented risk assessment, requiring a more comprehensive approach, which will be discussed in the next section.

3.2   Post-Attack Behavior Analysis

Despite the above risk assessment scheme, a detailed risk assessment of the UAS against cyber attack is still complicated due to the complex nature of UAS dynamics and the lack of a well-defined measure of attack severity. As a cyber–physical system (CPS), ensuring the security of the UAS requires an analysis of the interactions between both the cyber and physical components. However, this analysis is challenging since each UAS design is physically unique. The mass properties, propulsion system specifications, sensors, actuators, control systems, and aerodynamics of the UA all contribute to the system dynamics. Such unique dynamics of individual UAS modify the vulnerabilities of the UAS to cyber attack, making it difficult to evaluate the generic security aspects. Besides, the obtained risk values with respect to the security properties, that is, the confidentiality, integrity, and availability, cannot encompass the dynamical behavior of the compromised UAS. Although the UAS dynamics have not been previously studied in conjunction with cyber attacks in depth, there is some research that has investigated UAS flight envelope enforcement as a method of fail-safe recovery (Colgren and Johnson, 2001). Currently, 33% of all UAS failures are caused by the UAS exceeding its designed flight envelope (Wilson and Peters, 2009).

In seeking to address the vulnerabilities and to assess risks within the unified logical and physical dynamic model of the UAS, the Purdue Hybrid Systems Lab has developed a simulation testbed that can analyze the postattack behavior of the UAS as illustrated in Figure 7 (Goppert et al., 2014).

In order to configure the testbed, a UA description is first obtained using the wind tunnel, if a physical model is available, or from the USAF Digital DATCOM software (1979). Once the aerodynamics and mass properties of the UA are defined, JSBSim (2009) (a C++ flight dynamics model library) is used to simulate the UA. Next, ScicosLab (2011) (a software package similar to Simulink from MathWorks) interfaces with JSBSim. At this point, the block diagram environment within ScicosLab (Scicos) provides a mechanism for rapid analysis. The digital controller can then either be simulated in ScicosLab or implemented using the actual hardware of the UA. Implementing the controller in software is called a software-in-the-loop simulation (SIL), and using the actual hardware is called a hardware-in-the-loop simulation (HIL). Telemetry data from the UA can then be sent to ground station software to allow complete testing and validation of all user interactions with the UA.

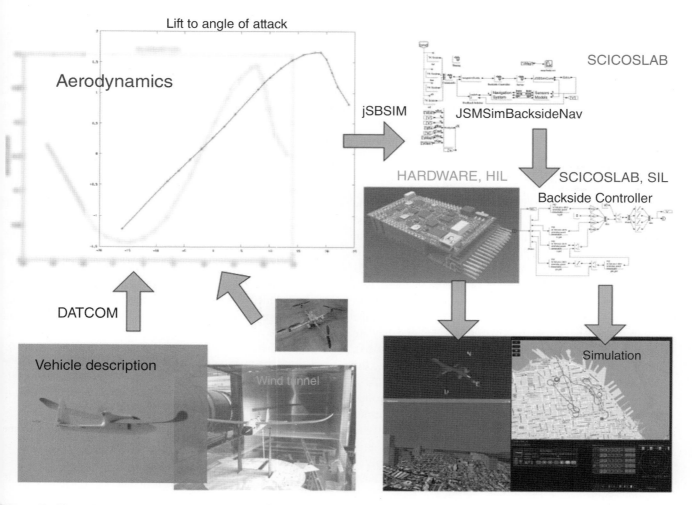

Figure 7. Simulation testbed in Purdue Hybrid System Lab (Goppert *et al.*, 2014).

Based on this simulation testbed, comprehensive UAS security risk assessment can be achieved through exhaustive testing on the UAS when attacked. As a way to motivate the different security measures used to quantify the severity of the attack, potential cyber attacks are classified by the intent of the attacker as follows:

- *Mission Obstruction:* In a mission obstruction attack, the objective of the attacker is to prevent the UA from completing the assigned mission objectives. There are several ways that this can be accomplished. For example, the UA can be delayed such that the time requirement of the mission is not met, or the UA can be caused to waste fuel or battery power so that the mission objectives are no longer feasible. Unpredictable errors could also be inserted into the navigation system in order to degrade the state awareness of the UA. Another feasible scenario is that the control system could be corrupted to the point that its sensors begin to perform poorly, introducing issues such as highly oscillatory physical motion. Some attacks leverage the collision avoidance system to obstruct the UA. By inserting a phantom aircraft in the path of the targeted UA, an attacker can cause the UA to deviate from its flight path in order to avoid collision. It should be noted that in a mission obstruction attack, the attacker does not have the ability to control the UA directly. If the attacker can control the UA directly, that would be considered a control acquisition attack.
- *Control Acquisition:* In a control acquisition attack, the objective of the attacker is to assume direct control of the UA. An example of this is the use of GPS spoofing to shift the flight path of the UA to suit the purposes of the attacker. For this type of attack, it may be possible for an attacker to have differing levels of control, that is, an attacker may be able to gain control of a UA's subsystems without gaining control of the entire UA. If the attacker is able to gain complete control of the UA, there is a possibility of a man-in-the-middle attack that refers to an attack where the attacker secretly relays and alters the communication between two parties who believe they are directly communicating with each other (Patange, 2013). In this attack, the attacker would send falsified data to the original controller to make it appear that the UA is behaving normally, when it is actually being controlled by the attacker. Such an attack is especially dangerous since an undetectable attack provides a clear advantage to the attacker.
- *UAS Destruction:* The attacker's intent may be simply to destroy the UA. It is possible that an attacker would have limited control over a few states of the UA so that they cannot perform a meaningful control acquisition attack,

but they are still capable of destroying the UA. For instance, if they have control of the altitude of the UA, they may command the UA to fly into the ground. Indeed, the primary area of danger, and thus the focus of this analysis for such an attack, will be the introduction of instability into the control and navigation system of the UA. Any instability in these critical systems will most likely result in a crash.

In order to determine whether or not a cyber attack has been successful, criteria for failure need to be established. Based on the identified attack intents described above, two failure modes can be posed that are described below. A specific metric to quantify the severity of an attack is the time elapsed till any of these failure criteria are reached, referred to as the time-till-failure.

- *Mission Envelope Failure:* The user may define parameters corresponding to the restrictions on the states of the UA, whose violations determine if the UA is outside the mission envelope. The parameters identified in this study are summarized in Table 1.
- *Flight Envelope Failure:* Flight envelope failure is defined as exceedance of some (or all) of the UA's states over the safe flight regime. This type of failure typically leads to destruction of the UA.

Employing these criteria, simulations can be performed using varying attack scenarios and magnitudes, and the results can be represented as a two-dimensional heat map of various attack combinations, as shown in Figure 8, which is capable of predicting the vulnerabilities of a UAS to cyber attacks. Specifically, the lower and middle shaded layers in the bottom portion and the top and middle shaded layers in the top portion of the figure represent the fastest failures, while the top shaded layer in the bottom portion and the bottom shaded layer in the top portion represent delayed or no failure within the simulation time window.

To avoid intensive computation and to gain better insight, analytical methods capable of discovering and analyzing the most severe attack scenarios were developed (Kwon and

Table 1. Mission envelope parameters

Mission theater	Geographic region to which the UA is to be confined (e.g., two-dimensional latitude–longitude coordinate)
Altitude window	Range of acceptable UA altitudes
Battery/fuel level	Amount of battery power and fuel that the UA must hold in reserve
Target window	Geographic region that the UA should stay during the specified time
Target time window	The time period during which the UA must be within the target window

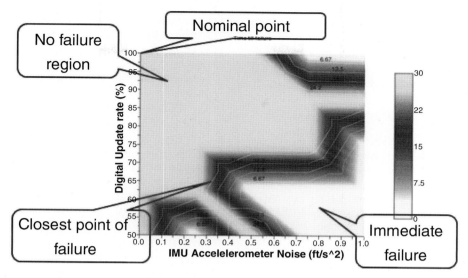

Figure 8. UAS risk assessment result example (Goppert *et al.*, 2014).

Hwang, 2013). Along with severity analysis, another type of analysis has been performed to provide some insight into which types of cyber attacks are more difficult to detect and how such attacks can be designed (Kwon, Liu, and Hwang, 2014).

3.3   UAS Cyber Threat Mitigation

Mitigation of cyber–physical threats demands novel UAS security considerations. For instance, cyber–physical integration for flight control must sense, process, and actuate, both correctly and in time. A security solution for guaranteeing correctness cannot be decoupled from its impacts on timeliness because a highly secured but delayed response may be ineffective and potentially unsafe. Furthermore, a security solution may not be able to use cryptographic solutions, for example, if solution overhead violates temporal or some physical resource constraints. Addressing such security considerations is perceived as difficult. Efficient cryptographic solution advances are available, but their applicability under UAS aviation performance requirements has not yet been explored. In that regard, promising cyber–physical security solution directions are indispensable. These include the joint leverage of capabilities, characteristics, and properties of cyberspace and the physical world. Furthermore, potential solution approaches can utilize system dynamics in the physical world, which inherently delays the physical impact of a cyber attack and can give a beneficial time window for applying threat mitigation. Solutions can also leverage intelligence related to cyberspace and UAS at

the right abstraction level, as well as use physical models and data to detect or predict cyber threats and vice versa.

4   CONCLUSION

In-depth UAS security issues against cyber threats have been discussed. Although advanced computer security continues to offer improved confidentiality, integrity, and availability in UAS data flow, it lacks an understanding of the complex interactions between cyber and physical components of the UAS. Thus, it is indispensable to consider a systems approach that closely integrates the cyber and physical behaviors of the UAS together with carefully designed measures of cyber attack severity. Within this framework, individual UAS components and their vulnerabilities are thoroughly investigated, and more comprehensive risk assessment against potential cyber threats can be done via proactive analysis and/or extensive simulation. The results of this assessment can leverage the reliability of the existing UAS security architecture, as well as provide the capability to detect or predict cyber threats, which could enhance safe operations of UAS in numerous applications.

REFERENCES

Arthur, C. (2009) SkyGrabber: The $26 Software Used by Insurgents to Hack into US drones. Technical Report, Guardian. Available at http://www.guardian.co.uk/technology/2009/dec/17/skygrabber-software-drones-hacked.

Bishop, M. (2004) *Introduction to Computer Security*, Addison-Wesley, ISBN 0-321-24744-2.

Colgren, R. and Johnson, T. (2001) Flight mishap prevention for UAVs. *IEEE Aerospace Conference*, pp. 2/647–2/656.

Flying Operations of Remotely Piloted Aircraft Unaffected by Malware (2011) U.S. Air Force Space Command Release 021011. Available at www.afspc.af.mil/news/story.asp?id=123275647 (retrieved 2014).

Goppert, J., Shull, A., Sathyamoorthy, N., Liu, W., Hwang, I., and Aldridge, H. (2014) Software/hardware-in-the-loop analysis of cyberattacks on unmanned aerial systems. *AIAA J. Aerosp. Inform. Syst.*, **11**(5), 337–343.

Hartmann, K. and Steup, C. (2013) The vulnerability of UAVs to cyber attacks – an approach to the risk assessment. *IEEE 5th International Conference on Cyber Conflict*, pp. 1–23.

Javaid, Y., Sun, W., Devabhaktuni, V. K., and Alam, M. (2012) Cyber security threat analysis and modeling of an unmanned aerial vehicle system. *IEEE Conference on Technologies for Homeland Security*, pp. 585–590.

JSBSim (2009) Web-link: http://jsbsim.sourceforge.net/ (released May 31, 2009).

Kerns, J., Shepard, D.P., Bhatti, , J.A., and Humphreys, T.E. (2014) Unmanned aircraft capture and control via GPS spoofing. *J. Field Robot.*, **11**(4), 617–636.

Kim, A., Wampler, B., Goppert, J., Hwang, I., and Aldridge, H. (2012) Cyber attack vulnerabilities analysis for unmanned aerial systems. *AIAA Conference on Infotech@Aerospace*.

Kwon, C. and Hwang, I. (2013) Analytical analysis of cyber attacks on unmanned aerial systems. *AIAA Conference on Guidance, Navigation, and Control*.

Kwon, C., Liu, W., and Hwang, I. (2014) Analysis and design of stealthy cyber attacks on unmanned aerial systems. *AIAA J. Aerosp. Inform. Syst.*, **11**(8), 525–539.

Patange, T. (2013) How to defend yourself against MITM or man-in-the-middle attack, Hacker space. Available at http:// hackerspace.kinja.com/how-to-defend-yourself-against-mitm-or-man-in-the-middl-1461796382.

Pelechrinis, K., Iliofotou, M., and Krishnamurthy, S. V. (2011) Denial of service attacks in wireless networks: the case of jammers. *IEEE Commun. Surv. Tutor.*, **13**(2), 245–257 (second quarter).

Poisel, R. (2003) *Modern Communications: Jamming Principles and Techniques*, Artech House Publisher.

Reed, T., Geis, J., and Dietrich, S. (2011) SkyNET: a 3G-enabled mobile attack drone and stealth botmaster. 5th Usenix Workshop on Offensive Technologies.

Rudniskas, D., Goraj, Z., and Stankunas, J. (2009) Security analysis of UAV radio communication system. *Aviation*, **13**(4), 116–121.

Sampigethaya, K. and Poovendran, R. (2013) Aviation cyber–physical systems: foundations for future aircraft and air transport. *Proc. IEEE*, **101**(8), 1834–1855.

ScicosLab (2011) Web-link: http://www.scicoslab.org/ (latest version 4.4.1, released April 8, 2011).

SkyGrabber software (2013) Web-link: http://www.skygrabber.com/en/index.php.

Snoopy (2014) Web-link: https://github.com/sensepost/Snoopy.

The USAF Stability and Control Digital Datcom (1979) USAF Technical Report AFFDL-TR-79-3032, April 1979.

Warner, J. S. and Johnston, R. G. (2003) *GPS Spoofing Countermeasures*, Los Alamos National Lab., TR LAUR-03-6163, Los Alamos, NM.

Wilson, J. and Peters, M. (2009) Automatic flight envelope protection for light general aviation aircraft. *IEEE/AIAA 28th Digital Avionics Systems Conference*.

Computer Virus Infects Drone Plane Command Center in US. Technical Report (2011) Associated Press, Guardian. Available at www.theguardian.com/technology/2011/oct/09/virus-infects-drone-plane-command.

Chapter 49
Social and Legal Issues*

C.E. "Noah" Flood
CAVU Global LLC, Purcellville, VA, USA

1 INTRODUCTION

Technological innovation and implementation typically precede legislation, regulation, and judicial rulings accommodating new technology. When a new technology is introduced, society reacts identifying social issues that may demand consideration, formulating new or additional regulations and laws. In addition to unmanned aircraft, the sensors that carried on board these vehicles will have a significant impact on evolving legal and social issues. In many nations, the legal and social impact of widespread unmanned aircraft operations and their evolving sensor packages is unknown.

Disclaimer: The information contained in this chapter provides general legal information. Nothing within is intended as legal advice, nor should be interpreted as legal advice. If legal advice is needed, please consult a legal professional with expertise within the appropriate jurisdiction.

Unmanned Aircraft Systems. Edited by Ella Atkins, Aníbal Ollero, Antonios Tsourdos, Richard Blockley and Wei Shyy.
© 2016 John Wiley & Sons, Ltd. ISBN: 978-1-118-86645-0.

Regulatory and legal systems do not follow predictable paths when addressing the introduction of new technologies. Some factors that influence this path include the culture of the regulatory agency chosen to regulate the new technology, societal response to the technology, legal precedent, and the political influence of interested parties. Early adopters of the technology will influence the International Civil Aviation Organization (ICAO) Standards and Recommended Practices (SARPs) that significantly influence the regulatory process of the remaining ICAO nations. As of this writing, following are some nations having regulations governing unmanned aircraft: Australia, Austria, Canada, Czech Republic, France, Germany, Italy, Japan, South Africa, Sweden, Switzerland, United Kingdom, and United States. The legal systems of individual countries will continue to address the civil and criminal implications of increasingly widespread operations by unmanned aircraft.

2 UNMANNED AIRCRAFT

Unmanned aircraft have operated for many decades in a military context. Most aspects of military operations fall outside the regulatory process. These aircraft do not necessarily meet certification requirements, nor their pilots licensing requirements. Flights of unmanned military aircraft in nonsegregated airspace are regulated. This is also true of state or public aircraft, which in addition to the military include customs, law enforcement, and other governmental functions. Civil use of unmanned aircraft beyond hobbyist with model aircraft is relatively recent.

In assessing the developing legal landscape related to unmanned aircraft, it is helpful to consider the legal backdrop of manned aircraft. Regulation and law related to manned aircraft have been well developed. Today and in the future, manned and unmanned aircraft will increasingly share the same airspace raising some similar legal and social issues. Myriad variations may evolve regarding unmanned aircraft as this technology has unique capabilities. Given commercially available unmanned aircraft are typically less expensive than manned aircraft, there is greater propensity for ubiquitous use.

The international legal foundation for today's regulation of manned aircraft is the Convention on International Civil Aviation (also known as the Chicago Convention). Signed in December 1944 and negotiated during the tumult of World War II, this document has well served international civil aviation. The three-paragraph preamble defines the goal and purpose of the Convention:

> Whereas the future development of international civil aviation can greatly help to create and preserve friendship and understanding among nations and peoples of the world, yet its abuse can become a threat to the general security; and
>
> Whereas it is desirable to avoid friction and promote that cooperation between nations and peoples upon which the peace of the world depends;
>
> Therefore, the undersigned governments having agreed on certain principles and arrangements in order that international civil aviation may be developed in a safe and orderly manner and that international air transport services may be established on the basis of equality of opportunity and operated soundly and economically;
>
> Have accordingly concluded this Convention to that end (Convention on International Civil Aviation, 1944, p. 1).

The ideals and concerns expressed in this preamble are still applicable today and apply to both manned and unmanned aircraft.

The Articles of the Convention outline the core principles and arrangements respected by ICAO's 191 member states (ICAO, 2015). Article 1 recognizes "that every State has complete and exclusive sovereignty over the airspace above its territory" (Convention on International Civil Aviation, 1944, p. 2). Article 3 distinguishes civil from state aircraft noting that "Aircraft used in military, customs and police services shall be deemed to be state aircraft." This Article also states that the Convention applies to civil aircraft and that "No state aircraft of a contracting State shall fly over the territory of another State or land thereon without authorization by special agreement or otherwise, and in accordance with the terms thereof" (Convention on International Civil Aviation, 1944, p. 2). Article 5 recognizes the right of nonscheduled flights "to make flights into or in transit non-stop across its territory and to make stops for non-traffic purposes without the necessity of obtaining prior permission, and subject to the right

of the State flown over to require landing" (Convention on International Civil Aviation, 1944, p. 4). Article 6 requires scheduled air services (scheduled airline operations) over or into a signatory State be conducted with "special permission or other authorization." In summary, the terms of the Chicago Convention apply to civil aircraft operations while requiring "special permission or other authorization" for state aircraft or aircraft involved in scheduled air services (Convention on International Civil Aviation, 1944, p. 5).

During World War I, the British and Americans conducted research into "remotely controlled pilotless aircraft" and an "aerial torpedo." These wartime research efforts did not field an unmanned aircraft for combat operations, but remind us of the often overlooked history of these vehicles (Keane and Carr, 2013, p. 559). This history and the potential civil applications for unmanned aircraft were not overlooked by the writers of the Convention on International Civil Aviation as witnessed by Article 8:

> No aircraft capable of being flown without a pilot shall be flown without a pilot over the territory of a contracting State without special authorization by that State and in accordance with the terms of such authorization. Each contracting State undertakes to insure that the flight of such aircraft without a pilot in regions open to civil aircraft shall be so controlled as to obviate danger to civil aircraft (Convention on International Civil Aviation, 1944, p. 5).

Article 37 of the Chicago Convention calls for contracting states "to collaborate in securing the highest practicable degree of uniformity in regulations, standards, procedures, and organization in relation to aircraft, personnel, airways and auxiliary services" with the goal of facilitating and improving air navigation. To effectively implement this Article, ICAO adopts "International Standards and Recommended Practices" in the following areas:

a) communications systems and air navigation aids, including ground marking;
b) characteristics of airports and landing areas;
c) rules of the air and air traffic control practices;
d) licensing of operating and mechanical personnel;
e) airworthiness of aircraft;
f) registration and identification of aircraft;
g) collection and exchange of meteorological information;
h) log books;
i) aeronautical maps and charts;
j) customs and immigration procedures;
k) aircraft in distress and investigation of accidents;
and such other matters concerned with the safety, regularity, and efficiency of air navigation as may from time to time appear appropriate (Convention on International Civil Aviation, 1944, pp. 16–17).

While SARPs for unmanned aircraft have not been published as of this date, ICAO has published the manual on Remotely Piloted Aircraft Systems (RPAS). This manual provides "guidance on technical and operational issues applicable to the integration of RPA [Remotely Piloted Aircraft] in non-segregated airspace and at aerodromes" (International Civil Aviation Organization, 2015, pp. 1–7).

This discussion of the Convention on International Civil Aviation provides a broad legal framework for the international operation of unmanned aircraft. It also introduces the ICAO Standards and Recommended Practices concept for use by civil aviation authorities, operators, manufacturers, and other stakeholders. These SARPs currently under development will serve as the basis for harmonizing the regulations, standards, and procedures for unmanned aircraft within the 191 ICAO member states.

The United States is regulating unmanned aircraft while ICAO SARPs are still under development. The United States was not the first nation to regulate and authorize use of civil unmanned aircraft, but will serve as an example of the regulatory process. In February 2006, the Federal Aviation Administration (FAA) established the Unmanned Aircraft Program Office, AIR 160. The foundation of the US regulatory process was the Small Unmanned Aircraft System (sUAS) Advisory Rulemaking Committee's (ARC's) work which started in 2008. Composed of members from government, industry, academia, and interest groups, the committee produced recommendations for Federal regulation of sUAS. These recommendations served as the basis for regulatory development. Regulatory recommendations were made in the following areas: model aircraft, operating rules, personnel, and aircraft and systems (sUAS ARC, 2009). By February 2012 with no unmanned aircraft regulations published, Congress provided specific guidance to the FAA by passing Title III, Subtitle B of the FAA Modernization and Reform Act of 2012 (FAA Modernization and Reform Act of 2012). Section 332 of this Act called for the integration of civil unmanned aircraft systems into America's National Airspace System (NAS) by September 30, 2015. This Section also called for the publication of a five-year "Roadmap" outlining how the Federal Aviation Administration (FAA) would introduce unmanned aircraft to the National Airspace System. Section 333 of the Act provided guidance for "expedited operational authorization" for use of civil unmanned aircraft until the regulations required under Section 332 were published. The first such exemption was issued to Astraeus Aerial for closed-set filming on July 23, 2014. Prior to this date, no commercial operation of unmanned aircraft was authorized within the United States. As of September 10, 2015, numerous commercial authorizations (1505) had been published permitting aerial mapping, data collection, inspections, filming, photography, surveying, and other airborne activities (FAA, 2015). The high demand for Section 333 exemptions forced the FAA to streamline their processes to reduce application backlog. Until the FAA issues final rules for sUAS, Section 333 exemptions remain the primary method for receiving authorization to conduct sUAS flights within the United States. Special Airworthiness Certification is a second more complex and costly flight authorization process.

It is unlikely that the FAA will meet its congressionally mandated deadline of September 30, 2015 for the publication of final rules on sUAS. However, the Notice of Proposed Rulemaking published on February 15, 2015 provides insight regarding the yet unpublished final rules. The primary efforts of the FAA's regulatory effort have been on sUAS defined as weighing less than 55 lb. Figure 1 provides a summary of the FAA's proposed rules.

Although hobby and recreational flying of sUAS are not regulated by the FAA, the agency recognizes the safety implications of such flying to other aircraft as well as people and property on the ground. It has entered into an education campaign with the sUAS industry to promote safe and responsible noncommercial unmanned flight. In addition to its "Model Aircraft Do's and Don'ts" campaign (Figure 1), the FAA is beta testing its B4UFLY smartphone application. "Many unmanned aircraft users today have little or no aviation experience, and some of them are flying where they could endanger manned aircraft. B4UFLY will give these flyers the tools and knowledge they need to operate safely."

"Key features of the B4UFLY app include:

- A clear "status" indicator that immediately informs operators about their current or planned location.
- Information on the parameters that drive the status indicator.
- A "Planner Mode" for future flights in different locations.
- Informative, interactive maps with filtering options.
- Links to other FAA UAS resources and regulatory information" (FAA News August 28, 2015).

3 CONSTITUTIONAL ISSUES

The Bill of Rights of the United States Constitution is composed of the first 10 Amendments of the Constitution. These Amendments address individual rights of citizens and protections from government. Cases involving Constitutional Law require that a federal or state governmental agency or body be involved. Constitutional Law does not apply between two or more private parties. Unmanned aircraft cases are likely to involve the First Amendment freedom of the press, the Fourth Amendment prohibition against unreasonable searches and seizures, and the Fifth Amendment declaration "nor shall private property be taken for public use, without just compensation" (Figure 2).

Under Section 333 of the FAA Modernization and Reform Act of 2012 several exemptions have been issued for operations related to journalism. On July 27, 2015, the FAA issued one such exemption to Cable News Network, Inc. (CNN).

Such exemptions reduce the likelihood of litigation under the First Amendment, but leave many operational questions highlighted later in this chapter.

Two legal issues with significant operational consequence are privacy and property rights. The legal community is still adjusting to the impact of unmanned aircraft on both property rights and privacy concerns. Although the operation of an

Overview of Small UAS Notice of Proposed Rulemaking

Summary of Major Provisions of Proposed Part 107	
The following provisions are being proposed in the FAA's Small UAS NPRM.	
Operational Limitations	• Unmanned aircraft must weigh less than 55 lbs. (25 kg). • Visual line-of-sight (VLOS) only; the unmanned aircraft must remain within VLOS of the operator or visual observer. • At all times the small unmanned aircraft must remain close enough to the operator for the operator to be capable of seeing the aircraft with vision unaided by any device other than corrective lenses. • Small unmanned aircraft may not operate over any persons not directly involved in the operation. • Daylight-only operations (official sunrise to official sunset, local time). • Must yield right-of-way to other aircraft, manned or unmanned. • May use visual observer (VO) but not required. • First-person view camera cannot satisfy "see-and-avoid" requirement but can be used as long as requirement is satisfied in other ways. • Maximum airspeed of 100 mph (87 knots). • Maximum altitude of 500 feet above ground level. • Minimum weather visibility of 3 miles from control station. • No operations are allowed in Class A (18,000 feet & above) airspace. • Operations in Class B, C, D and E airspace are allowed with the required ATC permission. • Operations in Class G airspace are allowed without ATC permission • No person may act as an operator or VO for more than one unmanned aircraft operation at one time. • No careless or reckless operations. • Requires preflight inspection by the operator. • A person may not operate a small unmanned aircraft if he or she knows or has reason to know of any physical or mental condition that would interfere with the safe operation of a small UAS. • Proposes a microUAS option that would allow operations in Class G airspace, over people not involved in the operation, provided the operator certifies he or she has the requisite aeronautical knowledge to perform the operation.

Operator Certification and Responsibilities	• Pilots of a small UAS would be considered "operators".
	• Operators would be required to:
	  o Pass an initial aeronautical knowledge test at an FAA-approved knowledge testing center.
	  o Be vetted by the Transportation Security Administration.

Figure 1. Overview of proposed small UAS notice of proposed rulemaking.

	o Obtain an unmanned aircraft operator certificate with a small UAS rating (like existing pilot airman certificates, never expires).
	o Pass a recurrent aeronautical knowledge test every 24 months.
	o Be at least 17 years old.
	o Make available to the FAA, upon request, the small UAS for inspection or testing, and any associated documents/records required to be kept under the proposed rule.
	o Report an accident to the FAA within 10 days of any operation that results in injury or property damage.
	o Conduct a preflight inspection, to include specific aircraft and control station systems checks, to ensure the small UAS is safe for operation.
Aircraft Requirements	• FAA airworthiness certification not required. However, operator must maintain a small UAS in condition for safe operation and prior to flight must inspect the UAS to ensure that it is in a condition for safe operation. Aircraft Registration required (same requirements that apply to all other aircraft).
	• Aircraft markings required (same requirements that apply to all other aircraft). If aircraft is too small to display markings in standard size, then the aircraft simply needs to display markings in the largest practicable manner.
Model Aircraft	• Proposed rule would not apply to model aircraft that satisfy all of the criteria specified in Section 336 of Public Law 112-95.
	• The proposed rule would codify the FAA's enforcement authority in part 101 by prohibiting model aircraft operators from endangering the safety of the NAS.

Figure 1. (*Continued*).

unmanned aircraft is the focus of property law, the onboard sensors may be the greater focus of privacy law.

A major factor in considering privacy concerns raised by unmanned aircraft is the type of onboard sensors. "Drones can be equipped with high-power cameras, thermal scanners, license plate readers, moving target indicators, LADAR (laser radar), LIDAR (light detection and ranging), and facial recognition software" (Matiteyahu, 2015, pp. 266–267). The United States Supreme Court has yet to address Fourth Amendment privacy issues with regard to unmanned aircraft. Existing privacy decisions may provide insight into future rulings regarding unmanned aircraft and the Fourth Amendment. The Fourth Amendment states

The right of the people to be secure in their persons, houses, papers, and effects, against unreasonable searches and seizures, shall not be violated, and no warrants shall issue, but upon probable cause, supported by oath or affirmation, and particularly describing the place to be searched, and the persons or things to be seized (U.S. Const. amend. IV, 1789).

Foundational to Fourth Amendment privacy law is *Katz v. United States* 389 U.S. 347 (1967). The Supreme Court decision in *Katz* cited a two-prong test for determining whether a Fourth Amendment violation occurred. Justice

Harlan stated, "My understanding of the rule that has emerged from prior decisions is that there is a twofold requirement, first that a person have exhibited an actual (subjective) expectation of privacy and, second, that the expectation be one that society is prepared to recognize as 'reasonable'." It is uncertain how the Court will rule regarding a person's expectation of privacy or the reasonableness of such expectation when unmanned aircraft are involved. Possible considerations are state legislation, government policy statements, prior court decisions, and public sentiment. Another consideration is identified in the Supreme Court decision in *Kyllo v. United States*.

We think that obtaining by sense enhancing technology any information regarding the interior of the home that could not otherwise have been obtained without physical "intrusion into a constitutionally protected area," *Silverman*, 365 U.S., at 512, constitutes a search – at least where (as here) the technology in question is not in general public use (*Kyllo v. United States*, 2001, p. 31).

The technology in *Kyllo* was a thermal imaging device. The unanswered question relating to unmanned aircraft is at what point are unmanned aircraft considered to be "in general public use"? The same question may independently apply to

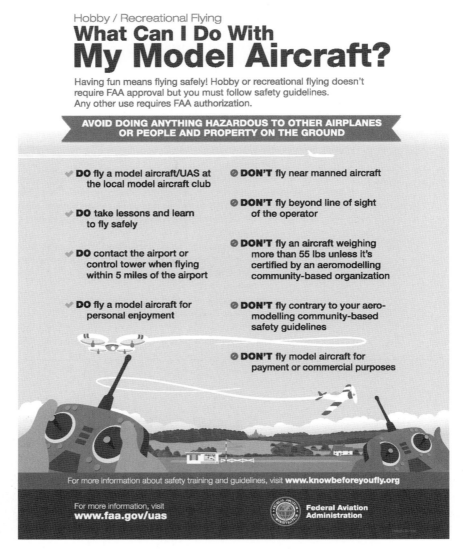

Figure 2. What can I do with my model aircraft?

any onboard sensors. It is also unknown whether the Court will focus on the platform (unmanned aircraft), the sensor(s), or both.

The Supreme Court decision in *United States v. Jones* did not involve an aircraft, but a Jeep. A global positioning system (GPS) tracker was used to track the vehicle of a criminal suspect's wife for 28 days. The Court ruled that placing the GPS tracker on the underside of the vehicle "encroached on a protected area" (*United States v. Jones*, 2012, p. 953). Perhaps more significant is a concurring opinion by Justice Sotomayor noting the ability of a GPS sensor to make "available at a relatively low cost such a substantial quantum of intimate information about any person" (*United States v. Jones*, 2012, p. 956). Questioning the

Constitutionality of such governmental information collection Justice Sotomayor states "I would ask whether people reasonably expect that their movements will be recorded and aggregated in a manner that enables the Government to ascertain, more or less at will, their political and religious beliefs, sexual habits, and so on" (*United States v. Jones*, 2012, p. 956). Time will determine if these concerns will be extended to the use of micro UAS or long endurance UAS and the plethora of sensors available or under development. State legislatures and Congress may also consider these concerns and chose to establish protections in laws addressing actions by private citizens and corporations.

In *United States v. Causby*, in examining the issue of a government taking under the Fifth Amendment (U.S. Const.

amend. V, 1789), the Court addressed a landowner's right to the airspace above his/her property.

> It is ancient doctrine that at common law ownership of the land extended to the periphery of the universe – *Cujusest solum ejus est usque ad coelum*. But that doctrine has no place in the modern world. The air is a public highway, as Congress has declared. Were that not true, every transcontinental flight would subject the operator to countless trespass suits. Common sense revolts at the idea. To recognize such private claims to the airspace would clog these highways, seriously interfere with their control and development in the public interest, and transfer into private ownership that to which only the public has a just claim (*United States v. Causby,* 1946, p. 261).

After recognizing the public interest in accommodating aviation's unimpeded use of navigable airspace, the Court states the rights of landowners in general terms:

> We have said that the airspace is a public highway. Yet it is obvious that if the landowner is to have full enjoyment of the land, he must have exclusive control of the immediate reaches of the enveloping atmosphere. Otherwise buildings could not be erected, trees could not be planted, and even fences could not be run . . . . The landowner owns at least as much of the space above the ground as he can occupy or use in connection with the land (*United States v. Causby,* 1946, p. 265).

The Court chose not to put a specific number on the height to which a landowner can enjoy the benefits of property ownership. The operation of sUAS has increased the strain between property owner rights and unmanned aircraft operators. With sUAS operations capped at an altitude of 400 ft, the continued growth of unmanned aircraft activity may demand a clearer delineation between the property owner rights and sUAS operator's rights.

4   ADDITIONAL ISSUES

Existing laws regarding trespass, nuisance, stalking, privacy, harassment, and liability will be utilized by courts in cases regarding unmanned aircraft systems. Characteristics unique to unmanned aircraft will result in judges adapting existing law when necessary. Legislatures may choose to make adaptations to some existing laws or create new laws specifically designed to address concerns related to unmanned aircraft.

In the first eight months of 2015, 156 bills related to unmanned aircraft have been debated within 45 of America's 50 states. "Nineteen states – Arkansas, Florida, Hawaii, Illinois, Louisiana, Maine, Maryland, Michigan, Mississippi, Nevada, New Hampshire, North Carolina, North Dakota, Oregon, Tennessee, Texas, Utah, Virginia, and West Virginia – have passed legislation" (National Conference of State Legislatures, 2015). The legislation passed covers a wide spectrum of social and legal issues. Among them are voyeurism, need for property owner consent to take an image when a reasonable expectation of privacy exists, commercial agriculture regulations, limitations related to unmanned aircraft use by law enforcement, prohibitions related to fishing, hunting, and trapping (National Conference of State Legislatures, 2015). The amount of state legislation related to unmanned aircraft systems has risen each year since 2013 and is likely to continue this trend for several years to come. Unmanned technology is still relatively new and legislatures, courts, and the FAA will continue to adapt as the technology evolves and new social issues arise.

Some legal considerations to consider before designing or operating an unmanned aircraft or an associated system include:

- Will operations be conducted where an expectation of privacy exists? (*Katz v. United States,*)
  - Is that expectation reasonable? (*Katz v. United States,* 1946)
- How invasive is the unmanned aircraft or associated sensor(s)? (*Katz v. United States,* 1946)
- Is the data collected in a way and over a period of time likely to limit the collection of nonrelevant personal information? (*United States v. Jones,* 2012)
- Is the technology related to the aircraft and/or sensors generally available to the public? (*Kyllo v. United States,* 2001)

While these questions are based on Constitutional legal decisions which require action by a governmental agency, the concepts involved may be considered by courts and legislatures hearing cases or enacting legislation involving private citizens and corporations.

The United States is at the beginning of an evolutionary process in the operation of unmanned aircraft. The focus is primarily on small unmanned aircraft and managing the risk to people (both airborne and on the ground) and property. According to the Section 333 exemptions issued to date and the FAA's proposed rules, small unmanned aircraft are required to operate no more than 400 ft above ground level (AGL) and remain five miles from airports. This is an attempt to keep sUAS clear of conflict with most manned aircraft traffic. However, with the proliferation of sUAS safety incidents are on the rise. The Consumer Electronics Association predicts unit sales of 700 000 consumer drones (unmanned aircraft) for 2015 in the United States (CEA website, 2015).

The proliferation of unmanned aircraft has resulted in a corresponding increase in safety and security-related incidents. Commercial airliners, general aviation, and military aircraft have experienced near misses with sUAS. In the first eight months of 2015, the FAA has received over 700 reports of unmanned aircraft interfering with the operation of manned aircraft. A Washington Post article details 12 incidents which occurred in seven states and Washington, DC, on Sunday, August 16, 2015 (Washington Post, August 20, 2015). It is likely that these numbers understate the number of incidents as small unmanned aircraft are only visible to a pilot for a short period of time and may not be seen by pilots during high workload periods such as takeoffs and landings.

Unmanned aircraft are a concern to the Secret Service after one crashed on the White House lawn, another flew nearby, and another hovered in the vicinity of the President while he played golf. A recent Department of Homeland Security Intelligence Assessment addresses concerns regarding the use of sUAS in support of terrorism (Washington Post, August 20, 2015).

Recent news articles report unmanned aircraft interfering with firefighting operations, flying into buildings, following a woman from a bar to her car, and private citizens shooting down unmanned aircraft (Washington Post, August 10, 2015) (Figure 3).

Government and industry are moving beyond educational efforts to ensure safe and legal sUAS flight operations. Taking advantage of a concept called geo-fencing, some

Figure 3. Drones near wildfires are not safe!

Figure 4. No fly zone.

manufacturers are limiting the ability of sUAS to operate above 400 ft, in the Special Flight Rules Area (SFRA) around Washington, DC, and in the vicinity of larger airports. A team at NASA Ames is leading the development of Unmanned Aircraft Systems Traffic Management (UTM). Their goal is to transition a prototype to the FAA along with recommendations for consideration. Ideally, this prototype will accommodate today's visual line-of-sight operations and beyond visual line-of-sight operations using a variety of technologies to ensure separation. Through education, geo-fencing, and traffic management, the number of no fly zones can be limited (Figure 4).

5 CONCLUSION

US Supreme Court Justice Samuel Alito provides prescient insight regarding the role technology can play with respect to social expectations:

> But technology can change those expectations. Dramatic technological change may lead to periods in which popular expectations are in flux and may ultimately produce significant changes in popular attitudes. New technology may provide increased convenience or security at the expense of privacy, and many people may find the tradeoff worthwhile. And even if the public does not welcome the diminution of privacy that new technology entails, they may eventually reconcile themselves to this development as inevitable" (*United States v.* Jones, 2012, 963).

The continued growth and development of unmanned aircraft systems and associated sensors will continue to influence social issues and the legal system's response. Continued monitoring of changes to regulations and laws

is highly recommended for those who are manufacturing and operating such systems.

REFERENCES

Consumer Electronics Association (2015) New Tech to Drive CE Industry Growth in 2015, Projects CEA's Midyear Sales and Forecasts Report [ONLINE] July 15. Available at https://www.ce.org/News/News-Releases/Press-Releases/2015-Press-Releases/New-Tech-to-Drive-CE-Industry-Growth-in-2015,-Proj.aspx (accessed September 10, 2015).

Convention on International Civil Aviation (Chicago Convention) (1944) ICAO Doc. 7300, opened for signature 7 December 1944, entered into force 5 March 1947.

Federal Aviation Administration (2015) B4UFLY, [ONLINE] August 28. Available at http://www.faa.gov/news/updates/?newsId=83608 (accessed September 10, 2015).

Federal Aviation Administration (2015) Authorizations Granted Via Section 333 Exemptions, 10th September. Available at http://www.faa.gov/uas/legislative_programs/section_333/333_authorizations/ (accessed September 10, 2015).

International Civil Aviation Organization (2015) *About ICAO*. [ONLINE] Available at http://www.icao.int/about-icao/Pages/default.aspx (accessed September 10, 2015).

International Civil Aviation Organization (2015) Manual on Remotely Piloted Aircraft Systems (RPAS) (Doc 10019), 1–7.

Katz v. United States, 389 US 347, 88S. Ct. 507, 1967, 361.

Keane, J. and Carr, S. (2013) A brief history of early unmanned aircraft. *Johns Hopkins APL Tech. Dig.*, **32**(3), 559–560 [ONLINE] Available at http://www.jhuapl.edu/techdigest/TD/td3203/index.htm (accessed September 10, 2015).

Kyllo v. United States, 533 US 27, 121S. Ct. 2038 (2001) 35.

Matiteyahu, T. (2015) Drone regulations and Fourth Amendment rights: the interaction of state drone statutes and the reasonable expectation of privacy. *Columbia J. Law Soc. Probl.*, **48**(2), [ONLINE] Available at http://www.columbia.edu/cu/jlsp/archive/issues/vol48issue2.html (accessed September 10, 2015).

National Conference of State Legislatures (2015) Current Unmanned Aircraft State Law Landscape, [ONLINE] August 26. Available at http://www.ncsl.org/research/transportation/current-unmanned-aircraft-state-law-landscape.aspx (accessed September 10, 2015).

Small Unmanned Aircraft System Advisory Rulemaking Committee (2009) Comprehensive Set of Recommendations for sUAS Regulatory Development, Washington, DC, 1st April.

United States v. Causby, 328 US 256, 66S. Ct. 1062 (1946) 261 & 265.

United States Constitution Amendment IV, Submitted for ratification 25 September (1789), ratification completed 15 December 1791.

United States Constitution Amendment V, Submitted for ratification 25 September (1789), ratification completed 15 December 1791.

United States H.R. 658 (112th): FAA Modernization and Reform Act of 2012, Title III, Subpart B, Sections 331–336. Signed by the President on Feb. 14 (2012).

United States v. Jones, 132S. Ct. 945, 565 US 945 (2012) 953 & 956.

Whitlock, C. (2015) FAA records detail hundreds of close calls between airplanes and drones. *The Washington Post.* [ONLINE] August 20. Available at https://www.washingtonpost.com/world/national-security/faa-records-detail-hundreds-of-close-calls-between-airplanes-and-drones/2015/08/20/5ef812ae-4737-11e5-846d-02792f854297_story.html (accessed September 9, 2015).

Whitlock, C. (2015) Rogue drones a growing nuisance across the U.S. *The Washington Post.* [ONLINE] August 10. Available at https://www.washingtonpost.com/world/national-security/how-rogue-drones-are-rapidly-becoming-a-national-nuisance/2015/08/10/9eae05d63c-3f61-11e5-8d45-d815146f81fa_story.html (accessed September 9, 2015).

Subject Index

Unmanned Aircraft Systems. Edited by Ella Atkins, Aníbal Ollero,
Antonios Tsourdos, Richard Blockley and Wei Shyy.
© 2016 John Wiley & Sons, Ltd. ISBN: 978-1-118-86645-0.

global positioning systems 221–222
guided weapons 218–223, 228
gyro characteristics 219–220
integration 221–222
local geographic navigation 307–308
mechanization 222, 307
missiles 218–223, 228
path-planning 218–223, 228
inertial navigation units (INU) 294, 305–309
inertial positioning & heading 242
inertial sensors 219
inertia moment equation 37–38
in-fleet collisions 618
information
    cooperative control 425–426
    filters 372–373
    fusion 293
    low-altitude flight 475–486
    multiple-UAVs 383, 389–390
    sensor fusion 293–312
infrared sensors 69, 303–304, 631, 633–634
inner-loop control 435, 438–441, 232–239
INS *see* inertial navigation systems
in-schedule dependency, swarms 399
insensitivity, state-space models 270–271
instantaneous task assignment 398–399
insurance
    absolutely liable 595–596
    criminal loss 592, 594
    federal regulatory compliance 594–595
    fraud losses 594
    indemnity 589–591
    indirect losses 592
    integration issues 589–597
    interest requirements 590–591
    key person loss 592
    legal indemnity 589–591
    liability 589–596
    losses, risk 589–596
    mission enablers 589–597
    national laws/regulations 595
    omnibus clauses 591
    personnel losses 594
    principle of subrogation 591–592
    privacy 596
    property loss 592
    property valuation 594
    regulatory compliance 594–595
    risk 589–596
    safety 593–596
    security 591
    state law/statutory requirements 594–595
    strict liability 595–596
    subrogation 591–592
    ultrahazardous activities 596

uninsurable activities 596
insured, definitions 591
integrated health monitoring 411–423
    see also integration
    abrupt fault detectors 411–416, 421–422
    actuators 411–414, 422
    automated threshold selection 418–422
    communications 411–413, 418
    Component-Level Fault Detection and Recovery 411–414
    control 412–413
    cooperation 411
    coordination 414
    decentralized systems 413–418
    dynamic coupling 416–422
    fault detection 411–422
    formation flight control 411–414, 416–418, 421–422
    multiple-UAVs 411–423
    nonabrupt fault detectors 411–412, 416–422
    observer robustness 413–416
    robust observer design 413–416
    sensors 411–416, 421–422
    signal redundancy 416–422
    simulations 421–422
    teams 411–412, 415
    threshold selection 412, 416–422
integration
    see also integrated health monitoring
    airspace access 329–330, 456–458, 470–473
    air traffic control 330–331
    challenges 13
    command and control 330–331
    commercial high altitude flight 470–473
    coordinating airspace requirements 329–330
    cyber security 637–650
    fail-safe systems 599–612
    fault detection 411–422
    global positioning systems 221–222, 173–174
    high altitude flight 470–473
    inertial navigation systems 221–222
    insurance 589–597
    legal issues 651–659
    National Airspace Systems 470–473
    navigation 221–222, 173–174
    policy documents, regulations 561–576
    privacy 596, 653–659
    regulations 456–458, 561–576
    relative navigation 173–174
    safety
        cyber security 637, 639–640, 647–649
        fail-safe systems 599–612
        insurance 593–596
        legal issues 651–653, 657–658
        level requirements 579–588
        sense-and-avoid systems 627–634
        social issues 651–652, 657